THE
BOOK OF
WAR

THE
BOOK OF
WAR

SUN-TZU
THE ART OF WARFARE

KARL VON CLAUSEWITZ
ON WAR

Series Introduction by Caleb Carr

Introduction by Ralph Peters

THE MODERN LIBRARY

NEW YORK

2000 Modern Library Paperback Edition

Introduction copyright © 2000 by Ralph Peters
Series Introduction copyright © 1999 by Caleb Carr
On War copyright © 1943 and renewed 1971 by Random House, Inc.
The Art of Warfare translation and new text copyright © 1993 by Roger T. Ames

On War was originally published by Random House, Inc., in 1943.
The Art of Warfare was originally published by Ballantine Books,
a division of Random House, Inc., in 1993.

MODERN LIBRARY and colophon are registered trademarks of Random House, Inc.

Grateful acknowledgment is made to Wing Tek Lum for permission to reprint
"Chinese Hot Pot" from *Expounding the Doubtful Points* (Bamboo Ridge Press) by
Wing Tek Lum. Copyright © 1987 by Wing Tek Lum.

LIBRARY OF CONGRESS CATALOGING-IN-PUBLICATION DATA
The book of war.
p. cm.
Includes index.
Contents: Sun-tzu The art of warfare & Karl von Clausewitz On war.
ISBN 978-0-375-75477-7
1. War. 2. Military art and science. I. Sun-tzu, 6th cent. B.C. Sun-tzu
ping fa. English. II. Clausewitz, Karl von, 1780–1831. Vom Kriege. English.
U21 .B58 2000
355.02—dc21 99-47953

Modern Library website address: www.modernlibrary.com

Printed in the United States of America

Introduction to the
Modern Library War Series

Caleb Carr

The term "military history" has always been a bit of a problem for me, as it has, I suspect, for many other students of the discipline. The uninitiated seem to have a prejudicial belief that those who study war are an exceedingly odd lot: men (few women enter the field) who at best have never outgrown boyhood and at worst are somewhat alienated, perhaps even dangerous, characters. Of course, much of this general attitude was formed during the sixties and early seventies (my own high school and college years), when an interest in the details of human conflict was one of the most socially ostracizing qualities a person could have. That tarnish has never quite disappeared: In our own day the popular belief that military historians are somehow, well, *off,* endures in many circles.

By way of counterargument let me claim that enthusiasts of military history are often among the most committed and well-read people one might hope to encounter. Rarely does an important work of military history go out of print; and those who know war well can usually hold their own in discussions of political and social history, as well. The reason for this is simple: The history of war represents fully half the tale of mankind's social interactions, and one cannot understand war without understanding its political and social underpinnings. (Conversely, one cannot understand political history or cultural development without understanding war.) Add to this the fact that military history very often involves tales of high adventure—peopled by extreme and fascinating characters and told by some of the best writers ever to take up a pen—and you have the actual secret of why the subject has remained so popular over the ages.

The new Modern Library War Series has been designed to both introduce the uninitiated to this, the real nature of military history, and to reacquaint the initiated with important works that they may have either forgotten or overlooked. For the sake of coherence, we have chosen to focus our four initial offerings on American military history specifically, in order to show how the study of war illuminates so many other aspects of a particular people's experience and character. Francis Parkman's *Montcalm and Wolfe,* for example, not only shows how very much about the psychology of pre-Revolutionary leaders one must understand in order to grasp the conflict known in North America as the French and Indian War, but is also the work of one of the great American prose stylists of the nineteenth century. Ulysses S. Grant's *Personal Memoirs* (which owe more than a little to the editorial efforts of one of Grant's champions, Mark Twain) contrasts the remarkable humility of their author with the overwhelmingly dramatic circumstances into which Fate flung him, and that he struggled so hard—in the end, successfully—to master. Theodore Roosevelt's *The Naval War of 1812,* too long neglected, was the first work to reveal the prodigious intellect, irrepressible character, and remarkably entertaining style of this future president, who (his father having spent most of the family fortune on charities) consistently made a good part of his income through writing. And finally we have *A Soldier's Story,* the memoirs of Omar Bradley, "the G.I. General," who, surrounded by a sea of prima donnas during World War II, never stopped quietly learning his trade, until he became, during the conquest of Germany in 1945, arguably the most progressive and important senior American commander in the European theater.

To read any or all of these books is to see that military history is neither an obscure nor a peculiar subject, but one critical to any understanding of the development of human civilization. That warfare itself is violent is true and unfortunate; that it has been a central method through which every nation in the world has established and maintained its independence, however, makes it a critical field of study. The fact that the personalities and stories involved in war are often so compelling is simply a bonus—but it is the kind of bonus that few academic disciplines can boast.

THE SEEKER AND THE SAGE

Ralph Peters

This book allies humankind's two most powerful works on warfare. Distant in time, space, and culture, Karl von Clausewitz and Sun-tzu offer dueling visions, with the Prussian appalled by fantasies of bloodless war and the Chinese crying that bloodless victory is the acme of generalship, and with Clausewitz anxious to increase military effectiveness, while Sun-tzu pleads, cleverly, for military restraint. Such discord assures their relevance to our time.

There is also plentiful agreement between Clausewitz's *On War* and Sun-tzu's *The Art of Warfare*, from their mutual vilification of heads of state who attempt to micro-manage distant battles to their similar emphasis on the key role of the commander. In the end—and I speak as a soldier, after decades of consideration—these two books complete each other, like a perfect couple formed of opposites. Between them, the two texts cover myriad aspects of the human experience of war—as well as reflecting the temperaments of their divergent civilizations. Clausewitz, the Western man, sought the grail of knowledge and found the pursuit endless, bottomless, and obsessive, while the Eastern sage who wrote down the sayings attributed to Sun-tzu polished what he knew until it shone. Each attained the universal, transcending personality and the particularity of experience. In the study of warfare, they have no peers, and these works remain the brightest lanterns we have to light our darkest endeavor.

The Western text embraces war's necessity, while the Eastern one despairs of its inevitability, but they are united by the recognition that the human remains at the heart of each combat encounter and every campaign. Each holds a flank in our approach to war: Clausewitz is the apostle of the relentless will, convinced there is no substitute for victory, while Sun-tzu seems a closet pacifist, wary of victory's hollowness. The first sought to sharpen the sword, the second to restrain it. The Prussian saw the power of the armed mass, while the Chinese pitied the suffering of the common man. Sun-tzu believed that the outcome of a campaign was predictable, but Clausewitz insisted that, although the odds can be improved, risk is inherent in warfare. This debate across millennia continues today, and placing these two works together highlights the strengths and weaknesses—and the inestimable value—of each book.

Each must be read. No cram notes will do, and summaries badly serve their genius. Clausewitz appears difficult, only to yield a hard, thrilling clarity; while Sun-tzu, a quick swallow, takes a lifetime to digest. One text is long, the other appealingly short. Both are inexhaustible.

———

Had I the skill, I would rescue Karl von Clausewitz from the admirers who have made of him a dry and reasoned thing. He was a man of fire, but acolytes have reduced him to ashes. His image is of the diligent staff officer: coldly avid, narrowly expert, and, for all his brilliance, a bit short of life. In reality, he was a courageous, driven man, struggling to contain his passions within Prussia's turgid hierarchies. His commitments were wholehearted, his abilities immense, and his sense of duty exemplary. Born on the doubtful fringes of the nobility, he experienced war at every level, found friends among the best men of his age, won the approval of a royal court—until he enraged its king—wrote love poetry, and married a higher-born woman after a courtship worthy of a nineteenth-century novel. When appeasement was the fashion, he ruined his career out of patriotism. He enjoyed a remarkably happy marriage and wrote a work so rich it may never be surpassed, yet he suffered disappointment and feared himself a failure. Above all, Clausewitz was a man of his times, and his times were the Romantic era in Germany.

Born in 1780 into a family clinging to threadbare claims of aristocracy, a relative arranged for his regimental apprenticeship to begin at age twelve. At thirteen, Clausewitz was a combat veteran. He participated in the frontier campaign against revolutionary France and the siege of Mainz (finely reported by Goethe, who was born a generation before Clausewitz and would outlive his fellow genius by a year). After the excitements of the field came garrison years, during which Clausewitz made the barracks his school. He read hungrily, with an appetite for a variety of intellectual dishes, and all his life he savored learning as only the autodidact can. Along with the current texts on military matters, he read philosophy, natural science, mathematics, and literature, developing an early fondness for Schiller, whose plays and histories, so vigorous and seductively written, exalted military valor and northern, Protestant virtue.

His superiors recognized his exceptional abilities, and at twenty Clausewitz was chosen to attend the *Allgemeine Kriegsschule,* Prussia's military college, in Berlin. By age twenty-three, his talent had won him a mentor and fatherly friend in Germany's greatest soldier of the age, Gerhard von Scharnhorst, and a place at court as adjutant to Prince August. He also met the woman he would marry, Countess Marie von Brühl, a favorite of Queen Luise, who valued depth and intellect in those around her.

Prussia was in ferment then, ruled by a bold, shining queen and a dullard king fitted for survival but not greatness. The revolution across the Rhine and Napoleon's dynamism had charged the atmosphere even where French policy met no approval. In Germany, philosophy, drama, and the applied arts had found their footing a generation before and now were on the march, while the new Romantic literature conquered the educated imagination. As with all disturbed ages, it was a period of explosive creativity. The classical tastes of the Enlightenment persisted among the old, but were subverted by the young. Literary giants were the idols of the day—Clausewitz's wife to be favored Goethe over her suitor's beloved Schiller, eventually winning her Karl over to the creator of Faust, Werther, and Wilhelm Meister. It is always a fine thing to be young, popular, and in love, but to be so in Berlin at the beginning of the new century must have gilded the bedazzlement.

The idyll ended in 1806. Napoleon's strategic vigor and speed shattered Prussia's brittle forces in the Jena-Auerstedt campaign, scattering the Prussian army with an ease that shocked Europe. The fighting ended for Clausewitz as he stood beside Prince August, surrounded in a swamp and covered in mud, with his battalion's last reserves of powder soaked through. As the French lined up their cannon on the high ground for the kill, the prince surrendered. In wretched, ruined uniforms, the prince and his adjutant were shunted from one French commander to another as trophies until they arrived in occupied Berlin and learned they would be prisoners in France.

For the prince and his subordinate, the terms of captivity proved gentle. Quartered not far from Paris, they had the liberty to visit on occasion, although Clausewitz sulked at the glories of Europe's new capital. Pride scalded, Clausewitz detested the French. Yet, he made the most of the opportunity to observe the world around him, writing compulsively in his attempt to understand how this coarse, self-obsessed nation could achieve such splendid victories (all his life, he used the act of writing to force himself to a deeper understanding). And he wrote a wealth of letters to his beloved, whose mother remained years away from consenting to the socially lopsided match. Devoted, the letters were a mix of heat and patience, of tenderness and intelligence. The lovers wrote of politics and longings, of important news and old friends, of books and dreams. Clausewitz described his efforts to appreciate painting as much as Marie did (he remained, all his life, a man whose tastes ran to ink on paper, not paint on canvas), and she wrote to soothe him in his anger and humiliation.

After a year, Prussia signed a degrading treaty and the prisoners began a slow journey homeward. En route, the prince took them on a detour to Schloss Coppet in Switzerland, to visit Madame de Staël, lioness of literary salons and author of romances unreadable today. Clausewitz liked her books—especially the sentimental novel *Corrine, ou l'Italie*—liked the woman, and enjoyed the "artsy" environment so much it gives cause to wonder what he might have done with his life had he not been consigned to the parade ground as a boy. While the prince flirted with another visitor, the couchable Madame Récamier, Clausewitz enjoyed long conversations with his hostess and her consort, August Wilhelm Schlegel, the German Romantic writer and philosopher, as well as with yet another guest, Johann Heinrich Pestalozzi, the Swiss educational theorist. Most revealing, Clausewitz was a hit in this exclusive milieu. Surrounded by the Alps—mountains were a passion new to the Romantic era—he wrote rhapsodic letters to his fiancée. For the first time in over a year, glints of joy shone through his penmanship.

But, back in Prussia, the gaiety was gone. A chastised nation and the broken remains of an army lacked bearings as much as they lacked prospects. Napoleon's victories had mounted—smashing the Austrians, crushing the Russians—until he seemed invincible, almost a cosmic force. The size of Prussia's army was restricted, and the French closely monitored government and military appointments through a web of spies and collaborators. Nonetheless, Scharnhorst—a genius of maneuver at court as on the battlefield—began to rebuild the Prussian military in the shadows. The great reformer's hour had come, and Clausewitz became his primary assistant. Although still only a captain, Clausewitz had long since earned the older man's trust. Scharnhorst once stated that he had only to speak a few words and the younger man could complete the thought and capture it on paper. Scharnhorst also wrote that, except for his own children, no one had been as close to him as Clausewitz.

Never in robust health, Clausewitz exhausted himself over his desk—there is far more ink than gunpowder in an officer's life, and in defeat ink must stand in for blood. He was determined that Prussia would rise again. His conception of war had begun to mature, and the quality of thought in those early writings is remarkable for an officer not yet thirty. Clausewitz had the gift that comes to few men of seeing clearly then drawing universal conclusions from the partiality of experience. Intuitively, he spotted relationships where others saw only disparate events and sensed the constants underlying the disruptions of his age. If his personal experience undeniably shaped his later work, it never deformed it. Among Scharnhorst's many virtues, the ability to spot an even greater talent than his own and to utilize it without jealousy was not the least important. He gave us Clausewitz.

King Friedrich Wilhelm III, who made an art of indecision, hedged between those subordinates who wished revenge and those who believed that Prussia's future lay with France. Meanwhile, Napoleon looked to the territories he had subjugated for fresh blood to lubricate new conquests. Bending to necessity, Friedrich Wilhelm appeased the French time and again. Then Napoleon began to prepare for the invasion of Russia, a campaign in which Prussian troops were to fight under the control of the *Grande Armée.*

It was too much for the newly promoted Major Clausewitz (and for dozens of other Prussian officers, as well). Leaving behind his adored Marie, whom he had married at last, he took off the blue *Königsrock* and accepted a commission from the czar as a lieutenant colonel. The defeat of Napoleon had become the central purpose of his life. But Clausewitz's yearning to contribute to the full extent of his abilities was stymied by his inability to speak Russian. For months he was shifted from staff to field unit to yet another staff, unable to find a position where his talents could emerge. It was a period of enormous frustration for the soldier who knew his skills were underutilized, but the variety of experiences proved beneficial to the military theorist. Those leaps from one post to another took him from the imperial staff down to brutal cossack raids, from rearguard cavalry actions to personal heroism on the field of Borodino. He witnessed the final destruction of the French army on the Beresina—a scene of utter savagery—and then his turn came at last. Clausewitz proved the pivotal man in negotiations with the Prussian general Yorck, whose formation was covering the French retreat. The Convention of Tauroggen withdrew Yorck's Prussians from the campaign—without the king's approval. It left the French with an open flank and Clausewitz in lasting disfavor at the Prussian court.

The king never forgave him. Although Prussia soon turned against Napoleon and took the field beside its Russian and Austrian allies, Friedrich Wilhelm felt that Clausewitz had betrayed him, first by putting on a Russian uniform, then through the impertinence of undercutting his policy. Clausewitz begged to be allowed to return to Prussian service, and his friends did what they could to help him trade his Russian green for Prussian blue again, but the king was adamant to the point of spite. He insisted that Clausewitz would have to prove himself, mightily, on the field of battle to have any hope of regaining a Prussian officer's patent.

Clausewitz *did* prove himself. Still in Russian uniform, he served as a liaison officer (but really as a deputy) to Scharnhorst on campaign. As Prussia rose again, the orders, directives, and assessments that flowed from Clausewitz's pen harnessed the power of a people in arms. This was a new world, in which regular military formations were reinforced by reserves drawn from a patriotic citizenry. A passion for nationhood that no Prussian king—not even Frederick the Great—had excited arose in response to French tyranny. It was a people's war on both sides now, and Scharnhorst and Clausewitz had been among the first outside of France to grasp the implications of the changing times. Even as Napoleon and his subordinates continued to win victories, the national cause strengthened and the Prussian military exhibited the resilience and determination that had been so lacking in the previous decade.

But all war brings tragedy. At the battle of Grossgörschen, in a desperate moment, the staff of Prussia's Army of Silesia plunged into the battle in an attempt to stave off defeat. Prussia's finest officers rode into the mêlée with sabers drawn—Blücher, Scharnhorst, Gneisenau, and Clausewitz. A bullet winged Blücher, Clausewitz received a flesh wound behind the ear from a French bayonet, and Scharnhorst was shot in the leg. One well-fired volley might have wiped out Prussia's military reformers. As it was, Scharnhorst refused to give his wound time to heal. He was a man whose sense of duty obliterated the self. And the general, reformer, and great soul, risen from the peasantry of Hannover, died of infection. When the king, warily, accepted Clausewitz back into Prussian service, it was a bittersweet return. Although Gneisenau—jovial, bright, and brave—assumed the role of mentor and friend to Clausewitz, their mutual affection could never fully replace the loss of Scharnhorst for the younger man. Clausewitz was the sort who does not need the applause of the masses but who craves the approval of a hero. Germany never produced a worthier hero in uniform than Scharnhorst.

With the blue coat came a disappointing assignment. Clausewitz was dispatched to a secondary theater in the north. He learned a great deal about economy of force operations and applied himself fully—as he always did—but he longed to be part of the great campaign that culminated in the Battle of Leipzig. From then on, Clausewitz remained on the edges of the battlefields whose names live on in history. He served well in the Waterloo campaign—but not on the crucial fields of Ligny or La Belle Alliance—and his laurels were fewer than those of lesser men. Denied his repeated requests to serve on the line, he was confined to the staff work that he could do better than anyone else—and that kept him out of sight of his rancorous king. Now forgotten, two of his brothers served with greater renown on the battlefields of the day than did the Clausewitz sibling we revere.

It is, finally, to the king of Prussia's credit that he allowed Clausewitz to serve on after the war's end—even promoting him to major general—but it is to his discredit that Friedrich Wilhelm so distrusted his subordinate that he "exiled" the brightest officer of the day to the Prussian military academy, not as director of the curriculum and trainer of officers but restricted to administrative and disciplinary matters. The commandant's job, by tradition, was used to sideline generals fallen out of favor; and new, potentially subversive ideas were no longer wanted. Confounding all the reformers of the period, a repressive twilight settled over the continent they had done so much to save. Clausewitz, the man of fire, was reduced to tending embers.

He felt his share of bitterness, but never surrendered his will. Only thirty-eight when he became head of the *Allgemeine Kriegsschule*, but with twenty-five years of military service behind him, he turned to his writing with renewed intensity. His duties were such that he could accomplish them in a few hours in the morning, and he spent the long afternoons on the papers that eventually would become *On War*. His greatest aide and friend was his wife, who not only copied his manuscripts but had the intellect to critique his work. Like the now-dead queen she had served in her youth, Marie von Clausewitz had more fire and intelligence than the mass of men given pride of place before her. Her husband's lifework is her work, too.

Clausewitz remained at his dreary post for more than a decade, grappling with ideas as single-mindedly as he had once struggled against Napoleon. He possessed what might be the most important virtue in a military man, the will to persist, to outlast the enemy. It has saved giants such as Frederick, Washington, and Giap; and although it could not prevent his bouts of depression, it kept Clausewitz from despair. For a dozen years, he fought to capture the complexity of warfare on the written page—an impossibility—and never brought his work to a conclusion. That he achieved as much success as he did is a monument to the human will.

Unexpectedly, his career underwent a late resurrection. He was assigned to an important post as inspector of artillery in Silesia, and he and his wife threw themselves into the provincial life of the Breslau garrison. Then, as Poland rose against its Russian oppressor, he was sent to serve as chief of staff to his old friend Gneisenau in the army of observation on the border of the czar's empire. The Russians drowned the rebellion in blood, and the Prussians never had to fire a shot.

But a fiercer enemy had crossed the frontier: cholera.

In the first of the great cholera epidemics that swept the world in the nineteenth century, Gneisenau, a living legend, died ignominiously. At first, it seemed that Clausewitz would be spared, and his splendid work in the field regained him favor at court, with praise even from old enemies. Depressed at the loss of his great friend, but with a brighter future before him, Clausewitz returned from the headquarters in Posen to his post in Breslau.

On a bare November day not long after his return, he sickened and died in the space of eight hours, leaving his book unfinished. With bitter symmetry, Germany's other great philosopher of the age, Hegel, died in the same epidemic.

Marie arranged for the publication of *On War* in its unfinished state. It has been the most fatally misunderstood book of the last two hundred years.

———

So much is known. But it seems to me there is more. Even the best biographies of Clausewitz have concentrated on the military man in his professional milieu or, sometimes, in his political context. But he was, indelibly, a man of his culture, as well—of the yearning, morbid atmosphere of German Romanticism. Placing the man in the intellectual context of his age reveals a new complexity in the writer and brings greater clarity to the work. This soldier was far more than his uniform.

The only philosopher referenced in most studies of Clausewitz is Kant, and then only to wonder whether Clausewitz actually read him or received his ideas watered down through a disciple who lectured in

Berlin. Yet, Kant—no Romantic—had at most a negligible influence upon Clausewitz, whose nature instinctively rejected Kant's belief that the essence of a thing was unknowable and embraced Hegel's conviction that human intelligence can penetrate to the essence. While Clausewitz would have been interested in Kant's theories on education, he would have dismissed Kant's political idealism as absurd. Their interests diverged as widely as did their conclusions. A range of thinkers and writers influenced Clausewitz—some directly and others by shaping the intellectual environment—but not Kant. These two Prussians shared only the regularity of habits that keeps the irregular mind from the abyss.

Clausewitz had plentiful opportunities to hear the great thinkers of his day in person—men such as Fichte, Hegel, Schelling, and the elder Schlegel (with whom he spoke at length in Switzerland). Their lectures were social occasions for the intelligentsia, of which Clausewitz was an enthusiastic member. Of even greater importance, his reading was broad and voracious. He was a man addicted to learning, and philosophy was the headiest drug. Nor could he have held his own in the court of Queen Luise—where he gained an early welcome—unless able to discuss the philosophy of the day, as well as poetry, fiction, and history, all of which blurred together in the Romantic era.

Even had Clausewitz been a blockheaded martinet—which this most thoughtful of soldiers was not—no man escapes the temper of his times. Just as today's military men are rendered inchoate by the march of technology, so Clausewitz, too, was a prisoner of his environment, if a more articulate one. Romantic thought and literature gave form to his ideas, just as his experience of war provided the content. For skeptics, there are concrete examples of the influence of Germany's Frederician- and Romantic-era philosophers on Clausewitz's writing. That they have gone unexplored says more about the narrowness of his biographers than about the man himself.

First, Clausewitz's concept of the trinity that decides the success or failure of policy—which we may crudely summarize as the relationship between the state, the army, and the population (reason, chance, and passion)—is fundamentally Romantic. Trinities pervade the thought of the period, with the trinity of *Geist, Seele, und Leib*—mind, soul, and body—encapsulating the Romantic view of Man and fundamental to the movement's philosophy, sciences, and literature. Here, Clausewitz appropriated an intellectual construct common to the intelligentsia of his day. He developed it wonderfully, but did not invent his trinity from thin air.

The Romantic era was also fascinated by dualities and opposites—male and female, day and night, good and evil, the *Doppelgänger* who pervade the popular ghost stories of the time—and reading German Romantic literature one starts to sense that the world is nothing but a series of bipolarities. Hence Clausewitz's identification of dualities in warfare: the relative strengths of the attack and the defense, war as a contest of opposing wills, the contrast between limited and absolute war, his juxtaposition of ends and means in war, and his insistence on the gulf between theory and practice. Again, he used the intellectual tools his age thrust upon him to build something new—but he did not create these tools.

Even his elaboration of the "center of gravity," or *Schwerpunkt*, concept owes less to Prussian military tradition than to Herder, the eve-of-the-Romantic-era philosopher whose discussion of the "center of gravity" of cultures clearly inspired the richness with which Clausewitz imbued the term (now unread, Herder is stunningly relevant to today's international dilemmas).

The Romantic era also created the cult of genius (perhaps its most damaging legacy). The titan of superhuman talents, ambitions, will, and achievements enraptured the chattering classes of the day, whether the colossus was Napoleon, bathing in oceans of blood, or the doomed poet Novalis, yearning after the blue flower, the impossible, the infinite. Clausewitz's ambition was to be a genius in the terms of his age, to take up an impossible challenge (*Don Quixote* was an especially popular book among the Romantics, who viewed its hero as nobly tragic more than comical). His attempt to capture the essence of war upon the page, to explain the complexity his own description of friction (*friktion*) in war renders boundless, was nothing less than heroic—and Clausewitz saw it as such. Novalis's statement, "What I want to do, I can do.... Nothing is impossible for mankind," might have served as the motto of Clausewitz's endeavor.

From *Faust* onward, the Promethean theme resounds through the literature—and behavior—of the day. Defiance and resistance set the tone. The Romantics sought to strive with gods, expecting their rewards from the effort itself rather than from ultimate success. All his life, Clausewitz was the man who reached for knowledge. His last, greatest effort, his struggle to contain war in words, was patently impossible.

I believe he knew it.

Perhaps he began in the belief that the task he set himself might be accomplished. On the surface, his work is rational and ordered—not unlike the persona he learned to offer the world around him. Like Goethe, Clausewitz was torn between the Classical, Enlightenment impulse and the stormy drives of the Romantic, between the scientific and the artistic. His was a split personality, riven between the noonday clarity of the drill field and the dark night of the soul, and demons lurked within the fitted uniform. Goethe defied the trend of his age, fleeing the Romantic wilderness for the safe Classical garden in his later years, but Clausewitz went the other way, beginning with the Enlightenment impulse to explain, only to be captured by the Romantic urge to understand. To borrow a line from Tennyson, an English Romantic of a later generation, Clausewitz sought "to follow knowledge like a sinking star, beyond the utmost bounds of human thought." Along the way, he saw that the mission he had set for himself was impossible. War could never be fully explained. It was the perfect subject for the Romantic age.

Clausewitz avoided finishing his work, sinking into his thoughts until they consumed him. Behind the disciplined exterior, he may have felt close to madness at times. To complete the book—to see it in print—would have been to lay his life's work open to criticism. But an unfinished work can only be criticized so far before we must ask if the author might not have altered this or clarified that, had the time been allotted him. Many a writer has failed to complete his masterwork—it is a disease of overweening ambition aggravated by cowardice. I believe that Clausewitz, the brave soldier, was afraid to finish his book and see it judged.

Even in leaving his book incomplete, Clausewitz was a child of his time. For the Romantics, the gorgeous fragment, not the finely rounded work, was the ultimate form in art. *Das Unvollendete*, the uncompleted, whether poem, symphony, or philosophy of war, was the true Romantic achievement.

Although his career had taken a late turn for the better, those who saw Clausewitz toward the end of his life described him as despondent and enervated. He loved his wife enduringly, but even the deepest mortal love could not sustain him. He had asked too much of himself and of the world. I believe his death by cholera, at age fifty-one, was a relief to him.

By placing Clausewitz among the high Romantics, we do not belittle his achievement. On the contrary, his work may be the most enduring text of the era, save only Goethe's *Faust*. While he likely would have scoffed if called a Romantic to his face, Clausewitz was, thoroughly, a creature of his times. And he might have been secretly pleased at the epithet. We cannot know with certainty. But we do know that no man sees himself objectively, and only history can place us. Despite ourselves, we are as the times are.

Indeed, the central character in *Faust* is timeless in his appeal because he is the intellectual's Everyman, blind to himself and desperate to escape the confines of mortality. He deals with the devil without to avoid the devil within. He seeks knowledge of the world to escape self-knowledge, and confuses excess with infinity. He must *act*, because *being* is terrifying. In the beginning, Clausewitz wrote to change an army; in the end, he wrote as devils chewed his soul.

No happy man thinks deeply or writes well.

———

Less is known of Sun-tzu. His life is the stuff of fables, not grounded by a single biographical fact. Said to be a general during China's "Warring States" period (453 to 221 B.C.) or earlier, Sun-tzu, or Sun Wu, has spurred more debate as to his existence as a historical figure than on the content of his thought. The single consistent story told about him, a nonsense about training court concubines for a military demonstration, does not ring true to anyone who has served either in a military or in a government. Even if he did exist, he did not write down the work we know as Sun-tzu's *Art of Warfare*. His pronouncements on campaigning, passed down, were recorded by other hands. The texts we have are laden with internal contradictions. Yet, "his" thought endures.

Sun-tzu pierces to the heart of warfare—and to the heart of man at war. He deals not only with the mass, but with the man, with psychology and instinct. He rightly sees human failings as the cause of most defeats, and gives a sense of sweat and death and truth. If Clausewitz wrote to understand, the man who first spoke the wisdom of Sun-tzu clearly knew more than he said. Here, knowledge is boiled down to its essence. I suspect this is not simply a matter of Chinese forms and conventions; rather, the speaker and the later chronicler of his speech, though separated by centuries, each understood enduring truths about rulers and generals: their time is always short, and their attention spans often limited. Sun-tzu was the original master of the sound bite, and he still quotes well today.

I do not read or speak Chinese, and cannot place Sun-tzu in the culture of his time as confidently as with Clausewitz. I can only approach him as a former soldier—and as an officer who was occasionally sentenced to work within a government. Still, some things seem obvious. Begin with what *is* known:

The Warring States period was long, brutal, and destructive. As with the tumultuous Romantic era more than two millennia later, it also bloomed with creativity—nothing inspires the arts to greatness like a sense of the blown-glass fragility of human life, and plague and slaughter spur the pen and brush. The times were, paradoxically, prosperous for many states, as well—war has always been good business. Yet, by the end, the destruction outweighed the progress, and the moral men of the day recoiled in horror. As in the Peloponnesian Wars of Greece, states were doomed or gutted.

The cost to the population—in famine, disease, dislocation, and slaughter—was high; and though a richer merchant class emerged, the lot of the common man or woman was bitter. Social order was maintained by savagery, but even vicious punishments could not keep states from decline. Earlier conventions of warfare were upset by burgeoning state ambitions and by innovations in the mechanics of warfare. As in the twentieth century, military capability carried within it the inevitability of use. The old order faltered and died, and warfare became total war. To those who loved tradition, who glorified a more pacific past, barbarism seemed the rule.

Undoubtedly, such a period produced talented commanders, some of whom were successful enough to command attention when they spoke and have their words passed on by their disciples. Perhaps Sun-tzu was one of these, or perhaps he was an idealization of them all, a distillation. In the end, it hardly matters: We have the work.

Yet, speculation is tempting. In the intelligence world, you develop a tactile feel for things, a honing of instincts, that allows you to construct models to fill in the gaps in what is known. Sometimes, you are wrong. Still, it seems to me there is a "likeliest" scenario for how this *Art of Warfare* emerged from the chaos of the Warring States.

Although archaeological finds have brought us closer to the original text, we may never know whether or not Sun-tzu lived. Does it matter? In the end, the chronicler may be the more interesting figure.

Toward the end of the Warring States period, in a landscape desolate here and newly prosperous there (not unlike northern Europe after the Thirty Years' War), a ruler probably turned to a trusted minister and asked him to gather and record the wisdom circulating in Sun-tzu's name for use in formulating strategy and conducting campaigns. Perhaps the minister himself, despairing of his times, undertook the project on his own, to urge the ruler to better his performance. He would have queried scholars and military experts, sending scribes to interview them, if the budget of the court allowed (the minister probably did not serve in a rich court, since there is so much stress on costs and economies in the text of the *Art of Warfare*). Gathering the many sayings that had appeared over the centuries, he would have pared them down to those he thought genuine—

and then discarded those he felt might lead his ruler down dangerous paths. For no minister of state ever passes the entire truth to his superior. Perhaps the courtier was charged only to gather the general military wisdom of the past for a young ruler anxious to make his mark on the world. Perhaps, to that end, he "discovered" Sun-tzu, a wise ancient—perfect for a culture that revered old wisdom above contemporary insight. Whatever the details, a close reading of the text convinces that the chronicler was not content with what he had gathered: before placing the work in the hands of his ruler, he tempered the military prescriptions and proscriptions with forged maxims that advanced his own view of the world.

The man who wrote this knowledge down was in despair. That much seems evident. The book speaks in two distinct voices (with other echoes, of which more later). Most of the chronicler's interpolations come early on, although there is some intermingling throughout the text. The muscle of the book is, of course, the advice on how to exploit terrain or how to deceive an enemy or when to attack or defend. This is the real Sun-tzu, the soldier. The minister of court, seeking to guide the ruler, appears when the head of state is warned of the terrible expense of making war or admonished to minimize costs and suffering by employing spies in place of battalions or, most famously, advised that "the highest excellence is to subdue the enemy's army without fighting at all." This is the chronicler and courtier speaking, the man who has seen enough of war in his lifetime to be sickened by it, and who, while recognizing how deeply the warrior impulse infects the human condition, longs for peace.

I find the higher greatness in the chronicler, although he often seems to me a fool. While we might long for a world in which "the expert in using the military subdues the enemy's forces without going into battle, takes the enemy's walled cities without launching an attack, and crushes the enemy's state without a protracted war," this rarely happens in reality. Warfare was, is, and will remain bloody and destructive. When the chronicler speaks, we may admire his virtue but wonder if his state perished because of his tomfoolery. Such chaste ploys are especially appealing and dangerous in our time, when heads of state no longer serve in uniform and long to believe in the myth of bloodless war. But if history suggests any pattern, it is that those who wait too long to raise their swords pay a greater price in blood and suffering than those given to timely and judicious action. And yet…the chronicler's repeated pleas for a more humane approach to warfare, for the sparing of lives and crops and cities whenever possible, must be heard by the generals who imagine that they might, finally, achieve Clausewitz's ideal of total war. Warfare that is excessive and indiscriminate in its violence only dissipates power, while inspiring resistance.

Likewise, the chronicler gives good advice when he tells the ruler or general to "provide for the captured soldiers and treat them well," but he lies when he insists that "there has never been a state that has benefited from an extended war." He was writing in an age when powerful states had expanded profitably over decades and centuries of warfare. Small states, with limited resources, may perish as a result of extended wars, and economically hollow states may decline; but, throughout history, a variety of states have prospered as a result of lengthy conflicts. We are dealing, ultimately, with a brilliant work of propaganda that aimed to lead a prince to wisdom.

The most difficult portions of the text to credit definitively to either the general or the minister are those dealing with intelligence and espionage, since both men would have known the value of those arts. Here, the minister probably supplemented observations culled from the general's legacy. The repeated insistence on the importance of knowing the enemy, of gathering information, and of deception and subversion are superior to Clausewitz's appreciation of intelligence affairs. In the end, Clausewitz is a philosopher, but a fighter: he wants to punch and even his defense is "a shield of blows." As soon as we labor through the quaintness of expression, *The Art of Warfare* offers the better advice on intelligence and operations in the shadows. While Clausewitz, in his most famous line, states that "war is a simple continuation of policy through other means," Sun-tzu's *Art of Warfare* is closer to the American position that warfare means that policy has failed. Clausewitz wants to win wars, while the chronicler of Sun-tzu wants to avoid them whenever possible, but to win them if they must be fought. Sun-tzu and his scribe both would rather bribe than march, and rather assassinate than bomb. They would have grasped the utility of the CIA immediately, while Clausewitz would have retained a measure of skepticism.

The blood-red core of *The Art of Warfare* clearly belongs to Sun-tzu the general—or to the various generals whose wisdom was collected by the chronicler. This is where we get the enduring wisdom of the man at arms. He stresses morale, here called the *tao*, or way; counsels the commander "in joining battle, seek the quick victory"; and, like his fellow soldier Clausewitz, stresses that the defense is inherently the stronger form of warfare (if *Blitzkrieg* or the technologies of our contemporary "Revolution in Military Affairs" appear to upset this maxim, recall that the German *Blitzkrieg* won campaigns but not wars and that postmodern Western militaries have yet to annihilate a single opponent). The military man describes much that is tactical and mechanical, from the role of fixing and flanking attacks—the "straightforward" and the "surprise"—to how to manage approach marches, but he also underscores the importance of avoiding too strict an adherence to formulas. He addresses the need for concentration of effort and economy of force operations, stating that "to be prepared everywhere is to be weak everywhere." In essential advice for our times, he demands that victory not be compromised through political irresolution: "to be victorious in battle and win the spoils, and yet fail to exploit your achievement, is disastrous." This is the lieutenant's handbook, and also the general's and the president's.

There is, unfortunately, a third hand—actually a series of hands—that adulterated *The Art of Warfare* we have received. Soon after the wisdom of Sun-tzu was recorded, revisions began. Various scholars and men of court inserted positions they wanted to dignify, or subtracted sections with which their recommended policies disagreed. The text has been heavily corrupted—the evidence is in the contradictions that arise from page to page. We are warned that the commander must share information with his subordinates, but then the general is cautioned "to blinker the ears and eyes of his officers and men, and to keep people ignorant." At one point, the commander is admonished not to throw his troops into hopeless situations, only to be told elsewhere that doing so will bring out the ferocity and will to survive in those he commands. He is advised to care for his men "as if they were his own beloved sons" shortly after he has been warned that "if he loves his people, he can be easily troubled and upset," because too much concern for the lives of his troops can "prove disastrous" (in fact, neither extreme is practical and the truth lies in the middle: the commander must cherish his men and protect them when he can, but he must be willing to send them to their deaths without hesitation when necessary). Likely, these passages are all corruptions of sounder military insights in the original lost to us.

It takes extraordinary power for a work to survive such bastardization as the *Art of Warfare* suffered over the centuries. The Bible is the supreme example of such a work, followed by Homer. That the work attributed to Sun-tzu, who may or may not have walked the earth, remains so relevant and immediate in so many of its parts is little short of miraculous. As the reader of this edition goes through the pages, he or she will likely do as I did, noting that a passage perfectly captures Napoleon or Stonewall Jackson, or Forrest, Montgomery, or MacArthur, and so on. It is intellectually breathtaking how many varieties of battle, how many command personalities, are captured in this small book. What could be more relevant for the digital age of warfare than the observation, "war is such that the supreme consideration is speed"? Or more vital to the asymmetrical, crudely barbaric conflicts that currently drain America's military resources and challenge our national will than the maxim, "the basic patterns of the human character must all be thoroughly investigated"? This small book is so encompassing that it almost seems to predict the emergence of air and space power when it says, metaphorically, "the expert on the attack strikes from out of the highest reaches of the heavens."

Although it deals with mankind's monstrous failings, *The Art of Warfare* is finally poetic—not in its subject matter or in the music of its language or even in its striking imagery, but in its economy and depth. As with poetry, it may be enjoyed at first reading, but, to be appreciated, it must be read more than once. And, like the greatest poems, it may be revisited for a lifetime, acquiring greater depth with each encounter.

This is a beautiful work on killing.

———

Will Clausewitz survive as long as Sun-tzu? He is the West's sole contender for such immortality in this field of study. And it is unlikely that we will see another entry in the race. In our time, Clausewitz is most often faulted for his failure to foresee the explosion of military technologies that would begin to revolutionize warfare only a generation after his death. The "missing technology" charge might be leveled at Sun-tzu, as well. And it would be as foolish as it is in the case of Clausewitz. Both men were blessed to live in ages in which the human factor remained the obvious key to understanding warfare. While even Sun-tzu's age saw the impact of new military technologies, neither writer was blinded by the dazzle of machinery—as our contemporary military theorists tend to be. The authors of *The Art of Warfare* and *On War* worked in a timeless laboratory of death, in which the factors of knowledge, morale, mass, and will were the primary chemicals. What might Sun-tzu have made of nuclear or precision-guided weapons? His chronicler—stunningly contemporary in his prejudices—might have rued the first and embraced the latter. But would he have seen through the ornaments to the essence? Likewise, had Clausewitz had the misfortune to survive into the age of the telegraph, repeating rifles, the railroad, and steam-driven ironclads, he might have missed the forest for the trees.

In this volume, we meet the two military minds who saw into that forest most deeply. They were not distracted by the transitory and ephemeral. If their souls were lesser, their works possess the timelessness of Shakespeare's—and both texts have been more influential, for better and worse, than any play. If, like Shakespeare, they are often quoted out of context or misquoted, that is the fate of genius. If their thought has been misused, it is our fault and not theirs.

RALPH PETERS is a retired Army officer and the author of a noted book on strategy, *Fighting for the Future: Will America Triumph?* His essays and commentaries on military affairs have been both controversial and influential, and have gained a worldwide audience. He is also a critically acclaimed novelist.

Contents

SUN-TZU
THE ART OF WARFARE

THE FIRST

ENGLISH TRANSLATION

INCORPORATING THE

RECENTLY DISCOVERED

YIN-CH'ÜEH-SHAN TEXTS

TRANSLATED,

WITH AN INTRODUCTION

AND COMMENTARY,

BY ROGER T. AMES

CONTENTS

SUN-TZU: THE ART OF WARFARE: A TRANSLATION

ACKNOWLEDGMENTS

These are exciting days for the study of classical China. Over the past few decades, students of early China, encouraged by the continuing discovery of textual materials lost for millennia, have been working to bring this culturally formative period into sharper focus.

Owen Lock, a China specialist and also editor-in-chief of Del Rey Books (an imprint of Ballantine Books), has followed these developments closely, and has been keenly aware of their significance for understanding the cultural origins of the longest continuous civilization in human history. *Classics of Ancient China,* the series of which this volume, *Sun-tzu: The Art of Warfare,* is a part, has been created by Owen as a means of bringing this revolution to the attention of a broad reading public. His careful attention to this book at every stage, his detailed comments on draft manuscripts, and his informed enthusiasm for the subject itself have made the project enjoyable and exciting from the outset, and I am most grateful to him.

Robert G. Henricks, with his translation of *Lao-tzu: Te-tao ching,* inaugurated this series, and by the sustained quality of his scholarship has set a high standard for us all. He, like Owen, read the manuscript, and gave me comments that have made it a better book. I have relied on another author in our series, Robin D. S. Yates, who has been ever generous with his advice on military technology.

In Beijing, I benefited from personal conversations and from the important publications of Wu Jiulong (Wu Chiu-lung) and Li Ling. In Shenyang, the consummate scholar Zhang Zhenze (Chang Chen-tse) shared his work and his warmth.

I would also like to thank Tian Chenshan of the Center for Chinese Studies at the University of Hawaii, who worked closely with me on the preparation of the critical Chinese text. Several of my colleagues gave their time and their thoughts in reading different generations of the manuscript: I am grateful to Michael Speidel, Tao T'ien-yi, Elizabeth Buck, and Daniel Cole.

The *Chuang-tzu* tells us that none of us walks alone; each of us is "a crowd," a "field of selves." D. C. Lau, Angus Graham, Yang Yu-wei, Eliot Deutsch, David L. Hall, Henry Rosemont, Jr., and Graham Parkes—and my family: Bonnie, Jason, Austin, and Cliff—have all crowded around as I collected myself and did this work, and I have spent a lot of time with each of them. If an expression of gratitude could ever be at once sincere and selfish, it is so for me on this occasion.

ROGER T. AMES
Honolulu

Preface to the
Ballantine Edition

THE "NEW" SUN-TZU

The *Sun-tzu*, or *"Master Sun,"* is the longest existing and most widely studied military classic in human history. Quite appropriately, it dates back to the Warring States period (c. 403–221 B.C.), a formative phase in Chinese civilization when contributions in literature and philosophy were rivaled in magnitude and sophistication only by developments in an increasingly efficient military culture.

Over the course of the preceding Spring and Autumn period (c. 722–481 B.C.), scores of small, semiautonomous states had joined in an ongoing war of survival, leaving in its wake only the dozen or so "central states" (*chung-kuo*) from which present-day "China" takes its actual Chinese-language name.[1] By the fifth century B.C., it had become clear to all contenders that the only alternative to winning was to perish. And as these rivals for the throne of a unified China grew fewer, the stakes and the brutality of warfare increased exponentially.

During this period, warfare was transformed from a gentlemanly art to an industry, and lives lost on the killing fields climbed to numbers in the hundreds of thousands. Itinerant philosophers toured the central states of China, offering their advice and services to the contesting ruling families. Along with the Confucian, Mohist, and the Legalist philosophers who joined this tour was a new breed of military specialists schooled in the concrete tactics and strategies of waging effective warfare. Of these military experts, history has remembered best a man named Sun Wu from the state of Wu, known honorifically as "Sun-tzu" or "Master Sun."

A major reason why Master Sun has remained such a prominent force in the military arts is the military treatise *Sun-tzu: The Art of Warfare* (*Sun-tzu ping-fa*), that came to be associated with his name early in the tradition. Over the centuries, a library of commentaries has accrued around the text, and it has been translated into many, if not most, of the world's major languages.

Although there are several popular English translations of the *Sun-tzu*, several of which are discussed below, there are reasons why a new translation and study of the text is necessary at this time. The *Sun-tzu* offered here in this *Classics of Ancient China* series differs markedly from previous editions in several important respects.

In 1972 a new text of the *Sun-tzu* was uncovered in an archaeological find in Shantung province, containing not only large sections of the thirteen-chapter work that has come down to the present day, but also portions of five lost chapters of the *Sun-tzu*. All of these materials, previously unavailable to the student of the *Sun-tzu* text, were entombed as burial items sometime between 140 and 118 B.C.

This archaeological discovery means several things.

The English translation of the thirteen-chapter core text contained in Part I of this book has been informed by a copy of the *Sun-tzu* over a thousand years older than those on which previous translations were based. Prior to the excavations at Yin-ch'üeh-shan, the most recent text on which translations could be based had been a Sung dynasty (960–1279) edition. The supplemental five chapters that have been translated below as Part II and which in length are about 20 percent of the thirteen-chapter core text, are entirely new, and provide us with additional insights into both the content and the structure of the original text.

Part III of this book contains another window on the *Sun-tzu* provided by traditional encyclopedic works and commentaries containing references to the *Sun-tzu* dating back as early as the first century A.D. In length, this section adds more than 2,200 characters—over a third of the thirteen-chapter text. The encyclopedic works as a genre were generally compiled by gathering citations from the classical texts around specific topics such as the court, animals, plants, omens, courtesans, and so on. One recurring encyclopedia topic has been warfare. Ancient commentaries written by scholars to explain classical works have also on occasion referred to the *Sun-tzu*. To this new, expanded text of *Sun-tzu* I have added Part III. It contains materials from the encyclopedias, and from some of the earliest commentaries, that have been ascribed directly to Master Sun. Now that we can have greater confidence that the *Sun-tzu* was a larger, more complex text, there is good reason to believe that at least some of these attributions are authentic. One factor that had previously brought these materials into question was a difference in style: The thirteen-chapter text is narrative prose while the encyclopedic citations are, by and large, in dialogue form. Now that we have confirmed "outer" commentarial chapters of the *Sun-tzu* that are also structured as dialogues and that share many stylistic features with the encyclopedic citations, our reasons for being suspicious are less compelling. Although the authenticity of any one of these passages is still impossible to determine, the general correspondence between passages found in the reconstructed *Sun-tzu* and those preserved in the encyclopedias suggest that many of the citations might well be from the lost portions of this text.

I also have included in Part III a few fragments from a 1978 archaeological find in Ch'ing-hai province dating from the first or second centuries B.C. Six of the strips uncovered refer explicitly to "Master Sun," suggesting some relationship with the *Sun-tzu*.

In addition to working from the earliest text of the *Sun-tzu* now available and translating newly recovered portions of the *Sun-tzu*, I have tried to underscore the philosophic importance of this early work. Most accounts of the *Sun-tzu* have tended to be historical; mine is cultural. In the Introduction that precedes the translations, I have attempted to identify those cultural presuppositions that must be consciously entertained if we are to place the text within its own world view. In our encounter with a text from a tradition as different from ours as is classical China's, we must exercise our minds and our imaginations to locate it within its own ways of thinking and living. Otherwise we cannot help but see only our own reflection appearing on the surface of Chinese culture when we give prominence to what is culturally familiar and important to us, while inadvertently ignoring precisely those more exotic elements that are essential to an appreciation of China's differences. By contrasting our assumptions with those of the classical Chinese world view, I have tried to secure and lift to the surface those peculiar features of classical Chinese thought which are in danger of receding in our interpretation of the text.

In addition to the role that philosophy plays in enabling us to distinguish the classical Chinese world view from our own, it has another kind of prominence. We must explain the intimate relationship in this culture between philosophy and warfare: We need to say why almost every one of the early Chinese philosophers took warfare to be an area of sustained philosophical reflection and how the military texts are themselves applied philosophy.

This edition of the *Sun-tzu* seeks to satisfy the needs of the China specialist as well as those of the generalist. To this end, a critical Chinese-language text of the *Sun-tzu* has been reconstructed from the available redactions of the work on the basis of the most authoritative scholarship available, and included with the translation. This critical text is based upon the collective judgment of China's leading scholars in military affairs. For the generalist who seeks a better understanding of the text within its broader intellectual environment, I have provided the aforementioned philosophic overview.

Given the long and eventful history of the *Sun-tzu* itself, its introduction to the English-speaking world has been very recent and rather undistinguished.[2] In spite of some illuminating criticisms by D. C. Lau (1965) about the quality of the Samuel B. Griffith translation (1963), one would still have to allow that Griffith's rendition of the *Sun-tzu* and his commentary on various aspects of the text was a quantum improvement over what had gone before and to date has been our best effort to capture the text for the English-speaking world. The first prominent English translation by Captain E. F. Calthrop (1908) was indeed so inadequate that the vitriolic and undignified assault that it provoked from the well-known sinologist and translator Lionel Giles, then an assistant curator at the British Museum, discolors the reasonable quality of Giles's own attempt. While the Giles translation of 1910 is somewhat compromised by his unrelenting unkindnesses to poor pioneering Calthrop, it is still a scholarly first run on a difficult text and has the virtue of including a version of the *Sun-tzu* in Chinese.

Not much happened in the half century between Giles and Griffith. The strength of Griffith's work is that it is the product of a mature and intelligent military man. Samuel B. Griffith rose to the rank of Brigadier General in the United States Marine Corps, and wrote extensively and well on military matters from the battle of Guadalcanal to the Chinese People's Liberation Army. The many practical insights provided by Griffith's commentary are invaluable, and the quality of his translation is superior to Giles's and to recent popular attempts such as the Thomas Cleary translation (1988), informed as the latter is by neither practical military wisdom nor scholarship.

Finally, I have used the occasion of this publication to introduce the reader to recent Chinese archaeological excavations—especially those beginning at Yin-ch'üeh-shan in 1972—in acknowledgment of the importance of these discoveries for rethinking the classical period in China. From these sites we have recovered a cache of textual materials, including everything from new redactions of extant classics to works that have been lost for thousands of years. In many ways, each of these excavations captures one historical moment from centuries long past, and allows us, with imagination, to step back and steal a glimpse of a material China that has not otherwise been available to us. And from these ancient relics and textual materials we are able to reconstitute one cultural site to test our theories and speculations about a world that is no more.

ARCHAEOLOGY: A REVOLUTION IN THE STUDY OF EARLY CHINA

For students of culture concerned with the formative period of Chinese civilization and its early development, the discovery of lost textual materials, reported to the world in China's archaeological journals since the resumption of their publication in 1972, has been nothing short of breathtaking. The texts that have been recovered are of several kinds.

One category of documents is that of extant texts. The texts in this grouping are important because they have been spared the mishandling of those perhaps well-intentioned but not always well-serving editors and scribes responsible for a two-thousand-year transmission. For example, the December 1973 excavation of the Ma-wang-tui Tomb #3 in Hunan dating from c. 168 B.C. has yielded us two editions of the *Lao-tzu*[3] that predate our earliest exemplars by five hundred years.

D. C. Lau's recently published revision of his *Lao-tzu* translation based on a study of the Ma-wang-tui texts is an effective demonstration of the value of these new documents in resolving textual problems that have plagued commentators for these two thousand years.[4] And with Robert G. Henrick's *Lao-tzu: Te-tao-ching* in this same *Classics of Ancient China* series, the careful textual work goes on. In addition to the *Lao-tzu* and the *Suntzu: The Art of Warfare* included here in this second volume of the series, portions of the *Book of Changes* (*I Ching*), *Intrigues of the Warring States* (*Chan-kuo ts'e*), and *The Spring and Autumn Annals of Master Yen* (*Yen-tzu ch'un-ch'iu*) have also been unearthed, and are undergoing the same kind of detailed analysis. It has even been reported that a partial text of the *Analects* was recovered at a Ting county site in Hopei province in a 1973 find, but at this writing the material has not yet been made available to foreign scholars.[5]

Another category of document that has been recovered is that of extant texts that have long been regarded by scholars as being apocryphal; that is, works of doubtful authorship and authority. Portions of the military treatises *Six Strategies* (*Liu-t'ao*) and *Master Wei-liao* (*Wei-liao-tzu*), found in the cache of military writings in Tomb #1 of Yin-ch'üeh-shan, belong to this group. Of course, the discovery of these texts in a tomb dating from c. 140 B.C. is ample evidence of their vintage. There is also the collection of *Master Wen* (*Wen-tzu*) fragments found at the Ting county site that, by virtue of the important differences between the authoritative recovered text and the altered received *Master Wen*, promise to relocate this work centuries earlier than previously thought.[6]

A third important classification of texts is works concerning astronomy and prognostication that have hitherto been entirely unknown to us. The *Wind Direction Divination* (*Feng-chiao-chan*) and *Portent and Omen Divination* (*Tsai-i-chan*) documents, and the calendrical register for 134 B.C. recovered from Yin-ch'üeh-shan Tombs #1 and #2 are examples of this kind of material.

A fourth category of textual materials is that of works we have known about by name, but which in substance have been lost to us for the better part of two millennia. Undoubtedly the most important finds in this category are the four treatises collectively referred to as the *Silk Manuscripts of the Yellow Emperor* (*Huang-ti po-shu*)—Ching-fa, Shih-liu-ching, Ch'eng and Tao-yüan—that precede the second copy of the *Lao-tzu* on the cloth manuscripts recovered from Ma-wang-tui #3,[7] and the *Sun Pin: The Art of Warfare*[8] found in Yin-ch'üeh-shan #1. Annotated translations of both of these works are in progress, and are scheduled to appear in this same *Classics of Ancient China* series.

———

In addition to these works that are new to us in their entirety, there are also lost portions of extant texts that themselves have been transmitted in some edited or otherwise abbreviated form. For example, the *Sun-tzu: The Art of Warfare* from Yin-ch'üeh-shan, in addition to containing over 2,700 characters of the received thirteen-chapter text, approximately one third of its total length, also includes five chapters of supplemental materials that we have not seen until now. In this same find, there are also some forty-two bamboo strips that look like lost portions of the *Master Mo* (*Mo-tzu*).

The value of these newly discovered documents for extending and clarifying our knowledge of early Chinese civilization cannot be exaggerated. And the prospects of new finds are very good indeed, especially since several important locations are already known to us—for example, the late-third century B.C. tomb of the First Emperor of the Ch'in dynasty. While work on these known sites proceeds slowly, with scholars awaiting those advances in technology necessary to maximize preservation of the contents of the tombs, many other finds are being uncovered by accident in unrelated construction projects. Because of the impact that this archaeological material is having and is bound to have on the scholarship of classical China, a continuing familiarity with developments in this area has become an essential element in the training of every China classicist. Having said this, the nature of the material, the painstaking work necessary to recover and analyze it, and the real possibility of new discoveries at any time makes the work available on these documents necessarily tentative. For this reason, the present book is and can only be a progress report—an update on one particularly important find. The mission of our *Classics of Ancient China* series is to continue to make the substance of these finds available to the Western reader.

THE EXCAVATION AT YIN·CH'ÜEH·SHAN

Of the various archaeological excavations published to date that have brought this new textual material to light, the two most important at this writing are the Western Han (202 B.C.–A.D. 8) tombs at Ma-wang-tui in Ch'ang-sha, Hunan, discovered in late 1973, and those at Yin-ch'üeh-shan near Lin-i city in Shantung. Portions of *Sun-tzu: The Art of Warfare*, the focus of this study, were recovered in the latter excavation in 1972.

After the initial find, the Yin-ch'üeh-shan Committee devoted some two years of research to the 4,942 bamboo strips and strip fragments on which the texts were written before making the preliminary results of this work known to the world in February 1974. For details of the early reports, a catalog of the contents of these tombs, and the best efforts of contemporary scholarship to date the tombs and identify the occupants, see the appendix.

Perhaps the most significant and exciting textual material uncovered in Tomb #1 is the additional text of the extant *Sun-tzu: The Art of Warfare* and the large portions of the long-lost *Sun Pin: The Art of Warfare*.

The contemporary archaeologist Wu Chiu-lung, in a 1985 revision of the earlier 1974 report, summarizes the overall content of the bamboo strips in the following more general terms.[9]

The Han strips from Tomb #1 can largely be divided into those of which we have extant traditional texts and those where the texts have been lost. Since the text provided by the Han strips and the extant text are often different, it is not always possible to keep the two categories clearly separate. In the first category of extant texts there are:

1. *Sun-tzu: The Art of Warfare* (*Sun-tzu ping-fa*) and five chapters of lost text
2. *Six Strategies* (*Liu-t'ao*)—fourteen segments
3. *Master Wei-liao* (*Wei-liao-tzu*)—five chapters
4. *Master Yen* (*Yen-tzu*)—sixteen sections

In the second category of lost texts, there are:

5. *Sun Pin: The Art of Warfare* (*Sun Pin ping-fa*)—sixteen chapters
6. *Obeying Ordinances and Obeying Orders* (*Shou-fa shou-ling*)—ten chapters
7. Materials on discussions of government and discussions on military affairs—fifty chapters
8. Materials on *yin-yang,* calendrics, and divination—twelve chapters
9. Miscellaneous—thirteen chapters

In addition, there are many leftover fragments, and the process of reconstruction goes on.

The 1985 first volume of the Yin-ch'üeh-shan Committee's anticipated three-volume set includes reconstructed texts for all of the documents 1–6 listed above; the remaining materials will be made available with the promised publication of volumes II and III.

From Tomb #2 we have a calendar for the first year of the *yüan-kuang* reign period (134 B.C.) of Emperor Wu (r. 141–87 B.C.) of the Western Han. It contains a total of thirty-two strips. The first strip records the year, the second strip lists the months, beginning with the tenth month and continuing until the following ninth month—a total of thirteen months. Strips three to thirty-two then record the days, listing the "stem and branches" designations for the first to the thirtieth day of each month. Together, these thirty-two strips constitute a complete calendar for the year.

There are varying opinions among scholars as to the dating of the texts themselves. From the archaeological evidence (see Appendix), we can estimate that Tomb #1 dates from between 140 and 118 B.C., and Tomb #2 dates from between 134 and 118 B.C. However, the dates at which the texts were transcribed would, of course, be earlier than the tombs in which they were buried, and the dates at which they were first compiled, earlier yet.

One potential clue as to the dates of the copied texts is the custom of avoiding the characters used in the emperor's name in texts transcribed during an emperor's reign. The Western Han, however, was not strict in its observance of such imperial taboos. The names of emperors Hui, Wen, and Wu all occur on the strips, and there are even instances of the less common characters of Empress Lü and Emperor Ching. The most that can be said is that these texts from Yin-ch'üeh-shan seem to observe the taboo on the first emperor of the Han dynasty, Liu Pang (r. 206–194 B.C.), avoiding the character *pang,* and using *kuo* (which also means "state") instead, with one exception in the supplemental strips of Chapter 4 of *Sun Pin: The Art of Warfare,* "T'ien-chi Inquires About Battlefield Defenses," which might have been an oversight.

The contemporary scholar Chang Chen-tse concludes that the strips must actually have been written during the dozen years Liu Pang was on the throne.[10] Other scholars are more cautious, insisting the taboos are inconclusive evidence. Wu Chiu-lung, for example, discounts the taboo factor, and instead compares the style of writing with other recent finds.[11] On this basis, he estimates the Yin-ch'üeh-shan texts were copied in the early years of the Western Han dynasty sometime during the period covered by the reigns of Emperor Wen (who ascended the throne in 179 B.C.), Emperor Ching, and the beginning years of Emperor Wu (who began his reign in 141 B.C.).

THE "ONE OR TWO
'MASTER SUNS' " DEBATE

Although fragmentary, the sizable portion (over one third) of the transmitted *Sun-tzu: The Art of Warfare* that was unearthed at Yin-ch'üeh-shan is the same in general outline as the received standard Sung dynasty edition: *Sun-tzu with Eleven Commentaries (Shih-i chia chu Sun-tzu).* This is significant because it demonstrates that by the time this text was copied sometime in the second century B.C., the thirteen-chapter "classic" of *Sun-tzu* had already become fixed as a text. Where the recovered text differs from the received Sung dynasty edition, it is usually more economical in its language. It frequently uses characters without their signifiers or with alternative signifiers, and homophonous loan characters have often been substituted for the correct forms—familiar features of those early writings that have been unearthed—suggesting perhaps a lingering resistance to the standardization of the characters promoted by the Ch'in dynasty (221–206 B.C.) some years earlier, and, further, the prominent role of oral transmission in the tradition. Where the Sung dynasty edition of the thirteen-chapter *Sun-tzu* is generally a fuller and more intelligible document, the opportunity to challenge problematic passages with a text dating from a full millennium earlier adds important new evidence for reconstructing a critical text. The other extraordinary value of the Yin-ch'üeh-shan text lies in the sixty-eight pieces constituting five partial chapters that had previously been lost. A version of one of these chapters, translated below in Part II as [An Interview with the King of Wu], was possibly reworked by Ssu-ma Ch'ien (c. 145–86 B.C.) into his biographical account of Master Sun. These additional sections are representative of the kinds of commentarial literature that would accrue over time around a classic once it had found its canonical form.

Sun Pin: The Art of Warfare, although only a partial and fragmentary text, still compares in length to the thirteen-chapter *Sun-tzu*. In the earliest published reports of the archaeological dig at Yin-ch'üeh-shan in 1975, the thirty fragmentary chapters identified as the *Sun Pin* contained some 8,700 characters. The *Sun-tzu*, for the sake of comparison, is approximately 6,000 characters. The revised sixteen-chapter text of the *Sun Pin* published in 1985 still provides sufficient detail to give us a reasonably clear picture of a work that has been known only by title for nearly two thousand years.

The fact that these two texts were recovered at the same time from the same tomb helps to resolve a question that has hovered over the two militarist treatises for centuries. Until this archaeological find, only the core chapters of one of the two texts had been available—the thirteen-chapter *Sun-tzu: The Art of Warfare*. Over the centuries all manner of speculation has arisen with respect to the authorship and even authenticity of the work, and particularly concerning its relationship to the second militarist text, *Sun Pin: The Art of Warfare*.

From the historical record, it is clear that scholars of the Han dynasty distinguished between the two military figures and their treatises, and that the debate among scholars as to whether there was one "Master Sun" or two (or one *Sun-tzu* text or two) is a post–Han dynasty phenomenon that arose after *Sun Pin: The Art of Warfare* was lost.

The *Historical Records* (*Shih-chi*), completed in 91 B.C., contains a biographical account that clearly separates Sun Wu (c. 544–496 B.C.), a contemporary of Confucius at the end of the Spring and Autumn period in the service of the state of Wu, and his descendent, Sun Pin (c. 380–316 B.C.), a contemporary of Mencius who flourished during the middle years of the fourth century B.C. in the employ of Ch'i.[12] In the biographies of these two persons who were separated by nearly two centuries, the *Historical Records* mentions both the *Sun-tzu: The Art of Warfare* in thirteen chapters (the same number of chapters as our extant Sung dynasty text), and *The Art of Warfare* attributed to Sun Pin. The latter text eventually disappeared from sight until portions of it were recovered in 1972.

Further, the "Record of Literary Works" (*Yi-wen chih*) of the *History of the Han Dynasty*, a catalog of the imperial library completed during the first century A.D., records the existence of two distinct texts:

1. *Sun-tzu of Wu: The Art of Warfare* in eighty-two chapters and nine
   scrolls of diagrams. Yen Shih-ku's (581–645) commentary states:
   "This refers to Sun Wu."
2. *Sun-tzu of Ch'i: The Art of Warfare* in eighty-nine chapters and four
   scrolls of diagrams. Yen Shih-ku comments: "This refers to Sun Pin."

In addition to this specific historical information, there are further references to the two figures in the late Warring States and Han dynasty corpus. In spite of the fact that these sources often refer to both men as
"Master Sun" ("Sun-tzu"), we are usually able to distinguish between
them. For example, in the *Intrigues of the Warring States* (*Chan-kuo ts'e*),
edited by Liu Hsiang (77–6 B.C.) in the late first century B.C., and throughout Ssu-ma Ch'ien's *Historical Records* as well, there are references to
"Master Sun" that, from context and historical situation, can only refer to
Sun Pin.[13] In the *Spring and Autumn Annals of Master Lü* (*Lü-shih ch'un-
ch'iu*), probably completed c. 240 B.C., reference is made specifically to Sun
Pin: "Sun Pin esteemed strategic advantage [*shih*]." This passage is glossed
by the late Eastern Han commentator, Kao Yu (fl. 205–212), who states:
"Sun Pin was a man of Ch'u [*sic*] who served as a minister in Ch'i working
out strategy. His eighty-nine chapters deal with the contingencies surrounding strategic advantage [*shih*]."[14]

From the many Ch'in and Han dynasty references to these two texts, it
would seem that at least until the end of the Eastern Han (A.D. 220), both
texts were extant and were clearly distinguished by scholars of the time.
Since the *History of the Later Han* (*Hou-Han-shu*), compiled over the third to
fifth centuries A.D., does not include a catalog of the court library, the next
logical place to expect a record of the two texts is the "Record of Classics
and Documents" (*Ching-chi chih*) in the *History of the Sui Dynasty* (*Sui-shu*)
compiled in the seventh century.[15] The total absence of any reference to
Sun Pin: The Art of Warfare in the *History of the Sui Dynasty* together with the
fact that Ts'ao Ts'ao (155–220), enthroned as King of the state of Wei
(220–265) during the Three Kingdoms period, makes no mention of it in
his commentary on the core thirteen chapters of *Sun-tzu: The Art of Warfare*, suggests rather strongly that *Sun Pin: The Art of Warfare* disappeared
sometime between the last years of the Eastern Han dynasty in the third
century, and the beginning of the Sui dynasty in the sixth century.

In spite of the many references and allusions to the two distinct texts
in the Ch'in and Han dynasty literature, from the southern Sung dynasty
(1127–1279) down to the present, prominent commentators such as Yeh
Shih, Ch'en Chen-sun, Ch'uan Tsu-wang, Yao Nai, Liang Ch'i-ch'ao, and

Ch'ien Mu have questioned both the authorship and the vintage of *Sun-tzu: The Art of Warfare*. Doubt concerning the historicity of Sun Wu was certainly reinforced by the fact that the *Commentary of Master Tso* (*Tso-chuan*), one of China's oldest narrative histories, which dates from the turbulent fourth century B.C. and which otherwise evidences great delight in recounting military events, never refers to him at all. Some of these later scholars have questioned the historicity of the strategist Sun Wu; others have claimed *Sun-tzu: The Art of Warfare* perhaps originated with Sun Wu, but was edited and revised by his mid-fourth-century descendant, Sun Pin. Some have even suggested that Ts'ao Ts'ao, canonized as the "Martial King," compiled *Sun-tzu: The Art of Warfare* on the basis of earlier works before appending his own commentary.

The unearthing of these two texts in the same Han dynasty tomb at Yin-ch'üeh-shan goes some way to resolving the "one or two 'Master Suns' " dispute. Firstly, there are unquestionably two distinct texts, both extant in the second century B.C. Secondly, the discovery supports the traditional opinion that there were in fact two "Master Suns"—Sun Wu and Sun Pin—and further lends credence to those historical records that offer such an opinion.

There is a real danger here, however, of pursuing the wrong questions and, in so doing, losing sight of what might be more important insights. We really must ask, for example: What do we mean by the *Sun-tzu* as a text, or even "Sun-tzu" as a historical person? The quest for a single text authored by one person and a preoccupation with historical authenticity is perhaps more a problem of our own time and tradition. There is a tendency on the part of the contemporary scholar to impose anachronistically our conceptions of "text" and "single authorship" on the classical Chinese artifact and, by doing so, to overlook the actual process whereby a text would come into being. This is a particular concern in dealing with cultures where oral transmission was a significant factor and in which authorship tended to be cumulative and corporate.

I am suggesting that works such as the *Sun-tzu* might have emerged more as a process than as a single event, and those involved in its authorship might well have been several persons over several generations. This Yin-ch'üeh-shan find reveals what I take to be a historical moment in the process. There is a redaction of the core thirteen-chapter *Sun-tzu* that certainly predates the imperial editing of the text undertaken by Liu Hsiang at the end of the first century B.C., and which corroborates the several early references to a thirteen-chapter work. The fact that Sun Wu is referred to honorifically as "Master Sun" (translating the *"tzu"* in "Sun-tzu" as "Master") is evidence the text was not written by Sun Wu himself, and is also an indication the text was compiled and transmitted by persons who held Sun Wu in high regard as a teacher and as an authority on military matters. We can be quite sure this thirteen-chapter document was not composed by Sun Wu, and was probably the product of some later disciple or disciples, probably several generations removed from the historical Sun Wu. The text itself is at the very least a secondhand report on what Master Sun had to say about military strategy.

In Part II, "The Questions of Wu" chapter refers directly to the events surrounding the dissolution of the state of Chin that climaxed in 403 B.C. to begin the Warring States period. Even though this chapter belongs to the "outer" text of the *Sun-tzu,* which we must assume to be later commentary, to have Master Sun rehearsing the incidents that followed from the collapse of Chin places this discussion well into the fourth century B.C. at the earliest. It is clearly an anachronism.

There is also a revealing discrepancy between the Sung dynasty edition of the *Sun-tzu* and the Han strips version that might be of some significance in dating the actual compilation of the text. The last paragraph of Chapter 13 in the Sung dynasty edition reads:

> Of old the rise of the Yin (Shang) dynasty was because of Yi Yin who served the house of Hsia; the rise of the Chou dynasty was because of Lü Ya who served in the house of Shang. Thus only those far-sighted rulers and their superior commanders who can get the most intelligent people as their spies are destined to accomplish great things.

The same passage in the Han strips version can be reconstructed as:

> [*The rise of the*] Yin (Shang) dynasty [*was because of Yi Yin*] who served the house of Hsia; the rise of the Chou dynasty was because of Lü Ya who served [*in the house of Shang*]; [*the rise of the state of...*] was because of Commander Pi who served the state of Hsing; the rise of the state of Yen was because of Su Ch'in who served the state of Ch'i. Thus only those far-sighted rulers and their superior [*commanders who can get the most intelligent people as their spies are destined to accomplish great things*].

While we have no information on the Commander Pi who served the state of Hsing, we do know that Su Ch'in was a Warring States military figure and statesman who, flourishing in the early years of the third century B.C., lived more than a century and a half after the historical Sun Wu.[16] Since Su Ch'in, in fact living a generation removed from Sun Pin, belongs to an era long after the historical Sun Wu, reference to him in this passage would, on the surface, suggest that the *Sun-tzu* is a text from the hand of a much later disciple or disciples. Alternatively (and this is the opinion of many, if not most, contemporary scholars), this passage in the Han strips text is a later interpolation.

In the introduction to his translation of the *Sun-tzu*, Samuel Griffith identifies several anachronistic references within the text itself that in sum push the date of the text well into the Warring States period: allusions to the scale of warfare, the professionalization of the soldier, the separation of aristocratic status and military rank, the deployment of shock and elite troops, the suggestion that rank-and-file troops as well as officers wore armor, the widespread use of metal currency, and so on.[17] For the most part, Griffith's arguments that *Sun-tzu* was compiled sometime in the period 400–320 B.C. are persuasive. The two references to the crossbow in the *Sun-tzu* that Griffith takes to be anachronistic, however, are probably not an issue. In their recent study of early military technology, Joseph Needham and Robin Yates have concluded that the crossbow was probably introduced into China by non-Han peoples in the middle Yangtze region as early as 500 B.C.[18]

On the basis of the Yin-ch'üeh-shan find, we can speculate that the eighty-two chapter *Sun-tzu*, a text including both the "inner" thirteen-chapter core and the "outer chapters" represented by fragments recovered in this archaeological dig, is assuredly a composite work—the product of many hands and many voices that accrued over an extended period of time. The role of oral transmission cannot be discounted. The nature and economy of the written text suggest its contents might have originally been discussion notes, copied down, organized, and edited by several generations of students, as was the case in the compilation of the *Analects of Confucius*. These materials were probably gathered together, collated, and subjected to a process of editorial refinement important for economical transmission—that is, a deletion of redundancies and marginally relevant references, the removal of historical detail that might bring the antiquity of the text into question, and so on. The main structural difference between the *Sun-tzu* and the *Analects of Confucius* is that the *Sun-tzu* is by and large organized thematically, while the order of the *Analects* is more random, with passages only sometimes being grouped—loosely—around a discernible theme or idea. The arrangement of the *Sun-tzu* is more linear, sequential, and thematic than the *Analects*, a characteristic increasingly in evidence in the texts compiled in the late fourth and third centuries B.C.

Another formal characteristic of the *Sun-tzu* that recommends a later rather than an earlier dating is the sustained dialogue structure of the newly recovered "outer" chapters. This distinctive feature suggests that these chapters were composed considerably later than the core thirteen-chapter text.

The overall congruencies between the Yin-ch'üeh-shan *Sun-tzu* and our received thirteen-chapter redactions suggest strongly that, by the time of the entombment of this text, the core text of *Sun-tzu* had already been edited into something closely resembling its present form, and thus had already become "fixed." Given the early Han dynasty date of this copy, this canonization of the *Sun-tzu* is what one would anticipate, following as it does the same pattern as other important pre-Ch'in works. In his examination of the *Lao-tzu*, D. C. Lau identifies the century between the writing of the *Master Han Fei* (c. 240 B.C.) and the compilation of the *Master of Huai Nan* (140 B.C.), as the time in which the *Lao-tzu* settled into its present form. Lau offers the following explanation for the congealing process that seems to have occurred at this particular historical juncture:

> It seems then that the text [the *Lao-tzu*] was still in a fluid state in the second half of the third century B.C. or even later, but by the middle of the second century B.C., at the latest, the text already assumed a form very much like the present one. It is possible this happened in the early years of the Western Han Dynasty. There is some reason to believe that in that period there were already specialist "professors" (*po shih*) devoted to the study of individual ancient works, including the so-called philosophers (*chu tzu*), as distinct from the classics (*ching*).... This would cause the text to become standardized....[19]

In 213 B.C., the Ch'in court at the urging of the Legalist counsellor, Li Ssu, decreed that all existing literature representing the writings of the various philosophical schools, and particularly the Confucian classics, be turned over to the governors of the commanderies to be burned. The "burning of the books," as this event has come to be called, might well have made the reclamation of the classical corpus a priority item for the newly established Han dynasty a few years later.

At Yin-ch'üeh-shan, in addition to the core thirteen-chapter text, however, representative fragments of five commentarial chapters were also found that are very different in structure and style. We can speculate that these extensions of the core chapters were probably appended by later generations in the Sun clan lineage (*chia*) to explain and elaborate what the passage of time had made increasingly unclear. These "outer" chapters of the text were again probably authored by the disciples and descendants of Sun Wu, but at some greater distance in time from the Master than the core chapters.

The central militarist (and later, Legalist) tenet that there are no fixed strategic advantages (*shih*) or positions (*hsing*) that can, in all cases, be relied upon to achieve victory, must be considered when we decide what kind of coherence we can expect from what was a growing body of work. Consistent with the stated principle that each situation must be taken on its own terms, different periods with different social, political, and material conditions would require different military strategies to be effective. The military philosophers, like any school that continued over time, would necessarily have to reflect changing historical conditions in the articulation of their doctrines. Reference to a specific historical site and occasion softens the otherwise more rigid demands of theoretical abstractions and categorical imperatives.

On the basis of the shared tenet that different circumstances require different strategies for success, we can make the claim that even where the *Sun-tzu* and the *Sun Pin* seem clearly to contradict each other, they are still entirely consistent. For example, *Sun-tzu* is explicit in discouraging the strategy of attacking walled cities:

> Therefore the best military policy is to attack strategies; the next to attack alliances; the next to attack soldiers; and the worst to assault walled cities. Resort to assaulting walled cities only when there is no other choice.[20]

Sun Pin, on the contrary, regards siege as a viable strategy.[21]

In what at present are regarded as supplemental chapters to the core text, the *Sun Pin* even recommends assaulting "female" fortifications.[22] The distinction between "male" and "female" fortifications is illustrated in the following terms:

> A walled fortification situated in the midst of a low-lying swamp which, even without high mountains or deep valleys around it, is still surrounded on all sides by crouching hills, is a male fortification, and cannot be attacked. [*A walled fortification in which*] the troops have access to fresh, flowing water [*has a vital water supply, and cannot be attacked*]. A walled fortification which has a deep valley in front of it and high mountains behind is a male fortification, and cannot be attacked. A walled fortification within which there is high ground while beyond its walls the land falls away is a male fortification and cannot be attacked. A walled fortification within which there are crouching hills is a male fortification, and cannot be attacked.

When troops on the march in setting up camp for the night are not in the vicinity of some source of water, their morale will flag and their purposes will be weakened, and they can be attacked. A walled fortification which has a deep valley behind it and no high mountains on its flanks is a weak fortification and can be attacked. [*An army camped*] on the ashes of scorched land is on dead ground, and can be attacked. Troops who have access only to standing pools of water have dead water, and can be attacked. A walled fortification situated in the midst of broad swamplands without deep valleys or crouching hills around it is a female fortification, and can be attacked. A walled fortification which is situated between two high mountains without deep valleys or crouching hills around it is a female fortification, and can be attacked. A walled fortification which fronts a high mountain and has a deep valley to the rear, which is high in front but falls away to the rear, is a female fortification, and can be attacked.

This seeming inconsistency between *Sun-tzu* and *Sun Pin* is understandable if we factor into our assessment developments in military technology that made siege more effective, and the development of walled cities as centers of wealth and commerce that made siege more profitable.

Chariots, ineffective against high walls, were a central military technology for Master Sun; a cavalry equipped with crossbows was an innovation important to Sun Pin. Do we conclude that we have competing opinions here, or can such seeming inconsistencies be adequately explained by the assertion, shared by both texts, that different situations require different strategies for success?

Somewhere in this process of the eighty-two-chapter *Sun-tzu* being composed, transcribed, edited, and transmitted to succeeding generations, the *Sun Pin* emerges as a second text that, while seeming to belong to the *Sun-tzu* lineage, at the same time achieved an increasingly significant degree of distinction and, in due course, independence. The differentiation of *Sun Pin* from the then still-growing *Sun-tzu* corpus was at least in part due to the military successes of Sun Pin himself that became an integral part of the historical record and set his textual materials off from the earlier *Sun-tzu*. Having achieved this relative independence, *Sun Pin: The Art of Warfare* then probably followed the pattern of the *Sun-tzu: The Art of Warfare* in first becoming "fixed" as a core text, and then accruing a commentarial tradition around itself. The *History of the Han Dynasty* reports that the *Sun Pin* comprised eighty-nine chapters, probably a mixture of "inner" core chapters and later commentarial appendixes.

The sixteen-chapter *Sun Pin* that has been reconstructed from the Yinch'üeh-shan find differs from the more consistently theoretical *Sun-tzu* by beginning from chapters that report on specific historical incidents. It then generalizes from these battles and strategy sessions to outline certain basic tenets of military theory. In the Yin-ch'üeh-shan Committee's first report on the *Sun Pin* (1974), it had reconstructed the *Sun Pin* text in thirty chapters. In the committee's 1985 review of these materials, one reason given for reducing the thirty-chapter text to sixteen is that some chapters that are not demonstrably *Sun Pin* might well belong to the "outer chapters" of the *Sun-tzu*. The line separating the two texts is, at best, often unclear. In fact, it is possible that the lineage of authors who contributed to the *Sun-tzu* might well have included Sun Pin himself, and some of the materials that came to constitute the *Sun Pin: The Art of Warfare* might have, at one time and in some form or another, been part of the "outer chapters" of the *Sun-tzu*. Indeed, the entire body of textual materials might, under different circumstances, have been revised and edited to constitute the one *Sun-tzu*. Instead, the materials were divided to become the two separate treatises on warfare, the *Sun-tzu* and the *Sun Pin*.

How else has the Yin-ch'üeh-shan archaeological dig shed light on the early years of the Han dynasty? In addition to the value of the Han strips in assessing the historicity of the classical corpus, they are an important resource for investigating the changing forms of written Chinese characters, especially during the early years in which the clerical form (*li shu*) was being institutionalized. The strips also offer up new loan characters, and new insights into rhyme patterns current in the formative period of Chinese civilization.

Perhaps the most important consequence of the Yin-ch'üeh-shan find is not the specific resolution of the "two Master Suns" debate, but a more general principle: That is, we must take the process of textual "growth" into account and give greater credence to the traditional dating of these early works. In addition to the *Sun-tzu* and *Sun Pin*, we have recovered portions of other texts previously dismissed as apocryphal. The fact that the fragments of the *Master Yen, Master Wei-liao,* and *Six Strategies* all have text very similar to the received redactions suggests a greater respect is due traditional claims of authenticity.

SUN WU AS A HISTORICAL PERSON

According to the biography in the *Historical Records*, the first comprehensive history of China completed in 91 B.C., Sun Wu was born in the state of Ch'i (in the area of present-day Shantung province) as a contemporary of Confucius (551–479 B.C.) at the end of the Spring and Autumn period, and came into the employ of King Ho-lu of Wu (r. 514–496 B.C.) as a military commander. He gained an audience with King Ho-lu who, after having read the thirteen chapters of the *Sun-tzu: The Art of Warfare*, summoned him to court. Putting Sun Wu to the test, the King requested that Sun Wu demonstrate his military skill by conducting a drill using the women of his court. An alternative version of this same story was reclaimed in the Yin-ch'üeh-shan dig, and has been translated below in Part II as "[An Interview with the King of Wu]." This must be one of the best-known anecdotes in Chinese military lore:

The King...dispatched 180 of his court beauties from the palace. Sun Wu divided them into two contingents, placed the King's two favorite concubines as unit commanders, and armed them all with halberds. He then instructed the women, "Do you know where your heart, your right and left hands and your back are?" The women replied, "We do indeed." "When I say 'Front'," he said, "face in the direction of your heart; when I say 'Left,' face in the direction of your left hand; when I say 'Right,' face in the direction of your right hand; when I say 'Back,' face in the direction of your back." The women agreed. Having set out the various drill commands, he then laid out the commander's broad-axe, and went through and explained his orders several times. Thereupon, he drummed for them to face right, but the women just burst into laughter.

Master Sun said, "Where drill orders are less than clear and the troops are not familiar enough with the commands, it is the fault of their commander." Again going through and explaining his orders several times, he then drummed for them to face left. Again the women just burst into laughter.

Master Sun addressed them, "Where the drill orders are less than clear and the troops are not familiar enough with the commands, it is the fault of their commander. But where they have already been made clear and yet are not obeyed, it is the fault of their supervising officers." He then called for the beheading of the right and left unit commanders.

The King, viewing the proceedings from his balcony, saw that Master Sun was in the process of executing his two favorite concubines, and was appalled. He rushed an attendant down to Master Sun with the command, "I am already convinced of the Commander's ability in the use of the military. If I don't have these two concubines, my food will be tasteless. It is my wish that you do not behead them."

Master Sun responded, "I have already received your mandate as Commander, and while I am in command of the troops, I am not bound by your orders." He thereupon beheaded the two unit commanders as an object lesson.

Appointing the next two in line as the new unit commanders, he again drilled them. Left, right, front, back, kneel, stand—at every turn the women performed with the precision of the square and compass, and did not dare to utter a sound. Master Sun thereupon sent a messenger to report to the King, "The troops have now been properly disciplined. Your Majesty can come down to inspect them. Do as you like with them—you can even send them through fire and water!"

The King of Wu replied, "The Commander may return to his chambers to rest. I have no desire to descend and review the troops."

Master Sun said, "The King is only fond of words, but has no stomach for their real application." At this, Ho-lu knew Master Sun's ability at military affairs, and ultimately made him his Commander. That Wu crushed the strong state of Ch'u to the west and occupied its capital at Ying, intimidated Ch'i and Chin to the north and rose to prominence among the various states, was in good measure due to Master Sun's military acumen.[23]

Elsewhere in his *Historical Records*, Ssu-ma Ch'ien records Sun Wu's counsel to King Ho-lu in the campaign against the state of Ch'u.[24] Following Sun Wu's advice, the state of Wu was able to occupy the Ch'u capital within six years. Evident from these historical reports is the fact that Sun Wu was not only a military tactician, but also a very capable strategist who was able to lead his state to victory.

Although the details of Sun Wu's life are for the most part lost, the place of his *Sun-tzu: The Art of Warfare* as *the* fundamental work in classical military literature is unassailable. The military chapters of the *Book of Lord Shang*, a Legalist text dating primarily from the third century B.C.,[25] are heavily indebted to material adapted from the *Sun-tzu*.[26] The *Master Hsün*'s "Debate on Warfare" treatise is in fact a very specific Confucian assault mounted by Master Hsün (c. 320–235 B.C.) against those prevailing military concepts and attitudes clearly drawn from the *Sun-tzu: The Art of Warfare*.[27] The Legalist Han Fei, a student of Master Hsün, reports on the popularity of the *Sun-tzu: The Art of Warfare* in a world that had been scorched with centuries of unrelenting military strife: "Everyone in the realm discusses military affairs, and every family keeps a copy of the *Master Wu* and the *Sun-tzu* on hand."[28] The "Military Strategies" treatise in the *Master of Huai Nan*, certainly one of the most lucid statements on early military ideas, evidences an intimate familiarity with the *Sun-tzu: The Art of Warfare* and builds upon it. From the centuries leading up to the founding of imperial China, over its two-millennia-long career, and during the decades of unprecedented military intensity in the twentieth century, the

Sun-tzu: The Art of Warfare has maintained its status as the world's fore-most classic on military strategy.

THE RECONSTRUCTED *SUN-TZU: THE ART OF WARFARE*

The 1985 volume of the Yin-ch'üeh-shan Committee's reconstruction of the *Sun-tzu* divides it into two parts. Part I includes the remnants of the thirteen-chapter edition (over 2,700 characters) with representative text from all of the chapters except Chapter 10, "The Disposition of the Terrain" (*ti-hsing*); Part II comprises five additional chapters unknown to us previously, one of which relates the story found in the *Historical Records* biography of Master Sun where Sun Wu disciplines the concubines of King Ho-lu of the state of Wu.

There are also six fragmentary segments of wood that, when pieced together, constitute a table of contents for the scrolls of bamboo strips containing the core *Sun-tzu* text.

From appearances, the bamboo manuscript was divided into two portions, with a table of contents and a character tally for each one. From what remains of the table of contents, we can still identify eight chapter titles of what, from all appearances, was a list of thirteen. This would suggest the table of contents of the *Sun-tzu* was the same then as our present Sung dynasty edition, although there seem to be discrepancies in the order of the chapters. The comparative similarity between the recovered text and the traditional text means that *Sun-tzu* was not edited into its present thirteen chapters by later commentators such as Ts'ao Ts'ao (155–220) or Tu Mu (803–852) as traditionally thought, but had this arrangement much earlier.

Part II's five newly recovered chapters, a total of over 1,200 characters or some additional 20 percent of the received text, have a commentarial relationship to the thirteen-chapter core. "The Questions of Wu" chapter records a dialogue between Master Sun and the King of Wu on the state of Chin and on governmental policies. Although this dialogue format is not found in the existing thirteen-chapter text, it is familiar from the long citations of *Sun-tzu* preserved in the T'ang dynasty (618–907) encyclopedic work on laws and institutions, the *T'ung-tien*, translated below in Part III of the present volume.

"The Yellow Emperor Attacks the Red Emperor" chapter begins with the "Master Sun said..." formula, and seems related in content to Chapter 9, "Deploying the Army" (*hsing chün*), which also alludes to the Yellow Emperor's victory over the emperors of the four quarters.

"[The Four Contingencies]" chapter further elaborates on sections of Chapter 8, "Adapting to the Nine Contingencies"; the fragments of "The Disposition [of the Terrain] II" seems related in content to Chapter 9, "Deploying the Army," and to Chapter 11, "The Nine Kinds of Terrain."

These chapters are all appended to the present text because, like much of the materials attributed to *Sun-tzu* recovered from other sources in the early corpus and the later encyclopedic works, they too elaborate on and explain the core thirteen-chapter text.

As we saw above, the "Record of Literary Works" (*Yi-wen chih*) of the *History of the Han Dynasty* lists in the category of "Military Strategists" the *Sun-tzu of Wu: The Art of Warfare* in eighty-two chapters and nine scrolls of diagrams. This certainly refers to a larger compilation than the familiar thirteen-chapter text. In the commentary of Chang Shou-chieh (fl. A.D. 737) to the biography of Master Sun in the *Historical Records,* he comments: "The Ch'i-lu of Juan Hsiao-hsü of the Liang dynasty (502–556) lists the *Sun-tzu: The Art of Warfare* in three scrolls. The thirteen-chapter text is the first scroll, and there are also a second and a third scroll."[29] It is possible that the last two scrolls were comprised of explanatory chapters that included among them the lost text recovered on these bamboo manuscripts.

The contemporary scholar Li Ling, in describing the compilation of the eighty-two-chapter *Sun-tzu,* compares it to the original inner and outer books of the *Mencius.*[30] Our present *Mencius* was edited by the Han dynasty commentator Chao Ch'i (d. A.D. 201), who expunged four "outer books" that he took to be the spurious work of a later age rather than the authentic work of Mencius. Ch'ing dynasty (1644–1911) collections have been made of passages attributed to Mencius but not contained in our present text, which might be remnants of those lost "outer" books.

Another analogous compilation is the *Master Kuan* (*Kuan-tzu*) (compiled c. 250 B.C.). Although more explicitly commentarial, the chronologically later "explanatory" (*chieh*) chapters of the *Master Kuan* serve a function similar to that of the "outer chapters" of the *Sun-tzu.*

It is most unlikely that the eighty-two-chapter *Sun-tzu* was one text by a single author. A plausible story is that the expository thirteen-chapter *Sun-tzu* differed substantially in date, content, and structure from the later outer books. Following the editing of the father-and-son Han dynasty bibliographers, Liu Hsiang (77–6 B.C.) and Liu Hsin (d. A.D. 23), the inner thirteen-chapter core and the outer chapters were brought together in the eighty-two-chapter text. The military strategist and scholar Ts'ao Ts'ao (155–220) wrote commentary only on the thirteen inner chapters, and subsequently, the outer chapters, supplementary to the inner chapters, were lost. Much of what has been preserved of the outer chapters recovered from the Yin-ch'üeh-shan (*Sun-tzu: Part II*, below) and from the later encyclopedic works and Ta-t'ung county archaeological finds (*Sun-tzu: Part III*, below) does differ in style and content from the thirteen inner chapters, although most of these materials bear a recognizable commentarial relationship.

ANALYSIS OF *SUN-TZU: THE ART OF WARFARE*

WISDOM AND WARFARE

Discussion of military affairs is pervasive in early Chinese philosophical literature. This in itself is a fair indication of the perceived importance of warfare as a topic of philosophic reflection in China, a concern that is not paralleled in Western philosophical literature. It is a seldom-advertised fact that many if not most of the classical Chinese philosophical works contain lengthy treatises on military thought: the *Master Mo, Master Hsün, Master Kuan,* the *Book of Lord Shang,* the *Spring and Autumn Annals of Master Lü,* the *Master of Huai Nan,* and so on. In addition, other central texts such as the *Analects, Mencius, Lao-tzu, Master Han Fei,* and the recently recovered *Silk Manuscripts of the Yellow Emperor* contain extended statements on military thought. In fact, in the imperial catalog included in the *History of the Han Dynasty,* the military writers are listed under the "philosophers" (*tzu*) classification.[31] It might be fair speculation to say that, in the philosophical literature of the classical period, a text would be perceived as less than complete if the conversation did not at some point turn to an extended discussion of military strategies and even tactics.

This abiding interest in military affairs is a particularly curious situation for a culture in which warfare is neither celebrated nor glorified, and in which military heroism is a rather undeveloped idea. When it comes to social status, the warrior in China did not have the benefit of having Greek and Roman forebears.[32] Even in those Chinese treatises that deal exclusively with military affairs, we generally find the same paternalistic concern for the welfare of the people familiar to us from the Confucian literature, and an explicit characterization of warfare as an always unfortunate last resort. There is no self-promoting militarism.

The question that emerges, then, is this: Given the general disparity in status between civil and martial virtue in the Chinese tradition, how do we explain the intimate, even interdependent relationship between the occupations of philosopher and warrior assumed by the early Chinese thinkers?

The military experience, early and late, was important in the culture. Armies up to the late Spring and Autumn period were still constituted by aristocratic families living in the vicinity of the capital, and ordinary people played a relatively minor role in the actual fighting. The merchant class was also largely excluded. The armies would be led personally by representatives of the ruling families and by high-ranking ministers of royal blood who would be educated from an early age in both civil and military arts. Even with Confucius, whose death in 481 B.C. usually marks the end of the Spring and Autumn period, it is clear from the profile preserved in the *Analects* by his disciples that he was trained for both a literary and military career.[33]

During the increasingly more frequent and brutal conflicts of the Warring States period, a real separation emerged between the civil and the military, with mercenaries from lower classes selling their talents to the highest bidder. Warfare moved from an honorable occupation to a profession,[34] and the numbers of those slaughtered on the battlefield and in the reprisals that sometimes followed increased from the hundreds to hundreds of thousands.[35]

The simple explanation for the relationship between philosophy and warfare is that military strategy, like any of the other "arts" (culinary, divinatory, musical, literary, and so on), can be used as a source of metaphors from which to shape philosophical distinctions and categories.[36] Further, military campaigns—particularly at that juncture in Chinese history when political survival was on the line—were a critical preoccupation in which the full range of human resources, including philosophical sensibilities, could be profitably applied. The resolutely pragmatic nature of classical Chinese philosophy resists any severe distinction between theory and application and, as a consequence, philosophizing in this culture is not merely theoretical—it entails practice, "doing." Hence warfare, to the extent that it is philosophical, is necessarily applied philosophy.

Such speculations are undoubtedly part of the answer. But is it simply that military practices can provide grist for philosophical reflection, and philosophy can be applied as some organizing apparatus for military action? Such surely is the case, but the relationship runs deeper. I want to suggest that beneath the rather obvious divergence in subject matter between the cultivation of wisdom in one's person and the cultivation of victory on the battlefield, there is an identifiable correlativity: There is a peculiarly Chinese model of "harmony" or achieved order (*ho*) both fundamental to and pervasive in the classical culture that is pursued by philosopher and military commander alike.

There is a more concrete way of reformulating this question about the intimate relationship between wisdom and warfare that underscores this shared sense of an achieved harmony. How can we explain the clear assumption in this classical Chinese culture that the quality of character which renders a person consummate and exemplary in the various roles of social, political, and cultural leader will also serve him equally well in the role of military commander? We might recall two relevant Confucian precepts:

1. The exemplary person is not a functionary (*ch'i*).
2. The exemplary person pursues harmony (*ho*), not sameness.[37]

What it means to be exemplary, then, is not determined by what function one serves or by what specific skills one possesses, but by one's character. The assumption is that persons of superior character will be exemplary in whatever occupation they turn their hand to—an assumption that is alive and well today. We need only recall the way in which cultural and political leaders are portrayed in the contemporary expression of the Chinese tradition. Mao Tse-tung, as a familiar recent example, was profiled for public view as a great statesman, a poet, a calligrapher, a military strategist, a philosopher, an economist—even an athlete swimming the Yangtze river. It is the ability of the leader to achieve "harmony," however it is defined, that is signatory of what it means to be a person of superior character, whether this harmony is expressed through communal leadership or through military prowess.

To understand the close relationship between warfare and philosophy in classical China, then, we must look to the dynamics of an underlying and pervasive conception of harmony (*ho*) that, for the classical Chinese world view, grounds human experience generally.

THE CLASSICAL CHINESE WORLD VIEW: THE UNCOMMON ASSUMPTIONS

In Chinese there is an expression, "We cannot see the true face of Mount Lu because we are standing on top of it." Although virtually all cultural traditions and historical epochs are complex and diverse, there are certain fundamental and often unannounced assumptions on which they stand that give them their specific genetic identity and continuities. These assumptions, extraordinarily important as they are for understanding the culture, are often concealed from the consciousness of the members of the culture who are inscribed by them, and become obvious only from a perspective external to the particular tradition or epoch. Often a tradition suspends within itself competing and even conflicting elements that, although at odds with one another, still reflect a pattern of importances integral to and constitutive of its cultural identity. These underlying strands are not necessarily or even typically logically coherent or systematic, yet they do have a coherence as the defining fabric of a specific and unique culture.

Within a given epoch, even where two members of a tradition might disagree in some very basic ways—the Confucian and the follower of Master Sun, for example—there are still some common assumptions more fundamental than their disagreements that identify them as members of that culture and have allowed meaningful communication, even where it is disagreement, to occur.

Looking at and trying to understand elements of the classical Chinese culture from the distance of Western traditions, then, embedded as we are within our own pattern of cultural assumptions, has both advantages and disadvantages. One disadvantage is obvious and inescapable. To the extent that we are unconscious of the difference between our own fundamental assumptions and those that have shaped the emergence of classical Chinese thought, we are sure to impose upon China our own presuppositions about the nature of the world, making what is exotic familiar and what is distant near. On the other hand, a clear advantage of an external perspective is that we are able to see with greater clarity at least some aspects of "the true face of Mount Lu"—we are able to discern, however imperfectly, the common ground on which the Confucian and the follower of Master Sun stand in debating their differences, ground that is in important measure concealed from them as unconscious assumptions.

While it is always dangerous to make generalizations about complex cultural epochs and traditions, it is even more dangerous not to. In pursuit of understanding, we have no choice but to attempt to identify and excavate these uncommon assumptions, and to factor them into our understanding of the Chinese tradition broadly, and in this instance, into our assessment of the Chinese art of warfare. The differences between the classical Chinese world view and those classical Greek, Roman, and Judaeo-Christian assumptions that dominate and ground Western traditions are fundamental, and can be drawn in broad strokes in the following terms.

SOME CLASSICAL WESTERN ASSUMPTIONS: A "TWO-WORLD" THEORY

We can call the world view that by the time of Plato and Aristotle had come to dominate classical Greek thinking a "two-world" theory. Later, with the melding of Greek philosophy and the Judaeo-Christian tradition, this "dualistic" mode of thinking became firmly entrenched in Western civilization as its dominant underlying paradigm. In fact, this way of thinking is so second nature to us in the Judaeo-Christian tradition that we do not have to be professional philosophers to recognize ourselves reflected in its outline. A significant concern among the most influential Greek thinkers and later the Christian Church Fathers was to discover and distinguish the world of reality from the world of change, a distinction that fostered both a "two-world theory" and a dualistic way of thinking about it. These thinkers sought that permanent and unchanging first principle that had overcome initial chaos to give unity, order, and design to a changing world, and which they believed makes experience of this changing world intelligible to the human mind. They sought the "real" structure behind change—called variously Platonic Ideas, natural or Divine law, moral principle, God, and so on—which, when understood, made life predictable and secure. The centrality of "metaphysics" in classical Greek philosophy, the "science" of these first principles, reflects a presumption that there is some originative and independent source of order that, when discovered and understood, will provide coherent explanation for the human experience.

There were many diverse answers to the basic question: What is the One behind the many? What is the unity that brings everything together as a "*uni*verse"? What—or Who—has set the agenda that makes human life coherent, and thus meaningful? For the Jewish prophets and scribes, and later for the Christian Church Fathers, it was the existence of the one transcendent Deity who through Divine Will overcame the formless void and created the world, and in whom truth, beauty, and goodness reside. It is this One who is the permanence behind change, and who unifies our world as a single-ordered "universe." It is this One who allows for objective and universal knowledge, and guarantees the truth of our understanding. Because this One is permanent and unchanging, it is more real than the chaotic world of change and appearances that it disciplines and informs. The highest kind of knowledge, then, is the discovery and contemplation (*theoria*) of what is in itself perfect, self-evident, and infallible. It is on the basis of this fundamental and pervasive distinction between a permanently real world and a changing world of appearance, then, that our classical tradition can be said to be dominated by a "two-world theory."

Another way of thinking about this "two-world" theory that has its origins in classical Greece begins from a fundamental separation between "that which creates" and "that which is created," "that which orders" and "that which is ordered," "that which moves" and "that which is moved." There is an assumption that there exists some preassigned design that stands independent of the world it seeks to order. The contrast between the real One—the First Cause, the Creator, the Good—and the less-real world of change, is the source of the familiar dualistic categories that organize our experience of the world: reality/appearance, knowledge/opinion, truth/falsity, Being/Non-being, Creator/creature, soul/body, reason/experience, cause/effect, objective/subjective, theory/practice, agent/action, nature/culture, form/matter, universal/particular, logical/rhetorical, cognitive/affective, masculine/feminine, and so on. What is common among these binary pairs of opposites is that the world defined by the first member is thought to stand independent of, and be superior to, the second. This primary world, defined in terms of "reality," "knowledge," and "truth," is positive, necessary, and self-sufficient, while the derivative world described by the second members as "appearance," "opinion," and "falsity" is negative, contingent, and dependent for its explanation upon the first. After all, it is reality that informs and explains what only appears to be the case, and allows us to separates the true from the false, fact from fiction. On the other hand, appearances are shadows—the false, the fictive. And like shadows, at best they are incidental to what is real; at worst, not only are they of no help to us in arriving at clear knowledge, they obscure it from us. Because the secondary world is utterly dependent on the first, we can say that the primary world is necessary and essential, the "Being" behind the "beings," and the secondary world is only contingent and passing. There is a fundamental discontinuity in this world view between what is real and what is less so.

It is because the first world determines the second that the first world is generally construed as the originative source—a creative, determinative principle, easily translatable into the Judaeo-Christian Deity, that brings both natural and moral order out of chaos. Hence, our early tradition tends to be both *cosmogonic*, meaning it assumes some original act of creation and initial beginning, and *teleological*, meaning it assumes some final purpose or goal, some design to which initial creation aspires. God created the world, and human life is made meaningful by the fact that God's creation has some design and purpose. It is from this notion of determinative principle that we tend to take explanation of events in the world to be linear and causal, entailing the identification of a premise behind a conclusion, a cause behind an effect, some agency behind an activity.

Perhaps a concrete example will help bring this dominant Western world view into clearer definition. The way in which we think about the human being serves this need because in many ways humanity is a microcosm of this "two-world" universe. In this tradition, we might generalize in the following terms. A particular person is a discrete individual by virtue of some inherent nature—a *psyche* or soul or mind—that guarantees a quality of reality and permanence behind the changing conditions of the body. The human being, as such, straddles the "two worlds" with the soul belonging to the higher, originative, and enduring world, and the body belonging to the realm of appearance. The soul, being the same in kind as the permanent principles that order the cosmos, has access to them through reason and revelation, and thus has a claim to knowledge. It is through the discovery of the underlying order that the universe becomes intelligible and predictable for the human being.

SOME CLASSICAL CHINESE ASSUMPTIONS: A "THIS-WORLD" VIEW

Turning to the dominant world view of classical China, we begin not from a "two-world" theory, but from the assumption that there is only the one continuous concrete world that is the source and locus of all of our experience. Order within the classical Chinese world view is "immanental"—indwelling in things themselves—like the grain in wood, like striations in stone, like the cadence of the surf, like the veins in a leaf. The classical Chinese believed that the power of creativity resides in the world itself, and that the order and regularity this world evidences is not derived from or imposed upon it by some independent, activating power, but inheres in the world. Change and continuity are equally "real."

The world, then, is the efficient cause of itself. It is resolutely dynamic, autogenerative, self-organizing, and in a real sense, alive. This one world is constituted as a sea of *ch'i*—psychophysical energy that disposes itself in various concentrations, configurations, and perturbations. The intelligible pattern that can be discerned and mapped from each different perspective within the world is *tao*—a "pathway" that can, in varying degrees, be traced out to make one's place and one's context coherent. *Tao* is, at any given time, both *what* the world is, and *how* it is. In this tradition, there is no final distinction between some independent source of order, and what it orders. There is no determinative beginning or teleological end. The world and its order at any particular time is self-causing, "so-of-itself" (*tzu-jan*). It is for this reason Confucius would say that "It is the person who extends order in the world (*tao*), not order that extends the person."[38] Truth, beauty, and goodness as standards of order are not "givens"—they are historically emergent, something done, a cultural product.

The "two-world" order of classical Greece has given our tradition a theoretical basis for *objectivity*—the possibility of standing outside and taking a wholly external view of things. Objectivity allows us to decontextualize things as "objects" in our world. By contrast, in the "this world" of classical China, instead of starting abstractly from some underlying, unifying, and originating principle, we begin from our own specific place within the world. Without objectivity, "objects" dissolve into the flux and flow, and existence becomes a continuous, uninterrupted process. Each of us is invariably experiencing the world as one perspective within the context of many. Since there is only this world, we cannot get outside of it. From the always unique place one occupies within the continuum of classical China, one interprets the order of the world around one as contrastive "thises" and "thats"—"this person" and "that person"—more or less proximate to oneself. Since each and every person or thing or event in the field of existence is perceived from some position or other, and hence is continuous with the position that entertains it, each thing is related to and a condition of every other. All human relationships are continuous from ruler and subject to friend and friend, relating everyone as an extended "family." Similarly, all "things," like all members of a family, are correlated and interdependent. Every thing is what it is at the pleasure of everything else. Whatever can be predicated of one thing or one person is a function of a network of relationships, all of which conspire to give it its role and to constitute its place and its definition. A father is "this" good father by virtue of the quality of the relationships that locate him in this role and the deference of "these" children and "that" mother, who all sustain him in it.

Because all things are unique, there is no strict notion of identity in the sense of some self-same identical characteristic that makes all members of a class or category or species the same. For example, there is no essential defining feature—no divinely endowed soul, rational capacity, or natural locus of rights—that makes all human beings equal. In the absence of such equality that would make us essentially the same, the various relationships that define one thing in relation to another tend to be hierarchical and contrastive: bigger or smaller, more noble or more base, harder or softer, stronger or weaker, more senior or more junior. Change in the quality of relationships between things always occurs on a continuum as movement between such polar oppositions. The general and most basic language for articulating such correlations among things is metaphorical: In some particular aspect at some specific point in time, one person or thing is "overshadowed" by another; that is, made *yin* to another's *yang*. Literally, *yin* means "shady" and *yang* means "sunny," defining in the most general terms those contrasting and hierarchical relationships that constitute indwelling order and regularity.

It is important to recognize the *interdependence* and correlative character of the *yin/yang* kind of polar opposites, and to distinguish this contrastive tension from the dualistic opposition implicit in the vocabulary of the classical Greek world we explored above, where one primary member of a set such as Creator stands *independent* of and is more "real" than the world He creates. The implications of this difference between dualism and polar contrast are fundamental and pervasive.

One such implication is the way in which things are categorized. In what came to be the dominant Western world view, categories were constituted analytically by an assumed formal and essential identity—all human beings who qualify for the category "human beings" are defined as having an essential *psyche* or soul. All just or pious actions share some essential element in common. The many diverse things or actions can be reduced to one essential identical feature or defining function.

In the dominant Chinese world view, "categories" (*lei*) are constituted not by "essences," but by analogy. One thing is associated with another by virtue of the contrastive and hierarchical relations that sets it off from other things. This particular human being evokes an association with other similar creatures in contrast with other less similar things, and hence gathers around itself a collection of analogous particulars as a general category. "This" evokes "that"; one evokes many. Coherence in this world, then, is not so much analytic or formally abstract. Rather it tends to be synthetic and constitutive—the pattern of continuities that lead from one particular phenomenon to some association with others. It is a "concrete" coherence that begins from the full consequence of the particular itself, and carries on through the category that it evokes.

If we were going to compare these two senses of "categorization," instead of "hammer, chisel, screwdriver, saw" being defined as "tool" by the assumption of some identical formal and abstract function, we are more likely to have a Chinese category that includes "hammer, nail, board, pound, blister, bandage, house, whitewash"—a category of "building a house" constituted by a perceived interdependence of factors in the process of *successfully* completing a given project. Where the former sense of category, defined by abstract essences, tends to be descriptive—what something "is"—the latter Chinese "category" is usually prescriptive and normative—what something "should be" in order to be successful.

The relative absence in the Chinese tradition of Western-style teleology that assumes a given "end" has encouraged the perception among Western historians that the Chinese, with libraries of carefully recorded yet seemingly random detail, are inadequate chroniclers of their own past. There seems to be little concern to recover an intelligible pattern from what seriously threatens to remain formless and meaningless. Jorge Luis Borges captures this Western perception in his well-known citation of "a certain Chinese encyclopedia" in which the category "animals" is divided into: 1) belonging to the Emperor, 2) embalmed, 3) tame, 4) suckling pigs, 5) sirens, 6) fabulous, 7) stray dogs, 8) included in the present classification, 9) frenzied, 10) innumerable, 11) drawn with a very fine camel-hair brush, 12) et cetera, 13) having just broken the water pitcher, and 14) that from a long way off look like flies.[39] From the perspective of the more rationalistic Western world view, the penalty the Chinese must pay for the absence of that underlying metaphysical infrastructure necessary to guarantee a single-ordered universe is what we take to be intelligibility and predictability. The compensation for this absence in the Chinese world is perhaps a heightened awareness of the immediacy and wonder of change, and one's complicity in it—the motive for revering the *Book of Changes* as the ultimate defining statement of the tradition, and as an apparatus for shaping a propitious world.

For the classical Greek philosophers, knowledge entailed the discovery and "grasping" of the defining "essence" or "form" or "function" behind elusively changing appearances. Hence the language of knowing includes "concept," "conceive," "comprehend." Reality is what is permanent, and hence its natural state is inertia. The paradigm for knowledge, then, is mathematics, and more specifically, geometry. Over the door of Plato's Academy was written: "Let none who have not studied geometry enter here." Visual and spatial language tends to predominate in the philosophical vocabulary, and knowledge tends to be understood in representational terms that are isomorphic and unambiguous—a true copy impressed on one's mind of that which exists externally and objectively.

In the classical Chinese model, knowledge is conceived somewhat differently. Form is not some permanent structure to be discovered behind a changing process, but a perceived intelligibility and continuity that can be mapped within the dynamic process itself. Spatial forms—or "things"— are temporal flows. "Things" and "events" are mutually shaping and being shaped, and exist as a dynamic calculus of contrasting foci emerging in tension with each other. Changing at varying degrees of speed and intensity, the tensions constitutive of things reveal a site-specific regularity and pattern, like currents in the water, sound waves in the air, or weather systems in the sky. Etymologically, the character *ch'i*—"the stuff of existence"—is probably acoustic, making "resonance" and "tensions" a particularly appropriate way of describing the relations that obtain among things. In contrast with the more static visual language of classical Greek thought typified by geometry, classical Chinese tends to favor a dynamic aural vocabulary, where wisdom is closely linked with communication—that keenness of hearing and those powers of oral persuasion that will enable one to encourage the most productive harmony out of relevant circumstances. Much of the key philosophic vocabulary suggests etymologically that the sage orchestrates communal harmony as a virtuoso in communicative action.

"Reason" is not a human faculty independent of experience that can discover the essences of things, but the palpable determinacy that pervades both the human experience and the world experienced. Reason is coherence—the pattern of things and functions. Rational explanation does not lie in the discovery of some antecedent agency or the isolation and disclosure of relevant causes, but in mapping out the local conditions that collaborate to sponsor any particular event or phenomenon. And these same conditions, once understood, can be manipulated to anticipate the next moment.

An important factor in classical Chinese "knowing" is comprehensiveness. Without an assumed separation between the source of order in the world and the world itself, causal agency is not so immediately construed in terms of relevant cause and effect. All conditions interrelate and collaborate in greater or lesser degree to constitute a particular event as a confluence of experiences. "Knowing" is thus being able to trace out and manipulate those conditions far or near that will come to affect the shifting configuration of one's own place. There is a direct and immediate affinity between the human being and the natural world so that no firm distinction is made between natural and man-made conditions—they are all open to cultivation and manipulation. In fact, it is because of the fundamental continuity between the human pattern and the natural pattern that all of the conditions, human and otherwise, that define a situation such as battle can be brought into sharp focus. In the absence of a severe animate/inanimate dualism, the battlefield with its complex of conditions is very much alive.

The inventory of philosophical vocabulary used in classical China to define this kind of "knowing" tends to be one of tracing out, unraveling, penetrating, and getting through. Knowing entails "undoing" something, not in an analytic sense to discover what it essentially "is," but to trace out the connections among its joints and sinews, to discern the patterns in things, and, on becoming fully aware of the changing shapes and conditions of things, to anticipate what will ensue from them. The underlying metaphor of "tracing a pattern" is implicit in the basic epistemic vocabulary of the tradition such as "to tread a pathway, a way" (*tao*), "to trace out, coherence" (*li*), "to figure, image, model" (*hsiang*), "to unravel, to undo" (*chieh*), "to penetrate" (*t'ung*), "to break through" (*ta*), "to name, to make a name, to inscribe" (*ming*), "to ritualize" (*li*), "to inscribe, markings, culture" (*wen*), and so on. In contrast with its classical Greek counterpart where "knowing" assumes a mirroring correspondence between an idea and an objective world, this Chinese "knowing" is resolutely participatory and creative—"tracing" in both the sense of etching a pattern and of following it. To know is "to realize," to "make real." The path is not a "given," but is made in the treading of it. Thus, one's own actions are always a significant factor in the shaping of one's world.

Because this emergent pattern invariably arises from within the process itself, the tension that establishes the line between one's own focus and one's field gives one a physical, psychological, social, and cosmological "skin"—a shape, a continuing, insistently particular identity. This dynamic pattern is reflexive in the sense that one's own dispositions are implicate in and affect the shaping of one's environment. One's own "shape" is constantly being reconstrued in tension with what is most immediately pressing in upon one and vice versa.

To continue the "person" example from our discussion of the classical Greek world view, generally in classical Chinese philosophy a particular person is not a discrete individual defined in terms of some inherent nature familiar in recent liberal democratic theory, but is a configuration of constitutive roles and relationships: Yang Ta-wei's father, An Lo-che's teacher, Kao Ta-jen's neighbor, a resident of Yung-ho village, and so on. These roles and relationships are dynamic, constantly being enacted, reinforced, and ideally deepened through the multiple levels of communal discourse: embodying (*t'i*), ritualizing (*li*), speaking (*yen*), playing music (*yüeh*), and so on. Each of these levels of discourse is implicit in every other, so there is a sense in which a person can be fairly described as a nexus of specific patterns of discourse. By virtue of these specific roles and relationships, a person comes to occupy a place and posture in the context of family and community. The human being is not shaped by some given design that underlies natural and moral order in the cosmos and that stands as the ultimate objective of human growth and experience. Rather, the "purpose" of the human experience, if it can be so described, is more immediate: to coordinate the various ingredients that constitute one's particular world here and now, and to negotiate the most productive harmony out of them. Simply put, it is to get the most out of what you've got here and now.

Creativity also has a different place in the classical Chinese world. Again, in gross terms, the preassigned design and ultimate purpose assumed in classical Western cosmology means that there is a large investment of creativity "up front" in the "birth" of a phenomenon—a condition reflected rather clearly in the preestablished "Ideas" of Plato, the "potentiality/actuality" distinction of Aristotle, or the Creator/creature dualism of the Judaeo-Christian tradition. For the classical Chinese world view, in the absence of an initial creative act that establishes a given design and a purpose governing change in the cosmos, the order and regularity of the world emerges from the productive juxtapositions of different things over the full compass of their existence. No two patterns are the same, and some dispositions are more fruitfully creative than others. For this reason, human knowledge is fundamentally performative—one "knows" a world not only passively in the sense of recognizing it, but also in the active shaping and "realizing" of it. It is the capacity to anticipate the patterned flow of circumstance, to encourage those dispositions most conducive to a productive harmony, and ultimately to participate in negotiating a world order that makes best advantage of its creative possibilities. Harmony is attained through the art of contextualizing.

A major theme in Confucius and in Confucianism alluded to earlier is captured in the phrase, "the exemplary person pursues harmony (*ho*), not sameness."[40] This Confucian conception of "harmony" is explained in the classical commentaries by appeal to the culinary arts. In the classical period, a common food staple was *keng*—a kind of a millet gruel in which various locally available and seasonal ingredients were brought into relationship with one another. The goal was for each ingredient—the cabbage, the squash, the bit of pork—to retain its own color, texture, and flavor, but at the same time to be enhanced by its relationship with the other ingredients. The key to this sense of harmony is that it begins from the unique conditions of a specific geographical site and the full contribution of those particular ingredients readily at hand—*this* piece of cabbage, *this* fresh young squash, *this* tender bit of pork—and relies upon artistry rather than recipe for its success. In the *Spring and Autumn Annals of Master Lü*, cooking as the art of contextualizing is described in the following terms:

> In combining your ingredients to achieve a harmony (*ho*), you have to use the sweet, sour, bitter, acrid and the salty, and you have to mix them in an appropriate sequence and proportion. Bringing the various ingredients together is an extremely subtle art in which each of them has its own expression. The variations within the cooking pot are so delicate and subtle that they cannot be captured in words or fairly conceptualized.[41]

The Confucian distinction between an inclusive harmony and an exclusive sameness has an obvious social and political application. There is a passage in the *Discourses of the States (Kuo-yü)*, a collection of historical narratives probably compiled around the fourth century B.C., which underscores the fertility of the kind of harmony that maximizes difference:

> Where harmony (*ho*) is fecund, sameness is barren. Things accommodating each other on equal terms is called blending in harmony, and in so doing they are able to flourish and grow, and other things are drawn to them. But when same is added to same, once it is used up, there is no more. Hence, the Former Kings blended earth with metal, wood, fire, and water to make their products. They thereby harmonized the five flavors to satisfy their palate, strengthened the four limbs to protect the body, attuned the six notes to please the ear, integrated their various senses to nourish their hearts and minds, coordinated the various sectors of the body to complete their persons, established the nine main visceral meridians to situate their pure potency, instituted the ten official ranks to organize and evaluate the bureaucracy... and harmony and pleasure prevailed to make them as one. To be like this is to attain the utmost in harmony. In all of this, the Former Kings took their consorts from other clans, required as tribute those products which distinguished each region, and selected ministers and counsellors who would express a variety of opinions on issues, and made every effort to bring things into harmony.... There is no music in a single note, no decoration in a single item, no relish in a single taste.[42]

A contemporary poet, Wing Tek Lum, reflects on the importation of this enduring Chinese sensibility to the new ways of immigrant life in his "Chinese Hot Pot":

My dream of America
is like *dá bìn lòuh*
with people of all persuasions and tastes
sitting down around a common pot
chopsticks and basket scoops here and there
some cooking squid and others beef
some tofu or watercress
all in one broth
like a stew that really isn't
as each one chooses what he wishes to eat
only that the pot and fire are shared
along with the good company
and the sweet soup
spooned out at the end of the meal.[43]

This "harmony" is not a given in some preassigned cosmic design, but it is the quality of the combination at any one moment created by effectively correlating and contextualizing the available ingredients, whether they be foodstuffs, farmers, or infantry. It is not a quest of discovery, grasping an unchanging reality behind the shadows of appearance, but a profoundly creative journey where the quality of the journey is itself the end. It is making the most of any situation.

In summary, at the core of the classical Chinese world view is the cultivation of harmony—a specifically "center-seeking" or "centripetal" harmony. This harmony begins from what is most concrete and immediate—that is, from the perspective of any particular human being—and draws from the outside in toward its center. Hence there is the almost pervasive emphasis on personal cultivation and refinement as the starting point for familial, social, political, and as we shall see, military order. A preoccupation in classical Chinese philosophy, then, is the cultivation of this centripetal harmony as it begins with oneself, and radiates outward. The cultivation of this radial harmony is fundamentally aesthetic. Just as Leonardo arranged those specific bits of paint to constitute the one and only Mona Lisa, so one coordinates those particular details that constitute one's own self and context, and in so doing seeks a harmony that maximizes their creative possibilities.

The Chinese world view is thus dominated by this "bottom-up" and emergent sense of order that begins from the coordination of concrete detail. It can be described as an "aestheticism," exhibiting concern for the artful way in which particular things can be correlated efficaciously to thereby constitute the ethos or character of concrete historical events and cultural achievements. Order, like a work of art, begins with always-unique details, from "this bit" and "that," and emerges out of the way in which these details are juxtaposed and harmonized. As such, the order is embedded and concrete—the coloration that differentiates the various layers of earth, the symphony of the morning garden, the striations in a wall of stone, the veins in the leaf of a plant, the wind piping through the orifices of the earth, the rituals and roles that constitute a communal grammar to give community meaning. Such an achieved harmony is always particular and specific—resistant to notions of formula and replication.

CENTRIPETAL HARMONY AND AUTHORITY

We begin from the premise in classical Chinese culture that human beings are irreducibly communal. The human being is a center of a radial pattern of roles and relationships. The question that emerges, then, is how do these overlapping yet disparate human "centers," having defined themselves as persons, families, and communities, come to be interrelated? And how is authority among them established and continued?

The answer: Authority is constituted as other centers are drawn up into one encompassing center and suspended within it through patterns of deference. This calculus of centers through their interplay produces a balancing centripetal center that tends to distribute the forces of its field symmetrically around its own axis. Authority has several parts. It resides in a role ("father," "commander," "ruler"), in the scope and quality of the extended pattern of relationships this role entails ("family members," "soldiers," "subjects"), and in the cultural tradition as it is conveyed within these relationships. Effective application of the cultural wealth of the tradition to prevailing circumstances through one's roles and relationships inspires deference and extends one's influence.

The analog to the hierarchical complex of relationships that make up a family or community can be found and illustrated in the political world by appeal to any number of concrete historical examples. Within the subcontinent that was Warring States China, the full spectrum of peoples—some paying their allegiance to traditional hereditary houses, some ruled by locally powerful warlords, others organized around religious doctrines, yet others governed by clan or tribal regulation—was suspended in the Han harmony, with each of them contributing in greater or lesser degree to the definition of Han culture. This political order was one in which all of the diversity and difference characteristic of the multiple, competing centers of the Warring States period was drawn up and suspended in the harmony of the Han dynasty. Moving from the radial extremes toward the center, the very disparate "zones" contributed to the imperial order in increasing degree to influence the authority at the center, shaping and bringing into focus the character of the social and political entity—its standards and values. Whatever constitutes the authority at the center is holographic. In this political example, the ruler derives his authority from having his field of influence implicate within him. He is the empire. The attraction of the center is such that, with varying degrees of success, it draws into its field and suspends within its web the disparate and diverse centers that constitute its world. It is the quality of these suspended centers in relationship to one another that defines the harmony of the field.

This same dynamic that defines Han culture politically can be discerned in its intellectual character. During the Warring States period, philosophical diversity flourished and schools of thought proliferated to become what the *Chuang-tzu* describes as "the doctrines of the Hundred Schools."[44] As the Han dynasty became established, the intellectual contest of the Hundred Schools gave way to a syncretic Confucianism-centered doctrine. This state ideology absorbed into itself (and in important degree concealed) the richness of what were competing elements, and out of this diversity articulated the philosophical and religious character of the period. The syncretism of Han dynasty Confucianism is harmony teased out of difference. This transition from diversity in the late Chou to coherent order in the Han is better expressed in the language of incorporation and accommodation than of suppression.

As the centripetal center of the Han court weakened in the second century A.D., and as the political order gradually dissolved into a period of disunity, disparate foci reasserted themselves, and what had been their contribution to a harmonious diversity became the energy of contest among them. What was a tightening centripetal spire in the early Han dynasty became a centrifugal gyre, disgorging itself of its now disassociated contents. It is not surprising that during this same period, there was a resurgence and interplay of competing philosophical schools and religious movements that reflected a contemporaneous disintegration of the centrally driven intellectual order. This is the familiar pattern of dynasty and interregnum repeated throughout the career of imperial China.[45]

Given the commitment to a centripetal sense of order pervasive at every level in the classical Confucian world view, a father or a magistrate or a commander or a ruler would derive his authority from being at the center, and having implicate within him the order of the whole. It is for this reason that "the exemplary person's errors are like an eclipse:

When he errs, everyone sees him; when he rights himself, everyone looks up to him."[46]

John Fairbank's essay, "The Grip of History on China's Leadership," makes a convincing argument that the social and political order of China under Mao Tse-tung was fully consistent with the tradition, from "the Chinese readiness to accept a supreme personality" to the phenomenon of a population continuing to struggle for proximity to the center.[47] It is by virtue of the supreme personality's embodiment of his world, as in the case of Mao Tse-tung, that he is able to lay claim to impartiality—his actions are not self-interested (*li*) but always appropriate (*yi*), accommodating the interests of all. Just as the traditional conception of Heaven (*t'ien*), encompassing within itself the world order, is credited with total impartiality, so the "Son of Heaven" (*t'ien-tzu*) with similar compass is devoid of a divisive egoism. As long as the center is strong enough to draw the deference and tribute of its surrounding spheres of influence, it retains its authoritativeness—that is, not only do these spheres willingly acknowledge this order, but actively participate in reinforcing it. Standing at the center, the ruler acts imperceptibly, a pole star that serves as a bearing for the ongoing negotiation of the human order while appearing to be unmoved and unmoving himself.[48]

WARFARE AS THE ART OF CONTEXTUALIZING

To return to the central contention, then, I want to suggest that the achieved harmony that we have identified as the goal of personal, social, and political cultivation in classical Confucianism is not limited to this school of thought or historical period, but is a signatory feature of the Chinese tradition more broadly construed. Centripetal harmony as the model of order operating in the classical Chinese world view is pervasive. To illustrate this, I want to juxtapose what for us might seem to be only marginally related concerns of personal cultivation and of effectiveness in battle in order to attempt to understand why concepts central to philosophy and to military affairs cannot be separated and, in fact, can only be fully explicated by appeal to one another. How then does this conception of achieved centripetal harmony figure into the military experience?

Beginning from the most general attitudes toward warfare in early China, John Fairbank makes the following observation:

Since the ideal of proper conduct was built into the Chinese concept of the cosmos, a rupture of this ideal threatened to break down the whole cosmic system. Consequently, the Chinese "right of rebellion" could not be asserted simply in the name of individual or corporate freedom against ruling class tyranny. It had to be asserted in the name of the system, alleging that the ruler had forfeited Heaven's mandate by not maintaining the social order adequately and correctly.... Rebels usually rose in the name of the social order, which was the great legitimizing myth of the state and the underlying moral sanction for all resort to warfare.[49]

Stated in another way, what makes any military action "appropriate" and "proper" (*yi*) as opposed to "self-seeking" (*li*) is the claim that it serves the quality of the sociopolitical order as a whole rather than any particular interest group within it. Those persons promoting military engagement must make their argument on the necessity of such action to revive and reshape the shared world order.

A note of explanation is needed to avoid a possible equivocation. The notion of sociopolitical order here is not justified as service to some universally applicable standard *independent* of oneself that sanctions conduct within its jurisdiction, as is the case where such service is devotion to the One True God, commitment to some doctrine of natural law, or respect for a universal Bill of Rights. Rather, it is a notion of sociopolitical order in which all orders are *interdependent* and mutually entailing, so that realization for oneself, one's family, one's community, and for one's state are codetermining and coextensive. .The "legitimizing myth" is symbiosis, where service to oneself and to one's community is the same. There is no "means/end" distinction that subordinates one's personal achievement to the social or political end, or vice versa. Hence, any assertion on behalf of any part is always an assertion on behalf of the whole. And by the same token, any protest is ultimately self-referential—a criticism of an order in which one's self is a constitutive factor.

Perhaps an analogy that might be illustrative here is the relationship that exists between any particular note in a symphonic performance and the symphony as a whole. There is a sense in which the value and meaning of each note can only be understood within the context of the entire symphony. In these terms, then, each note has the entire symphony implicate within it. At the same time, the symphony is only available through one note at a time as particular perspectives on the symphony, and the only sense of "objective" vantage point from which the entire symphony can be entertained lies in the presumption that each note appropriately executed serves the interests of the symphony as a whole.

The qualification on "order" that needs to be introduced here, then, is that even righteous war in service to the social order as a whole is invariably pursued from some particular perspective within the whole—some claim to authority that occupies or seeks to occupy the center. At the same time, it is impartial and "objective" in that it claims as one perspective to represent all interests. Military action, then, is generally seen as an attunement on the existing order from within—ideally it is always responsive, always punitive, always pro-social.

There is a deep and abiding association in the Chinese world between the execution of punishments and of warfare. In both instances, the central authority is acting in the interests of the whole to define the sociopolitical order at its boundaries. The character used for "punishment" (*hsing*) is homophonous and often used interchangeably with the character meaning "to shape," and carries with it a strong sense of drawing a line and configuring a defined order by excluding those who are antisocial, usually by amputating something or disfiguring them, and thus, quite literally, reshaping them. Similarly, warfare frequently occurs on the borders as a final effort to define what belongs within one's circle and what lies beyond it. There is an obvious cognate relationship between the characters "to order" (*cheng*), "to govern" (*cheng*), and "to dispatch a punitive expedition" (*cheng*). Warfare is an attempt to redefine sociopolitical order.

STRATEGIC ADVANTAGE (*SHIH*)

The key and defining idea in the *Sun-tzu: The Art of Warfare* is *shih* (pronounced like the affirmative, "sure").[50] Although I have translated *shih* consistently as "strategic advantage," it is a complex idea peculiar to the Chinese tradition, and resists easy formulaic translation.[51] In fact, an understanding of *shih* entails not only the collation of an inventory of seemingly alternative meanings, but also a familiarity with those presuppositions outlined above that make the classical Chinese world view so very different from our own. *Shih*, like ritual practices and role playing (*li*), speaking (*yen*), playing music (*yüeh*), and embodying (*t'i*), is a level of discourse through which one actively determines and cultivates the leverage and influence of one's particular place.

In studying the Chinese corpus, one consults dictionaries that encourage us to believe that many if not most of the characters such as *shih* have "multiple" alternative meanings from which the translator, informed by the context, is required to select the most appropriate one. This approach to the language, so familiar to the translator, signals precisely the problem that I have worried over in the introductory comments about alternative world views. The irony is that we serve clarity in highlighting what makes sense in our own conceptual vocabulary only to bury the unfamiliar implications that, in themselves, are the most important justification for the translation.

I would suggest that with the appearance of any given character in the text, with varying degree of emphasis, the full seamless range of its meaning is introduced. And our project as interpreters and translators is to negotiate an understanding and rendering that is sensitive to this full undifferentiated range of meaning. In fact, it is this effort to reconstitute the several meanings as an integrated whole and to fathom how the character in question can carry what for us might well be a curious, often unexpected, and sometimes even incongruous combination of meanings that leads us most directly to a recognition of difference.

For example, the character *shen* does not *sometimes* mean "human spirituality" and *sometimes* "divinity." It always means both and, moreover, it is our business to try and understand philosophically how it *can* mean both. Given the prominence of transcendent Deity in our tradition, human beings do not generally get to be gods. Reflection on the range of meaning represented by *shen* reveals that gods in the Chinese world are by and large dead people—they are ancestors who have embodied, enriched, and transmitted the cultural tradition. They are cultural heroes, as in the case of Confucius, who do the work of our transcendent Deity by establishing the enduring standards of truth, beauty, and goodness. Culturally productive ancestors such as Confucius are not *like* gods—they are precisely what the word "gods" conveys in this alternative tradition. Gods are historical, geographical, and cultural. They grow out of the ground, and when neglected, fade and die. Such gods have little to do with the notion of a transcendent Creator Deity that has dominated the Western religious experience, and unless we are sensitive to the "this-world" presuppositions that ground the classical Chinese world view, we stand the risk of willy-nilly translating Chinese religiousness into our own.

The key militarist idea, *shih*, is as complex as *shen*, "spirituality/divinity," and, fortunately for our grasp of the tradition, as revealing of the underlying sense of order. We must struggle to understand how *shih* can combine in one idea the following cluster of meanings:

1. "aspect," "situation," "circumstances," "conditions"
2. "disposition," "configuration," "outward shape"
3. "force," "influence," "momentum," "authority"
4. "strategic advantage," "purchase"

In defining *shih* or any of the other key ideas, the military texts such as the *Sun-tzu* and *Sun Pin* do not rely solely or even primarily upon the currency of abstract concepts and theoretical programs. Rather, these texts, emerging as they do out of concrete historical experience, tend to communicate through the mediums of image, historical allusion, and analogy. What constitutes evidence and makes things clear in the text is often an effectively focused image, not a theory; an inexpressible and inimitable experience, not an argument; an evocative metaphor, not a logically demonstrated truth. The style, then, respects the priority of the unique particular—a defining characteristic of emergent harmony. It resists the abstractness of universalizing principles in favor of the concrete image. The aphoristic statements seek, with the assistance of a sympathetic reading, to make points rather than lay down categorical imperatives. The readers, on their part, are required to generate a specific set of circumstances that make the assertions meaningful and important.

This claim that image has an important role is illustrated in the putative origins of the term *shih* itself. The *Sun Pin* states:

> Thus, animals not equipped with natural "weapons" have to fashion them for themselves. Such were the contributions of the sages.... Yi created the bow and crossbow and derived the notion of strategic advantage (*shih*) from them.... How do we know that the notion of strategic advantage is derived from the bow and crossbow? An archer shoots from between shoulder and chest and kills a soldier over a hundred paces away who does not even know where the bolt came from. Hence it can be said: the bow and crossbow exemplify strategic advantage (*shih*).[52]

The images used to express *shih* in these texts are many, each of them focused in such a way as to suggest some specific area in its range of meaning. Round boulders and logs avalanching down a precipitous ravine and cascading water sending boulders bobbing about underscore the sense of fluidity and momentum.[53] The taut trigger on the drawn crossbow emphasizes timing and precision.[54] The bird of prey swooping down to knock its victim out of the air stresses the agility that gives one full control over one's own movement, the coordination of this movement and that of one's target, and the resolutely aggressive posture one assumes throughout. The drawn crossbow locates one well beyond the range of the enemy. The scales tilting in one's favor highlights the logistical advantage of one's position relative to one's enemy.[55] The "sudden striker" snake suggests the flexibility and the total preparedness that turns defense into offense.[56]

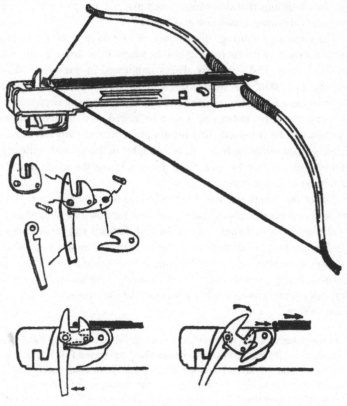

Line drawing of a crossbow and its trigger device dating from the Warring States period (403–221 B.C.) unearthed in a 1952 excavation at Ch'ang-hsia Tomb #138

Lifting the discussion from the metaphorical to a more theoretical level, the first point that can be made is *shih* (like immanental order generally) begins from the concrete detail. It begins from a recognition that the business of war does not occur as some independent and isolated event, but unfolds within a broad field of unique natural, social, and political conditions. These conditions and the relations that exist among them are ever changing. Further, although the changes that occur within any local field of conditions are always unique to it, they proceed according to a general pattern that can not only be anticipated, but can be manipulated to one's advantage. It is the changing configuration of these specific conditions that determines one's place at any point in time, and gives one a defining disposition and "shape."

The constantly shifting "disposition" of any thing or event is constituted in tension with environing others, where their dispositions condition one's own. The enemy is always implicated in one's own shifting position. The "skin" that defines one's "inner/outer" circle and separates one from the enemy also conjoins one to him, making any change mutual and pervasive. If he moves, one is thereby moved. And more importantly, if one moves, he is moved. This presumption of continuity between self and other means that each focus is holographic in the sense that the entire field is implicate in every one. Each position brings the whole into focus from its own unique perspective.

One of the "supplemental chapters" to the *Sun Pin: The Art of Warfare* provides us with a cosmological explanation for how *shih* functions. This treatise on the complementarity of "straightforward and surprise operations" begins from a description of how, in an immanental cosmos, change is always movement between polar opposites on a continuum. This explanation of change is fundamental to the classical Chinese world view, and although probably most familiar to us from the Taoist sources, is pervasive in the culture.[57]

> In the pattern of the heavens and the earth: when something has reached its extreme, it then returns; when something has waxed full, it then wanes. This is exemplified by [*the sun and the moon*]. Flourishing and fading succeed each other. This is exemplified in the succession of the four seasons. Something prevails only to then be prevailed over. This is exemplified in the succession of the five phases (*wu hsing*). Life and death succeed one another. This is exemplified in the life cycle of the myriad things. Capacity and incapacity succeed each other. This is exemplified in the maturation process of the myriad life-forms. And that while some things are had in surplus, there is a deficiency in others—this is exemplified in the dynamics of shapes or dispositions (*hsing*), and strategic advantages (*shih*).[58]

It is because change is always movement between polar opposites that the fluid dispositions that obtain among phenomena can be described in the language of *yin-yang* "sunny/shady" contrasts. Since these contrasts can only be explained by reference to each other, they are correlative (as opposed to dualistic) opposites. The vocabulary used to express military insights in these treatises depends heavily upon a cluster of just such correlations: us/enemy (*wo/ti*); aggressor/defender (*chu/k'o*); attack/defend (*kung/shou*); many/few (*chung/kua*); strong/weak (*ch'iang/jo*); courage/

timidity (*yung/ch'ieh*), intimate/distant (*chi/shu*); full/empty (*ying/hsü*); slow/fast (*hsü/chi*); movement/stillness (*tung/ching*); rested/exhausted (*yi/lao*); order/disorder (*chih/luan*); viable/fatal (*sheng/ssu*); victory/defeat (*sheng/pai*); surprise/straightforward (*ch'i/cheng*); advance/retreat (*chin/t'ui*); and so on. This correlative vocabulary reflects the assumption that any situation definable on a continuum can be manipulated into its polar opposite: Order can be teased out of disorder, courage can be stoked out of timidity, largeness can be conjured out of smallness, victory can be lifted out of defeat. As the *Sun-tzu* observes:

> Disorder is born from order; cowardice from courage; weakness from strength. The line between disorder and order lies in logistics (*shu*); between cowardice and courage, in strategic advantage (*shih*); and between weakness and strength, in strategic positioning (*hsing*).[59]

All determinate situations can be turned to advantage. The able commander is able to create differentials and thus opportunities by manipulating his position and the position of the enemy. By developing a full understanding of those factors that define one's relationship with the enemy, and by actively controlling and shaping the situation so that the weaknesses of the enemy are exposed to one's acquired strength, one is able to ride the force of circumstances to victory.

> All things and events that have a distinguishing shape or disposition can be named, and all things that can be named can be prevailed over. Thus, because the sages would use that characteristic in which any one thing excels to prevail over all other things, they were always successful in whatever they did.[60]

This general observation, of course, has an immediate military application:

> Battle then is simply one disposition trying to prevail over another. All distinguishable dispositions can be prevailed over. The problem lies in knowing which disposition will enable one to prevail. The changing calculus of dispositions that can lead one thing to prevail over another is as inexhaustible as all that happens between the heavens and earth. These dispositions that can lead one thing to prevail over another could not be fully described if you were to write on all of the bamboo that could be cut from the states of Ch'u and Yüeh. Such dispositions are, in all cases, using that characteristic in which a particular thing excels to prevail over other things.[61]

While the military strategist can articulate general principles concerning the nature of change and how to manipulate it to one's own advantage, a real limitation on what can be said arises from the fact that each situation is site-specific; it is local and unique, and must be dealt with on its own terms.

> But you will never find a winning characteristic of one particular disposition that will enable you to prevail in all situations. The need to figure out a disposition is in principle the same, but what disposition will actually prevail is always different.[62]

A central theme of both the *Sun-tzu* and the *Sun Pin* is the need for flexibility and negotiation in dealing with the specific conditions that make each situation particular.

> In the business of war, there is no invariable strategic advantage (*shih*) which can be relied upon at all times.[63]

In fact, a fundamental insight into nature that one must accord with in prosecuting military affairs is the irrepressibility of change itself. One must find security by revising and redefining one's own strength by immediate yet unannounced responsiveness to the enemy's shifting position.

> Thus an army does not have fixed strategic advantages (*shih*) or an invariable position (*hsing*). To be able to take the victory by varying one's position according to the enemy's is called being inscrutable (*shen*). Thus of the five phases (*wu hsing*), none is the constant victor; of the four seasons, none occupies a constant position; the days are both short and long; the moon waxes and wanes.[64]

The able commander does not resist the rhythm of change, but, finding its pulse, translates defining conditions into correlative terms as a means of controlling the situation, anticipating the enemy's movements, and making his victory inevitable.

> Thus, the expert at warfare can infer the enemy's weaknesses from observing his strengths, and can infer his surpluses from observing his deficiencies. He can see the victory as clearly as the sun or moon, and can grasp it as certainly as water douses fire.[65]

If we allow that there are several different ways in which we can look at *shih*, it enables us to bring its cluster of meanings together. When looked at spatially from outside of one's own "skin," *shih* is that set of conditions that is defining of one's situation. It is one's context in relationship to one-self. When looked at from an internal perspective, *shih* is one's own place and posture relative to one's context. When looked at temporally, taking into account the full calculus of dispositions, *shih* is the tension of forces and the momentum that brings one position in immediate contact with another. And, of course, what brings these various dimensions of meaning together is the acknowledgment that, in this classical Chinese world view, the spatial (*yü*) and the temporal (*chou*) are themselves correlatives that require reference to each other for explanation. In fact, the combination of these two terms as "space-time" means "cosmos" (*yü-chou*) in the classical language.

STRATEGIC ADVANTAGE (*SHIH*)
AND STRATEGIC POSITIONING (*HSING*)

What is the difference, then, between strategic advantage (*shih*) and strategic positioning (*hsing*)? As D. C. Lau has indicated, in the *Sun-tzu* there are passages in which these two terms are used as near-synonyms.[66] This is because *shih* overlaps with *hsing* in having the connotation of physical position—not position as specific location, but rather as a fluid disposition ever responsive to context. Where *hsing* is limited to the tangible and determinate shape of physical strength, *shih* includes intangibles such as morale, opportunity, timing, psychology, and logistics. Effective strategic positioning (*hsing*) creates a situation where we can use "the undivided whole to attack his one,"[67] "weigh in a full hundredweight against a few ounces,"[68] and "use many to strike a few"[69]—that is, to win the war before joining battle.[70] Strategic advantage (*shih*), by contrast, is the full concentrated release of that latent energy inherent in one's position, physical and otherwise.

The military treatise in the Han dynasty work, the *Master of Huai Nan*, describes in some detail those several factors implied by *shih* that go beyond one's physical position:

> When the commander is full of courage and regards the enemy with contempt, when his troops are steeled in their resolve and delight in the prospect of battle, when the determination of his troops, countless in number, outstrips the skies, when their fury is like a tempest and their battle cries ring out like thunder, when utterly committed they thoroughly intimidate the enemy—this is called a morale advantage (*ch'i shih*).
>
> A narrow crossing in the mountain gorges, a well-known obstruction in high and mountainous terrain, a snaking and coiling pathway, the summit

of a rise, a road that spirals like a ram's horn, a bottleneck through which there is entry but no retreat, a point at which one man holds a thousand enemy at bay—this is called a terrain advantage (*ti shih*).

Capitalizing fully on the enemy's fatigue, his ill-preparedness and disorder, his hunger and thirst, and his exposure to the elements, to press in upon him where he has lost his footing, and to give him no quarter where he is most vulnerable—this is called an opportunity advantage (*yin shih*).[71]

ACCORDING WITH THE ENEMY (*YIN*)

Another elusive notion essential to an understanding of classical Chinese philosophy in general, and militarist thought specifically, is *yin*, conventionally translated "to avail oneself of," "to make the best of," "to rely upon." Every situation has its "give and take," and, as such, can be parlayed into an opportunity. The basic meaning of *yin* is responsiveness to one's context: to adapt oneself to a situation in such a manner as to take full advantage of the defining circumstances, and to avail oneself of the possibilities of the situation in achieving one's own purposes.

Yin requires sensitivity and adaptability. Sensitivity is necessary to register the full range of forces that define one's situation, and, on the basis of this awareness, to anticipate the various possibilities that can ensue. Adaptability refers to the conscious fluidity of one's own disposition. One can only turn prevailing circumstances to account if one maintains an attitude of readiness and flexibility. One must adapt oneself to the enemy's changing posture as naturally and as effortlessly as flowing water winding down a hillside:

> As water varies its flow according to (*yin*) the fall of the land, so an army varies its method of gaining victory according to (*yin*) the enemy.[72]

Yin means feeding your army from the enemy's fields;[73] *yin* means taking advantage of inflammable materials in the vicinity of the enemy's camp;[74] *yin* means shifting your posture so adroitly and imperceptibly that, from the enemy's perspective, you are inscrutable.[75]

When this notion of *yin* is applied to espionage, it designates a "local" spy—the enemy's own countrymen in our employ. It means using the enemy against himself.

AN ATTITUDE TOWARD WARFARE

Both the *Sun-tzu* and its literary descendant, the *Sun Pin*, are military treatises that share a fundamental distaste for warfare. Warfare always constitutes a loss. As the *Sun-tzu* observes, "If one is not fully cognizant of the evils of waging war, he cannot be fully cognizant either of how to turn it to best account."[76] It is on this principle that the *Sun Pin* claims that "a distaste for war is the most basic principle of the True King."[77] This being the case, "you must go to war only when there is no alternative."[78] At times, however, virtuous government is not enough to maintain social and political order, and it can become necessary to resort to arms.[79] This unfortunate reality is tempered by the assertion that even military victory is "defeat" in the sense that it requires an expenditure of a state's manpower and resources. As the *Sun-tzu* states, "To win a hundred victories in a hundred battles is not the highest excellence; the highest excellence is to subdue the enemy's army without fighting at all."[80] Similarly, the *Sun Pin* insists that "even ten victories out of ten, while evidencing an able commander, is still a source of national misfortune."[81] For this reason, war is justifiable only when all possible alternatives have been exhausted, and must be entertained with the utmost seriousness. The first line of the *Sun-tzu* declares: "War is a vital matter of state."[82] The first priority is the avoidance of warfare if at all possible. Once, however, a commitment has been made to a military course of action, the project becomes to achieve victory at the minimum cost. The able commander's first concern is to guarantee the integrity of his own forces: "He must use the principle of keeping himself intact to compete in the world."[83] After all, "invincibility depends on oneself."[84] The ruler commissions the able commander as a means of achieving victory with minimum loss. From the perspective of his cultivated humanity, he, like his ruler, regards warfare as a losing proposition that must be approached with the utmost caution and gravity, and with absolute control.

A second characteristic of the able commander is that he is active rather than reactive—he takes the offense and controls the situation: The expert in battle moves the enemy, and is not moved by him.[85] Such control is evident where defense itself is always offense:

> Do not depend on the enemy not coming; depend rather on being ready for him. Do not depend on the enemy not attacking; depend rather on having a position that cannot be attacked.[86]

Always maintaining the offense requires precision: "War is such that the supreme consideration is speed."[87] Speed is certainly defined in terms of timing:

> In advancing he cannot be resisted because he bursts through the enemy's weak points; in withdrawing he cannot be pursued because, being so quick, he cannot be caught.[88]

Moreover, speed in the sense of a short duration of battle is also desirable:

> In joining battle, seek the quick victory.... I have heard tell of a foolish haste, but I have yet to see a case of cleverly dragging on the hostilities. There never has been a state that has benefited from an extended war.[89]

Any deviation from this attitude represents military adventurism, and is outrightly condemned: "...one who takes pleasure in military affairs shall ultimately perish, while one who seeks to profit from victory shall incur disgrace."[90]

The fundamental question the *Sun-tzu* seeks to respond to is how does the enlightened ruler achieve victory at the minimum cost? The answer, then, is the ruler must give free rein to the consummate military commander.

THE EXEMPLARY COMMANDER

The emphasis in the *Sun-tzu* placed on the effective selection of military personnel reflects a fundamental assumption in the tradition. The first and foremost defining feature of the consummate military commander is that he must be an exemplary person (*chün tzu*), and must ply his military skills from a foundation of superior character. In this respect, the military commander is like any other officer in the service of the state. His ability to achieve great things within the parameters of his office—his efficacy—is a function of his cultivated humanity rather than any specific set of skills:

> A commander who advances without any thought of winning personal fame and withdraws in spite of certain punishment, whose only concern is to protect his people and promote the interests of his ruler, is the nation's treasure.[91]

What it means to be a person of exemplary character is defined in the text in the standard Confucian "virtue" vocabulary of "wisdom, integrity, humanity, courage, and discipline."[92] A commander defined in such holistic terms "is the side-guard on the carriage of state. Where this guard is in place, the state will certainly be strong...."[93] It·is by virtue of his status as an exemplary person, and the consonance this gives him with the tradition as embodied in his ruler, that the commander has sufficient authority on the battlefield to place him at the center of the centripetal field. It is from this particular perspective in the hierarchy, then, that he sets about the configuring of an optimal harmony.

The exemplary commander in the context of warfare stands as the self-organizing center, where the chaos of battle, far from interfering with order, feeds into and stimulates it. For the *Sun-tzu,* "the commander who understands war is the final arbiter of people's lives, and lord over the security of the state."[94] The first condition of effective command is that this commander must have complete control of the campaign, unchallenged even by the authority of the ruler at home: "The side on which the commander is able and the ruler does not interfere will take the victory."[95] The *Sun-tzu* is both explicit and emphatic on this point: "There are ... commands from the ruler not to be obeyed."[96]

The reason why, in this model, the commander must have sole control over his localized area is because an effective harmony must be pursued through the coordination of the immediate constituent elements, unmediated by some distant and undoubtedly less informed perspective:

> Thus, if the way (*tao*) of battle guarantees you victory, it is right for you to insist on fighting even if the ruler has said not to; where the way (*tao*) of battle does not allow victory, it is right for you to refuse to fight even if the ruler has said you must.[97]

If the military command has to take its orders from a source that cannot possibly be apprised of the full circumstances, the situation is tantamount to making decisions without a comprehensive understanding of the battlefield.

FOREKNOWLEDGE (*CHIH*)

The nature and effectiveness of the commander's "wisdom" requires comment. The commander must understand that any set of circumstances is the consequence of a dynamic process of organically related, mutually determining conditions. All of the characteristics of the able military commander follow from this insight into the interdependent nature of circumstances. He is aware that conditions constituting a situation are correlative and interdependent, and that what affects any one situation in this process to some greater or lesser degree has an effect on the whole field of conditions.

The complex of forces that define a battle situation are organically related, and hence, in spite of analogies that might be made with other seemingly similar engagements, each event must be respected for its particularity. Familiar patterns can be unpredictable because slight variations, when magnified through this organic relationship, can have massive consequences; minute fluctuations can amplify into dramatic changes. The uniqueness of each and any situation makes globalization precarious, and forces the commander to take each engagement on its own terms.

Complex systems such as battle conditions are rich in information—information that must be acquired immediately. The commander's wisdom must be funded by direct access to persons who serve him as eyes on the site-specific conditions, and who enable him to anticipate the outcome. To be reliable, information must be firsthand. The commander in defining the configuration of his forces treats his spatial form as a temporal flow. There is thus an important relationship between intelligence and timing. Once the specific time has past, information loses its strategic function and importance, and at best retains only historical value.

The effective gathering and dissemination of information can constitute an additional, albeit intangible, battle line:

> Intelligence is of the essence in warfare—it is what the armies depend upon in their every move.[98]

Ideally, effective intelligence provides clear discernment of the enemy's situation and a full concealment of one's own:

> If we can make the enemy show his position (*hsing*) while concealing ours from him, we will be at full force where he is divided.[99]

Such intelligence is by its immediacy distinguished from other apparent sources of information, such as the application of historical precedents, or revelatory knowledge gained from divination practices:

> Thus the reason the farsighted ruler and his superior commander conquer the enemy at every move, and achieve successes far beyond the reach of the common crowd, is foreknowledge. Such foreknowledge cannot be had from ghosts and spirits, educed by comparison with past events, or verified by astrological calculations. It must come from people—people who know the enemy's situation.[100]

Even though *Sun-tzu* advocates seeking victory from strategic advantage rather than from one's men, he also makes it clear that it is only by select-

ing the right person that one is able to exploit the strategic advantage.[101]
The commander must spare no cost in finding the right person and in acquiring reliable intelligence:

> If, begrudging the outlay of ranks, emoluments, and a hundred pieces of gold, a commander does not know the enemy's situation, his is the height of inhumanity. Such a person is no man's commander, no ruler's counsellor, and no master of victory.[102]

Front-line reconnaissance must be fortified by covert operations. Of particular importance is the selection of operatives and saboteurs for espionage.

> Thus only those farsighted rulers and their superior commanders who can get the most intelligent people as their spies are destined to accomplish great things.[103]

It is because the commander's wisdom is resolutely performative that it is foreknowledge—it creates the victory. This wisdom entails a cognitive understanding of those circumstances that bear on the local situation, an awareness of possible futures, the deliberate selection of one of these futures, and the capacity to manipulate the prevailing circumstances and to dispose of them in such a way as to realize the desired future. The emphasis here is on the commander's access to intelligence acquired directly from the specific situation, and his capacity to thus control events.

Given that warfare is always defeat, the commander in pursuing the best possible outcome seeks to disarm the enemy without ever joining him on the battlefield:

> ... the expert in using the military subdues the enemy's forces without going to battle, takes the enemy's walled cities without launching an attack, and crushes the enemy's state without a protracted war.[104]

The *Sun-tzu* defines military wisdom in terms of sober and methodical deliberation and planning. Where at all possible, the commander attempts to defeat the enemy with this careful planning rather than military might:

> ... the best military policy is to attack strategies; the next to attack alliances; the next to attack soldiers; and the worst to assault walled cities.[105]

In any case, the commander never enters a battle where there is any question of defeat. Victory must be a predetermined certainty:

...the victorious army only enters battle after having first won the victory, while the defeated army only seeks victory after having first entered the fray.[106]

As a consequence, the able commander is not the one who is celebrated for daring and courage, for his victory requires neither:

He whom the ancients called an expert in battle gained victory where victory was easily gained. Thus the battle of the expert is never an exceptional victory, nor does it win him reputation for wisdom or credit for courage. His victories in battle are unerring. Unerring means that he acts where victory is certain, and conquers an enemy that has already lost.[107]

The foreknowledge required to be in complete control of events is gained by acquiring complete information, by anticipating the ensuing situations, and by going over and scoring the battle strategy in a formal exercise:

It is by scoring many points that one wins the war beforehand in the temple rehearsal of the battle....[108]

This is a somewhat obscure passage, which seems to describe "mock battles" acted out in advance. But it likely refers to the practice of assessing relative battlefield strengths by identifying a set of relevant categories, and then using counting rods or some similar device to indicate an advantage on one side or the other, enabling one to thus predict the outcome.[109]

Since all *yin*-like correlations can only be fully fathomed by reference to *yang*, understanding the local situation completely entails understanding both sides of all correlative pairs:

...the deliberations of the wise commander are sure to assess jointly both advantages and disadvantages. In taking full account of what is advantageous, he can fulfill his responsibilities; in taking full account of what is disadvantageous, his difficulties become resolvable.[110]

It is as important to keep information from the enemy as it is to acquire information about him. In the absence of information, the enemy has only unconcentrated force that is dissipated across the lines of attack:

If our army is united as one and the enemy's is fragmented, in using the undivided whole to attack his one, we are many to his few.[111]

The integrity of one's position conceals the details of his battle configuration from the enemy's view, and makes one impenetrable:

> The ultimate skill in taking up a strategic position (*hsing*) is to have no form (*hsing*). If your position is formless (*hsing*), the most carefully concealed spies will not be able to get a look at it, and the wisest counsellors will not be able to lay plans against it.[112]

Another way to achieve this desired "formlessness" is through deceit. Deceit is used to become "one" by reconciling correlations. If one is close but seems far to the enemy, his distance is indeterminate. If one is slow but seems fast to the enemy, his speed is indeterminate.

> Warfare is the art (*tao*) of deceit. Therefore, when able, seem to be unable; when ready, seem unready; when nearby, seem far away; and when far away, seem near. If the enemy seeks some advantage, entice him with it. If he is in disorder, attack him and take him. If he is formidable, prepare against him. If he is strong, evade him. If he is incensed, provoke him. If he is humble, encourage his arrogance. If he is rested, wear him down. If he is internally harmonious, sow divisiveness in his ranks. Attack where he is not prepared; go by way of places where it would never occur to him you would go.[113]

The consummate commander is able to achieve and retain control of a military situation in a way analogous to an able ruler's control of the civil situation and a farmer's control over his crops: by a thorough understanding of the conditions determining the situation and the manipulation of these circumstances to his chosen end:

> He who knows the enemy and himself
> Will never in a hundred battles be at risk.[114]

Introduction to the

Translations

In translating the core thirteen chapters of the *Sun-tzu: The Art of Warfare* that comprise Part I of this volume, I have relied upon the *Sun-tzu chiao-shih* edited by Wu Chiu-lung et al. and published in 1990. It reflects the judgment of a group of China's most prominent scholars presently working on the reconstruction of the military texts and is informed by a detailed knowledge of the recent archaeological finds. I have followed this work for the Chinese text with a few typographical corrections.

In translating the five additional chapters recovered in the Yin-ch'üeh-shan dig that comprise Part II, I have followed the authoritative *Yin-ch'üeh-shan Han-mu chu-chien* (Bamboo strips recovered from the Han tombs at Silver Sparrow mountain) Collection, Vol. I prepared by the Yin-ch'üeh-shan Han-mu chu-chien cheng-li hsiao-tsu (Committee for the Reconstruction of the Yin-ch'üeh-shan Han strips) and published by Wen-wu Publishing House in 1985.

For the encyclopedic materials I have translated in Part III, I worked from the appendixes of Yang Ping-an's *Sun-tzu hui-chien* (1986) and Huang K'uei's *Sun-tzu tao-tu* (1989) which are based on the Ch'ing dynasty collections of Pi I-hsün, *Sun-tzu hsü-lu* (*Citations from Sun-tzu*), and Wang Jen-chün, *Sun-tzu i-wen* (unpublished text preserved in the archives of the Shanghai Library). I have then checked these citations against authoritative editions of the encyclopedias and corrected them accordingly (see Bibliography). These Ch'ing dynasty collections have been augmented from the cache of strips dating from the late Western Han dynasty discovered in 1978 in Tomb #115 of the Sun family compound in Ta-t'ung county, Ch'ing-hai province. Of the sixty-odd fragments reported in the *Wen-wu* (*Cultural Relics* 1981:2) description of this find, six strips had "Master Sun" on them, suggesting some relationship with the *Sun-tzu*. The contemporary scholar Li Ling (1983) rejects any suggestion that these strips are lost text of the *Sun-tzu* or some related military treatise, as was suggested in first reports of this find in *Wen-wu*. He argues that these strips are works on military regulations that cite the *Sun-tzu*.

The disorderly and corrupt condition of the bamboo strips and the fact that there has not always been an extant text that can be used for comparison has made the project of arranging these strips and reconstructing an intelligible text from them a task fraught with difficulties. Given the necessary amount of speculation involved in reassembling the strips, conclusions can often be no more than tentative. There is a good possibility, for example, that material from texts other than the original *Sun-tzu* has crept into the reconstructed text.

Even where it is clear that certain strips belong together in a given chapter, the position of the chapter relative to the other chapters cannot always be determined with any confidence. Further, there are some strips that for one reason or another seem to belong to a given chapter, but are devoid of any further context. Where these fragments make sense and add to our understanding of the chapter, they have been translated and appended separately at the end of the chapter. Otherwise, they have been omitted.

Chapter titles that have been added at the discretion of the Yin-ch'üeh-shan Committee are provided in square brackets. Where it is apparent that the bamboo strips contained in any one chapter, in spite of missing characters or strips, constitute a continuous passage, they are translated accordingly. Where there is a break in a passage, the translation is also broken at this point. If the text is interrupted with lacunae, this is indicated by ellipses. Where what is missing can be restored from context with some degree of confidence, a translation is provided in italics within square brackets.

ROGER T. AMES is one of the leading interpreters of Chinese philosophy in America today. He received his Ph.D. from the School of Oriental and African Studies, University of London in 1978, under the supervision of Professor D. C. Lau. He is presently Professor of Comparative Philosophy and Director of the Center for Chinese Studies at the University of Hawai'i. He edits the journal *Philosophy East & West*, and is Executive Editor of *China Review Interna-*

tional. His major publications include *The Art of Rulership: Studies in Ancient Chinese Political Thought* (1983); *Thinking Through Confucius* (with David L. Hall) (1987); *Nature in Asian Traditions of Thought: Essays in Environmental Philosophy* (edited with J. Baird Callicott) (1989); *Interpreting Culture Through Translation: A Festschrift in Honor of D. C. Lau* (edited with Ng Mausang and Chan Sin-wai) (1991).

SUN-TZU:
PART I

THE
THIRTEEN-CHAPTER
TEXT

計 篇

孫子曰：兵者，國之大事也。死生之地，存亡之道，不可不察也。故經之以五，效（校）之以計而索其請（情）：一曰道，二曰天，三曰地，四曰將，五曰法。道者，令民與上同意也，故可與之死，可與之生而不詭也。天者，陰陽、寒暑、時制也。地者，高下、遠近、險易、廣狹、死生也。將者，知（智）、信、仁、勇、嚴也。法者，曲制、官道、主用也。凡此五者，將莫不聞，知之者勝，不知者不勝。故校之以計，而索其情。曰：主孰有道？將孰有能？天地孰得？法令孰行？兵衆孰强？士卒孰練？賞罰（罰）孰明？吾以此知勝負矣。

將聽吾計，用之必勝，留之；將不聽吾計，用之必敗，去之。計利以聽，乃為之勢，以佐其外。勢者，因利而制權也。兵者，詭道也。故能而示之不能，用而視（示）之不用，近而視（示）之遠，遠而視（示）之近。利而誘之，亂而取之，實而備之，强而避之，怒而撓（撓）之，卑而驕之，佚而勞之，親而離之。攻其無備，出其不意。此兵家之勝，不可先傳也。

夫未戰而廟筭勝者，得筭多也；未戰而廟筭不勝者，得筭少也。多筭勝，少筭不勝，而況於无筭乎！吾以此觀之，勝負見矣。

CHAPTER 1

ON ASSESSMENTS

Master Sun said:

War[115] is a vital matter of state. It is the field on which life or death is determined and the road that leads to either survival or ruin, and must be examined with the greatest care.

Therefore, to gauge the outcome of war we must appraise the situation on the basis of the following five criteria, and compare the two sides by assessing their relative strengths. The first of the five criteria is the way (*tao*), the second is climate, the third is terrain, the fourth is command, and the fifth is regulation.

The way (*tao*) is what brings the thinking of the people in line with their superiors. Hence, you can send them to their deaths or let them live, and they will have no misgivings one way or the other.

Climate is light and shadow, heat and cold, and the rotation of the seasons.[116]

Terrain refers to the fall of the land,[117] proximate distances, difficulty of passage, the degree of openness, and the viability of the land for deploying troops.

Command is a matter of wisdom, integrity, humanity, courage, and discipline.

And regulation entails organizational effectiveness, a chain of command, and a structure for logistical support.

All commanders are familiar with these five criteria, yet it is he who masters them who takes the victory, while he who does not will not prevail.

Therefore, to gauge the outcome of war we must compare the two sides by assessing their relative strengths. This is to ask the following questions:

Which ruler has the way (*tao*)?
Which commander has the greater ability?
Which side has the advantages of climate and terrain?
Which army follows regulations and obeys orders more strictly?
Which army has superior strength?
Whose officers and men are better trained?
Which side is more strict and impartial in meting out rewards and punishments?

On the basis of this comparison I know who will win and who will lose.

If you heed my assessments, dispatching troops into battle would mean certain victory, and I will stay. If you do not heed them, dispatching troops would mean certain defeat, and I will leave.[118]

Having heard what can be gained from my assessments, shape a strategic advantage (*shih*) from them to strengthen our position. By "strategic advantage" I mean making the most of favorable conditions (*yin*) and tilting the scales in our favor.

Warfare is the art (*tao*) of deceit. Therefore, when able, seem to be unable; when ready, seem unready; when nearby, seem far away; and when far away, seem near. If the enemy seeks some advantage, entice him with it. If he is in disorder, attack him and take him. If he is formidable, prepare against him. If he is strong, evade him. If he is incensed, provoke him. If he is humble, encourage his arrogance. If he is rested, wear him down. If he is internally harmonious, sow divisiveness in his ranks. Attack where he is not prepared; go by way of places where it would never occur to him you would go. These are the military strategist's calculations for victory—they cannot be settled in advance.

It is by scoring many points that one wins the war beforehand in the temple rehearsal of the battle; it is by scoring few points that one loses the war beforehand in the temple rehearsal of the battle. The side that scores many points will win; the side that scores few points will not win, let alone the side that scores no points at all. When I examine it in this way, the outcome of the war becomes apparent.[119]

CHAPTER 2

ON WAGING BATTLE

Master Sun said:

The art of warfare is this:[120]

For an army of one thousand fast four-horse chariots, one thousand four-horse leather-covered wagons, and one hundred thousand armor-clad troops, and for the provisioning of this army over a distance of a thousand *li*,[121] what with expenses at home and on the field, including foreign envoys and advisors, materials such as glue and lacquer, and the maintenance of chariots and armor, only when you have in hand one thousand pieces of gold for each day can the hundred thousand troops be mobilized.

In joining battle, seek the quick victory. If battle is protracted, your weapons will be blunted and your troops demoralized. If you lay siege to a walled city, you exhaust your strength. If your armies are kept in the field for a long time, your national reserves will not suffice. Where you have blunted your weapons, demoralized your troops, exhausted your strength and depleted all available resources, the neighboring rulers will take advantage of your adversity to strike. And even with the wisest of counsel, you will not be able to turn the ensuing consequences to the good.

Thus in war, I have heard tell of a foolish haste, but I have yet to see a case of cleverly dragging on the hostilities. There has never been a state that has benefited from an extended war. Hence, if one is not fully cognizant of the evils of waging war, he cannot be fully cognizant either of how to turn it to best account.

作　戰　篇

孫子曰：凡用兵之法，馳車千駟，革車千乘，帶甲十萬，千里饋糧（糧），則內外之費，賓客之用，膠漆之材，車甲之奉，日費千金，然後十萬之師舉矣。其用戰也貴勝，久則頓（鈍）兵挫銳，攻城則力屈，久暴師則國用不足。夫鈍兵挫銳，屈力殫貨，則諸侯乘其弊而起，雖有知（智）者，不能善其後矣。故兵聞拙速，未覩巧之久也。夫兵久而國利者，未之有也。故不盡知用兵之害者，則不能盡知用兵之利也。善用兵者，役不再籍，糧不三載，取用於國，因糧（糧）於敵，故軍食可足也。國之貧於師者：遠師者遠輸，遠輸則百姓貧。近市（師）者貴賣，貴賣則財竭，財竭則急於丘役。屈力中原，內虛於家，百姓之費十去其七。公家之費，破車罷馬，甲胄矢弩，戟楯矛櫓，丘牛大車，十去其六。故智將務食於敵，食敵一鍾，當吾二十鍾，萁秆一石，當吾二十石。故殺適（敵）者，怒也，取敵之利者，貨也。故車戰，得車十乘已上，賞其先得者，而更其旌旗，車雜而乘之，卒善而養之，是胃（謂）勝敵而益強。故兵貴勝，不貴久。故知兵之將，民之司命，國家安危之主也。

The expert in using the military does not conscript soldiers more than once or transport his provisions repeatedly from home. He carries his military equipment with him, and commandeers (*yin*) his provisions from the enemy. Thus he has what he needs to feed his army.

A state is impoverished by its armies when it has to supply them at a great distance. To supply an army at a great distance is to impoverish one's people. On the other hand, in the vicinity of the armies, the price of goods goes up. Where goods are expensive, you exhaust your resources, and once you have exhausted your resources, you will be forced to increase district exactions for the military. All your strength is spent on the battlefield, and the families on the home front are left destitute. The toll to the people will have been some 70 percent of their property; the toll to the public coffers in terms of broken-down chariots and worn-out horses, body armor and helmets, crossbows and bolts, halberds and bucklers, lances and shields, draft oxen and heavy supply wagons will be some 60 percent of its reserves.

Therefore, the wise commander does his best to feed his army from enemy soil. To consume one measure of the enemy's provisions is equal to twenty of our own; to use up one bale of the enemy's fodder is equal to twenty of our own.

Killing the enemy is a matter of arousing the anger of our men; snatching the enemy's wealth is a matter of dispensing the spoils.[122] Thus, in a chariot battle where more than ten war chariots have been captured, reward those who captured the first one and replace the enemy's flags and standards with our own. Mix the chariots in with our ranks and send them back into battle; provide for the captured soldiers and treat them well. This is called increasing our own strength in the process of defeating the army.

Hence, in war prize the quick victory, not the protracted engagement. Thus, the commander who understands war is the final arbiter of people's lives, and lord over the security of the state.

謀攻篇

孫子曰：凡用兵之法，全國為上，破國次之；全軍為上，破軍次之；全旅為上，破旅次之；全卒為上，破卒次之；全伍為上，破伍次之。是故百戰百勝，非善之善者也；不戰而屈人之兵，善之善者也。

故上兵伐謀，其次伐交，其次伐兵，其下攻城。攻城之法，為不得已，修（修）櫓轒轀，具器械，三月而後成，距闉，有（又）三月而後已。將不勝其忿而蟻附之，殺士三分之一，而城不拔者，此攻之災（災）也。故善用兵者，屈（屈）人之兵而非戰也，拔人之城而非攻也，毀人之國而非久也，必以全爭於天下，故兵不頓而利可全，此謀攻之法也。

故用兵之法：十則圍之，五則攻之，倍則戰之，敵則能分之，少則能守之，不若則能避之。故小敵之堅，大敵之擒也。

夫將者，國之輔也，輔周則國必強，輔隙則國必弱。故君之所以患於軍者三：不知軍之不可以進而謂之進，不知軍之不可以退而謂之退，是謂縻軍。不知三軍之事，而同三軍之政，則軍士惑矣；不知三軍之權，而同三軍之任，則軍士疑矣。三軍譬（既）惑且疑，則諸侯之難至矣，是謂亂軍引勝。

故知勝有五：知可以戰與不可以戰者勝，識衆寡之用者勝，上下同欲者勝，以虞侍（待）不虞者勝，將能而君不御者勝。此五者，知勝之道也。

故曰：知皮（彼）知己，百戰不殆；不知彼而知己，一勝一負；不知彼不知己，每戰必殆。

CHAPTER 3

Planning the Attack

Master Sun said:

The art of warfare is this:

It is best to keep one's own state intact; to crush the enemy's state is only a second best. It is best to keep one's own army, battalion, company, or five-man squad intact; to crush the enemy's army, battalion, company, or five-man squad is only a second best.[123] So to win a hundred victories in a hundred battles is not the highest excellence; the highest excellence is to subdue the enemy's army without fighting at all.

Therefore, the best military policy is to attack strategies; the next to attack alliances; the next to attack soldiers; and the worst to assault walled cities.

Resort to assaulting walled cities only when there is no other choice. To construct siege screens and armored personnel vehicles and to assemble all of the military equipment and weaponry necessary will take three months, and to amass earthen mounds against the walls will take another three months. And if your commander, unable to control his temper, sends your troops swarming at the walls, your casualties will be one in three and still you will not have taken the city. This is the kind of calamity that befalls you in laying siege.

Therefore, the expert in using the military subdues the enemy's forces without going to battle, takes the enemy's walled cities without launching an attack, and crushes the enemy's state without a protracted war. He must use the principle of keeping himself intact to compete in the world. Thus, his weapons will not be blunted and he can keep his edge intact. This then is the art of planning the attack.[124]

Therefore the art of using troops is this:

When ten times the enemy strength, surround him; when five times, attack him; when double, engage him; when you and the enemy are equally matched, be able to divide him;[125] when you are inferior in numbers, be able to take the defensive; and when you are no match for the enemy, be able to avoid him. Thus what serves as secure defense against a small army will only be captured by a large one.[126]

The commander is the side-guard on the carriage of state.[127] Where this guard is in place, the state will certainly be strong; where it is defective, the state will certainly be weak.

There are three ways in which the ruler can bring grief to his army:[128]

To order an advance, not realizing the army is in no position to do so, or to order a retreat, not realizing the army is in no position to withdraw—this is called "hobbling the army."

To interfere in the administration of the army while being ignorant of its internal affairs will confuse officers and soldiers alike.

To interfere in military assignments while being ignorant of exigencies will lose him the confidence of his men.

Once his army has become confused and he has lost the confidence of his men, aggression from his neighboring rulers will be upon him. This is called sowing disorder in your own ranks and throwing away the victory.

Therefore there are five factors in anticipating which side will win:

The side that knows when to fight and when not to will take the victory.

The side that understands how to deal with numerical superiority and inferiority in the deployment of troops will take the victory.

The side that has superiors and subordinates united in purpose will take the victory.

The side that fields a fully prepared army against one that is not will take the victory.

The side on which the commander is able and the ruler does not interfere will take the victory.

These five factors are the way (*tao*) of anticipating victory.

Thus it is said:

He who knows the enemy and himself
Will never in a hundred battles be at risk;

He who does not know the enemy but knows himself
Will sometimes win and sometimes lose;
He who knows neither the enemy nor himself
Will be at risk in every battle.[129]

形　篇

孫子曰：昔之善戰者，先爲不可勝，以侍（待）適（敵）之可勝。不可勝

在適（敵）。故善戰者，能爲不可勝，不能使適（敵）必可勝。故曰：勝可智（知），

而不可爲。不可勝者，守也；可勝者，攻也。守則不足，攻則不勝。善守者，臧（藏）於

九地之下，善攻者，動於九天之上，故能自葆（保）而全勝也。見勝不過衆人之所

知，非善之善者也；戰勝而天下曰善，非善之善者也。故舉秋毫不爲多力，見日

月不爲明目，聞雷霆不爲蔥（聰）耳。古之所胃（謂）善戰者，勝於易勝者也。故善

戰者之勝也，無奇〔勝〕，無智名，無勇功。故其戰勝不貸（忒）不貸（忒）者，其

所錯（措）必勝，勝已敗者也。故善戰者，立於不敗之地，而不失敵之敗也。是故，

勝兵先勝而後求戰，敗兵先戰而後求勝。善用兵者，脩（修）道而保法，故能爲勝

敗正。法：一曰度，二曰量，三曰數，四曰稱，五曰勝。地生度，度生量，量生數，

數生稱，稱生勝。故勝兵若以洫（鎰）稱朱（銖），敗兵若以朱（銖）稱洫（鎰）。稱勝

者之戰民也，若決積水於千仞（仞）之谿者，形也。

CHAPTER 4

STRATEGIC POSITIONS
(HSING)[130]

Master Sun said:

Of old the expert in battle would first make himself invincible and then wait for the enemy to expose his vulnerability. Invincibility depends on oneself; vulnerability lies with the enemy.[131] Therefore the expert in battle can make himself invincible, but cannot guarantee for certain the vulnerability of the enemy. Hence it is said:

Victory can be anticipated,
But it cannot be forced.

Being invincible lies with defense; the vulnerability of the enemy comes with the attack.[132] If one assumes a defensive posture, it is because the enemy's strength is overwhelming; if one launches the attack, it is because the enemy's strength is deficient.[133] The expert at defense conceals himself in the deepest recesses of the earth; the expert on the attack strikes from out of the highest reaches of the heavens. Thus he is able to both protect himself and to take the complete victory.

To anticipate the victory is not going beyond the understanding of the common run; it is not the highest excellence. To win in battle so that the whole world says "Excellent!" is not the highest excellence. Hence, to lift an autumn hair is no mark of strength; to see the sun and moon is no mark of clear-sightedness; to hear a thunder clap is no mark of keen hearing. He whom the ancients called an expert in battle gained victory where victory was easily gained. Thus the battle of the expert is never an exceptional victory, nor does it win him reputation for wisdom or credit for courage.[134] His victories in battle are unerring.[135] Unerring means that he acts where victory is certain, and conquers an enemy that has already lost.

Therefore, the expert in battle takes his stand on ground that is unassailable, and does not miss his chance to defeat the enemy. For this reason, the victorious army only enters battle after having first won the victory, while the defeated army only seeks victory after having first entered the fray.[136]

The expert in using the military builds upon the way (*tao*) and holds fast to military regulations,[137] and thus is able to be the arbiter of victory and defeat.[138]

Factors in the art of warfare are: First, calculations; second, quantities; third, logistics; fourth, the balance of power; and fifth, the possibility of victory. Calculations are based on the terrain, estimates of available quantities of goods are based on these calculations, logistical strength is based on estimates of available quantities of goods, the balance of power is based on logistical strength, and the possibility of victory is based on the balance of power.

Thus a victorious army is like weighing in a full hundredweight against a few ounces, and a defeated army is like pitting a few ounces against a hundredweight.[139] It is a matter of strategic positioning (*hsing*) that the army that has this weight of victory on its side, in launching its men into battle, can be likened to the cascading of pent-up waters thundering through a steep gorge.[140]

CHAPTER 5

STRATEGIC ADVANTAGE (SHIH)

Master Sun said:

In general, it is organization[141] that makes managing many soldiers the same as managing a few. It is communication with flags and pennants[142] that makes fighting with many soldiers the same as fighting with a few. It is "surprise" (*ch'i*) and "straightforward" (*cheng*) operations that enable one's army to withstand the full assault of the enemy force[143] and remain undefeated.[144] It is the distinction between "weak points" and "strong points" that makes one's army falling upon the enemy a whetstone being hurled at eggs.

Generally in battle use the "straightforward" to engage the enemy and the "surprise" to win the victory. Thus the expert at delivering the surprise assault is as boundless as the heavens and earth, and as inexhaustible as the rivers and seas.[145] Like the sun and moon, he sets only to rise again; like the four seasons, he passes only to return again.

There are no more than five cardinal notes, yet in combination, they produce more sounds than could possibly be heard; there are no more than five cardinal colors, yet in combination, they produce more shades and hues than could possibly be seen; there are no more than five cardinal tastes, yet in combination, they produce more flavors than could possibly be tasted. For gaining strategic advantage (*shih*) in battle, there are no more than "surprise" and "straightforward" operations, yet in combination, they produce inexhaustible possibilities. "Surprise" and "straightforward" operations give rise to each other endlessly just as a ring is without a beginning or an end.[146] And who can exhaust their possibilities?

勢 篇

孫子曰：凡治眾如治寡，分數是也；鬥眾如鬥寡，形名是也；三軍之眾，可使畢受適（敵）而无敗者，奇正是也。兵之所加，如以段（碬）投卵者，虛實是也。凡戰者，以正合，以奇勝。故善出奇者，無窮如天地，不謁（竭）如江河。冬（終）而復始，日月是也；死而復生，四時是也。聲不過五，五聲之變不可勝聽也；色不過五，五色之變不可勝觀也；味不過五，五味之變不可勝嘗也。戰勢不過奇正，奇正之變不可勝窮也。奇正環（還）相生，如環之毋（無）端，孰能窮之？激水之疾，至於漂石者，勢也；鷙鳥之擊，至於毀折者，節也。是故善戰者，其勢險，其節短。勢如彍弩，節如發機。紛紛紜紜，鬥亂而不可亂也；渾渾沌沌，形圓而不可敗也。亂生於治，怯（怯）生於患（勇），弱生於強。治亂，數也；患（勇）怯（怯），勢（勢）也；強弱，形也。故善動適（敵）者，刑（形）之，適（敵）必從之；予之，敵必取之。以此動之，以卒侍（待）之。故善戰者，求之於執（勢），不責於人，故能擇人而任勢。任勢者，其戰人也，如轉木石；木石之生（性）：安則静，危則動，方則止，圓則行。故善戰人之勢，如轉圓石於千仞之山者，勢也。

That the velocity of cascading water can send boulders bobbing about is due to its strategic advantage (*shih*). That a bird of prey when it strikes[147] can smash its victim to pieces is due to its timing. So it is with the expert at battle that his strategic advantage (*shih*) is channeled and his timing is precise. His strategic advantage (*shih*) is like a drawn crossbow and his timing is like releasing the trigger. Even amidst the tumult and the clamor of battle, in all its confusion, he cannot be confused. Even amidst the melee and the brawl of battle, with positions shifting every which way, he cannot be defeated.

Disorder is born from order; cowardice from courage; weakness from strength. The line between disorder and order lies in logistics (*shu*); between cowardice and courage, in strategic advantage (*shih*); and between weakness and strength, in strategic positioning (*hsing*). Thus the expert at getting the enemy to make his move shows himself (*hsing*), and the enemy is certain to follow. He baits the enemy, and the enemy is certain to take it. In so doing,[148] he moves the enemy, and lies in wait for him with his full force.

The expert at battle seeks his victory from strategic advantage (*shih*) and does not demand it from his men. He is thus able to select the right men and exploit the strategic advantage (*shih*).[149] He who exploits the strategic advantage (*shih*) sends his men into battle like rolling logs and boulders. It is the nature of logs and boulders that on flat ground, they are stationary, but on steep ground, they roll; the square in shape tends to stop but the round tends to roll. Thus, that the strategic advantage (*shih*) of the expert commander in exploiting his men in battle can be likened to rolling round boulders down a steep ravine thousands of feet high says something about his strategic advantage (*shih*).[150]

虛實篇

孫子曰：凡先處戰地而侍（待）敵者失（佚），後處戰地而趨戰者勞。故善戰者，致人而不致於人。能使適（敵）人自至者，利之也；能使適（敵）人不得至者，害之也。故敵佚能勞之、飽能飢之、安能動之者，出其所必趨也。行於無人之地也；攻而必取者，攻其所不守也；守而必固者，守其所必攻也。故善攻者，適（敵）不知其所守；善守者，適（敵）不知其所攻。微乎微乎，至於無形。神乎神乎，至於無聲，故能爲適（敵）之司命。進而不可御者，衝其虛也；退而不可追者，速而不可及也。故我欲戰，適（敵）雖高壘深溝，不得不與我戰者，攻其所必救也；我不欲戰，畫地而守之，適（敵）不得與我戰者，乖其所之也。故刑（形），則我專（專）而適（敵）分；我專（專）爲壹，適（敵）分而爲十，是以十攻其壹也。則我衆而適（敵）寡，能以衆擊寡者，則吾之所與戰者約矣。吾所與戰之地不可知，不可知，則適（敵）所備者多，敵所備者多，則吾所與戰者寡矣。故備

CHAPTER 6

WEAK POINTS AND STRONG POINTS

Master Sun said:

Generally he who first occupies the field of battle to await the enemy will be rested; he who comes later and hastens into battle will be weary. Thus the expert in battle moves the enemy, and is not moved by him. Getting the enemy to come of his own accord is a matter of making things easy for him; stopping him from coming is a matter of obstructing him. Thus being able to wear down a well-rested enemy, to starve one that is well-provisioned, and to move one that is settled, lies in going by way of places where the enemy must hasten in defense.[151]

To march a thousand *li* without becoming weary is because one marches through territory where there is no enemy presence. To attack with the confidence of taking one's objective is because one attacks what the enemy does not defend. To defend with the confidence of keeping one's charge secure is because one defends where the enemy will not attack.[152] Thus against the expert in the attack, the enemy does not know where to defend, and against the expert in defense, the enemy does not know where to strike.

So veiled and subtle,
To the point of having no form (*hsing*);
So mysterious and miraculous,
To the point of making no sound.
Therefore he can be arbiter of the enemy's fate.

前則後寡，備後則前寡，備左則右寡，備右則左寡，无所不備，則无所不寡。寡者，備人者也；衆者，使人備己者也。故知戰之地，知戰之日，則可千里而戰；不知戰地，不知戰日，則左不能救右，右不能救左，前不能救後，後不能救前，而（况）遠者數十里，近者數里乎！以吾度之，越人之兵雖多，亦奚益於勝哉（哉）？故曰：勝可爲也。適（敵）雖（雖）衆，可使無所（鬥）。故策之而知得失之計，作之而知動靜之理，形之而知死生之地，角之而知有餘不足之處。故刑（形）兵之極，至於无刑（形）。无刑（形），則深間不能規（窺），知（智）者不能謀。因刑（形）而錯勝於衆，衆不能知；人皆知我所以勝之形，而莫知吾所以制勝之形。故其戰勝不復，而應刑（形）於無窮。夫兵刑（形）象水，水之行，辟（避）高而趨下，兵之勝，辟（避）實而擊虛。水因地而制行，兵因敵而制勝。故兵无成埶（勢），无恒刑（形）。能因敵變化而取勝者，謂之神。故五行无常勝，四時無常立（位），日有短長，月有死生。

In advancing he cannot be resisted because he bursts through the enemy's weak points; in withdrawing he cannot be pursued because, being so quick, he cannot be caught.

Thus, if we want to fight, the enemy has no choice but to engage us, even though safe behind his high walls and deep moats, because we strike at what he must rescue. If we do not want to fight, the enemy cannot engage us, even though we have no more around us than a drawn line, because we divert him to a different objective.

If we can make the enemy show his position (*hsing*) while concealing ours from him, we will be at full force where he is divided.[153] If our army is united as one and the enemy's is fragmented, in using the undivided whole to attack his one, we are many to his few. If we are able to use many to strike few, anyone we take the battle to will be in desperate circumstances.[154]

The place we have chosen to give the enemy battle must be kept from him. If he cannot anticipate us, the positions the enemy must prepare to defend will be many. And if the positions he must prepare to defend are many, then any unit we engage in battle will be few in number.

Thus if the enemy makes preparations by reinforcing his numbers at the front, his rear is weakened; if he makes preparations at the rear, his front is weakened; if he makes them on his left flank, his right is weakened; if he makes them on his right flank, his left is weakened. To be prepared everywhere is to be weak everywhere.

One is weak because he makes preparations against others; he has strength because he makes others prepare against him.

Thus if one can anticipate the place and the day of battle, he can march a thousand *li* to join the battle. But if one cannot anticipate either the place or the day of battle, his left flank cannot even rescue his right, or his right his left; his front cannot even rescue his rear, or his rear his front. How much more is this so when your reinforcements are separated by at least a few *li*, or even tens of *li*.

The way I estimate it, even though the troops of Yüeh are many, what good is this to them in respect to victory?[155] Thus it is said: Victory can be created. For even though the enemy has the strength of numbers, we can prevent him from fighting us.

Therefore, analyze the enemy's battle plan to understand its merits and its weaknesses; provoke him to find out the pattern of his movements; make him show himself (*hsing*) to discover the viability of his battle position; skirmish with him to find out where he is strong and where he is vulnerable.

The ultimate skill in taking up a strategic position (*hsing*) is to have no form (*hsing*).[156] If your position is formless (*hsing*), the most carefully concealed spies will not be able to get a look at it, and the wisest counsellors will not be able to lay plans against it. I present the rank and file with victories gained through (*yin*) strategic positioning (*hsing*), yet they are not able to understand them. Everyone knows the position (*hsing*) that has won me victory, yet none fathom how I came to settle on this winning position (*hsing*). Thus one's victories in battle cannot be repeated—they take their form (*hsing*) in response to inexhaustibly changing circumstances.

The positioning (*hsing*) of troops can be likened to water: Just as the flow of water avoids high ground and rushes to the lowest point, so on the path to victory avoid the enemy's strong points and strike where he is weak.[157] As water varies its flow according to (*yin*) the fall of the land, so an army varies its method of gaining victory according to (*yin*) the enemy.

Thus an army does not have fixed strategic advantages (*shih*) or an invariable position (*hsing*).[158] To be able to take the victory by varying one's position according to (*yin*) the enemy's is called being inscrutable (*shen*).[159]

Thus, of the five phases (*wu hsing*), none is the constant victor; of the four seasons, none occupies a constant position; the days are both short and long; the moon waxes and wanes.[160]

CHAPTER 7

ARMED CONTEST

Master Sun said:

The art of using troops is this: In the process of the commander's receiving his orders from the ruler, assembling his armies, mobilizing the population for war, and setting up his camp facing the enemy, there is nothing of comparable difficulty to the armed contest itself. What is difficult in the armed contest is to turn the long and tortuous route into the direct, and to turn adversity into advantage. Thus, making the enemy's road long and tortuous, lure him along it by baiting him with easy gains. Set out after he does, yet arrive before him. This is to understand the tactic of converting the tortuous and the direct.

Armed contest can be both a source of advantage and of danger.[161] If you mobilize your entire force to contend for some advantage, you arrive too late; if you abandon your base camp to contend for advantage, your equipment and stores will be lost. For this reason, if an army were to stow its armor and set off in haste, and stopping neither day nor night, force-march at double time for a hundred *li* to contend for some advantage, its commanders would all be taken, its strongest men would be out in front, the exhausted ones would lag behind, and as a rule only one tenth of its strength would reach the target.

Were it to travel fifty *li* at such a pace to contend for some advantage, the commander of the advance force would be lost, and as a rule only half of its strength would reach the target. Were it to travel thirty *li* at such a pace to contend for some advantage, only two thirds of its strength would reach the target. For this reason, if an army is without its equipment and

軍爭篇

孫子曰：凡用兵之法，將受命於君，合軍聚衆，交和而舍，莫難於軍爭。軍爭之難者，以迂爲直，以患爲利。故迂其途而誘之以利，後人發，先人至，此知迂（迂）直之計者也。故軍爭爲利，軍爭爲危。舉軍而爭利則不及，委軍而爭利則輜重捐。是故卷（卷）甲而趨，日夜不處倍道兼行，百里而爭利，則擒三軍將，勁者先，罷者後，其法十一而至。五十里而爭利，則厥（蹶）上軍將，其法半至。三十里而爭利，則三分之二至。是故軍毋（無）輜重則亡，無糧食則亡，无委責（積）則亡。故不知諸侯之謀者，不能豫交，不知山林、險阻、沮澤之刑（形）者，不能行軍，不用鄉（向）道（導）者，不能得地利。故兵以詐立，以利動，以分合爲變者也。故其疾如風，其徐如林，侵掠如火，不動如山，難知如陰，動如雷震。掠鄉分衆，廓地分利，懸（懸）權而動。先知汙（迂）直之計者勝，此軍爭之法也。軍政曰：言不相聞，故爲金鼓，視不相見，故爲旌旗。夫金鼓旌旗者，所以一民之耳目也。民濟（既）槫（專）壹，則勇者不得獨進，怯者不得獨退。此用衆之法也。故夜戰多金鼓，晝戰多旌旗。故三軍可奪氣，將軍可奪心。是故朝氣銳，晝氣惰，暮氣歸。故善用兵者，辟（避）其兌（銳）氣，擊其惰歸，此治氣者也。以治待亂，以靜待譁，此治心者也。以近待遠，以失（佚）待勞，以飽侍（待）飢，此治力者也。無邀正正之旗，勿擊堂堂之陳（陣），此治變者也。故用兵之法：高陵勿向，背丘勿逆，佯北勿從，銳卒勿攻，餌兵勿食，歸師勿遏，圍師必闕，窮寇勿迫，此用兵之法也。

stores, it will perish; if it is without provisions, it will perish; if it is without its material support, it will perish.

[*Therefore, unless you know the intentions of the rulers of the neighboring states, you cannot enter into preparatory alliances with them; unless you know the lay of the land—its mountains and forests, its passes and natural hazards, its wetlands and swamps—you cannot deploy the army on it; unless you can employ local scouts, you cannot turn the terrain to your advantage.*][162]

Therefore, in warfare rely on deceptive maneuvers to establish your ground, calculate advantages in deciding your movements, and divide up and consolidate your forces to make your strategic changes.

Thus, advancing at a pace, such an army is like the wind; slow and majestic, it is like a forest; invading and plundering, it is like fire; sedentary, it is like a mountain; unpredictable, it is like a shadow; moving, it is like lightning and thunder.

In plundering the countryside, divide up your numbers;[163] in extending your territory, divide up and hold the strategic positions; weigh the pros and cons before moving into action.

He who first understands the tactic of converting the tortuous and the direct will take the victory. This is the art of armed contest.[164]

The Book of Military Policies[165] states: It is because commands cannot be heard in the din of battle that drums and gongs are used; it is because units cannot identify each other in battle that flags and pennants are used. Thus, in night battle make extensive use of torches and drums, and in battle during the day make extensive use of flags and pennants.[166] Drums, gongs, flags, and pennants are the way to coordinate the ears and eyes of the men.[167] Once the men have been consolidated as one body, the courageous will not have to advance alone, and the cowardly will not get to retreat alone.[168] This is the art of employing large numbers of troops.

An entire enemy army can be demoralized, and its commander can be made to lose heart.[169] Now, in the morning of the war, the enemy's morale is high; by noon, it begins to flag; by evening, it has drained away.[170] Thus the expert in using the military avoids the enemy when his morale is high, and strikes when his morale has flagged and has drained away. This is the way to manage morale.

Use your proper order to await the enemy's disorder; use your calmness to await his clamor. This is the way to manage the heart-and-mind.

Use your closeness to the battlefield to await the far-off enemy; use your well-rested troops to await his fatigued; use your well-fed troops to await his hungry. This is the way to manage strength.

Do not intercept an enemy that is perfectly uniform in its array of banners; do not launch the attack on an enemy that is full and disciplined in its formations. This is the way to manage changing conditions.

Therefore, the art of using troops is this:

Do not attack an enemy who has the high ground; do not go against an enemy that has his back to a hill; do not follow an enemy that feigns retreat; do not attack the enemy's finest; do not swallow the enemy's bait; do not obstruct an enemy returning home; in surrounding the enemy, leave him a way out; do not press an enemy that is cornered. This is the art of using troops. (465 characters)[171]

CHAPTER 8

ADAPTING TO THE NINE CONTINGENCIES (*PIEN*)[172]

Master Sun said:

The art of using troops is this: When the commander receives his orders from the ruler, assembles his armies, and mobilizes the population for war,[173] he should not make camp on difficult terrain;[174] he should join with his allies on strategically vital intersections;[175] he should not linger on cut-off terrain;[176] he should have contingency plans on terrain vulnerable to ambush;[177] and he should take the fight to the enemy on terrain from which there is no way out.[178] There are roadways not to be traveled,[179] armies not to be attacked,[180] walled cities not to be assaulted,[181] territory not to be contested,[182] and commands from the ruler not to be obeyed.[183]

Thus, a commander fully conversant with the advantages to be gained in adapting to these nine contingencies will know how to employ troops; a commander who is not, even if he knows the lay of the land, will not be able to use it to his advantage. One who commands troops without knowing the art of adapting to these nine contingencies, even if he knows the five advantages,[184] will not be able to get the most from his men.

For this reason, the deliberations of the wise commander are sure to assess jointly both advantages and disadvantages. In taking full account of what is advantageous, he can fulfill his responsibilities; in taking full account of what is disadvantageous, his difficulties become resolvable.

For this reason, to subjugate neighboring states, use the threat of injury; to keep them in service, drive them on; to lure them out, use the prospect of gain.

The art of using troops is this:[185]

九變篇

孫子曰：凡用兵之法，將受命於君，合軍聚眾，圮地無舍，衢（衢）地合交，絕地無留，圍地則謀，死地則戰。途有所不由，軍有所不擊，城有所不攻，地有所不爭，君命有所不受。故將通於九變之利者，知用兵矣。將不通於九變之利者，雖知地形，不能得地之利矣。治兵不知九變之術，雖知五利，不能得人之用矣。是故，智者之慮，必雜於利害。雜於利，而務可信也；雜於害，而患可解也。是故，屈諸侯者以害，役諸侯者以業，趨諸侯者以利。故用兵之法：無恃其不來，恃吾有以待也；無恃其不攻，恃吾有所不可攻也。故將有五危：必死，可殺也；必生，可虜也；忿速，可侮也；潔廉，可辱也；愛民，可煩也。凡此五者，將之過也，用兵之災也。覆軍殺將，必以五危，不可不察也。

Do not depend on the enemy not coming; depend rather on being ready for him. Do not depend on the enemy not attacking; depend rather on having a position that cannot be attacked.

There are five traits that are dangerous in a commander:[186] If he has a reckless disregard for life, he can be killed; if he is determined to live at all costs, he can be captured; if he has a volatile temper, he can be provoked; if he is a man of uncompromising honor, he is open to insult; if he loves his people, he can be easily troubled and upset. These five traits are generally faults in a commander, and can prove disastrous in the conduct of war. Since an army's being routed and its commander slain is invariably the consequence of these five dangerous traits, they must be given careful consideration.

行軍篇

孫子曰：凡處軍、相敵，絕山依谷，視生處高，戰隆無登，此處山之軍也。絕水必遠水，客絕水而來，勿迎之於水內，令半濟而擊之，利；欲戰者，無附於水而迎客；視生處高，無迎水流，此處水上之軍也。絕斥澤，惟亟去無留；若交軍於斥澤之中，必依水草而背眾樹，此處斥澤之軍也。平陸處易，而右背高，前死後生，此處平陸之軍也。凡此四軍之利，黃帝之所以勝四帝也。

凡軍好高而惡下，貴陽而賤陰，養生而處實，軍无百疾，是謂必勝。陵丘隄防，必處其陽而右倍（背）之。此兵之利，地之助也。上雨，水沫至，止涉，待其定也。絕天澗、天井、天牢、天羅、天陷、天隙，必亟去之，勿近也。吾遠之，敵近之，吾迎之，敵背之。軍旁有險阻、潢井、葭葦（葦）、山林、蘙薈（薈）者，必謹覆索之，此伏姦之所處也。敵近而靜者，恃其險也；遠而挑戰者，欲人之進也。其所居易者，利也。眾樹動者，來也；眾草多障者，疑也。鳥起者，伏也；獸駭者，覆也。塵高而銳者，車來也；卑而廣者，徒來也；散

CHAPTER 9

DEPLOYING THE ARMY

Master Sun said:

In positioning your armies and assessing the enemy:

Pass through the mountains keeping to the valleys; pitch camp on high ground facing the sunny side; and joining battle in the hills, do not ascend to engage the enemy.[187] This is positioning an army when in the mountains.

Crossing water, you must move to distance yourself from it. When the invading army crosses water in his advance, do not meet him in the water. It is to your advantage to let him get halfway across and then attack him. Wanting to join the enemy in battle, do not meet his invading force near water. Take up a position on high ground facing the sunny side that is not downstream from the enemy. This is positioning an army when near water.

Crossing salt marshes, simply get through them in all haste and without delay. If you engage the enemy's force on the salt marshes, you must take your position near grass and water and with your back to the woods. This is positioning an army when on salt marshes.

On the flatlands, position yourself on open ground, with your right flank backing on high ground, and with dangerous ground in front and safe ground behind.[188] This is positioning an army when on flatlands.

Gaining the advantageous position for his army in these four different situations was the way the Yellow Emperor defeated the emperors of the four quarters.[189]

而條達者，薪來也；少而往來者，營軍也；辭卑（卑）而益備者，進也；辭强而進驅（驅）者，退也。輕車先出，居其側者，陳也；無約而請和者，謀也；奔走而陳兵者，期也；半進半退者，誘也。杖而立者，飢也；汲役先飲者，渴也；見利而不進者，勞也；鳥集者，虛也；夜呼（呼）者，恐也；軍擾（擾）者，將不重也；旌旗動者，亂也；吏怒者，倦也；粟馬肉食，軍無懸甀（甀），不反（返）其舍者，窮寇也；諄諄翕翕，徐言入人者，失衆也；數賞者，窘也；數罰者，困也；先暴而後畏其衆者，不精之至也；來委謝者，欲休息也；兵怒而相迎，久而不合，又不相去，必謹察之。夫唯無慮而易敵者，必擒於人。

卒未親附而罰之，則不服，不服則難用也；卒已親附而罰不行，則不可用也。故合之以文，濟（齊）之以武，是謂必取。令素行以教其民，則民服；令素不行以教其民，則民不服。令素行者，與衆相得也。

Generally speaking, an army prefers high ground and dislikes the low, prizes the sunny side and shuns the shady side, seeks a place in which food and water are readily available and ample to supply its needs, and wants to be free of the numerous diseases. These conditions mean certain victory. Encountering rises, hills, embankments, and dikes, you must position yourself on the sunny side and on your right flank have your back to the slope. This is an advantage for the troops, and is exploiting whatever help the terrain affords.

When it is raining upstream and churning waters descend, do not try to cross, but wait for the waters to subside.[190]

Encountering steep river gorges, natural wells, box canyons, dense ground cover, quagmires, or natural defiles,[191] quit such places with haste. Do not approach them. In keeping our distance from them, we can maneuver the enemy near to them; in keeping them to our front, we can maneuver the enemy to have them at his back.

If the army is flanked by precipitous ravines, stagnant ponds, reeds and rushes, mountain forests, and tangled undergrowth, these places must be searched carefully and repeatedly, for they are where ambushes are laid and spies are hidden.

If the enemy is close and yet quiet,
He occupies a strategic position;
If he is at a distance and yet acts provocatively,
He wants us to advance.
Where he has positioned himself on level ground,
He is harboring some advantage;
If there is movement in the trees,
He is coming;
If there are many blinds in the bushes,
He is looking to confuse us;
If birds take to flight,
He is lying in ambush;
If animals stampede in fear,
He is mounting a surprise attack;
If the dust peaks up high,
His chariots are coming;
If the dust spreads out low to the ground,
His infantry is coming;
If the dust reaches out in scattered ribbons,
His firewood details have been dispatched;
If a few clouds of dust come and go,

He is making camp.
If his envoys are modest of word yet he continues to increase his readiness for war,
He will advance;
If his language is belligerent and he advances aggressively,
He will withdraw;
If his light chariots move out first
And take up position on the flanks,
He is moving into formation;
If he has suffered no setback and yet sues for peace,[192]
He is plotting;
If he moves rapidly with his troops in formation,[193]
He is setting the time for battle;
If some of his troops advance and some retreat,
He is seeking to lure us forward.
If the enemy soldiers lean on their weapons,
They are hungry;
If those sent for water first drink themselves,
They are thirsty;
If there is an advantage to be had yet they do not advance to secure it,
They are weary;
Where birds gather,
The enemy position is unoccupied;
Where there are shouts in the night,
The enemy is frightened;
Where there are disturbances in the ranks,
The enemy commander is not respected;
Where their flags and pennants are shifted about,
The enemy is in disorder;
Where his officers are easily angered,
The enemy is exhausted.
Where the enemy feeds his horses grain and his men meat,
And where his men no longer bother to hang up their water vessels,
Or return to camp,
The now-desperate enemy is ready to fight to the death.
Where, hemming and hawing,
The enemy commander speaks to his subordinates in a meek and halting voice,
He has lost his men.
Meting out too many rewards
Means the enemy is in trouble,
And meting out too many punishments
Means he is in dire straits.

The commander who erupts violently at his subordinates,
Only then to fear them,
Is totally inept.
When the enemy's emissary comes with conciliatory words
He wants to end hostilities.

When an angry enemy confronts you for an extended time, without either joining you in battle or quitting his position, you must watch him with the utmost care.

In war it is not numbers that give the advantage. If you do not advance recklessly, and are able to consolidate your own strength, get a clear picture of the enemy's situation, and secure the full support of your men, it is enough. It is only the one who has no plan and takes his enemy lightly who is certain to be captured by him. If you punish troops who are not yet devoted to you, they will not obey, and if they do not obey, they are difficult to use. But once you have their devotion, if discipline is not enforced, you cannot use them either. Therefore, bring them together by treating them humanely and keep them in line with strict military discipline. This will assure their allegiance.

If commands are consistently enforced in the training of the men, they will obey; if commands are not enforced in their training, they will not obey. The consistent enforcement of commands promotes a complementary relationship between the commander and his men.

地形篇

孫子曰：地形有通者，有挂者，有支者，有隘者，有險者，有遠者。我可以往，彼可以來，曰通。通形者，先居高陽，利糧道，以戰則利。可以往，難以返，曰挂。挂形者，敵無備，出而勝之；敵有備，出而不勝，難以返，不利。我出而不利，彼出而不利，曰支。支形者，敵雖利我，我無出也，引而去之，令敵半出而擊之，利。隘形者，我先居之，必盈之以待敵；若敵先居之，盈而勿從，不盈而從之。險形者，我先居之，必居高陽以待敵；若敵先居之，引而去之，勿從也。遠形者，勢均，難以挑戰，戰而不利。凡此六者，地之道也，將之至任，不可不察也。故兵有走者，有弛者，有陷者，有崩者，有亂者，有北者。夫勢均，以一擊十，曰走。卒強吏弱，曰弛。吏強卒弱，曰陷。大吏怒而不服，遇敵懟

CHAPTER 10

THE TERRAIN

Master Sun said:

Kinds of terrain include the accessible, that which entangles, that which leads to a stand-off, the narrow pass, the precipitous defile and the distant.

Terrain that both armies can approach freely is called accessible. On accessible terrain, the army that enters the battle having been first to occupy high ground on the sunny side and to establish convenient supply lines, fights with the advantage.

Terrain that allows your advance but hampers your return is entangling. On entangling ground, if you go out and engage the enemy when he is not prepared, you might defeat him. But when the enemy is prepared, if you go out and engage him and fail to defeat him, you will be hard-pressed to get out, and will be in trouble.

Terrain that when entered disadvantages both our side and the enemy is ground that will lead to a stand-off. On this kind of terrain, even if the enemy tempts us out, we must not take the bait, but should quit the position and withdraw. Having lured the enemy halfway out, we can then strike to our advantage.

With the narrow pass, if we can occupy it first, we must fully garrison it and await the enemy. Where the enemy has occupied it first, if he garrisons it completely, do not follow him, but if he fails to, we can go after him.

With the precipitous defile, if we can occupy it first, we must take the high ground on the sunny side and await the enemy. Where the enemy has occupied it first, quit the position and withdraw, and do not follow him.

而自戰，將不知其能，曰崩。將弱不嚴，教道不明，吏卒無常，陳兵縱橫，曰亂。

將不能料敵，以少合衆，以弱擊強，兵無選鋒，曰北。凡此六者，敗之道也；將之至任，不可不察也。

夫地形者，兵之助也。料敵制勝，計險易遠近，上將之道也，知此而用戰者必勝，不知此而用戰者必敗。

故戰道必勝，主曰無戰，必戰可也；戰道不勝，主曰必戰，無戰可也。故進不求名，退不避罪，惟民是保，而利合於主，國之寶也。

視卒如嬰兒，故可與之赴深谿；視卒如愛子，故可與之俱死。厚而不能使，愛而不能令，亂而不能治，譬如驕子，不可用也。

知吾卒之可以擊，而不知敵之不可擊，勝之半也。知敵之可擊，而不知吾卒之不可以擊，勝之半也。知敵之可擊，知吾卒之可以擊，而不知地形之不可以戰，勝之半也。

故知兵者，動而不迷，舉而不窮，故曰：知彼知己，勝乃不殆；知地知天，勝乃可全。

When the enemy is at some distance, if the strategic advantages of both sides are about the same, it is not easy to provoke him to fight, and taking the battle to him is not to our advantage.

Now these are the six guidelines (*tao*) governing the use of terrain. They are the commander's utmost responsibility, and must be thoroughly investigated.

In warfare there is flight, insubordination, deterioration, ruin, chaos, and rout.[194] These six situations are not natural catastrophes but the fault of the commander.

Where the strategic advantages of both sides are about the same, for an army to attack an enemy ten times its size will result in flight.

If the troops are strong but the officers weak, the result will be insubordination.

If the officers are strong but the troops weak, the result will be deterioration.

If your ranking officers are angry and insubordinate and, on encountering the enemy, allow their rancor to spur them into unauthorized engagements so that their commander does not know the strength of his own forces, the result will be ruin.

If the commander is weak and lax, his instructions and leadership unenlightened, his officers and troops undisciplined, and his military formations in disarray, the result will be chaos.

If the commander, unable to assess his enemy, sends a small force to engage a large one, sends his weak troops to attack the enemy's best, and operates without a vanguard of crack troops, the result will be rout.

These are six ways (*tao*) to certain defeat. They are the commander's utmost responsibility, and must be thoroughly investigated.

Strategic position (*hsing*) is an ally in battle. To assess the enemy's situation and create conditions that lead to victory, to analyze natural hazards and proximate distances—this is the way (*tao*) of the superior commander.[195] He who fights with full knowledge of these factors is certain to win; he who fights without it is certain to lose.

Thus, if the way (*tao*) of battle guarantees you victory, it is right for you to insist on fighting even if the ruler has said not to; where the way (*tao*) of battle does not allow victory, it is right for you to refuse to fight even if the ruler has said you must.

Hence a commander who advances without any thought of winning personal fame and withdraws in spite of certain punishment, whose only concern is to protect his people and promote the interests of his ruler, is the nation's treasure. Because he fusses over his men as if they were infants, they will accompany him into the deepest valleys; because he fusses over his men as if they were his own beloved sons, they will die by his side. If he is generous with them and yet they do not do as he tells them, if he loves them and yet they do not obey his commands, if he is so undisciplined with them that he cannot bring them into proper order, they will be like spoiled children who can be put to no good use at all.

To know our troops can attack and yet be unaware that the enemy is not open to attack, reduces our chances of victory to half; to know the enemy is open to attack and yet be unaware that our own troops cannot attack, reduces our chances of victory again to half; to know the enemy is open to attack and our troops can attack, and yet be unaware that the terrain does not favor us in battle, reduces the chances of victory once again to half.

Thus when one who understands war moves, he does not go the wrong way, and when he takes action, he does not reach a dead end.

Hence it is said:

Know the other, know yourself,
And the victory will not be at risk;
Know the ground, know the natural conditions,
And the victory can be total.[196]

CHAPTER 11

THE NINE KINDS OF TERRAIN

Master Sun said:

In the art of employing troops, the kinds of terrain include scattering terrain, marginal terrain, contested terrain, intermediate terrain, the strategically vital intersection, critical terrain, difficult terrain, terrain vulnerable to ambush, and terrain from which there is no way out.

Where a feudal ruler does battle within his own territory, it is a terrain that permits the scattering of his troops.

Where one has penetrated only barely into enemy territory, it is marginal terrain.

Ground that gives us or the enemy the advantage in occupying it is contested terrain.

Ground accessible to both sides is intermediate terrain.

The territory of several neighboring states at which their borders meet is a strategically vital intersection. The first to reach it will gain the allegiance of the other states of the empire.

When an army has penetrated deep into enemy territory, and has many of the enemy's walled cities and towns at its back, it is on critical terrain.

Mountains and forests, passes and natural hazards, wetlands and swamps, and any such roads difficult to traverse constitute difficult terrain.

Ground that gives access through a narrow defile, and where exit is tortuous, allowing an enemy in small numbers to attack our main force, is terrain vulnerable to ambush.

九地篇

孫子曰：用兵之法，有散地，有輕地，有爭地，有交地，有衢（衢）地，有重地，有圮地，有圍地，有死地。諸侯自戰其地者，爲散地。我可以往，彼可以來者，爲交地。諸侯之地三屬，先至而得天下之衆者，爲衢（衢）地。入人之地深，倍（背）城邑多者，爲重地。山林、險阻、沮澤，凡難行之道者，爲圮地。所由入者隘，所從歸者迂，彼寡可以擊吾之衆者，爲圍地。疾戰則存，不疾戰則亡者，爲死地。是故散地則無戰，輕地則無止，爭地則無攻，交地則無絕，衢（衢）地則合交，重地則掠，圮地則行，圍地則謀，死地則戰。所謂古之善用兵者，能使敵人前後不相及，衆寡不相恃，貴賤不相救，上下不相收，卒離而不集，兵合而不齊。合於利而動，不合於利而止。敢問：適（敵）衆以正（整）將來，侍（待）之若何？曰：先奪其所愛，則聽矣。兵之情主速，乘人之不及，由不虞之道，攻其所不戒也。

凡爲客之道：深入則專，主人不克。掠於饒野，三軍足食，謹養而勿勞，併氣積力，運兵計謀，爲不可測。投之無所往，死且不北，死，焉不得士人盡力。兵士甚陷則不懼，無所往則固，入深則

Ground on which you will survive only if you fight with all your might, but will perish if you fail to do so, is terrain with no way out.

This being the case, do not fight on scattering terrain; do not stay on marginal terrain; do not attack the enemy on contested terrain; do not get cut off on intermediate terrain; form alliances with the neighboring states at strategically vital intersections; plunder the enemy's resources on critical terrain; press ahead on difficult terrain; devise contingency plans on terrain vulnerable to ambush; and on terrain from which there is no way out, take the battle to the enemy.

The commanders of old said to be expert at the use of the military were able to ensure that with the enemy:

His vanguard and rearguard could not relieve each other,
The main body of his army and its special detachments could not support each other,
Officers and men could not come to each other's aid,
And superiors and subordinates could not maintain their lines of communication.
The enemy forces when scattered could not regroup,
And when their army assembled, it could not form ranks.

If it was to the advantage of these expert commanders, they would move into action; if not, they would remain in place. Suppose I am asked: If the enemy, in great numbers and with strict discipline in the ranks, is about to advance on us, how do we deal with him? I would reply: If you get ahead of him to seize something he cannot afford to lose, he will do your bidding.

War is such that the supreme consideration is speed. This is to take advantage of what is beyond the reach of the enemy, to go by way of routes where he least expects you, and to attack where he has made no preparations.[197]

The general methods of operation (*tao*) for an invading army are:

The deeper you penetrate into enemy territory, the greater the cohesion of your troops, and the less likely the host army will prevail over you.

Plunder the enemy's most fertile fields, and your army will have ample provisions.

Attend to the nourishment of your troops and do not let them get worn down; lift their morale and build up their strength.

Deploy your troops and plan out your strategies in such a way that the enemy cannot fathom your movements.

拘，不得已則鬥。是故不修而戒，不求而得，不約而親，不令而信，禁祥去疑，至死無所之。吾士無餘財，非惡貨也，無餘命，非惡壽也。令發之日，士坐者涕霑襟，臥者涕交頤。投之無所往者，諸歲（劌）之勇也。故善用兵者，譬如率然；率然者恆山之蛇也，擊其首則尾至，擊其尾則首至，擊其中則首尾俱至。敢問：兵可使如率然乎？曰：可。夫吳人與越人相惡也，當其同周（舟）而濟，其相救也，如左右手。是故方馬埋輪，未足恃也；齊勇若一，政之道也；剛柔皆得，地之理也。故善用兵者，攜手若使一人，不得已也。將軍之事，靜以幽，正以治。能愚士卒之耳目，使民無知；易其事，革其謀，使民無識；易其居，迂其途，使民不得慮。帥與之期，如登高而去其梯；帥與之深入諸侯之地，而發其機，若驅群羊，驅而往，驅而來，莫知所之。聚三軍之眾，投之於險，此謂將軍之事也。九地之變，屈伸之利，人情之理，不可不察也。凡爲客之道，深則專，淺則散。去國越境而師者，絕地也；四徹者，衢地也；入深者，重地也；入淺者，輕地也；背固前隘者，圍地也；無所往者，死地也。是故散地，吾將壹其志；輕地，吾將使之屬；爭地，吾將趨其

Throw your troops into situations from which there is no way out, and they will choose death over desertion. Once they are ready to die, how could you get less than maximum exertion from your officers and men?

> Even where your troops are in the most desperate straits,
> They will have no fear,
> And with nowhere else to turn,
> They will stand firm;
> Having penetrated deep into enemy territory,
> They are linked together,
> And if need be,
> They will fight.
> For this reason, with no need of admonishment, they are vigilant;[198]
> Without compulsion, they carry out their duties;
> Without tying them down, they are devoted;
> With no need for orders, they follow army discipline.
> Proscribe talk of omens and get rid of rumors,
> And even to the death they will not retreat.

Our soldiers do not have an abundance of wealth, but it is not because they despise worldly goods; their life expectancy is not long, but it is not because they despise longevity. On the day these men are ordered into battle, those sitting have tears soaking their collars, and those lying on their backs have tears crossing on their cheeks. But throw them into a situation where there is no way out and they will show the courage of any Chuan Chu or Ts'ao Kuei.[199]

Therefore, those who are expert at employing the military are like the "sudden striker." The "sudden striker" is a snake indigenous to Mount Heng.[200] If you strike its head, its tail comes to its aid; if you strike its tail, its head comes to its aid; if you strike its middle, both head and tail come to its aid.[201]

Suppose I am asked: Can troops be trained to be like this "sudden striker" snake? I would reply: They can. The men of Wu and Yüeh hate each other. Yet if they were crossing the river in the same boat and were caught by gale winds, they would go to each other's aid like the right hand helping the left.

For this reason, it has never been enough to depend on tethered horses and buried chariot wheels.[202] The object (*tao*) of military management is to effect a unified standard of courage. The principle of exploiting terrain is to get value from the soft as well as the hard.[203] Thus the expert in using

後，交地，吾將謹其守；衢地，吾將固其結；重地，吾將繼其食；圮地，吾將進其途，圍地，吾將塞其闕；死地，吾將示之以不活。故兵之情，圍則禦，不得已則鬪，過則從。是故，不知諸侯之謀者，不能豫交；不知山林、險阻、沮澤之形者，不能行軍。不用鄉（向）道（導）者，不能得地利。四五者，一不智（知），非王霸之兵也。夫王霸之兵，伐大國，則其衆不得聚；威加於敵，則其交不得合。是故，不爭天下之交，不養天下之權，信己之私，威加於敵。故其城可拔，其國可隳（隳）。施无法之賞，懸无正（政）之令，犯三軍之衆，若使一人。犯之以事，勿告以言，犯之以害，勿告以利。投之亡地然後存，陷之死地然後生。夫衆陷於害，然後能爲勝敗。故爲兵之事，在於順詳敵之意，并敵一向，千里殺將，此謂巧能成事者也。是故，正（政）舉之日，夷關折符，無通其使，厲於郎（廊）廟之上，以誅其事。敵人開闔，必亟入之。先其所愛，微與之期。踐墨隨敵，以決戰事。是故始如處女，敵人開戶，後如脫兔，敵不及拒。

the military leads his legions as though he were leading one person by the hand. The person cannot but follow.

As for the urgent business of the commander:

He is calm and remote, correct and disciplined. He is able to blinker the ears and eyes of his officers and men, and to keep people ignorant. He makes changes in his arrangements and alters his plans, keeping people in the dark.[204] He changes his camp, and takes circuitous routes, keeping people from anticipating him. On the day he leads his troops into battle, it is like climbing up high and throwing away the ladder. He leads his troops deep into the territory of the neighboring states and releases the trigger.[205] Like herding a flock of sheep, he drives them this way and that, so no one knows where they are going. He assembles the rank and file of his armies, and throws them into danger.

This then is the urgent business of the commander.

The measures needed to cope with the nine kinds of terrain, the advantages that can be gained by flexibility in maneuvering the army, and the basic patterns of the human character must all be thoroughly investigated.

The general methods of operation (*tao*) for an invading force are:

The deeper you penetrate into enemy territory, the greater the cohesion of your troops; the more shallow the penetration, the more easily you are scattered. When you quit your own territory and lead your troops across the border, you have entered cut-off terrain. When you are vulnerable on all four sides, you are at a strategically vital intersection. When you have penetrated deep into enemy territory, you are on critical terrain; when you have penetrated only a short distance, you are on marginal terrain. When your back is to heavily secured ground, and you face a narrow defile, you are on terrain vulnerable to ambush. When you have nowhere to turn, you are on terrain with no way out.

Therefore, on terrain where the troops are easily scattered, I would work to make them one of purpose; on marginal terrain, I would keep the troops together; on contested terrain, I would pick up the pace of our rear divisions;[206] on intermediate terrain, I would pay particular attention to defense;[207] at a strategically vital intersection, I would make sure of my alliances;[208] on critical terrain, I would maintain a continuous line of provisions;[209] on difficult terrain, I would continue the advance along the road; on terrain vulnerable to ambush, I would block off the paths of access and retreat; on terrain from which there is no way out, I would show our troops my resolve to fight to the death.

Thus the psychology of the soldier[210] is:

Resist when surrounded,
Fight when you have to,
And obey orders explicitly when in danger.

Unless you know the intentions of the rulers of the neighboring states,
you cannot enter into preparatory alliances with them;[211] unless you know
the lay of the land (*hsing*)—its mountains and forests, its passes and natural hazards, its wetlands and swamps—you cannot deploy the army on it;
unless you can employ local scouts, you cannot turn the terrain to your
advantage.[212] If an army is ignorant of even one of these several points, it
is not the army of a king or a hegemon.[213]

When the army of a king or hegemon attacks a large state, it does not
allow the enemy to assemble his forces; when it brings its prestige and influence to bear on the enemy, it prevents his allies from joining with him.
For this reason, one need not contend for alliances with the other states in
the empire or try to promote one's own place vis-à-vis these states. If you
pursue your own program, and bring your prestige and influence to bear
on the enemy, you can take his walled cities and lay waste to his state.

Confer extraordinary rewards and post extraordinary orders, and you
can command the entire army as if it were but one man. Give the troops
their charges, but do not reveal your plans; get them to face the dangers,
but do not reveal the advantages.[214] Only if you throw them into life-and-
death situations will they survive; only if you plunge them into places
where there is no way out will they stay alive. Only if the rank and file
have plunged into danger can they turn defeat into victory.

Therefore, the business of waging war lies in carefully studying the designs of the enemy.[215]

Focus your strength on the enemy
And you can slay his commander at a thousand *li*.

This is called realizing your objective by your wits and your skill.

For this reason, on the day a declaration of war is made, close off the
passes, destroy all instruments of agreement, and forbid any further contact with enemy emissaries. Rehearse your plans thoroughly in the ances-
tral temple and finalize your strategy. When the enemy gives you the
opening, you must rush in on him. Go first for something that he cannot
afford to lose, and do not let him know the timing of your attack. Revise
your strategy according to the changing posture of the enemy to determine the course and outcome of the battle.[216] For this reason,

At first be like a modest maiden,
And the enemy will open his door;
Afterward be as swift as a scurrying rabbit,
And the enemy will be too late to resist you.

火攻篇

孫子曰：凡火攻有五：一曰火人，二曰火積（積），三曰火輜，四曰火庫，五曰火隊。行火必有因，因必素具。發火有時，起火有日。時者，天之燥也。日者，月在箕、壁、翼、軫也。凡此四宿者，風起之日也。凡火攻，必因五火之變而應之。火發於內，則早應之於外。火發其兵靜而勿攻，極其火央，可從而從之，不可從而止之。火可發於外，無寺（待）於內，以時發之。火發上風，無攻下風。晝風久，夜風止。凡軍必知有五火之變，以數守之。故以火佐攻者明，以水佐攻者強。水可以絕，不可以奪。夫戰勝攻取，而不修其功者，凶，命曰費留。故曰：明主慮之，良將修之，非利不動，非得不用，非危不戰。主不可以怒而興軍，將不可以慍（慍）而戰。怒可復喜，慍（慍）可復悅，亡國不可以復存，死者不可以復生。

故明主慎之，良將警之，此安國全軍之道也。

CHAPTER 12

INCENDIARY ATTACK

Master Sun said:

There are five kinds of incendiary attack: The first is called setting fire to personnel; the second, to stores; the third, to transport vehicles and equipment; the fourth, to munitions; the fifth, to supply installations.

In order to use fire there must be some inflammable fuel (*yin*), and such fuel must always be kept in readiness.[217] There are appropriate seasons for using fire, and appropriate days that will help fan the flames. The appropriate season is when the weather is hot and dry; the appropriate days are those when the moon passes through the constellations of the Winnowing Basket, the Wall, the Wings, and the Chariot Platform.[218] Generally these four constellations mark days when the winds rise.

With the incendiary attack, you must vary your response to the enemy according to (*yin*) the different changes in his situation induced by each of the five kinds of attack. When the fire is set within the enemy's camp, respond from without at the earliest possible moment. If in spite of the outbreak of fire, the enemy's troops remain calm, bide your time and do not attack. Let the fire reach its height, and if you can follow through, do so. If you cannot, stay where you are. If you are able to raise a fire from outside, do not wait to get inside, but set it when the time is right. If the fire is set from upwind, do not attack from downwind. If the wind blows persistently during the day, it will die down at night.[219] In all cases an army must understand the changes induced by the five kinds of incendiary attack, and make use of logistical calculations to address them.

He who uses fire to aid the attack is powerful;
He who uses water to aid the attack is forceful.[220]
Water can be used to cut the enemy off,
But cannot be used to deprive him of his supplies.[221]

To be victorious in battle and win the spoils, and yet fail to exploit your achievement, is disastrous. The name for it is wasting resources.

Thus it is said:

The farsighted ruler thinks the situation through carefully;
The good commander exploits it fully.
If there is no advantage, do not move into action;
If there is no gain, do not deploy the troops;
If it is not critical, do not send them into battle.

A ruler cannot mobilize his armies in a rage; a commander cannot incite a battle in the heat of the moment.[222] Move if it is to your advantage; bide your time if it is not. A person in a fit of rage can be restored to good humor and a person in the heat of passion can be restored to good cheer, but a state that has perished cannot be revived, and the dead cannot be brought back to life. Thus the farsighted ruler approaches battle with prudence, and the good commander moves with caution. This is the way (*tao*) to keep the state secure and to preserve the army intact.

CHAPTER 13

USING SPIES

Master Sun said:

In general, the cost to the people and to the public coffers to mobilize an army of 100,000 and dispatch it on a punitory expedition of a thousand *li* is a thousand pieces of gold per day. There will be upheaval at home and abroad, with people trekking exhausted on the roadways and some 700,000 households kept from their work in the fields. Two sides will quarrel with each other for several years in order to fight a decisive battle on a single day. If, begrudging the outlay of ranks, emoluments, and a hundred pieces of gold, a commander does not know the enemy's situation, his is the height of inhumanity. Such a person is no man's commander, no ruler's counsellor, and no master of victory.

Thus the reason the farsighted ruler and his superior commander conquer the enemy at every move, and achieve successes far beyond the reach of the common crowd, is foreknowledge. Such foreknowledge cannot be had from ghosts and spirits, educed by comparison with past events, or verified by astrological calculations. It must come from people—people who know the enemy's situation.

There are five kinds of spies that can be employed:[223] local (*yin*) spies, inside agents, double agents, expendable spies, and unexpendable spies. When the five kinds of spies are all active, and no one knows their methods of operation (*tao*), this is called the imperceptible web,[224] and is the ruler's treasure.

Local spies are the enemy's own countrymen in our employ.

Inside agents are enemy officials we employ.

用間篇

孫子曰：凡興師十萬，出征千里，百生（姓）之費，公家之奉，日費千金，內外騷動，怠於道路，不得操事者，七十萬家。相守數年，以爭一日之勝，而愛爵祿百金，不知適（敵）之請（情）者，不仁之至也，非民之將也，非主之佐也，非勝之注（主）也。故明君賢將，所以動而勝人，成功出於衆者，先知也。先知者，不可取於鬼神，不可象於事，不可驗於度，必取於人，知敵之情者也。故用間有五：有鄉間、有內間、有反間、有死間、有生間。五間俱起，莫知其道，是謂神紀，人君之葆（寶）也。鄉間者，因其鄉人而用之。內間者，因其官人而用之。反間者，因其敵間而用之。死間者，爲誑事於外，令吾間知之，而傳於敵間也。生間者，反報也。故三軍之親，莫親於間，賞莫厚於間，事莫密於間。非聖不能用間，非仁不能使間，非微妙不能得間之實。間與所告者皆死。凡軍之所欲擊，城之所欲攻，人之所欲殺，必先知其守將、左右、謁者、門者、舍人之姓名，令吾間必索知之。必索敵人之間來間我者，因而利之，導而舍之，故反間可得而用也。因是而知之，故鄉間、內間可得而使也；因是而知之，故死間爲誑事，可使告敵；因是而知之，故生間可使如期。五間之事，主必知之，知之必在於反間，故反間不可不厚也。昔殷之興也，伊摯在夏；周之興也，呂牙在殷。故唯明君賢將，能以上智爲間者，必成大功，此兵之要，三軍之所恃而動也。

Double agents are enemy spies who report to our side.

Expendable spies are our own agents who obtain false information we have deliberately leaked to them, and who then pass it on to the enemy spies.

Unexpendable spies are those who return from the enemy camp to report.

Thus, of those close to the army command, no one should have more direct access than spies,[225] no one should be more liberally rewarded than spies, and no matters should be held in greater secrecy than those concerning spies.

Only the most sagacious ruler is able to employ spies; only the most humane and just commander is able to put them into service; only the most sensitive and alert person can get the truth out of spies.

So delicate! So secretive! There is nowhere that you cannot put spies to good use. Where a matter of espionage has been divulged prematurely, both the spy and all those he told should be put to death.

In general terms, whether it is armies we want to attack, walled cities we want to besiege, or persons we want to assassinate, it is necessary to first know the identities of the defending commander, his retainers, counsellors, gate officers, and sentries. We must direct our agents to find a way to secure this information for us.

It is necessary to find out who the enemy has sent as agents to spy on us. If we take care of them (*yin*) with generous bribes, win them over and send them back,[226] they can thus be brought into our employ as double agents. On the basis of what we learn from (*yin*) these double agents, we can recruit and employ local and inside spies. Also, from (*yin*) this information we will know what false information to feed our expendable spies to pass on to the enemy. Moreover, on what we know from (*yin*) this same source, our unexpendable spies can complete their assignments according to schedule. The ruler must have full knowledge of the covert operations of these five kinds of spies. And since the key to all intelligence is the double agent, this operative must be treated with the utmost generosity.

Of old the rise of the Yin (Shang) dynasty was because of Yi Yin who served the house of Hsia; the rise of the Chou dynasty was because of Lü Ya who served in the house of Shang.[227] Thus only those farsighted rulers and their superior commanders who can get the most intelligent people as their spies are destined to accomplish great things. Intelligence is of the essence in warfare—it is what the armies depend upon in their every move.

SUN-TZU:
PART II

TEXT RECOVERED FROM THE YIN-CH'ÜEH-SHAN HAN DYNASTY STRIPS[228]

吳問

吳王問孫子曰：六將軍分守晉國之地，孰先亡？孰固成？孫子曰：范、中行是

(氏)先亡。孰爲之次？智是(氏)爲次。孰爲之次？韓、巍(魏)趙毋失

其故法，晉國歸焉。吳王曰：其說可得聞乎？孫子曰：可。范、中行是(氏)制田，

以八十步爲嬈(畹)，以百六十步爲畛，而伍稅之。其□田陝(狹)，置士多，伍稅

之，公家富。公家富，置士多，主喬(驕)臣奢，冀功數戰，故曰先【亡】。……公家

富，置士多，主喬(驕)臣奢冀功數戰，故爲范、中行是(氏)次。韓、巍(魏)制田，

以百步爲嬈(畹)，以二百步爲畛，而伍稅【之】。其□田陝(狹)，其置士多，伍稅

之，公家富。公家富，置士多，主喬(驕)臣奢，冀功數戰，故爲智是(氏)次。趙

是(氏)制田，以百廿步爲嬈(畹)，以二百卌步爲畛，公无稅焉。公家貧，其置

士少，主僉臣收，以御富民，故曰固國。晉國歸焉。吳王曰：善。王者之道，□□

厚愛其民者也。二百八十四

CHAPTER 1

THE QUESTIONS OF WU

The King of Wu asked Master Sun: "Which of the six commanders[229] who divided up the territory of the state of Chin were the first to perish? And which succeeded in holding on to their lands?"

Master Sun replied: "Fan and Chung-hang were the first to perish."

"Who was next?"

"Chih was next."

"And who was next?"

"Han and Wei were next. It was because Chao did not abandon the traditional laws that the state of Chin turned to him."

The King of Wu said: "Could you explain this to me?" Master Sun replied: "Indeed. In regulating the measurement of land area, Fan and Chung-hang took eighty square paces as a *yüan*, and took one hundred and sixty square paces as a *chen*, and then made five households their basic tax unit.[230] The land area was small and the officials in office were many. With five households as the basic tax unit, the public coffers prospered. With the public coffers prospering and the officials in office many, the ruler became arrogant and his ministers wasteful. And in pursuit of great exploits they embarked on frequent wars. Thus I say they were the first to perish.

"[*In regulating the measurement of land area, Chih took ninety square paces as a* yüan *and took one hundred and eighty square paces as a* chen, *and then made five households his basic tax unit. The land area was also small and the officials in office were also many. With five households as the basic tax unit, the public coffers prospered.*] With the public coffers prospering and the officials in office many, the ruler became arrogant and his ministers wasteful. And in pursuit of great exploits they embarked on frequent wars. Thus I say he was the next to perish after Fan and Chung-hang.

"In regulating the measurement of land area, Han and Wei took one hundred square paces as a *yüan*, and took two hundred square paces as a *chen*, and then made five households their basic tax unit. The land area was again small and the officials in office were again many. With five households as the basic tax unit, the public coffers prospered. With the public coffers prospering and the officials in office many, the ruler became arrogant and his ministers wasteful. And in pursuit of great exploits they embarked on frequent wars. Thus I say they were the next to perish after Chih.

"In regulating the measurement of land area, Chao took one hundred and twenty square paces as a *yüan*, and took two hundred and forty square paces as a *chen*, and so there were no new taxes forthcoming for the public coffers. With the public coffers empty and the officials in office few, the ruler was frugal and his ministers humble in their management of what was a prosperous people. Thus I say that he held on to his lands, and the whole state of Chin turned to him."

The King of Wu said: "Excellent! The way (*tao*) of the True King is [*that he should*] love the people generously." (284 characters)

CHAPTER 2

[THE FOUR CONTINGENCIES]²³¹

[*There are roadways not to be traveled, armies not to be attacked,*] walled cities not to be assaulted, territory not to be contested, and commands from the ruler [*not to be obeyed*].

That there are roadways not to be traveled refers to a roadway where if we penetrate only a short distance we cannot bring the operations of our vanguard into full play, and if we penetrate too deeply we cannot link up effectively with our rearguard. If we move, it is not to our advantage, and if we stop, we will be captured. Given these conditions, we do not travel it.

That there are armies not to be attacked refers to a situation in which the two armies make camp and face off. We estimate we have enough strength to crush the opposing army and to capture its commander. Taking the long view, however, there is some surprise advantage (*shib*) and clever dodge he has, so his army...its commander. Given these conditions, even though we can attack, we do not do so.

That there are walled cities not to be assaulted refers to a situation in which we estimate we have enough strength to take the city. If we take it, it gives us no immediate advantage, and having gotten it, we would not be able to garrison it. If we are [*lacking*] in strength,²³² the walled city must by no means be taken. If in the first instance we gain advantage, the city will surrender of its own accord; and if we do not gain advantage, it will not be a source of harm afterward. Given these conditions, even though we can launch an assault, we do not do so.

That there is territory not to be contested refers to mountains and

四 變

徐(途)有所不由,軍有所不擊,城有所不攻,地有所不爭,君令有所不行,徐(途)
之所不由者,曰:淺入則前事不信,深入則後利不接(接)。動則不利,立則囚。如
此者,弗由也。軍之所不毄(擊)者,曰:兩軍交和而舍,計吾力足以破其軍,獲其
將。遠計之,有奇執(勢)巧權於它,而軍……□將。如此者,軍唯(雖)可毄(擊),
弗毄(擊)也。城之所不攻者,曰:計吾力足以拔之,拔之而不及利於前,得之而
後弗能守。若力[不]足,城必不取。及於前,利得而城自降,利不得而不爲害於
後。若此者,城唯(雖)可攻,弗攻也。地之所不爭者,曰:山谷水□无能生者,□
□□,而□□……虛。如此者,弗爭也。君令有所不行者,君令有反此四變者,則
弗行也。□□□□□□□□□□□行也。事……變者,則智(知)用兵矣。

gorges ... that are not able to sustain life ... vacant. Given these conditions, do not contest it.

That there are commands from the ruler not to be obeyed means that if the commands of the ruler are contrary to these four contingencies, do not obey them ... obey them. Where affairs ... contingencies, one understands how to use troops.

黃帝伐赤帝

黃帝伐赤帝

孫子曰:【黃帝南伐】赤帝,【至於□□】,戰於反山之原,右陰,順術,倍(背)衝,大威(滅)有之。【□年休民,執(熟)穀,赦罪。東伐□帝,至於襄平,戰於平□,【右】陰,順術,倍(背)衝,大威(滅)有之。□年休民,執(熟)穀,赦罪。

北伐黑帝,至於武隧,戰於□□,右陰,順術,【倍衝,大威有之。□年休民,執穀,赦罪】。西伐白帝,至於武剛,戰於□□,右陰,順術,倍衝,大威有之。已勝

四帝,大有天下,暴者……以利天下,天下四面歸之。湯之伐桀也,【至於□□】,戰於薄田,右陰,順術,倍(背)衝,大威(滅)有之。武王之伐紂,至於戮遂,戰牧之野,右陰,順術,【倍衝,大威】有之。一帝二王皆得天之道、□之□、民之請(情),

故……

CHAPTER 3

THE YELLOW EMPEROR ATTACKS THE RED EMPEROR[233]

Master Sun said: [*The Yellow Emperor to the south attacked the Red Emperor, penetrated as far as...*] and did battle on the steppes of Mount Pan.[234] Advancing with the *yin* conditions on his right, following the roadway, and keeping his back to strategic ground[235] he exterminated the enemy and annexed his territory. For [*...years*] he gave his people respite, allowed the grains to ripen, and gave amnesty to the criminals.

Then to the east he attacked the [*Green*] Emperor, penetrated as far as Hsiang-p'ing, and did battle at P'ing.... Advancing with the *yin* conditions [*on his right*], following the roadway, and keeping his back to strategic ground, he exterminated the enemy [*and annexed his territory. For...*] years he gave his people respite, allowed the grains to ripen, and gave amnesty to the criminals.

Then to the north he attacked the Black Emperor, penetrated as far as Wu-sui, and did battle at...Advancing with the *yin* conditions on his right, following the roadway, [*and keeping his back to strategic ground, he exterminated the enemy and annexed his territory. For...years he gave his people respite, allowed the grains to ripen, and gave amnesty to the criminals*].

Then to the west he attacked the White Emperor, penetrated as far as Wu-kang, and did battle at [*...Advancing with the* yin *conditions on his right, following the roadway, and keeping his back to strategic ground, he exterminated the enemy and annexed*] his territory. Having defeated the four emperors he ruled over all under heaven. The violent...for the advantage of the empire, and the people under heaven from all four directions turned to him.

When King T'ang of Shang attacked King Chieh of Hsia, [*he penetrated as far as*...], and did battle at Po-t'ien. Advancing with the *yin* conditions on his right, following the roadway, and keeping his back to strategic ground, he exterminated the enemy and annexed his territory.

When King Wu of Chou attacked King Chou of Shang, he penetrated as far as Shu-sui, and did battle on the fields of Mu. Advancing with the *yin* conditions on his right, following the roadway, [*and keeping his back to strategic ground, he exterminated the enemy*] and annexed his territory.

This one emperor and these two kings all realized the way (*tao*) of heaven,... the basic nature of the people. Thus...

CHAPTER 4

THE DISPOSITION
[OF THE TERRAIN] II[236]

...in the disposition of terrain, east is left and west is [*right*]...

...head, on terrain that is flat, use the left, and the army...

...is terrain from which there is no way out. Places that produce grasses...

...the ground is hard, do not...

...[*natural*] net, natural well, and natural prison...[237]

...This is called an important advantage. If it is in front, this is called a concealed guard; if it is on the right, it is called a natural fortification; if it is on the left, it is called...

...what dwells on high is said to be the Constant Hall,...is said to be...

...follow, water on the left is said to be advantageous, and water on the right is said to be pent up...

...when the army enters into formation, regardless of the time of day, on the right flank to its back should be rises and hills, and on its left flank to its front should be waters and marshes. Those who follow...

...the measures needed to cope with the nine kinds of terrain, and the basic patterns of the human character must all be [*thoroughly investigated.*][238]

地形 二

【□】地刑（形）東方爲左，西方爲【右】……

……首，地平用左，軍

……地也。交□水□

……者，死地也。産草者□……

……地剛者，毋□□□也□……

【天】離、天井、天宛□……

……是胃（謂）重利。前之，是胃（謂）獸守。右之，是胃（謂）天固。左之，是胃

（謂）……

……所居高曰建堂，□曰□【□】□遂左水曰利，右水曰積……

……□五月度□地，七月□……

……三軍出陳（陣），不問朝夕，右負丘陵，左前水澤，順者……

……九地之法，人請（情）之里（理），不可不□□……

CHAPTER 5

[AN INTERVIEW WITH
THE KING OF WU]²³⁹

[EDITOR'S NOTE: This passage is similar in content to the story recorded in the *Historical Records* biography of Sun Wu, where he has an interview with King Ho-lu of Wu and where he applies his military arts to discipline the court beauties. See the section entitled "Sun Wu as a Historical Person" in the Introduction above.

In comparing these two accounts, to the extent that they overlap, the language in the *Historical Records* biography is more summary, and also more polished. The account preserved in the archaeological dig is more didactic, portraying Sun Wu as a serious-minded teacher addressing King Ho-lu of Wu in a way reminiscent of Mencius's lecturing King Hui of Liang in the opening chapter of the *Mencius.* The surface difference between Sun Wu and Mencius is that Sun Wu espouses a positive position on "advantage" where Mencius condemns it. Sun Wu makes a distinction between pursuing "advantage" in warfare that redounds to the national good, and treating the brutality of warfare as a kind of royal blood sport. Mencius, on the other hand, rejects "advantage" as an unworthy consideration for a ruler when morality should be his first and major concern. Both Sun Wu and Mencius at a more fundamental level would agree that warfare, taking its toll in lives and property, is invariably a national sacrifice, and must be pursued only as a last and unavoidable resort.

We can speculate that the *Historical Records* might be a revised version of an earlier and more primitive account similar to the one preserved here.

【見吳王】

……□於孫子之館，曰：「不穀好□□□□□□□兵者與（歟）？孫……

乎？不穀之好兵□□之□□也，適之好之也。孫……

也。兵，□【也】，非戲也。□□□君王以好與戲問之，外臣不敢對。蓋（闔）廬曰：不穀未

聞道也，不敢趣之利與……□孫子曰：唯君王之所欲，以貴者可也，賤者可也，

婦人可也。試男於右，試女於左，□□□……曰：不穀順（願）以婦人。孫子

曰：婦人多所不忍，臣請□……畏，有何悔乎？孫子曰：然則請得宮□……

之國左後壨圉之中，以爲二陳（陣）□□……曰：陳（陣）未成，不足見也。及

已成……□不辟（辭）其難。君曰：若（諾）。孫子以其御爲□參乘爲司

空，告其御、參乘曰：□□……婦人而告之曰：知女（汝）右手？……之。知女

（汝）心？曰：知之。知女（汝）北（背）？曰：知之。……左手。……

從女（汝）心。□不從令者也。七周而澤（釋）之，鼓而前

之，【三告而】五申之，鼓而前之，婦人亂而笑。□□胃（謂）女（汝）前，

申之，鼓而前之，婦人亂而笑。……金而坐之，有（又）三告而五

司馬與□司空而告之曰：兵法曰……三告而五申之者三矣，而令猶不行。孫子乃召其

也。兵法曰：弗令弗聞，君將之罪也；已令已申，卒長之罪

也。□請謝之。孫子曰：君□……引而員（圓）之，

員（圓）中規，引而方之，方中巨（矩）。……

蓋（闔）廬六日不自□□□□……

□□□孫子再拜而起曰：道得矣。□□□長遠近習此教也，以爲恒命。此

素教也，將之道也。民……□莫貴於威。 威行於衆，嚴行於吏，三軍信其將畏

（威）者，乘其適（敵）。

而用之，□□□得矣。 若□十三扁（篇）所……

【十】三扁（篇）所明道言功也，誠將聞□……

至日中請令……

也，君王居臺上而侍（待）之，臣……

【孫】子曰：「唯……

□□□之孫子曰：「外內貴賤得矣。」孫……

□而試之□得□……

【孫】子曰：「古（姑）試之，得而用之，無不□……

人主也。 若夫發令而從，不聽者誅□……

□也。 請合之於□□□之於……

陳（陣）已成矣，教□□聽……

□不穀請學之。」爲終食而□……

將軍□不穀不敢不□……

The references to the "thirteen chapters" and *The Art of Warfare* indicate that both versions of this story are later than the core text.]

...to Master Sun's guest house, the King of Wu asked: "I am fond of...the use of the military?" Again he inquired: "Sun...? My fondness for using the military...is using them properly and being fond of them."

Master Sun replied: "Using the military is to gain the advantage; it is not a matter of being fond of it. Using the military is to...; it is not a matter of sport. If Your Majesty wants to ask about war in terms of fondness and sport, I dare not reply."

King Ho-lu said: "I have never been told about the way of warfare; I dare not go after advantage and..."

Master Sun responded: "It is only important that it be what Your Majesty wants to do. We can use noble persons, we can use common folk, we can use your court ladies. We will train the men on the right and the ladies on the left...."

[*King Ho-lu*] said: "I want to use my court ladies."

Master Sun replied: "Many of the court ladies lack the stamina. I would rather use..."

[*King Ho-lu replied:*] "...awe, what is there to regret?"

Master Sun said: "In that case, then from your palace please let me have..., to go to the outer hunting park to the east side of the capital..., and get them to form two lines...."

[*Master Sun*] said: "When they have not yet been drilled in their formations, they are not ready for parade. Once they have been drilled,...cannot excuse their difficulties. [*Could Your Majesty please go to the high balcony and wait there for us?...at midday I will ask for Your orders. Once they have been drilled in their formations, following commands...not...not difficult.*]"[240]

The ruler replied: "I consent."

Master Sun used his chariot driver as [*his major*] and his arms bearer as field officer, and instructed the driver and the arms bearer, saying: "..."

[*Master Sun then turned to*] the court ladies and instructed them, saying: "Do you know which is your right hand?"

"[*We know*] it, [*they replied*]."

"Do you know your heart?"

"We know it," they replied.

"Do you know your back?"

"We know it," they replied.

[*"Do you know which is your left hand?"*
"We know it," they replied.]

... "[*When I tell you 'Left,' follow the direction of*] your left hand. When I tell you 'Front,' follow the direction of your heart. When I tell you [*'Back,' follow the direction of your back. When I tell you 'Right,' follow the direction of your right hand.*] ... it is your life. [*But my commands are to be obeyed. Those who do not obey will be executed* ...]²⁴¹ ... those who do not obey commands. Having circled seven times, fall out. On hearing the drums, advance ..."

Going through and explaining his commands several times, he then drummed for them to advance, but the court ladies being all out of place, [*they broke into laughter*] ... Striking the gong, he had them kneel. Again going through and explaining his commands several times, he drummed for them to advance, but the court ladies were all out of place, and broke into laughter. Three times he went through and explained his commands, but still his orders were not carried out. Master Sun then summoned his major and his field commander, and told them: "It says in *The Art of Warfare:* If one does not order them, or if one's orders are not understood, it is the fault of the commander. If one has already issued orders and has explained them, it is the fault of the field officers. *The Art of Warfare* also says: In rewarding the good, begin from the lowliest; in punishing ..."

[*King Ho-lu said:*] "... please excuse them."

Master Sun replied: "The ruler ..."

... [*Master Sun said:*] "Now if you direct them to assume a circular formation, their circle will satisfy the compass; if you direct them to assume a square formation, their square will satisfy the set square."

... for six days King Ho-lu did not ...

... Master Sun, bowing several times, arose and said: "You now know the way of warfare ... far and near practice this doctrine, and take it as their constant rule. This unadorned doctrine is the way of the commander. For the masses ... nothing is more exalted than authority. If the commander acts with authority over his men and enforces discipline among his officers, the entire army will have faith in his authority, and will conquer the enemy."

... and use it, ... will get. As the thirteen chapters ...

... the way elucidated and the attainments spoken of in the thirteen chapters is really what the commander told ...

... Master Sun said: "In the meantime we will try them, and if they get it we will use them. There is nothing that is not ...

... and try them ... getting ...

... to Master Sun, and said: "We have those within and without and the noble and the base." Master Sun ...

"... I, your ruler, would like to practice it." For the duration of the meal ...

"... the commander, I would not dare to not ...

Sun-tzu:
Part III

Text Recovered from
Later Works

[EDITOR'S NOTE: The textual materials that comprise Part III are passages recovered from later encyclopedic works and commentaries. I have grouped these passages into "chapters," sometimes on the basis of a shared theme, and sometimes because they come from the same source or archaeological site. The specific reference for each passage is indicated in the endnotes.]

CHAPTER 1

A CONVERSATION BETWEEN
THE KING OF WU AND SUN WU[242]

(i)*

吳王問孫武曰：散地，士卒顧家，不可與戰，則必固守不出。敵攻我小城，掠吾田野，禁吾樵採，塞吾要道，待吾空虛而急來攻，則如之何？武曰：敵人深入吾都，多背城邑，士卒以軍爲家，專志輕鬭。吾兵在國，安土懷生，以陳則不堅，以鬭則不勝，當集人衆，聚穀蓄帛，保城備險，遣輕兵絕其糧道。彼挑戰不得，轉輸不至，野無所掠，三軍困餒，因而誘之，可以有功。若與戰，必因勢。勢者，依險設伏，無險則隱於天陰暗昏霧，出其不意，襲擊懈怠。

---

* The topic of this passage, "scattering terrain," is defined in Chapter 11 above: "Where a feudal ruler does battle within his own territory, it is a terrain that permits the scattering of his troops." Master Sun is explicit in his warning, "On scattering terrain do not fight."

The King of Wu enquired of Sun Wu, saying: "If on 'scattering terrain' my officers and men are thinking of their homes, and cannot engage the enemy in battle, we must consolidate our defenses, and not go out against him. If the enemy then attacks our smaller walled cities, plunders our fields and meadowlands, prevents us from gathering our crops, blocks off our main thoroughfares, and, waiting until we have nothing left, attacks us in earnest, then what are we to do?"

Sun Wu replied: "If the enemy has penetrated deep into our territory, many of our walled cities and towns will be to his back. His officers and men will take the army as family, and with a united resolve, will think nothing of going to battle. But our troops are fighting on the home front, are comfortable on their native soil and have a great love of life. If you deploy them in a defensive position, they are not solid; if you send them into battle, they will not win.

"You should assemble a large number of troops, lay in ample provisions, stockpile cloth, fortify your walled cities, and guard the strategic passes. Dispatch light infantry to cut off the enemy's supply lines. If he tries to provoke an engagement do not give it to him. His supply wagons will not get through, and the countryside will have nothing left for him to pillage. With his whole army in the grips of hunger, you can succeed in drawing him into a blind. If you engage him in battle, you must make the most of strategic advantage (*shih*). Making the most of strategic advantage (*shih*) means occupying the key passes and lying in ambush. Where there is no such terrain, hide in the shadows and the mist, go by way of places where it would never occur to him you would go, and attack him when he is off his guard."

(ii)*

吳王問孫武曰：吾至輕地，始入敵境，士卒思邊，難進易退，未背險阻，三軍恐懼，大將欲進，士卒欲退，上下異心。而敵盛守，修其城壘，整其車騎，或當吾前，或擊吾後，則如之何？孫武曰：軍入敵境，敵人固壘不戰，士卒思歸，欲退且難，謂之輕地。當選驍騎伏要路，我退敵追，來則擊之。軍在輕地，士卒未專以入爲務，無以戰固。故無近其名城，無由其通路，設疑佯惑，示若將去。乃選驍騎，銜枚先入，掠其牛馬六畜。三軍見得進，乃不懼。分吾良卒，密有所伏，敵人若來，擊之勿疑；若其不至，捨之而去。

The King of Wu enquired of Sun Wu, saying: "We have reached marginal terrain, and have begun to press into enemy territory. Officers and men alike are thinking of the return home; it is hard to advance, and so easy to withdraw. With no passes or natural hazards to their backs, the

* The topic of this passage, "marginal terrain," is defined in Chapter 11 above: "Where one has penetrated only barely into enemy territory, it is marginal terrain." Master Sun is explicit in his warning, "Do not stay on marginal terrain."

armies are fearful. The commanders want to advance, and their officers and men want to withdraw; superiors and subordinates are of two minds. Moreover, the enemy is amply defended. He has reinforced his walled cities and fortifications, and strengthened his chariot and mounted detachments. With some of his forces blocking our front and others attacking us from the rear, what are we to do?"

Sun Wu replied: "[*A situation in which our troops have entered enemy territory, the enemy is secure behind his walls and does not bring the battle to us, our officers and men are thinking of the return home, and for us to withdraw would be difficult indeed is called occupying marginal terrain. We should select our elite mounted troops and place them in ambush on the main thoroughfare. As we withdraw, the enemy will give us chase, and when they reach us, we attack them.*][243] When an army is on marginal terrain, the officers and men are not one in spirit, they are doing what they must only because they are on enemy ground, and are going into battle without the heart for it. Therefore, do not approach the enemy's major walled cities, and do not advance on his main thoroughfares. Set up decoys and feign confusion, and give the enemy the impression we are about to quit our position. Then select our elite mounted troops, and send them on ahead into enemy territory under a cloak of silence to seize cattle, horses, and livestock. When our armies see the spoils, they will be ready to advance without fear. Separate off our best troops and lay them secretly in ambush. If the enemy comes, attack him in full fury; if he does not, break camp and quit the position."

(iii)*

又問曰：爭地，敵先至，據要保利，簡兵練卒，或出或守，以備我奇，則如何？　武曰：爭地之法，讓之者得，求之者失。敵得其處，慎勿攻之。引而佯走，建旗鳴鼓，趣其所愛，曳柴揚塵，惑其耳目；分吾良卒，密有所伏，敵必出救，人欲我與，人棄吾取，此爭先之道。若我先至而敵用此術，則選吾銳卒，固守其所，輕兵追之，分伏險阻，敵人還鬭，伏兵旁起，此全勝之道也。

The King of Wu enquired of Sun Wu, saying: "The enemy has been first to reach contested terrain, has taken up key strategic positions and has secured the advantageous ground. In an effort to check our mounted detachments, he then dispatches some of his select troops and crack offi-

---

* The topic of this passage, "contested terrain," is defined in Chapter 11 above: "Ground that gives us or the enemy the advantage in occupying it is contested terrain." Master Sun is explicit in his warning, "Do not attack the enemy on contested terrain."

cers to attack us while others are kept defending their position. What are we to do?"

Sun Wu replied: "The principle governing contested terrain is that if you let the enemy have it, you can get it, but if you try to get it, you will lose it. If the enemy has occupied the contested terrain, move carefully and do not attack him. Feign retreat and withdraw. Set up the flags and sound the drums, and hasten to the enemy's most vital points. Drag brush behind the troops and raise the dust to confuse the ears and eyes of the enemy. Separate off our best troops and lay them secretly in ambush. The enemy must come out to the rescue. What he wants we give him, and what he abandons we take. This is the way of contested terrain.

"If we are first to arrive, and the enemy tries to use this strategy on us, select out our finest troops and reinforce the defenses of our position, and send our light infantry in pursuit. Deploy a detachment to lay in ambush in some difficult stretch of terrain, and when the enemy comes out to meet our pursuing force, our concealed troops launch an attack from both sides. This is the way to take the complete victory."

(iv)*

又問曰：武曰：交地，吾將絕敵，令不得來，必全吾邊城，修其所備，深絕通道，固其阨塞。若不先圖，敵人已備，彼可得來，而吾不可往，衆寡又均，則如之何？武曰：既吾不可以往，彼可以來，吾分卒匿之，守而易怠，示其不能。敵人且至，設伏隱廬，出其不意也。

The King of Wu enquired of Sun Wu, saying: "On intermediate terrain, we want to cut off the enemy line and prevent him from advancing. We must preserve our border walled cities intact and fortify their de-

* The topic of this passage, "intermediate terrain," is defined in Chapter 11 above: "Ground accessible to both sides is intermediate terrain." Master Sun is explicit in his warning, "Do not get cut off on intermediate terrain."

fenses, make a deep cut in the main road and reinforce our hazards and blockades. What if we have not planned in advance and the enemy is already prepared, so he can advance at will and yet we cannot get away? Where the numerical strength of our armies is about the same, what are we to do?"

Sun Wu replied: "Since we cannot leave and yet the enemy can come at will, we deploy a detachment and secrete the men in ambush. We are vigilant in our defenses, but give the enemy the impression we are not up to battle. Then when the enemy arrives, our troops in ambush will appear from hiding places where it would never occur to the enemy they would."

(v)*

又問曰：衢地必先，吾道遠，發後，雖馳車驟馬，至不能先，則如之何？武
曰：諸侯參屬，其道四通，我與敵相當，而傍有國。所謂先者，必重幣輕使，
約和傍國，交親結恩，兵雖後至，眾以屬矣。簡兵練卒，阻利而處，親吾軍
事，實吾資糧，令吾車騎出入瞻候。我有眾助，彼失其黨，諸國犄角，震鼓
齊攻，敵人驚恐，莫知所當。

---

* The topic of this passage, the "strategically vital intersection," is defined in Chapter 11 above: "The territory of several neighboring states at which their borders meet is a strategically vital intersection. The first to reach it will gain the allegiance of the other states of the empire." Master Sun is explicit in his injunction, "Form alliances with the neighboring states at strategically vital intersections."

f Wu enquired of Sun Wu, saying: "The strategically vital
ust be reached before the enemy forces, but our road is long
gotten under way after the enemy. If, even with racing our
galloping our horses, we cannot possibly reach the intersec-
he enemy, what are we to do?"

replied: "The territories of our neighboring rulers border on
and our thoroughfares go in all four directions. Our military
about the same as the enemy's, but there are other neighboring
ved. What is meant by arriving first is we must send lavish gifts
voys and effect alliances with our neighboring states, so that re-
s are intimate and there is mutual good will. Even if our armies
r the enemy, we are more numerous by virtue of these alliances.
our select troops and crack officers to check enemy operations
the upper hand. People sympathetic to our troops will provide us
full complement of supplies and provisions, and will act as look-
r our chariots and mounted troops in their comings and goings.
we have abundant support, the enemy will have lost all those who
have sided with him. The neighboring states will be one flank in our
front, the sound of our drums will rock the heavens, and we will at-
as one. The enemy will be alarmed, and will not know how to re-
d."

(vi)*

又問曰：吾引兵深入重地，多所踰越，糧道絕塞，設欲歸還，勢不可過。欲食於敵，持兵不失，則如之何？……武曰：凡居重地，士卒輕勇，轉輸不通，則掠以繼食。下得粟帛，皆貢於上，多者有賞，士無歸意。若欲還出，切爲戒備，深溝高壘，示敵且久，敵疑通途，私除要害之道，乃令輕車衘枚而行，塵埃氣揚，以牛馬爲餌。敵人若出，鳴鼓隨之，陰伏吾士，與之中期。內外相應，其敗可知。

The King of Wu enquired of Sun Wu, saying: "Our forces have pushed deep into critical terrain and have passed by many of the enemy's cities and towns. Our supply lines have been cut off and stopped. If we try to go

---

* The topic of this passage, "critical terrain," is defined in Chapter 11 above: "When an army has penetrated deep into enemy territory, and has many of the enemy's walled cities and towns at its back, it is on critical terrain." Master Sun is explicit in his injunction, "Plunder the enemy's resources on critical terrain."

back now, there is no way we will make it. If we try to feed off of the enemy, he is sure to put up a fight. What then are we to do?"

Sun Wǔ replied: "Generally when an army has occupied critical terrain, the officers and men rely on courage in pressing ahead. If the supply lines are broken, they plunder to provision themselves. If the rank and file get grain or cloth, it is all handed over to the superiors. When many receive rewards, the men will have no thought of going back. If we intend to launch another attack, we must be thoroughly prepared with deep ditches and high barriers, giving the enemy the impression it will be a long and protracted battle. If the enemy doubts our capacity to move on his roads, he himself will recall his troops from guarding vital arteries. Under a cloak of silence we can then dispatch a detachment of light chariots at the quick. Under the cover of a cloud of dust, we can use horses and cattle to bait him. If the enemy sends his troops out, sound the drums and go after him. Conceal our troops, and when the enemy has walked into the ambush, fall on him from all sides. His defeat is assured."

(vii)*

又問曰：吾人圯地，山川險阻，難從之道，行久卒勞，敵在吾前而伏吾後，營居吾左而守吾右，良車曉騎，要吾隘道，則如之何？ 武曰：先進輕車，去軍十里，與敵相候，接期險阻。 或分而左，或分而右，大將四觀，擇空而取，皆會中道，倦而乃止。

The King of Wu enquired of Sun Wu, saying: "We have entered difficult terrain and, with the mountains and rivers, passes and natural hazards, the road is hard to follow. We have been pressing on for a long time,

---

* The topic of this passage, "difficult terrain," is defined in Chapter 11 above: "Mountains and forests, passes and hazards, wetlands and swamps, and any such roads hard to traverse constitute difficult terrain." Master Sun is explicit in his injunction, "Press ahead on difficult terrain."

and our troops are exhausted. The enemy occupies the ground ahead, and has also set an ambush behind us. He has established camp to the left of our forces, and has set up defenses against our right flank. His fine chariots and elite mounted troops threaten our precarious route. What are we to do?"

Sun Wu replied: "First dispatch our light chariots to advance about ten *li* in front of the main force to keep an eye on the enemy. Prepare to engage the enemy in battle amid the passes and natural hazards of this difficult terrain. Divert the troops to the left and to the right. On the signal of the high command, select vulnerable targets and take them, with all of the men regrouping back at the main road. Break off the operation once the troops are exhausted."

(viii)*

又問曰：吾入圍地，前有強敵，後有險難，敵絕我糧道，利我走勢，敵鼓噪不進，以觀吾能，則如之何？ 武曰：圍地之宜，必塞其闕，示無所往，則以軍爲家，萬人同心，三軍齊力。并炊數日，無見火烟，故爲毀亂寡弱之形。敵人見我，備之必輕。告勵士卒，令其奮怒，陣伏良卒，左右險阻，擊鼓而出。敵人若當，疾擊務突，前闘後拓，左右犄角。又問曰：敵在吾圍，伏而深謀，示我以利，縈我以旗，紛紛若亂，不知所之，奈何？ 武曰：千人操旆，分塞要道，輕兵進挑，陣而勿搏，交而勿去，此敗謀之法。

---

* The topic of this passage, "terrain vulnerable to ambush," is defined in Chapter 11 above: "Ground that gives access through a narrow defile and where exit is tortuous, allowing an enemy in small numbers to attack our main force, is terrain vulnerable to ambush." Master Sun is explicit in his injunction, "Devise contingency plans on terrain vulnerable to ambush."

The King of Wu enquired of Sun Wu, saying: "We have entered terrain vulnerable to ambush. Directly in our path is a formidable enemy, and to our back are natural hazards and rough terrain. The enemy has cut off our supply lines, and wants us to think our best advantage lies in flight. He sounds his drums and raises a hue and cry, yet does not advance on us, trying to gauge our battle strength. What are we to do?"

Sun Wu replied: "On terrain vulnerable to ambush, we must seal off the passes. If we show the men there is nowhere to go, they will take their fellows-at-arms as family, everyone will be united in spirit, and the entire army will fight as one. Prepare several days' provisions at once, but do not let the enemy see the fire and smoke, thus creating the impression that our forces are run-down, disorderly, and few in number. The enemy forces will take our measure, and in preparing against us are sure to think we are of little consequence. Arouse the officers and men, and rally them to rise up in fury against the enemy. Detail our superior fighting men in attack formation and in ambush. With defiles and natural hazards on all sides, sound the battle drums and launch the attack. If the enemy forces offer resistance, lash out at them suddenly and in full fury. Those on the front line carry the fight and those behind buttress them, working together to ram the enemy position."

The King of Wu again enquired: "The enemy has fallen into our ambush, but takes cover and plans his strategy carefully. He offers us some concessions, encircles us with his standards, and mills about as though his ranks are in disorder. We do not know what to make of it. What are we to do?"

Sun Wu replied: "Dispatch a thousand men to take care of the standard bearers, send a detachment to block off the main arteries, and send the light chariots ahead to harass the enemy. Deploy our main force in battle formation, but do not pounce on him. Join him in battle and do not withdraw. This is the way to defeat his strategy."

(ix)

吳王問孫武曰：吾師出境，軍於敵人之地。敵人大至，圍我數重，欲突以出，四塞不通。欲勵士激眾，使之投命潰圍，則如之何？武曰：深溝高壘，示為守備，安靜勿動，以隱吾能。告令三軍，示不得已。殺牛燔車，以饗吾士。燒盡糧食，填夷井灶，割髮捐冠，絕去生慮。將無餘謀，士有死志。於是砥甲礪刃，并氣一力，或攻兩旁，震鼓疾譟，敵人亦懼，莫知所當。銳卒分行，疾攻其後。此是失道而求生。故曰：困而不謀者窮，窮而不戰者亡。吳王曰：若吾圍敵，則如之何？武曰：山峻谷險，難以踰越，謂之窮寇。擊之之法，伏卒隱廬，開其去道，示其走路，求生透出，必無鬥意。因而擊之，雖眾必破。

The King of Wu enquired of Sun Wu, saying: "Our army has moved across our own borders and has entered enemy territory. The enemy arrives in force, and throws a cordon around us several times over. We want to break through his lines and escape, but the enemy has blocked off the roadways in all directions. We want to arouse the officers and inflame the rank and file so our men are willing to sacrifice their lives in bursting through the blockade. What are we to do?"

Sun Wu replied: "Gouge out deep ditches and pile up high barriers, showing the enemy we are prepared to defend our ground. Lie still and motionless, thereby concealing our strength from the enemy. Solemnly inform the entire army that our situation is a last-ditch fight to the death. Slaughter the oxen and burn the wagons to feast our troops, cook up all of the remaining provisions, and fill in and flatten our wells and cooking holes. Shave your head, throw away your official cap, and give up any thought of living. The commander has no further strategies; the officers and men are armed with their death resolve. At this, wipe down the armor, sharpen the blades, unite the men in spirit and strength, and launch the attack on two flanks. With the thundering of our drums and our ferocious battle cries, the enemy will be terrified, and will not know how to stop us. Divide our crack troops into two divisions to smash through enemy lines and launch a stinging attack on his rear lines. This is what is called snatching life from a disaster of our own making. Thus it is said:

> To fail to think fast when surrounded by the enemy
> is to have your back pressed to the wall;
> And to fail to take the battle to the enemy when
> your back is to the wall is to perish.

King Wu again enquired: "What do we do if it is we who have surrounded the enemy?"

Sun Wu replied: "Our mountains and valleys, high crags and defiles, are difficult for the enemy to traverse. His predicament is called the invader with his back to the wall. As for how to attack him: Conceal our troops in unlikely hiding places, and give the enemy a way out so he thinks there is a road to safety. He will pass through the corridor in an effort to save himself, and is sure to have no heart for battle. Take this opportunity and attack him, and even though he may be more numerous, you are sure to smash him."

(x)*

又問曰：吾在死地，糧道已絕，敵伏吾險，進退不得，則如之何？武曰：燔吾蓄積，盡我餘財，激士勵衆，使無生慮。鼓呼而衝，進而勿顧，決命爭強，死而須鬭，若敵在死地，士卒氣勇，欲擊之法：順而勿抗，陰守其利，絕其糧道，恐有奇伏，隱而不睹，使吾弓弩，俱守其所。

* The topic of this passage, "terrain from which there is no way out," is defined in Chapter 11 above: "Ground on which you will survive only if you fight with all your might, but will perish if you fail to do so, is terrain with no way out." Master Sun is explicit in his injunction, "On terrain from which there is no way out, take the battle to the enemy."

The King of Wu enquired of Sun Wu, saying: "We occupy terrain from which there is no way out, and our supply lines have already been cut. The enemy ambushes us on the rough terrain, and we can neither advance nor retreat. What are we to do?"

Sun Wu replied: "Put our stores to the torch and use up whatever goods we have left. Inflame the officers and incite the rank and file so they have no thought of living. With war drums and battle cries mounting to the heavens, advance on the enemy without looking back. Enter the fray having resolved to win or die, being fully aware that the only alternative to death is to do what is needed in the struggle.

"If it is the enemy who is on terrain from which there is no way out, and the morale and courage of his officers and men is at its height, the way to attack him is this: Be responsive to the enemy's moves and do not take him head on. Secretly deploy troops to safeguard our advantages, cut off the enemy's supply lines, and watch out for the surprise ambush. Go into hiding where we cannot be seen, dispatch the archers and crossbowmen, and have them all hold their ground."

(xi)

吳王問孫武曰：敵人保據山險，擅利而處之，糧食又足，挑之則不出，乘間則侵掠，爲之奈何？ 武曰：分兵守要，謹備勿懈。潛探其情，密候其怠。 以利誘之，禁其樵牧。久無所得，自然變改。待離其固，奪其所愛。敵據險隘，我能破之也。

The King of Wu enquired of Sun Wu, saying: "The enemy occupies the mountains and passes, and constantly uses his terrain advantage against us. He moreover has all he needs of supplies and provisions, and though we harass him he does not come out. And as soon as he sees an opening, he breaks through and pillages. What can we do about this?"

Sun Wu replied: "Divide up and deploy the army to safeguard our critical points, and prepare against the enemy thoroughly and with utmost vigilance. Covertly explore the enemy's situation, and wait in readiness for the least sign of negligence. Try to tease him out with seeming opportunities, and put an end to his herding and gathering so that for an extended period he brings in nothing. He will change his posture of his own accord. Wait for him to leave his stronghold, and then snatch what he covets the most. Though the enemy might occupy strategic passes and terrain, we are still able to smash him."

CHAPTER 2

SUN WU DISCUSSES
THE COMMANDER

孫武兵書云：軍井未達，將不言渴，軍灶未炊，將不言飢。

The military treatise of Sun Wu says: "Before the army's watering hole has been reached, the commander does not speak of thirst; before the fires have food on them, the commander does not speak of hunger."[244]

(ii)

孫子曰：將者：智也，仁也，敬也，信也，勇也，嚴也。是故智以折敵，仁以附眾，敬以招賢，信以必賞，勇以益氣，嚴以一令。故折敵則能合變，眾附則思力戰，賢智則陰謀利，賞罰必則士盡力，氣勇益則兵威令自倍，威令一則惟將所使。

Master Sun said: "The traits of the true commander are: wisdom, humanity, respect, integrity, courage, and dignity. With his wisdom he humbles the enemy, with his humanity he draws the people near to him, with his respect he recruits men of talent and character, with his integrity he makes good on his rewards, with his courage he raises the morale of his men, and with his dignity he unifies his command. Thus, if he humbles his enemy, he is able to take advantage of changing circumstances; if the people are close to him, they will be of a mind to go to battle in earnest; if he employs men of talent and wisdom, his secret plans will work; if his rewards and punishments are invariably honored, his men will give their all; if the morale and courage of his troops is heightened, they will of themselves be increasingly martial and intimidating; if his command is unified, the men will serve their commander alone."[245]

(iii)

孫子兵法曰：人效死而上能用之，雖優游暇譽，令猶行也。

Sun-tzu: The Art of Warfare states: "Where men are committed to fight to the death, their superiors are able to make good use of them. Even when they are taking it easy and are at their leisure, commands will still be carried out."[246]

(iv)

孫子兵法云：貴之而無驕，委之而不專，扶之而無隱，危之而不懼。故良將之動也，猶璧玉之不可污也。

Sun-tzu: The Art of Warfare states: "Exalt him and he is not arrogant; commission him and he does not act autocratically; support him and he does not conspire; threaten him and he is not afraid. Thus the actions of the able commander are as incorruptible as a jade insignium."[247]

(v)

孫子兵法秘要云：良將思計如飢，所以戰必勝，攻必取也。

Sun-tzu: The Secret Essentials of the Art of Warfare[248] states: "Because the able commander plans and calculates like a hungry man, he is invincible in battle and unconquerable in the attack."[249]

(vi)

孫子兵法云：非文無以平治，非武無以治亂。善用兵者有三略
焉：上略伐智，中略伐義，下略伐勢。

Sun-tzu: A Discussion of the Art of Warfare[250] states: "It takes a person of
civil virtue to bring peace to the empire; it takes a person of martial virtue
to quell disorder in the land. The expert in using the military has three
basic strategies that he applies: The best strategy is to attack the enemy at
the level of wisdom and experience; the second is to expose the injustice
of the enemy's claims; and the last is to attack the enemy's battle position
(*shih*)."[251]

(vii)

孫子曰：將者，勇，智，仁，信。

Master Sun said: "The traits of the true commander are: courage, wisdom, humanity, and integrity."[252]

(viii)

孫子曰：將必擇其福厚者。

Master Sun said: "The commander will surely choose those who are most fortunate."[253]

CHAPTER 3

Sᴜɴ Wᴜ Dɪsᴄᴜssᴇs Dᴇᴘʟᴏʏɪɴɢ ᴛʜᴇ Aʀᴍʏ

(i)

孫子曰：天隙之地，丘墓故城，兵不可處。

Master Sun said: "On marching through terrain with natural defiles, grave mounds, and the ruins of old walls, the army cannot tarry."[254]

(ii)

孫子兵法曰：林木翳薈，草樹蒙籠。

Sun-tzu: *The Art of Warfare* states: "The forests lie thick and tangled, the vegetation is lush and overgrown."[255]

(iii)

孫子曰：故曰：深草蓊穢者，所以逃遁也；深谷險阻者，所以止禦車騎也；隘塞山林者，所以少擊眾也；沛澤杳冥者，所以匿其形也。

Master Sun said: "Therefore it is said: Terrain covered with thick brush and lush foliage is used for escape and for hiding; ground marked with deep valleys, defiles, and natural hazards is used to ward off chariots and mounted troops; narrow passes and mountain forests are used for the few to attack the many; terrain covered with marshy jungle and dark thickets is used to conceal one's position."[256]

(iv)

孫子曰：凡地多陷曲，曰天井。

Master Sun said: "Lowlands covered with quagmires and labyrinths, are called natural wells."[257]

CHAPTER 4

THE PROGNOSTICATIONS
OF SUN-TZU[258]

(i)

孫子占曰：三軍將行，其旌旗從容以向前，是爲天送，必亟擊之，得其大將。三軍將行，其旌旗墊然若雨，是爲天霑，其帥失。三軍將陣，雨甚，是爲浴師，勿用陣戰。三軍方行，大風飄起於軍前，右周中其師，得糧。絕軍，其將亡，右周中其師，得糧。雲其上而赤，勿用陣，先陣戰者，莫復其跡。三軍將戰，有雲其上而赤，勿用陣，先陣戰者，莫復其跡。三軍將行，其旌旗亂於上，東西南北，無所主方，其軍不還。

The Prognostications of Sun-tzu says: "The combined army is about to set off. When the standards and banners are unfurled, they flutter in the direction the army is to go. This means Heaven is sending it on its way. It must strike quickly and will capture the enemy's high command.

The combined army is about to set off, and the standards and banners droop limply as though rain-soaked. This means Heaven has opened up a deluge on them, and its officers will be lost.

The combined army is about to set off, and the standards and banners flap around every which way on their staffs, without blowing in any particular direction. This army will not return.

The combined army is about to assume battle formation, and it rains in torrents. This is an army awash. It should not go to battle in formation.

The combined army is about to enter battle. Clouds gather above that are flaming red in color. Do not use battle formation in engaging the enemy. The first one to deploy in battle formation will not retrace his steps.

The combined army has just set off. Strong winds blow up in front of the troops. If the winds sweep to the right and cut off the advancing forces, the army's commanders will perish; if the winds sweep to the right behind the troops, the army will capture provisions."[259]

(i)

孫子稱司雲氣，非雲非煙非霧，形似禽獸，客吉，主人忌。

Master Sun said of those cloudlike vapors that govern a situation that they are neither cloud nor smoke nor mist. Where they take the shape of birds or animals, it is auspicious for the aggressor and a bad omen for the defending forces.[260]

CHAPTER 5

SUN WU DISCUSSES THE "EIGHT-DIVISION FORMATION"[261]

(i)

孫子八陣，有蘋車之乘。

Master Sun's "eight-division formation" includes the armored personnel-style chariots.[262]

(ii)

孫子兵法曰：長陣爲甄。

The *Sun-tzu: The Art of Warfare* states: "The extended battle formation deploys winged flanks."[263]

(iii)

八陣圖曰：以後爲前，以前爲後，四頭八尾，觸處爲首，敵衝其中，首尾俱救。

The *Eight-Division Formation Diagrams*[264] states: "In deploying in this formation, make the rear the front line, and the front line the rear. It has four heads and eight tails, so wherever the enemy strikes is its head. And when the enemy bursts through the lines, the head and tail can both come to the rescue."[265]

THE CLASSIC OF THE
THIRTY-TWO RAMPARTS[266]

(i)

移軍移旗，以順其意。銜枚而陣，分師而伏。後至先擊，以戰則克。

Redeploy the army and redistribute the banners in response to the enemy's intentions. Move the troops under a cloak of silence into their battle formation, and lay detachments in ambush. If the enemy is last to arrive at the battlefield, be first to launch the attack. If you use this battle strategy, you will defeat him.

CHAPTER 7

HAN DYNASTY BAMBOO STRIPS FROM TA-T'UNG COUNTY[267]

(i)

孫子曰：夫十三篇……

Master Sun said: "As for these thirteen chapters…"[268]

(ii)

相勝奈何？孫子曰：

"How do we take the victory?" Master Sun replied, "…

(iii)

軍鬭令

孫子曰：能當三□……

Cited in *Battle Ordinances*, Master Sun says: "To be able to face three ..."

(iv)

合戰令

孫子曰：戰貴齊成，以□□

Cited in *Ordinances for Joining the Enemy in Battle*, Master Sun says: "On the battlefield exalt concerted achievement, thereby ..."

(v)

□令孫子曰：軍行患車轄之，相……

Cited in [...] *Ordinances*, Master Sun says: "As the army advances concern yourself if the war chariots break ranks, and ... each other ..."

(vi)

子曰：軍患陣不堅，陣不堅則前破，而□……

The Master said: "In deploying the troops concern yourself if the formation is not solid, for if the formation is not solid, the front line will be crushed, and..."

CHAPTER 8

MISCELLANEOUS

(i)

孫子兵法曰：其鎮如岳，其淳如淵。

The *Sun-tzu: The Art of Warfare* states: "Stable, it is like a mountain peak; at rest, it is like a deep abyss."[269]

(ii)

孫子曰：強弱長短雜用。

Master Sun said: "Weak and strong, short and long, are mixed together in their use."[270]

(iii)

又曰：遠則用弩，近則用兵，兵弩相解也。

[Master Sun] went on to say: "At a distance, use your crossbow; at close quarters, use your hand weapons. Hand weapons and crossbow are of mutual aid."[271]

(iv)

又曰：以步兵十人擊騎一匹。

[Master Sun] went on to say: "Use a ratio of ten infantry to each mounted soldier in attacking."[272]

(v)

孫子曰：金城湯池而無粟者，太公、墨翟不能守之。

Master Sun said: "A city might have walls of iron and be surrounded by moats of boiling water, but if it is inadequately provisioned, even a Chiang T'ai-kung or a Mo Ti would be unable to defend it."[273]

APPENDIX

BACKGROUND TO THE EXCAVATION
AT YIN·CH'ÜEH·SHAN

Lin-i is a city and prefecture some 120 miles southwest of Ch'ing-tao (Tsingtao) in Shantung province. The mountain, Meng-shan, stands to the north of Lin-i city. To the south of the city, there are flat farmlands which the I river traverses from north to south. The prefecture of Lin-i takes its name from the fact that in its eastern reaches, it "meets" (*lin*) the I river.

Prevailing opinion has located Lin-i in what was, during the Warring States period, the southern portion of the state of Ch'i—the state that Sun Pin served as military adviser. It is also very near to the border of what was Confucius' home state of Lu. Approximately two thirds of a mile south of the old city wall, there are two small rises that run east and west, the eastern hill being called Chin-ch'üeh-shan (Gold Sparrow mountain) and the western hill, Yin-ch'üeh-shan (Silver Sparrow mountain).

In April 1972, during a construction project, two major finds dating from the Western Han dynasty (202 B.C.–A.D. 9), designated Tomb #1 and Tomb #2, were discovered at Yin-ch'üeh-shan. They were excavated under the direction of the Shantung Provincial Museum. Three institutions took responsibility for the find: The Institute of Scientific Technology for the Preservation of Artifacts, The Shantung Provincial Museum, and the Forbidden Palace Museum. From 1972 to 1974, a team of scholars

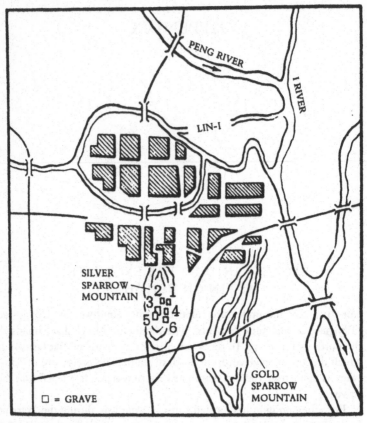

The Yin-ch'üeh-shan excavation site at Lin-i city in Shantung province

including Lo Fu-i, Ku T'ieh-fu, and Wu Chiu-lung did initial reconstruction work on the texts found in the tombs. In 1974, the preliminary results of the find were published under the name of the Committee for the Reconstruction of Yin-ch'üeh-shan Han Dynasty Bamboo Strips (hereafter, the Yin-ch'üeh-shan Committee). The committee, having focused first on the texts of the *Sun-tzu* and *Sun Pin,* completed a preliminary editing and annotation of all of the strips. The translations in this book are based on the ongoing published work of this committee.

The crypts in both tombs are rectangular pits dug out of the rock. Tomb #1, running north and south, is 3.14 meters long and 2.26 meters wide; Tomb #2, also running north and south, is 2.91 meters long and 1.96

meters across. Tomb #1 ranges from 2 to 3 meters in depth, and Tomb #2 is from 3.5 to 4 meters deep. Over the course of time, the upper covering of both of the tombs suffered breakage and water collected within them.

The tombs are divided lengthwise into two sections, one section as a crypt to contain a coffin, and the other to contain the various burial effects. The coffin crypts are intact, and between the pit and the wooden coffin, there is a very fine grayish clay. The structure of the coffin crypts in both tombs is approximately the same: Tomb #1 is 2.64 meters long, 1.76 meters wide, and 1 meter deep; Tomb #2 is 2.41 meters long, 1.56 meters wide, and .88 meters deep. In each of the coffin areas, there are skeletal remains that have decomposed to the extent that sex is impossible to determine, but the extension and direction of the corpse can be discerned. Both tombs contain a wealth of burial effects (see diagrams).

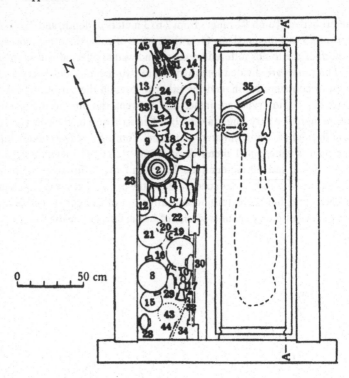

Diagram of Tomb #1 at Yin-ch'üeh-shan

1–4	earthenware pots	27–30	lacquered ear-shaped cups
5, 6	earthenware bowls	31	wooden vessel
7–10	earthenware tripods	32	wooden gameboard
11, 12	earthenware containers	33–34	wooden canes
13	cocoon-shaped earthenware pot with four legs	35	lacquered pillow
		36	lacquered toilet case
14–16	earthenware jars	37	coarse wooden comb
17–20	earthenware figurines	38–40	fine wooden combs
21	earthenware vessel with cover	41	comb stand
		42	bronze mirror
22–23	earthenware containers	43	bamboo basket
24–25	lacquered plate with colored ornamentation	44	chestnuts
		45	35 *pan-liang* (half-tael) coins
26	wine goblet		

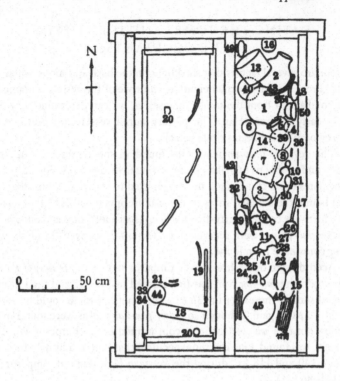

Diagram of Tomb #2 at Yin-ch'üeh-shan

1	earthenware jar with characters	22	38 *pan-liang* (half-tael) coins
2	earthenware jar	23–25	lacquered bowls with
3–4	earthenware pots with		colored ornamentation
	colored ornamentation	26	small lacquered bowl
5–6	earthenware containers with	27–32	lacquered ear-shaped cups
	colored ornamentation	33	coarse wooden comb
7–8	earthenware tripods with	34	fine wooden comb
	colored ornamentation	35–38	small wooden heads
9–12	earthenware figurines with	39–40	lacquered bowls
	colored ornamentation	41	lacquered ladle
13	lacquered toilet case	42	wooden gate
14	lacquered tube	43	broken piece of a lacquered
15	oval wooden container		circular box
16	remains of a wooden owl	44	lacquered toilet case
17	wooden cane	45	plain lacquered bowl
18	wooden pillow	46–47	lacquered ear-shaped cups
19	wooden handle		with colored ornamentation
20	small lacquered container	48–49	lacquered ear-shaped cups
21	bronze mirror	50	bronze pot

DATING THE TOMBS AND IDENTIFYING
THE OCCUPANTS

There are several factors that shed light on the basic questions: when and who? From the shape, ornamentation, and style of the vessels contained in the tombs, and from the tombs themselves, we can determine they date from early Western Han. The pottery burial vessels still preserve the legacy of Warring States ritual vessels.

The seventy-four ancient coins found in the tombs are important clues, and enable us to make a closer estimate of the dates. Wu Chiu-lung speculates these coins might have been used to secure the thin silk cords that bound the bamboo strips of the manuscripts together.[274] From Tomb #1, thirty-six coins—thirty-five *pan-liang* (half-tael) coins and one *san-chu* (three-*chu*) coin—were recovered. In Tomb #2, thirty-eight of the *pan-liang* coins have been found.

According to the "Chronicle of Emperor Wu" in the *History of the Han Dynasty*,[275] the *san-chu* coin was first minted in 140 B.C. and then discontinued shortly thereafter in 136 B.C. Thus, neither tomb could be earlier than 140 B.C. Again, although so many *pan-liang* coins were found in the tombs, not one *wu-chu* (five-*chu*) coin minted under Emperor Wu from 118 B.C. was found. One can thus speculate that both tombs date from between 140 and 118 B.C. during the Western Han reign of Emperor Wu (r. 141–86 B.C.).

This date can be further refined, at least in the case of Tomb #2. In addition to the *pan-liang* coins, a calendar was recovered that begins in 134 B.C. and covers a thirteen-month period. This narrows the date of Tomb #2 to 134–118 B.C.

As to *who* was buried in these tombs, engraved on the bottom of two lacquer cups from Tomb #1 in clerical script (*li-shu*) are the two characters, "Ssu-ma." From its appearance we can determine it is not the imprint of the craftsman who made the vessel, but rather the surname of the owner engraved onto the vessel at some later date. "Ssu-ma," then, was probably the name of the occupant of Tomb #1. Ssu-ma could also have been this person's official title. The *History of the Han Dynasty* states that "the militarist school possibly derives from the ancient office of *Ssu-ma*, the royal office of military preparations."[276] According to this same source, "for seventy years from the beginning of the reign of Emperor Wu," there was the custom of "taking the name of one's office as one's surname."[277] Taking into account the collection of specifically military books found in the tomb together with this custom, it has been speculated that the occupant or his family had taken this military office as a surname. However, it is perhaps more likely that a person would engrave his name than the title of his office on to a cup, and by this time, "Ssu-ma" had already become a popular surname. If "Ssu-ma" was the name of the occupant's office, it is also likely it was the surname of the occupant.[278]

Burial effects were customarily things in daily use or of some particular value. From the large collection of military works found in Tomb #1, we can assume the occupant was someone who knew about the art of warfare or who was connected with the military in some way. The absence of weapons of any kind might suggest this interest stopped somewhat short of the occupant's being an active military man.

On the shoulder of an earthenware jar found in Tomb #2 there are the four characters *Shao-shih shih tou*—"ten pecks of the Shao family." This would suggest the occupant of Tomb #2 was surnamed Shao.[279]

The sum of this rather limited evidence has led to speculation that the occupants of Tombs #1 and #2, dated 140–118 B.C. and 134–118 B.C. respectively, were husband and wife, perhaps surnamed Ssu-ma and Shao. Even though the actual burials date from this period, it should be remembered that the texts themselves may have been copied considerably earlier.

The texts with which we are concerned were discovered among the lacquerware, pottery, bronzeware, coins, and various other burial effects. The bamboo strips constituting these early documents were found in both tombs.

THE FIRST PUBLISHED REPORTS

After the initial find, the Yin-ch'üeh-shan Committee devoted some two years of research to the 4,942 strips and fragments before making the preliminary results of this research known to the world in the February 1974 issue of the academic journal *Cultural Relics* (*Wen-wu*). In addition to the partial texts of the *Sun-tzu: The Art of Warfare* and the *Sun Pin: The Art of Warfare*, portions of the following texts were tentatively identified and published:[280]

1. *Six Strategies* (*Liu-t'ao*)—illustrated *Wen-wu* (*Cultural Relics*) (hereafter, *WW*) 1974:2; transcribed in *WW* 1977:2, pp. 21ff.—54 pieces
2. *Master Wei-liao* (*Wei-liao-tzu*)—illustrated *WW* 1972:2; transcribed in *WW* 1974:2, pp. 30ff.—36 pieces

3. *Master Kuan* (*Kuan-tzu*)—shown *WW* 1974:2 and 1976:12; transcribed in *WW* 1976:12, pp. 36ff.—10 pieces
4. *Master Yen* (*Yen-tzu*)—shown *WW* 1974:2–112 pieces
5. *Master Mo* (*Mo-tzu*)—shown *WW* 1974:2–1 piece (plus 42 additional pieces resembling lost chapters)

To a greater or lesser extent these partial texts all seem to deal with military affairs. Another group of textual materials, although primarily dealing with *yin-yang* theory and prognostication, also relates in some way to military affairs. There is speculation that these materials might in some part be divination texts from the lost *Miscellaneous Prognostications of Sun-tzu: The Art of Warfare*,[281] or perhaps treatises on divination originally included in *Sun Pin: The Art of Warfare*. These materials include:

6. *Yin-Yang of Master Ts'ao* (*Ts'ao-shih yin-yang*)—24 pieces
7. *Wind Direction Divination* (*Feng-chiao-chan*)—51 pieces
8. *Portent and Omen Divination* (*Tsai-i-chan*)—53 pieces
9. unknown textual material concerning divination—82 pieces

There is also one veterinary text dealing with the physiognomic examination of dogs:

10. *The Classic on Examining Dogs* (*Hsiang-kou-ching*)—11 pieces

From Tomb #2 was unearthed the oldest and most complete ancient calendrical record we have to date, which, as such, has considerable value for the study of calendrical methods during the ancient period:

11. *A Calendrical Record for the First Year of the yüan-kuang Reign Period of Emperor Wu of Han: 134 B.C.* (*Han Wu-ti yüan-kuang yüan-nien li-p'u*)—32 pieces

Perhaps the most significant and exciting textual material uncovered in Tomb #1 is the additional text of the extant *Sun-tzu: The Art of Warfare* and the portions of the long-lost *Sun Pin: The Art of Warfare*. The initial description of these materials in a report of the Yin-ch'üeh-shan Committee published as *Yin-ch'üeh-shan Han-mu chu-chien* I in July 1975[282] contains the following information:

12. *Sun-tzu: The Art of Warfare* (*Sun-tzu ping-fa*)—196 pieces: 11 complete ones, 122 partial slips, and 63 fragments, totaling 3,160 charac-

ters. Fragments from all 13 chapters of the received text have been identified, together with 68 pieces from what seem to be 6 previously unknown chapters.

13. *Sun Pin: The Art of Warfare* (*Sun Pin ping-fa*)—364 pieces: 187 complete ones, 109 partial slips, and 68 fragments, totalling 8,700 characters.

Considering for comparison that the *Lao-tzu* is approximately 5,000 characters in length and the thirteen-chapter *Sun-tzu* is about 6,000, the supplement to the *Sun-tzu* of over 1,200 characters and the newly recovered *Sun Pin*, though it has been revised downward to about half of this initial 8,700 figure, are substantial documents.

The updated report of the Yin-ch'üeh-shan Committee is summarized in the Introduction.

THE BAMBOO STRIP MANUSCRIPTS AND THEIR DATES

In ancient China, "books" were generally written on bamboo, wood, or silk. From the extant classical records, it would seem that bamboo strips were already in use during the Shang dynasty (traditionally, 1751–1112 B.C.). The character *ts'e,* a graph depicting symbolically "rolls of strips," is frequently seen on the oracle-bone inscriptions that date from the Shang dynasty.

Certain material factors have contributed to a situation in which, after two thousand years, these strips have been recovered in a still legible condition. The strips in these tombs were buried quite deep and out of the sunlight, and hence kept at a low and relatively constant temperature. The tombs themselves were carved out of the rock from the top down, and, as a consequence, were relatively easy to seal tightly. The fine, gray glutinous clay must have functioned as a relatively water-tight sealant. The combination of these factors certainly has had a bearing on the condition of the strips. Still, after such a long time immersed in the muddy water that eventually did leak in, the strips were much discolored, had become very fragile, and are now easily fragmented.

Most of the strips came from Tomb #1. A total of more than 7,500 strips and fragments were found in the northern corner of the burial-effects pit between the lacquerware and pottery. Few of these are whole; most are fragments; many have only one or two characters. Of the 7,500 strips, then, 4,942 constitute the working core of the reconstructed texts.

Physically, the complete strips divide into long and short strips. The long ones are about 27.5 cm with most having a width of 0.5 to 0.7 cm and a thickness of 0.1 to 0.2 cm. Almost all of the texts are copied on the long strips. Only the divination texts seem to be found on the shorter ones. All of these shorter strips are fragmentary, but the estimated length is about 18 cm with a width of about 0.5 cm. The bamboo strips were originally bound into scrolls or *ts'e* and tied with cords, but the cords rotted away long ago, and hence were not in place to preserve the order when the strips were recovered. Most of the long strips were originally joined with three cords, with those at the top and bottom about 2 cm from the ends, and one in the center. The shorter strips and some of the long strips were joined with only two cords, with the cords at the top and bottom being about 2 cm from the ends.

Tomb #2 produced a total of 32 bamboo strips found in the southeastern corner with other burial effects. These strips constitute the calendrical register for the first year of the *yüan-kuang* reign period, the seventh year of Emperor Wu of the Han dynasty (134 B.C.). They are about 69 cm long, 1 cm wide, and 0.2 cm thick, and were joined together with three cords.

The characters on the strips from Tomb #1 are all in clerical script (*li-shu*), which had become the standard style with the unification of Chinese states under the state of Ch'in in 221 B.C. Given that the strips belong to an early period in the standardization of the clerical script (*li-shu*), they are a substantial resource for investigating developments in the construction of the character and the style of writing during the process of transition from the seal-style (*chuan-shu*) to the clerical-style script. They were written with a brush using black ink. The immediate observation that some of the characters are well formed and symmetrical while others are more crude leads to the assumption that they are not the product of one hand. Because they were not written by one person or at one time, there is some variation in the written form of the characters and in the length of the lines of text. In the written form of the characters, for example, the strips divide into the two large categories of standard and cursive, with each of these including several different hands. Most complete strips have over thirty characters, but some of those written with characters close together have over forty, while those written with some space between the characters have only twenty-odd. There is not always internal consistency within the texts. The same text can include several different hands and may vary in the length of lines.

The basic unit of text is the chapter (*p'ien*). Some of the chapter titles are written on the back of the first strips, some are written on strips by themselves, and some are written at the end of the chapter. This variation can be explained if we understand the physical structure of the classical Chinese "scroll," or *ts'e*. Most *ts'e* would have the last strip of text as its core, and then be rolled up from left to right (from the end of the text to the beginning) with the characters facing inward to protect them. Once a *ts'e* was rolled up in this manner, the outer surface of the first strip in the text could then be used for recording the title of the first chapter on the scroll as a way of facilitating easy reference.

In some cases a scroll would contain only the one chapter with the title written on the outer surface of the first strip, that is, the back of the first strip. But if it contained more than one chapter, typically the title of the first chapter in the scroll would be written on the outer surface of the first strip, and the titles of the remaining chapters would be written either on separate strips at the beginning of each chapter, or at the end of the chapters. Several of the shorter chapters of the reconstructed Yin-ch'üeh-shan bamboo scrolls have the titles on both the back of the first strip and at the end, while some of the others just have the title at the end. The former were probably the first chapters in a scroll, while the latter chapters were contained somewhere within the scroll.

For example, on the strips that constitute "The Eight-Division Formation" and "Terrain as Treasure" chapters of the *Sun Pin: The Art of Warfare*, there is a similarity in the form of the characters and in the length of the lines, suggesting that they belong to the same scroll. But "The Eight-Division Formation" chapter has its title on both the back of the first strip and at the end of the chapter. This chapter then was probably the first chapter of the scroll. "Terrain as Treasure" has its title at the end, but not on the back of the first strip, and hence was probably a chapter following "The Eight-Division Formation" within the scroll.

In the Yin-ch'üeh-shan bamboo scrolls, each chapter that has a title on an individual strip has no title on either the back of the first strip or at the end. For example, the fragments of the thirteen chapters of the *Obeying Ordinances* (*Shou-fa*) are like this. We can surmise that in these bamboo scrolls, those chapters that have their titles on individual strips, and those that have them on the back of the first strip, did not belong to the same scroll.

In addition to the bamboo strips found in Tomb #1, several slats of wood were found with what seems to be a listing of the titles of the chapters on them. A wooden slat was probably bound to the outside of each scroll as a table of contents.

Li Hsüeh-ch'in describes four wooden slats that list the chapter titles: #1 has the thirteen core chapters of *Sun-tzu: The Art of Warfare*; #2 has the thirteen chapters of the *Obeying Ordinances and Orders* (*Shou-fa shou-ling*); #4 has the *Yin-Yang of Master Ts'ao* (*Ts'ao-shih yin-yang*). Wooden slat #3 has only a few characters remaining, but at least two lines, "Fatal Weaknesses of the Commander" (*chiang-pai*) and "Common Mistakes of the Military" (*Ping chih heng-shih*), seem to refer to chapters that have been included in the supplemental Part II of *Sun Pin: The Art of Warfare*.[283] The style of the characters is similar to those strips that we can be sure belong to *Sun Pin: The Art of Warfare*. Some of the other titles on this slat seem to refer to discussions on government. This does not preclude the possibility that slat #3 was originally part of the table of contents for *Sun Pin: The Art of Warfare*. As Li Hsüeh-ch'in observes, the other military texts—*Sun-tzu, Six Strategies*, and *Master Wei-liao*—all include discussions of government.[284] The fact that there is no fuller table of contents for the *Sun Pin: The Art of Warfare*, however, makes the process of reconstructing this particular text more difficult

NOTES

References to *Sun Pin: The Art of Warfare* are to the *Yin-ch'üeh-shan Han-mu chu-chien* Collection I, prepared by Yin-ch'üeh-shan Han-mu chu-chien cheng-li hsiao-tsu (Committee for the Reconstruction of the Yin-ch'üeh-shan Han strips) and published in 1985. For those chapters originally included in the *Sun Pin* but excluded from the 1985 publication, I refer to the *Sun Pin ping-fa* published by Yin-ch'üeh-shan Han-mu chu-chien cheng-li hsiao-tsu (Committee for the Reconstruction of the Yin-ch'üeh-shan Han strips) in 1975.

1. The Chinese expression for China, *"chung-kuo,"* often translated as "Middle Kingdom," dates back to pre-imperial days, and refers to a plurality of "central states," not one unified political entity. The English name, "China," is reportedly taken from the state of Ch'in, which was the ultimate victor in establishing a unified empire in 221 B.C.

2. For a short history of the *Sun-tzu* in European languages, see Appendix III in Samuel B. Griffith (1963), *"Sun-tzu* in Western Languages."

3. I retain the untranslated title *Lao-tzu* throughout for two reasons. Firstly, there is an ambiguity with respect to its meaning: It could reasonably be translated as either "Master Lao" or "the old Master." Secondly, there is the recognition factor.

4. D. C. Lau (1982).

5. *Wen-wu (Cultural Relics)* 1981.8:11–13. Ting-hsien Tomb #40 dating from the Han dynasty has yielded bamboo strips totaling approximately half of the received *Analects*. Although the *Analects* was one of the earliest texts to find its canonical form, several alternative versions circulated until it was edited into its present form by Ho Yen (A.D. 190–249). It is reported that there are some important differences be-

tween this Han dynasty version and our present text. The sectioning of passages is often not the same, and if we take grammatical particles into account, there are variants in almost every passage.

6. The *Wen-wu* (*Cultural Relics*) 1981.8:11–12 report on the *Master Wen* recovered from the Ting county tomb sheds important new light on the *Master Wen*. Reasons for taking the Han strips text as more authoritative than the received text are several. Firstly, of course, is its incontrovertible presence in a tomb that can be dated back to the Han dynasty.

Secondly, the structure of the Han strips text is consistent with the *Master Wen* as described in the library catalog of the Han dynasty court. The *History of the Han Dynasty* lists "*Master Wen* in nine chapters" in the "Record of Literary Works" (*Yi-wen chih*) of the *History of the Han Dynasty*, a catalog of the imperial library completed during the first century A.D. A commentary is appended, identifying Master Wen as a disciple of Lao-tzu who lived as a contemporary of Confucius, and who was asked questions by King P'ing of Chou. Some passages of the Han strips text are similar in content to the received Master Wen; some of them are entirely different. On the passages in the Han strips text that parallel the received *Master Wen*, the occurrences of "Master Wen" have been changed in the received text to read "Lao-tzu," and Master Wen has become the student asking the questions. On the Han strips, consistent with the description in the *History of the Han Dynasty* account, King P'ing asks the questions, and Master Wen answers them.

Thirdly, it would seem likely that the text would be named for Master Wen the teacher rather than Master Wen the disciple.

The discovery of the *Master Wen* from the Ting county tomb not only gives this text new status as an indisputably ancient work, but further, for the history of early Chinese thought, adds new textual material that has previously been unknown to us.

7. A great deal of scholarly attention is being invested in these four documents. Robin Yates is presently working on an annotated translation that is forthcoming in this same *Classics of Ancient China* series. These documents have, in recent scholarship, notably by R. P. Peerenboom, been used to articulate a definition of "Huang-Lao" Taoism, a specific mixture of Taoist and Legalist thought that emerged in the early Han dynasty. This clarification is particularly important, because many scholars continue to use the "Huang-Lao" category as a catch-all for any and all early Han philosophical works that make reference to Taoism. See Peerenboom (1993).

8. I am working on an annotated translation of *Sun Pin: The Art of Warfare* that will appear in this same *Classics of Ancient China* series.

9. Wu Chiu-lung (1985):9.

10. Chang Chen-tse (1984):2 (preface).

11. Wu Chiu-lung (1985):13.
12. *Historical Records* (1959): Chapter 65.
13. See *Intrigues of the Warring States* (1920):4/6b; cf. Crump (1979):154; 7/8b, 380; and *Historical Records* (1959):279, 1845–1846, 1962, 2343, 2351, 3300.
14. See *Lü-shih ch'un-ch'iu* (1935):802.
15. *History of the Sui Dynasty* (1973):1012–1013.
16. Su Ch'in is frequently found in the list of commoners who became high ministers of state. Although Su Ch'in figures frequently in the *Historical Records* and the *Intrigues of the Warring States* that chronicle events of this period, there has been considerable speculation concerning the historicity of these accounts. See J. I. Crump (1979):13–15. For traditional views on Su Ch'in's dates, see Ch'ien Mu (1956):306 who, following Ssu-ma Ch'ien, places him as having died in 321 B.C.

The discovery of the *Documents of the Warring States Strategists* (*Chan-kuo tsung-heng-chia shu*) in Tomb #3 at Ma-wang-tui in 1973 has provided us with textual materials not available to China's earliest historians that are enabling scholars to revise significantly the historical profile of Su Ch'in, and the dates of his life. This text, dating from 195–188 B.C., was initially thought to be a version of the *Intrigues of the Warring States*, but since the *Intrigues of the Warring States* was not compiled by Liu Hsiang until the late first century B.C., further study has shown that the *Documents of the Warring States Strategists* was copied from at least three independent sources. See Blanford (1991):198n12. It consists of twenty-seven passages, ten of which in some form appear in Liu Hsiang's *Intrigues of the Warring States* and eight of which appear in Ssu-ma Ch'ien's *Historical Records*. Discounting the overlap between these two records, sixteen passages are new.

In his 1976 article, "Precious Historical Materials Unseen by Ssu-ma Ch'ien," T'ang Lan asserts that Ssu-ma Ch'ien never saw any firsthand historical materials recounting the thoughts and ideas of Su Ch'in and, as a consequence, confused the order of persons and events, introduced various errors concerning these same events, and even invented some episodes. Chronologically, Ssu-ma Ch'ien inadvertently pushed those events of Su Ch'in's life that occurred at the beginning of the third century B.C. back into the end of the fourth century B.C. T'ang Lan suggests that the *Historical Records* account of Su Ch'in is more like a historical novel than a chronicle of events. It is clear that these newly recovered documents will be of inestimable value in sorting out the sequence of events during the key era when the Warring States period was drawing toward its climax, and in reevaluating the role played by Su Ch'in.
17. Samuel B. Griffith (1963):3–11.
18. See Joseph Needham *et al.* and Robin Yates (forthcoming). Jerry Norman and Mei Tsu-lin (1976):293–294 make a case for an Austroasiatic origin

of the crossbow and the term for it, *nu*, on philological and historical grounds, suggesting that the Chinese acquired this weapon from the proto-Tai and Vietnamese sometime in the third or fourth century B.C.

19. D. C. Lau (1982): 134.
20. *Sun-tzu* 3:111.
21. *Sun Pin* (1985):50–51.
22. *Sun Pin* (1975):115–116, "Male and Female Fortifications."
23. *Historical Records* (1959): 2161–2162.
24. *Historical Records* (1959):1466.
25. See J. J. L. Duyvendak (1928).
26. See particularly Chapter 10, "The Method of Warfare"; also, Chapters 11 and 12. Duyvendak (1928):244–252.
27. See John Knoblock (1990):211–234.
28. *Han Fei-tzu so-yin* (1982):49.12.25. See Burton Watson (1964):110.
29. See *Historical Records* (1964):2162.
30. See Li Ling (1983):552–553.
31. Pan Ku (1962):1731. The military writers are listed under the Taoist school.
32. See John Fairbank in Frank A. Kierman, Jr. (1974):7; H. G. Creel (1970) Chapter 10, especially pp. 247–257.
33. See for example *Analects* 9/26, 13/29, 13/30, and 16/7.
34. See Mark Lewis (1990): chapters 2 and 3 for a thorough discussion of this changing situation in classical China.
35. See Lai Hai-tsung (1984) for a discussion of the changing role of the military in early China.
36. The concept *shih*, translated as "strategic advantage," for example, can be traced in the Legalist, Confucian, and even Taoist philosophical sources back to a specifically militarist notion of battle advantage. See Roger T. Ames (1983): Chapter 2.
37. See *Analects* 2/12 and 13/23. See also 1/12, 2/14, and 15/22. An explication of this one passage is really the central theme of Hall and Ames (1987), *Thinking Through Confucius.*
38. *Analects* 15/29.
39. Michel Foucault's *The Order of Things: An Archaeology of the Human Sciences* is a self-conscious response to Borges's category. See Foucault (1973):xv.
40. *Analects* 13/23.
41. Hsü Wei-yü (1935):540.
42. *Discourses of the States* (*Kuo-yü*) 16/4a–b.
43. Wing Tek Lum (1987):105.
44. Graham (1981):275.
45. See Paul Demiéville's essay, "Philosophy and Religion from Han to Sui" in Denis Twitchett and Michael Loewe (1986).

46. *Analects* 19/21.
47. See John Fairbank (1987):83–94.
48. *Analects* 2/1 and 15/5.
49. John Fairbank, "Introduction: Varieties of the Chinese Military Experience" in Frank A. Kierman, Jr. (1974):8.
50. *Shih* is written as *chih* in the ancient texts.
51. For an exploration of the later political implications of this term, see Ames (1983):Chapter 3 *passim*.
52. *Sun Pin* (1985):62–63.
53. *Sun-tzu* 5:120–121.
54. *Sun-tzu* 5:120.
55. *Sun-tzu* 1:104.
56. *Sun-tzu* 11:158.
57. See for example *Lao-tzu* 36, 40, 50, 57, and 78. This same understanding of the nature of change can be found in such disparate sources as the *Book of Changes* and the Confucian *Analects*.
58. *Sun Pin* (1975): 121.
59. *Sun-tzu* 5:120.
60. *Sun Pin* (1975):121.
61. *Sun Pin* (1975):121.
62. *Sun Pin* (1975):121.
63. *Sun Pin* (1985):48. See also *Sun-tzu* 6:126–127.
64. *Sun-tzu* 6:127.
65. *Sun Pin* (1975):122.
66. See Lau (1965):332–333 and Ames (1983):67.
67. *Sun-tzu* 6:125.
68. *Sun-tzu* 4:116.
69. *Sun-tzu* 6:125.
70. *Sun-tzu* 1:105.
71. *Master of Huai Nan* (1968):15/8a–b.
72. *Sun-tzu* 6:127.
73. *Sun-tzu* 2:108.
74. *Sun-tzu* 12:165.
75. *Sun-tzu* 6:127.
76. *Sun-tzu* 2:107–108.
77. *Sun Pin* (1985):58.
78. *Sun Pin* (1985):59.
79. *Sun Pin* (1985):48.
80. *Sun-tzu* 3:111.
81. *Sun Pin* (1985):59.
82. *Sun-tzu* 1:103.
83. *Sun-tzu* 3:111–112.
84. *Sun-tzu* 4:115.

85. *Sun-tzu* 6:123.

86. *Sun-tzu* 8:136.

87. *Sun-tzu* 11:157.

88. *Sun-tzu* 6:125.

89. *Sun-tzu* 2:107.

90. *Sun Pin* (1985):48.

91. *Sun-tzu* 10:150.

92. *Sun-tzu* 1:103. See also *Sun-tzu* III:2:ii, p. 226.

93. *Sun-tzu* 3:112.

94. *Sun-tzu* 2:109.

95. *Sun-tzu* 3:113. Compare *Sun Pin* (1985):58:
    There are five conditions that ensure constant victory: to have the full confidence of one's ruler and full authority over the army will lead to victory;...

96. *Sun-tzu* 8:135.

97. *Sun-tzu* 10:150.

98. *Sun-tzu* 13:171.

99. *Sun-tzu* 6:125.

100. *Sun-tzu* 13:169.

101. *Sun-tzu* 5:120.

102. *Sun-tzu* 13:169.

103. *Sun-tzu* 13:171.

104. *Sun-tzu* 3:111.

105. *Sun-tzu* 3:111.

106. *Sun-tzu* 4:116.

107. *Sun-tzu* 4:116.

108. *Sun-tzu* 1:105.

109. See D. C. Lau (1965):331–332.

110. *Sun-tzu* 8:135.

111. *Sun-tzu* 6:125.

112. *Sun-tzu* 6:126.

113. *Sun-tzu* 1:104–105.

114. *Sun-tzu* 3:113.

115. The basic meaning of the character *ping* translated here as "war" is "arms" or "weapons." By extension in different contexts it means "soldiers," "army," and "war" itself.

116. The Han strips text has an additional phrase here:
    Complying with or defying these conditions will determine military victory.

117. I have followed Wu Chiu-lung (1990) in adding this phrase from the Han strips.

118. *Chiang* can be read as "commander" rather than as a particle indicating futurity. An alternative translation, then, would be:

If a commander heeds my assessments, to employ him is certain victory. Keep him. If a commander does not heed my assessments, to employ him is certain defeat. Dismiss him.

119. I have adapted this passage from D. C. Lau's review of Samuel B. Griffith's translation. Lau (1965):332 points out that the key to understanding this passage is *suan:* "counters or, more precisely, counting rods which are used to form numbers.... In the final calculation, the number of rods is totted up, and for the side that has scored more rods victory is predicted."

120. The character *fan*, conventionally translated "in general" or "generally speaking," is taken care of here and elsewhere through the text by treating this kind of introductory phrase as a caption. See D. C. Lau (1965):323–324.

121. A thousand *li* amounts to several hundred miles.

122. I read the *ku* here—often translated inferentially as "therefore"—as simply a passage marker. This is a familiar usage of *ku* in the classical corpus, and there is a real danger if we insist on a formulaic translation that suggests linear inference where there is none.

123. There is an ambiguity in this passage. D. C. Lau (1965):334–335 uses contemporaneous texts to construct a persuasive argument that the *"ch'üan"* as it occurs in *ch'üan kuo* and the other parallel binomials is a technical term meaning "to preserve intact." I have followed his reading of this passage. The more standard interpretation would be:

> It is best to preserve the enemy's state intact; to crush the enemy's state is only a second best. It is best to preserve the enemy's army, battalion, company, or five-man squad intact; to crush the enemy's army, battalion, company, or five-man squad is only a second best.

Mark Lewis (1990):116 has this reading. Another possible reading would be:

> It is best to preserve a state intact, and only second best to crush it; it is best to preserve an army, battalion, company, or five-man squad intact, and only second best to crush it.

124. I follow D. C. Lau (1965):334–335 in his reading of *"chüan"* as a technical term here.

125. I follow the arguments of Wu Chiu-lung *et al.* (1990):42 in reversing these two phrases, which originally read:

> ... when double, divide him; when you and the enemy are equally matched, be able to engage him ...

126. Many of the commentators reconstrue the grammar here in imaginative ways to arrive at the alternative reading:

> Thus, if a small force simply digs in, it will fall captive to a larger enemy,

I have opted for a more literal translation that seems to make perfect sense.

127. Mark Lewis (1990):109 identifies this guard as "the pole mounted on the side of a chariot to keep it from overturning."

128. See note 122 above for *ku* as a passage marker.

129. Several texts have "in every battle will be defeated," but both the rhyme and the meaning recommend "be at risk (*tai*)" over "be defeated (*pai*)."

130. See D. C. Lau (1965):332–333, for a discussion of the various meanings of *hsing*, here translated "strategic positions." See also Ames (1983):Chapter 3 for a discussion on the relationship between *hsing* and *shih*, "strategic advantage/political purchase."

131. Compare the *Master of Huai Nan* (1968):15/14a.

132. Many of the commentators have an alternative reading of this passage: "He who cannot win takes the defensive; he who can win attacks."

133. I follow the word order on the Han strips here. All of the received texts have an alternative reading:
    > If one assumes a defensive posture, it is because one's strength is deficient; if one launches the attack, it is because one's strength is more than enough.

    Both versions make good sense. The Han strips version is familiar in the Han dynasty histories; see, for example, the *History of the Han Dynasty*, Pan Ku (1962):2981, and the *History of the Later Han Dynasty*, Fan Yeh *et al.* (1965):650. However, the version found in the received texts also occurs in the *History of the Later Han Dynasty*, Fan Yeh *et al.* (1965):2305, and the commentary of Ts'ao Ts'ao (155–220) on the *Sun-tzu* is based upon the received text version. It would appear the change in the transmission of *Sun-tzu* had already been made in the later Han dynasty. See Yin-ch'üeh-shan Committee (1985):8n7.

134. I am following the Han strips version here. The Sung edition reads:
    > Therefore, in the battle victory for the expert neither wins him reputation for wisdom nor credit for courage.

135. See note 122 above for *ku* as a passage marker.

136. Compare the *Master of Huai Nan* (1968):15/6b.

137. The "way" is defined in *Sun-tzu* 1:103 in the following terms:
    > The way (*tao*) is what brings the thinking of the people in line with their superiors. Hence, you can send them to their deaths or let them live, and they will have no misgivings one way or the other.

    "Regulation" is also defined in the same chapter 1:103:
    > And regulation entails organizational effectiveness, a chain of command, and a structure for logistical support.

138. I follow the Han strips version here. There is corroboration for it in *Lao-tzu* 47 and *Kuan-tzu* 2:75–130. The alternative reading is "... and thus is able formulate policies governing the outcome of the war."

139. It is difficult to be certain about these classical measures because of regional and temporal disparities, but the units referred to in this passage probably indicate about a 600 to 1 differential.

140. The Han strips text has two drafts of this particular chapter that, each written in a different hand, can be separated into "A" and "B" texts. I follow both "A" and "B" versions of the Han strips text here in substituting the phrase "the army that has this weight of victory on its side" for "the victorious army" in the received texts. Compare the *Master of Huai Nan* (1968):15/11b.

141. This passage reads literally:
     ...it is the dividing and counting of numbers that makes managing many soldiers the same as managing a few.

142. The expression here is literally "forms and names" rather than "flags and pennants." I follow Robin Yates (1988):220–222 who makes his argument on the basis of the following passage which he translates from the *Mo-tzu:*

     The standard method/procedure for defending cities is: Make gray-green flags for wood; make red flags for fires; make yellow flags for firewood and fuel; make white flags for stones; make black flags for water; make bamboo flags for food; make gray goshawk flags for soldiers that fight to the death; make tiger flags for mighty warriors; make double rabbit flags for brave [?] soldiers; make youth flags for fourteen-year-old boys; make grasping arrow flags for women; make dog flags for crossbows; make forest flags for *ch'i* halberds; make feather flags for swords and shields; make dragon flags for carts; make bird flags for cavalry. In general, when the name of the flag that you are looking for is not in the book, in all cases use its form and name to make [the design on the flag].

     See also Robin Yates (1980):387–390.

     Yates argues persuasively that many expressions such as "forms and names (*hsing-ming*)" that were to become central political terms in the fourth and third centuries B.C. originated with often more concrete military terminology in centuries prior to it. "Forms and names" in later Legalist doctrine came to mean "accountability," where one's actual performance would be carefully weighed against what one had promised as a result. I have made the same connection between military terminology and later political vocabulary with respect to "strategic advantage (*shih*)"—see Ames (1983):65–107.

143. I follow the Han strips text here that corroborates the commentarial emendation.

144. These terms are commonly translated as "regular" and "irregular," but this does not capture their correlativity. In military language, "regular"

and "irregular" conjures forth a distinction between "regular army" and "irregular militia." The distinction here is between what can be anticipated by the enemy and what catches him off guard—the element of surprise. Importantly, it might be an otherwise "regular" action that surprises an enemy using guerrilla tactics.

145. The Han strips version has "rivers and seas," where the Sung text has "rivers and streams." It is their flowing that makes the rivers and seas inexhaustible.

146. All of the received texts have "like going around a ring without beginning or end," but this seems to entail a later emendation.

147. I follow Wu Chiu-lung *et al.* (1990):75 in this reading. The alternative more common version is:

> ... that the velocity of a bird of prey can even smash its victim to pieces is due to its timing.

148. I follow the Han strips text here. The alternative is "He tempts the enemy..."

149. I take the *ku* at the beginning of this passage as a passage marker. See note 122 above.

150. Compare the *Master of Huai Nan* (1968):15:11b.

151. I am following Yang Ping-an (1985):73–75 and Wu Chiu-lung *et al.* (1990):85 in reconstructing this passage on the basis of the Han strips, the received text, and the encyclopedic works. This reading is more consistent with the demands of context. D. C. Lau (1965):321, prior to the discovery of the Han text, argued for the unemended received text that would translate:

> Thus being able to wear down a well-rested enemy, to starve one that is well-provisioned, and to move one that is settled lies in going by way of territory that the enemy does not make for, and making for places where it never occurred to him you would go.

152. Compare *Sun Pin* (1985):51.

153. Compare the *Master of Huai Nan* (1968):15/15a, and *Lao-tzu* 68.

154. The order of this passage is reversed and somewhat corrupt in the Han strips. Yang Ping-an (1985):79 makes the point that in Sun Wu's own world, his state of Wu compared with Ch'u and Yüeh could be fairly described as small and weak.

155. I am following the Han strips version here. Most of the other redactions read "what good is this to them in determining the outcome?"

156. The Han strips does not have *ku* here—commonly translated as "therefore." Even if we retain it, it only functions as a passage marker.

157. I am following the Han strips text here. The Sung edition has:

> ... just as it is the disposition (*hsing*) of water to avoid high ground and rush to the lowest point, so in dispositioning (*hsing*) troops, avoid the enemy's strong points and strike where he is weak.

158. I follow the Han strips redaction here. Many of the traditional texts have:
> Thus an army does not have constant strategic advantages (*shih*); water does not assume a constant form (*hsing*).

This change might have been made to avoid the character *heng* (translated here as "invariable") that was the given name of Emperor Wen of the Han (r. 179–157 B.C.), and hence taboo. The presence of *heng* in the Han strips might suggest the text was copied before Emperor Wen came to the throne, although during the Western Han, the taboo on the emperor's name was not always strictly observed.

159. The Han strips text has:
> To be able to transform oneself according to the enemy is called being inscrutable.

I translate *shen*—"spiritual, godlike, divine"—as "inscrutable" here, taking this sense of *shen* from the *Chou I* (*Book of Changes*) (1978):41/*hsi shang*/5: "What cannot be fathomed with *yin* and *yang* is called *shen*."

160. The Han strips text concludes with a round dot and an additional two characters, *shen yao*—"the essentials of inscrutability." This could be an alternative chapter title, or possibly a reader's summary of the contents of the chapter.

161. I read *ku* here as a passage marker. See note 122 above.

162. This passage seems out of place. Yang Ping-an suggests it belongs to the following chapter, and has been mistakenly interpolated here. The same passage appears again in *Sun-tzu* 11:161, where it also seems out of context.

163. An alternative reading for this phrase is:
> Use flags and pennants to divide up your numbers...

164. Although this phrase seems out of place here, its location is corroborated by the Han strips text.

165. The Han strips redaction begins this passage with "for this reason" while most other versions do not. Most commentators take this expression *"chün cheng"* to refer to a lost text entitled *The Book of Military Policies*, similar to a text called *Military Annals* (*chün chih*) cited in the *Tso-chuan* (Duke Hsi 28 and Duke Hsüan 12).

166. I have followed the arguments of the Yin-ch'üeh-shan Committee (1985):16 in retaining the order of the Han strips text in reorganizing this passage, inserting this sentence here instead of its location a few lines later in the received texts.

167. On the basis of the Han strips text, I have replaced *jen* with "the men" (*min*) here and below.

168. The *Master of Huai Nan* (1968):15/9b–10a has a similar passage:
> Thus the skilled commander in using the troops makes them of one mind and unifies their strength, so that the courageous will not have to advance alone, and the cowardly will not get to retreat alone.

169. I read *ku* here as a passage marker rather than as "therefore." See note 122 above. Several redactions omit it altogether.

170. Yang Ping-an interprets the "day" metaphor here to refer to the progress of battle. Hence, his reading would be:

    ... at the outset of battle, the enemy's morale is high; as the battle continues, it begins to flag; by battle's end, it has drained away.

    D. C. Lau's interpretation of the text is persuasive. See Lau (1965):321–322.

171. This tally appears on the last strip. The Sung edition has a total of 477 characters, which is close.

172. Literally, this chapter is entitled "Nine Contingencies (*chiu pien*)." There has been a debate among the commentators whether or not to take the number "nine" literally, given that this passage seems to enumerate ten contingencies. Some suggest that as the largest primary number, "nine" should be construed in this text as "all of the various."

    I take "nine" literally. The "[Four Contingencies]" chapter recovered in the Yin-ch'üeh-shan find is a commentary on this chapter. Having elaborated on four of the contingencies, it then states:

    ... if the commands of the ruler are contrary to these four contingencies, do not obey them.

    Thus it excludes this concluding summary proscription from its tally of four. If we do the same here and take the phrase "and commands from the ruler not to be obeyed" as a summary proscription, we arrive at the tally of nine. Other examples of this usage are "nine heavens (*chiu t'ien*)" and "nine kinds of terrain (*chiu ti*)."

173. The opening passage of Chapter 7 is identical, and commentators speculate that it has been erroneously interpolated here in the process of transmission. The uncharacteristic brevity of this chapter would suggest some substantial textual problem.

174. This particular kind of terrain, *yi ti*, is defined in *Sun-tzu* 11:153 in the following terms:

    Mountains and forests, passes and natural hazards, wetlands and swamps, and any such roads difficult to transverse constitute difficult terrain.

175. This kind of terrain, *ch'ü ti*, is defined in *Sun-tzu* 11:153:

    The territory of several neighboring states at which their borders meet is a strategically vital intersection. The first to reach it will gain the allegiance of the other states of the empire.

    See D. C. Lau (1965):328 for a discussion of this passage.

176. This type of terrain, *chüeh ti*, is defined in *Sun-tzu* 11:160:

    When you quit your own territory and lead your troops across the border, you have entered cut-off terrain.

    See D. C. Lau (1985):327–328.

177. The kind of terrain called *wei ti* is described in *Sun-tzu* 11:153–155:
    Ground that gives access through a narrow defile, and where exit is tortuous, allowing an enemy in small numbers to attack our main force, is terrain vulnerable to ambush.

178. This kind of terrain, *szu ti*, is described in *Sun-tzu* 11:155:
    Terrain in which you will survive only if you fight with all your might, but perish if you fail to do so, is terrain with no way out.
    See D. C. Lau (1965):328.

179. In the "[Four Contingencies]" chapter recovered in the Yin-ch'üeh-shan find and translated below (Part II:2 p. 179), it elaborates on this passage:
    That there are roadways not to be traveled refers to a roadway where if we penetrate only a short distance we cannot bring the operations of our vanguard into full play, and if we penetrate too deeply we cannot link up effectively with our rearguard. If we move, it is not to our advantage, and if we stop, we will be captured. Given these conditions, we do not travel it.

180. In the "[Four Contingencies]" chapter recovered in the Yin-ch'üeh-shan find and translated as Part II:2 p. 179, it elaborates on this passage:
    That there are armies not to be attacked refers to a situation in which the two armies make camp and face off. We estimate we have enough strength to crush the opposing army and to capture its commander. Taking the long view, however, there is some surprise advantage (*shih*) and clever dodge he has, so his army...its commander. Given these conditions, even though we can attack, we do not do so.

181. In the "[Four Contingencies]" chapter recovered in the Yin-ch'üeh-shan find and translated as Part II:2 p. 179–180, it elaborates on this passage:
    That there are walled cities not to be assaulted refers to a situation in which we estimate we have enough strength to take the city. If we take it, it gives us no immediate advantage, and having gotten it, we would not able to garrison it. If we are [*lacking*] in strength, the walled city must by no means be taken. If in the first instance we gain advantage, the city will surrender of its own accord; and if we do not gain advantage, it will not be a source of harm afterwards. Given these conditions, even though we can launch an assault, we do not do so.

182. In the "[Four Contingencies]" chapter recovered in the Yin-ch'üeh-shan find and translated as Part II:2 p. 180, it elaborates on this passage:
    That there is territory not to be contested refers to mountains and gorges...that are not able to sustain life.... vacant. Given these conditions, do not contest it.

183. In the "[Four Contingencies]" chapter recovered in the Yin-ch'üeh-shan find and translated as Part II:2 p. 180, it elaborates on this passage:
    That there are commands from the ruler not to be obeyed means that

if the commands of the ruler are contrary to these four contingencies, do not obey them.... obey them. Where affairs... contingencies, one understands how to use troops.

184. It is not clear what the "five advantages" are, although the commentaries are ready to speculate that they refer to five contingencies listed at the beginning of this chapter. The shortness of this chapter suggests that much of the original text is missing, and perhaps with it, a more conclusive explanation.

185. I read *ku* ("therefore") here as a passage marker. See note 122 above.

186. I read *ku* ("therefore") here as a passage marker. See note 122 above.

187. This passage is problematic. I am following most of the commentaries with this reading. Alternative interpretations of this passage are (1) "do not ascend high ground that stands alone" (because you can then be surrounded), and (2) "descend rather than ascend to engage the enemy."

188. Many of the commentators simply read this as "low ground to your front and high ground to your rear," and while this is generally true, it is also to simplify the insight. If your position forces the enemy to have mountainous terrain to his rear, you thereby disadvantage him by cutting off his retreat. If you have an exit to your rear that can be easily defended in the case of retreat, this is to give yourself advantage at the rear.

189. This passage reads literally, "the four emperors," but the reference has been unclear in the legend surrounding the Yellow Emperor, symbolic ancestor of the Han peoples. Commentators have speculated that "four Emperors" should read "four regions," or "four armies." In the newly recovered chapter, "The Yellow Emperor Attacks the Red Emperor," it states:

> [to the south he attacked the Red Emperor] ... to the east he attacked the [Green] Emperor... to the west he attacked the White Emperor... to the north he attacked the Black Emperor... and having defeated the four Emperors, he ruled over all under heaven.

Hence "the four Emperors" is not an error, but might refer to the ancestors of those peoples occupying the territory in each of the four directions.

Compare an alternative account of these battles attributed to Chiang-tzu in the *T'ai-p'ing yü-lan* (1963):79/369–370.

190. Although this passage seems out of place, it does occur here in the Han strips text. Several commentators suggest that it has been erroneously interpolated here from the passage near the beginning of this chapter where it describes "positioning an army when near water."

191. Compare *Sun Pin* (1985):61.

192. A popular alternative reading of this phrase is "if he has no treaty yet sues for peace," but it would seem unnecessary to sue for peace if a treaty had in fact already been completed.

193. The Sung text has "military vehicles" instead of "troops," but I follow the Han strips text here.

194. I have not translated the *ku* at the beginning of this passage as "therefore," taking it to be simply a passage marker. See note 122 above.

195. Compare *Sun Pin* (1985):51.

196. I follow Sun Hsing-yen (1965) in emending this passage for the rhyme. The unemended text reads:
    Know the natural conditions, know the ground, and the victories will be inexhaustible.

197. I am relying on D. C. Lau (1965):321 here.

198. I follow the Han strips here in omitting "his troops," thereby preserving the pattern of four character rhymed phrases.

199. Ts'ao Kuei was a native of Lu who in 681 B.C., disregarding his own life, succeeded in recovering lands lost to the state of Ch'i by grabbing Duke Huan of Ch'i and holding him at knifepoint. Chuan Chu was an assassin in the state of Wu who disregarded the certainty of his own death to take the life of King Liao of Wu in 515 B.C. These stories are recounted in *Historical Records* (1959):2515–2518.

200. In the received texts, Mount Heng is written as Mount Ch'ang, but given that the Han strips text has *"heng," ch'ang* was probably substituted for *heng* by scribes following the convention of avoiding the given name of the emperor—in this case, Emperor Wen of the Han (r. 179–157 B.C.). See note 158 above.

201. See *Intrigues of the Warring States* (1920):7/33b; cf. Crump (1979):412–413.

202. This passage suggests that the commander takes such measures as tying up the legs of the horses and rendering the chariots inoperable to show his troops there is no retreat, and to make plain his resolve to fight to the death.

203. A common alternative reading of this ambiguous phrase is:
    The principle of exploiting terrain is to get value from both your shock units and your weaker troops.
    That is, a commander can maximize his effectiveness by coordinating the quality of his troops with the features of the terrain.

204. The received texts have "keeping the enemy [literally, "others" (*jen*)] in the dark," but I follow the Han strips text here that has "keeping people (*min*) in the dark." It would appear that throughout this passage, scribes made this substitution to avoid the personal name of Li Shih-min, first emperor of the T'ang.

205. Most of the received texts include an additional phrase here:
    He sets fire to his boats and smashes his cooking pots.
    I follow the Han strips text in omitting it.

206. The Han strips version reads "On contested terrain, I would not allow them to remain."

207. The Han strips version reads "...on intermediate terrain, I would make sure of my alliances."

208. The Han strips version reads "...at a strategically vital intersection, I would pay particular attention to reliability."

209. The Han strips version reads "...on critical terrain, I would hurry up our rear divisions."

210. The Han strips text reads "Therefore, the psychology of the [feudal] lords is:..."

211. I treat the *shih ku*—conventionally, "for this reason"—as a passage marker. See note 122 above.

212. This passage occurs almost word for word in Chapter 7:130, but seems as out of place there as it does here.

213. I follow the Han strips text in reversing the order of hegemon and king to restore what is a familiar classical expression.

214. I follow the Han strips version of the text here, which seems more consistent with the passage that follows. The received texts read "...get them to make the gains, but do not reveal the dangers."

215. There is an alternative interpretation of this passage, but it is not responsible to the syntax of the text:
    Therefore, the business of waging war lies in the pretence of accommodating the designs of the enemy.

216. The text at this point is clearly corrupt, and any translation can only be tentative.

217. I follow the Han strips version here, which repeats "fuel" in the second phrase rather than "smoke" or "sparks."

218. The Winnowing Basket (four stars), the Wall (two stars), the Wings (twenty-two stars), and the Chariot Platform (four stars) are four of the twenty-eight constellations called "lunar mansions" (*hsiu*), equatorial divisions that constitute segments of the celestial sphere. See Needham (1970):229–283.

219. Yang Ping-an (1986):192 recommends emending this text to read:
    If the wind blows in the daytime, follow through; if it blows at night, do not.
    His argument is that the unemended interpretation of this passage is not consistent with what follows.

220. I read the *ku*—conventionally, "therefore"—that begins this sentence as a passage marker. See note 122 above.

221. Yang Ping-an (1986):193 suggests an emendation that would reinforce the parallel structure:
    Water can be used to cut the enemy off; Fire can be used to deprive him of his supplies.

222. Compare *Lao-tzu* 68.

223. I read *ku*—conventionally, "therefore"—as a passage marker. See note 122 above.

224. I follow the similar passage in *Lao-tzu* 14 in interpreting this phrase.

225. I have emended this on the basis of the Han strips text. In the received Sung text, it reads:

> Thus, in the operations of the combined forces, no one should have more direct access than spies, . . .

226. I follow Yang Ping-an here, but the more popular reading with the traditional commentators is:

> It is necessary to search out the enemy's agents who have been sent to spy on us, take care of them with generous bribes, and provide them with a place to live.

227. The same passage in the Han strips version can be reconstructed as:

> [The rise of the] Yin (Shang) dynasty [*was because of Yi Yin*] who served the house of Hsia; the rise of the Chou dynasty was because of Lü Ya who served [*in the house of Shang*]; [*the rise of the state of. . .*] was because of Commander Pi who served the state of Hsing; the rise of the state of Yen was because of Su Ch'in who served the state of Ch'i. Thus only those farsighted rulers and their superior [*commanders who can get the most intelligent people as their spies are destined to accomplish great things.*].

The implication here is that Yi Yin was a minister of Chieh, the classically diabolical last ruler of the Hsia, and led troops against the throne to install T'ang, the first ruler of the Shang dynasty. Lü Ya was a minister of the equally evil Chou, the last ruler of the Shang dynasty, and was instrumental in overthrowing his sovereign and founding the Chou dynasty. These are examples of historical figures who, having inside information, could effectively topple the power at the center.

The mention of Su Ch'in here, a historical figure who lived many generations after Sun Wu, introduces the problem of dating the *Sun-tzu*. See the discussion in note 16, above.

228. I have followed the Yin-ch'üeh-shan Committee (1985) for the order of these additional chapters.

229. "Six commanders" refers to the six high ministers of Chin: Han, Chao, Wei, Fan, Chung-hang, and Chih-po.

230. The *yüan* and the *chen* are classical units of land measure. The point here is that a commander's tenure as ruler was related in inverse proportion to the taxes exacted from his people. Each household was allowed a fixed parcel of land, and the commander whose "square foot" was the smallest could exact the most taxes. Fan and Chung-hang taxed their people at 150 percent the rate of Chao.

231. This section elaborates four of the contingencies discussed in Chapter

8, "Adapting to the Nine Contingencies," above. See note 172, which suggests an explanation for why five situations are addressed as "four contingencies."

232. This passage is unclear on the Han strips, and the Yin-ch'üeh-shan Committee's reconstruction is only tentative. If an army's strength is insufficient, it would seem too obvious that siege as a strategy must be ruled out.

233. See the reference to the Yellow Emperor's attack on the Four Emperors in Chapter 9, "Deploying the Army," above. This fragmentary chapter seems to be a later commentary on "Deploying the Army."

234. In other sources, the battle site is identified as "the fields of the Pan Springs." See, for example, Ssu-ma Ch'ien (1959):3.

235. The terminology in this chapter is somewhat obscure. Some commentators suggest this passage uses terminology associated with the prognostications of the Yin-yang Five Phases School that arose under Tsou Yen (c. 305–240 B.C.)—"right/left," "yin/yang," "facing/to the back," "complying/going against." According to this school, all natural phenomena and events in the processes of the world are defined in terms of their place relative to the changing conditions of their context. The description of the conquest of the Yellow Emperor also seems to refer to the "five processes" (wu hsing). The Master Han Fei 19.1.43 refers rather impatiently to such beliefs:

> Thus it is said: Divining to the gods and spirits on tortoise shells is not going to give you the victory; calculating your position "right or left," "facing or to the back" is not going to decide the battle. There is no greater stupidity than, in spite of the irrelevance of such factors, to still rely upon them.

The Master Wei-liao (1977):1–4 has a similar reference:

> Master Wei-liao replied: "The Yellow Emperor's dispatching the military to punish the disorderly and showing his generosity in taking care of the people had nothing to do with calculating the auspicious days, the yin and yang relations or his relative direction described in the Heavenly Almanac. The Yellow Emperor was the Yellow Emperor because of his ability to use people, and nothing else."

However, the Historical Records (1959):2617 has the following reference to taking practical advantage of the terrain:

> The Art of Warfare recommends that the right flank of the army have its back to the mountains and hills, and the left flank have the water and swamplands to its front.

This Historical Records description is certainly consistent with the language of Chapter 9, "Deploying the Army," which describes the advantages of maintaining high ground to the rear of the right flank and keeping to the sunny side, and also refers explicitly to the conquest of the Yellow Emperor over the four quarters. In addition, *Sun-tzu* is rather practical in tone, and unsympathetic to divinatory "revelation." In Chapter 13 above it states explicitly:

> Thus the reason the farsighted ruler and his superior commander conquer the enemy at every move and achieve successes far beyond the reach of the common crowd is foreknowledge. Such foreknowledge cannot be had from ghosts and spirits, educed by comparison with past events, or verified by astrological calculations. It must come from people—people who know the enemy's situation.

If we insist that the terminology here reflects the influence of a Yin-yang Five Phases School, it would mean a relatively late date not only for this fragmentary commentarial chapter, but for the thirteen-chapter *Sun-tzu* itself.

236. Of the fragments assigned to this chapter, I have tentatively translated those strips which provide us with additional content.

237. Compare Chapter 9:141 above, and *Sun Pin* (1985):61. "The natural net is dense forest and underbrush, the natural well is a box canyon, and the natural prison is terrain that is closed on all sides."

238. This is similar to a passage in Chapter 11:159, "The Nine Kinds of Terrain."

239. I follow the order of the 1985 Yin-ch'üeh-shan Han strips edition. See Hattori Chiharu (1987) for a revised ordering of these strips. I have tentatively translated those remaining fragments where they provide us with additional content—for example, the explicit references to the "thirteen-chapter" text.

240. I have followed the reconstruction of Hattori Chiharu (1987) in inserting this passage from the remaining fragments at this point in the text.

241. I have followed the reconstruction of Hattori Chiharu (1987) in inserting this passage from the remaining fragments at this point in the text.

242. These eight passages have been reconstructed from treatise 159 of a T'ang dynasty encyclopedic work in two hundred books, the *T'ung-tien*, compiled by Tu Yu (735–812). See *T'ung-tien* (1988):4076–4079.

243. The *T'ung-tien* (1988):4076 text does not have the first portion of this paragraph. It is restored on the basis of the (Sung) Ho Yen-hsi commentary. See Sun Hsing-yen (1965):226.

244. See the K'ung Ying-ta (574–648), (1931):57/4a commentary to *Tso-chuan* (Duke Ai 1).

245. See Wang Fu's (76–157) *Ch'ien-fu-lun* (1928):5/8b (Advice to the Commander). It is difficult to tell what is being cited from *Sun-tzu*, and what is Wang Fu's own commentary. The first sentence adds one additional trait ("respect") of the commander to a list found in Chapter 1:103 above of the original thirteen-chapter text. But what follows this one sentence appears to be Wang Fu's own elaboration.

246. See the Li Shan (d.689) commentary to *Wen-hsüan* (1931):9/97.
247. See *Pei-t'ang shu-ch'ao* 115/2a.
248. This title does not appear in any of the early court catalogs, and might be an apocryphal text from some later period.
249. See *Pei-t'ang shu-ch'ao* 115/3a.
250. This title does not appear in any of the early court catalogs, and might be an apocryphal text from some later period. Pi I-hsün (1937):10 states that in later times, many descendants of Sun Wu wrote on military affairs. This work might well have been the product of one of them.
251. See the *Pei-t'ang shu-ch'ao* 116/1a.
252. See *T'ai-p'ing yü-lan* (1963):273/4b. Compare the list of traits in Chapter 1:103.
253. Cited in Yang Ping-an (1986):216.
254. Cited in Huang K'uei (1989):251. In chapter 9:141 above, it states that in "encountering... natural defiles, quit such places with haste. Do not approach them. In keeping our distance from them, we can maneuver the enemy near to them; in keeping them to our front, we can maneuver the enemy to have them at his back."
255. See the Li Shan commentary to *Wen-hsüan* (1931):3/89. Chapter 9:000 above has a similar passage that states:

    If the army is flanked by precipitous ravines, stagnant ponds, reeds and rushes, mountain forests and tangled undergrowth, they must be searched carefully and repeatedly, for these are the places where ambushes are laid and spies are hidden.

256. See *T'ung-tien* (1988):4074.
257. In Chapter 9:141 it states: "Encountering... natural wells,... quit such places with haste." This passage is cited in Pi I-hsün (1937):8 as coming from *T'ai-p'ing yü-lan*.
258. The court catalog of the *History of the Sui Dynasty* has the entry, "*The Miscellaneous Prognostications of Sun-tzu* in four scrolls." *Sun-tzu* Chapter 11:158 explicitly rejects prognostication as a positive source of military information: "proscribe talk of omens and get rid of rumors, and even to the death they will not retreat." This makes any direct attribution of this work problematic.
259. See *T'ai-p'ing yü-lan* (1963): 328/3b.
260. See *T'ai-p'ing yü-lan* (1963):8/7a.
261. *Sun Pin* (1985):60 describes this deployment in a chapter entitled "The Eight-Division Formation":

    Master Sun Pin said, "When putting the eight-division formation into battle operation, turn whatever advantages the terrain permits to account, and adapt the formation to meet these conditions. Divide your main body in three with each of these detachments having a vanguard force, and each vanguard having reinforcements to its rear. All should move only upon command. Commit one detachment to the fray while holding the other two in reserve. Use one detachment to actually assault the enemy, and the other two to consolidate your gains. Where the enemy is weak and in disarray, commit your elite troops first to gain a quick advantage. But where he is strong and tight in formation, commit your weaker troops first to bait him. Divide the chariots and cavalry that will be used in combat into three detachments: one on either flank and one at the rear. On flat and easy ground, make greater use of the war chariots; on rugged terrain use more cavalry; on terrain that is sheer and closes in on both sides, use more crossbowmen. Taking into account both the rugged and the easy terrain, you must distinguish between safe ground and terrain from which there is no way out. And you must occupy the safe ground yourself while attacking the enemy where he has no way out."

262. Pi I-hsün (1937):9 attributes this passage incorrectly to the Cheng Hsüan (A.D. 127–200) commentary to the *Chou-li*. I do not know where he found it. Such chariots used screens to conceal the occupants and protect them from bolts and other projectiles.

263. See the Li Shan commentary to *Wen-hsüan* (1931):9/93.

264. The commentary to the court catalog of the *History of the Sui Dynasty* (*Sui-shu*) (1973):1012 records that the *Eight-Division Formation Diagrams of Sun-tzu* in one scroll has been lost.

265. This passage is cited in the Chang Yü commentary to Chapter 11 of the Sung edition of *Sun-tzu with Eleven Commentaries* (*Shih-i chia chu Sun-tzu*) (1978):274.

266. There is no record of this text in the court catalogs until the *Former History of the T'ang Dynasty*, which might suggest its vintage.

267. In addition to the Han dynasty strips found in the Yin-ch'üeh-shan tombs, another cache of strips dating from the late Western Han dynasty was discovered in 1978 in Tomb #115 of the Sun family compound in Ta-t'ung county, Ch'ing-hai province, and reported in the *Wen-wu* (*Cultural Relics*) 1981 no. 2. Six strips had "Master Sun" on them, suggesting some relationship with the *Sun-tzu*.

268. This corroborates the received "thirteen-chapter" text referred to both in the Yin-ch'üeh-shan Han strips and the *Historical Records*.

269. See the Li Shan commentary to *Wen-hsüan* (1931):9/99. It is reminiscent of Chapter 7:130 above:

> Thus, advancing at a pace, such an army is like the wind; slow and majestic, it is like a forest; invading and plundering, it is like fire; sedentary, it is like a mountain; unpredictable, it is like a shadow; moving, it is like lightning and thunder.

270. Cited from the *T'ung-tien* in Pi I-hsün (1937):9.

271. Cited from the *T'ung-tien* in Pi I-hsün (1937):9.

272. Cited from the *T'ung-tien* in Pi I-hsün (1937):9.
273. See *Feng-su t'ung-yi* (1980):403. If "Master Sun" here refers to Sun Wu, the mention of Mo Ti (Master Mo) is an anachronism, and makes the relationship of this passage to the historical Sun Wu suspect.
274. See Wu Chiu-lung (1985):12.
275. *History of the Han Dynasty* (*Han-shu*) (1962):156, 1164.
276. *History of the Han Dynasty* (*Han-shu*) (1962):1762.
277. *History of the Han Dynasty* (*Han-shu*) (1962):1135–1136.
278. I am grateful to my colleague Tao Tien-yi for helping me think this through.
279. See Lo Fu-i (1974):35.
280. Compare Michael Loewe (1977).
281. See *History of the Sui Dynasty* (1973):1013.
282. Yin-ch'üeh-shan Han-mu chu-chien cheng-li hsiao-tsu (1975b).
283. In the first assessment of the Yin-ch'üeh-shan find in 1975, the *Sun Pin* was reconstructed in thirty chapters. After another decade of study, the committee in its 1985 publication reconsidered this attribution, and reduced the number of chapters to fifteen of these original thirty, and added one new one to make a total of sixteen chapters. The fifteen chapters excluded from the *Sun Pin* in the 1985 publication can be regarded as "supplemental."
284. See Li Hsüeh-ch'in's preface to Li Ching (1990):4 (preface).

BIBLIOGRAPHY OF WORKS CITED

Ames, Roger T. *The Art of Rulership: A Study in Ancient Chinese Political Thought.* Honolulu: University of Hawaii Press, 1983.

Blanford, Yumiko F. "A Textual Approach to *Zhanguo zonghengjia shu:* Methods of Determining the Proximate Original Word among Variants" in *Early China* 16, 1991.

Calthrop, Captain E. F. *The Book of War.* London: John Murray, 1908.

Carson, Michael F. (comp.). *A Concordance to Lü-shih ch'un-ch'iu,* 2 volumes. San Francisco: Chinese Materials Center, Inc., 1985.

Chan-kuo-ts'e. Peking: Shang-wu shu-chü. Ssu-pu ts'ung-k'an edition, 1920.

Chang Chen-tse. *Sun Pin ping-fa chiao-li.* Peking: Chung-hua shu-chü, 1984.

Chang-sun Wu-chi (d. 659), (comp.). *History of the Sui Dynasty (Sui-shu).* Peking: Chung-hua shu-chü, 1973.

Ch'i Kuang. *Sun Wu ping-fa chu-shih.* Peking: Pei-ching ku-chi ch'u-pan-she, 1988.

Ch'ien Mu. *Hsien-Ch'in chu-tzu hsi nien* (revised edition). Hong Kong: Hong Kong University Press, 1956.

Chih Wei-ch'eng (editor). *Sun-tzu ping-fa shih-cheng.* Peking: Chung-kuo shu-tien, 1988.

Chou I (Book of Changes). Reprint of Harvard–Yenching Institute Sinological Index Series. Taipei: Nan-yü ch'u-pan-she, 1978.

Chu Chun. *Sun-tzu ping-fa shih-yi.* Peking: Hai-ch'ao ch'u-pan-she, 1988.

Cleary, Thomas. *The Art of War: Sun-tzu.* Boston: Shambhala, 1988.

Creel, H. G. *The Origins of Statecraft in China,* Vol. I. Chicago: Chicago University Press, 1970.

Crump, James (trans.). *Chan-kuo ts'e [Intrigues of the Warring States].* San Francisco: Chinese Materials Center, Inc. Second revised edition, 1979.

Discourses of the States (Kuo-yü). Shanghai: Chung-hua shu-chü. Ssu-pu pei-yao edition, 1928.

Duyvendak, J. J. L. (trans.). *The Book of Lord Shang.* London: Arthur Probsthain, 1928.

Fairbank, John. *China Watch.* Cambridge, MA: Harvard University Press, 1987.

Fan Yeh *et al.*, (comp.). *History of the Later Han Dynasty (Hou-Han-shu).* Peking: Chung-hua shu-chü, 1965.

Foucault, Michel. *The Order of Things: An Archaeology of the Human Sciences.* New York: Vintage Books, 1973.

Fu Chen-lun. *Sun Ping ping-fa shih-chu.* Ssu-ch'uan: Pa-shu shu-she, 1986.

Giles, Lionel. *Sun Tzu on the Art of War.* London: Luzac & Co., 1910.

Graham, A. C. *Chuang-tzu: The Inner Chapters.* London: George Allen & Unwin, 1981.

Griffith, Samuel B. *Sun Tzu: The Art of War.* Oxford: Oxford University Press, 1963.

Guisso, Richard W., and Stanley Johannesen (eds.). *Women in China: Current Directions in Historical Scholarship.* Youngstown, NY: Philo Press, 1981.

Hall, David L., and Roger T. Ames. *Thinking Through Confucius.* Albany, NY: State University of New York Press, 1987.

Han Fei-tzu so-yin. Peking: Chung-hua shu-chü, 1982.

Hattori, Chiharu. *Sun-tzu ping-fa chiao-chieh.* Peking: Chün-shih k'o-hsüeh ch'u-pan-she, 1989.

Henricks, Robert G. *Lao-tzu Te-tao Ching: A New Translation Based on the Recently Discovered Ma-wang-tui Texts.* New York: Ballantine Books, 1989.

Hou Yin-chang. *Sun Pin ping-fa ch'ien-shuo.* Peking: Chieh-fang-chün ch'u-pan-she, 1986.

Hsiao T'ung (501–531) (comp.). *Wen-hsüan.* Shanghai: Shang-wu yin-shu-kuan, 1931.

Hsü P'ei-ken and Wei Ju-lin. *Sun Pin ping-fa chu-shih.* Taipei: Liming wen-hua shih-yeh kung-szu, 1976.

Hsün-tzu. Harvard–Yenching Institute Sinological Index Series, Supplement 22. Peking: Harvard-Yenching, 1950.

Hsü Wei-yü (ed.). *Lü-shih ch'un-ch'iu.* Taipei: Shih-chieh shu-chü, 1935 (1955 reprint).

Huang K'uei. *Sun-tzu tao-tu.* Ch'eng-tu: Pa-shu shu-she ch'u-pan, 1989.

Kanaya Osamu. *Sonbin heiho.* Tokyo: Toho shuten, 1976.

Kierman, Frank A., Jr. (editor). *Chinese Ways in Warfare.* Cambridge, MA: Harvard University Press, 1974.

Knoblock, John. *Xunzi: A Translation and Study of the Complete Works,* Vol. II, Books 7–16. Stanford: Stanford University Press, 1990.

Kuan-tzu. *A Concordance to the Kuan-tzu.* Compiled by Wallace Johnson. Taipei: Ch'eng-wen ch'u-pan-she, 1970.

K'ung Ying-ta (574–648). *Tso-chuan chu-su* (*Commentary on the Tso-chuan*). Shanghai: Chung-hua shu-chü. Ssu-pu pei-yao edition, 1931.

Kuo Hua-jo. *Sun-tzu shih-chu.* Shanghai: Shang-hai ku-chi ch'u-pan-she, 1984.

Lai Hai-tsung. *Chung-kuo wen-hua yü Chung-kuo ping.* Taipei: Li-jen shu-chü, 1984 (Taiwan reprint).

Lau, D. C. "Some Notes on the *Sun-tzu.*" *Bulletin of the School of Oriental and African Studies* 1965:28:2:318–335.

———— (trans.). *Chinese Classics: Tao Te Ching.* Hong Kong: Chinese University Press, 1982.

———— (trans.). *Chinese Classics: Mencius.* Hong Kong: Chinese University Press, 1984.

Le Blanc, Charles. *Huai-nan-tzu: Philosophical Synthesis in Early Han Thought.* Hong Kong: Hong Kong University Press, 1985.

Lei Hai-tsung. *Chung-kuo wen-hua yü Chung-kuo te ping.* Taipei: Li-jen shu-chü, 1984.

Lewis, Mark Edward. *Sanctioned Violence in Early China.* Albany, NY: State University of New York Press, 1990.

Li Ching. *Ch'i Sun-tzu fa-chieh.* Peking: Chung-kuo shu-tien, 1990.

Li Fang (925–996) (comp.). *T'ai-p'ing yü-lan.* Peking: Chung-hua shu-chü, 1963.

Li Ling. "Ch'ing-hai Ta-t'ung-hsien shang Sun-chia-chai Han-chien hsing-chih hsiao-i [A brief discussion of the nature of the Han dynasty strips recovered in the Ch'ing-hai Ta-t'ung county Sun family compound dig]" in *K'ao-ku* (*Archaeology*) 1983:6:549–553.

Liu An (editor). *Huai Nan-tzu* (*Master of Huai Nan*). Taipei: Yi-wen yin-shu-kuan. Ssu-pu ts'ung-k'an edition, 1968.

Liu Chung-p'ing. *Wei-liao-tzu chin-chu chin-shih* (trans. and commentary). Taipei: Commercial Press, 1975.

Liu Hsiang (attributed). *Intrigues of the Warring States* (Chan-kuo ts'e). Peking: Shang-wu shu-chü. Ssu-pu ts'ung-k'an edition, 1920.

Liu Hsin-chien (editor). *Sun Pin ping-fa: hsin-pien chu-shih.* Honan: Honan University Press, 1989.

Lo Fu-i. "Lin-i Han-chien kai-shu." *Wen-wu* (*Cultural Relics*), 1974:2.

Loewe, Michael. "Manuscripts found recently in China: A preliminary survey." *T'oung Pao* Vol. 63, no. 2–3:99–136, 1977.

Lum, Wing Tek. *Expounding the Doubtful Points.* Honolulu: Bamboo Ridge Press, 1987.

Mencius (*Meng-tzu*). Harvard–Yenching Institute Sinological Index Series, Supplement 17. Peking: Harvard-Yenching, 1941.

Needham, Joseph. *Science and Civilisation in China,* Vol. III. Cambridge: Cambridge University Press, 1970.

Needham, Joseph, *et al.,* and Robin D. S. Yates. *Science and Civilisation in China* Vol. 5, Part VI. Cambridge: Cambridge University Press (forthcoming).

Nei Meng-ku ta-hsüeh chung-wen-hsi. *Sun Pin ping-fa yen-chiu*. Hu-ho-hao-t'e: Nei Meng-ku ta-hsüeh chung-wen-hsi, 1978.

Norman, Jerry, and Tsu-lin Mei. "The Austroasiatics in Ancient South China: Some Lexical Evidence" in *Monumenta Serica* Vol. XXXII, 1976, 274–301.

Pan Ku (32–92). *History of the Han Dynasty (Han-shu)*. Peking: Chung-hua shu-chü, 1962.

Peerenboom, R.P. *Law and Morality in Ancient China. The Silk Manuscripts of Huang-Lao.* Albany, NY: State University of New York Press, 1993.

Pi I-hsün. *Sun-tzu hsü-lu (Citations from Sun-tzu)*. Shanghai: Shang-wu yin-shu-kuan, 1937.

————— (trans.). *Kuan-tzu: A Repository of Early Chinese Thought.* Hong Kong: Hong Kong University Press, 1965.

Rickett, W. A. (trans.). *Guanzi: Political, Economic, and Philosophical Essays from Early China.* Princeton, New Jersey: Princeton University Press, 1985.

Ssu-ma Ch'ien *et al. Historical Records (Shih-chi)*. Peking: Chung-hua shu-chü, 1959.

Sun Hsing-yen. *Sun-tzu shih-chia chu*. Taipei: Shang-wu yin-shu-kuan, 1965.

Sun-tzu shih-i-chia chu. Taipei: reprint of the Central Library's Sung edition in the *Chung-kuo tzu-hsüeh ming-chu chi-ch'eng* series, 1978.

T'ang Lan. "Ssu-ma Ch'ien so mei-you chien-kuo-te chen-kuei shih-liao" in *Chan-kuo tsung-heng-chia shu*. Ma-wang-tui Han-mu po-shu cheng-li hsiao-tsu (ed.). Peking: Wen-wu ch'u-pan she, 1976.

T'ao Han-chang. *Sun-tzu ping-fa kai-lun*. Peking: Chieh-fang-chün ch'u-pan-she, 1989.

Teng Tse-tsung. *Sun Pin ping-fa chu-shih*. Peking: Chieh-fang-chün ch'u-pan-she, 1986.

Tu Yu (735–812). *T'ung-tien*. Peking: Chung-hua shu-chü, 1988.

Twitchett, Denis, and Michael Loewe (editors). *The Cambridge History of China, Vol. I: The Ch'in and Han Empires 221 B.C.–A.D. 220.* Cambridge: Cambridge University Press, 1986.

Wang Fu (76–157) (comp.). *Ch'ien-fu-lun*. Shanghai: Chung-hua shu-chü. Ssu-pu pei-yao edition, 1928.

Wang Jen-chün. *Sun-tzu i-wen* (unpublished text preserved in the archives of the Shanghai Library).

Wang Yin-chih. *Ching-chuan shih-tsu*. Hong Kong: T'ai-p'ing shu-chü, 1966.

Watson, Burton (trans.). *Han Fei Tzu: Basic Writings*. New York: Columbia University Press, 1964.

Wei Cheng. *History of the Sui Dynasty (Sui-shu)*. Peking: Chung-hua shu-chü, 1966.

Wu Chiu-lung. *Yin-ch'üeh-shan Han-chien shih-wen*. Peking: Wen-wu ch'u-pan-she, 1985.

Wu Chiu-lung *et al.* (editors). *Sun-tzu chiao-shih*. Chün-shih k'o-hsüeh ch'u-pan-she, 1990.

Yang Kuan. "Ma-wang-tui po shu *Chan-kuo tsung-heng-chia shu* te shih-liao chia-chih" in *Wen-wu* (*Cultural Relics*) 1975.2.

Yang Ping-an. *Sun-tzu hui-chien*. Chung-chou: Chung-chou ku-chi ch'u-pan-she, 1986.

Yates, Robin D. S. *The City Under Siege: Technology and Organization as Seen in the Reconstructed Text of the Military Chapters of Mo-tzu.* Unpublished Ph.D. dissertation, Harvard University, 1980.

Yates, Robin D. S. "New Light on Ancient Chinese Military Texts: Notes on Their Nature and Evolution, and the Development of Military Specialization in Warring States China." *T'oung Pao*, 1988:64:211–248.

Yin-ch'üeh-shan Han-mu chu-chien cheng-li hsiao-tsu (Committee for the Reconstruction of the Yin-ch'üeh-shan Han strips). (1975a). *Sun Pin ping-fa*. Peking: Wen-wu ch'u-pan-she, 1975(a).

Yin-ch'üeh-shan Han-mu chu-chien cheng-li hsiao-tsu (Committee for the Reconstruction of the Yin-ch'üeh-shan Han strips). *Yin-ch'üeh-shan Han-mu chu-chien* I, 10 *ts'e* (1–4 *Sun-tzu: The Art of Warfare;* 5–10 *Sun Pin: The Art of Warfare*). Peking: Wen-wu ch'u-pan-she, 1975(b).

Yin-ch'üeh-shan Han-mu chu-chien cheng-li hsiao-tsu (Committee for the Reconstruction of the Yin-ch'üeh-shan Han strips) *Yin-ch'üeh-shan Han-mu chu-chien* Collection I. Peking: Wen-wu ch'u-pan-she, 1985.

Ying Shao (fl. 189–220). *Feng-su t'ung-yi* (*Comprehensive Meaning of Customs*). Peking: T'ien-chin jen-min ch'u-pan-she, 1980.

Yü Shih-nan (558–638) (comp.). *Pei-t'ang shu-ch'ao.* Blockprint edition re-cut from traced Sung edition, 1888.

INDEX

Karl von Clausewitz

On War

TRANSLATED BY

O. J. MATTHIJS JOLLES

OF THE UNIVERSITY OF MILITARY STUDIES,

THE UNIVERSITY OF CHICAGO

With Notes by the Author

CONTENTS

BOOK III
OF STRATEGY IN GENERAL

BOOK IV
THE ENGAGEMENT

BOOK V
MILITARY FORCES

BOOK VI
DEFENSE

BOOK VII
THE ATTACK

Two Notes by the Author

I regard the first six books, of which a fair copy has now been made, as only a rather formless mass which will be thoroughly revised once more. In this revision the two kinds of war will everywhere be more clearly kept in view. Thus all ideas will acquire a more distinct meaning, a definite direction and a more detailed application. These two kinds of war are: the one in which the object is *the overthrow of our adversary,* whether we wish to destroy him politically or merely to disarm him and thus force him to accept whatever peace we will; and the other in which we want *merely to make some conquests on the frontier of his country,* either to retain them or to avail ourselves of them as useful bargaining points in settling the terms of peace. The two kinds must, no doubt, continue to blend into one another, but the wholly different characters of the two efforts must everywhere prevail, and things which are irreconcilable must be kept in their separate categories.

Apart from this difference existing in fact between wars, we must further expressly and exactly establish the point of view, no less necessary in practice, from which war is regarded as *nothing but the continuation of state policy with other means.* This point of view, everywhere maintained, will bring much more unity into our investigation, and everything will be easier to disentangle. Although this point of view will chiefly find its application in Book VIII, nevertheless it must be fully explained in Book I and also contribute to the revision of the first six Books. By such a revision the first six Books will be freed from many a piece of dross; many a fissure and gap will be closed; and many a generality can be converted into more definite thoughts and forms.

Book VII, "On the Attack," for which the rough drafts of the several chapters have already been prepared, is to be regarded as a reflection of Book VI. It is to be revised at once in accordance with the more definite points of view just stated, so that it will need no further revision, but may rather serve as a model in the revision of the first six Books.

For Book VIII, "On the Plan of a War," that is, generally, on the arranging of a whole war, several chapters have been drafted, which, however, cannot even be regarded as real materials but are merely a rough survey of the whole subject, for the purpose of finding, in the very process itself, a clear idea of what are the important points. This purpose they have fulfilled, and I propose after finishing Book VII to proceed at once to the final revision of Book VIII, in which the two points of view above stated will be principally brought out and should simplify everything and also at the same time raise it to a higher intellectual level. I hope in this Book to iron out various wrinkles in the heads of strategists and statesmen and at all events to show what are the points at issue and what in the case of a war is specially to be taken into consideration.

If by the revision of Book VIII I have cleared up my ideas, and the main features of war have been properly established, it will then be the easier for me to infuse the same spirit into the first six Books and there, too, to make those features everywhere visible. Not till then, therefore, shall I undertake the revision of the first six Books.

Should I be interrupted in this work by an early death, what exists of it may certainly be described as merely a hotchpotch of ideas, which, being exposed to ceaseless misunderstandings, will give rise to a multitude of hasty criticisms. For in these matters everyone thinks that whatever comes into his head when he takes up a pen is quite good enough to be said and printed, and holds it to be as far beyond doubt as that two and two make four. If he would give himself the trouble, as I have done, to think over the subject for years and years and always compare his conclusions with the actual history of war, he would certainly be more cautious with his criticism.

But in spite of their unfinished form, I nevertheless believe that an unprejudiced reader, anxious for truth and conviction, will not fail in the first six Books to appreciate the fruits of many years of thought and a diligent study of war, and perhaps will find in them the leading ideas from which a revolution in the theory of war might proceed.

Berlin, July 10, 1827

—

[Besides this "Note" there was also found among the author's papers the following unfinished statement.]

The manuscript on the conduct of war on a great scale which will be found after my death, can, in its existing state, only be regarded as a collection of materials out of which a theory of war on a great scale was to be constructed. With most of it I am not yet satisfied, and Book VI is to be considered a mere experiment; I would have rewritten it entirely and sought another solution.

But the leading features which predominate in these materials I hold to be the right ones in the view of war. They are the fruits of a comprehensive study with constant reference to practical life, and in constant remembrance of what experience and association with distinguished soldiers have taught me.

Book VII was to cover the attack, the subjects of which have been lightly sketched; Book VIII, the plan of a war. In the latter I should have dealt more particularly with the political and human side of war.

The first chapter of Book I is the only one which I regard as finished. It will, at any rate, serve to give the whole the direction which I wished to maintain throughout.

The theory of war on a great scale, or what is called strategy, presents extraordinary difficulties. It may perhaps be said that on its different subjects very few people have clear ideas—ideas, that is to say, which are logically traced back to their underlying necessity. In action most men follow a mere instinctive judgment, which hits the mark more or less successfully, according as they have in them more or less of genius.

So have all great commanders acted, and their greatness and their genius lay partly in the fact that this instinct of theirs always proved correct. So too it will always be with action, and for this that instinct is amply sufficient. But when it is a matter not of acting ourselves, but of persuading others in council, then it is a question of clear ideas, of demonstrating the internal connection of things. And because so little progress has yet been made in this direction, most councils are a futile bandying of words, in which either everyone sticks to his own opinion or a superficial agreement out of mutual consideration leads to a middle way, fundamentally devoid of all value.

Clear ideas, therefore, on these matters are not without utility. Moreover, the mind of man has quite universally a tendency toward clarity and a need of establishing necessary connections.

The great difficulties presented by such a philosophical construction of the art of war, and the many very unsuccessful attempts that have been made at it, have reduced most people to declaring that such a theory is not possible, for it is a question of things which no permanent law can include. We should agree with this opinion and give up all attempt at a theory if there were not a whole multitude of propositions which can be made perfectly clear without difficulty. Of such are the following: defense is the stronger form with the negative object, and attack the weaker form with the positive object; great successes help to determine the small; strategic effects can, therefore, be referred to certain centers of gravity; a demonstration is a weaker application of force than a real attack, and therefore demands special conditions; victory consists not merely in the conquest of the battlefield, but in the destruction of physical and moral forces and this is usually attained only in the pursuit after the battle is won; success is always at its greatest where a battle has been fought and won, and the changing over from one line and direction to another can only, therefore, be regarded as a necessary evil; an enveloping movement can only be justified by a general superiority or by the superiority of our lines of communication and retreat to those of the enemy; flank positions, therefore, are also conditioned by the same circumstances; every attack weakens as it advances.

Author's Preface

That the conception of the scientific does not solely or chiefly consist of a system, with its course of instruction all complete, needs no explanation today. In the present exposition, no trace of system is to be found on the surface, and instead of a course of instruction all complete, it is nothing but a collection of materials.

Its scientific character lies in the attempt to investigate the essence of the phenomena of war and to show their connection with the nature of the things of which they are composed. Nowhere have we tried to avoid logical conclusion, but when its thread runs altogether too fine, the author has preferred to sever it, and depend upon the corresponding phenomena of experience. As many plants only bear fruit if their stems do not shoot too high, so in the practical arts the theoretical leaves and blossoms must not be allowed to grow too high, but must be kept close to experience, their proper soil.

Undoubtedly it would be a mistake to try from the chemical constituents of the grain of wheat to investigate the form of the ear it produces; we have only to go into the field to see the ears all complete. Investigation and observation, philosophy and experience must never despise or exclude each other; each goes bail for the other. The propositions of this book rest the short span of their internal necessity either upon experience or on the conception of war itself as an external point and therefore do not lack their buttresses.[1]

---

[1] That this is not the case with many military writers, especially such as wished to deal with war scientifically, is shown by the many instances in which the pros and the cons get so tangled up in their arguments that not even, as with the two lions, the tails are left.

It is not impossible, perhaps, to write a systematic theory of war both logical and wide in scope. Ours, up to the present, is far from being either. Not to mention their unscientific spirit, in the attempt to make their systems consistent and complete, many such works are stuffed with banalities, commonplaces and twaddle of every kind. If anyone wants a faithful picture of them, let him read Lichtenberg's extract from a Fire Regulation:

> If a house is on fire, it is above all things necessary to try to protect the right wall of the house standing to the left, and the left wall of the house standing to the right respectively. For if, for instance, we proposed to protect the left wall of the house standing to the left, the right wall of the house lies to the right of the left wall, and consequently, as the fire also lies to the right of this wall and the right wall (for we have assumed that the house lies to the left of the fire) so the right wall lies nearer to the fire than the left, and the right wall of the house could catch fire, if it were not protected, before the fire reached the left wall, which is protected. Consequently something might catch fire which is not protected, and indeed sooner than anything else would catch fire, even if not protected; consequently the former must be left alone and latter must be protected. In order to impress the matter thoroughly on one's mind, one may merely note: if the house lies to the right of the fire, it is the left wall, and if the house lies to the left, it is the right wall.

In order not to frighten away the intelligent reader with such commonplaces and make what little is good unpalatable by dilution with water, the author has preferred to offer in the form of small nuggets of pure metal what many years of thought about war, association with shrewd observers acquainted with war and many an experience of his own have suggested to him and established in his mind. It is thus that the chapters of this book, outwardly but loosely connected, have come into being. Nevertheless, it is hoped they do not lack internal consistency. Perhaps before long a greater brain will appear, which will give us, instead of these isolated nuggets, the whole in one casting of pure metal without dross.

BOOK I

ON THE
NATURE OF WAR

CHAPTER 1

What Is War?

1. INTRODUCTION

We propose to consider, first, the several elements of our subject, then its several parts or divisions, and, finally, the whole in its internal connection. Thus we proceed from the simple to the complex. But in this subject more than in any other it is necessary to begin with a glance at the nature of the whole, because here more than elsewhere the part and the whole must always be considered together.

2. DEFINITION

We shall not begin here with a clumsy, pedantic definition of war, but confine ourselves to its essence, the duel. War is nothing but a duel on a larger scale. If we would combine into one conception the countless separate duels of which it consists, we would do well to think of two wrestlers. Each tries by physical force to compel the other to do his will; his immediate object is to overthrow his adversary and thereby make him incapable of any further resistance.

War is thus an act of force to compel our adversary to do our will.

Force, to meet force, arms itself with the inventions of art and science. It is accompanied by insignificant restrictions, hardly worth mentioning, which it imposes on itself under the name of international law and usage, but which do not really weaken its power. Force, that is to say, physical force (for no moral force exists apart from the conception of a state and law), is thus the *means*; to impose our will upon the enemy is the *object*. To achieve this object with certainty we must disarm the enemy, and this disarming is by definition the proper aim of military action. It takes the place of the object and in a certain sense pushes it aside as something not belonging to war itself.

3. THE USE OF FORCE THEORETICALLY WITHOUT LIMITS

Now philanthropic souls might easily imagine that there was an artistic way of disarming or overthrowing our adversary without too much bloodshed and that this was what the art of war should seek to achieve. However agreeable this may sound, it is a false idea which must be demolished. In affairs so dangerous as war, false ideas proceeding from kindness of heart are precisely the worst. As the most extensive use of physical force by no means excludes the co-operation of intelligence, he who uses this force ruthlessly, shrinking from no amount of bloodshed, must gain an advantage if his adversary does not do the same. Thereby he forces his adversary's hand, and thus each pushes the other to extremities to which the only limitation is the strength of resistance on the other side.

This is how the matter must be regarded, and it is a waste—and worse than a waste—of effort to ignore the element of brutality because of the repugnance it excites.

If the wars of civilized nations are far less cruel and destructive than those of the uncivilized, the reason lies in the social condition of the states, both in themselves and in their relations to one another. From this condition, with its attendant circumstances, war arises and is shaped, limited and modified. But these things do not themselves belong to war; they already exist. Never in the philosophy of war itself can we introduce a modifying principle without committing an absurdity.

Conflict between men really consists of two different elements: hostile feeling and hostile intention. We have chosen the latter of these two elements as the distinguishing mark of our definition because it is the more general. We cannot conceive the most savage, almost instinctive, passion of hatred as existing without hostile intention, whereas there are many hostile intentions accompanied by absolutely no hostility, or, at all events, no predominant hostility, of feeling. Among savages intentions inspired by emotion prevail; among civilized peoples those prescribed by intelligence. But this difference lies not in the intrinsic nature of savagery and civilization, but in their accompanying circumstances, institutions, and so forth. It does not necessarily, therefore, exist in every case, but only prevails in the majority of cases. In a word, even the most civilized nations can be passionately inflamed against one another.

From this we see how far from the truth we should be if we ascribed war among civilized men to a purely rational act of the governments and conceived it as continually freeing itself more and more from all passion, so that at last there was no longer need of the physical existence of armies, but only of the theoretical relations between them—a sort of algebra of action.

Theory was already beginning to move in this direction when the events of the last war[1] taught us better. If war is an act of force, the emotions are also necessarily involved in it. If war does not originate from them, it still more or less reacts upon them, and the degree of this depends not upon the stage of civilization, but upon the importance and duration of the hostile interests.

If, therefore, we find that civilized peoples do not put prisoners to death or sack cities and lay countries waste, this is because intelligence plays a greater part in their conduct of war and has taught them more effective ways of applying force than these crude manifestations of instinct.

The invention of gunpowder and the advances continually being made in the development of firearms, in themselves show clearly enough that the demand for the destruction of the enemy, inherent in the theoretical conception of war, has been in no way actually weakened or diverted by the advance of civilization.

So we repeat our statement: War is an act of force, and to the application of that force there is no limit. Each of the adversaries forces the hand of the other, and a reciprocal action results which in theory can have no limit. This is the first reciprocal action that we meet and the first extreme.

(First reciprocal action)

4. THE AIM IS TO DISARM THE ENEMY

We have said that the disarming of the enemy is the aim of military action, and we shall now show that, theoretically, at all events, this is necessarily so.

---

[1] The war with Napoleon.

If our opponent is to do our will, we must put him in a position more disadvantageous to him than the sacrifice would be that we demand. The disadvantages of his position should naturally, however, not be transitory, or, at least, should not appear to be so, or our opponent would wait for a more favorable moment and refuse to yield. Every change in his position that will result from the continuance of military activity, must thus, at all events in theory, lead to a position still less advantageous. The worst position in which a belligerent can be placed is that of being completely disarmed. If, therefore, our opponent is to be forced by military action to do our will, we must either actually disarm him or put him in such a condition that he is threatened with the probability of our doing so. From this it follows that the disarming or the overthrow of the enemy—whichever we choose to call it—must always be the aim of military action.

Now war is not the action of a live force upon a dead mass—absolute non-resistance would be no sort of war at all—but always the collision of two live forces with each other, and what we have said of the ultimate aim of military action must be assumed to apply to both sides. Here, then, is again reciprocal action. So long as I have not overthrown my adversary I must fear that he may overthrow me. I am no longer my own master, but he forces my hand as I force his. This is the second reciprocal action, which leads to the second extreme.

(Second reciprocal action)

5. UTMOST EXERTION OF FORCES

If we want to overthrow our opponent, we must proportion our effort to his power of resistance. This power is expressed as a product of two inseparable factors: *the extent of the means at his disposal* and *the strength of his will.* The extent of the means at his disposal would be capable of estimation, as it rests (though not entirely) on figures, but the strength of the will is much less so and only approximately to be measured by the strength of the motive behind it. Assuming that in this way we have got a reasonably probable estimate of our opponent's power of resistance, we can proportion our efforts accordingly and increase them so as to secure a preponderance or, if our means do not suffice for this, as much as we can. But our opponent does the same; and thus a fresh competition arises between us which in pure theory once more involves pushing to an extreme. This is the third reciprocal action we meet and a third extreme.

(Third reciprocal action)

6. MODIFICATIONS IN PRACTICE

In the abstract realm of pure conceptions the reflective mind nowhere finds rest till it has reached the extreme, because it is with an extreme that it has to do—a conflict of powers left to themselves and obeying no law but their own. If, therefore, we wanted from the mere theoretical conception of war to deduce an absolute aim which we are to set before ourselves and the means we are to employ, these continuous reciprocal actions would land us in extremes which would be nothing but a play of fancies produced by a scarcely visible train of logical hair-splitting. If, adhering closely to the absolute, we proposed to get round all difficulties with a stroke of the pen and insist with logical strictness that on every occasion we must be prepared for the extreme of resistance and meet it with the extreme of effort, such a stroke of the pen would be a mere paper law with no application to the real world.

Assuming, too, that this extreme of effort were an absolute quantity that could easily be discovered, we must nevertheless admit that the human mind would hardly submit to be ruled by such logical fantasies. In many cases the result would be a futile expenditure of strength which would be bound to find a restriction in other principles of statesmanship. An effort of will would be required disproportionate to the object in view and impossible to call forth. For the will of man never derives its strength from logical hair-splitting.

Everything, however, assumes a different shape if we pass from the abstract world to that of reality. In the former everything had to remain subject to optimism and we had to conceive both one side and the other as not merely striving toward perfection but also attaining it. Will this ever be so in practice? It would if:

1. war were a wholly isolated act, which arose quite suddenly and had no connection with the previous course of events,

2. if it consisted of a single decision or of several simultaneous decisions,

3. if its decision were complete in itself and the ensuing political situation were not already being taken into account and reacting upon it.

7. WAR IS NEVER AN ISOLATED ACT

With reference to the first of these three points we must remember that neither of the two opponents is for the other an abstract person, even as regards that factor in the power of resistance which does not depend on external things, namely, the will. This will is no wholly unknown quantity:

what it has been today tells us what it will be tomorrow. War never breaks out quite suddenly, and its spreading is not the work of a moment. Each of the two opponents can thus to a great extent form an opinion of the other from what he actually is and does, not from what, theoretically, he should be and should do. With his imperfect organization, however, man always remains below the level of the absolute best, and thus these deficiencies, operative on both sides, become a modifying influence.

8. WAR DOES NOT CONSIST OF ONE BLOW WITHOUT DURATION

The second of the three points gives occasion for the following observations:

If the issue in war depended on a single decision or several simultaneous decisions, the preparations for that decision or those several decisions would naturally have to be carried to the last extreme. A lost opportunity could never be recalled; the only standard the real world could give us for the preparations we must make would, at best, be those of our adversary, so far as they are known to us, and everything else would once more be relegated to the realm of abstraction. But if the decision consists of several successive acts, each of these with all its attendant circumstances can provide a measure for those which follow, and thus here, too, the real world takes the place of the abstract, and modifies, accordingly, the trend to the extreme.

Every war, however, would necessarily be confined to a single decision or several simultaneous decisions if the means available for the conflict were all brought into operation together or could be so brought into operation. For an adverse decision necessarily diminishes these means, and if they have all been used up in the first decision, a second really becomes unthinkable. All acts of war which could follow would be essentially part of the first and really only constitute its duration.

But we have seen that in the preparations for war the real world has already taken the place of the mere abstract idea, and an actual standard that of a hypothetical extreme. Each of the two opponents, if for no other reason, will therefore in their reciprocal action stop short of the extreme effort, and their resources will thus not all be called up together.

But the very nature of these resources and of their employment makes it impossible to put them all into operation at one and the same moment. They consist of *the military forces proper, the country* with its superficial extent and its population, and *the allies.*

The country with its superficial extent and its population, as well as being the source of all military forces proper, is also in itself an integral part of the factors operative in war, if only with that part which provides the theater of war or has a marked influence upon it.

Now all movable military resources can very well be put into operation simultaneously, but not all the fortresses, rivers, mountains, inhabitants, and so forth—in a word, the whole country, unless it is so small as to be wholly embraced by the first act of war. Furthermore, the co-operation of the allies does not depend upon the will of the belligerents, and from the very nature of political relations, it frequently does not come into effect or become active till later, for the purpose of restoring a balance of forces that has been upset.

That this part of the means of resistance, which cannot be brought into operation all at once, in many cases is a much larger part of the whole than at first sight we should think; and that consequently it is capable of restoring the balance of forces even when the first decision has been made with great violence and that balance has thus been seriously disturbed, will be more fully explained later. At this point it is enough to show that to make all our resources available at one and the same moment is contrary to the nature of war. Now in itself this could furnish no ground for relaxing the intensity of our efforts for the first decision, because an unfavorable issue is always a disadvantage to which no one will purposely expose himself, because even if the first decision is followed by others, the more decisive it has been, the greater will be its influence upon them. But the possibility of a subsequent decision is something in which man's shrinking from excessive effort causes him to seek refuge, and thus for the first decision his resources are not concentrated and strained to the same degree as they would otherwise have been. What either of the two opponents omits from weakness becomes for the other a real, objective ground for relaxing his own efforts, and thus, through this reciprocal action, the trend to the extreme is once more reduced to a limited measure of effort.

9. THE RESULT OF A WAR IS NEVER ABSOLUTE

Lastly, the final decision of a whole war is not always to be regarded as an absolute one. The defeated state often sees in it only a transitory evil, for which a remedy can yet be found in the political circumstances of a later day. How greatly this also must modify the violence of the strain and the intensity of the effort is obvious.

10. THE PROBABILITIES OF REAL LIFE
TAKE THE PLACE OF THE EXTREME AND
ABSOLUTE DEMANDED BY THEORY

In this way the whole field of war ceases to be subject to the strict law of forces pushed to the extreme. If the extreme is no longer shunned and no longer sought, it is left to the judgment to determine the limits of effort, and this can only be done by deduction according to the *laws of probability* from the data supplied by the phenomena of the real world. If the two adversaries are no longer mere abstractions but individual states and governments, if the course of events is no longer theoretical but one that is determined according to its own laws, then the actual situation supplies the data for ascertaining what is to be expected, the unknown that has to be discovered.

From the character, the institutions, the situation and the circumstances of the adversary, each side will draw its conclusions, in accordance with the laws of probability, as to what the action of the other will be and determine its own accordingly.

11. THE POLITICAL OBJECT NOW
COMES FORWARD AGAIN

At this point a subject, which in Section 2 we had dismissed, now once more insists on claiming our consideration: namely, *the political object of the war.* The law of the extreme, the intention of disarming the enemy and overthrowing him, had up to now, so to speak, more or less swallowed it up. As this law loses its force, and this intention falls short of its aim, the political object of the war once more comes to the front. If all we have to consider is a calculation of probabilities starting from definite persons and circumstances, the political object as the original motive must be an essential factor in this process. The smaller the sacrifice we demand from our adversary, the slighter we may expect his efforts to be to refuse it to us. The slighter, however, his effort, the smaller need our own be. Furthermore, the less important our political object, the less will be the value we attach to it and the readier we shall be to abandon it. For this reason also our own efforts will be the slighter.

Thus the political object as the original motive of the war will be the standard alike for the aim to be attained by military action and for the efforts required for this purpose. It cannot be in itself an absolute standard, but, as we are dealing with real things and not with mere ideas, it will be the standard relative to the two contending states. One and the same political object can in different nations, and even in one and the same nation at different times, produce different reactions. We can therefore allow the political object to serve as a standard only in so far as we bear in mind its influence on the masses which it is to affect. So the character of these masses must be considered. It is easy to see that the result may be quite different, according as the action is strengthened or weakened by the feeling of the masses. In two nations and states such tensions, and such a mass of hostile feelings, may exist that a motive for war, very trifling in itself, still can produce a wholly disproportionate effect—a positive explosion.

This holds good for the efforts which the political object can call forth in the two states, and for the aim it can assign to military action. Sometimes it can itself become this aim, for example, if it is the conquest of a certain province. Sometimes the political object will not itself be suited to provide the aim for military action, and in such cases one must be chosen of such a kind as will serve as an equivalent for it and can take its place in the conclusion of peace. But in this case also due consideration for the character of the states concerned is always presupposed. There are circumstances in which the equivalent must be much more considerable than the political object, if the latter is to be attained by it. The greater the indifference of the masses and the less serious the tensions that on other grounds also exist in the two states and their relations, the more dominant as a standard, and decisive in itself, will the political object be. There are cases in which it is, almost by itself, the deciding factor.

Now if the aim of the military action is an equivalent for the political object, that action will in general diminish as the political object diminishes. The more this object comes to the front, the more will this be so. This explains how, without self-contradiction, there can be wars of all degrees of importance and energy, from a war of extermination down to a mere state of armed observation. But this leads us to a question of another kind, which we have still to analyze and answer.

12. A SUSPENSION OF MILITARY ACTION NOT EXPLAINED BY ANYTHING YET SAID

However insignificant the political claims made on either side, however weak the means employed and however trifling the aim to which military action is directed, can this action ever for a moment be suspended? This is a question that goes deep into the essence of the matter.

Every action requires for its accomplishment a certain time, which we call its duration. This may be longer or shorter, according as the person acting is more or less quick in his movements.

About this we shall not here trouble ourselves. Everyone does his business in his own fashion; but the slow person does not do it more slowly because he wants to spend more time on it but because by his nature he needs more time, and if he were to make greater haste, he would do it less well. This time, therefore, depends on subjective causes and belongs to the actual duration of the action.

If we now allow to every action in war its duration, we must admit, at all events at first sight, that every expenditure of time in excess of this duration, that is to say, every suspension of military action, seems to be absurd. In this connection we must always remember that the question is not of the progress of one or other of the two opponents, but of the progress of the military action as a whole.

13. THERE IS ONLY ONE CAUSE THAT CAN SUSPEND ACTION, AND THIS SEEMS ALWAYS TO BE POSSIBLE ON ONE SIDE ONLY

If two parties have armed themselves for the conflict, a hostile motive must have caused them to do so. So long then as they remain under arms, so long, that is, as they do not make peace, this motive must be present and can only cease to act with either of the two opponents for one sole reason, namely, *that he wants to wait for a more favorable moment for action.* Now, it is obvious that this reason can only be present on one of the two sides, because by its very nature it becomes the opposite on the other. If it is to the interest of the one commander to act, it must be to the interest of the other to wait.

A complete equilibrium of forces can never produce a suspension of action, for in such a suspension he who has the positive aim—that is, the assailant—would necessarily retain the initiative.

But if we chose to conceive the equilibrium as such that he who has the positive aim, and therefore the stronger motive, has at the same time the smaller forces at his disposition, so that the equation would arise from the product of motives and forces, we should still have to say that if no change in this condition of equilibrium is to be foreseen, both sides must make peace. But if a change is to be foreseen, it will be in favor of one side only, and for that reason the other will necessarily be moved to action. We see that the idea of an equilibrium cannot explain a suspension of hostilities, but all it amounts to is the waiting for a more favorable moment. Let us assume, therefore, that of two states one has a positive aim, the conquest, for instance, of one of the adversary's provinces to be used as a counter in the settlement of peace. After this conquest his political object is attained, the need for action ceases and he can take rest. If his adversary is prepared to acquiesce in this result, he must make peace; if not, he must act. If it is thought now that in four weeks' time he will be in a better condition to do so, then he has sufficient grounds for postponing his action.

But from that moment the duty of action seems to fall logically upon his opponent, in order that no time be allowed to the vanquished to prepare for action. In all this, it is, of course, assumed that each side has a complete knowledge of the circumstances.

14. THUS A CONTINUITY WOULD BE INTRODUCED INTO MILITARY ACTION FORCING EVERYTHING AGAIN TO A CLIMAX

If this continuity of military action actually existed, everything would again be driven by it to the extreme. For apart from the fact that such ceaseless activity would give a greater bitterness to the feelings and impart to the whole a higher degree of passion and a greater elemental force, there would also arise through the continuity of action a more inevitable sequence of events and a less disturbed causal connection between them. Each action would in consequence become more important and thus more dangerous.

But we know that military action seldom or never has this continuity, and that there are many wars in which action fills by far the smallest part of the time occupied, and inaction all the rest. This cannot possibly be always an anomaly. Suspension of military action must be possible, that is to say, not a contradiction in itself. That this is so, and why, we will now show.

15. HERE, THEREFORE, A PRINCIPLE OF POLARITY IS BROUGHT INTO EVIDENCE

By supposing the interests of the one commander to be always diametrically opposed to those of the other, we have assumed a true *polarity*. We propose later on to devote a special chapter to this principle, but for the present must make one observation upon it.

The principle of polarity only holds good if it is conceived in one and the same thing, in which the positive and its opposite, the negative, exactly destroy one another. In a battle each of the two parties wishes to win; that is true polarity, for the victory of the one destroys that of the other. But if we are speaking of two different things which have a common relation external to themselves, it is not the things but their relations that have the polarity.

16. ATTACK AND DEFENSE ARE THINGS DIFFERENT IN KIND AND OF UNEQUAL FORCE. POLARITY THEREFORE IS NOT APPLICABLE TO THEM

If there were only one form of war, namely, the attack of the enemy, therefore no defense; in other words if the attack were distinguished from the defense merely by the positive motive, which the one has and the other has not, but the methods of the fight were always one and the same, in such a fight every advantage to the one side would be an equal disadvantage to the other and true polarity would exist.

But military activity takes two separate forms, attack and defense, which, as we shall later on explain in detail, are very different and of unequal strength. Polarity lies therefore in that to which they both bear a relation, namely, the decision, but not in attack or defense itself. If one commander wishes to postpone the decision, the other must wish to hasten it, but, of course, only in the same form of conflict. If it is to A's interest not to attack his opponent at once but four weeks hence, it is to B's interest to be attacked by him at once and not four weeks hence. Here is a direct opposition; but it does not follow therefrom that it is to B's interest to attack A at once. That is obviously something quite different.

17. THE EFFECT OF POLARITY IS OFTEN DESTROYED BY THE SUPERIORITY OF THE DEFENSE TO THE ATTACK. THIS EXPLAINS THE SUSPENSION OF MILITARY ACTION

If the form of defense, as we shall hereafter show, is stronger than that of attack, the question arises whether the advantage of a deferred decision is as great for the one side as that of the defense is for the other. When it is not, it cannot by means of its opposite outweigh the latter and so influence the course of military action. We thus see that the impulsive force which lies in the polarity of interests may be lost in the difference between the strength of the attack and that of the defense, and thereby becomes ineffectual.

If, therefore, the side for which the present is favorable is too weak to be able to dispense with the advantage of the defensive, it must resign itself to facing a less favorable future. For it may still be better to fight a defensive battle in the unfavorable future than an offensive one in the present, or than to make peace. Now as we are convinced that the superiority of the defense (rightly understood) is very great and much greater than may appear at first sight, a very large proportion of the periods of suspended action which occur in war are thereby explained, without our being necessarily involved in a contradiction. The weaker the motives to action are, the more they will be swallowed up and neutralized by this difference between attack and defense. The more frequently, therefore, will military action be brought to a standstill, as, indeed, experience teaches.

18. A SECOND CAUSE LIES IN THE IMPERFECT KNOWLEDGE OF THE SITUATION

But there is still another cause that can stop military action, and that is imperfect knowledge of the situation. No commander has accurate personal knowledge of any position but his own; that of his adversary is only known to him by uncertain reports. He can make a mistake in his judgment of them and in consequence of this mistake believe that the initiative lies with his opponent when it really lies with himself. This want of knowledge could, it is true, just as often occasion untimely action as untimely inaction and would in itself no more contribute to delay than to hasten military action. Still it must always be regarded as one of the natural causes that, without involving an internal contradiction, may bring military action to a standstill. If, however, we reflect how much more we are inclined and induced to estimate the strength of our opponent too high rather than too low, because it lies in human nature to do so, we must also admit that imperfect knowledge of the situation must in general greatly contribute to putting a stop to military action and modifying the principles on which it is conducted.

The possibility of a standstill introduces into military action a new modification by diluting it, so to speak, with the element of time, halting danger in its stride and increasing the means for restoring a lost balance of forces. The greater the tensions out of which the war has sprung and the greater in consequence the energy with which it is waged, the shorter will be these periods of inaction; the weaker the hostile feeling, the longer will they be. For stronger motives increase the power of the will, and this, as we know, is always a factor in the product of our forces.

19. FREQUENT PERIODS OF INACTION REMOVE WAR STILL FURTHER FROM THE REALM OF EXACT THEORY AND MAKE IT STILL MORE A CALCULATION OF PROBABILITIES

But the more slowly military action proceeds and the longer and more frequent the periods of inaction, so much the more readily can a mistake be repaired, the bolder the commander will thus become in his assumptions, and the more readily will he at the same time remain below the extreme demanded by theory and build everything upon probability and conjecture. So the more or less leisurely course of military action allows more or less time for what the nature of the concrete situation in itself already demands, namely, a calculation of probabilities in accordance with the given circumstances.

20. SO ONLY THE ELEMENT OF CHANCE IS NOW LACKING TO MAKE OF WAR A GAMBLE, AND IN THIS ELEMENT IT IS LEAST OF ALL DEFICIENT

We see from the foregoing how much the objective nature of war makes it a calculation of probabilities. It now needs but one single element more to make of it a gamble, and that element it certainly does not lack—the element of *chance*. There is no human activity that stands in such constant and universal contact with chance as does war. Thus together with chance, the accidental and, with it, good luck play a great part in war.

21. THROUGH ITS SUBJECTIVE AS WELL AS THROUGH ITS OBJECTIVE NATURE WAR BECOMES A GAMBLE

If we now glance at the *subjective nature* of war, that is, at those qualities with which it must be carried on, it must strike us as still more like a gamble. The element in which the activity of war moves is danger; but, in danger, which is the most superior of all moral qualities? It is *courage*. Now courage is certainly quite compatible with prudent calculation, but courage and calculation are nevertheless things different in kind and belonging to different parts of the mind. On the other hand, daring, reliance on good fortune, boldness and foolhardiness are only manifestations of courage, and all these efforts of the spirit seek the accidental because it is their proper element.

We thus see that from the very first the absolute, the so-called theoretical, faculty finds nowhere a sure basis in the calculations of the art of war. From the outset there is a play of possibilities and probabilities, of good and bad luck, which permeates every thread, great or small, of its web and makes war, of all branches of human activity, the most like a game of cards.

22. HOW THIS BEST ACCORDS WITH THE HUMAN MIND IN GENERAL

Although our intellect always feels itself urged toward clarity and certainty, our mind still often feels itself attracted by uncertainty. Instead of threading its way with the intellect along the narrow path of philosophical investigation and logical deduction, in order, almost unconsciously, to arrive in spaces where it finds itself a stranger and where all familiar objects seem to abandon it, it prefers to linger with imagination in the realm of chance and luck. Instead of being confined, as in the first instance, to meager necessity, it revels here in the wealth of possibilities. Enraptured thereby, courage takes to itself wings, and thus daring and danger become the element into which it flings itself as a fearless swimmer flings himself into the stream.

Shall theory leave it here and move on, self-satisfied, to absolute conclusions and rules? In that case it is of no practical use. Theory must also take into account the human element and accord a place to courage and boldness and even to foolhardiness. The art of war has to do with living and with moral forces; from this it follows that it can nowhere attain the absolute and certain; there remains always a margin for the accidental just as much with the greatest things as with the smallest. As on the one side stands this accidental element, so on the other courage and self-confidence must step forward and fill up the gap. The greater the courage and self-confidence, the larger the margin that may be left for the accidental. Courage and self-confidence are thus principles absolutely essential for war. Consequently theory must only lay down such rules as allow free scope for these necessary and noblest of military virtues in all their degrees and variations. Even in daring there is still wisdom and prudence as well, only they are estimated by a different standard of value.

23. YET WAR STILL REMAINS A SERIOUS MEANS FOR A SERIOUS OBJECT. MORE PARTICULAR DEFINITIONS OF IT

Such is war, such the commander who conducts it, and such the theory that rules it. But war is no pastime, no mere passion for daring and winning, no work of a free enthusiasm; it is a serious means to a serious end. All that it displays of that glamour of fortune, all that it assimilates of the thrills of passion and courage, of imagination and enthusiasm, are only particular properties of this means.

The war of a community—of whole nations and particularly of civilized nations—always arises from a political condition and is called forth by a political motive. It is, therefore, a political act. Now if it were an act complete in itself and undisturbed, an absolute manifestation of violence, as we had to deduce it from its mere conception, it would, from the moment it was called forth by policy, step into the place of policy and, as something quite independent of it, set it aside and follow only its own laws, just as a mine, when it is going off, can no longer be guided into any other direction than that given it by previous adjustments. This is how the thing has hitherto been regarded even in practice, whenever a lack of harmony between policy and the conduct of war has led to theoretical distinctions of this kind. But it is not so, and this idea is radically false. War in the real world, as we have seen, is no such extreme thing releasing its tension in a single discharge; it is the operation of forces which do not in every case develop in exactly the same way and the same proportion but which at one moment rise to a pitch sufficient to overcome the resistance which inertia and friction oppose to them, while at another, they are too weak to produce any effect. War is, therefore, so to speak, a regular pulsation of violence, more or less vehement and consequently more or less quick in relaxing tensions and exhausting forces—in other words, more or less quickly leading to its goal. But it always lasts long enough to exert, in its course, an influence upon that goal, so that its direction can be changed in this way or that—in short, long enough to remain subject to the will of a guiding intelligence. Now if we reflect that war has its origin in a political object, we see that this first motive, which called it into existence, naturally remains the first and highest consideration to be regarded in its conduct. But the political object is not on that account a despotic lawgiver;

it must adapt itself to the nature of the means at its disposal and is often thereby completely changed, but it must always be the first thing to be considered. Policy, therefore, will permeate the whole action of war and exercise a continual influence upon it, so far as the nature of the explosive forces in it allow.

24. WAR IS A MERE CONTINUATION OF POLICY BY OTHER MEANS

We see, therefore, that war is not merely a political act but a real political instrument, a continuation of political intercourse, a carrying out of the same by other means. What now still remains peculiar to war relates merely to the peculiar character of the means it uses. The art of war in general and the commander in each particular case can demand that the tendencies and designs of policy shall be not incompatible with these means, and the claim is certainly no trifling one. But however powerfully it may react on political designs in particular cases, still it must always be regarded as only a modification of them; for the political design is the object, while war is the means, and the means can never be thought of apart from the object.

25. DIVERSITY IN THE NATURE OF WARS

The greater and the more powerful the motives for war, the more they affect the whole existence of the nations involved, and the more violent the tension which precedes war, so much the more closely will war conform to its abstract conception. The more it will be concerned with the destruction of the enemy, the more closely the military aim and the political object coincide, and the more purely military, and the less political, war seems to be. But the weaker the motives and the tensions, the less will the natural tendency of the military element, the tendency to violence, coincide with the directives of policy; the more, therefore, must war be diverted from its natural tendency, the greater is the difference between the political object and the aim of an ideal war, and the more does war seem to become political.

But that the reader may not form false conceptions, we must here remark that by this natural tendency of war we only mean the philosophical, the strictly logical tendency, and by no means that of the forces actually engaged in conflict, to the point where, for instance, all emotions and passions of the combatants should be reckoned as included. These too, it is true, might in many cases be excited to such a pitch that they could with difficulty be kept confined to the political road; but in most cases such a contradiction will not arise, because the existence of such strong emotions will imply the existence also of a great plan in harmony with them. If the plan is directed only to a trifling object, the emotional excitement of the masses will be so slight that they will always be rather in need of being pushed on than of being held back.

26. ALL WARS MAY BE REGARDED AS POLITICAL ACTS

To return to our main subject: Though it is true that in one kind of war policy seems entirely to disappear, while in another it very definitely comes to the front, we can nevertheless maintain that the one kind is as political as the other. For if we regard policy as the intelligence of the personified state, we must include among the combinations of circumstances which its calculations have to take into account that in which the nature of all the circumstances postulates a war of the first kind. It is only if we understand the term "policy" not as a comprehensive knowledge of the situation but the conventional idea of a cautious, crafty, even dishonest cunning, averse to violence, that the latter kind of war could belong to it more than does the former.

27. CONSEQUENCES OF THIS VIEW FOR THE UNDERSTANDING OF MILITARY HISTORY AND FOR THE FOUNDATIONS OF THEORY

We see, therefore, in the first place that in all circumstances we have to think of war not as an independent thing, but as a political instrument. And only by taking this point of view can we avoid falling into contradiction with the whole of military history. This alone opens the great book to intelligent appreciation. In the second place, this same point of view shows us how wars must differ according to the nature of their motives and of the circumstances out of which they arise.

Now the first, the greatest and the most decisive act of the judgment which a statesman and commander performs is that of correctly recognizing in this respect the kind of war he is undertaking, of not taking it for, or wishing to make it, something which by the nature of the circumstances it cannot be. This is, therefore, the first and most comprehensive of all strategic questions. Later on, in the chapter on the plan of a war, we shall examine it more closely.

For the moment we content ourselves with having brought our subject to this point and thereby fixed the main point of view from which war and the theory of war must be regarded.

28. RESULT FOR THEORY

War is, therefore, not only a veritable chameleon, because in each concrete case it changes somewhat its character, but it is also, when regarded as a whole, in relation to the tendencies predominating in it, a strange trinity, composed of the original violence of its essence, the hate and enmity which are to be regarded as a blind, natural impulse, of the play of probabilities and chance, which make it a free activity of the emotions, and of the subordinate character of a political tool, through which it belongs to the province of pure intelligence.

The first of these three sides is more particularly the concern of the people, the second that of the commander and his army, the third that of the government. The passions which are to blaze up in war must be already present in the peoples concerned; the scope that the play of courage and talent will get in the realm of the probabilities of chance depends on the character of the commander and the army; the political objects, however, are the concern of the government alone.

These three tendencies, which appear as so many lawgivers, lie deep in the nature of the subject and at the same time vary in magnitude. A theory which insisted on leaving one of them out of account, or on fixing an arbitrary relation between them, would immediately fall into such contradiction with reality that through this alone it would forthwith necessarily be regarded as destroyed.

The problem, therefore, is that of keeping the theory poised between these three tendencies as between three centers of attraction.

How this difficult problem can be solved in the most satisfactory way, we propose to investigate in the book dealing with the theory of war. In any case this definition of the conception of war becomes for us the first ray of light that falls upon the foundations of theory, and will for the first time separate its main features and enable us to distinguish them.

CHAPTER 2

END AND MEANS IN WAR

Having in the previous chapter ascertained the complex and variable nature of war, we shall now occupy ourselves in considering what influence this has upon the means and the end in war.

If, first of all, we ask what is the aim to which the whole war must be directed so as to be the proper means for attaining the political object, we find that this is just as variable as are the political object and the particular circumstances of the war.

If we begin by keeping once more to pure theory, we are bound to say that the political object of war really lies outside of war's province; for if war is an act of violence to compel the enemy to do our will, then in every case everything would necessarily and solely depend on overthrowing the enemy, that is to say, of disarming him. This object, which is deduced from pure theory but to which in reality a large number of cases nearly approximate, we shall first of all examine in the light of this reality.

Later on, in the plan of a war, we shall consider more closely what disarming a state means, but we must here at once distinguish between three things, which as three general categories include everything else. They are the *military forces*, the *country* and the *will of the enemy*.

The *military forces* must be destroyed, that is to say, put into such a condition that they can no longer continue to fight. We take this opportunity to explain that in what follows the expression "destruction of the enemy's military forces" is to be understood only in this sense.

The *country* must be conquered, for from the country fresh military forces could be raised.

But even if both of these things have been done, the war, that is to say the hostile tension and the activity of hostile agencies, cannot be regarded as ended so long as the *will* of the enemy is not subdued also, that is, until his government and his allies have been induced to sign a peace or his people to submit. For though we are in full possession of his country, the conflict may break out again in the interior or through assistance from his allies. No doubt this may also happen after the peace, but this only shows that not every war admits of a complete decision and settlement. But even when this is the case, the conclusion of a peace in itself always extinguishes a quantity of sparks, which would have quietly gone on smouldering, and the tensions slacken because the minds of those inclined to peace, of whom in every nation and in all circumstances there is always a great number, turn wholly away from the idea of resistance. However that may be, we must always regard the end as attained with the peace and the business of war as finished.

Of the three things above enumerated, the military forces are meant to defend the country. So the natural order is that these should first be destroyed; then the land should be conquered; and as a result of these two successes and of the strength which we shall then still possess, the enemy should be induced to make peace. Usually the destruction of the enemy's military forces takes place by degrees, and in a corresponding measure the conquest of the country immediately follows. The two things usually react upon one another, the loss of the provinces helping to weaken the military forces. But this order is by no means necessary, and on that account it is also not always followed. The enemy forces, even before they have been noticeably weakened, may retreat to the opposite side of the country, even right into foreign territory. In this case the greater part of the country, or even the whole, is therefore conquered.

But this object of war in the abstract, this last means for attaining the political object, in which all others are to be included, namely, the *disarming of the enemy*, by no means universally occurs in practice nor is it a necessary condition to peace. It can, therefore, in no wise be set up in theory as a law. There are innumerable instances of peace treaties which have been concluded before either of the parties could be regarded as disarmed, even indeed before the balance of strength was so much as noticeably altered. And, what is more, if we look at the actual cases, we must admit that in a whole class of them the overthrow of the enemy would be a futile playing with ideas—the cases, namely, in which the enemy is distinctly the stronger.

The reason why the object of war deduced from theory is not always suited to real war lies in the difference between the two with which we have been occupied in the previous chapter. According to pure theory, a war between states of noticeably unequal strength would appear to be an absurdity and therefore impossible. The inequality in physical strength would have to be at most no greater than could be neutralized by moral strength, and that would not go far in Europe in our present social condition. If, therefore, we have seen wars take place between states of unequal power that is because war in reality is often very far removed from our original theoretical conception of it.

There are two things which in practice can take the place of the impossibility of further resistance as motives for making peace. The first is the improbability of success, the second an excessive price to pay for it.

As we have seen in the previous chapter, war must from first to last free itself from the strict law of internal necessity and resign itself to a calculation of probabilities. This is always so much the more the case, the more it is adapted thereto by the circumstances out of which it has sprung; that is to say, the slighter are the motives for it and the existing tensions. This being so, it is quite conceivable that out of this calculation of probabilities, even the motive for making peace may arise. A war need not, therefore, always be fought out until one of the parties is overthrown, and we may suppose that when the motives and tensions are weak, a slight, scarcely perceptible probability is in itself enough to move that side to which it is unfavorable to give way. Now, were the other side convinced of this beforehand, he would, naturally, strive for this probability only, instead of first going out of his way to attempt to effect a complete overthrow of the enemy.

Still more general in its effect on the decision to make peace is the consideration of the expenditure of force already made and further required. As war is no act of blind passion, but is dominated by the political object, therefore the value of that object determines the measure of the sacrifices by which it is to be purchased. This will be the case not only as regards the extent of these sacrifices, but also their duration. As soon, therefore, as the expenditure of force becomes so great that the political object is no longer equal in value, this object must be given up, and peace will be the result.

We see, therefore, that in wars in which the one side cannot completely disarm the other, the motives to peace will rise and fall on both sides according to the probability of future success and the expenditure of force required. If these motives were equally strong on both sides, they would meet in the middle of their political difference. What they gain in strength on the one side they should lose on the other. As long as the sum of them added together is sufficient, peace will result, but naturally to the advantage of the side that had the weakest motives thereto.

At this point we purposely pass over for the moment the difference which the *positive* or *negative* character of the political object must necessarily produce in practice. Although, as we shall hereafter show, this is of the highest importance, we must here keep to a still more general point of view, because the original political intentions change very much in the course of the war and may at last become totally different, *just because they are partly determined by the successes and by the probable results.*

The question of how the probability of success can be influenced now arises. In the first place, naturally, by the same means as those used for the overthrow of the enemy, namely, the destruction of his military forces and the conquest of his provinces; though neither of these is quite the same in this connection as it would be if used with that object. If we attack the enemy's forces, it is a very different thing whether we intend to follow up the first blow with a succession of others, until the whole force is destroyed, or whether we mean to content ourselves with one victory in order to shatter the enemy's feeling of security, to give him a feeling of our superiority, and so to instill into him apprehensions about the future. If this is our intention, we only go so far in the destruction of his forces as is sufficient for that purpose. In like manner, the conquest of the enemy's provinces is quite a different measure if the object of it is not the overthrow of the enemy. If this were our object, the destruction of his forces would be the really effective action and the taking of the provinces only the consequence of it. To take them before his forces have been shattered would always have to be regarded as only a necessary evil. On the other hand, if our purpose is not the overthrow of the enemy forces and if we are convinced that the enemy does not seek but fears to bring matters to a bloody decision, the taking of a weak or entirely undefended province is an advantage in itself, and if this advantage is great enough to make the enemy apprehensive about the final result, then it may also be regarded as a shorter road to peace.

But now we come upon yet another particular means of influencing the probability of success without defeating the enemy's armed forces, namely, such enterprises as have an immediate bearing upon policy. If there are any enterprises which are particularly suited to breaking up the enemy's alliances or making them ineffective, to winning new allies for ourselves, to stimulating political activities in our favor, and so forth, then it is easy to conceive how much these can increase the probability of success and become a much shorter way to our object than the defeat of the enemy's armed forces.

The second question is how to influence the enemy's expenditure of strength, that is to say, how to raise for him the price of success.

The enemy's expenditure of strength lies in the *wastage* of his forces, consequently in the *destruction* of them on our part, and in the *loss of provinces*, consequently the *conquest* of them by us.

That each of these terms varies in meaning and that the operation it designates differs in character, according to the object it has in view, will here again upon closer examination be self-evident. That the differences as a rule will be only slight should not cause us perplexity, for in practice when the motives are weak, the finest shades of difference are often decisive in favor of this or that method of applying force. We are only concerned here to show that, certain conditions being supposed, other ways of attaining our object are possible, and that they are neither self-contradictory and absurd, nor even an error.

Besides these two means, there are three other special ways of directly increasing the enemy's expenditure of force. The first is *invasion,* that is, *the occupation of the enemy's territory, not with a view to keeping it,* but in order to levy contributions upon it or even to devastate it. The immediate object here is neither the conquest of the enemy's territory nor the defeat of his armed forces, but merely to *do him damage in a general way.* The second way is to direct our enterprises preferably to the points at which we can do the enemy most harm. Nothing is easier to conceive than two different directions in which our forces may be employed, the first of which is much to be preferred if our object is to defeat the enemy, while the other is more advantageous if there is and can be no question of defeating him. According to the customary way of speaking, the first would be considered the more military course, the second the more political. But from the highest point of view, both are equally military, and each only effective if it suits the given conditions. The third way, by far the most important from the number of cases to which it applies, is the *wearing out* of the enemy. We choose this expression not merely to give a verbal definition, but because it represents the thing exactly and is not so figurative as at first sight appears. The idea of wearing out in a struggle implies *a gradual exhaustion of the physical powers and the will by the long continuance of action.*

Now, if we want to outlast the enemy in the continuance of the struggle, we must content ourselves with as small objects as possible, for naturally a great object demands a greater expenditure of forces than a small one; but the smallest object that we can propose to ourselves is pure resistance, that is, a combat without any positive intention. In this case, therefore, our means will be relatively at their maximum and the result therefore most certainly assured. How far now can this negative mode of proceeding be carried? Clearly not to absolute passivity, for mere endurance would cease to be a combat; but resistance is an active thing and by it so many of the enemy's forces are to be destroyed that he must give up his intention. That alone is what we aim at in each single act, and therein resides the negative character of our intention.

No doubt this negative intention in its single act is not so effective as a positive one would be in the same direction, assuming that it succeeded; but there is just this difference in its favor that it succeeds more easily than the positive and thus offers greater certainty. What it loses in effectiveness in its single act, it must recover with time, that is, with the duration of the struggle, and thus this negative intention, which constitutes the essence of pure resistance, is also the natural means for outlasting the enemy in the duration of the struggle, that is to say, for tiring him out.

Herein lies the origin of the difference between *offensive* and *defensive*, which dominates the whole province of war. We cannot, however, here pursue this subject further than to observe that from this negative intention are to be deduced all the advantages and all the stronger forms of combat which are on the side of the defensive and in which, therefore, that philosophico-dynamic law establishing a constant relation between the magnitude and the certainty of success is realized. We shall resume the consideration of all this hereafter.

If, therefore, the negative intention, that is, the concentration of all means in pure resistance, affords a superiority in combat, and if this is sufficient to *balance* whatever preponderance the enemy may have, then the mere *duration* of the combat will suffice gradually to bring the loss of force on the side of the enemy to the point at which his political object can no longer be an adequate equivalent; at which point, therefore, he must give up the contest. We thus see that this method, the tiring out of the enemy, characterizes the great number of cases in which the weaker wants to offer resistance to the stronger.

Frederick the Great in the Seven Years' War would never have been in a position to defeat the Austrian monarchy, and if he had tried to do so after the fashion of a Charles XII, he would inevitably have been brought to ruin. But after his skilful use of a wise economy of his forces had for seven years shown the powers allied against him that the expenditure of forces on their side would far exceed what they had at first imagined, they made peace.

We see then that there are many ways to our object in war; that the defeat of the enemy is not in every case necessarily involved; that the destruction of the enemy's military forces, the conquest of enemy provinces, the mere occupation, the mere invasion of them, enterprises aimed directly at political relations, and lastly a passive expectation of the enemy's onset—that all these are means which, each in itself, may be used to subdue the enemy's will, according as the peculiar circumstances of the case lead us to expect more from the one or the other. We can still add to these a whole class of objects, as shorter ways of gaining our aim, which we might call arguments *ad hominem*. In what field of human intercourse do sparks of personality that overleap all material circumstances fail to appear? And least of all, surely, can they fail to appear in war, where the personalities of the combatants play so important a part, in the cabinet and in the field. We confine ourselves to pointing this out, as it would be pedantry to attempt to classify them. Including these, we may say that the number of possible ways of attaining the object in view rises to infinity.

To avoid underestimating the value of these various shorter ways to our aim, either reckoning them as merely rare exceptions or holding that the difference they make in the conduct of war is insignificant, we have only to bear in mind the diversity of political objects which may cause a war, or to measure with a glance the distance that separates a death-struggle for political existence from a war which a forced or tottering alliance makes a matter of disagreeable duty. Between the two, innumerable gradations occur in practice. If we reject one of these gradations, we might with equal right reject them all, that is to say, lose sight of the real world entirely.

In general, that is the substance of the aim that has to be pursued in war; let us now turn to the means.

There is only one means: combat. However diversified this may be in form, however widely it may be removed from a rough venting of hatred and animosity in hand-to-hand encounter, whatever number of things may introduce themselves which are not actual combat, still it is always implied in the conception of war that all the effects manifested in it must have their origin in combat.

That this must always be so even in the greatest diversity and complication of reality can be proved in a very simple manner. All that occurs in war takes place through military forces; but where military forces, that is to say, armed men, are used, the idea of combat must necessarily underlie everything.

All, therefore, that relates to the military forces, and, thus, all that appertains to their creation, maintenance, and employment, belongs to warfare.

Creation and maintenance are obviously only the means, while employment is the object.

Combat in war is not a combat of individual against individual, but an organized whole made up of many parts. In this great whole we may distinguish units of two kinds: the one determined by the subject, the other by the object. In an army the mass of combatants ranges itself always into an order of new units, which, again, form members of a higher order. The combat of each of these members also forms, therefore, a more or less distinct unit. Furthermore, the purpose of the combat—and therefore its object—makes it a unit.

Now, to each of the units in the combat we attach the name of engagement.

If the idea of combat lies at the foundation of every employment of armed forces, then the employment of armed forces in general is nothing more than the determining and arranging of a certain number of engagements.

Every military activity, therefore, necessarily relates to the engagement, either directly or indirectly. The soldier is levied, clothed, armed, trained, sleeps, eats, drinks, and marches *merely to fight at the right place and the right time.*

If, therefore, all the threads of military activity terminate in the engagement, we shall also be able to grasp them all when we settle the arrangement of the engagements; only from this arrangement and its execution do the effects proceed, never directly from the conditions preceding them. Now, in the engagement all the activity is directed to the destruction of the enemy or rather of his *ability to fight,* for this is inherent in the conception of an engagement. The destruction of the enemy's armed forces is, therefore, always the means to attain the object of the engagement.

This object may likewise be the mere destruction of the enemy's armed forces; but that is not by any means necessary, and it may be something quite different. Whenever, for instance, as we have shown, the defeat of the enemy is not the only means to attain the political object, whenever there are other things which may be pursued as an aim in war, then it follows of itself that these things may become the object of particular acts of war, and, therefore, also the object of engagements.

But even those engagements, which as subordinate acts are in the strict sense devoted to the defeat of the enemy's armed forces, need not have the destruction of them as their immediate object.

If we think of the complex organization of a great armed force, of the quantity of details that come into play when it is employed, we can understand that the combat of such a force must also acquire a complex organization, with parts subordinated one to the other and acting in correlation. There may and must arise for single parts a number of objects which are not themselves the destruction of the enemy's armed forces, and while they certainly contribute to increase that destruction, do so only indirectly. If a battalion is ordered to drive the enemy from a hill, or a bridge, for instance, as a rule the occupation of this position is the real object, and the destruction of the enemy's forces there only the means or secondary matter. If the enemy can be driven away by a mere demonstration, the object is attained all the same; but this hill or bridge will only be occupied in order thereby to cause a greater destruction of the enemy's armed forces. If this is the case on the field of battle, much more must it be so in the whole theatre of war, where not merely one army is opposed to the other, but one state, one nation, one country, to the other. Here the number of possible relations, and consequently of combinations, must be greatly multiplied, the diversity of arrangements increased, and by the gradation of objects, each subordinate to the other, the first means further removed from the final object.

It is, therefore, for many reasons possible that the object of an engagement is not the destruction of the enemy's forces, that is of the forces immediately opposed to us, but this appears only as a means. In all such cases, however, it is no longer a question of making this destruction complete, for the engagement is here nothing but a *trial* of strength. It has in itself no value but only that of its result, that is to say, of its decision.

But in cases where the strengths are very unequal, a measure of them can be obtained by mere estimation. In such cases the engagement will not take place, but the weaker force will at once give way.

If the object of an engagement is not always the destruction of the enemy's forces therein engaged, and if its object can often be attained as well without the engagement taking place at all, merely by the estimated result of it and the circumstances to which this gives rise, we can understand how whole campaigns can be carried on with great activity without the actual engagement playing any notable part in them.

That this may be so, military history proves by a hundred examples. In how many of such cases the bloodless decision was justified, that is to say, did not involve a self-contradiction, and whether some of the reputations therein gained would stand criticism, we shall leave undecided, for all that concerns us now is to show the possibility of such a course of events in war.

We have only one means in war—the engagement; but this means by the multiplicity of the ways in which it can be employed leads us into all the various paths of which the multiplicity of its objects allows, so that we seem to have achieved nothing. But this is not the case, for from this unity of means proceeds a thread which we follow with our eye as it runs through the whole web of military activity and which, indeed, holds it together.

But we have considered the destruction of the enemy's forces as one of the objects which may be pursued in war and left undecided what importance should be assigned to it in comparison to the other objects. In single instances it will depend on circumstances, and as a general principle we have left its value undetermined. Once more we are brought back to this point, and we shall learn to understand the value which must necessarily be accorded to it.

The engagement is the sole effective activity in war; in the engagement the destruction of the enemy forces opposed to us is the means to the end. It is so even if the engagement does not actually take place, because at all events there lies at the root of the decision the assumption that this destruction is to be regarded as beyond doubt. It follows, therefore, that the destruction of the enemy's forces is the foundation-stone of all action in war, the ultimate support of all combinations, which rest upon it like the arch on its abutments. All action, therefore, takes place on the assumption that if the decision by force of arms which lies at its foundation should actually take place, it would be a favorable one. The decision by arms is, for all operations in war, great and small, what cash payment is in bill transactions. However remote these relations may be, however seldom the settlements may take place, they must eventually be fulfilled.

If a decision by arms lies at the foundation of all combinations, it follows that our opponent by means of a fortunate decision by arms can make any one of these impracticable, not merely if it is the decision on which our combination directly rests, but also by means of any other, provided only it be of sufficient importance. For every important decision by arms—that is, destruction of the enemy's forces—reacts upon all preceding it, because, like a fluid, they tend to bring themselves to a level.

Thus, the destruction of the enemy's forces always appears as the superior and more effectual means, to which all others must give way.

It is, however, only when there is an assumed equality in all other conditions that we can ascribe to the destruction of the enemy's forces a higher efficacy. It would, therefore, be a great mistake to conclude that a blind dash must always gain the victory over cautious skill. An unskilful dash would lead to the destruction not of the enemy's forces but of our own and cannot, therefore, be what we mean. The higher efficacy belongs not to the *means* but to the *end*, and we are only comparing the effect of one realized end with the other.

If we speak of the destruction of the enemy's force, we must expressly point out that nothing obliges us to confine this idea to the mere physical force. On the contrary, the moral force is necessarily implied as well, because in fact both are interwoven with each other even in the most minute details and therefore cannot be separated. In connection with the inevitable effect which has been referred to as a great act of destruction—a great victory—upon all other decisions by arms, it is just the moral element that is most fluid, if we may use that expression, and that distributes itself the most easily through all the parts. Against the superior value which the destruction of the enemy's forces has over all other means stands the expense and the risk this involves, and only to avoid these are other methods employed.

That the means in question must be costly stands to reason, for, other things being equal, the wastage of our own forces is always the greater, the more our aim is directed to the destruction of those of the enemy.

The risk of this means lies in the fact that the greater efficacy which we seek recoils in the event of failure, upon ourselves, and thus brings more disastrous consequences.

Other methods are, therefore, less costly when they succeed, less risky when they fail; but this necessarily implies the condition that they are only opposed by similar ones, that is, that the enemy employs the same methods. For if the enemy should choose the method of a great decision by arms, our own method *would just by that very fact be changed against our will into a similar one.* Thus all then depends on the issue of the act of destruction; now it is obvious that, other things being equal, in this act we must be in all respects at a disadvantage, because our intentions and methods had been partly directed to other things, which was not the case with the enemy. Two different objects of which one is not part of the other exclude each other, and therefore a force which is applied to attain the one cannot at the same time serve the other. If, therefore, one of two belligerents is determined to take the way of great decisions by arms, he has a high probability of success as soon as he is certain that the other does not want to take it but seeks a different object; and anyone who sets before himself any such other object can reasonably do so only on the assumption that his adversary has as little intention as he has himself of seeking great decisions by arms.

But what we have here said of another direction of intentions and forces relates only to other *positive objects* which we may propose to ourselves in war besides the destruction of the enemy's forces, and not by any means to pure resistance, which may be adopted with a view thereby to exhaust the enemy's strength. In mere resistance the positive intention is wanting, and therefore in this case our forces cannot be directed to other objects, but can only be confined to defeating the intentions of the enemy.

We have now to consider the negative side of the destruction of the forces of the enemy, that is to say, the preservation of our own. These two efforts always go together, as they react upon each other; they are integral parts of one and the same intention, and we have only to examine what effect is produced when one or the other has the predominance. The endeavor to destroy the enemy's forces has a positive object and leads to positive results, of which the final aim would be the defeat of the enemy. The preservation of our own forces has a negative object, and thus leads to the defeat of the enemy's intentions, that is to say, to pure resistance, of which the ultimate aim can be nothing else than to prolong the duration of the contest so that the enemy exhausts himself in it.

The effort with a positive object calls into existence the act of destruction; the effort with the negative object awaits it.

How far this waiting should and may be carried we shall enter upon more particularly in the theory of attack and defense, at the origin of which we again find ourselves. Here we must content ourselves with saying that the waiting must be no mere passive endurance, and that in the action bound up with it the destruction of the enemy forces engaged in the conflict may be the aim just as well as anything else. It would thus be a great error in fundamental principles to suppose that the consequence of the negative effort must be that we are precluded from choosing the destruction of the enemy's forces as our object, but must prefer a bloodless decision. The preponderance of the negative effort may certainly lead to that, but only at the risk of its not being the most suitable method, a question which depends on totally different conditions, resting not with ourselves but with our opponent. This other, bloodless way can thus in no way be considered as the natural means of satisfying our predominating anxiety to preserve our own forces. On the contrary, in cases in which such a course did not suit the circumstances we should be much more likely to bring them to utter ruin. Very many generals have fallen into this error and been brought to ruin by it. The only necessary effect resulting from the preponderance of the negative effort is the delay of the decision, so that the defender takes refuge, as it were, in waiting for the decisive moment. The consequence of this is usually the *putting back of the action* both in time and, so far as space is connected with it, also in space, so far as circumstances permit. If the moment has arrived when this could no longer be done without overwhelming disadvantage, the advantage of the negative effort must be considered as exhausted and then comes forward, unchanged, the effort for the destruction of the enemy's force, which was only kept back by a counterpoise but never discarded.

We have thus seen in the foregoing reflections that in war there are many ways to its aim, that is, to the attainment of the political object; but that the only means is the engagement and that consequently everything is subject to a supreme law: which is the *decision by arms;* that when this is actually claimed by the enemy, such an appeal can never be refused, and that, therefore, a belligerent who proposes to take another way must be sure that his opponent will not make this appeal, or he will lose his case before this supreme tribunal; that thus, in a word, the destruction of the enemy's forces seems among all the objects which can be sought in war always to be that which overrules all else.

What may be achieved in war by combinations of another kind we shall only learn in the sequel, and, naturally, only by degrees. We content ourselves here with acknowledging in general their possibility as something pointing to the deviation of practice from theory and to the influence of particular circumstances. But we cannot avoid showing here at once that the *bloody solution of the crisis,* the effort to destroy the enemy's force, is the first-born son of war. When political objects are unimportant, motives slight and the tensions of the forces small, a cautious commander may skilfully try all sorts of ways by which, without great crises and bloody solutions, he may twist himself into a peace through the characteristic weaknesses of his opponent in the field and in the cabinet. We have no right to blame him if the assumptions on which he acts are well founded and promise success; but still we must require him to remember that it is a slippery path he is treading on which the god of war may surprise him, and to keep his eye always upon the enemy that he may not have to defend himself with a dress rapier if that enemy takes up a sharp sword.

The consequences of the nature of war, how ends and means act in it, how in the deviations of practice it departs now more and now less from its original, strict conception, fluctuating backward and forward yet always remaining under that strict conception as under a supreme law—all this we must keep in view and bear constantly in mind in the consideration of each of the succeeding subjects if we wish rightly to understand their true relations and proper importance, and not become incessantly involved in the most glaring contradictions with reality and at last with ourselves.

CHAPTER 3

THE GENIUS FOR WAR

Every special activity, if it is to be pursued with a certain perfection, demands special qualifications of intellect and temperament. When these are of a high degree of distinction and manifest themselves by extraordinary achievements, the mind to which they belong is distinguished by the term "genius."

We know very well that this word is used with meanings which vary very greatly, both in their application and their nature, and that in the case of many of these meanings it is a very difficult task to distinguish the essence of genius. But as we do not profess to be either a philosopher or a grammarian, we shall take leave to keep to a meaning usual in ordinary speech and to understand by "genius" a very superior mental capacity for certain activities.

We wish to stop for a moment at this faculty and dignity of the mind in order to indicate more closely its justification and become more closely acquainted with the content of the conception. But we cannot dwell on genius that has obtained its title through a very superior talent, on *genius* properly so called, for that is a conception that has no defined limits. What we must do is generally to bring into consideration all the combined tendencies of the mind and soul toward military activity, and these we may then regard as the *essence of military genius*. We say "combined," for military genius consists of being not just a single capacity bearing on war, as for instance, courage, while other capacities of mind and soul are lacking or take a direction useless for war, but in being a *harmonious combination of powers*, in which one or other may predominate, but none must be in opposition.

Were every combatant required to be more or less endowed with military genius, our armies would probably be very weak; for precisely because it implies a *special* trend of mental and moral powers, it can but rarely be found when the mental and moral powers of a people are employed and trained on so many sides. But the fewer different activities a people has, and the more the military activity predominates in it, the more prevalent in that people must military genius also be found to be. This, however, determines merely its prevalence and by no means its degree, for the latter depends on the general state of the people's mental and moral development. If we look at a wild, warlike people, we find a warlike spirit in individuals much more common than among civilized peoples, for in the former almost every warrior possesses it, while in the latter there is a whole multitude of persons who are only roused by the pressure of necessity and not at all by inclination. But among uncivilized peoples we never find a really great general, and extremely seldom what may be called a military genius, because that demands a development of the intellectual powers which an uncivilized people cannot have. That civilized peoples can also have a more or less warlike tendency and development is obvious, and the more this is the case, the more frequently also will the military spirit be found in individuals in their armies. As this now coincides with the higher degree of civilization, such peoples always provide the most brilliant example of military achievement, as the Romans and the French have shown. The greatest names in these and in all other peoples renowned in war always belong only to epochs of higher civilization.

From this we may at once infer the importance of the share that intellectual powers have in superior military genius. We shall now examine it more closely.

War is the province of danger, and therefore *courage* above all things is the first quality of a warrior.

Courage is of two kinds: first, courage in presence of danger to the person, and next, courage in the presence of responsibility, whether before the judgment seat of an external authority, or before that of the internal authority which is conscience. We speak here only of the first.

Courage in the presence of danger to the person, again, is of two kinds. First, it may be indifference to danger, whether proceeding from the way the individual is constituted, from contempt of death, or from habit; in any of these cases it is to be regarded as a permanent condition.

Secondly, courage may proceed from positive motives, such as ambition, patriotism, enthusiasm of any kind. In this case courage is not so much a permanent condition as an emotion, a feeling.

We can understand that the two kinds act differently. The first kind is more certain, because, having become a second nature, it never deserts a man; the second often leads him further. Firmness belongs more to the first; boldness to the second. The first leaves the intellect cooler; the second raises its power at times, but also often bewilders it. The two combined make up the most perfect form of courage.

War is the province of physical exertion and suffering. In order not to be overcome by these a certain strength of body and soul is required, which, whether natural or acquired, produces indifference to them. With these qualifications, under the guidance of mere common sense, a man is already a good instrument for war; and these are the qualifications so commonly to be met among wild and half-civilized peoples. If we go further in the demands that war makes upon its votaries, we find intellectual qualifications predominating. War is the province of uncertainty; three-fourths of the things on which action in war is based lie hidden in the fog of a greater or less uncertainty. Here then, first of all, a fine and penetrating intellect is called for to feel out the truth with instinctive judgment.

An average intellect may occasionally hit upon this truth by chance; an extraordinary courage may on other occasions repair a blunder; but the majority of cases, the average result, will always bring to light defective intelligence.

War is the province of chance. In no other sphere of human activity has such a margin to be left for this intruder, because none is in such constant contact with it on every side. It increases the uncertainty of every circumstance and deranges the course of events.

Owing to this uncertainty of all reports and assumptions and to these continual incursions of chance, the person acting in war constantly finds things different from his expectations. This inevitably has an influence upon his plan, or, at all events on the expectations connected with this plan. If this influence is so great as to render the predetermined designs entirely useless, as a rule new designs must be substituted for them; but at the moment, the necessary data for this are often wanting, because in the course of action, circumstances press for immediate decision and allow no time for a fresh look round, often not even enough for careful consideration. But it much more frequently happens that the correction of our premises and the knowledge of chance events that have intruded is not enough to overthrow our design entirely but only to shake it. Our knowledge of the circumstances has increased, but our uncertainty has not been diminished thereby, but intensified. The reason for this is that we do not gain all these experiences simultaneously, but by degrees, because our decisions are unceasingly assailed by them, and our mind, if we may say so, must always be "under arms."

Now if it is to get safely through this continual conflict with the unexpected, two qualities are indispensable: in the first place, an intellect which even in the midst of this intensified obscurity is not without some traces of inner light which lead to the truth, and next, courage to follow this faint light. The first is figuratively denoted by the French expression, *coup d'oeil*, the second is *resolution*.

As engagements are the feature in war to which attention was first and chiefly directed, and as in engagements time and space are important elements and were still more so in the period when cavalry with its rapid decisions was the chief arm, the idea of rapid and correct decision originated in the first instance from the estimation of these two elements, and to denote this idea an expression was adopted which applies only to correct judgment by eye. Many teachers of the art of war have in consequence also given it that limited meaning. But it is undeniable that all sound decisions formed in the moment of execution soon came to be understood by the expression, as, for instance, recognizing the right point for attack, etc. It is, therefore, not only the physical but, more frequently, the mental eye that is meant in *coup d'oeil*. Naturally, the expression, like the thing, is always more in its place in the field of tactics, yet it cannot be excluded from that of strategy either, inasmuch as in that too rapid decisions are often necessary. If we strip this conception of the too figurative and limited elements given to it by that expression, it amounts simply to the rapid hitting upon a truth which to the eye of the ordinary mind is either not visible at all or only becomes so after long examination and reflection.

Resolution is an act of courage in a single instance, and, if it becomes a characteristic trait, a habit of the mind. But here we do not mean courage in facing physical danger, but courage in facing responsibility, therefore to a certain extent in facing moral danger. This has often been called *courage d'esprit*, on the ground that it springs from the intellect, but it is not on that account an act of the intellect but one of feeling. Mere intellect is not quite courage, for we often see the cleverest people devoid of resolution. The intellect must first, therefore, awaken the feeling of courage to be maintained and supported by it, because in emergencies of the moment man is governed more by his feelings than by his thoughts.

We have assigned to resolution the office of removing the torments of doubt and the dangers of hesitation when there are no sufficient motives for guidance. Colloquial usage, it is true, has no scruple in applying the word "resolution" even to the mere propensity for daring, bravery, boldness or temerity. But when a man has sufficient motives, whether they be subjective or objective, true or false, we have no reason to speak of his resolution, for, in doing so, we put ourselves in his place and throw into the scale doubts that did not exist with him at all.

Here there is no question of anything but of strength and weakness. We are not pedantic enough to quarrel with colloquial usage about this little misapplication; our observation is only intended to remove unjustified objections.

Now this resolution, which overcomes the state of doubting, can only be called forth by the intellect, and in fact by a quite special direction of it. We maintain that the mere union of superior intelligence and the necessary feelings is not enough to constitute resolution. There are persons who possess the keenest perception for the most difficult problem and who also are not lacking in courage to accept grave responsibilities, yet in difficult cases cannot come to a resolution. Their courage and their intelligence stand apart, do not give each other a hand, and on that account do not produce resolution as a result. Resolution springs only from an act of the intellect, which makes evident the need for daring and determines thereby the will. This quite special direction of the intellect, which conquers every other fear in man together with the fear of wavering or hesitating, is what makes up resolution in strong minds. Therefore, men of little intelligence can never be resolute, in our sense of the word. In difficult situations they may act without hesitation, but they do so then without reflection, and a man who acts without reflection is not, of course, torn asunder by any doubts. Such a course of action may even, now and then, turn out to be correct: but we say, now as before, it is the average result which indicates the existence of military genius. Should our assertion seem strange to anyone because he knows many a resolute hussar officer who is no deep thinker, we must remind him that the question here is of a special direction of the intellect and not of a capacity for deep meditation.

We believe, therefore, that resolution is indebted to a special direction of the intellect for its existence, a direction which belongs to a strong mind rather than to a brilliant one. In corroboration of this genealogy of resolution, we may add that there have been many instances of men who had shown the greatest resolution in an inferior rank and have lost it in a higher position. While on the one hand they see the necessity of coming to a resolution, on the other they realize the dangers of a wrong decision, and as they are not familiar with the things they are concerned with, their intelligence loses its original force, and they become only the more timid the more they become aware of the danger of the irresolution which holds them spellbound, and the more they have formerly been in the habit of acting on the spur of the moment.

From the *coup d'oeil* and resolution we are naturally led to speak of its kindred quality, *presence of mind,* which must play a great part, in a region of the unexpected like war, for it is indeed nothing but a supreme instance of the conquest of the unexpected. As we admire presence of mind in a telling repartee to anything said unexpectedly, so we admire it in a quickly found resource at a moment of sudden danger. Neither the repartee nor the resource need be in themselves extraordinary if only they meet the case; for that which as a result of mature reflection would be nothing unusual, and therefore insignificant in its impression upon us, may, as an instantaneous act of the intelligence, give us pleasure. The expression "presence of mind" certainly denotes very fitly the readiness and rapidity of the help rendered by the intelligence.

Whether this noble quality of a man is to be ascribed more to the special quality of his intelligence or to the steadiness of his emotional balance, depends on the nature of the case, although neither of the two can ever be entirely wanting. A telling repartee is more the work of a ready wit; a telling counterstroke in sudden danger implies above all else steadiness of emotional balance.

If we take a comprehensive view of the four components of the atmosphere in which war moves, *danger, physical effort, uncertainty* and *chance,* it is easy to understand that a great moral and mental force is needed to advance with safety and success in this baffling element, a force which, according to the different modifications arising out of circumstances, we find historians and chroniclers of military events describing as *energy, firmness, staunchness, strength of mind and character.* All these manifestations of the heroic nature might be regarded as one and the same force of will, modified according to circumstances; but, nearly related as these things are to one another, still they are not one and the same, and it is desirable for us to distinguish at least a little more closely these moral qualities and their relation to one another.

In the first place, to get our ideas clear it is essential to observe that the weight, burden, resistance, or whatever it may be called, by which that force of the soul in the person acting is called forth, is only in a very small measure the enemy's activity, the enemy's resistance, the enemy's action directly. The enemy's activity only affects the general directly in the first place in relation to his person, without affecting his action as commander. If the enemy resist for four hours instead of two, the commander is four hours instead of two in danger. This is a consideration which clearly diminishes in importance, the higher the rank of the commander. What is it for one in the position of commander-in-chief? It is nothing.

Secondly, the enemy's resistance has a direct effect on the commander through the loss of means he incurs in a more prolonged resistance, and the responsibility connected with that loss. It is at this point, through these anxious considerations, that his force of will is first put to the test and called forth. Still we maintain that this is far from being the heaviest burden he has to bear, for he has to settle it only with himself. But all the other effects of the enemy's resistance act upon the combatants under his command, and through them react upon him.

As long as his men are full of good courage and fight with zeal and spirit, the commander seldom has an opportunity of displaying great force of will in the pursuit of his object. But as soon as difficulties arise—and this can never fail to happen when great results are to be achieved—then things no longer move on of themselves like a well-oiled machine, but the machine itself begins to offer resistance, and to overcome this, the commander must have great force of will. By this resistance we must not exactly suppose disobedience and contradictions, though these are frequent enough with single individuals; it is the general impression of the dissolution of all physical and moral forces and the heartrending sight of the bloody sacrifice which the commander has to contend with in himself, and then in all others who directly or indirectly transfer to him their impressions, feelings, anxieties, and efforts. As the forces in one individual after another die away and can no longer be excited and maintained by his own will, the whole inertia of the mass gradually rests its weight on the will of the commander. By the spark in his breast, by the light of his spirit, the spark of purpose, the light of hope, must be kindled afresh in all others. Only in so far as he is equal to this, does he control the masses and remain their master. When that ceases and his own courage is no longer strong enough to revive the courage of all others, then the masses drag him down to them into the lower region of animal nature which shrinks from danger and knows not shame. These are the burdens under which the military commander has to bear up if he wishes to do great deeds. They increase with the size of the masses under him, and, therefore, if the forces in question are to continue equal to the weight upon his shoulders, they must rise in proportion to the height of his rank.

Energy in action expresses the strength of the motive by which the action is called forth, whether the motive has its origin in a conviction of the intellect or in an impulse of feeling. The latter, however, can hardly ever be absent where great force is to be shown.

Of all the noble feelings which fill the human heart in the fierce stress of battle, none, we must admit, is so powerful and constant as the soul's thirst for honor and renown, which the German language treats so unfairly and strives to depreciate by two unworthy associations, in *Ehrgeiz* (greed of honor) and *Ruhmsucht* (hankering after glory). No doubt it is especially in war that the abuse of these proud aspirations of the soul were bound to bring forth the most shocking outrages upon the human race, but by their origin these feelings are certainly to be counted among the noblest that belong to human nature, and in war they are the animating spirit which gives the gigantic body a soul. Although other feelings may become more general in their influence, and many of them—such as love of country, devotion to an idea, revenge, enthusiasm of every kind—may seem to stand higher, the thirst for honor and renown still remains indispensable. Those other feelings may rouse the great masses in general, and inspire them with loftier feelings, but they do not give the leader the desire to aim higher than his fellows, which is an essential requisite in his position if he is to achieve in it anything noteworthy. They do not, like ambition, make the individual military act the special property of the leader, which he then strives to use to the best advantage; in which he plows with toil and sows with care that he may reap plentifully. It is these aspirations, however, shared by all commanders from the highest to the lowest—this sort of energy, this spirit of emulation, this incentive—that more than anything else quicken the efficiency of an army and make it successful. And now as to that which specially concerns the head of all, we ask: Has there ever been a great commander destitute of ambition, or is such a phenomenon even so much as conceivable?

Firmness denotes the resistance of the will with reference to the force of a single blow, *staunchness* with reference to duration. Close as the analogy between the two is, and often as the one is used in place of the other, still there is a notable difference between them which cannot be mistaken, inasmuch as firmness against a single powerful impression may have its root in the mere strength of a feeling, but staunchness must be supported rather more by the intelligence. For the longer an action lasts, the more deliberate it becomes, and from this deliberation staunchness partly derives its power.

If we now turn to *strength of mind or of character*, then the first question is: What are we to understand thereby?

Obviously it is not violence in expressions of feeling, or proneness to strong emotion, for that would be contrary to all the usage of language, but the power of listening to reason even in the midst of the most intense excitement, in the storm of the most violent emotions. Should this power depend on strength of intellect alone? We doubt it. The fact that there are men of outstanding intellect who cannot control themselves would, it is true, prove nothing to the contrary for we might say that it perhaps requires an intellect of a powerful rather than of a comprehensive nature; but we believe we are nearer to the truth if we assume that the power of submitting oneself to the control of the intellect, even in moments of the most violent excitement of the feelings, the power that we call *self-command*, has, itself, its root in the heart. It is in point of fact another feeling, which in men of stout heart balances the excited emotions without destroying them; and it is only through this balance that the mastery of the intellect is secured. This counterpoise is nothing but a feeling of the dignity of man, that noblest pride, that innermost need of the soul, always to act as a being endowed with judgment and intelligence. We may therefore say that a stout heart is one which does not lose its balance even under the most violent excitement.

If we cast a glance at the variety to be observed in men in respect of their emotional side, we find, first, some people who have very little excitability and are called phlegmatic or indolent.

Secondly, some very excitable persons, whose feelings, however, never exceed a certain strength, and who are therefore known as sensitive but gentle.

Thirdly, those who are very easily roused, whose feelings blaze up quickly and violently like gunpowder, but do not last.

Fourthly, and lastly, those who cannot be moved by slight causes, and who generally are not moved suddenly but gradually, but whose feelings become very powerful and are much more lasting. These are men with strong passions, lying deep and hidden.

This difference between men in respect of their emotional constitution lies probably close on the confines of the physical forces active in the human organism, and belongs to that amphibious organization which we call the nervous system and which seems akin on the one side to matter, on the other to spirit. We, with our feeble philosophy, have nothing further to seek in this dark field. But it is important for us to spend a moment over the effect which these differences in character have upon action in war, and to see how far a great strength of character is to be expected from them.

Indolent men cannot easily be thrown off their balance, but we certainly cannot say there is strength of character where there is a total lack of any manifestation of force. It is not, however, to be denied that such men have a certain one-sided efficiency in war, just on account of their steadfast balance. They are often lacking in the positive motive for action. That is to say, in driving force, and consequently in activity, but they are not likely to spoil anything.

The peculiarity of the second class is that they are easily excited to act on trifling grounds, but in great matters they are easily overwhelmed. Men of this kind show great activity in helping an individual in distress, but by the distress of a whole nation they are only depressed, not roused to action.

In war such men are not lacking either in activity or in balance, but they will not accomplish anything great, unless a very powerful intelligence supplies the motives for it. But it is seldom that a very strong, independent intelligence is combined with such temperaments.

Excitable, inflammable feelings are in themselves little suited for practical life, and therefore they are not very fit for war. They have the advantage, it is true, of strong impulses, but these do not last. If, however, the excitability of such men takes the direction of courage or ambition, they may often be very useful in inferior positions in war, simply because the action in war over which commanders in inferior positions have control is usually of shorter duration. Here one courageous resolution, one outpouring of the forces of the soul, will often suffice. A brave attack, a soul-stirring charge, is the work of a few moments, whilst a brave contest on the battlefield is the work of a day, and a campaign the work of a year.

Owing to the rapid movement of their feelings, it is doubly difficult for men of this description to keep their emotional balance; they therefore frequently lose their heads, and that, for the conduct of war, is the worst of their defects. But it would be contrary to experience to maintain that men of very excitable temperament could never be strong, never, that is to say, be capable of keeping their balance even under the strongest excitement. Why should not the feeling for their own dignity exist in them, for as a rule their nature is of the nobler sort? This feeling is seldom wanting in them, but it has not time to become effective. After an outburst they are mostly overcome by a feeling of humiliation. If through education, through keeping watch on themselves, and through experience, they have sooner or later learned to be on their guard against themselves, so as in moments of wild excitement to become conscious betimes of the coun-teracting force within their own breasts, they too can be capable of great strength of soul.

Lastly there are the men who are difficult to move, but on that account are moved deeply, men who stand in the same relation to the preceding as red heat to a flame. It is they who are best fitted, by means of their titanic force to move the gigantic masses by which we may figuratively represent the difficulties of action in war. The working of their feelings is like the movement of great masses, which, though slower, is yet the more irre-sistible.

Although such men are not so likely to be suddenly surprised by their feelings and carried away so as to be afterwards ashamed of themselves, like the preceding, still it would be once more contrary to experience to believe that they could never lose their balance, or be overcome by blind passion. This, on the contrary, will always happen as soon as the noble pride of self-control fails, or whenever it has not sufficient weight. We see examples of this most frequently in great men belonging to uncivilized peoples, where the scanty cultivation of the intellect always favors the predominance of passion. But even among the most civilized classes of the civilized peoples life is full of examples of this kind—of men carried away by the violence of their passions, as the poacher in the Middle Ages, chained to a stag, was carried away through the forest.

We therefore say once more that a strong mind is not one that is merely capable of strong emotions, but one that under stress of the strongest emotions keeps its balance, so that in spite of the storms within the breast, judgment and conviction can act with perfect freedom, like the needle of the compass on a storm-tossed ship.

By the term *strength of character,* or simply *character,* is denoted tenacity of conviction, whether that conviction be the result of our own or of others' judgment and whether it is based upon principles, opinions, mo-mentary inspirations, or any other product of the intelligence. But this kind of firmness cannot, it is true, manifest itself if the judgments them-selves are subject to frequent change. This frequent change need not be the result of an external influence. It may arise from the continuous activ-ity of our own intelligence, but in that case, certainly, it indicates an un-steadiness peculiar to that intelligence. Obviously it will not be said of a man who changes his views every moment that he has character, however much such changes may proceed from himself. Only those men, therefore, can be said to have this quality whose conviction is very constant, either because it is deeply rooted, clear and little liable in itself to alteration, or because, as in the case of indolent men, there is a lack of mental activity, and therefore a lack of motive for alteration; or lastly, because an explicit act of will, derived from an imperative maxim of the intelligence refuses, up to a certain point, any change of opinions.

Now in war, owing to the many and powerful impressions to which the mind is exposed, and to the uncertainty of all knowledge and all judgment, more things occur to distract a man from the road he has entered upon, to make him doubtful of himself and others, than in any other human activity.

The harrowing sight of suffering and danger easily leads to feeling gaining ascendancy over intellectual conviction; and in the twilight which surrounds everything it is so hard for a judgment to be clear and profound that to change it is more intelligible and more pardonable. It is at all times only conjecture or guesses at truth that we have to act upon. This is why differences of opinion are nowhere so great as in war, and the stream of impressions acting counter to one's own convictions never ceases to flow. Even the greatest intellectual impassivity is scarcely proof against them, because the impressions are too strong and vivid and are always also at the same time directed at the emotions.

When the judgment is clear and profound, none but general principles and views of action governing it from a higher standpoint can be the result; and on them the opinion on the particular case immediately under consideration lies, as it were, at anchor. But to hold fast to these results of previous reflection, in opposition to the stream of opinions and phenomena which the present brings with it, is just the difficulty. Between the particular case and the principle there is often wide space which cannot always be traversed on a visible chain of conclusions, and where a certain belief in oneself is necessary and a certain amount of scepticism is serviceable. Here often nothing will help us but an imperative maxim which, independent of reflection, controls it; the maxim is, in all doubtful cases to adhere to our first opinion and not to give it up until a clear conviction forces us to do so. We must firmly believe in the superior authority of well-tried maxims, and not let the vividness of momentary phenomena make us forget that the truth of these is of an inferior stamp. By this preference which in doubtful cases we give to our previous convictions, and by adherence to them, our actions acquire that stability and consistency which make up what we call character.

It is easy to see how greatly a well-balanced temperament promotes strength of character; that, too, is why men of great moral strength generally have a great deal of character.

Strength of character leads us to a degenerate form of it—*obstinacy*.

It is often very difficult in concrete cases to say where the one ends and the other begins; on the other hand, it does not seem difficult to determine the difference in the abstract.

Obstinacy is not a fault of the intellect; we use the term as denoting resistance to our better judgment, and that cannot be located, without involving us in a contradiction, in the intellect, which is the capacity of judgment. Obstinacy is a *fault of temperament*. This inflexibility of will and impatience of contradiction find their origin only in a particular kind of egotism, which sets above every other pleasure that of governing itself and others solely by its own caprice. We would call it a form of vanity if it were not, of course, something better; vanity is satisfied with the appearance but obstinacy rests upon the enjoyment of the thing.

We say, therefore, strength of character becomes obstinacy as soon as resistance to an opposing judgment proceeds not from a better conviction or reliance upon a higher principle, but from a feeling of opposition. If this definition, as we have already admitted, is of little practical assistance, still it will prevent obstinacy from being considered merely strength of character intensified, while it is something essentially different—something which, it is true, lies close to it and borders upon it, but is at the same time so little an intensification of it that there are very obstinate men who from lack of intelligence have little strength of character.

Having in these high attributes of a great military commander made ourselves acquainted with those qualities in which temperament and intellect co-operate, we now come to a peculiarity of military activity which may perhaps be considered the most influential though it is not the most important, and which only demands mental capacity without regard to temperamental qualities. It is the connection which exists between war and terrain, that is to say, country or ground.

This connection is, in the first place, one that is constantly present, so that it is quite impossible to conceive of an operation of war on the part of our organized army taking place otherwise than in a definite space; it is, secondly, of the most decisive importance, because it modifies and at times completely alters the operation of all forces; thirdly, while, on the one hand, it often concerns the most minute features of locality, on the other, it may embrace the widest expanses of country.

In this manner the relation which war has to country and ground, gives to its action a very peculiar character. If we think of other human activities which have a relation to these objects—of horticulture, agriculture, house building, hydraulic works, mining, hunting and forestry—they are all confined to very limited spaces, which can all be soon explored with sufficient exactness. But the commander in war must commit the business on which he is engaged to a space which is his partner in it, which his eyes cannot survey, which the keenest zeal cannot always explore, and with which, owing to the constant changes taking place, he can seldom become properly acquainted. The enemy, it is true, is generally in the same situation; still, in the first place, the difficulty, though common to both, is not the less a difficulty, and he who by talent and training masters it will have a great advantage on his side; secondly, this equality of difficulty only occurs in general, not necessarily in a particular case, in which as a rule one of the two combatants (the defender) knows much more about the locality than the other.

This very peculiar difficulty must be overcome by a mental gift of a peculiar kind called *sense of locality*, a term which is too narrow. It is the capacity for quickly forming for oneself a correct geometrical representation of any given piece of country and consequently of correctly and easily finding at any time one's position in it. This is obviously an act of imagination. The perception, no doubt, is formed partly by the physical eye, partly by the intellect, which by means of judgments derived from knowledge and experience supplies what is wanting, and out of the fragments visible to the physical eye forms a whole. But that this whole should present itself vividly to the mind, should become a picture, a map drawn in the brain, that this picture should be permanent, that the details should not always be falling apart again—all this can only be effected by the mental faculty we call imagination. If some great poet or painter should feel hurt at our attributing to his goddess such an office, if he shrugs his shoulders at the notion that a smart gamekeeper is on that account to be credited with a first-rate imagination, we readily grant that we are speaking here only of a very limited application of the term and of its employment in a truly menial office. But, however slight this service, still it must be the work of that natural gift, for if that gift is altogether wanting, it would become difficult to form a clear, coherent picture of things as if they were before our eyes. That a good memory is of great assistance in this, we readily allow, but whether memory is then to be considered as an independent faculty of the mind, or whether it is just that capacity for forming pictures that fixes these things better in the memory, we must leave undecided, the more readily because it seems really difficult to think of these two mental faculties as in many respects separated from each other.

That practice and intelligent judgment have much to do with it is not to be denied. Puysegur, the famous Quartermaster-General of the famous

General Luxemburg, used to say that at first he had little confidence in himself in this respect, because he noticed that if he had to fetch the *parole* from a distance, he always lost his way.

Scope for the exercise of this talent naturally increases with height in rank. If the hussar, or the rifleman, in command of a patrol must be able easily to locate his position on highways and byways, and needs few landmarks for the purpose and only a limited gift of observation and imagination, the commander-in-chief must rise to a knowledge of the general geographical features of a province and of a country, must have always vividly before his eyes the direction of the roads, rivers and mountains, without at the same time being able to dispense with the limited sense of locality. No doubt, for objects in general, information of all kinds, maps, books, and memoirs, are a great help to him, and for details, the assistance of his staff; but it is nevertheless certain that a considerable talent for quickly and clearly grasping the features of a country lends to his action an easier and firmer step, saves him from a certain mental helplessness, and makes him less dependent on others.

If this capacity then is to be ascribed to imagination, it is at the same time almost the only service which military activity demands from that erratic goddess, whose influence is otherwise rather more harmful than useful.

We think we have now passed in review those manifestations of the powers of mind and soul which military activity requires from human nature. Everywhere intelligence appears as an essential co-operative force and thus we can understand how the work of war, plain and simple though it appears, can never be conducted with distinguished success by people without distinguished intellectual powers.

If we have reached this view, we need no longer look upon such a natural thing as the turning of an enemy's position, which has been done a thousand times, and a hundred other such feats, as the result of a great mental effort.

Certainly, one is accustomed to regard the plain, efficient soldier as the very opposite of the men of reflection, the men whose heads are full of inventions and ideas, the brilliant spirits who dazzle us with every educational adornment. This antithesis is also by no means devoid of actuality; but it does not indicate that the efficiency of the soldier consists merely in his courage and that it does not also demand a certain special energy and efficiency of brain to be more than what is called a good fighter. We must continue to insist that there is nothing more common than to hear of men losing their energy on being raised to a higher position, to which their abilities are no longer equal; but we must also continue to remind our readers that we are speaking of outstanding achievements, such as give renown in the branch of their profession to which they belong. Each grade of command in war therefore creates its own standard of mental qualities required and of fame and honor.

An immense gulf lies between a commander-in-chief, that is to say, a general in supreme command of a whole war or theater of war, and his Second in Command, for the simple reason that the latter is subject to a much more detailed guidance and supervision and consequently confined to a much smaller sphere of independent mental activity. This is why common opinion sees no room for distinguished intellectual activity except in high places, and thinks ordinary intelligence is sufficient for all beneath; why, indeed, people are rather inclined to discover in a subordinate general, who has grown gray in the service and in whom his one-sided activities have produced an unmistakable poverty of mind, a sort of stupidity, and with all respect for his bravery, to laugh at his simplicity. It is not our object to gain for these brave men a better lot; that would contribute nothing to their efficiency and little to their happiness. We only wish to represent things as they are, and to warn against the error of supposing that a mere bravo without intelligence can do distinguished service in war.

As we consider that, even in the humblest positions, the leader who is to attain distinction must have distinguished talents and that the higher his rank, the higher the quality of these talents must be, it naturally follows that we take a quite different view of those who occupy with credit the place of Second in Command of an army; and their seeming simplicity of character, by comparison with a universal genius, a business man mighty with his pen, or a statesman in conference, should not lead us astray as to their practical intelligence. It happens, indeed, sometimes that men carry over with them into a higher position credit they have won in a lower, without deserving it in the higher position; and then, if they are not much employed, and therefore run no risk of exposing themselves, the judgment does not distinguish so clearly what kind of credit is due to them. Thus such men are often the occasion of a low estimate being formed of a personality, which can in certain positions nevertheless be a brilliant one.

For each rank, from the lowest upward, in order to render distinguished services in war, a particular genius is required. But the title of genius, history and the judgment of posterity usually confer only on those minds that have shone in the highest rank, that of Commander-in-Chief. The reason is that here, in point of fact, the demand on mental and moral qualities is, of course, much greater.

To conduct a whole war or its great acts, which we call campaigns, to a brilliant end, requires a keen insight into state policy in its higher relations. The conduct of the war and the policy of the state here coincide and the general becomes at the same time the statesman.

Charles XII is denied the name of a great genius because he could not make the power of his sword subservient to a higher judgment and wisdom, could not attain by it a glorious object. That title is denied to Henry IV (of France) because he did not live long enough to influence the conditions of several states by his military prowess and to have experience in that higher field in which noble feelings and a chivalrous character are less effective in mastering an enemy than in overcoming internal dissension.

In order that the reader may appreciate all that a general must comprehend and correctly divine in one glance, we refer to the first chapter. We say that the general becomes a statesman, but he must not cease to be the general. On the one hand, he must comprehend in one glance all the political conditions; on the other, he knows exactly what he can do with the means at his disposal.

The diversity and undefined limits of all relations in war bring a great number of factors into consideration. Most of these factors can only be estimated according to the laws of probability, and, therefore, if the person acting did not divine everything with the glance of a mind that in all circumstances intuitively senses the truth, a confusion of views and considerations would arise, out of which his judgment would be totally unable to see its way. In this sense Bonaparte was quite right when he said that many of the decisions a general has to make would furnish a problem of mathematical calculation not unworthy of the powers of a Newton or an Euler.

Of the higher powers of the mind those here required are a sense of unity and a judgment raised to a marvelous pitch of mental vision which in its range touches upon and sets aside a thousand half-obscure ideas which an ordinary intellect only brings to light with great effort and over which it would exhaust itself. But this higher activity of mind, this glance of genius, would still not become a matter of historical importance unless it were supported by those qualities of temperament and character of which we have treated.

Truth alone is but a very weak motive for action, and hence there is always a great difference between cognition and volition, between knowing what to do and being able to do it. Man always gets his strongest impulses to action through his emotions, and his most powerful backing, if we may be permitted the expression, from those amalgamations of temperament and intelligence which we have learned to recognize as resolution, firmness, staunchness and strength of character.

If, however, this exalted activity of heart and brain in the general did not manifest itself in the final success of his work, and were only accepted on faith, it would rarely become a matter of historical importance.

All that becomes known of the course of events in war is usually very simple, and has a great sameness in appearance. No one on the mere narration of such events has any perception of the difficulties encountered and overcome. It is only now and again, in the memoirs of generals or of those in their confidence, or on the occasion of some special historical investigation to which an event has been subjected, that a portion of the many threads composing the whole web is brought to light. Most of the deliberations and mental conflicts which precede the execution of great schemes are purposely concealed because they affect political interests, or the recollection of them is accidentally lost because they are looked upon as mere scaffolding which has to be removed on the completion of the building.

If now, in conclusion, without venturing upon a closer definition of the higher powers of the soul, we still admit a distinction in the intellectual faculty itself according to the common ideas as they have been fixed in language, and if we then ask what kind of intellect is most closely associated with military genius, then a glance at the subject as well as at experience will tell us that searching, rather than creative, minds, comprehensive minds rather than such as pursue one special line, cool, rather than fiery, heads are those to which in time of war we should prefer to trust the welfare of our brothers and children, the honor and safety of our country.

CHAPTER 4

ON DANGER IN WAR

Usually before we have learned what danger really is, we form an idea of it which is more attractive than repulsive. In the intoxication of enthusiasm to fall upon the enemy at the charge, who cares then about bullets and men falling? To hurl ourselves, with eyes a few moments shut, into the face of chill death, uncertain whether we or others shall escape him, and all this close on the golden goal of victory, close to the refreshing fruit for which ambition thirsts—can this be difficult? It will not be difficult, and still less will it appear so. But such moments, which, however, are not the work of a single pulse-beat, as they are supposed to be, but rather, like doctors' draughts, must be taken diluted and spoiled by time—such moments, we say, are but few.

Let us accompany the novice to the battlefield. As we approach, the thunder of the cannon becoming plainer and plainer is soon accompanied by the howling of shot, which now attracts the attention of the inexperienced. Balls begin to strike the ground close to us, in front and behind. We hasten to the hill where stands the general and his numerous staff. Here the close striking of the cannon balls and the bursting of shells are so frequent that the seriousness of life forces itself through the youthful picture of imagination. Suddenly someone we know falls. A shell strikes among the crowd and causes some involuntary movements; we begin to feel that we are no longer perfectly at ease and collected, and even the bravest is at least to some degree distracted. Now, a step farther into the battle raging before us, as yet almost like a scene in a theater, to the nearest general of division. Here ball follows ball, and the noise of our own guns increases the confusion. From the general of division to the brigadier. He, a man of acknowledged bravery, keeps carefully behind a hill, a house, or some trees—a sure sign of increasing danger. Grape rattles on the roofs of the houses and in the fields; cannon balls roar in all directions over us, and already there is a frequent whistling of musket balls. A step farther toward the troops, to that sturdy infantry which has been for hours holding its ground under fire with indescribable steadiness. Here the air is filled with the hissing of balls, which announce their proximity by a short, sharp noise as they pass within an inch of the ear, the head, the breast. To add to all this, compassion strikes our beating heart with pity at the sight of the maimed and fallen.

On none of these different stages of increasing danger will the novice set foot without feeling that the light of reason neither moves here through the same mediums nor is refracted in the same way as when it is engaged in thought. He must, indeed, be a very extraordinary man who under stress of these first impressions does not lose the capacity of making a prompt decision. It is true that habit soon blunts these impressions; in half an hour we begin to be more indifferent in greater or less degree, to everything that is going on around us; but the ordinary man never attains complete coolness and natural elasticity of mind. So we perceive that here again ordinary qualities will not suffice—a thing which becomes all the truer the wider the sphere of activity which has to be filled. Enthusiastic, stoical, natural bravery and overmastering ambition or long familiarity with danger—much of all this must be there if all the effects produced in this baffling medium are not to fall far short of that which in the study may appear only the ordinary standard.

Danger in war belongs to its friction. A correct idea of it is necessary for true understanding, and for that reason it is mentioned here.

CHAPTER 5

ON PHYSICAL EFFORT IN WAR

If no one were allowed to pass an opinion on the events of war except at a moment when he is benumbed by frost, suffocating with heat and thirst or overwhelmed by hunger and fatigue, we should certainly have even fewer judgments that would be correct objectively; but they would at least be so subjectively, that is, they would contain the exact relation between the person giving the judgment and the object. We recognize this when we see how depreciatory, how spiritless and humble is the judgment passed on the results of untoward events by those who were eye-witnesses of them, especially while they have been involved in them. This, in our opinion, is a criterion of the influence which physical effort exercises and of the consideration that must be given to it in forming a judgment.

Among the many things in war for the value of which no objective standard can be fixed, physical effort must above all be included. Provided that it is not wasted, it is a factor in the efficiency of all forces and no one can say precisely how far it can be carried. But the remarkable thing is that just as only a strong arm enables the archer to tighten up his bowstring, so it is only a strong spirit that can be expected to exact the utmost from the forces of his army. For it is one thing if an army, in consequence of great misfortunes, surrounded with danger, falls all to pieces like a collapsing wall, and can only find safety in the utmost effort of its physical strength and quite another when a victorious army, borne on only by feelings of pride, is led by its chief exactly wherever he wills it to be led. The very effort that in the former case could at most excite our pity, in the latter could not but fill us with admiration, because it is so much more difficult to sustain.

To the inexperienced eye, this brings to light one of those things which put fetters, as it were, in darkness on the movements of the mind and wear away secretly the forces of the soul.

Although here the question is strictly of the effort required by a general from his army, by a leader from his subordinates, and thus of the courage to demand it and of the art of maintaining it, still the physical effort of the leader and of the general himself must not be overlooked. Having brought the analysis of war conscientiously up to this point, we must also take account of the importance of this remaining sediment.

We have spoken here of physical effort, chiefly because, like danger, it belongs to the fundamental causes of friction, and because its indefinite quantity makes it like an elastic body, the friction of which is known to be difficult to calculate.

To prevent the misuse of these considerations and of this examination of the conditions which aggravate the difficulties of war, nature has given our judgment a sure guide in our feelings. Just as an individual cannot with advantage refer to his personal deficiencies if he is insulted and ill-treated, but may well do so if he has successfully averted the insult or brilliantly avenged it, so no general and no army will improve a disgraceful defeat by depicting the danger, the distress, and the effort, which would immensely enhance the glory of a victory. Thus our feeling, which is after all only a higher kind of judgment, forbids us to do what seems an act of justice to which our judgment would be inclined.

CHAPTER 6

INFORMATION IN WAR

By the word "information" we denote all the knowledge which we have of the enemy and his country; therefore, in fact, the foundation of all our plans and actions. Let us consider the nature of this foundation, its unreliability and uncertainty, and we shall soon feel what a dangerous edifice war is, how easily it may fall to pieces and bury us in its ruins. Although it is a maxim in all books that we should only trust information which is certain, and that we must always be suspicious, this is only a miserable book comfort, belonging to that philosophy in which writers of systems and compendiums take refuge for want of anything better to say.

A great part of the information obtained in war is contradictory, a still greater part is false, and by far the greatest part somewhat doubtful. What is required of an officer in this case is a certain power of discrimination, which only knowledge of men and things and good judgment can give. The law of probability must be his guide. This is no trifling difficulty even with the first plans, those which are made in the study and as yet outside the actual sphere of war, but it is enormously increased when in the turmoil of war itself one report follows hard upon another. It is then fortunate if these reports in contradicting each other produce a sort of balance and themselves arouse criticism. It is much worse for the inexperienced when chance does not render him this service, but one report supports the other, confirms it, magnifies it, continually paints the picture in new colors, until necessity in urgent haste forces from us a decision which will soon be discovered to be folly, all these reports having been lies, exaggeration, errors, etc., etc. In a few words, most reports are false, and the timidity of men gives fresh force to lies and untruths. As a general rule, every-one is more inclined to believe the bad than the good. Everyone is inclined to magnify the bad in some measure, and although the perils thus re-ported subside like the waves of the sea, yet like them they rise again with-out any apparent cause. Firm in reliance on his own better convictions, the leader must stand fast like the rock on which the wave breaks. The role is not an easy one; he who is not by nature of a buoyant disposition or has not been trained and his judgment strengthened by experience in war may let it be his rule to do violence to his own inner conviction, and incline from the side of fear to the side of hope. Only by that means will he be able to maintain a true balance. This difficulty of seeing things correctly, which is one of the greatest sources of friction in war, makes things appear quite different from what was expected. The impression of the senses is stronger than the force of the ideas resulting from deliberate calculation, and this goes so far that probably no scheme of any importance has ever been executed without the commander in the first moments of its execu-tion having had to conquer fresh doubts in his mind. Ordinary men, who follow the suggestions of others, generally, therefore, become undecided on the field of action; they think they have found the circumstances dif-ferent from what they had expected, all the more so, indeed, since here again they give way to the suggestions of others. But even the man who has made his own plans, when he comes to see things with his own eyes, easily loses faith in his former opinion. Firm reliance upon himself must make him proof against the apparent pressure of the moment; his first conviction will in the end prove true, when the foreground of scenery which fate pushes on to the stage of war with its exaggerated shapes of danger is drawn aside and the horizon extended. This is one of the great gulfs that separate *conception* from *execution*.

CHAPTER 7

FRICTION IN WAR

As long as we have no personal knowledge of war we cannot conceive where the difficulties of the matter lie, nor what genius and the extraordinary mental and moral qualities required in a general really have to do. Everything seems so simple, all the kinds of knowledge required seem so plain, all the combinations so insignificant, that in comparison with them the simplest problem in higher mathematics impresses us with a certain scientific dignity. But if we have seen war, all becomes intelligible. Yet it is extremely difficult to describe what brings about this change and to put a name to this invisible and universally operative factor.

Everything is very simple in war, but the simplest thing is difficult. These difficulties accumulate and produce a friction of which no one can form a correct idea who has not seen war. Suppose a traveler who, toward the end of a day's journey, is thinking of accomplishing two more stages— a very small matter of four or five hours with post-horses on a high road. He arrives now at the stage before the last, finds no horses or bad ones, then a hilly country, roads out of repair; it is getting pitch dark and he is glad after many difficulties to have reached the next stage and find there some miserable accommodation. So in war, through the influence of innumerable trifling circumstances, which on paper cannot properly be taken into consideration, everything depresses us and we come far short of our mark. A powerful, iron will overcomes this friction; it crushes the obstacles, but at the same time the machine along with them. We shall often meet with this result. Like an obelisk toward which the principal streets of a town converge, the proud will of a strong spirit stands prominent and commanding in the middle of the art of war.

Friction is the only conception which in a fairly general way corresponds to that which distinguishes real war from war on paper. The military machine, the army and all belonging to it, is fundamentally simple, and appears on this account easy to manage. But let it be borne in mind that no part of it consists of one piece, that it is all composed of individuals, each of whom maintains his own friction in all directions. Theoretically it sounds very well: the commander of a battalion is responsible for the execution of the order given; and as the battalion by its discipline is cemented together into one piece, and the chief must be a man of recognized zeal, the beam turns on an iron pin with little friction. But it is not so in reality, and all that is exaggerated and false in the conception manifests itself at once in war. The battalion always remains composed of a number of men, of whom, if chance so wills, the most insignificant is able to cause delay or some irregularity. The danger that war brings with it, the physical efforts which it demands, intensify the evil so greatly that they must be regarded as its most considerable causes.

This enormous friction which is not concentrated, as in mechanics, at a few points, is, therefore, everywhere brought into contact with chance, and thus produces incidents quite impossible to foresee, just because it is to chance that to a great extent they belong. One such instance of chance, for example, is the weather. Here, fog prevents the enemy from being discovered in time, a gun from firing at the right moment, a report from reaching the general; there, rain prevents one battalion from arriving at all, and another from arriving at the right time, because it has had to march perhaps eight hours instead of three, the cavalry from charging effectively because it is stuck fast in heavy ground.

These few details are only by way of illustration and in order that the reader may be able to follow the author in the matter, for otherwise whole volumes might be written on these difficulties. To give a clear conception of the host of small difficulties to be contended with in war, we might exhaust ourselves in illustrations. To avoid the risk of being tiresome, the few we have given will suffice.

Action in war is movement in a resistant medium. Just as a man immersed in water is unable to perform with ease and regularity even the most natural and simplest of movements, that of walking, so in war, with ordinary powers one cannot keep even the line of mediocrity. This is the reason why the correct theorist is like a swimming master, who teaches on dry land movements which are required in the water, which must appear ludicrous and exaggerated to those who forget about the water. This is also why theorists who have never plunged in themselves, or who cannot deduce any generalization from their experiences, are unpractical and even absurd, because they only teach what everyone knows—how to walk.

Further every war is rich in individual phenomena. It is in consequence an unexplored sea, full of rocks which the mind of the general may sense but which he has never seen with his eyes and round which he now must steer in dark night. If a contrary wind also springs up, that is, if some great chance event declares against him, then the most consummate skill, presence of mind and effort are required, while to a distant observer everything seems to be going like clockwork. The knowledge of this friction is a chief part of that often boasted experience of war which is required in a good general. It is true that he is not the best general in whose mind this knowledge fills the largest space and who is most overawed by it (this constitutes the class of over-anxious generals, of whom there are so many amongst the experienced); but a general must be aware of it that he may overcome it, where this is possible, and that he may not expect a degree of precision in his operations which just because of this friction is impossible. Besides, it can never be learned theoretically; and if it could, there would still be wanting that practiced judgment which we call instinctive, and which is always more necessary in a field full of innumerable small and diversified objects than in great and decisive cases in which we deliberate with ourselves and with others. Just as the instinctive judgment which has become almost a habit makes a man of the world speak and act and move only as befits the occasion, so only the officer experienced in war will always, in great and small matters, at every pulsation, so to speak, of the war, decide and determine suitably to the occasion. Through this experience and practice the thought comes into his mind of itself: this is the right thing and that not. And thus he will not easily place himself in a position to expose a weakness, a thing which if it frequently occurs in war shakes all the foundations of confidence and is extremely dangerous.

It is friction, therefore, or what is here so called, which makes that which appears easy in reality difficult. As we proceed we shall often meet with this subject again, and it will then become plain that, besides experience and a strong will, there are still many other rare qualities of mind required to make a distinguished general.

CHAPTER 8

CONCLUDING REMARKS ON BOOK I

Those things which enter as elements into the atmosphere of war and make it a resistant medium for every activity, we have designated under the terms of danger, physical effort, information, and friction. In their hindering effects they may thus be included again in the collective idea of a general friction. Now is there, then, no oil capable of diminishing this friction? Only one, and that is not always available at the will of the commander or his army. It is the habituation of the army to war.

Habit gives strength to the body in great efforts, to the mind in danger, to the judgment against first impressions. By its means a precious self-possession is gained throughout every rank, from the hussar and rifleman up to the general of division, which makes the commander-in-chief's work easier.

As the human eye in a dark room dilates its pupil, draws in the little light there is, by degrees imperfectly distinguishes objects and at last sees them quite accurately, so it is in war with the experienced soldier, while the novice only encounters pitch-dark night.

No general can give his army habituation to war, and maneuvers (peace exercises) furnish but a weak substitute for it, weak in comparison with real experience in war, but not weak in relation to other armies in which even these maneuvers are limited to mere mechanical exercises of routine. So to arrange the maneuvers in peace time as to include some of these causes of friction, in order that the judgment, circumspection, even resolution, of the separate leaders may be exercised, is of much greater value than those believe who do not know the thing from experience. It is of immense importance that the soldier, high or low, whatever be his rank, should not see for the first time in war those phenomena of war which, when seen for the first time, astonish and perplex him. If he has only met them once before, even by that he is half acquainted with them. This applies even to physical efforts. They must be practiced not so much to accustom the body to them, but the mind. In war the young soldier is very apt to regard unusual efforts as the consequence of serious faults, mistakes and embarrassments in the conduct of the whole, and on that account to become doubly depressed and despondent. This will not happen if he has been prepared for it beforehand in peace maneuvers.

Another less comprehensive but still very important means of gaining habituation to war in time of peace is to invite into the service officers of foreign armies who have had experience in war. Peace seldom reigns over all Europe, and never in all quarters of the world. A state which has been long at peace should therefore always try to secure some officers from these theaters of war—though only, of course, such as have done good service—or to send there some of its own, that they may get a lesson in war.

However small the number of officers of this description may appear in proportion to the mass of an army, still their influence is very clearly felt. Their experience, the bent of their mind, the development of their character, influence their subordinates and comrades; and besides that, even if they cannot be placed in positions of superior command, they may always be regarded as men acquainted with the country, who may be consulted in many special cases.

BOOK II

THE THEORY
OF WAR

CHAPTER 1

BRANCHES OF THE ART OF WAR

War in its literal meaning is combat, for combat alone is the efficient principle in the manifold activity which in a wide sense is called war. But combat is a trial of strength of the moral and physical forces by means of the latter. That the moral cannot be omitted is self-evident, for the state of the mind does have the most decisive influence on the forces employed in war.

The need of combat very soon led men to special inventions to turn the advantage in it in their own favor. In consequence of these, combat has undergone very many changes; but in whatever way it is conducted its conception remains unaltered, and combat constitutes war.

The inventions have been, first of all, weapons and equipments for the individual combatants. These have to be provided and the use of them learned before war begins. They are devised in accordance with the nature of the combat and thus are governed by it; but evidently the fashioning of them is a different thing from the combat itself; it is only a preparation for the combat not the conduct of it. That arms and equipment are not an essential part of the concept of combat is clear, because mere wrestling is also combat.

Combat has determined everything appertaining to arms and equipment, and these in turn modify the combat. There is therefore a reciprocal relation between the two.

Nevertheless, combat itself remains still a quite special form of activity, the more so because it moves in a quite special element, the element of danger.

If, then, there is anywhere a necessity for drawing a line between two different activities, it is here; and in order to see clearly the practical importance of this idea, we need only call to mind how often the greatest personal fitness in one field has turned out to be nothing but the most useless pedantry in the other.

It is also in no way difficult to separate, in treatment, one activity from the other, if we regard the armed and equipped forces as given means. In order to use these forces profitably we need know nothing but their main results.

In its proper sense, therefore, the art of war is the art of making use of the given means in combat, and we cannot give it a better name than "the conduct of war." On the other hand, in a wider sense all activities, of course, which exist for the sake of war—the whole process of creating armed forces, that is, the levying, arming, equipping and training of them—belong to the art of war.

To make a sound theory it is essential to separate these two activities, for it is easy to see that if every art of war insisted on starting with the organization of the armed forces and on training them for the conduct of war according to the requirements of the latter, it could only be applicable in practice to the few cases in which the actually existing forces were exactly suitable to those requirements. If, on the other hand, we wish to have a theory which fits the majority of cases and for none is quite inapplicable, it must be founded upon the great majority of the means of making war in common use, and with these, too, only upon their most important results.

The conduct of war, therefore, is the arrangement and conduct of combat. If this combat were a single act, there would be no necessity for any further subdivision. But combat is composed of a more or less large number of single acts, each complete in itself, which we call engagements (as we have shown in Book I, Chapter 1) and which form new units. From this two different activities spring: *individually arranging and conducting* these single engagements and *combining* them with one another to attain the object of the war. The former is called *tactics*, the latter, *strategy*.

This division into tactics and strategy is now in fairly general use and everyone knows tolerably well under which head to place any single fact, without being clearly aware of the grounds on which the division is made. But when such divisions are blindly adhered to in practice, there must be some deep reason for them. We have sought for this reason, and we can say that it is just the usage of the majority which has brought us to it. On the other hand, we must consider the arbitrary, irrelevant definitions of the idea sought out by some writers as non-existent in common usage.

According to our classification, therefore, tactics teaches *the use of the armed forces in engagements,* and strategy *the use of engagements to attain the object of the war.*

How the idea of the single or independent engagement is more closely defined, and on what conditions this unity is dependent, we shall not be able to make quite clear till we examine the engagement more closely. We must for the present content ourselves with saying that in relation to space, that is, in the case of simultaneous engagements, the unity extends just so far as the *personal command,* but in relation to time, that is, in successive engagements, until the crisis which occurs in every engagement is entirely over.

That doubtful cases may occur here, cases in which several engagements can also be regarded as one, will not suffice to overthrow the principle of classification which we have adopted, for it shares that peculiarity with all principles of classification when applied to real things, which, though different, always pass into one another by gradual transitions. There can, of course, in this way be single instances of action which may just as well, and without any change of our point of view, be reckoned as belonging to tactics as to strategy, very extended positions, for example, which resemble a chain of posts, the dispositions made for certain crossings of rivers and so forth.

Our classification concerns and covers only *the use of the armed forces.* But now there are in war a number of activities, subservient to it but still different from it, related to it sometimes more closely and sometimes less. All these activities relate to *the maintenance of the armed forces.* As the creation and training of these forces precedes their use, so their maintenance is inseparable from that use and a necessary condition of it. But, strictly considered, all activities related thereto are always to be regarded as preparations for combat. Of course, these activities, being very closely connected with action, run through the whole fabric of war, occurring alternately with the use of the forces. We have, therefore, the right to exclude them as well as the other preparatory activities from the art of war in its restricted sense—from the conduct of war properly so called; and we are obliged to do so if we wish to fulfil the first task of any theory: the separation of things that are unlike. Who would include in the conduct of war proper the whole catalogue of things like subsistence and administration? These things, it is true, stand in a constant reciprocal relation to the use of the troops, but they are something essentially different from it.

We have said, Book I, Chapter 2, that while the combat or the engagement is defined as the sole directly effective activity, the threads of all the others are included in it, because in it they end. By this we meant to indicate that the object is thereby assigned to all the others, which they then seek to attain in accordance with the laws peculiar to themselves. Here we must explain ourselves more fully on this subject.

The subjects of the remaining activities, exclusive of the engagement, are of very various kinds.

In one respect one part still belongs to combat itself, is identical with it, while in another respect it serves for the maintenance of the armed forces. The other part belongs exclusively to the maintenance and has only, in consequence of its reciprocal action, a conditioning influence on combat through its results.

The subjects which in one respect belong to the engagement itself are *marches, camps* and *quarters,* for all three imply different situations in which troops may be, and whenever we think of troops, the idea of an engagement must always be present.

The other subjects which belong only to maintenance are: *provisioning, care of the sick* and *the supply and repair of arms and equipment.*

Marches are wholly identical with the use of the troops. The act of *marching in the engagement,* generally called maneuvering, it is true, does not amount to the actual use of weapons, but it is so closely and necessarily connected with it that it forms an integral part of what we call an engagement. But, outside of the engagement, the march is nothing but the execution of a strategic plan. By this plan is settled *when, where, and with what forces* to give battle; and the march is the only means by which that can be carried out.

Outside of the engagement, the march is, therefore, an instrument of strategy, but it is not only, on that account, an affair of strategy, but also, because the force that carries it out at every moment implies a possible engagement, its carrying out is governed by tactical as well as strategic laws. If we prescribe to a column the route on this side of a river or range of mountains, this is a strategic measure, for in it lies the intention, if during the march an engagement should become necessary, of offering battle to the enemy on this side rather than on that.

But if a column, instead of following the road in the valley, advances along the heights parallel to it, or for convenience of marching divides itself into several columns, then these are tactical measures, for they relate to the manner in which we wish to use our forces in an engagement, if one occurs.

The particular order of march is in constant relation to readiness for engagement and, therefore, tactical in its nature. For it is nothing but the first, preliminary disposition for the engagement that might occur.

As the march is the instrument by which strategy distributes its effective elements, the engagements, and these often count only by their results and not by the actual course they take, it cannot fail to happen that in the consideration of them the instrument has often been put in the place of the effective element. Thus we speak of a decisive, skilfully designed march and mean the way the engagement was fought to which it led. This substitution of one idea for another is too natural and the conciseness of the expression too desirable to be rejected, but it is always only a condensed sequence of ideas, and in using it we must not forget to bear in mind its proper meaning if we do not want to fall into error.

It is such an error to attribute to strategical combinations a power independent of tactical results. Marches and maneuvers are combined, the object attained, yet there is no question of any engagement—the conclusion drawn being that there are means of overcoming the enemy even without an engagement. Not till later shall we be able to show the whole magnitude of this error, so prolific in consequences.

But although a march can be regarded absolutely as an integral part of the combat, still certain things are connected with it which do not belong to the combat and therefore are neither tactical nor strategic. To these belong all arrangements which are concerned merely with the comfort of the troops, with the construction of bridges and roads, and so forth. These are only prerequisites; in many circumstances they may come very near to the use of troops and almost be identical with it, as in the building of a bridge under the eyes of the enemy; but in themselves they are always extraneous activities, the theory of which does not form part of the theory of the conduct of war.

Camps, by which we mean every disposition of troops in concentration, therefore ready to fight, as distinguished from quarters, are places of rest and therefore of recuperation. At the same time they also imply a strategic decision to offer battle on the spot where they are pitched, while the manner in which they are taken up already indicates the outline of the engagement, a condition from which every defensive engagement starts. They are thus essential parts of both strategy and tactics.

Quarters take the place of camps for the better refreshment of the troops. Like camps, they belong to strategy in respect of their position and extent, to tactics in respect of their internal organization which aims at readiness for battle.

In addition to the recuperation of the troops, camps and quarters usually, of course, have another object: the covering, for instance, of a piece of country or the holding of a position. But they may very well have only that first object. We must not forget that the objects which strategy pursues may be extremely varied, for everything that looks like an advantage may be the object of an engagement, and the preservation of the instrument with which war is conducted must very frequently become the object of particular strategic combinations.

If, therefore, in such a case strategy ministers only to the preservation of the troops, we do not on that account find ourselves, so to speak, in a strange country, for we are still dealing with the use of the armed force, because every disposition of that force upon any point whatever of the theater of war is such a use.

But if the maintenance of the troops in camps or quarters calls forth activities which are not a use of the armed forces as such, like the building of huts, the pitching of tents, subsistence and sanitary service in camp or quarters, that is part neither of strategy nor of tactics.

Even entrenchments, the site and preparation of which are quite obviously part of the order of battle, and therefore matters of tactics, do not belong to the theory of the conduct of war in respect of the *carrying out of their construction*. The knowledge and skill required for such work must already be present in a trained force. The technique of the engagement takes them for granted.

Among the things which belong to the mere maintenance of the armed force, because no part of them is identical with the engagement, the subsistence of the troops stands nevertheless nearest to it, because it must be in daily operation and for every individual. Thus it is that it completely permeates military action in the parts constituting strategy—we say "constituting strategy," because in one particular engagement the feeding of the troops will very rarely have an influence strong enough to modify the plan of it, although this still remains quite conceivable. The care for the subsistence of the forces will therefore come into reciprocal action chiefly with strategy, and there is nothing more common than for the main strategic outlines of a campaign or a war to be to some extent laid down out of consideration for this subsistence. But however frequently this consideration may be taken into account and however important it may be, the provision of subsistence for the troops still remains an essentially different activity from the use of the troops, and the former only influences the latter by its results.

The other branches of administrative activity which we have mentioned stand much further apart from the use of the troops. The care of the sick and wounded, highly important as it is for the well-being of an army, directly affects it only in a small portion of the individuals composing it, and therefore has only a weak and indirect influence upon the use of the rest. The replacing and repairing of arms and equipment, except in so far as through the organization of the forces it is a continuous activity inherent in them, takes place only periodically and therefore seldom affects strategic plans.

We must, however, here guard ourselves against a misunderstanding. In individual cases these matters may be really of decisive importance. The distance of hospitals and depots of munitions may very easily be considered the sole ground for very important strategic decisions. We do not wish either to contest that point or to underestimate its importance. But we are at present occupied not with the concrete facts of an individual case, but with abstract theory. We therefore maintain that such an influence is too rare to give the theory of sanitary measures and the supply of munitions and arms an importance in the theory of the conduct of war such as to make it worth while to include the different methods and systems which the above theories may furnish, together with their results, in the same way as is certainly necessary in regard to the subsistence of troops.

If we once more review the result of our reflections, then the activities belonging to war are divided into two principal classes: such as are only *preparations for war* and such as are *the war itself.* This division must therefore also be made in theory.

The kinds of knowledge and skill involved in the preparations for war will be concerned with the creation, training and maintenance of all the armed forces. What general name shall be given them is a matter we leave open, but it is clear that among them are included artillery, fortification, so-called elementary tactics, the whole organization and administration of the armed forces and all similar matters. But the theory of war itself is occupied not with perfecting these means but with their use for the object of the war. It needs only the results of them, that is to say the knowledge of the principal properties of the means it has taken over. This we call the "Art of War" in a limited sense, or the "Theory of the Conduct of War" or the "Theory of the Employment of the Armed Forces"—all of them denoting for us the same thing.

This theory will therefore deal with the engagement as the real combat, and with marches, camps and quarters as matters more or less identical with it. The subsistence of the troops will only come into consideration like *other given circumstances* in respect of its results, not as an activity belonging to the theory itself.

This art of war, in its more limited sense, is in its turn divided into tactics and strategy. The former occupies itself with the form of the individual engagement, the latter with its use. Both are concerned with the circumstances of marches, camps and quarters only through the engagement, and these things will be tactical or strategic, according as they relate to the form or to the significance of the engagement.

No doubt there will be many readers who will consider superfluous this careful separation of two things lying so close together as tactics and strategy, because it has no direct effect on the conduct itself of war. One would certainly have to be a great pedant to expect to find on the field of battle direct effects of a theoretical distinction.

But the first business of every theory is to clear up conceptions and views which have been mixed up and, we may say, very much entangled and confused. Only when we have come to an understanding about terms and conceptions, may we hope to advance clearly and easily in the discussion of the things they stand for and be sure that author and reader will always see things from the same point of view. Tactics and strategy are two activities mutually permeating each other in time and space, but are also essentially different activities, the inherent laws and mutual relations of which cannot be intelligible to the mind at all until a clear conception of the nature of each activity is established.

He to whom all this is nothing must either repudiate all theoretical consideration or never have had his intelligence annoyed by the confused and confusing ideas, resting on no fixed point of view and leading to no satisfactory result, sometimes dull, sometimes fantastic, sometimes floating in empty generalities, which we have so often to listen to and read on the proper conduct of war because a spirit of scientific investigation has as yet rarely dealt with the subject.

CHAPTER 2

On the Theory of War

1. AT FIRST BY THE ART OF WAR WAS UNDERSTOOD ONLY THE PREPARATION OF THE ARMED FORCES

Formerly by the terms "art of war" or "science of war" nothing was understood but the totality of those branches of knowledge and skill which are concerned with material things. The contrivance and preparation and use of weapons, the construction of fortifications and entrenchments, the organization of the army and the mechanism of its movements were the subject of these branches of knowledge and skill, and the object of them all was the description of an armed force suitable for war. Here one had to do with concrete things and with a one-sided activity, which at bottom was nothing but an activity gradually rising from manual work to a refined mechanical art. The relation of all this to combat was very much that of the art of the sword cutler to that of the fencer. Of the employment of it in the moment of danger and in a constant state of reciprocal action, or of the actual movements of thought and courage in the direction prescribed to them, there was as yet no question.

2. WAR FIRST APPEARS IN THE ART OF SIEGES

In the art of sieges was first to be seen something of the conduct of war itself, something of the movement of the mind to which these material things are entrusted, but for the most part only in so far as that mind quickly embodied itself in new material objects like approaches, trenches, counter-approaches, batteries, etc., and as each step it took showed itself in such a result. It was only the thread that was needed whereon to string these material creations. Since in this sort of war it is almost in these things alone that the mind finds expression, this kind of approach was then more or less adequate.

3. THEN TACTICS TRIED TO FIND ITS WAY IN THE SAME DIRECTION

Later on tactics tried to impose upon the mechanism of its combinations the character of an arrangement universally valid and founded on the peculiar properties of the instrument. This certainly leads to the battlefield, but not to a free activity of the mind. On the contrary, with an army reduced to an automaton by rigid formation and order of battle and put in motion by the mere word of command, its activity was intended to proceed like clockwork.

4. THE REAL CONDUCT OF WAR ONLY MADE ITS APPEARANCE INCIDENTALLY AND INCOGNITO

The conduct of war, properly so called, the free employment of available means previously prepared—free, that is to say, in the sense of being adaptable to the most particular needs—could not, it was thought, be a subject for theory, but must be left to natural talents alone. By degrees, as war passed from the hand-to-hand encounters of the Middle Ages into a more regular and composite form, stray reflections on this matter did also, it is true, obtrude themselves upon the minds of men, but for the most part they appeared only in memoirs and narratives, incidentally and, as it were, incognito.

5. REFLECTIONS ON MILITARY EVENTS MADE THE WANT OF A THEORY FELT

As these reflections became more and more numerous and history assumed a more and more critical character, urgent need arose for a basis of principles and rules, whereby the controversy which had so naturally arisen in military history, the conflict of opinions, could be brought to some kind of an end. This whirl of opinions, revolving round no central point and obeying no perceptible laws, could not but be very distasteful to the human mind.

6. ENDEAVORS TO ESTABLISH A POSITIVE THEORY

There arose, therefore, an endeavor to establish principles, rules and even systems for the conduct of war. Thus a positive end was set up, without keeping properly in view the innumerable difficulties which the conduct of war presents in this connection. The conduct of war has, as we have shown, no fixed limits in any direction. Every system, every theoretical construction, however, possesses the limiting nature of a synthesis, and the result is an irreconcilable opposition between such a theory and practice.

7. LIMITATION TO MATERIAL OBJECTS

Writers on theory felt the difficulty of the subject soon enough and considered themselves justified in avoiding it once more by confining their principles and systems to material things and a one-sided activity. They wished, as in the sciences dealing with the preparation for war, to arrive at perfectly certain and positive results and thus, therefore, only to take into consideration what could be made a matter of calculation.

8. SUPERIORITY IN NUMBERS

Superiority in numbers, being a material thing, was chosen from all other factors in the product of victory, because by combinations of time and space it could be brought under mathematical laws. It was thought possible to consider it apart from all other circumstances, by supposing these to be equal on both sides and, consequently, to neutralize one another. That would have been quite right if the intention had been to do so temporarily for the purpose of studying this one factor in its relations; but to do so permanently—to consider superiority in numbers as the sole law, and to see the whole secret of war in the formula: *to bring up superior numbers to a certain point at a certain time*—was a restriction absolutely untenable against the force of reality.

9. SUBSISTENCE OF THE TROOPS

In one theoretical treatment the attempt was made to systematize yet another material element, making the subsistence of the troops, based on a certain assumed organic character of the army, the supreme arbiter in the higher conduct of war.

In this way definite figures were certainly arrived at, but figures which rested on a multitude of quite arbitrary assumptions and which could not stand the test of experience.

1O. BASE

An ingenious author tried to concentrate in a single conception, that of a *base*, a whole host of things, among which even some relations with mental and moral forces found their way in as well. The list comprised the subsistence of the army, the keeping up of its numbers and equipment, the security of its communications with the home country, and finally the security of retreat in case it should become necessary. First, he tried to substitute this conception of a base for all these separate functions, and then again for the base itself to substitute the magnitude of the base, and finally the angle which the armed forces make with this base; and all this merely in order to arrive at a purely geometrical result, which is quite worthless. This last is, in fact, unavoidable if we reflect that none of these substitutions could be made without doing violence to the truth and leaving out some of the things which were still included in the earlier conception. The conception of a base is for strategy a real need, and to have conceived it is meritorious; but to make such a use of it as we have indicated is quite inadmissible and could only lead to one-sided conclusions, which have forced these theorists even into an absolutely absurd direction, to a belief, namely, in the superior efficacy of the enveloping form of attack.

11. INTERIOR LINES

As a reaction against this false tendency, another geometrical principle, namely, that of the so-called interior lines, has been elevated to the throne. This principle rests on a sound foundation, on the truth that the engagement is the only effectual means in war, but, nevertheless, just on account of its purely geometrical nature, it is nothing but another instance of one-sided theory which was never able to govern real life.

12. ALL THESE ATTEMPTS ARE OBJECTIONABLE

All these attempts at theory can be regarded as advances in the domain of truth only in so far as they are analytical; in so far as they are synthetical, they must in their precepts and rules be regarded as utterly useless.

They strive after determinate quantities, while in war all is undetermined, and the calculation has to be made with quantities entirely variable.

They direct the attention only upon material quantities, while military action is permeated throughout by immaterial forces and effects.

They consider the action on one side only, while war is a constant reciprocal action, between the one side and the other.

13. THEIR RULES TAKE NO ACCOUNT OF GENIUS

All that was unattainable by this paltry wisdom of a treatment that neglected all elements but one lay outside the precincts of science. It was, according to them, the field of genius, which *raises itself above all rules.*

Alas for the warrior who was to crawl about in this beggarly realm of rules, which are too bad for genius, over which it can set itself as superior and over which it can also at all events make merry! What genius does, must be the finest of all rules, and theory cannot do better than show how and why it is so.

Alas for the theory, which sets itself in opposition to mental and moral forces! It cannot make up for this contradiction by any humility, and the humbler it is, the sooner will ridicule and contempt drive it out of real life.

14. THE DIFFICULTY OF THEORY AS SOON AS MENTAL AND MORAL QUANTITIES COME INTO CONSIDERATION

Every theory becomes infinitely more difficult from the moment that it touches on the province of mental and moral quantities. Architecture and painting know exactly where they stand so long as they have only to deal with matter; there is no dispute about mechanical and optical construction. But as soon as mental and moral effects begin to operate, as soon as mental and moral impressions and feelings are to be produced, the whole set of rules dissolves into vague ideas.

The art of medicine, for the most part, deals only with physical phenomena; it has to do with the animal organism, which is subject to perpetual changes and is never quite the same for two moments. This makes its task very difficult and places the judgment of the physician above his knowledge; but how much more difficult is the case if a mental and moral effect comes in as well, and how much higher do we set the physician of the soul!

15. THE MENTAL AND MORAL QUANTITIES
CANNOT BE EXCLUDED IN WAR

But now the activity in war is never directed solely against matter; it is always at the same time directed against the mental and moral force which gives life to this matter, and to separate the one from the other is impossible.

But the mental and moral quantities are only visible to the inner eye, and this is different in each person, and often different in the same person at different times.

As danger is the general element in which everything moves in war, it is chiefly courage, the feeling of our own strength, that influences our judgment in different ways. It is, so to speak, the crystal lens through which all images pass before reaching the intelligence.

And yet we cannot doubt that these things must acquire a certain objective value, if merely through being experienced.

Everyone knows the moral effects of a surprise, of an attack in flank or rear. Everyone thinks less of the enemy's courage as soon as he turns his back, and everyone ventures much more in pursuit than when pursued. Everyone judges his opponent by his reputed talents, by his age and experience, and acts accordingly. Everyone casts a critical glance at the spirit and morale of his own and the enemy's troops. All these and similar effects in the province of man's mental and moral nature have been proved by experience and are constantly recurring. They therefore warrant our reckoning them in their kind as real quantities. And what should become of a theory which wanted to leave them out of consideration?

But these truths certainly need to be authenticated by experience. No theory, no general, should have anything to do with psychological and philosophical sophistries.

16. PRINCIPAL DIFFICULTY OF A THEORY
OF THE CONDUCT OF WAR

In order to comprehend clearly the difficulty of the problem involved in a theory of the conduct of war, and thence to deduce the necessary character of such a theory, we must take a closer view of the chief characteristics which make up the nature of military action.

17. FIRST CHARACTERISTIC: MENTAL AND MORAL FORCES AND EFFECTS

(Hostile Feeling)

The first of these characteristics consists of the mental and moral forces and effects.

Combat is in its origin the expression of *hostile feeling*, but in our great combats, which we call wars, the hostile feeling certainly often becomes merely a hostile *intention*, and there is usually at least no hostile feeling of individual against individual. Nevertheless, the combat never comes off without such feelings becoming active. National hatred, which is seldom lacking in our wars, becomes a more or less powerful substitute for personal hostility of individual against individual. But where this also is wanting, and, at first, no animosity existed, a hostile feeling is kindled by the combat itself. An act of violence which anyone commits upon us by order of his superior will excite in us the desire to retaliate and be revenged on him sooner than on the superior power at whose command the act was done. This is human—animal, if you will—but it is a fact. In theory we are very apt to look upon the combat as an abstract trial of strength, an isolated phenomenon in which the feelings have no part. This is one of the thousand errors which theories deliberately commit, because they never see the consequences of them.

Besides that excitation of feelings arising from the nature of the combat itself there are others also which do not essentially belong to it, but from their kindred nature easily unite with it—ambition, desire to dominate, enthusiasms of every kind, and so forth.

18. THE IMPRESSIONS OF DANGER

(Courage)

Finally the combat gives birth to the element of danger, in which all the activities of war must live and move, like the bird in the air or the fish in the water. The effects of danger, however, all pass on to the emotions, either directly, and thus instinctively, or through the intelligence. The effect, in the first case, would be a desire to escape from the danger, and, if that cannot be done, fear and anxiety. If this effect does not take place, then it is *courage* which acts as a counterpoise to that instinct. Courage, however, is by no means an act of the intelligence, but likewise a feeling, like fear; the latter is directed to physical preservation, courage to moral preservation. Courage is a nobler instinct. But because it is so, it cannot be used like a lifeless instrument, which produces its effects in a degree precisely predetermined. Courage is, therefore, no mere counterpoise to danger in order to neutralize this in its effects, but a special quantity in itself.

19. EXTENT OF THE INFLUENCE OF DANGER

But in order to estimate correctly the influence of danger upon leaders in war, we must not limit its sphere to the physical danger of the moment. It dominates the leader, not only by threatening him personally, but also by threatening all those entrusted to him; not only at the moment in which it is actually present, but also through the imagination at all other moments related to the present; and lastly, not only directly by itself, but also indirectly by the responsibility which makes it bear with tenfold weight on the leader's mind. Who could advise, or resolve upon, a great battle, without feeling his mind more or less wrought up and paralyzed by the danger and responsibility which such a great act of decision carries in itself? We may say that action in war, in so far as it is real action and not mere endurance, is never quite out of the sphere of danger.

20. OTHER EMOTIONAL FORCES

If we consider these emotional forces that are excited by hostility and danger as peculiar to war, we do not, therefore, exclude from it all others that accompany man on his life's journey. Here too they will often enough find room. We may say, it is true, that in this serious duty of life, many a petty play of the passions is silenced; but that holds good only of those in the lower ranks, who, hurried from one state of exertion and danger to another, lose sight of all else in life, *become unused to deceit*, because it is of no avail with death, and so attain that soldierly simplicity of character which has always been the best and most characteristic quality of the military profession. In the higher ranks it is otherwise, for the higher a man's rank, the more he has to look about him. There interests arise on every side and a manifold activity of passions, good and evil. Envy and nobility of mind, pride and humility, anger and tenderness—all may appear as active forces in the great drama.

21. QUALITY OF MIND

The mental qualities of a leader, next to his moral qualities, are likewise of great importance. From an imaginative, extravagant, inexperienced mind other things may be expected than from a cool and powerful intellect.

22. FROM THE DIVERSITY OF MENTAL AND MORAL INDIVIDUALITIES ARISES THE DIVERSITY OF WAYS LEADING TO THE END IN VIEW

This great diversity in mental and moral individuality, the influence of which must be considered as principally felt in the higher ranks, because it increases as we go upward, is what principally produces the diversity of ways, noticed in the first book, of attaining our end. It is this also that gives to the play of probability and luck such an unequal share in determining the course of events.

23. SECOND QUALITY: QUICK REACTION

The second quality in a soldier is quick reaction and the reciprocal action that springs from it. We do not speak here of the difficulty of estimating such reaction, for that is included in the difficulty, already mentioned, of dealing with mental and moral qualities as quantities. What we have in mind is the fact that reciprocal action revolts against all regularity. The effect which any measure produces upon the enemy is the most individual of all the items that figure among the data for action. But every theory must keep to classes of phenomena and can never assimilate a really individual case; that must everywhere be left to judgment and talent. It is, therefore, natural that in a business such as war, which so frequently in its plan built on general circumstances, is upset by unexpected, individual events, more must generally be left to talent, and less use can be made in it of a theoretical *guide* than in any other.

24. THIRD QUALITY

Lastly, the great uncertainty of all data in war is a characteristic difficulty, because all action must be directed, to a certain extent, in a mere twilight, which in addition not unfrequently—like fog and moonlight—gives to things exaggerated size and a grotesque appearance.

What this feeble light denies to clear vision, talent must divine, or it must be left to luck. It is therefore once more talent or even the favor of fortune in which, for lack of objective knowledge, we must put our trust.

25. POSITIVE SYSTEM OF RULES IS IMPOSSIBLE

This being the nature of the subject we must admit that it would be a sheer impossibility by means of an edifice of positive rules to provide the art of war with a scaffolding, as it were, which should give the leader support on all sides. In all those cases in which he is thrown upon his talents, the leader would find himself outside of this edifice of rules and in opposition to it. However many-sided its construction might be, the same result would ensue of which we have already spoken. Talent and genius would act beyond the law, and theory would become an opposite to reality.

26. OPENINGS FOR THE POSSIBILITY OF A THEORY

(THE DIFFICULTIES ARE NOT EVERYWHERE EQUALLY GREAT)

Two ways out of this difficulty are open to us.

In the first place, what we have said of the nature of military action in general does not apply in the same manner to the action of all ranks. In the lower ranks the courage of self-sacrifice is more required, but the difficulties which the intelligence and the judgment meet are infinitely less. The field of events is much more confined. Ends and means are fewer in number. The data are more distinct, for the most part even contained in things actually visible. But the higher we ascend, the more the difficulties increase, until in the commander-in-chief they reach their climax, so that with him almost everything must be left to genius.

But also, a division of the subject, *according to the intrinsic nature of its elements,* shows that the difficulties are not everywhere the same, but diminish the more the effects manifest themselves in the material world, and increase the more they pass over into the mental and moral, and become motives that determine the will. On that account it is easier by theoretical rules to determine the order, plan and conduct of an engagement than the use to be made of the engagement itself. In the engagement physical weapons clash together, and though mental and moral elements cannot be absent, yet matter must be allowed its rights. But in the effects of the engagement, when the material results become motives, we have to do solely with mental and moral elements. In a word: it is much easier to offer a theory for *tactics* than for *strategy.*

27. THEORY MUST BE OF THE NATURE OF OBSERVATION, NOT OF RULES FOR ACTION

The second opening for the possibility of a theory is to take the point of view that it need not be a body of positive rules, that is to say, that it need not be a *guide* to action. Whenever an activity has, for the most part, continually to do with the same things, with the same ends and means, although with small differences and a corresponding variety of combinations, these things must be capable of becoming an object of observation by reason. Such observation, however, is the most essential part of every *theory* and quite properly lays claim to that name. It is an analytical investigation of the subject; it leads to an exact acquaintance with the subject, and if brought to bear on experience, which in our case would be military history, to a thorough familiarity with it. The more nearly it attains this last object, the more it passes over from the objective form of knowledge to the subjective form of power; and so much the more, therefore, will it prove effective in cases where the nature of the matter admits of no other decision than that of talent; it will have an effect upon talent itself. If theory investigates the things that make up war, if it separates more distinctly that which at first sight seems confused, if it explains fully the properties of the means, if it shows their probable effects, if it clearly defines the nature of the ends in view, if it sheds the light of a deliberate, critical observation over the whole field of war—then it has achieved the main object of its task. It then becomes a guide to whoever wishes to become familiar with war from books; it everywhere lights up for him the road, facilitates his progress, educates his judgment, and keeps him from going astray.

If an expert spends half his life in the endeavor to clear up an obscure subject in all its details, he will probably know more about it than a person who seeks to master it in a short time. Theory, therefore, exists in order that each person need not have to clear the ground and toil through it afresh, but may find it cleared and put in order. It should educate the mind of the future leader in war, or rather guide him in his self-instruction, but not accompany him to the field of battle. Just as a sensible tutor guides and helps a youth's intellectual development without, on that account, keeping him in leading-strings for all the rest of his life.

If principles and rules result of themselves from the observations that theory institutes, if the truth of itself crystallizes into these forms, then theory will not oppose this natural law of the mind. It will rather, if the arch ends in such a keystone, bring it out more prominently; but it does this only to satisfy the philosophical law of thought, in order to show distinctly the point to which the lines all converge, not in order to construct from it an algebraical formula for use upon the battlefield. For even these principles and rules serve more to determine in the reflecting mind the leading outlines of its accustomed movements than, like signposts, to point the way for it to take in execution.

28. WITH THIS POINT OF VIEW THEORY BECOMES POSSIBLE AND CEASES TO BE IN CONTRADICTION TO PRACTICE

This point of view makes possible a satisfactory theory of the conduct of war, that is to say, a theory which will be useful and never in opposition to reality, and it will only depend on intelligent handling so completely to reconcile it with practice that between theory and practice there will no longer be that absurd difference, which an unintelligent theory, divorced from sound common sense, has often produced, but which narrow-mindedness and ignorance have just as often used as a pretext for letting themselves continue in their congenital ineptitude.

29. THEORY, THEREFORE, CONSIDERS THE NATURE OF ENDS AND MEANS. END AND MEANS IN TACTICS

Theory has, therefore, to consider the nature of means and ends.

In tactics the means are the trained armed forces which are to carry on the combat. The end is victory. How this idea can be more accurately defined will be better explained later on in considering the engagement. Here we content ourselves by specifying the withdrawal of the enemy from the field of battle as the sign of victory. By means of this victory strategy gains the object which it assigned to the engagement and which constitutes its real significance. This significance has indeed a certain influence on the nature of the victory. A victory which is intended to weaken the enemy's forces is a different thing from one which is designed merely to put us in possession of a position. The significance of an engagement can, therefore, have a noticeable influence on the planning and conduct of it; consequently it will also be a subject of consideration in tactics.

30. CIRCUMSTANCES WHICH ALWAYS ACCOMPANY THE EMPLOYMENT OF THE MEANS

As there are certain circumstances which constantly accompany the engagement and have more or less influence upon it, these must also be taken into consideration in the employment of the armed forces.

These circumstances are the locality of the engagement (terrain), the time of day and the weather.

31. LOCALITY

Locality, which we prefer to reduce to the idea of country and ground could, strictly speaking, be without any influence at all if the engagement took place on a completely level and uncultivated plain.

In districts consisting of steppes such a case might occur, but in the cultivated districts of Europe it is almost a figment of the imagination. Therefore between civilized nations an engagement on which country and ground have no influence is scarcely conceivable.

32. TIME OF DAY

The time of day influences the engagement by the difference between day and night; but the influence naturally extends further than merely to the limits of these divisions, as every engagement has a certain duration, and great battles a duration of many hours. In the planning of a great battle, it makes an essential difference whether it begins in the morning or the afternoon. Many battles, however, are, of course, fought in which the question of the time of day is quite immaterial, and in the majority of cases its influence is but trifling.

33. WEATHER

Still more rarely has the weather any decisive influence, and it is mostly only through fogs that it plays a part.

34. ENDS AND MEANS IN STRATEGY

For strategy the victory, that is the tactical success, is primarily only a means, and the things which should lead directly to peace are its ultimate object. The employment of its means to this object is likewise accompanied by circumstances which have more or less influence on it.

35. CIRCUMSTANCES WHICH ATTEND THE EMPLOYMENT OF THE MEANS OF STRATEGY

These circumstances are country and ground, the former including the territory and inhabitants of the whole theater of war; next the time of day, and the time of the year as well; lastly, the weather, particularly any unusual state of it, severe frost, etc.

36. THESE FORM NEW MEANS

By combining these things with the result of an engagement, strategy gives this result—and therefore the engagement—a special significance, *assigns to it a special object.* But in so far as this object is not that which is to lead directly to peace and is thus a subordinate one, it is also to be regarded as a means. Thus in strategy we may regard successful engagements, or victories, with all their different significances, as means. The conquest of a position is such a success of an engagement applied to terrain. But it is not only the different engagements with their special ends that are to be regarded as means. Whenever a deeper insight shows itself in the combination of the engagements to secure a common end, that also is to be regarded as a means. A winter campaign is such a combination applied to the time of year.

There are left over, therefore, as objects only those things which are conceived as leading *directly* to peace. Theory investigates all these ends and means according to the nature of their effects and interrelations.

37. THE ENDS AND MEANS TO BE INVESTIGATED STRATEGY TAKES ONLY FROM EXPERIENCE

The first question is: How does strategy arrive at an exhaustive enumeration of these things? If a philosophical investigation were to lead to an absolute result, it would tangle itself up in all the difficulties which exclude logical necessity from the conduct of war and the theory of it. So it turns to experience and directs its attention to those precedents which military history already has to show. In this way it will certainly be a limited theory, which only fits the circumstances as military history presents them. But this limitation is from the first inevitable, because, in every case, what theory says of things, it must either have abstracted from military history, or at all events compared with that history. Besides, such limitation is in any case more theoretical than real.

One great advantage of this method is that theory cannot lose itself in subtleties, hair-splittings and chimeras, but must remain practical.

38. HOW FAR THE ANALYSIS OF THE MEANS SHOULD BE CARRIED

Another question is: How far should theory go in its analysis of the means? Obviously, only so far as the different components present themselves for consideration in use. The range and effect of different weapons are very important to tactics; their construction, although these effects result from it, is a matter of complete indifference. For the conduct of war is not the production of powder and cannon out of a given quantity of charcoal, sulphur and saltpeter, of copper and tin; the given quantities for the conduct of war are arms in a finished state and their effects. Strategy makes use of maps without troubling itself about triangulations; it does not inquire what the institutions of a country must be and how a people must be educated and governed to give the best results in war, but it takes these things as it finds them in the community of European states and points out where very different conditions have a notable influence on war.

39. GREAT SIMPLIFICATION OF THE KNOWLEDGE REQUIRED

That in this manner the number of subjects for theory is greatly simplified and the knowledge required for the conduct of war greatly limited is easy to see. The very numerous varieties of expert knowledge and skill which minister to military activity in general, and which are necessary before a fully equipped army can take the field, coalesce into a few major groups before they arrive at the point of attaining in war the final end of their activity, as the streams of a country unite into rivers before they fall into the sea. Only with these activities which pour directly into the sea of war has the student who wishes to direct their course to make himself familiar.

40. THIS EXPLAINS HOW QUICKLY GREAT GENERALS ARE FORMED AND WHY THE GENERAL IS NOT A MAN OF LEARNING

This result of our investigation is in fact so necessary that any other could not but have made us distrustful of its accuracy. Only thus is explained how so often men have made their appearance with great success in war and indeed in the higher ranks, even in supreme command, whose pursuits had previously been of a totally different nature; how indeed the most distinguished generals have never risen from the very learned, or really erudite, class of officers, but have been mostly men who, from the circumstances of their position, could not have attained any great amount of knowledge. On that account those who have considered it necessary, or even merely useful, to begin the education of a future general by instruction in all details have always been ridiculed as absurd pedants. It is easy to show that such a course will do him harm, because the human mind is formed by the kinds of knowledge imparted to it and the direction given to its ideas. Only what is great can make it great; the little can only make it little, if the mind itself does not reject it as something repugnant to it.

41. FORMER CONTRADICTION

Because this simplicity of the knowledge required in war was disregarded, while this knowledge was jumbled up with the whole ruck of subordinate kinds of knowledge and skill that minister to it, the obvious contradiction into which it fell with the facts of the real world could only be reconciled by ascribing everything to genius, which needs no theory and for which theory was not supposed to be written.

42. FOR THIS REASON ALL USE OF KNOWLEDGE WAS DENIED, AND EVERYTHING ASCRIBED TO NATURAL TALENTS

People with whom common sense prevailed realized what an enormous distance remained to be filled up between a genius of the highest order and a learned pedant. They became in a way free-thinkers, rejected all belief in theory, and held the conduct of war to be a natural function of man, which he performs more or less well according as he has brought with him into the world more or less talent in that direction. It cannot be denied that these were nearer to the truth than those who attached value to a false knowledge; at the same time it is easy to see that such a view is itself but an exaggeration. No activity of the human intelligence is possible without a certain abundance of ideas, but these, at all events for the most part, are not innate but acquired, and make up its knowledge. The only question, therefore, is of what kind these ideas should be; and we think we have answered it when we say that for war they should be directed on those things with which man in war is immediately concerned.

43. THE KNOWLEDGE REQUIRED MUST VARY WITH THE RANK

Inside this field of military activity, the knowledge required must be different according to the position of the leader. It has to be directed on less important and more limited objects if he holds an inferior position, upon greater and more comprehensive, if his position is higher. There are commanders-in-chief who would not have shone at the head of a cavalry regiment, and *vice versa*.

44. THE KNOWLEDGE IN WAR IS VERY SIMPLE, BUT NOT, AT THE SAME TIME, VERY EASY

But although the knowledge in war is very simple, that is to say, directed to so few subjects, and embracing these only in their final results, to put it into practice is not, at the same time, very easy. Of the difficulties to which action in war is subject generally, we have already spoken in Book I; we here pass over those which can only be overcome by courage, and we maintain that the proper activity of the intelligence is also only simple and easy in the lower positions, but increases in difficulty with increase of rank, and in the highest position, that of commander-in-chief, is to be reckoned among the most difficult things of which the mind of man is capable.

45. THE NATURE OF THIS KNOWLEDGE

The commander of an army need not be either a learned student of history or a publicist, but he must be familiar with the higher affairs of state; he must know and be able to judge correctly of traditional tendencies, the interests at stake, the questions at issue, and the leading personalities. He need not be a subtle observer of men, a delicate dissector of human character, but he must know the character, the ways of thinking and habits, and the characteristic strong and weak points of those whom he is to command. He need not understand anything of the construction of a wagon, or of the harnessing of a battery horse, but he must know how to calculate accurately the march of a column, under different circumstances, according to the time it requires. These are kinds of knowledge which cannot be extorted by an apparatus of scientific formulas and machinery; they are only to be gained by the exercise of an accurate judgment in the observation of things in life and of a special talent for their apprehension.

The necessary knowledge for a high position in military activity is thus distinguished by the fact that by observation, that is, by study and reflection, it can only be gained through a special talent, which as an intellectual instinct knows how to extract from the phenomena of life only their essence, as bees do the honey from the flowers. This instinct can also be gained by experience of life as well as by study and reflection. Life will never with its rich teaching produce a Newton or an Euler, but it may well produce the higher powers of calculation possessed by a Condé or a Frederick.

It is, therefore, not necessary that, in order to vindicate the intellectual dignity of military activity, we should resort to untruth and silly pedantry. There never has been a great and distinguished commander of mean intelligence, but very numerous are the instances of men who, after serving with the greatest distinction in inferior positions, remained below mediocrity in the highest, from insufficiency of intellectual capacity. That even among those holding the position of commander-in-chief a distinction can be made, according to the degree of their authority, is a matter of course.

46. THEORETICAL KNOWLEDGE MUST BECOME PRACTICAL SKILL

Now we have yet to consider one condition which for the knowledge of the conduct of war is more necessary than for any other, which is that it must become wholly part of oneself and almost wholly cease to be something objective. In almost all other arts and occupations in life the person acting can make use of truths which he has only once learned, but in the spirit and sense of which he no longer lives, and which he extracts from dusty books. Even truths which he daily handles and uses may remain something quite external to himself. If the architect takes up a pen to determine by a complicated calculation the strength of a buttress, the truth found as a result is no emanation from his own mind. First he has had laboriously to find the data and then submit these to a mental operation, the rule for which he did not discover and the necessity of which at the moment he is only partly conscious of, but which he applies, for the most part, mechanically. But it is never so in war. The mental reaction, the ever-changing form of things, makes it necessary for the person acting to carry in himself the whole mental apparatus of his knowledge and be able, anywhere and at any moment, to produce from himself the decision required. The knowledge must, therefore, by being thus completely assimilated with his own mind and life, be transformed into a real skill. That is the reason why with men distinguished in war, it seems so easy, and why everything is ascribed to natural talent; we say natural talent to distinguish it from that which is formed and matured by observation and study.

We think that by these reflections we have explained the problem of a theory of the conduct of war, and indicated how it may be solved.

Of the two fields into which we have divided the conduct of war, tactics and strategy, the theory of the latter contains unquestionably, as before observed, the greater difficulties, because the first is almost entirely limited to a circumscribed field of things, but the latter, in respect of the objects leading directly to peace, opens into an undefined region of possibilities. But since essentially it is only the commander-in-chief who has to keep these ends in view, the part of strategy in which he moves is also particularly subject to this difficulty.

Theory, therefore, in strategy, and especially where it embraces the highest achievements, will stop much sooner than it does in tactics at the mere consideration of things and content itself with helping the soldier to that insight into things which, blended with his whole thought, makes his course easier and surer, and never forces him into opposition with himself in order to obey an objective truth.

CHAPTER 3

ART OF WAR OR SCIENCE OF WAR

1. USAGE IS STILL UNSETTLED

(PRACTICAL SKILL AND THEORETICAL KNOWLEDGE. SCIENCE, WHEN MERE KNOWLEDGE, ART, WHEN PRACTICAL SKILL, IS THE OBJECT.)

The choice between these terms seems to be still unsettled, and no one seems really to know on what grounds it is to be decided, simple though the matter is. We have already said elsewhere that knowledge is something different from practical skill. The two are so different that they should not easily be mistaken for each other. Practical skill cannot properly be contained in a book, and therefore "art"[1] should never be in the title of a book. But because we have accustomed ourselves to lump together the branches of knowledge required for the practice of an art (which branches may separately be pure sciences) under the name of "theory of art," or simply "art," it is consistent to maintain this ground of distinction and call everything art when a creative skill is the object—the art, for instance of building; and science when mere knowledge is the object—as in mathematics, for instance, and astronomy. That in every theory of art separate, entire sciences may be included is obvious and should not confuse us. But it is

---

[1] In German, *Kunst* (art) belongs to *können* (to be able).—Ed.

further worth noting that there also is no science quite without art. In mathematics, for example, the use of arithmetic and algebra is an art, but that is only one of many instances. The reason is that, however plain and perceptible the difference may be between knowledge and skill in the compounds which result from the combination of different branches of human knowledge, yet it is difficult in man himself clearly to trace the line of demarcation between them.

2. DIFFICULTY OF SEPARATING COGNITION FROM JUDGMENT

(ART OF WAR)

All thinking is indeed art. Where the logician draws the line, where the premises, which are the results of cognition, stop and judgment begins, there art begins. But more than this: even cognition by the mind is again a judgment and consequently art, and, finally, so too is cognition by the senses. In a word, if it is as impossible to imagine a human being possessing merely the faculty of cognition without judgment as it is to imagine the reverse, art and knowledge can never be completely separated from each other. The more these subtle elements of light embody themselves in the outward forms of the world, the more widely separated their realms become; and now, once more: where creation and production is the object, there is the domain of art; where investigation and knowledge is the goal, there science reigns. After all this, it is obvious that it is more fitting to speak of "art of war" than of "science of war."

So much for this, because we cannot do without these conceptions. But now we come forward with the assertion that war is neither an art nor a science in the proper sense, and that it is just the setting out from that starting-point of ideas which has led to a false direction being taken, and which has caused war to be put on a par with other arts and sciences, and has led to a host of erroneous analogies.

This has indeed been felt before now, and on that account it was maintained that war is a handicraft; but there was more lost than gained by that, for a handicraft is only an inferior art, and, as such, subject to more definite and rigid laws. In point of fact, the art of war did go on for some time in the spirit of a handicraft, namely, in the days of the *condottieri*. But it took this direction not for internal but external reasons, and military history shows how unnatural and unsatisfactory it was.

3. WAR IS A FORM OF HUMAN INTERCOURSE

We say therefore that war belongs not to the province of the arts and sciences but to that of social existence. It is a conflict of great interests which is settled by bloodshed, and only in that is it different from other conflicts. It would be better, instead of comparing it to any art, to compare it to trade, which is also a conflict of human interests and activities; and it is much more like politics, which again, for its part, may be regarded as a kind of trade on a large scale. Furthermore, politics is the womb in which war is developed, in which its outlines lie hidden in a rudimentary state, like the qualities of living creatures in their embryos.

4. DIFFERENCE

The essential difference consists in this: that war is an activity of the will exerted not, like the mechanical arts, upon dead matter, nor, like the human mind and emotions, in the fine arts, upon a living, but still passive and yielding object, but upon a living and reacting object. How little the categories of arts and sciences are applicable to such an activity strikes us at once; and we can understand at the same time how the constant seeking and striving after laws, similar to those which can be evolved from the dead world of matter, could not but lead to constant errors. And yet it is just the mechanical arts that people have wanted to take as a model for constructing an art of war. To take the fine arts as such a model was out of the question, because these themselves still too sorely lack laws and rules, and those hitherto tried have invariably been acknowledged as insufficient and narrow, and been perpetually undermined and washed away by the current of opinions, feelings and customs.

Whether such a conflict of living elements as arises and is settled in war is subject to general laws, and whether these can provide a useful guide to action, will be partly investigated in this book. But this much is self-evident: that this, like every other subject which does not exceed our powers of understanding, may be lighted up and made more or less clear in its inner relations by an inquiring mind; and that alone is sufficient to realize the idea of a *theory*.

CHAPTER 4

METHODISM

To explain clearly the idea of method and of what we shall call "methodism," which plays so great a part in war, we must be allowed to cast a hasty glance at the logical hierarchy by means of which, as if by regularly constituted official authorities, the world of action is governed.

Law, the most general conception valid both for cognition and action, has plainly something subjective and arbitrary in its literal meaning, and nevertheless expresses exactly that on which we and the things external to us are dependent. As a subject of cognition, *law* is the relation of things and their effects to one another; as a subject of the will, it is a determination of action, and is then equivalent to *command* or *prohibition.*

Principle is likewise such a law for action, except that it has not the formal, definite meaning that law has, but is only the spirit and sense of law, so as to leave the judgment more freedom of application when the diversity of the real world cannot be apprehended under the definite form of a law. As the judgment must itself find reasons to explain cases to which the principle is not applicable, the principle becomes in that way a real aid or guiding star for the person acting.

Principle is objective when it is the result of objective truth, and consequently of equal value for all men; it is subjective and then generally called a *maxim,* that is, a self-chosen rule of conduct, if there are subjective relations in it and if, therefore, it has a positive value only for the person who makes it for himself.

Rule is frequently taken in the sense of *law,* and then means the same as *principle,* for we say "no rule without exceptions," but we do not say "no law without exceptions," a sign that with *rule* we reserve to ourselves more freedom of application.

In another meaning *rule* is the means used to recognize a deeper-lying truth in a particular nearer-lying mark, in order to attach to this particular mark the law of action applying to the whole truth. Of this kind are all the rules for playing games, all abridged modes of procedure in mathematics, etc.

Regulations and *instructions* are determinations of action which deal with a number of minor circumstances, which would be too numerous and too insignificant for general laws, but which help to indicate the way more clearly.

Lastly *method*, mode of procedure, is a constantly recurring way of proceeding chosen out of several possible ways; and by methodism we mean the determining of action not by general principles or individual regulations, but by methods. When this is done, the cases dealt with by such a method must necessarily be assumed to be alike in their essential features. As they cannot all be alike, then the point is that at least as many as possible should be; in other words that method should be based on the most probable cases. Methodism is thus founded not upon definite, particular premises, but upon the average probability of analogous cases; and its ultimate tendency is to set up an average truth, the constant and uniform application of which soon acquires something of the nature of a mechanical skill, which in the end does the right thing almost unconsciously.

The idea of law in relation to cognition can, for the conduct of war, well be dispensed with, because the complex phenomena of war are not so regular, and the regular phenomena are not so complex, that we should gain anything more by this conception than by the simple truth. And where a simple conception and simple language are sufficient, to resort to the complex becomes affected and pedantic. The idea of law in relation to action cannot be used by the theory of the conduct of war, because owing to the variation and diversity of the phenomena there is in it no determination of such a general nature as to deserve the name of a law.

But principles, rules, regulations and methods are conceptions indispensable to a theory of the conduct of war, in so far as that theory leads to positive instruction, because in instruction the truth can only crystallize into such forms.

As tactics is the branch of the conduct of war in which theory can most often arrive at positive instruction, these conceptions will appear in it most frequently.

Not to use cavalry against unbroken infantry except in case of necessity, not to use firearms until the enemy is within their effective range, in an engagement to spare the forces as much as possible for the end—these are tactical principles. None of them can be applied absolutely in every case, but they must always be present to the mind of the commander, in order that the benefit of the truth contained in them may not be lost in cases where that truth can be applied to them.

If from the unusual cooking in the enemy's camp we infer that he is about to move, or if the intentional exposure of troops in an engagement indicates a feint attack, then this way of discerning the truth is called a rule, because from a single visible circumstance is inferred the purpose it serves.

If it is a rule to attack the enemy with renewed vigor as soon as he begins to limber up his artillery in the engagement, then with this particular fact is linked a course of action which is aimed at the general situation of the enemy as divined from it, namely, that he is about to break off the engagement, that he is beginning to draw off his troops, and is capable neither of a serious resistance while thus drawing off, nor, as when making his retreat, of successfully avoiding the enemy.

Regulations and *methods* are brought into the conduct of war by the theories of preparation for war, in so far as disciplined troops are inoculated with them as active principles. The whole body of instructions for formations, drill and field service are regulations and methods. In the drill instructions the former predominate, in the field-service instructions, the latter. To these things the real conduct of war is linked; it takes them over, therefore, as given modes of procedure, and as such they must appear in the theory of the conduct of war.

But for those activities retaining freedom in the employment of these forces there can be no regulations, that is, no definite instructions, just because they exclude freedom of action. Methods, on the other hand, are a general way of executing duties as they arise, based, as we have said, on average probability. As a governing body of principles and rules, carried through to application, they may certainly appear in the theory of the conduct of war, provided they are not represented as something different from what they are, not as the absolute and necessary binding laws of action (systems), but as the best of the general forms which can be used or suggested as shorter ways in place of individual decision.

The frequent application of methods will also be seen to be most essential and unavoidable in the conduct of war, if we reflect how many actions proceed on mere conjectures or in complete uncertainty. Measures in war must always be calculated on a certain number of possibilities. One side is prevented from learning all the circumstances which influence the dispositions of the other. Even if these circumstances which influence the decisions of the one were really known, there is not sufficient time for the other to carry out all the necessary counteracting measures, owing to their extent and complexity. There are numberless trifling circumstances belonging to any single event and requiring to be taken into account along with it, and there is no means of doing so but to infer the one from the other and base our arrangements only upon what is general and probable. Finally, owing to the number of officers increasing as we descend the scale of rank, the lower the sphere of action, the less must be left to the correct insight and trained judgment of the individual. When we reach those ranks where we can look for no other knowledge than that which service regulations and experience afford, we must meet them half way with routine methods bordering on those regulations. This will serve both as a support to their judgment and a barrier against those extravagant and erroneous views which are especially to be dreaded in a sphere where experience is so costly.

Aside from its indispensability, we must also recognize in methodism a positive advantage; which is that by practicing its constantly recurring forms, a readiness, precision and certainty are attained in the leading of the troops which diminish the natural friction and make the machine move more easily.

Method will therefore be the more generally used and become the more indispensable, the lower in rank are the persons acting; while upward its use will diminish until in the highest positions it entirely ceases. For this reason it is more in place in tactics than in strategy.

War in its highest aspects consists not of an infinite number of little events, which are analogous to one another in spite of their diversities and therefore by a better or worse method would be better or worse controlled, but of separate, great, decisive events which must be dealt with individually. It is not a field of stalks, which with a better or worse scythe is better or worse mown without regard to the shape of the single stalks, but it consists of great trees to which the axe must be laid with judgment, according to the particular nature and inclination of each separate trunk.

How far up the admissibility of methodism in military action extends is naturally determined not really according to rank but according to things; and it affects the highest positions in a lesser degree only because these positions have the most comprehensive subjects of activity. A permanent order of battle, a permanent formation of advance guards and outposts are examples of routine methods by which a general ties not only his subordinates' hands, but also for certain cases his own. They may, it is true, have been devised by himself and be adapted by him according to circumstances, but they may also be a subject of theory, in so far as they are based on the general characteristics of troops and weapons. On the other hand, any routine method for drawing up plans for a war or a campaign and delivering them ready-made, as if by a machine, would be absolutely worthless.

As long as there exists no tolerable theory, that is, no intelligent treatment of the conduct of war, methodism—routine methods—must encroach even on the higher spheres of activity, for the men who are employed in them have not always been able to educate themselves through study and through contact with the higher walks of life. In the unpractical and contradictory discussions of the theorists and critics they cannot find their way; their sound common sense spontaneously rejects them; and thus they bring no knowledge with them but that of experience. Consequently in those cases which admit of a free, individual treatment and require it, they also readily make use of the means that experience offers them, that is, they imitate the methods of procedure characteristic of the greatest generals, whereby what we have called methodism arises of itself. If we see Frederick the Great's generals always coming forward with the so-called oblique order of battle, the generals of the French Revolution always using turning movements with a long extended line of battle, and Bonaparte's lieutenants rushing to the attack with the bloody energy of concentrated masses, we recognize in the recurrence of the procedure what is obviously a borrowed method, and thus see that methodism may extend to regions bordering on the highest. Should an improved theory facilitate the study of the conduct of war, and educate the mind and judgment of the men who are rising to the higher commands, then methodism will also no longer reach so high, and so much of it as is to be considered indispensable will then at least be deduced from theory itself and not be the product of mere imitation. However excellently a great general does things, there is always something subjective in the way he does them; and if he has a certain manner, a good deal of his individuality is contained in it, which does not always agree with the individuality of the person who imitates that manner.

At the same time it would neither be possible nor right to banish subjective methodism, or manner, completely from the conduct of war; it is rather to be regarded as a manifestation of that influence which the general character of a war has upon its separate events, and which, if theory has not been able to foresee and take account of it can only be satisfied in this way. What is more natural than that the war of the French Revolution had its own way of doing things? And what theory could ever have included that peculiar method? The trouble is that such a manner, originating from a special case easily outlives its day, because it continues while circumstances imperceptibly change. That is what theory should prevent by lucid and rational criticism. When in the year 1806 the Prussian generals, Prince Louis at Saalfeld, Tauentzien on the Dornberg near Jena, Grawert before, and Rüchel behind Kappelldorf, all threw themselves into the open jaws of destruction in the oblique order of Frederick the Great, and managed to ruin Hohenlohe's army in a way that no other army has ever been ruined on the actual field of battle—all this was due not merely to a manner which had outlived its day, but to the most downright stupidity to which methodism has ever led.

CHAPTER 5

CRITICISM

The influence of theoretical truths upon practical life is always exerted more through criticism than through rules for practice. Criticism is the application to actual events of theoretical truth, and so not only brings the latter nearer to life but also accustoms the intelligence more to these truths through the constant repetition of their applications. Consequently we think it necessary, next to the point of view of theory, to establish that of criticism.

From the simple narration of a historical event which merely places things one by the side of the other and at most touches on their most immediate causal connections, we distinguish the *critical* narration.

In this *critical* narration, three different activities of the intelligence can be present.

First, there is the historical discovery and establishment of doubtful facts. This is historical investigation proper and has nothing in common with theory.

Second, there is the tracing of the effect from its causes. This is *critical investigation proper.* It is indispensable to theory, for everything that in theory is to be established, supported, or even only explained, through experience, can only be settled in this way.

Third, there is the testing of the means employed. This is *criticism proper,* which contains praise and blame. This is where theory is of service to history or rather, to the teaching to be derived from it.

In these two last, strictly critical, parts of historical study, everything depends on tracing things to their final elements, that is, to truths that are beyond doubt, and not, as so often happens, stopping half way at some arbitrary assumption or hypothesis, and going no further.

As respects the tracing of an effect to its causes, an insuperable external difficulty is often encountered in the true causes being quite unknown. In none of the circumstances of life does this occur so frequently as in war, where the events are seldom fully known and still less the motives, which are either purposely suppressed by the persons who acted on them, or, when they were of a very transient and accidental character, may also be lost to history. Consequently, critical narration must, for the most part, go hand in hand with historical investigation, and even so, there often remains such a disparity between cause and effect that history is not justified in regarding the effects as necessary consequences of the causes known. In this case, therefore, gaps must necessarily occur; that is to say, we get historical results from which no teaching can be extracted. All that theory can demand is that the investigation should be rigidly conducted up to these gaps and there suspend all its demands. Real trouble only arises if what is known has, at all costs, to suffice to explain the results, and thus a false importance is given to it.

Besides this difficulty, critical investigation also encounters another very serious intrinsic one in the fact that effects in war seldom proceed from one simple cause, but from several joint causes, and that therefore it is not enough in a candid and impartial spirit to trace back the series of events to its beginning, but it is then still necessary to assign to each of the contributing causes its due weight. This leads, therefore, to a closer investigation of their nature, and thus a critical investigation may lead us into what is the proper field of theory.

The critical *consideration*, that is, the examination of the means, leads to the question: Which are the effects peculiar to the means applied, and were these effects intended by the person acting?

The effects peculiar to the means lead to the investigation of their nature, and thus again into the field of theory.

We have seen that in criticism all depends upon attaining truths which are beyond doubt; that is to say, not stopping at arbitrary propositions which are not valid for others and to which other perhaps equally arbitrary assertions are then opposed, so that there is no end to *pros* and *cons* and the whole is without result and therefore without instruction.

We have seen that both the investigation of causes and the testing of means lead into the field of theory, that is, into the field of universal truth not derived solely from the individual case under examination. If there is a serviceable theory, critical investigation will appeal to what has there been settled, and at that point the investigation may stop. But where no such theoretical truth is to be found, the investigation must be pushed on to the ultimate elements. If this necessity occurs often, it must lead the historian into greater and greater detail. He then has his hands full, and it is almost impossible for him to dwell on every point with the deliberation required. The consequence is that, to set limits to his examination, he stops at arbitrary assertions, which though they would not really be arbitrary for him, yet remain so for everyone else, because they are neither self-evident nor demonstrated.

A serviceable theory is therefore an essential foundation for criticism, and without the assistance of a reasonable theory it is impossible for criticism to reach the point at which it chiefly begins to be instructive, that is to say, to be a demonstration convincing and *sans replique* (unanswerable).

But it would be a visionary hope to believe in the possibility of a theory which took care of every abstract truth and only left to criticism the task of putting the individual case under its appropriate law. It would be a ridiculous piece of pedantry to lay down as a rule for criticism that it must always halt and turn round on reaching the boundaries of sacred theory. The same spirit of analytical investigation which is the origin of theory must also guide the critic in his work; and it can and may, therefore, happen that he strays over into the domain of theory and goes on to elucidate for himself those points which are of particular importance to him. Criticism, on the contrary, may much more probably entirely fail to attain its object if it becomes a mechanical application of theory. All positive results of theoretical investigation, all principles, rules and methods are the more lacking in universality and absolute truth, the more they become positive rules for practice. They exist to offer themselves for use as required, and it must always be left for judgment to decide whether they are suitable or not. Such results of theory must never be used in criticism as standard rules or norms, but in the same way as the person acting is to use them, merely as aids to judgment. If in tactics it is a settled thing that, in the general order of battle, cavalry should be placed behind infantry and not in line with it, still it would be folly on that account to condemn every deviation from this arrangement. Criticism must investigate the reasons for the deviation, and it is only if these are inadequate that it has a right to appeal to what is established by theory. If, further, it is settled in theory that a divided attack diminishes the probability of success, it would be unreasonable, whenever a divided attack and an unsuccessful issue occur together, to regard the latter as the consequence of the former without further investigation as to whether that is really the case. And it would be equally unreasonable, when a divided attack is successful, to infer from it, on the contrary, the fallacy of what theory asserts. The investigating spirit of criticism refuses to allow either. Criticism, therefore, is based essentially upon the results of the analytical investigation accomplished by theory. What has been settled by theory does not need to be established afresh by criticism, and it is settled in theory in order that criticism may find it already established.

This task of criticism, to investigate what effect has been produced by a cause and whether a means employed has been the right one to attain its end, will be easy if cause and effect, end and means, lie near together.

If an army is surprised and therefore cannot make a regular and intelligent use of its powers and resources, the effect of the surprise is not doubtful. If theory has settled that in a battle an enveloping attack leads to a greater, but less certain, success, then the question is whether he who employs the enveloping attack has principally had in view the magnitude of the success as his object. In that case the means was rightly chosen. But if his desire was to make his success more certain, and if this expectation was based not on the particular circumstances but on the general nature of the enveloping attack, then he mistook the nature of that means and committed an error, as has happened a hundred times before.

Here the work of military investigation and testing is easy and it always will be so where we confine ourselves to the immediate effects and ends. We can do this exactly as we please, provided we regard things apart from their connection with the whole, and only look at them as thus abstracted.

But in war, as generally in the world, there is a connection between everything that belongs to a whole; and, in consequence, every cause, however small it may be, must in its effects influence the whole of the rest of the war, and modify in some degree the final result, however slight that degree may be. In like manner, every means must exert its influence up to the attainment of the final end.

We can, therefore, trace the effects of a cause as far as they are still worth noticing, and in like manner we can test a means not only for its immediate end, but also test this end itself as a means for a higher end, and thus ascend along the chain of ends, each subordinated to the one above it, until we come to one which requires no testing because its necessity is indubitable. In many cases, especially if it is a question of great and decisive measures, our examination will have to extend to the final end: that which is directly to bring about peace.

It is evident that in thus ascending, at every new station that we reach, we get a new point of view for the judgment, so that the very means which from the immediate point of view appears advantageous, when regarded from a higher one, must be rejected.

The search for the causes of phenomena and the testing of the means according to the ends they serve, must always in the critical consideration of a piece of history go hand in hand, for only the search for the cause brings us to the things which deserve to be made a subject of testing.

This attempt to follow the chain of causation up and down involves considerable difficulties, for the farther from an event the cause that we are seeking lies, the greater must be the number of other causes which must at the same time be kept in view, allowed for with reference to the share they may have had in shaping events, and eliminated; because the higher a phenomenon stands in the chain of causation, the more numerous are the separate forces and circumstances by which it is conditioned. If we have ascertained the causes of a lost battle, we have certainly also ascertained part of the causes of the consequences which this lost battle had for the whole war. But we only ascertain a part, for to the final result effects of other causes will, according to circumstances, more or less contribute.

The same multiplicity of things to be dealt with presents itself in the testing of the means, as our points of view become successively higher; for the higher the ends, the more numerous the means employed to attain them. The final end of the war is pursued by all the armies simultaneously, and we must therefore also take into consideration all that has been accomplished of this or could have been accomplished.

It is obvious that this may sometimes lead into a wide field of inquiry in which it is easy to lose our way and in which difficulty prevails, because a host of assumptions must be made about things which have not actually happened but which were probable and on that account cannot possibly be left out of our consideration.

When Bonaparte in 1797 at the head of the Army of Italy advanced from the Tagliamento against the Archduke Charles, he did so with the intention of forcing him to a decision before the reinforcements which the Archduke expected from the Rhine had reached him. If we look only at the immediate decision, the means were well chosen. And the result proved it, for the Archduke was still so weak that he made no more than an attempt at resistance on the Tagliamento. When he saw his adversary too strong and resolute, he abandoned to him the field of battle and the passes into the Norican Alps. Now what could Napoleon have been aiming at with this happy success? To penetrate into the heart of the Austrian Empire himself, to facilitate the advance of the Rhine armies under Moreau and Hoche and get into close communication with them? This was the position taken by Bonaparte, and from this point of view he was right. But if criticism now places itself at a higher point of view, namely, that of the French Directory, which body was able to see and must have seen that the campaign on the Rhine would not be opened till six weeks later, then the advance of Napoleon over the Norican Alps can only be regarded as an extravagant piece of bravado, for if the Austrians had drawn largely upon their Rhine armies to reinforce their army in Styria so as to enable the Archduke to fall upon the Army of Italy, not only would that army have been routed but the whole campaign lost. This consideration which forced itself upon Napoleon at Villach, induced him to sign the armistice of Leoben with so much readiness.

If the critic takes a still higher position and if he knows that the Austrians had no reserves between the army of the Archduke Charles and Vienna, then we see that Vienna became threatened by the advance of the Army of Italy.

Suppose that Bonaparte had known that the capital was thus uncovered and had also known that he still retained in Styria this decisive superiority in numbers over the Archduke; then his hurried advance against the heart of the Austrian States would no longer have been without purpose, for its value depended only on the value the Austrians set upon preserving Vienna. If that was so great that, rather than lose it, they would accept the conditions of peace which Bonaparte was ready to offer them, the threat to Vienna was to be regarded as his ultimate aim. If Bonaparte had for some reason known this, criticism can stop there, but if he were still uncertain about it, criticism must take a still higher position and ask what would have followed if the Austrians had abandoned Vienna and retired further into the vast stretches of their dominions still left to them. But it is easy to see that this question cannot possibly be answered without bringing into consideration the probable course of events between the Rhine armies on both sides. In view of the decided superiority of numbers on the side of the French—130,000 to 80,000—there could, it is true, have been little doubt as to the result, but then again the question arose: What use would the Directory make of a victory? Would they follow up their success to the opposite frontiers of the Austrian monarchy, to the complete breaking up, therefore, or overthrow of that power, or would they be satisfied with the conquest of a considerable portion to serve as a guaranty for peace? The probable result in each case must be estimated, in order to come to a conclusion as to the probable choice of the Directory. Suppose that the result of these considerations had been that the French forces would have been much too weak for the complete overthrow of the Austrian monarchy, so that the attempt would have of itself completely reversed the situation, and that even the conquest and occupation of a considerable part of it would have placed the French in a strategic position to which their forces were probably unequal; then that result was bound to influence their judgment of the position of the Army of Italy and induce that army to lower its expectations. And it was this, no doubt, that induced Bonaparte, even when he could see at a glance the helpless condition of the Archduke, to sign the peace of Campo Formio, which imposed no greater sacrifices on the Austrians than the loss of provinces which, even after the most fortunate of campaigns, they would not have reconquered. But the French could not even have reckoned on the moderate treaty of Campo Formio, and therefore could not have made it the object of their bold advance, if two questions had not had to be considered. The first was what value the Austrians would have attached to each of the above-mentioned results; whether, notwithstanding the probability of a satisfactory result for them in either of the two cases, it would have been worth their while to make the sacrifices involved in them, that is, in the continuation of the war, when they could be spared those sacrifices by a peace on terms not too humiliating. The second question was whether the Austrian Government would seriously weigh the final possible results of their continued resistance and would not allow themselves to be disheartened by the impression of their present reverses.

The consideration which forms the subject of the first question is no piece of idle subtlety but of such decisive practical importance that it comes up whenever a plan for pushing things to the last extremity is under discussion, and this is what most frequently prevents such plans from being put into execution.

The second question is just as necessary, for war is waged not with an abstract opponent, but with a real one, who must always be kept in view. And we may be sure that the bold Bonaparte was not unaware of this point of view, that is to say of the confidence which he placed in the terror his sword inspired. It was this same confidence that in 1812 led him to Moscow. There it left him in the lurch. The terror inspired by him had already somewhat worn off in the gigantic struggles in which he had been engaged. In 1797 it was, of course, still fresh, and the secret of the strength of a resistance pushed to the last extremity was not yet discovered. But none the less, even in 1797, his boldness would have led him to a negative result, if, as already said, with a sort of presentiment he had not chosen the moderate peace of Campo Formio as a way of escape.

Here we must bring this examination to a close. It will have sufficed to show by an example the wide range, the diversity and the difficulty which a critical examination may present, if we ascend to the ultimate ends, that is to say, if we speak of measures of an important and decisive character, the influence of which must necessarily extend to them. This examination will reveal that, besides a theoretical insight into the subject, natural talent must also have a great influence upon the value of a critical examination, for on this it will chiefly depend to throw light upon the connection of things, to distinguish those which are essential from the innumerable interrelations of events.

But talent will be required in yet another way. Critical consideration is not merely an examination of the means actually employed, but of all the means possible, which, therefore, must first be discovered and specified; and we are certainly not in a position to censure any particular means unless we are able to specify a better. Now however small the number of possible combinations may be in most cases, still it must be admitted that to point out those which have not been used is not a mere analysis of actual things, but a spontaneous creation which cannot be foreseen, depending upon the fertility of the mind.

We are far from seeing a field for great genius in a case in which everything can be traced back to a very few practically possible and very simple combinations. We find it exceedingly ridiculous, as so often has been done, to regard the turning of a position as a discovery indicating great genius, but none the less this act of spontaneous creation is necessary and the value of critical examination is essentially determined by it.

When Bonaparte on July 30, 1796, decided to raise the siege of Mantua in order to march against the advancing Wurmser and with his whole force to beat in detail his columns, separated by the Garda Lake and the Mincio, this appeared the surest way to the attainment of brilliant victories. These victories actually followed and were afterward again repeated with the same means and with still more brilliant success on the attempt to relieve the fortress being again renewed. We hear only one opinion on these achievements, that of unmixed admiration.

At the same time, Bonaparte could not adopt this course on July 30th without quite giving up the idea of the siege of Mantua, because it was impossible to save the siege train and this could not be replaced by another in this campaign. In fact the siege was converted into a mere blockade. The city, which, if the siege had continued, would have soon fallen, held out for six months in spite of Bonaparte's victories in the open field.

Criticism has generally regarded this as an evil that was quite unavoidable, because critics have not been able to suggest any better way of resistance. Resistance to a relieving army within lines of circumvallation had fallen into such disrepute and contempt that it appears to have entirely escaped consideration as a means. And yet in the time of Louis XIV that measure was so often used with success that it can only be called a freak of fashion if a hundred years later it did not occur to anyone that it might at least be taken into account with the rest. If this possibility had been admitted, a closer investigation of the circumstances would have shown that 40,000 of the best infantry in the world under Bonaparte, behind strong lines of circumvallation around Mantua, had so little to fear from the 50,000 Austrians coming to the relief under Wurmser that it was very unlikely that even so much as an attempt at an attack would have been made upon their lines. We shall not seek here to establish this point, but we believe enough has been said to show that this means was entitled to consideration. Whether in the action Bonaparte himself thought of this means, we do not wish to decide. No trace of it is to be found in his memoirs and the other printed sources. None of the later critics thought of it, because such a measure had passed entirely out of their field of vision. The merit of calling this means to mind is not great, for we have only to free ourselves from the arrogance of a freak of fashion to think of it. But nevertheless it is necessary to think of it in order to take it into account and compare it with the means which Bonaparte did employ. Whatever the result of the comparison may prove to be, it is a comparison that criticism should not omit.

When Bonaparte, in February, 1814, turned away from Blücher's army which he had defeated in the engagements of Etogues, Champ-Aubert, Montmirail, etc., so as to throw himself again upon Schwarzenberg and beat his troops at Montereau and Mormant, everyone was full of admiration because Bonaparte, just by throwing his concentrated force first upon one opponent and then upon another, made a brilliant use of the mistake made by the Allies in advancing with their forces divided. That these brilliant strokes in all directions failed to save him was generally considered to be at least no fault of his. No one has yet asked the question: What would the result have been if instead of turning again upon Schwarzenberg he had gone on hammering at Blücher and pursued him to the Rhine? We are convinced that a complete reversal of the campaign would have taken place and that the army of the Allies, instead of marching to Paris, would have retired behind the Rhine. We do not ask others to share our conviction, but when this alternative has once been mentioned no expert will doubt that criticism had to take notice of it with the rest.

In this case the means of comparison lay also much nearer than in the former. It has been equally overlooked because a biased tendency was blindly followed and there was no impartiality of judgment.

From the necessity of assigning a better means in place of one that has been condemned, a kind of criticism has arisen which is almost the only one in use, and which contents itself with merely pointing out a supposedly better procedure without adducing the real proof of it. The consequence is that some are not convinced, that others do exactly the same thing, and that then a controversy arises which affords no basis for discussion. All military literature teems with this sort of thing.

The proof we demand is always necessary when the advantage of the means put forward is not so evident as to leave no room for doubt, and it consists in investigating each of the two means on its own merits and comparing it with the end in view. If the matter has been traced back in this way to simple truths, the controversy must finally cease or at all events new results are obtained, while in the other way of proceeding the pros and the cons always completely destroy one another.

Should we, for example, not rest content with assertion in the case before mentioned and wish to prove that the persistent pursuit of Blücher would have been better than the turning on Schwarzenberg, we should rely on the following simple truths:

1. In general it is more advantageous to continue our blows in one and the same direction than to strike in different directions, because striking in different directions involves loss of time, and, furthermore, because when the moral force has already been weakened by considerable losses, fresh successes are easier to gain; in that way, therefore, no part of the preponderance already gained is left unused.

2. Because Blücher, although weaker than Schwarzenberg, was still, on account of his enterprising spirit, the more important adversary; in him, therefore lay the center of gravity, which draws everything else with it in its direction.

3. Because the losses that Blücher had sustained amounted to a defeat and had given Bonaparte such a preponderance over him that his retreat to the Rhine could scarcely be a matter of doubt, because on this line no reinforcements of any consequence existed.

4. Because no other possible success would have seemed so terrible or appeared to the imagination in such gigantic proportions; an immense advantage in dealing with an irresolute, timorous staff such as Schwarzenberg's notoriously was. What losses the Crown Prince of Württemberg had suffered at Montereau, and Count Wittgenstein at Mormant, Prince

Schwarzenberg must have known well enough. What sort of misfortunes, on the other hand, Blücher would have experienced on his entirely detached and disconnected line from the Marne to the Rhine would have only reached him by the avalanche of rumor. The desperate movement which Bonaparte made upon Vitry at the end of March, to try what sort of effect the threat of a strategic envelopment would have on the Allies, was obviously based on the principle of striking terror, but under quite different circumstances, after he had been defeated at Laon and Arcis, and Blücher with 100,000 men was at Schwarzenberg's side.

There are people, no doubt, who will not be convinced by these arguments, but at all events, they cannot retort by saying that "while Bonaparte threatened Schwarzenberg's base by advancing to the Rhine, Schwarzenberg at the same time threatened Paris, Bonaparte's base," because we wished to show by the reasons given above that Schwarzenberg would never have thought of marching on Paris.

With regard to the example quoted by us from the campaign of 1796 we should say: Bonaparte looked upon the plan which he adopted as the safest way to beat the Austrians. Even if it had been that, the object which would be attained by it would still be an empty military glory, which could scarcely have had a perceptible influence on the fall of Mantua. The way that we would have chosen was, in our opinion, much more certain to prevent the relief of Mantua; but even if we held that, as the French general thought, it was not so, and we preferred to look upon the certainty of success as slighter, the question would once more become one of balancing a more probable, but almost useless, and therefore slight, success in the one case, against a not altogether probable, but much greater, success in the other. If the matter is presented in this way, boldness would have had to declare itself in favor of the second solution, which is exactly the reverse of what a superficial view of the affair would lead us to believe. Bonaparte certainly did not have the less bold intention, and we may be sure that he had not made the nature of the case so clear to himself and realized its consequences as we have learned from experience to know them.

Naturally the critic, in considering the means, must often appeal to military history, as experience is of more value in the art of war than all philosophical truth. But this historical evidence is, no doubt, subject to its own conditions, of which we shall treat in a special chapter; and unfortunately these conditions are so seldom fulfilled that reference to history generally serves only to increase the confusion of ideas.

We have still a most important subject to consider, which is: how far criticism in passing judgments on a particular event is permitted, or in duty bound, to make use of its superior view of things and therefore of what the results have established, or when and where it is obliged to leave these things out of consideration in order to put itself exactly in the place of the person acting.

If criticism wants to praise or blame the person acting, it must certainly put itself exactly in his place; that is to say, it must collect all he knew and all the motives on which he acted, and on the other hand disregard all that he could not, or did not, know, that is, above all things, the result that followed. But that is only an end which we can strive toward but never quite attain, for the state of things out of which an event arises never lies before the eyes of the critic exactly as it did before those of the person acting. A multitude of minor circumstances which could have influenced his decision have been completely lost, and many a subjective motive has never been brought to light. Such motives can only be learned from the memoirs of the person acting or of his very intimate friends, and in such memoirs these things are often handled very vaguely or even purposely misrepresented. Criticism must, therefore, always forgo much of what was present to the mind of the person acting.

On the other hand, it is still harder for it to forgo what it knows too much of. This is only easy in respect of accidental circumstances—circumstances, that is to say, which were not necessarily related to the situation but have got mixed up with it. But in all essential matters it is extremely hard and never completely attainable.

Let us take the result first. If it did not proceed from accidental circumstances, it is almost impossible that the knowledge of it should not influence our judgment of the circumstances from which it did proceed, for we see these circumstances in the light of it, and, to some extent, only through it do we get our knowledge of them and our estimate of their importance. Military history with all its events is a source of instruction for criticism itself, and it is only natural that criticism should throw upon things the same light that it has derived from the consideration of the whole. If, therefore, in many cases it had to intend entirely to deny itself that light, it still would never wholly succeed in doing so.

But this happens not merely in respect of the result, that is, of what does not come in till later, but also in respect of what was already existing, that is, of the data which determine the action. Criticism will in most cases have more of these in its possession than had the person acting. Now it might be supposed that it was easy to rule them completely out, and yet it is not so. The knowledge of preceding and simultaneous circumstances rests not merely upon definite information, but upon a large number of conjectures or assumptions. There is, indeed, hardly any information respecting things not purely accidental, which has not been preceded by an assumption or conjecture, which will take the place of authentic information, if the latter remains missing. Now is it conceivable that criticism at a later time, which has before it as facts all the preceding and concurrent circumstances, should not allow itself to be thereby prejudiced when it asks itself what portion of the unknown circumstances it would have held to be probable at the moment of action? We maintain that in this case, as in the case of the results and for the same reason, it is impossible to disregard all these things completely.

If therefore the critic wishes to praise or blame any single act, he will only succeed to a certain degree in putting himself in the position of the person acting. In very many cases he will be able to do so to an extent sufficient for practical purposes, but in some cases not at all, a fact of which we must not lose sight.

But it is neither necessary nor desirable that criticism should completely identify itself with the person acting. In war, as in all things demanding skill, there is a certain natural aptitude required which we call virtuosity. This can be either great or small. In the first case it may easily be superior to that of the critic, for what critic would claim to possess the virtuosity of a Frederick or a Bonaparte! Therefore if criticism is not to abstain altogether from offering an opinion where eminent talent is concerned, it must be allowed to make use of the advantage which its enlarged horizon affords. Criticism cannot, therefore, check a great general's solution of his problem with the same data, like a sum in arithmetic, but must first, from the result, from the way he was invariably confirmed by events, recognize with admiration what was due to the higher activity of his genius, and first learn to see as established fact the essential connection which the glance of genius instinctly perceived.

But for even the smallest act of virtuosity it is necessary that criticism should take a higher point of view, in order that, richly provided as it is with objective reasons for decision, it should be as little subjective as possible, and that the limited mind of the critic should not make itself the standard by which to judge.

The superior position of criticism, its praise and blame pronounced in accordance with a complete knowledge of the circumstances, have in themselves nothing that offends our feelings; they only do so when the critic pushes himself forward and speaks in a tone as if all the wisdom which he had obtained by his exhaustive knowledge of the event under consideration were his own special talent. Gross as this deception may be, it is one that vanity readily perpetrates and that is naturally annoying to others. Still oftener, though no such arrogant self-exaltation is intended by the critic, it is ascribed to him by the reader unless he expressly guards himself against it, and then and there he is at once charged with lack of critical judgment.

Therefore when criticism points out an error of a Frederick or a Bonaparte, that does not mean that the critic would not have made it. He could, indeed, concede that in the place of these generals he might have made much greater errors, but these he knows from the general connection of events, and he demands from the sagacity of the general in question that he ought to have seen them.

This is, therefore, an opinion formed from the connection of events, and therefore also *from the result*. But there is another quite different effect of the result itself upon the judgment, that is, if it is used quite simply as evidence for or against the soundness of a measure. This may be called *judgment according to the result*. Such a judgment appears at first sight absolutely worthless, and yet again it is not absolutely so.

When Bonaparte marched to Moscow in 1812, everything depended on whether, through the taking of the capital and the events preceding it, he would be able to force the Emperor Alexander to make peace as he had done after the battle of Friedland in 1807, and as he had forced the Emperor Francis to do in 1805 and 1809 after Austerlitz and Wagram. For if Bonaparte did not obtain a peace at Moscow, there was no alternative for him but to return—that is, nothing but a strategic defeat. We shall leave out of the question what Bonaparte had done to reach Moscow and whether in his advance he had not missed many possible opportunities of inducing the Emperor Alexander to decide on peace. We shall also exclude all consideration of the disastrous circumstances which attended his retreat, and which had, perhaps, their origin in the general conduct of the campaign. The question will always remain the same, for however much more brilliant the result of the campaign up to Moscow might have been, still there always remained an uncertainty whether the Emperor Alexander would be frightened by it into making peace. And even if the retreat had contained within itself no such seeds of disaster, it could never have been anything else than a great strategic defeat. If the Emperor Alexander had agreed to a peace which was disadvantageous to him, the campaign would have ranked with those of Austerlitz, Friedland and Wagram. But these campaigns also, if they had not led to peace, would in all probability have ended in similar catastrophes. Whatever, therefore, of force, skill, and wisdom the conqueror of the world applied to the task, this last "question addressed to fate" remained always the same. Shall we then discard the campaigns of 1805, 1807, 1809, and say on account of the campaign of 1812 that they were acts of imprudence, that their success was against the nature of things, and that in 1812 strategic justice at last found vent for itself against blind fortune? That would be an unwarrantable conclusion, a very arbitrary judgment of which half the proof would necessarily remain lacking, because no human eye can trace the thread of the necessary connection of events up to the decision of the conquered princes.

Still less can we say that the campaign of 1812 deserved the same success as the others, and that the reason why it turned out otherwise lies in something that was unnatural, for we cannot regard the firmness of Alexander as such.

What can be more natural than to say that in the years 1805, 1807, 1809, Bonaparte judged his opponents correctly, and that in 1812 he blundered? On the former occasions, therefore, he was right, on the last, wrong, and in both cases we must admit that the justification for our opinion lies in the *result*.

All action in war, as we have already said, is aimed only at probable, not at certain results. Whatever is lacking in certainty must always be left to fate, or fortune, call it which you will. We may demand that what is so left should be as little as possible, but only in relation to the particular case— that is, as little as in this particular case is possible, but not that the case in which the uncertainty is the least should always have to be preferred. That would be an enormous error, as will follow from all our theoretical views. There are cases in which the greatest daring is the greatest wisdom.

Now in everything which must be left to chance by the person acting, his personal merit, and therefore his responsibility as well, seems to be completely set aside. Nevertheless we cannot suppress an inward feeling of satisfaction whenever our expectation is realized, and if it has been disappointed, we are conscious of a mental discomfort. No more than this is to be implied in our judgment of a measure as right or wrong, if we derive that judgment merely from the result of the measure, or rather, if we find it in that result.

But it cannot be denied that the satisfaction which success, the dissatisfaction which failure, gives our mind, rests on the vague feeling that between a success ascribed to luck and one ascribed to the genius of the person acting there is a subtle connection, invisible to the mind's eye, and the supposition gives us pleasure. What tends to confirm this idea is that our sympathy increases and becomes a more definite feeling, if the success and failure are often repeated in the case of the same person. Thus it becomes intelligible how luck in war takes on a much nobler character than does luck at play. In general, when a fortunate warrior does not in some other way chill the interest we feel in him, we shall take pleasure in following him in his career.

Criticism, therefore, after having weighed all that comes within the sphere of human reckoning and conviction, will let the result be the standard of judgment for that part where the deep, mysterious interrelation of things does not take shape in visible phenomena, and will protect this quiet judgment of a higher authority from the tumult of crude opinions, on the one hand, while, on the other, it rejects the gross abuse which might be made of that highest tribunal.

This verdict of the result must therefore always bring forth that which human sagacity cannot discover; and it will thus be chiefly required for the powers and workings of the mind, partly because these least admit of a reliable judgment being formed of them, and partly because their close connection with the will allows them to influence it the more easily. When fear or courage precipitates a decision, there is no longer anything objective to decide between them, and in consequence nothing whereby sagacity and calculation could once more arrive at the probable result.

We must now allow ourselves to make a few observations on the instrument of criticism, that is, the language which it uses, because that is in a certain way closely connected with action in war, for critical examination is nothing but the deliberation which should precede such action. We therefore think it very essential that the language of criticism should have the same character which that of deliberation in war must have, for otherwise it would cease to be practical, and, for criticism, no access to life would be provided.

In considering the theory of the conduct of war we have said that it should educate the mind of the leader in war, or rather that it should guide his education; that it is not intended to furnish him with positive instructions and systems which he could use as instruments of the mind. But if in war to judge of a case before us the construction of scientific aids is never necessary or even so much as admissible, if truth does not enter there in systematic shape and is never found in an indirect way but directly by the unaided vision of the mind, so too it must be in critical examination.

It is true, as we have seen, that in all cases in which it would be too complicated a matter to establish the real nature of the circumstances, criticism must rely on the truths that theory has established on the point.

But just as in war the person acting obeys these theoretical truths, more because he has absorbed the spirit of them into his own than because he regards them as an external, inflexible law, criticism also should make use of them not as an external law or an algebraical formula the truth of which has absolutely no need of being demonstrated afresh on each application, but should always let this truth shine through, leaving to theory only the more detailed and circumstantial proof. It thus avoids a mysterious, obscure phraseology and takes its way in simple speech, and with a clear, that is, an ever visible, chain of ideas.

Certainly this cannot always be completely attained, but it must always be the aim in critical exposition. Such exposition must use complex forms of cognition as little as possible and never employ the construction of scientific aids as if it were an apparatus that had truth in itself, but should accomplish everything by natural and free insight.

But this pious endeavor, if we may use the expression, has unfortunately seldom hitherto prevailed in critical examinations; most of them have rather been guided by a certain vanity to a pompous display of ideas.

The first fault we constantly encounter is a clumsy, wholly inadmissible application of certain one-sided systems as of a veritable code of laws. But it is never difficult to show the one-sidedness of such a system, and nothing more is needed to reject once and for all its judicial verdict. We have to do here with a definite object, and as the number of possible systems after all can be but small, they are in themselves also only the lesser evil.

Much more serious is the disadvantage which lies in the pompous retinue of technical terms, scientific expressions and metaphors which these systems carry in their train and which, like a disorderly rabble, or the camp-followers of an army detached from its leader, trails around in all directions. Any critic who has not yet risen to a complete system, either because none pleases him or he has not yet got so far as to make himself master of one, wants at least occasionally to apply a fragment of one as one would apply a ruler, to show the blunders committed by a general. Most of them are incapable of reasoning at all without using as a support here and there some shred of scientific military theory. The smallest of these shreds, consisting in mere scientific words and metaphors, are often nothing more than ornamental flourishes of critical narration. Now, naturally, all technical and scientific expressions which belong to a system lose their propriety, if they ever had any, as soon as they are torn from that system to be used as general axioms, or as tiny crystals of truth which have more power of demonstration than simple speech.

Thus it has come to pass that our theoretical and critical books, instead of being simple, straightforward treatises, in which the author at least always knows what he says and the reader what he reads, are brimful of these technical terms, which form dark points of intersection where author and reader part company. But they are often something much worse still, being nothing but hollow shells without any kernel. The author himself has no clear perception of what he means and contents himself with vague ideas, which if expressed in plain language would be unsatisfactory even to himself.

A third fault of criticism is the *misuse of historical examples*, and a display of great reading or learning. What the history of the art of war is, we have already said, and we shall further develop our views on examples and military history in general in special chapters. A fact merely touched upon in a very cursory manner may be used to support the most opposite views, and three or four such facts of the most heterogeneous description, brought together out of the most distant lands and remote times and heaped up, generally distract and bewilder the judgment without demonstrating anything; for when exposed to the light, they turn out to be only trumpery rubbish, used to show off the author's learning.

But what can be gained for practical life by such obscure, partly false, confused, arbitrary conceptions? So little is gained that theory, on account of them, has always been a true antithesis of practice, and frequently a subject of ridicule to those whose soldierly qualities in the field were above question.

But it is impossible that this could have come about if theory, in simple language and by natural treatment of those things which constitute the conduct of war, had merely sought to establish just so much as admits of being established; if, avoiding all false pretensions and irrelevant display of scientific forms and historical parallels, it had kept close to the subject and gone hand in hand with those who must conduct affairs in the field by their own natural insight.

CHAPTER 6

ON EXAMPLES

Examples from history make everything clear, and in addition they afford the most convincing kind of proof in the empirical fields of knowledge. This applies more to the art of war than to anything else. General Scharnhorst, whose compendium is the best ever written on actual war, declares historical examples to be the most important thing in this subject, and makes admirable use of them. Had he survived the war in which he fell, he would have given us a still finer proof of the observant and enlightening spirit with which he dealt with all experience.

But such use of historical examples is rarely made by theoretical writers. Rather, the way in which they make use of them is for the most part calculated not only to leave the intellect dissatisfied, but also to offend it. We therefore think it important to give special consideration to the correct use and the abuse of examples.

Unquestionably the branches of knowledge which lie at the foundation of the art of war belong to the empirical sciences. For although they are mostly derived from the nature of things, still for the most part we can only get to know this nature itself from experience. Besides that, however, the practical application is modified by so many circumstances that the effects can never be completely discerned from the mere nature of the means.

The effects of gunpowder, that great agent in our military activity, were learned through experience only, and up to this hour experiments are continually in progress in order to investigate them more fully. That an iron ball, which by means of powder has been given a velocity of 1,000 feet in a second, smashes every living thing which it touches in its course is, no doubt, obvious. We do not need experience to tell us that. But in determining this effect, how many hundreds of attendant circumstances are concerned, some of which can only be learned by experience! And the physical is not the only effect which we have to take into account; it is the moral effect which we are in search of, and there is no other way of learning and estimating it but by experience. In the Middle Ages, when firearms had just been invented, their physical effect, owing to their imperfect construction, was naturally but trifling compared to what it is now, but their moral effect was much greater. One must have actually seen the steadiness of one of those masses taught and led by Bonaparte in his career of conquest, under the heaviest and most unintermittent cannonade, in order to understand what troops, hardened by long practice in danger, can do, when a superabundance of victory has brought them to act on the noble rule of demanding from themselves the uttermost. To mere imagination, it would never be credible. On the other hand, it is well known that there are troops in the European armies even today, such as Tartars, Cossacks, Croatians, who would easily be dispersed by a few cannon shots.

But no empirical field of knowledge, consequently also no theory of the art of war, can always accompany its truths with historical proof; to some extent it would also be difficult to illustrate each individual instance from experience.

If one finds in war that a certain means has shown itself very effective, it is repeated; one copies the other, the thing becomes a regular fashion, and in this manner it comes into use, supported by experience, and takes its place in theory, which contents itself with appealing to experience in general in order to indicate its origin, but not in order to prove its truth.

But it is quite different if experience is to be used to supersede a means in use, to establish a doubtful one, or to introduce a new one; then particular examples from history must be cited as proofs.

Now if we consider the use of a historical example more closely, four points of view can readily be distinguished.

First, it may be used merely as an *explanation* of an idea. In every abstract discussion it is very easy to be misunderstood, or not be intelligible at all; when an author is afraid of this, an exemplification from history serves to throw on his idea, the light which is lacking and to ensure his being intelligible to his reader.

Second, it may serve as an *application* of an idea, because by means of an example there is an opportunity of showing the action of those minor circumstances which cannot all be apprehended together with it in any general expression of an idea; for in that consists, indeed, the difference between theory and experience. In both these cases we are dealing with true examples; the two that follow are concerned with historical proof.

Third, one can make special reference to historical fact, in order to support what has been advanced. This suffices in all cases where one wishes to prove the *mere possibility* of a phenomenon or effect.

Fourth, and finally, from the circumstantial presentation of a historical event, and from the comparison of several of them, we may deduce some theory, which then has its true *proof* in this testimony itself.

For the first of these purposes all that is generally required is a cursory mention of the case, as it is only used from one point of view. Even historical correctness is a secondary consideration; a case invented might also serve the purpose as well; only historical examples are always to be preferred because they bring the idea which they illustrate nearer to practical life itself.

The second use presupposes a more circumstantial presentation of events, but historical correctness is again of secondary importance, and in respect to this point the same is to be said as in the first case.

For the third purpose the mere quotation of an undoubted fact is generally sufficient. If it is asserted that fortified positions may fulfil their object under certain conditions, it is only necessary to mention the position of Bunzelwitz in support of the assertion.

But if by the narrative of a historical case, a general truth is to be demonstrated, then everything in the case bearing on the assertion must be analyzed accurately and minutely; it must, so to speak, be reconstructed carefully before the eyes of the reader. The less effectively this can be done the weaker the proof will be, and the more necessary it will be to make up for the demonstrative power which is lacking in the single case by citing a large number of cases, because we have a right to suppose that the more minute details which we were unable to mention, neutralize each other in a certain number of cases in respect to their effects.

If we want to prove by experience that cavalry is better placed behind, than in a line with, infantry; that it is very dangerous without a decided preponderance of numbers to attempt an enveloping movement, with widely separated columns, either on a field of battle or in the theater of war—that is, either tactically or strategically—then in the first of these cases it would not be sufficient to specify some defeats in which the cavalry was on the flanks and some victories in which the cavalry was in the rear of the infantry; and in the latter of these cases it is not sufficient to refer to the battles of Rivoli and Wagram, to the attack of the Austrians on

the theater of war in Italy, in 1796, or of the French upon the German theater of war in the same year. The way in which these forms of battle position and attack essentially contributed to the bad outcome in the
individual instances must be shown by closely tracing the circumstances
and single events. Then it will appear how far such forms are to be condemned, a point which it is very necessary to show, for a total condemna-
tion would in any event be inconsistent with truth.

It has been already admitted that when a detailed account of facts is
impossible, the demonstrative power which is deficient may to a certain
extent be supplied by the number of cases quoted; but it cannot be denied
that this is a dangerous expedient, and one which has been much abused.
Instead of one example expounded in great detail, three or four are just
touched upon, and thus the appearance may be given of a convincing
proof. But there are matters where a whole dozen of cases brought forward prove nothing, when, that is, these matters are of frequent occurrence and therefore a dozen other cases with an opposite result might just
as easily be brought forward on the other side. If any one names a dozen
lost battles in which the defeated party attacked in separated columns, we
can cite a dozen that have been gained in which the same order was used.
It is evident that in this way no result could be obtained.

Upon carefully considering these different circumstances, it will be
seen how easily examples may be misapplied.

An occurrence which, instead of being carefully reconstructed in all its
parts, is superficially mentioned, is like an object seen at a great distance,
presenting the same appearance on every side, and in which the position
of its parts cannot be distinguished. Such examples have really served to
support the most contradictory opinions. To some, Daun's campaigns are
models of restraint. To others, they are nothing but examples of timidity
and lack of resolution. Bonaparte's passage of the Norican Alps in 1797
may appear as the noblest resolution, but also as an act of sheer recklessness. His strategic defeat in 1812 may be represented as the consequence
either of an excess, or of a deficiency, of energy. All these opinions have
been expressed, and it is easy to see that they might well arise, because
each opinion has interpreted in a different way the connection of events.
At the same time these antagonistic opinions cannot be reconciled with
each other, and therefore one of the two must necessarily be false.

Much as we are obliged to the excellent Feuquières for the numerous
examples introduced in his memoirs—partly because a great number of
historical incidents have thus been preserved which would otherwise have
been lost, and partly because he was the first to bring theoretical, that is,

abstract, ideas into a very useful connection with practical life, in so far as the cases brought forward may be regarded as explaining and more closely defining what is theoretically asserted—yet, in the opinion of impartial readers of our own times, he has hardly attained the object he for the most part proposed to himself: that of proving theoretical principles by historical examples. For although he sometimes relates occurrences with great minuteness, still he falls very short of showing that the deductions drawn necessarily proceed from the internal connection of these events.

Another evil resulting from the superficial notice of historical events is that some readers have not sufficient knowledge or memory of them to be able even to grasp the author's meaning; so that there is nothing left for them but either to accept blindly what he says or remain without any convictions at all.

It is indeed extremely difficult to reconstruct or unfold historical events before the eyes of a reader in such a way as is necessary in order to be able to use them as proofs; for the writer mostly lacks no less the means than the time or the space to do this. But we maintain that, when it is our object to establish a new or doubtful opinion, one single event, thoroughly analyzed, is far more instructive than ten which are superficially treated. The great evil of this superficial treatment is not that the writer presents his story with the unjustified claim that he wants to prove something by it, but that he himself has never been properly acquainted with the events, and that from this sort of slovenly, frivolous treatment of history, a hundred false views and attempts at the construction of theories arise which would never have made their appearance if the writer had looked upon it as his duty to deduce conclusively from the strict connection of events everything new which he has to offer and seeks to prove from history.

When we are convinced of these difficulties in the use of historical examples, and at the same time of the necessity of demanding them, we shall also be of the opinion that the latest military history must always be the most natural field from which to select examples, so long as that history is only sufficiently well known and well digested.

It is not only that more remote periods are related to different circumstances, and therefore to a different conduct of war, and that consequently their events are less instructive to us either theoretically or practically; but it is also natural that military history, like every other, gradually loses a number of small traits and details which it originally still had to exhibit, loses more and more in color and life, like a faded or darkened picture, so that at last only the large masses and leading features chance to remain and thus acquire undue proportions.

If we look at the state of the present conduct of war, we must say that the wars since that of the Austrian Succession are almost the only ones which, at least as far as armament is concerned, have still a considerable similarity to the present, and which, notwithstanding the many changes which have taken place in great and small circumstances, are still close enough to modern wars to afford us considerable instruction. It is quite different with the war of the Spanish Succession, as the use of firearms was not yet so well developed, and cavalry was still the most important arm. The farther we go back, the less useful military history becomes, as it becomes so much the more meager and barren of detail. The least usable and most barren history must necessarily be that of the old world.

But this uselessness is certainly not an absolute one; it relates only to those subjects which depend on a knowledge of minute details, or on those things in which the method of conducting war has changed. Although we know very little about the tactics in the battles of the Swiss against the Austrians, the Burgundians and French, still we find in them unmistakable evidence that they were the first in which the superiority of a good infantry over the best cavalry was displayed. A general glance at the time of the *condottieri* teaches us how the whole method of conducting war is dependent on the instrument used, for at no period have the forces used in war had so much the characteristics of a special instrument and been so totally separated from the rest of the political and civil life. The remarkable way in which the Romans in the second Punic War attacked Carthage in Spain and Africa, while Hannibal was still in Italy unconquered, can be a most instructive subject to study, as the general relations of the states and armies on which the success of this indirect resistance rested are sufficiently well known.

But the farther things descend into particulars and deviate in character from the merest generalities, the less can we look for examples and experiences from very remote periods, for we have neither the means of judging properly of analogous events, nor can we apply them to our completely different means.

Unfortunately, however, it has always been the fashion with historical writers to talk about events of ancient times. How great a share vanity and charlatanism may have had in this, we do not wish to decide, but in most cases we fail to discover any honest intention and earnest endeavor to instruct and convince, and we can therefore only regard such allusions as embellishments to fill up gaps and hide defects.

It would be an immense service to teach the art of war entirely by historical examples, as Feuquières proposed to do; but it would be fully the work of a lifetime, if we reflect that he who undertakes it must first qualify himself for the task by a long personal experience in actual war.

Whoever, stirred by inner powers, wishes to undertake such a task, let him prepare himself for his pious undertaking as for a long pilgrimage; let him sacrifice his time, let him shrink from no exertion, fear no temporal power and might, and rise above all feelings of personal vanity and false shame, in order, according to the French code, to speak the truth, the whole truth and nothing but the truth.

BOOK III

OF STRATEGY IN GENERAL

CHAPTER 1

STRATEGY

The conception of strategy has been defined in Book II, Chapter 1. Strategy is the use of the engagement to attain the object of the war. It has properly only to do with the engagement, but the theory of it must consider, at the same time, the agent of its proper activity, namely, the armed forces, both in themselves and in their chief relations, the engagement being determined by these and in turn exercising upon them its immediate effects. The engagement itself must be studied in relation both to its possible results and to the mental and moral forces which are most important in the use of it.

Strategy is the use of the engagement to attain the object of the war. It must therefore give an aim to the whole military action, which aim must be in accordance with the object of the war. In other words, strategy maps out the plan of the war, and to the aforesaid aim it affixes the series of acts which are to lead to it; that is, it makes the plans for the separate campaigns and arranges the engagements to be fought in each of them. As all these are matters, which to a great extent can only be determined on suppositions, some of which do not materialize, while a number of other decisions pertaining to details cannot be made beforehand at all, it is self-evident that strategy must take the field with the army in order to arrange particulars on the spot, and to make the modifications in the general plan which incessantly become necessary. Strategy can, therefore, never for a moment take its hand from the work.

That this, at least as far as the whole is concerned, has not always been the view taken is evident from the former custom of keeping strategy in the cabinet and not with the army. Such a thing is permissible only if the cabinet remains so close to the army that it can be regarded as its chief headquarters.

Theory will, therefore, follow strategy in this plan, or more properly speaking, it will cast light on things both in themselves and in their relations to one another and emphasize the little that there is of principle or rule.

If we recall from the first chapter of Book I how many matters of the highest importance war touches upon, we will realize that a consideration of all these presupposes a rare mental grasp.

A prince or general, who knows how to organize his war exactly according to his object and means, who does neither too much nor too little, furnishes thereby the greatest proof of his genius. But the effects of this genius do not manifest themselves so much in the invention of new modes of action, which might strike the eye immediately, as in the successful final result of the whole. It is the exact fulfilment of silent suppositions, the quiet harmony of the entire action, which we should admire, and which makes itself known only in the total result.

The inquirer, who starting from the total result does not perceive that harmony, is one who is apt to seek for genius where it is not and cannot be.

The means and forms which strategy uses are in fact so extremely simple, so well known by their constant repetition, that it only appears ridiculous to sound common sense to hear critics so frequently speaking of them with high-flown emphasis. Turning a flank, which has been done a thousand times, is regarded by one as a mark of the most brilliant genius, by another as a proof of the most profound penetration, indeed even of the most comprehensive knowledge. Can there be in the academic world any excesses more absurd?

It is still more ridiculous if, in addition to this, we reflect that the very same critics, in accordance with the most common opinion, exclude all moral quantities from theory, and will not allow it to be concerned with anything but the material forces, so that everything is confined to a few mathematical relations of equilibrium and preponderance, of time and space, and a few lines and angles. If it were nothing more than this, then out of such a miserable business a scientific problem could scarcely be formulated even for a schoolboy.

But let us admit that there is no question at all here of scientific formulas and problems. The relations of the material things are all very simple. The comprehension of the moral forces which come into play is more difficult. Still, even in respect to these, it is only in the highest branches of strategy that intellectual complications and a great diversity of quantities and relations are to be looked for. At this point strategy borders on politics and statesmanship, or rather it becomes both itself, and, as we have observed before, these have more influence on how much or how little is to be done than on how it is to be executed. Where the latter is the principal question, as in the single acts of war both great and small, the mental and moral quantities are already reduced to a very small number.

Thus, then, in strategy everything is very simple, but not on that account very easy. Once it is determined from the conditions of the state what war shall and can do, then the way thereto is easy to find; but to follow that way straight forward, to carry out the plan without being obliged to deviate from it a thousand times by a thousand varying influences, requires, besides great strength of character, great clearness and steadiness of mind. Out of a thousand men who are remarkable, some for intellect, others for penetration, others again for boldness or strength of will, perhaps not one will combine in himself all those qualities which raise him above mediocrity in the career of a general.

It may sound strange, but for all who know war in this respect it is a fact beyond doubt, that much more strength of will is required to make an important decision in strategy than in tactics. In the latter we are carried away by the moment; a commander feels himself borne along in a whirl-pool, against which he dares not contend without the most destructive consequences, he suppresses the rising doubts, and boldly ventures further. In strategy, where all moves much more slowly, there is much more room allowed for our own doubts and those of others, for objections and remonstrances, consequently also for untimely regrets; and since in strategy we do not see things with our own eyes as we do at least half of them in tactics, but everything must be conjectured and assumed, the convictions produced are less powerful. The result is that most generals, when they should act, are stuck fast in false doubts.

Now let us cast a glance at history—upon Frederick the Great's campaign of 1760, which is famed for its fine marches and maneuvers; a perfect masterpiece of strategic skill, as critics tell us. Are we then to be beside ourselves with admiration, because the king first sought to turn Daun's right flank, then his left, then again his right, and so forth? Are we to see profound wisdom in this? No, that we cannot, if we are to decide naturally and without affectation. Rather we must admire above all the sagacity of the king, that while pursuing a great object with very limited means, he undertook nothing beyond his powers, and *just enough* to gain his object. His sagacity as a general is visible not only in this campaign, but throughout all the three wars of the great king.

To bring Silesia into the safe harbor of a well-guaranteed peace was his object.

At the head of a small state, which was like other states in most things, and only ahead of them in some branches of administration, he could not become an Alexander, and, as a Charles XII, he would only, like him, have ended in disaster. We find, therefore, in the whole of his conduct of war, a restrained power, always well balanced, and never wanting in vigor, which at the critical moment rises to astonishing deeds, and at the next moment swings quietly on, accommodating itself to the play of the most subtle political influences. Neither vanity, thirst for glory, nor vengeance could make him deviate from his course, and this course alone brought him to a fortunate termination of the contest.

How little justice these few words are able to do to that aspect of the great general's genius! Only if we carefully observe the extraordinary outcome of this war, and trace the causes which led to this outcome, will the conviction be borne in upon us that nothing but the king's keen insight brought him safely through all perils.

This is one feature in that great commander which we admire in the campaign of 1760—and in all others, but in this especially, because in none did he keep the balance against such a superior hostile force, with such small sacrifice.

Another feature concerns the difficulty of execution. Marches to turn a flank, right or left, are easily mapped out; the idea of always keeping a small force well concentrated in order to be able to meet the scattered enemy on equal terms at any point and of multiplying a force by rapid movement is as easily conceived as expressed. The discovery of it, therefore, cannot excite our admiration, and with respect to such simple things, it suffices to admit that they are simple.

But let a general try to imitate Frederick the Great in these things. Long afterward authors, who were eye-witnesses, have spoken of the danger, indeed of the imprudence of the king's camps, and doubtless, at the time he pitched them, the danger appeared three times as great as afterward.

It was the same with his marches, under the eyes, nay, often under the cannon of the enemy's army. These camps were taken up, these marches made, because in Daun's mode of procedure, in his method of drawing up his army, in his sense of responsibility and in his character, Frederick found that security which made his camps and marches daring, but not reckless. But it required the king's boldness, determination and strength of will to see things in this light, and not to be led astray and intimidated by the danger of which thirty years after people still wrote and spoke. Few generals in this situation would have believed these simple strategic means to be practicable.

Then again there was another difficulty of execution, namely, that the king's army in this campaign was constantly in motion. Twice the army marched by wretched by-roads, from the Elbe into Silesia, behind Daun and pursued by Lascy (beginning of July and beginning of August). At every moment it had to be prepared for battle, and its marches had to be organized with a degree of skill which necessarily resulted in equally great exertion. Although attended and delayed by thousands of wagons, still its subsistence was extremely scanty. In Silesia, for eight days before the battle of Liegnitz, it was involved in constant night marches, and forced to defile alternately right and left along the enemy's front; this cost great exertion and entailed great privations.

Is it to be supposed that all this could have been done without producing great friction in the machine? Can the mind of a commander produce such movements as easily as the hand of a land surveyor manipulates the astrolabe? Does not the sight of the sufferings of their hungry, thirsty comrades pierce the hearts of the commander and his generals a thousand times? Must not the complaints and doubts concerning these reach his ear? Has an ordinary man the courage to demand such sacrifices, and would not such efforts unavoidably demoralize the army, break down discipline, and, in short, undermine its military virtue, if strong confidence in the greatness and infallibility of the commander did not compensate for all? Here, therefore, it is that we should pay respect; it is these miracles of execution which we should admire. But it is impossible to realize all this in its full weight without a foretaste of it through experience. For the person, who knows war only from books and from the drill-ground, none of this paralyzing effect on action really exists; we beg him, therefore, to accept from us on faith and trust all that he is unable to supply from any personal experiences of his own.

By means of this illustration we intended to give greater clarity to the course of our ideas, and in closing this chapter we hasten to say that in our presentation of strategy we shall describe those individual aspects of it which appear to us as the most important, be they of a material or a mental and moral nature. We shall proceed from the simple to the complex, and conclude with the inner connection of the whole act of war, in other words, with the plan for a war or campaign.[1]

[1] In an earlier manuscript of Book II the following passages endorsed by the author himself are to be found: "To be used for chapter 1 of Book III." The projected revision of that chapter not having been made, the passages referred to are introduced here in full.

By the mere disposition of armed forces at one point, an engagement there is made possible, but does not always actually take place. Is that possibility to be regarded as a reality and therefore as a real thing? Certainly. It is so by virtue of its consequences, and these effects, whatever they may be, can never be wanting.

1. POSSIBLE ENGAGEMENTS ARE ON ACCOUNT OF THEIR CONSEQUENCES TO BE REGARDED AS REAL ONES

If a detachment is sent to cut off the retreat of a flying enemy, and the enemy surrenders without further resistance, it is to the engagement which is offered to him by the detachment so sent that his decision is due.

If a part of our army occupies an enemy province which was undefended, and thus deprives the enemy of considerable means with which he might reinforce his own army, it is only through the engagement which this detachment causes the enemy to expect in the event of his proposing to recover the lost province that we remain in possession of it.

In both cases, therefore, the mere possibility of an engagement has produced consequences, and has therefore entered the category of real things. Suppose that in these cases the enemy had opposed our troops with others superior in force, and thus forced ours to give up their object without an engagement, then, no doubt, our plan has failed, but the engagement which we offered to the enemy at this point has not been without effect, for it has attracted the enemy forces. Even if the entire undertaking has amounted to a loss for us, it cannot be said that those positions, those possible engagements, have been without effect. These effects are then similar to those of a lost engagement.

In this manner we see that the destruction of the enemy's military forces and the overthrow of the enemy's power are accomplished only through the effects of the engagement, be it that the engagement actually takes place or that it is merely offered and not accepted.

2. THE TWOFOLD OBJECT OF THE ENGAGEMENT

But these effects are also twofold, namely, direct and indirect. They are of the latter type, if other things intervene and become the object of the engagement—things which in themselves cannot be regarded as the destruction of the enemy's forces, but which are only supposed to lead up to it, indirectly, no doubt, but with so much the greater force. The possession of provinces, cities, fortresses, roads, bridges, magazines, etc., may be the *immediate* object of an engagement, but never the ultimate one. Things of this description must only be regarded as a means of gaining greater superiority, in order that the engagement may finally be offered to the opponent in such a way that it will be impossible for him to accept it. Therefore all these things are only to be regarded as intermediate steps, that is, as guides to the effective principle, but never as that principle itself.

3. EXAMPLES

In 1814, by the capture of Bonaparte's capital the object of the war was attained. The political divisions which had their roots in Paris became effective, a violent cleavage caused the power of the Emperor to collapse. Nevertheless it is necessary to regard this from the point of view that hereby Bonaparte's military force and his power of resistance were suddenly reduced, that the superiority of the Allies was proportionately increased, and that now any further resistance became impossible. It was this impossibility which produced the peace with France. If we suppose the military forces of the Allies to have been proportionately reduced at this moment through the influence of external causes, the superiority vanishes, and the entire effect and importance of the capture of Paris disappears also.

We have gone through this chain of argument in order to show that this is the natural and only true view of the thing from which its importance is derived. It constantly leads back to the question: What at any given moment of the war or campaign will be the probable result of the great and small engagements which the two sides might offer one another? In the consideration of a plan for a campaign or war, only this question is decisive as to the measures which are to be taken from the outset.

4. WHEN THIS VIEW IS NOT TAKEN, THEN A FALSE VALUE IS GIVEN TO OTHER THINGS

If we do not accustom ourselves to look upon war, and upon a single campaign in war as a chain, composed of nothing but engagements, of which one is always the cause of the other; if we adopt the idea that the capture of certain geographical points, the occupation of undefended provinces is in itself something; then we are very likely to regard it as an advantage, which can be picked up in passing; and if we look at it so, and not as a link in the whole series of events, we do not ask ourselves whether this possession may not lead to greater disadvantages later. How often we find this mistake recurring in the history of war! We might say that, just as in commerce the merchant cannot set apart and place in security gains from one single transaction, so in war a single advantage cannot be separated from the result of the whole. Just as the former must always operate with the whole sum of his means, so in war only the final total will decide whether any particular item is profit or loss.

But if the mind's eye is always directed upon the series of engagements, so far as it can be perceived beforehand, then it is fixed upon the direct road to its goal, and thereby the movement of our strength acquires that rapidity, that is to say, our volition and action acquire that energy which the occasion demands and which is not disturbed by extraneous influences.

CHAPTER 2

ELEMENTS OF STRATEGY

The causes which condition the use of the engagement in strategy may be conveniently divided into elements of different kinds, namely, the moral, physical, mathematical, geographical and statistical elements.

The first class includes all that are called forth by mental and moral qualities and effects; to the second belong the magnitude of the military force, its composition, the proportion of arms, etc.; to the third, the angle of the lines of operation, the concentric and eccentric movements in so far as their geometrical nature acquires any value in the calculation; to the fourth, the influence of terrain, such as commanding points, mountains, rivers, woods, roads; lastly, to the fifth, all the means of supply, etc. The fact that for the moment one thinks of these elements separately has its advantage in that it gives clarity to our ideas and helps us to estimate the higher or lower value of the different classes as we pass onward. For when we consider them separately, many of them spontaneously lose their borrowed importance; we feel, for instance, quite clearly, that the value of a base of operations, even if we wished to look at nothing but the position of the line of operations, depends much less even in that simple form on the geometrical element of the angle which they form with one another than on the nature of the roads and the country through which they pass.

But to treat strategy according to these elements would be the most unfortunate idea possible, for these elements are generally manifold, and intimately connected with each other in every single operation of war. We should be lost in the most soulless analysis, and as in a nightmare we should for ever be seeking in vain to erect an arch which would connect this base of abstractions with facts belonging to the real world. May heaven protect every theorist from such an undertaking! We shall keep to the world of complex phenomena, and not pursue our analysis further than is necessary on each occasion to give distinctness to the idea which we wish to impart, and which has come to us, not by a speculative investigation, but through the impression made by the realities of war in their totality.

CHAPTER 3

MORAL QUANTITIES

We must return again to this subject which is touched upon in Book I, Chapter 3, because the moral quantities are among the most important subjects in war. They are the spirits which permeate the whole sphere of war. They attach themselves sooner or later and with greater affinity to the will which sets in motion and guides the whole mass of forces, and they unite so to speak with it in one whole, because it is itself a moral quantity. Unfortunately, they seek to withdraw from all book knowledge, for they can neither be measured in figures nor grouped into classes, and require to be both seen and felt.

The spirit and other moral qualities of an army, a general or a government, public opinion in provinces in which the war is proceeding, the moral effect of a victory or of a defeat—these are things which in themselves vary greatly in their nature, and which, according as they stand with regard to our object and our circumstances, may also have a very different kind of influence.

Although little or nothing can be said about these things in books, they still belong to the theory of the art of war, as much as everything else which constitutes war. For once more I must here repeat that it is a miserable philosophy if, according to the old manner, we establish rules and principles regardless of all moral quantities, and then, as soon as these quantities make their appearance, we begin to count the exceptions, which we thereby after a fashion scientifically formulate, that is, make them the rule; or if we resort to an appeal to genius, which is above all rules, whereby we are given to understand that rules were not only made for fools, but also must themselves be really folly.

Even if the theory of war could actually do no more than recall these things to memory, showing the necessity of allowing to the moral quantities their full value, and of always taking them into consideration, still by so doing it would have included in its borders this sphere of immaterial forces, and by establishing that point of view, it would have condemned beforehand every one who would endeavor to justify himself before its judgment seat by the mere physical conditions of forces.

Further for the sake of all other so-called rules, theory cannot banish the moral quantities from its scope, because the effects of the physical and moral forces are completely fused, and are not to be broken down like a metal alloy by a chemical process. In every rule relating to the physical forces, theory must bear in mind at the same time the share which the moral quantities can have in the matter, if it is not to be misled into categorical propositions, which are at times too timid and limited, at times too dogmatic and broad. Even the most uninspired theories have unconsciously had to stray over into this moral kingdom, for, as an example, the effects of a victory can never be wholly explained without taking into consideration the moral impressions. And therefore most of the subjects which we shall go through in this book are composed half of physical, half of moral, causes and effects, and we might say the physical are almost no more than the wooden handle, whilst the moral are the noble metal, the real, brightly polished weapon.

The value of the moral quantities, and their frequently incredible influence, are best exemplified by history, and this is the noblest and most genuine nourishment which the mind of the general can extract from it. In this connection it is to be observed, that the seeds of wisdom which are to bear fruit in the mind are sown not so much by demonstrations, critical examinations and learned treatises, as by sentiments, general impressions and single, flashing sparks of intuition.

We might go through the most important moral phenomena in war, and with all the care of a diligent professor try what we could impart about each, either good or bad. But in such a method one lapses too readily into the commonplace and ordinary, while the real spirit in the analysis quickly vanishes, and one drops, without being aware of it, into telling things which everybody knows. We prefer, therefore, here still more than elsewhere, to remain incomplete and rhapsodical, content to have drawn attention to the importance of the subject in a general way, and to have indicated the spirit in which the views advanced in this book have been formed.

CHAPTER 4

THE CHIEF MORAL POWERS

They are as follows: *the talents of the commander, the military virtue of the army, its national feeling.* Which of these has a greater value no one can determine in a general way. For it is very difficult to say anything at all concerning their strength, and still more so to compare the strength of one with that of another. The best plan is not to undervalue any of them, a fault which human judgment is inclined to in its whimsical vacillation, tending now to the one side now to the other. It is better to adduce sufficient evidence from history of the undeniable efficacy of these three things.

It is true, however, that in modern times the armies of European states have arrived very much at a par in respect to discipline and training. The conduct of war has—as philosophers would say—so naturally developed, thereby becoming a kind of method, common as it were to almost all armies, that even on the commander's side we can no longer reckon on the application of special devices in the more limited sense (such as Frederick the Second's oblique order). Hence it cannot be denied that, as matters now stand, greater scope is afforded for the influence of national spirit and of the habituation of an army to war. A long peace may again alter all this.

The national spirit of an army (enthusiasm, fanatical zeal, faith, opinion) displays itself most in mountain warfare, where everyone, down to the common soldier, is left to himself. On this account, mountains are the best campaigning grounds for general levies.

Technical skill in an army, and that well-tempered courage which holds the ranks together as if they had been cast in a mold, show to the best advantage in an open plain.

The talent of a general has most scope in a broken, undulating country. In mountains he has too little command over the separate parts, and the direction of all gets beyond his powers; in open plains it is simple and does not exhaust those powers.

According to these unmistakable elective affinities plans should be formulated.

CHAPTER 5

MILITARY VIRTUE OF AN ARMY

This is distinguished from mere bravery, and still more from enthusiasm for the cause for which the war is fought. The first is of course a necessary constituent part of it, but just as bravery, which is ordinarily a natural gift, can also arise in a soldier, as member of an army, from habit and training, so with him it must also have a different direction from that which it has with other men. It must lose that impulse to unbridled activity and manifestation of force which is its characteristic in the individual, and submit itself to demands of a higher kind, such as, obedience, order, rule and method. Enthusiasm for the cause gives life and greater fire to the military virtue of an army, but does not constitute a necessary part of it.

War is a special profession. However general its relation may be, and even if all the male population of a country capable of bearing arms were to practice it, war would still continue to be different and separate from the other activities which occupy the life of man. To be imbued with the spirit and essence of this profession, to train, to rouse, to assimilate into our system the powers which should be active in it, to apply our intelligence to every detail of it, to gain confidence and expertness in it through exercise, to go into it heart and soul, to pass from the man into the role which is to be assigned to us in it—that is the military virtue of an army in the individual.

However carefully we may seek to conceive of the citizen and the soldier, as existing in one and the same individual, however much we may look upon wars as national affairs, and however far our ideas may depart from those of the *condottieri* of former days, never will it be possible to do away with the individuality of the professional routine. And if this cannot be done, then those who belong to the profession, and as long as they belong to it, will always look upon themselves as a kind of guild, in the regulations, laws and customs of which the spirit of war is predominantly expressed. And so it is in fact. Even with the most decided inclination to look at war from the highest point of view, it would be very wrong to depreciate this corporative spirit, this *esprit de corps,* which can and must exist more or less in every army. This corporative spirit forms, so to speak, the bond of union between the natural forces which are active in what we have called military virtue. The crystals of military virtue form more easily upon the corporative spirit.

An army which preserves its customary formations under the heaviest fire, which is never shaken by imaginary fears, and resists with all its might any that are well founded, which, proud in the feeling of its victories, never loses its sense of obedience, its respect for and confidence in its leaders, even in the midst of the disaster of defeat; an army with its physical powers strengthened in the practice of privations and exertion, like the muscles of an athlete; an army which looks upon all its toils as the means to victory, not as a curse which rests on its standards, and which is always reminded of its duties and virtues by the short catechism of one single idea, namely, the honor of its arms—such an army as this is imbued with the true military spirit.

Soldiers may fight bravely like the Vendeans, and effect great things like the Swiss, the Americans or the Spaniards, without developing this military virtue. A commander may also be successful at the head of standing armies, like Eugene and Marlborough, without enjoying the benefit of its assistance. Therefore, we must not say that a successful war cannot be imagined without it. We draw special attention to this point, in order to give greater individuality to the conception presented here, so that our ideas may not dissolve into a vague generalization and we may not think that military virtue is in the end the one and only thing. This is not so. Military virtue in an army appears as a definite moral power which we can abstract, and the influence of which we can therefore estimate as an instrument, the strength of which may be calculated.

Having thus characterized it, we shall see what can be said about its influence, and about the means of gaining this influence.

Military virtue is everywhere for the parts what the genius of the commander is for the whole. The general can only direct the whole, not each part, and where he cannot direct the part, there military spirit must be its leader. A general is chosen by the reputation of his excellent qualifications, the more distinguished leaders of large masses by careful examination; but this examination decreases as we descend the scale of rank, and in just the same measure we may count less and less upon individual talents; but what is wanting in this respect, military virtue must supply. It is just this part that is played by the natural qualities of a people mobilized for war: *bravery, flexibility, powers of endurance and enthusiasm*. These properties may therefore be substituted for military virtue, and vice versa, from which it may be deduced that:

1. Military virtue is a quality of standing armies only, and they require it the most. In national uprisings and in war natural qualities, which develop more rapidly in these, are substituted for it.
2. Standing armies opposed to standing armies can more easily dispense with it than a standing army opposed to a national uprising, for in that case, the troops are more scattered, and the parts left more to themselves. But where an army can be kept concentrated, the genius of the general plays a greater role and makes up for what is lacking in the spirit of the army. Generally, therefore, military virtue becomes all the more necessary, the more the theater of operations and other circumstances complicate the war and scatter the forces.

From these truths the only lesson to be derived is that if an army is deficient in this quality, every endeavor should be made to simplify the operations of the war as much as possible or to double the attention paid to other points of the military system and not to expect from the mere name of a standing army that which only the thing itself can give.

The military virtue of an army is, therefore, one of the most important moral powers in war, and where it has been lacking, we either see it replaced by one of the others, such as the superior genius of the general or the enthusiasm of the people, or we find results which are not commensurate with the effort made. How many a great thing this spirit, this sterling worth of an army, this refinement of ore into gleaming metal, has already accomplished we see in the history of the Macedonians under Alexander, the Roman legions under Caesar, the Spanish infantry under Alexander Farnese, the Swedes under Gustavus Adolphus and Charles XII, the Prussians under Frederick the Great, and the French under Bonaparte. We would have deliberately to shut our eyes against all historical proof, if we refused to admit that the wonderful successes of these generals and their greatness in situations of extreme difficulty were only possible with armies raised by military virtue to a higher power of efficiency.

This spirit can only arise from two sources, and these can produce it only by working together. The first is a series of wars and successful results; the other is the practice of often working an army to the last ounce of its strength. Only in this effort does the soldier learn to know his powers. The more a general is in the habit of demanding from his troops, the more certain he is that his demands will be satisfied. The soldier is as proud of hardship overcome as he is of danger surmounted. Therefore it is only in the soil of incessant activity and effort that this germ will thrive, but also only in the sunshine of victory. Once it has developed into a strong tree, it will resist the fiercest storms of misfortune and defeat, and even the sluggish inactivity of peace, at least for a time. Therefore, it can only be created in war, and under great generals, but no doubt it may last at least for several generations, even under generals of moderate capacity, and also through considerable periods of peace.

Between this extended and ennobled *esprit de corps* in a handful of scar-covered, war-hardened veterans and the self-esteem and vanity of a standing army which is held together merely by the bond of service regulations and a drill book there is no comparison. A certain grim severity and strict discipline may prolong the life of military virtue, but cannot create it. These things retain, nevertheless, a certain value, but must not be overrated. Order, smartness, good will, also a certain degree of pride and high morale, are qualities of an army trained in times of peace which are to be valued, but which, however, cannot stand alone. The whole maintains the whole, and as with glass too quickly cooled, a single crack breaks the whole mass. Above all, the highest spirit in the world changes only too easily with the first misfortune into depression, and one might say into a kind of gasconade of fear, the French *sauve qui peut*. Such an army can only achieve something through its leader, never by itself. It must be led with double caution, until by degrees, in victory and hardship, its strength becomes adequate to its task. Beware then of confusing the spirit of an army with its morale.

CHAPTER 6

BOLDNESS

The place and part which boldness plays in the dynamic system of forces, where it stands opposite to foresight and discretion, has been stated in the chapter on the certainty of success, in order to show thereby that theory has no right to restrict it on the pretext of her legislative power.

But this noble buoyancy, with which the human soul raises itself above the most formidable dangers, is to be regarded as a separate active agent in war. In fact, in what branch of human activity should boldness have a right to citizenship if not in war?

From the camp follower and the drummer boy up to the general, it is the noblest of virtues, the true steel which gives the weapon its edge and luster.

Let us admit in fact, that it has even special prerogatives in war. Over and above the result of the calculation of space, time and quantity, we must allow a certain percentage to boldness, which it derives from the weakness of the enemy, whenever it shows itself superior. It is, therefore, a truly creative power. This is not difficult to demonstrate even philosophically. As often as boldness encounters hesitation, the probability of success is of necessity in its favor, because the very state of hesitation is already a loss of equilibrium. It is only when it encounters cautious foresight—which we may say is just as bold, in every case just as strong and powerful as itself—that it is at a disadvantage; such cases, however, rarely occur. Out of the whole multitude of cautious men a considerable majority are so out of timidity.

In the great mass, boldness is a force, the special cultivation of which can never become detrimental to other forces, because the great mass is bound to a higher will by the framework and organization of the order of battle and of the service, and therefore is guided by an intelligent power which is not its own. Boldness is therefore here only like a spring held down until released.

The higher the rank the more necessary it is that boldness should be accompanied by a reflective mind, that it may not be a mere blind outburst of passion to no purpose; for with increase of rank it becomes less and less a matter of self-sacrifice and more a matter of the preservation of others, and the good of the whole. What service regulations, as a kind of second nature, prescribe for the great mass must be prescribed to the general by reflection, and in his case individual boldness in a single act may easily become an error. But nevertheless it is a fine error and must not be regarded in the same light as any other. Happy the army in which an untimely boldness frequently manifests itself; it is an exuberant growth but it indicates a rich soil. Even foolhardiness, that is boldness without an object, is not to be despised; fundamentally it is the same energy of temperament, only exercised in a kind of passion without any co-operation of the intellectual faculties. It is only where boldness rebels against obedience, when it forsakes with contempt a definitely higher authority, that it must be repressed as a dangerous evil, not on its own account but on account of the act of disobedience, for there is nothing in war which is of greater importance than obedience.

Given an equal degree of intelligence, a thousand times more is lost in war through anxiety than through boldness. We need only say this in order to be assured of our reader's approval.

Fundamentally the intervention of a reasonable object should make boldness easier, and therefore lessen its intrinsic merit, and yet the very reverse is the case.

The intervention of lucid thought or even more so the supremacy of mind deprives the emotional forces of a great part of their violence. On that account boldness becomes *more infrequent the higher we ascend the scale of rank,* for although insight and intelligence may not increase with rank, nevertheless objective quantities, circumstances, relations and considerations from without are forced upon commanders in their different stations so much and so strongly that the burden upon them increases all the more, the more that on their own insight decreases. This, in so far as war is concerned, is the chief foundation of the truth of the French proverb:—

"Tel brille au second qui s'éclipse au premier."

Almost all generals whom history presents to us as merely having attained to mediocrity, and as wanting in decision when in supreme command, are men who had distinguished themselves in the lower ranks by boldness and resolution.

We must make a distinction in those motives to bold action which arise from the pressure of necessity. Necessity has its degrees of intensity. If it lies near at hand, if in the pursuit of his object the person acting is driven into great dangers in order to escape others equally great, then the only thing we can admire is his resolution, which, however, still has its value. If a young man to show his skill in horsemanship leaps across a deep abyss, then he is bold; if he makes the same leap pursued by a troop of head-chopping Janissaries he is only resolute. But the farther removed the necessity from the action and the greater the number of circumstances which the mind has to traverse in order to realize it, so much the less does it disparage boldness. If Frederick the Great, in the year 1756, regarded war as inevitable, and could only escape destruction by forestalling his enemies, it was necessary for him to begin the war himself, but at the same time it was certainly very bold, for few men in his position would have resolved to do so.

Although strategy is only the province of commanders-in-chief or generals in the higher positions, still boldness in all other branches of the army is as little a matter of indifference to it as the other military virtues. With an army which emanates from a bold people, and in which the spirit of boldness has always been nourished, other things may be undertaken than with one which is a stranger to this virtue. For that reason we have mentioned boldness in connection with the army. But our particular subject is the boldness of the general, and yet we have not much to say about it after having described this military virtue in a general way to the best of our ability.

The higher we ascend in positions of command, the more will mind, intellect and insight predominate in activity, and the more, therefore, will boldness, which is a property of temperament, be thrust into the background. For that reason we find it so rarely in the highest positions; but it is then all the more worthy of admiration. Boldness, directed by a predominating intelligence, is the stamp of the hero: this boldness does not consist in venturing directly against the nature of things, in a downright violation of the laws of probability, but in the forceful support of that higher calculation which genius, with its instinctive judgment, has run through with lightning speed, and but half consciously, when it makes its choice. The more boldness lends wings to the mind and insight, so much the higher these will reach in their flight, and so much the more comprehensive will be the vision, and the more correct the result; but of course only on the assumption that with greater objects greater dangers are associated. The ordinary man, not to speak of the weak and irresolute, arrives at a correct result in so far as such is possible without living experience, while pursuing, at most, an imaginary activity in his study, far away from danger and responsibility. Let danger and responsibility surround him from every direction, and he loses his perspective and if he retains this in any measure by the influence of others, still he would lose his power of *decision*, because in that point no one can help him.

We think then that it is impossible to imagine a distinguished general without boldness; that is to say, that no man can become one who is not born with this strength of temperament, which we therefore regard as the first requisite for such a career. How much of this innate strength, developed and molded through education and the circumstances of life, is left when the man has attained a high position is the second question. The greater this power remains, the stronger is the soaring of genius, the higher its flight. The risk becomes ever greater, but the object of it grows accordingly. Whether its lines emanate and get their direction from a distant necessity, or whether they converge to the keystone of a building which ambition has designed, whether it is a Frederick or an Alexander who acts is much the same as regards the critical view. If the latter alternative excites the imagination more because it is bolder, the former satisfies the intellect more, because it has in it more inherent necessity.

Now, however, we have still to consider one very important circumstance.

The spirit of boldness may be in an army, either because it is in the people, or because it has been created in a successful war conducted by bold generals. In the latter case, however, it will be lacking in the beginning.

Nowadays there is hardly any other means of educating the spirit of a people in this respect than war alone and that under bold leadership. Only this can counteract that effeminacy of feeling and that inclination to seek the enjoyment of comfort which drag down a people in conditions of increasing prosperity and heightened commercial activity.

A nation may hope to have a firm position in the political world only if national character and habituation to war mutually support each other in constant reciprocal action.

CHAPTER 7

PERSEVERANCE

The reader expects to hear of angles and lines, and finds, instead of these inhabitants of the scientific world, only people of common life, such as he meets every day in the street. And yet the author cannot make up his mind to be a hair's breadth more mathematical than the subject seems to him to require, and he is not afraid of the astonishment which the reader might show.

In war more than anywhere else in the world things happen differently from what we had expected, and look differently when near from what they did at a distance. With what serenity the architect can watch his work gradually rising and taking the shape of his plan! The doctor, although much more at the mercy of inscrutable agencies and contingencies than the architect, still knows enough of the forms and effects of his means. In war, on the other hand, the commander of a great mass finds himself in a constant surge of false and true information, of mistakes committed through fear, through negligence, through thoughtlessness, of acts of disobedience to his orders, committed either from mistaken or correct views, from ill will, a true or false sense of duty, indolence or exhaustion, of accidents which no mortal could have foreseen. In short, he is the victim of a hundred thousand impressions, of which the most have an intimidating, the fewest an encouraging tendency. By long experience in war, the instinct of readily appreciating the value of these incidents is acquired; high courage and strength of character withstand them, as the rock resists the beating of the waves. He who would yield to these impressions would never carry out any of his undertakings, and on that account *perseverance* in the course decided upon, so long as the most decisive reasons against it are not forthcoming, is a very necessary counterpoise. Further, there is hardly any glorious enterprise in war which was not achieved by endless effort, pains and privations; and as here the physical and moral weaknesses of human nature are ever disposed to yield, only a great strength of will, which manifests itself in steadfastness admired by present and future generations, can conduct us to our goal.

CHAPTER 8

SUPERIORITY IN NUMBERS

This is in tactics, as well as in strategy, the most general principle of victory, and we shall first examine it from this general point of view. For that purpose we venture to offer the following exposition:

Strategy determines the point where, the time when and the military force with which the battle is to be fought. By this threefold determination it has therefore a very essential influence on the issue of the engagement. If tactics has fought the engagement, if the result is there, let it be victory or defeat, strategy makes such use of it as can be made in accordance with the ultimate object of the war. This object is naturally often a very distant one; seldom does it lie quite close at hand. A series of other objects are subordinated to it as means. These objects, which are at the same time means to a higher end, in practice may be of various kinds; even the ultimate aim of the whole war is a different one in almost every case. We shall acquaint ourselves with these things according as we come to know the separate subjects with which they come in contact, and it is not our intention here to embrace the whole subject by a complete enumeration of them, even if that were possible. Consequently, we are not considering the use of the engagement, for the present.

Those things, through which strategy has an influence on the issue of the engagement, inasmuch as it determines the engagement (to a certain extent decrees it), are not so simple either that they could be comprehended in a single investigation. As strategy appoints time, place and strength, it can do so in practice in many ways, each of which influences in a different manner the outcome of the engagement as well as its success. Therefore we shall acquaint ourselves with this only by degrees, that is, through the subjects which more closely determine practice.

If we strip the engagement of all modifications which it may undergo according to its purpose and the circumstances from which it proceeds, if finally we set aside the value of the troops, because that is a given quantity, there remains only the bare conception of the engagement, that is, a combat without form, in which we distinguish nothing but the number of the combatants.

This number will therefore determine victory. Now from the number of abstractions we have had to make in order to arrive at this point, it follows that superiority in numbers is only one of the factors which produce victory; that therefore far from having obtained everything or even the principal thing with superiority in numbers, we have perhaps obtained very little by it, according as the circumstances involved happen to vary.

But this superiority has degrees, it may be imagined as twofold, threefold or fourfold, and every one realizes that by increasing in this way, it must overpower everything else.

In this connection we grant that superiority in numbers is the most important factor in the result of an engagement, only it must be sufficiently great to counter-balance all other concurrent circumstances. The direct consequence of this is that the greatest possible number of troops should be brought into action at the decisive point of the engagement.

Whether then these troops suffice or not, we have done in this respect all that our means permitted. This is the first principle in strategy, and in the general way that it is stated here, it can be applied just as well to Greeks and Persians, or Englishmen and Mahrattas, as to Frenchmen and Germans. But let us turn our attention to the military conditions in Europe, in order to arrive at some more definite idea on this subject.

Here we find armies much more alike in equipment, organization and practical skill of every kind. Only a passing difference still exists in the military virtue of the army and in the talent of the general. If we go through the military history of modern Europe, we find no example of a Marathon.

Frederick the Great beat 80,000 Austrians at Leuthen with about 30,000 men, and at Rossbach with 25,000 some 50,000 allies; these are, however, the only instances of victories gained against an enemy double, or more than double in number. We cannot very well cite the battle which

Charles XII fought at Narva. For the Russians were at that time hardly to be regarded as Europeans, and furthermore even the principal circumstances of this battle are too little known. Bonaparte had at Dresden 120,000 against 220,000, the odds being, therefore, not even twice his own number. At Kollin, Frederick the Great did not succeed, with 30,000 against 50,000 Austrians, neither did Bonaparte in the battle of Leipzig, where he was fighting with 160,000 men against 280,000, the superiority of the enemy therefore being far from double.

From this we may infer, that it is very difficult in the present state of Europe, for the most talented general to gain a victory over an enemy double his strength. Now if we see double numbers prove such a weight in the scale against the greatest generals, we may be sure that, in ordinary cases, in small as well as great engagements, an important superiority of number, which, however, need not be more than twice as many, will be sufficient to ensure the victory, however disadvantageous other circumstances may be. Of course, one may conceive of a mountain pass, in which even a tenfold superiority would not suffice to overpower the enemy, but in such a case we cannot speak of an engagement at all.

We think, therefore, that in our own circumstances, as well as in all similar ones, strength at the decisive point is a matter of capital importance, and that this, in most cases, is definitely the most important thing of all. The strength at the decisive point depends on the absolute strength of the army, and on the skill with which it is employed.

The first rule would therefore be to enter the field with an army as strong as possible. This sounds very much like a truism, but it really is not one.

In order to show that for a long time the strength of the military forces was by no means regarded as a vital matter, we need only observe that in most and even in the more detailed histories of the wars of the eighteenth century, the strength of the armies is either not given at all, or only incidentally, and in no case is any special value set upon it. Tempelhoff in his history of the Seven Years' War is the earliest writer who mentions it regularly, but nevertheless he does it only very superficially.

Even Massenbach, in his manifold critical observations on the Prussian campaigns of 1793–1794 in the Vosges, talks a great deal about hills and valleys, roads and footpaths, but never says a syllable about the strength on either side.

Another proof lies in a wonderful notion which haunted the heads of many critical writers, according to which there was a certain size of an army which was the best, a normal quantity, beyond which excessive forces were burdensome rather than useful.[1]

Lastly, there are a number of instances to be found in which all the available forces were not really used in the battle, or in the war, because superiority in numbers was not considered to have that importance which in the nature of things belongs to it.

If we are thoroughly imbued with the conviction that with a considerable superiority in numbers everything possible can be gained by force, this clear conviction cannot fail to react on the preparations for the war, in order that we may take the field with as many troops as possible, and either ourselves obtain the preponderance, or at least guard against one on the part of the enemy. So much for what concerns the absolute force with which the war is to be conducted.

The measure of this absolute force is determined by the government; and although with this determination the real military activity begins, and though it forms an essential part of the strategy of the war, still in most cases the general who is to command these forces in the war must regard their absolute strength as a given quantity, whether it be that he had no voice in determining it, or that circumstances prevented a sufficient expansion being given to it.

There remains nothing, therefore, but to produce, even where an absolute superiority is not attainable, a relative one at the decisive point, by making skilful use of what we have.

The calculation of space and time appears as the most essential thing in this matter. This has caused people to regard this subject of strategy as one which embraces nearly the whole art of using military forces. Indeed, some have gone so far as to ascribe to great generals in strategy and tactics a mental organ peculiarly adapted to this purpose.

But the co-ordination of time and space, although it everywhere lies at the foundation of strategy, and is, so to speak, its daily bread, is nevertheless neither the most difficult of its tasks nor the most decisive.

If we take an unprejudiced glance at military history, we shall find that the instances in which mistakes in such a calculation have proved the cause of serious losses are, at least in strategy, very rare. But if the conception of a skilful correlation of time and space is to explain every in-

---

[1] Tempelhoff and Montalembert first occur to us in this connection, the former in a passage of his first part, page 148, the latter in his correspondence with reference to the Russian plan of operations for 1759.

stance of a resolute and active commander beating several of his opponents with one and the same army (Frederick the Great, Bonaparte) by means of rapid marches, then we confuse ourselves unnecessarily with conventional language. To make ideas clear and profitable, it is necessary that things should always be called by their right names.

Correct judgment of their opponents (Daun, Schwarzenberg), audacity to oppose them for a short time with a small force only, energy in prolonged marches, boldness in sudden assaults, the intensified activity which great souls acquire in the moment of danger, these are the grounds of such victories. And what have these to do with the ability correctly to coordinate two such simple things as time and space?

But even that repercussion of forces, in which the victories at Rossbach and Montmirail give the impulse to victories at Leuthen and Montereau, and upon which great generals on the defensive have often relied, is still, if we wish to be clear and exact, only a rare occurrence in history.

Much more frequently the relative superiority—that is, the skilful massing of superior forces on the decisive point—has its foundation in the correct appreciation of such points, in the appropriate direction which by that means has been given to the forces from the outset, and in the resolution which is required if we have to sacrifice the unimportant in favor of the important—that is, to keep our forces concentrated in an overpowering mass. In this respect, Frederick the Great and Bonaparte are particularly characteristic.

With this we feel that we have allotted to superiority in numbers the importance which is due it. It is to be regarded as the fundamental idea, and is always to be sought before anything else, and as far as possible.

But to regard it on this account as a necessary condition of victory would be a complete misconception of our exposition. In the conclusion to be drawn from it there is nothing more than the value which we should attach to numerical strength in the engagement. If that strength is made as great as possible, then it has conformed with the principle, and only a view of the general situation decides whether or not the engagement is to be avoided for lack of sufficient force.

CHAPTER 9

THE SURPRISE

From the subject of the preceding chapter—the general endeavor to attain a relative superiority—there follows another endeavor which must consequently be just as general in its nature; this is the *surprise* of the enemy. It lies more or less at the foundation of all undertakings, for without it superiority at the decisive point is really not conceivable.

Surprise becomes, therefore, the means to the attainment of numerical superiority; but is also to be regarded as an independent principle in itself, on account of its moral effect. When it is successful to a high degree, confusion and broken courage in the enemy's ranks are the consequences, and there are sufficient examples great and small to show how these multiply a success. We are not now speaking of the actual raid, which belongs to the chapter on attack, but of the endeavor by measures in general, and especially by the distribution of forces, to surprise the enemy, which is just as conceivable in defense, and which in tactical defense particularly is a chief point.

We say: surprise lies at the foundation of all undertakings without exception, only in very different degrees, according to the nature of the undertaking and other circumstances.

This difference, indeed, begins with the characteristics of the army and its commander, even with those of the government.

Secrecy and rapidity are the two factors in this product. Both presuppose great energy in the government and the commander-in-chief, and a high sense of military duty on the part of the army. With effeminacy and loose principles it is vain to count on a surprise. But general, indeed indispensable, as this endeavor is, and true as it is that it will never be wholly ineffective, it is still none the less true that it seldom succeeds to a *remarkable* degree, and that this follows from the nature of the thing itself. We should form an erroneous conception, therefore, if we believed that by this means above all others there is much to be attained in war. Theoretically it promises a great deal; in execution it generally bogs down in the friction of the whole machine.

In tactics the surprise is much more at home, for the very natural reason that all times and distances are shorter. In strategy, therefore, it will be the more feasible, according as its measures lie nearer to the province of tactics, and the more difficult the higher they lie toward the province of policy.

The preparations for a war usually require several months; the assembly of an army at its principal positions requires generally the establishment of depots and magazines and considerable marches, the direction of which can be guessed soon enough.

It is therefore very seldom that one state surprises another by a war, or by the general direction of its forces. In the seventeenth and eighteenth centuries, when war was very much concerned with sieges, it was a frequent aim, and quite a peculiar and important chapter in the art of war, to surround a strong place unexpectedly, but even that succeeded only rarely.

On the other hand, with things that can be done in a day or two, a surprise is much more conceivable, and it is often not difficult, therefore, to steal a march, and thereby seize a position, a point of country, a road, etc. But it is evident that what surprise gains in this way in easy execution, it loses in its effectiveness, just as this effectiveness increases in the other direction. Whoever believes that he may connect great results with such surprises on a small scale—as, for example, the gain of a battle, the capture of an important magazine—believes in something which, no doubt, is quite conceivable, but for which there is no warrant in history; for there are, on the whole, very few instances where anything great has resulted from such surprises. From this we may justly conclude that there are inherent difficulties in the matter.

Certainly, whoever consults history on such points must not depend on certain show pieces of historical critics, on their wise dicta and self-complacent pomp of technical terms, but must face the facts themselves squarely. There is, for instance, a certain day in the campaign in Silesia, 1761, which, in this respect, has attained a kind of notoriety. It is the 22nd of July, the day on which Frederick the Great stole from Laudon the march to Nossen, near Neisse, by which, as is said, the junction of the

Austrian and Russian armies in Upper Silesia became impossible, and, therefore, a period of four weeks was gained for the king. Whoever reads about this event carefully in the principal histories,[1] and considers it impartially, will never find this significance in the march of the 22nd of July; and generally in the whole argument on this subject, which has become so popular, he will see nothing but contradictions, whereas in the proceedings of Laudon, in this renowned period of maneuvers, he will see much that was objectless. How could one, with a thirst for truth and clear conviction, accept such historical evidence?

When we expect great effects from the principle of surprise in the course of a campaign, we think of great activity, rapid resolutions and forced marches as the means of producing them. That these things, however, even when they are present in a high degree, will not always produce the desired effect we see in examples given by two generals, who may be considered to have had the greatest talent in the use of these means, Frederick the Great and Bonaparte. The former when from Bauzen he fell so suddenly on Lascy in July, 1760, and turned against Dresden, gained nothing by the whole of that intermezzo, but rather placed his affairs in a condition notably worse, as the fortress of Glatz fell in the meantime.

In 1813, Bonaparte twice turned suddenly from Dresden against Blücher, to say nothing of the invasion of Bohemia from Upper Lusatia, and both times without attaining the desired object. They were a beating of the air, which only cost him time and force, and might have placed him in a dangerous position at Dresden.

A very successful surprise, therefore, in this field also, does not proceed from the mere activity, energy and resolution of the commander. It must be favored by other circumstances. But by no means do we deny that there can be success; we only wish to connect it with the necessity of favorable circumstances, which, of course, do not occur very frequently, and which the commander can seldom bring about.

Those very generals, named above, afford a striking example of this. We take first Bonaparte in his famous enterprise against Blücher's army in 1814, when, separated from the main army, it was marching down the Marne. A two days' march to surprise the enemy could hardly have given greater results. Blücher's army, extended over a distance of three days' march, was beaten in detail, and suffered a loss equal to that of defeat in a major battle. This was completely the effect of surprise, for if Blücher had thought of such a close possibility of an attack from Bonaparte he would

---

[1] Tempelhoff: *Der Veteran Friedrich der Grosse.*

have organized his march quite differently. The result is to be attributed to this mistake on the part of Blücher. Bonaparte, of course, did not know these circumstances, and so, as far as he was concerned, it was an intervention of lucky chance.

It is the same with the battle of Liegnitz in 1760. Frederick the Great gained this fine victory by changing during the night a position which he had just before taken up. Laudon was thereby completely surprised, and the result was the loss of seventy pieces of artillery and 10,000 men. Although Frederick the Great had at this time adopted the principle of moving backward and forward, in order thereby to make a battle impossible, or at least to disconcert the enemy's plans, still the change of position on the night of the 14th–15th was not made exactly with that intention, but, as the king himself says, because the position of the 14th did not please him. Here, too, therefore, chance played a large part. Without this happy coincidence of the attack and the change of position in the night, and the inaccessible country, the result would not have been the same.

Also in the higher and highest province of strategy there are some instances of surprises having important results. We shall only cite the brilliant marches of the Great Elector against the Swedes from Franconia to Pomerania and from the Mark (Brandenburg) to the Pregel, in the campaign of 1757, and the famous passage of the Alps by Bonaparte in 1800. In the latter case an army gave up its whole theater of war by a capitulation, and in 1757 another army was very near giving up its theater of war and itself as well. Lastly, as an instance of wholly unexpected war, we may adduce the invasion of Silesia by Frederick the Great. Great and sweeping are the successes in all these cases, but such events are not common in history if we do not confuse with them cases in which a state, for want of activity and energy (Saxony in 1756, and Russia in 1812), has not completed its preparations in time.

Now there still remains an observation which concerns the core of the matter. A surprise can only be effected by that party *which gives the law to the other*, and he who takes the right action *gives the law*. If we surprise the adversary by a wrong measure, then instead of getting good results, we may have to endure a strong counterattack; in any case the adversary need not trouble himself much about our surprise, for he finds in our mistake the means of avoiding the evil. As the offensive includes much more positive action than the defensive, so the surprise is, of course, much more in place in an attack, but by no means exclusively so, as we shall see later on. Mutual surprises by the offensive and defensive, therefore, may occur, and then whoever has best hit the nail on the head would necessarily carry the day.

It ought to be thus, but practical life does not keep to this line so exactly, and that for a very simple reason. The moral effects of a surprise often convert the worst case into a good one for the side which enjoys their assistance, and does not allow the other to come to a proper decision. Here, more than anywhere else, we have in mind not only the chief commander, but each individual one, because the surprise has the very peculiar effect of violently loosening the bond of unity, so that the individuality of each separate leader readily comes to light.

Much depends here on the general relation in which the two parties stand to each other. If the one side through a general moral superiority is able to intimidate and outdo the other, then it will be able to use the surprise with greater success, and even achieve good results where properly it should come to ruin.

CHAPTER 10

STRATAGEM

Stratagem presupposes a concealed intention, and is, therefore, opposed to straightforward, simple, that is, direct dealing, just as wit is opposed to direct proof. It has therefore nothing in common with means of persuasion, of self-interest, of force, but has a great deal to do with deceit, because that likewise conceals its intention. It is even itself a deceit, when all is said and done, but still it differs from what is commonly called deceit, and in this respect: that there is no direct breach of word. He who employs stratagem allows the person himself, whom he wishes to deceive, to commit the errors of intelligence, which at last coalescing into *one* effect, suddenly change the nature of things before his eyes. We may therefore say that as wit is a sleight-of-hand with ideas and conceptions, so stratagem is a sleight-of-hand with actions.

At first sight it appears as if strategy had not without justification derived its name from stratagem, and that, with all the real and apparent changes which in its long history war has undergone since the time of the Greeks, this term still points to its real nature.

If we leave to tactics the actual delivery of the blow, the engagement itself, and look upon strategy as the art of using with skill the means thereto, then, besides the forces of temperament, such as burning ambition which is always pressing like a spring, a strong will which yields with difficulty, and so forth, there seems no subjective gift of nature so suited to guide and inspire strategic activity as stratagem. The general tendency to surprise, treated in the preceding chapter, points to this conclusion, for there is a degree of stratagem, be it ever so small, which lies at the foundation of every attempt to surprise.

But however much we feel a desire to see those acting in war outdo each other in sly activity, skilfulness and stratagem, still we must admit that these qualities manifest themselves but little in history, and have rarely been able to work their way to the surface from the mass of events and circumstances.

The reason for this is quite easily seen, and it is almost identical with the subject matter of the preceding chapter.

Strategy knows no other activity than the arrangement of engagements together with the measures which relate to it. Unlike ordinary life, it is not concerned with transactions which consist merely of words—that is, in statements, declarations, etc. But it is chiefly these means, which do not cost much, wherewith the person using stratagem takes people in.

That which is similar to this in war—such as, plans and orders given merely for the sake of appearances, false reports purposely given out to the enemy—is usually of so little effect in the field of strategy that it is only resorted to in particular cases which arise spontaneously. Therefore it cannot be regarded as a free activity emanating from the person acting.

But to carry out such measures as the arrangement of engagements to such an extent that they will make an impression on the enemy requires a considerable expenditure of time and forces; of course, the greater the impression to be made, the greater the expenditure. Because we are not usually willing to make the sacrifice required, very few so-called demonstrations in strategy have the desired effect. In fact, it is dangerous to use large forces for any length of time merely for the sake of appearances, because there is always the risk of its being done in vain, and then these forces are lacking at the decisive point.

The person acting in war is always aware of this sober truth, and therefore has no desire for the game of crafty agility. The bitter earnestness of necessity usually forces us into direct action, so that there is no room for that game. In a word, the pieces on the strategical chessboard are lacking in that agility which is the element of stratagem and cunning.

The conclusion we draw is that a correct and penetrating eye is a more necessary and more useful quality for a general than stratagem, although that also does no harm as long as it does not exist at the expense of qualities of temperament, which is only too often the case.

But the weaker the forces become which are under the command of strategy, so much the more they become adapted for stratagem, so that to the very weak and small, for whom no prudence, no sagacity is any longer sufficient, at the point where all art seems to forsake them, stratagem offers itself as a last resource. The more desperate their situation and the more everything concentrates into one single, desperate blow, the more readily stratagem comes to the aid of their boldness. Relieved of all further calculations, freed from all later penalty, boldness and stratagem may intensify each other, and thus concentrate at one point an infinitesimal glimmering of hope into a single ray, which may likewise serve to kindle a flame.

CHAPTER 11

ASSEMBLY OF FORCES IN SPACE

The best strategy is *always to be very strong*, first of all generally, then at the decisive point. Therefore, apart from the effort which creates the army and which does not always proceed from the general, there is no more imperative and no simpler law for strategy than to *keep the forces concentrated*. Nothing is to be separated from the main army unless called away by some pressing object. We stand firm on this criterion, and regard it as a guide to be depended upon. The reasonable grounds on which a separation of forces may be made we shall learn by and by. Then we shall also see that this principle cannot produce the same general results in every war, but that these results differ according to the means and end.

It seems incredible, and yet it has happened hundreds of times, that troops have been divided and separated merely on account of vague adherence to traditional fashion, without any clear perception of the reason.

If the concentration of the whole force is acknowledged as the norm, and every division and separation as an exception which must be justified, not only will that folly be completely avoided, but also many an erroneous ground for separating troops will be eliminated.

CHAPTER 12

ASSEMBLY OF FORCES IN TIME

We have here to deal with a conception which, when it comes in contact with active life, spreads many kinds of deceptive illusions. We, therefore, consider a clear definition and development of the idea to be necessary, and we hope to be allowed another short analysis.

War is the impact of opposing forces upon each other, from which it follows as a matter of course that the stronger not only destroys the other, but carries it along in its movement. This fundamentally admits of no successive operation of forces, but makes the simultaneous application of all forces intended for the impact appear a primary law of war.

So too it really is, but only so far as the struggle also really resembles a mechanical impact. But when it consists in a lasting mutual interaction of destructive forces, then we can of course imagine a successive action of them. This is the case in tactics, principally because firearms form the basis of all tactics, but also for other reasons as well. If in an engagement with firearms, 1,000 men are used against 500, then the amount of losses is the sum of the losses of the enemy's forces and our own. One thousand men fire twice as many shots as 500, but more shots will strike the 1,000 than the 500 because it is to be assumed that they stand in closer order than the other. If we were to suppose the number of hits to be double, then the losses on each side would be equal. From the 500 there would be, for example, 200 disabled, and out of the 1,000 likewise the same; now if the 500 had kept another body of equal number in reserve entirely out of fire, then both sides would have 800 effective men; but of these, on the one side, there would be 500 men quite fresh, fully supplied with ammunition, and in their full vigor; on the other side only 800, all equally disorganized, in want of sufficient ammunition and weakened in physical force. The assumption that the 1,000 men merely on account of their greater number would lose twice as many as 500 would have lost in their place is of course not correct; therefore the greater loss which the side suffers that has placed half of its force in reserve must be regarded as a disadvantage. Further it must be admitted, that in most cases the 1,000 men could in the first moment gain the advantage of being able to drive their opponent out of his position and force him to withdraw. Now, whether these two advantages are equivalent to the disadvantage of finding ourselves with 800 men to a certain extent disorganized by the engagement, opposed to an enemy who is at least not materially weaker in numbers and who has 500 entirely fresh troops, is a question which cannot be decided by means of further analysis; we must here rely on experience, and there will scarcely be an officer with some experience in war who will not in most cases assign the advantage to that side which has the fresh troops.

In this way it becomes evident how the employment of too many forces in an engagement may be disadvantageous; for whatever advantages the superiority may give in the first moment, we may have to pay dearly for it in the next.

But this danger goes only as far as the disorder, the state of disintegration and weakness goes, in a word, up to the crisis which every engagement brings with it even for the victor. So long as this weakened state lasts, the appearance of a proportionate number of fresh troops is decisive.

But where this disintegrating effect of victory ceases, and therefore only the moral superiority remains which every victory gives, it is no longer possible for fresh troops to repair the losses; they would only be carried along in the general movement. A beaten army cannot be led back to victory a day after by means of a strong reserve. Here we find ourselves at the source of a most essential difference between tactics and strategy.

The tactical results, the results within the engagement, and before its close, lie for the most part within the limits of that period of disintegration and weakness. But the strategic result, that is to say, the result of the engagement as a whole, of the accomplished victory, be it small or great, lies outside the limits of that period. It is only when the results of partial engagements have combined into an independent whole that the strategic success is accomplished, but then the state of crisis is over, the forces have resumed their original form and have only been weakened to the extent of their actual losses.

The consequence of this difference is that tactics can make a successive use of forces, strategy only a simultaneous one.

If I cannot, in tactics, decide everything by the first success, if I have to fear the next moment, it naturally follows that I employ only so much of my force for the success of the first moment as appears necessary for that object, and keep the rest beyond the reach both of fire and of hand-to-hand fighting, in order to be able to oppose fresh troops to fresh, or with such to overcome those that are exhausted. But it is not so in strategy. Partly, as we have just shown, it has not so much reason to fear a reaction after an accomplished success, because with that success the crisis comes to an end; partly because not all the forces strategically employed are necessarily weakened. Only so much of them as have been tactically in conflict with the enemy's force, that is, engaged in a partial engagement, are weakened by it; consequently, unless tactics has expended them uselessly, only so much as is unavoidably necessary, but by no means all that is strategically in conflict with the enemy. Corps which, on account of the general superiority in numbers, have either been little engaged or not at all, whose mere presence has contributed to the decision, are after the decision the same as they were before, and as fit for use for new enterprises as if they had been entirely inactive. How greatly such corps, which constitute our superiority, may contribute to the total success is self-evident; indeed, it is not difficult to see how they may even diminish considerably the loss of the forces on our side engaged in tactical conflict.

If, therefore, in strategy the loss does not increase with the number of troops employed, but is on the contrary often even diminished by it, and if, as a natural consequence, the decision in our favor is, by that means, the more certain, then it naturally follows that we can never employ too many forces, and consequently also that those which are at hand for action must be employed simultaneously.

But we must vindicate this proposition on another ground. Hitherto we have only spoken of the combat itself; it is the real activity in war. But men, time and space, which appear as the agents of this activity, must also be taken into account, and the effects of their influence brought into consideration as well.

Fatigue, exertion and privation constitute in war a special agent of destruction, not essentially belonging to combat, but more or less inseparably bound up with it, and indeed one which especially belongs to strategy. They no doubt exist in tactics as well, and perhaps there in the highest degree; but since the duration of the tactical acts is shorter, the effects of exertion and privation can come but little into consideration in them. But in strategy, on the other hand, where time and space are on a larger scale, their influence is not only always noticeable, but very often quite decisive. It is not at all uncommon for a victorious army to lose many more by sickness than on the field of battle.

If, therefore, we look at this sphere of destruction in strategy in the same manner as we have considered that of fire and hand-to-hand fighting in tactics, then we may well imagine that everything exposed to it will, at the end of the campaign or of any other strategic period, be weakened, which makes the arrival of a fresh force decisive. We might therefore conclude that there is a motive in the latter case as well as the former to strive for the first success with as few forces as possible, in order to reserve this fresh force for the last.

In order accurately to evaluate this conclusion, which, in numerous instances of actual practice, will have a great appearance of truth, we must direct our attention to the separate ideas which it contains. In the first place, we must not confuse the idea of mere reinforcement with that of fresh unused troops. There are few campaigns at the end of which an increase of force would not be highly desirable for the conqueror as well as the conquered, and indeed would seem decisive; but that is not the point here, for that increase of force would not be necessary if the force had been so much larger at the beginning. It would, however, be contrary to all experience to suppose that an army coming fresh into the field is to be esteemed higher in point of moral value than an army already in the field, just as a tactical reserve is indeed more to be valued than a body of troops which has already suffered severely in engagement. Just as an unfortunate campaign lowers the courage and moral force of an army, a successful one raises their value in this respect. In the majority of cases, therefore, these influences balance one another, and then there remains over and above as clear gain the habituation to war. Moreover, here we must look rather at successful than at unsuccessful campaigns, because when the course of the latter can be foreseen with greater probability, the forces are lacking anyhow, and, therefore, the reservation of a part of them for future use is out of the question.

This point being settled, there remains the question: Do the losses which a force sustains through fatigue and privation increase in proportion to the size of the force, as is the case in an engagement? And to that we must answer "No."

Fatigue results for the most part from the dangers which more or less pervade every moment of the act of war. To encounter these dangers at all points, to go forward with assurance on our course of action, is the object of a great number of activities which constitute the tactical and strategic service of the army. This service is the more difficult the weaker the army, and the easier as the army's numerical superiority over that of the enemy increases. Who can doubt this? A campaign against a much weaker enemy will therefore cost less fatigue than against one just as strong or stronger.

So much for fatigue. It is somewhat different with privations; they consist chiefly of two things, want of food and want of shelter for the troops, either in quarters or in comfortable camps. The greater the number of men in one place, the greater, of course, these two deficiencies will be. But does not the superiority in force afford also the very best means of spreading out and finding more room, and therefore more means of subsistence and shelter?

If Bonaparte, in his advance into Russia in 1812, concentrated his army in great masses upon one single road in a manner never heard of before, and thus caused privations equally unparalleled, we must ascribe it to his principle that it is impossible to be too strong at the decisive point. Whether in this instance he did not strain the principle too far is a question which would be out of place here. But it is certain that, if he had made a point of avoiding the hardship thus brought about, he would only have had to advance on a wider front. Room was not lacking for the purpose in Russia, and in very few cases will it be lacking elsewhere. Therefore, this cannot serve as a proof that the simultaneous employment of very superior forces was bound to produce greater weakness. But now, suppose that wind and weather and the inevitable fatigues of war, had produced a diminution even in that part of the army, which as a supplementary force could have been reserved in any case for later use. Then in spite of the relief provided by such a force for the whole, we are still obliged to take a comprehensive general view of the whole situation, and therefore ask: Will this diminution of force suffice to counterbalance the gain in forces, which we, through our superiority in numbers, may be able to make in more ways than one?

But there still remains a most important point to be mentioned. In a limited engagement, we can without much difficulty roughly determine the force required to obtain a major result which we are contemplating, and, consequently, we can also determine what would be superfluous. In strategy this is practically impossible, because the strategic success has no such well-defined object and no such circumscribed limits as the tactical. Thus what can be looked upon in tactics as an excess of forces must be regarded in strategy as a means of extending the success, if an opportunity is offered. With the magnitude of the success the percentage of profit increases at the same time, and in this way the superiority in numbers may soon reach a point which the most careful economy of forces could never have attained.

By means of his enormous numerical superiority, Bonaparte succeeded in reaching Moscow in 1812, and in taking that central capital. Had he by means of this superiority, in addition to that, succeeded in completely annihilating the Russian Army, he would, in all probability, have concluded a peace in Moscow which in any other way was less attainable. This example is used only to explain the idea, not to prove it, which would require a circumstantial demonstration, for which this is not the place.

All these reflections have had reference merely to the idea of a successive employment of forces, and not to the conception of a reserve properly so called, which they do indeed take into account, but which, as we shall see in the following chapter, is connected with other ideas.

What we desired to establish here is that whereas in tactics the military force through the mere duration of its actual employment suffers a diminution of power, and time, therefore, appears as a factor in the result, this is not the case in strategy in an essential way. The destructive effects which time also produces upon the forces in strategy are partly diminished through the bulk of these forces, partly made good in other ways, and, therefore, in strategy it cannot be the object to make time an ally for its own sake by bringing troops successively into action.

We say "for its own sake," for on account of other circumstances which it brings about but which are different from it, the value time can have, indeed must necessarily have, for one of the two parties, is quite another thing, anything but indifferent or unimportant, and will be the subject of consideration later.

The law which we have been seeking to set forth is, therefore, that all forces which are available and destined for a strategic object should be *simultaneously* applied to it; and this application will be so much the more complete the more everything is compressed into one act and into one moment.

But there is nevertheless in strategy an after-pressure and a successive action which, as a chief means toward the ultimate success, is the less to be overlooked. It is the continual development of new forces. This is also the subject of another chapter, and we only refer to it here in order to prevent the reader from having something in view of which we have not been speaking.

We now turn to a point which is very closely connected with what we have been considering and the settlement of which will cast full light on the whole—we mean the *strategic reserve*.

CHAPTER 13

STRATEGIC RESERVE

A reserve has two objects which are clearly distinct from each other, namely, first, to prolong and renew the combat, and secondly, use in case of unforeseen events. The first object implies the usefulness of a successive application of forces, and on that account cannot occur in strategy. Cases in which a corps is sent to a point which is about to fall are obviously to be placed in the category of the second object, as the resistance which has to be offered here had not been sufficiently foreseen. A corps, however, which is merely intended to prolong the combat, and with that object in view is placed in rear, would only be placed out of reach of fire, but under the command and at the disposition of the commanding officer in the engagement, and accordingly would be a tactical and not a strategic reserve.

But the need for a force ready for unforeseen events may also occur in strategy, and consequently there can also be a strategic reserve, but only where unforeseen events are imaginable. In tactics, where the enemy's measures are generally ascertained only by direct sight, and where they may be concealed by every wood, every small valley of undulating ground, we must naturally always be prepared, more or less, for the possibility of unforeseen events, in order to strengthen, subsequently, those points which prove to be too weak, and, in fact, to modify generally the disposition of our troops, so as to make it correspond better to that of the enemy.

Such cases must also happen in strategy, because the strategic act is directly linked to the tactical. In strategy also many a measure is adopted only in consequence of what is actually seen, of uncertain reports arriving from day to day, or even from hour to hour, and, lastly, from the actual results of the engagements. It is, therefore, an essential condition of strategic command that, according to the degree of uncertainty, forces must be kept in reserve for later use.

In the defensive generally, but particularly in the defense of certain sections of ground, like rivers, hills, etc., this, as is well known, has to be done constantly.

But this uncertainty diminishes in proportion as the strategic activity departs from the tactical, and ceases almost altogether in those regions where it borders on politics.

The direction in which the enemy leads his columns to battle can be perceived by actual sight only; where he intends to cross a river is learned from a few preparations which are revealed shortly before; the side from which he will invade our country is usually announced by all the newspapers before a pistol shot has been fired. The greater the scale of the measure the less it is possible to surprise by it. Time and space are so considerable, the circumstances out of which the action proceeds so public and so little subject to change that the result is either known in good time or can be discovered with certainty.

On the other hand, the use of a reserve in this province of strategy, provided one were actually available, will also be always less effective the more the measure tends to be of a general nature.

We have seen that the decision of a partial engagement is nothing in itself, but that all partial engagements only find their complete solution in the decision of the total engagement.

But even this decision of the total engagement has only a relative importance of many different gradations, according as the force over which the victory has been gained forms a more or less large and important part of the whole. The lost battle of a corps may be repaired by the victory of an army. Even the lost battle of an army may not only be counterbalanced by the gain of a more important one, but could be converted into a fortunate event (the two days of Kulm, August 29 and 30, 1813). No one can doubt this; but it is quite clear that the weight of each victory (the successful issue of each total engagement) is the more independent the more important the part conquered, and that consequently the possibility of repairing the loss by subsequent events diminishes in the same proportion. In another place we shall have to examine this in greater detail; it suffices for the present to have drawn attention to the unquestionable existence of this progression.

If we now add lastly to these two considerations the third, namely, that if the successive use of forces in tactics always shifts the main decision to the end of the whole act, the law of the simultaneous use of the forces in strategy, on the contrary, lets the main decision (which need not be the final one) take place almost always at the beginning of the great action, then in these three conclusions we have sufficient grounds to find strategic reserves more and more superfluous, more and more useless, more and more dangerous, the more general their purpose.

The point where the idea of a strategic reserve begins to become untenable is not difficult to determine: it lies in the *main decision*. All the forces must be used for the main decision, and every reserve (active force available) which is only intended for use after that decision is absurd.

If, therefore, tactics has in its reserves the means of not only meeting unforeseen dispositions on the part of the enemy, but also of repairing that which never can be foreseen, namely, the result of the engagement, should that be unfortunate, strategy on the other hand must, at least as far as the main end is concerned, renounce the use of these means. As a rule, it can only repair the losses which occur at one point by advantages gained at another, in a few cases by moving troops from one point to another. The idea of preparing for such reverses in advance by placing forces in reserve must never be entertained in strategy.

We have pointed out as an absurdity the idea of a strategic reserve which is not to co-operate in the main decision, and as this is so beyond any doubt, we should not have been led into such an analysis as we have made in these two chapters, were it not that, in the disguise of other conceptions, it looks somewhat better, and so frequently makes its appearance. One person sees in it the acme of strategic sagacity and caution; another rejects it, and with it the idea of any reserve, consequently even of a tactical one. This confusion of ideas passes into real life, and if we wish to see a memorable instance of it we have only to recall that Prussia in 1806 left a reserve of 20,000 men cantoned in the Mark (Brandenburg), under Prince Eugene of Württemberg, which could not reach the Saale in time to be of any use, and that another force of 25,000 men belonging to this power remained in East and South Prussia, intended only to be put on a war footing afterward as a reserve.

After these examples we cannot be accused of having been fighting with windmills.

CHAPTER 14

ECONOMY OF FORCES

The road of reason, as we have said, seldom allows itself to be reduced by principles and opinions to a mere mathematical line. There always remains a certain margin. It is the same in all the practical arts of life. For the lines of beauty there are no abscissae and ordinates; circles and ellipses are not brought into being by means of their algebraical formulae. The person acting in war, therefore, must at one moment trust himself to the delicate instinctive judgment which, founded on natural sagacity and educated by reflection, almost unconsciously hits upon the right course; at another he must simplify the law by reducing it to leading distinctive features, which form his rules; and at yet another, the established routine must become the standard to which he adheres.

As one such simplified distinctive feature or mental aid we look upon the principle of watching continually over the co-operation of all forces, or, in other words, of keeping constantly in view that no part of them should ever be idle. Whoever has forces in places where the enemy does not give them sufficient employment, whoever has part of his forces on the march—that is, allows them to lie idle—while the enemy's are fighting, is a bad manager of his forces. In this sense there is a waste of forces, which is even worse than their inappropriate use. If there must be action, then the first necessity is that all parts should act, because the most inappropriate activity still keeps employed and destroys a portion of the enemy's forces, whilst troops completely inactive are for the moment quite neutralized. Obviously this idea is connected with the principles contained in the last three chapters. It is the same truth, but seen from a somewhat more comprehensive point of view and condensed into a single conception.

CHAPTER 15

GEOMETRICAL ELEMENT

The length to which the geometrical element or form in the disposition or military forces in war can be carried as a governing principle, we see in the art of fortification, where geometry manages almost everything, great or small. Also in tactics it plays a great part. It is the basis of tactics in the narrower sense of the theory of moving troops. In field fortification, as well as in the theory of positions, and of their attack, its angle and lines rule like law-givers who have to decide the contest. Many things here were at one time misapplied, and others were mere trifling. Still, however, in the tactics of the present day, in which in every engagement the aim is to envelop the enemy, the geometrical element has attained anew a great influence, though in a very simple, but constantly recurring, application. Nevertheless, in tactics, where everything is more mobile, where the moral forces, individual traits, and chance are more influential than in a war of sieges, the geometrical element can never attain to the same degree of supremacy as in the latter. Still less is its influence in strategy. Here also, to be sure, the formations in the disposition of troops, the shape of countries and states is of great influence, but the geometrical element is not decisive, as in the art of fortification, and not nearly so important as in tactics. The manner in which this influence manifests itself can only be shown later on at those points where it makes its appearance, and deserves consideration. Here we wish rather to direct attention to the difference which exists between tactics and strategy in this matter.

In tactics time and space quickly dwindle to their absolute minimum. If a body of troops is attacked in flank and rear by the enemy, a point is soon reached at which retreat no longer is possible; such a position is very close to an absolute impossibility of continuing the fight; the army must therefore extricate itself from it, or avoid getting into it. All expedients aiming at this are thus from the very start very effective, chiefly on account of the apprehensions they cause the enemy as to the consequences. This is why the geometrical disposition of the forces is such an important factor in the result.

In strategy this is only faintly reflected, on account of the greater spaces and times involved. We do not fire from one theater of war to another; and often weeks and months pass before a strategic movement designed to surround the enemy can be executed. Further, the distances are so great that the probability of hitting the right point at last, even with the best arrangements, remains but small.

In strategy therefore the scope for such expedients, that is of the geometrical element, is much smaller, and for the same reason the effect of an advantage actually gained for the moment at any point is much greater. Such advantage has time to show all its effects before it is disturbed or quite neutralized by any counteracting apprehensions. We therefore do not hesitate to regard it as an established truth that in strategy more depends on the number and the magnitude of the victorious engagements than on the form of the great features in which they are connected.

A view just the reverse has been a favorite theme of modern theory, because a greater importance was supposed to be thus given to strategy. The higher functions of the mind were seen in strategy. It was thought thereby to ennoble war, and, as it was said—by a new substitution of ideas—to make it more scientific. We hold it to be one of the principal uses of a complete theory to expose such vagaries, and as the geometrical element is the fundamental idea from which they usually proceed, we have expressly stressed this point.

CHAPTER 16

ON THE SUSPENSION OF
ACTION IN WARFARE

If one considers war as an act of mutual destruction, we must of necessity imagine both parties as generally making some progress; but at the same time, as regards each existing moment, we must with almost the same necessity suppose the one party to be waiting, and only the other actually advancing, for circumstances can never be absolutely the same on both sides, or continue to be so. In time a change must ensue, from which it follows that the present moment is more favorable to one side than to the other. Now if we suppose that both commanders have a full knowledge of this circumstance, then the one has a motive for action, which at the same time is a motive for the other for waiting. According to this it cannot be in the interest of both to advance at the same time, nor can waiting be in the interest of both at the same time. This mutual exclusion of the same object is not deduced here from the principle of general polarity, and therefore is not a contradiction of the assertion in Book I, Chapter 1, but originates from the fact that here actually the same thing becomes the decisive motive for both commanders, namely, the probability of improving or impairing their position by future action.

But even if we suppose the possibility of a perfect equality of circumstances in this respect, or if we take into account that through imperfect knowledge of their mutual position such an equality may appear to the two commanders to exist, still the difference of political objects does away with this possibility of suspension. One of the parties must of necessity be assumed politically to be the aggressor, because no war could originate from defensive intentions on both sides. But the aggressor has the positive object, the defender merely a negative one. To the first then belongs the positive action, for it is only by that means that he can attain the positive object; in cases where both parties are in precisely similar circumstances, the aggressor is thus called upon to act by virtue of his positive object.

From this point of view, a suspension in the act of war, strictly speaking, is in contradiction to the nature of the thing; because two armies, like two incompatible elements, must destroy one another unremittingly, just as fire and water can never put themselves in equilibrium, but act and react upon one another, until one disappears entirely. What would we say of two wrestlers who remained clasped in a mutual grip for hours without making a movement? Action in war, therefore, like that of a clock which is wound up, should go on running down in constant motion. But wild as is the nature of war, it still wears the chains of human weakness, and the contradiction we see here, viz., that man seeks and creates dangers which, at the same time, he fears, will astonish no one.

If we cast a glance at military history in general, we find so much the opposite of an incessant advance toward the aim, that *suspension* and *inactivity* is quite obviously the *normal condition* of an army in the midst of war, *action*, the exception. This ought almost to raise a doubt as to the correctness of the conception which we have formed. But if military history leads to this doubt when the bulk of its events are taken into account, the latest series of these events redeems our position. The war of the French Revolution shows too plainly its reality, and only proves too clearly its necessity. In this war and especially in the campaigns of Bonaparte, the conduct of war attained that unlimited degree of energy which we have represented as its natural, elemental law. This degree is therefore possible, and if it is possible then it is necessary.

How could any one in fact justify in the eyes of reason the expenditure of forces in war, if action was not the object? The baker only heats his oven if he has bread to put into it; horses are harnessed to the carriage only if we intend to drive; why then make the enormous effort of a war if we intend to engender nothing by it but similar efforts on the part of the enemy?

So much in justification of the general principle. Now we turn to its modifications, as far as they lie in the nature of the thing and do not depend on special cases.

There are three causes to be noticed here, which appear as inherent counterbalances and prevent the too rapid or the uninterrupted movement of the wheel-work.

The first, which produces a constant tendency to delay and thereby becomes a retarding influence, is the natural timidity and want of resolution in the human mind, a kind of gravity in the moral world, which, however, is produced not by attractive, but by repellent, forces, that is to say, by dread of danger and responsibility.

In the elemental flame of war, ordinary natures appear heavier; the impulses must therefore be stronger and more frequently repeated if the motion is to be a continuous one. The mere conception of the object for which arms have been taken up is seldom sufficient to overcome this resistant force, and if a warlike enterprising spirit is not at the head, who feels himself in war in his natural element, as much as a fish in the ocean, or if there is not the pressure from above of some great responsibility, suspension will be the rule of the day, and progress the exception.

The second cause is the imperfection of human understanding and judgment, which is greater in war than anywhere else, because a person hardly knows exactly his own position from one moment to another, and can only conjecture on slight grounds that of the enemy, which is concealed. This often gives rise to the case of both parties looking upon one and the same object as advantageous for them, while in reality the interest of one must preponderate; thus then each may think he acts wisely by waiting for another moment, as we have already said in Book I, Chapter 1.

The third cause, which like a cog-wheel gears into the machinery producing from time to time a complete suspension, is the greater strength of the defensive. A may feel too weak to attack B, from which it does not follow that B is strong enough for an attack on A. The addition of strength, which the defensive gives is not only lost by assuming the offensive, but, in addition to that, passes to the enemy just as, figuratively expressed, the difference of $a + b$ and $a - b$ is equal to $2b$. Therefore it may so happen that both parties, at one and the same time, not only feel themselves too weak to attack, but also are so in reality.

Thus in the midst of the art of war itself, anxious sagacity and the apprehension of too great danger find convenient standpoints from which to assert themselves and tame the elemental violence of war.

However, these causes, without being forced, can hardly explain the long suspensions that undertakings suffered in earlier wars, which were stirred up by no great cause and in which inactivity consumed nine-tenths of the time that the troops remained under arms. This phenomenon is to be traced principally to the influence which the demands of the one party, and the conditions and feeling of the other, exercise on the conduct of war, as has been already observed in the chapter on the essence and object of war.

These things may obtain such a preponderating influence as to make of war a half-hearted affair. A war is often nothing more than an armed neutrality or a menacing attitude to support negotiations or a moderate attempt to gain some small advantage and then await the result or a disagreeable obligation to an alliance which is fulfilled in the most niggardly way possible.

In all these cases in which the impulse given by interest is slight, and the principle of hostility feeble, in which there is no desire to do much to the opponent, and also not much to fear from him; in short, where no powerful motives press and urge, cabinets will not risk much in the game; hence this tame mode of carrying on war, in which the hostile spirit of real war is kept in fetters.

The more war becomes in this manner a half-hearted affair, so much the more its theory becomes destitute of the necessary abutments and buttresses for reasoning; the necessary is constantly diminishing, the accidental constantly increasing.

Nevertheless, in this kind of warfare, there will also be a certain shrewdness; indeed, its action is perhaps more diversified, and of wider range than in the other. The game of hazard with rouleaux of gold pieces seems changed into a game of commerce with pennies. And on this field, where the conduct of war spins out the time with a number of small flourishes, with skirmishes at outposts, half in earnest, half in jest, with long maneuverings which end in nothing, with positions and marches, which afterward are called scientific only because their infinitesimally small causes have been forgotten and common sense can make nothing of them—here on this very field many theorists find the art of war at home. In these feints, parades, half- and quarter-thrusts of former wars, they find the aim of all theory, the supremacy of mind over matter, and modern wars appear to them mere savage fisticuffs, from which nothing is to be learned, and which must be regarded as mere retrograde steps toward barbarism. This opinion is as frivolous as the objects to which it relates. Where great forces and great passions are wanting, it is, of course, easier for an adroit shrewdness to display its dexterity. But is the management of great forces, the piloting in dashing waves and tempest, not in itself a higher exercise of the moral and intellectual faculties? Is not that kind of conventional sword-play included and implicit in the other mode of conducting war? Does it not bear the same relation to it as the motions upon a ship to the motion of the ship itself? Truly it can take place only under the tacit condition that the adversary does no better. And can we tell how long he may choose to respect those conditions? Did not the French Revolution fall upon us in the midst of the fancied security of our old system of war, and drive us from Chalons to Moscow? And did not Frederick the Great in like manner surprise the Austrians reposing in their old military traditions and make their monarchy tremble? Woe to the cabinet which, with a policy of half measures and a fettered military system, comes upon an adversary who, like the rude element, knows no other law than that of his intrinsic strength. Every deficiency in activity and effort is then a weight in the scales in favor of the enemy. Then it is not so easy to change from the fencing posture into that of an athlete, and a slight blow is often sufficient to throw the whole to the ground.

The result of all the causes just mentioned is that the hostile action of a campaign does not progress by a continuous, but by an intermittent, movement, and that, therefore, between the separate bloody actions, there is a period of watching, during which both parties fall into the defensive, and also that usually a higher object causes the principle of aggression to predominate on one side, and thus allows it in general to remain in an advancing position, whereby its proceedings become to some degree modified.

CHAPTER 17

ON THE CHARACTER OF
MODERN WAR

The attention which must be paid to the character of modern war has a great influence upon all plans, especially the strategic.

All conventional methods have been upset by Bonaparte's luck and boldness, and first-rate powers have been annihilated with almost a single blow. The Spaniards by their stubborn resistance have shown what the general arming of a nation and insurgent measures on a great scale can effect, in spite of weakness and looseness in details. Russia, by the campaign of 1812, has taught us that an empire of great dimensions cannot be conquered (which might have been easily known before), and, secondly, that the probability of final success does not in all cases diminish in the same measure as battles, capitals and provinces are lost (which was formerly an irrefragable principle with all diplomatists and made them always ready to enter at once into some bad temporary peace). Russia has proved that, on the contrary, a nation is often strongest in the heart of its own country, when the enemy's offensive power has exhausted itself, and she has shown us with what enormous force the defensive then springs over to the offensive. Prussia (1813), furthermore, has shown that sudden efforts can increase an army sixfold by means of the militia, and that this militia is just as fit for service abroad as in its own country. These events have all shown what an enormous factor the heart and the sentiment of a nation may be in its total political and military strength, and since governments have found out all these additional aids, it is not to be expected that they will let them lie idle in future wars, whether it be that danger threatens their own existence, or fervent ambition drives them on.

That a war which is waged with the whole weight of the national power on each side must be organized according to other principles than those in which everything was calculated according to the relations of standing armies to each other, it is easy to perceive. Standing armies once resembled fleets, the land force resembled the sea force in its relations to the remainder of the state, and from that the art of war on land had in it something of naval tactics, which it has now quite lost.

CHAPTER 18

TENSION AND REST

THE DYNAMIC LAW OF WAR

We have seen in the sixteenth chapter of this book (p. 440), how, in most campaigns, much more time used to be spent in suspension and inactivity than in action. Now, although, as observed in the preceding chapter, we see quite a different character in the present form of war, still it is certain that real action will always be interrupted by more or less long pauses, and this leads to the necessity of our examining more closely the nature of these two phases of war.

If there is a suspension of action in war, that is, if neither party wants anything positive, there is rest, and consequently equilibrium, but of course an equilibrium in the widest sense, in which not only the moral and physical military forces, but all circumstances and interests, are taken into account. As soon as one of the two parties aims at a new positive object and takes active steps to attain it, if only by preparations, and as soon as the adversary opposes this, a tension of forces is created; this lasts until the decision is made—that is, until one party either gives up his object or the other has conceded it to him.

This decision—the foundation of which always lies in the effect of the combinations of engagements which originate from both sides—is followed by a movement in one or the other direction.

When this movement has exhausted itself, either through the difficulties which had to be mastered in overcoming its own internal friction or through newly intervening counterpoises, then either a state of rest sets in again or a new tension and decision, and then a new movement, in most cases in the opposite direction.

This theoretical distinction between equilibrium, tension and motion is more essential for practical action than may at first sight appear.

In a state of rest and of equilibrium various kinds of activity may prevail that result from mere accidental causes, and have no great change for their object. Such an activity may include important engagements—even main battles—but in that case it is of quite a different nature, and on that account mostly operates in a different way.

If a state of tension exists, the effects of the decision are always greater, partly because a greater force of will and a greater pressure of circumstances manifest themselves therein, partly because everything has been prepared and arranged for a great movement. The decision in such cases resembles the effect of a mine well sealed and tamped, while an event, in itself perhaps just as great, occurring in a state of rest, is more or less like a mass of powder puffed away in the open air.

Moreover, the state of tension must, of course, be conceived as existing in different degrees of intensity, and it may therefore approach the state of rest by so many gradations that in the last of them there is very little difference between the two.

The most essential profit which we derive from these reflections is the conclusion that every measure which is taken during a state of tension is more important and more effective than the same measure would have been in a state of equilibrium, and that this importance increases immensely in the highest degrees of tension.

The cannonade of Valmy decided more than the battle of Hochkirch.

In a tract of country which the enemy abandons to us because he cannot defend it, we can settle ourselves quite differently than we would if the retreat of the enemy was only made with the view to a decision under more favorable circumstances. Against a strategic attack in course of execution, a faulty position, a single false march, may be of decisive consequences; while in a state of equilibrium such errors must be of a very glaring kind to excite the activity of the enemy at all.

Most bygone wars, as we have said before, consisted, for the greater part of the time, in this state of equilibrium, or at least in such slight tensions with long intervals between them, and so weak in their effects that the events which occurred in them were seldom of great consequence; often they were theatrical performances in honor of a royal birthday

(Hochkirch), often a mere satisfaction of military honor (Kunersdorf), or of the personal vanity of the commander (Freiberg).

That a commander should thoroughly understand these circumstances, that he should have an instinct to act in the spirit of them, we hold to be highly requisite, and we have experienced in the campaign of 1806 how very much this is sometimes lacking. In that tremendous tension, when everything was pressing toward a supreme decision, and that alone, with all its consequences, should have occupied the whole soul of the commander, measures were proposed and even partly carried out (such as the reconnaissance toward Franconia), which in a state of equilibrium could at most have produced a kind of gentle oscillation. Over these confusing schemes and views, absorbing the activity of the army, the really necessary measures, which alone could save it, were lost.

But this theoretical distinction which we have made is also necessary for further progress in the construction of our theory, because all that we have to say on the relation of attack and defense, and on the execution of this two-sided action, concerns the state of crisis in which the forces are to be found during tension and movement, and because all the activity which can take place during the state of equilibrium will be regarded and treated only as a corollary. For that crisis is the real war and this state of equilibrium only its reflection.

Book IV

THE ENGAGEMENT

CHAPTER 1

INTRODUCTORY

Having in the foregoing book examined the subjects which may be regarded as the effective elements of war, we shall now turn our attention to the engagement as the real activity in warfare, which, by its physical and psychological effects, embraces sometimes more simply, sometimes in a more complex manner, the object of the whole war. In this activity and in its effects those elements must again appear.

The construction of the engagement is tactical in its nature; we only glance at it here in a general way in order to get acquainted with it in its aspect as a whole. In practice the more immediate objects give to every engagement a characteristic form; these more immediate objects we shall not discuss until later. But such peculiarities are, in comparison to the general characteristics of an engagement, mostly only insignificant, so that most of them are much like one another, and, therefore, in order to avoid constant repetition of that which is general, we are compelled to look into it here, before taking up the subject of its more special application.

In the first place, we shall give in the next chapter, in a few words, the characteristics of the modern battle in its tactical course, because this lies at the foundation of our conceptions of the engagement.

CHAPTER 2

CHARACTER OF THE
MODERN BATTLE

From the concepts which we have accepted of tactics and strategy, it follows, as a matter of course, that if the nature of the former is changed, that change must have an influence on the latter. If the tactical phenomena in the one case have an entirely different character from that in the other, then the strategic phenomena must have it also, if they are to remain consistent and reasonable. Therefore it is important to describe a main battle in its modern form before we advance with the study of its employment in strategy.

What do we usually do now in a great battle? We place ourselves quietly in great masses arranged next to one another and behind one another. We deploy only a relatively small portion of the whole, and let it fight it out in a musketry duel which lasts for hours, and which is interrupted now and again and pushed hither and yon by separate small thrusts from charges at the double and bayonet and cavalry attacks. When this line has gradually exhausted its warlike fire in this manner, and there remains nothing more than the ashes, it is withdrawn and replaced by another.

In this manner the battle, with moderated violence, burns slowly away like wet powder, and if the veil of night commands it to stop, because no one can see any longer, and no one chooses to run the risk of blind chance, then an account is taken by each side respectively of the masses remaining, which can still be called effective, that is, which have not yet completely collapsed like extinct volcanoes; account is taken of the ground gained or lost, and of the security of the rear; these results combined with the special impressions as to bravery and cowardice, sagacity and stupidity, which are thought to have been perceived in ourselves and in our opponents, are collected into one single total impression, out of which then springs the resolution to quit the field or to renew the engagement next morning.

This description, which is not intended to be a finished picture of a modern battle, but only to give its general tone, holds good for the offensive and defensive, and the special traits which are lent by the object proposed, the country, etc., may be introduced into it without materially changing this tone.

But modern battles are not so by accident; they are so because the parties are more or less on the same level as regards military organization and the art of war, and because the violence of war, enkindled by great national interests, has broken through artificial limits and been led into its natural paths. Under these two conditions battles will always preserve this character.

This general idea of the modern battle will be useful to us in the sequel in more than one place, if we want to estimate the importance of the individual co-efficients of strength, country, etc. It is only for general, great and decisive engagements, and such as approach them, that this description holds good; minor ones have changed their character in the same direction also, but less than great ones. The proof of this belongs to tactics; we shall, however, have an opportunity later of making this subject clearer by giving a few particulars.

CHAPTER 3

THE ENGAGEMENT IN GENERAL

The engagement is the real warlike activity, everything else only ministers to it; let us therefore take an attentive look at its nature.

Engagement is combat, and in this the object is the destruction or the overcoming of the opponent; the opponent in a particular engagement, however, is the military force which stands opposed to us.

This is the simple conception; we shall return to it, but before we can do that, we must insert a series of others.

If we conceive of the state and its military force as a unit, then the most natural idea is to think of war also as one great single engagement, and in the simple conditions of savage peoples it is indeed not much different. But our wars are made up of a number of simultaneous and consecutive engagements, great and small, and this splitting of the activity into so many individual actions is due to the great diversity of the circumstances out of which with us war arises.

In point of fact, the ultimate object of our wars, the political object, is not always an entirely simple one; and even if it were, still the action is bound up with such a number of conditions and considerations that the object can no longer be attained by one single great act, but only through a number of greater or smaller acts, which are bound up into a whole. Each of these separate activities is therefore a part of a whole, and has consequently a special object by which it is bound to this whole.

We have already said that every strategic action can be reduced to the idea of an engagement, because it is an employment of the military force, at the basis of which there always lies the idea of the engagement. We may therefore reduce every military activity in the province of strategy to the unity made up of individual engagements, and occupy ourselves with the object of these alone. We shall get acquainted with these special objects only by degrees, as we come to speak of the causes which produce them. Here we content ourselves with saying that every engagement, great or small, has its own special object, which is subordinated to the whole. If this is the case, the destruction and conquest of the enemy is only to be regarded as the means of gaining this object, as it unquestionably is.

But this result is true only in its form, and important only because of the connection which the ideas have with one another; and it is precisely in order to rid ourselves of it that we have sought it out.

What does overcoming the enemy mean? Invariably nothing but the destruction of his military force, whether it be by death or wounds or any other means, whether it be completely or only to such a degree that he no longer wishes to continue the combat. Therefore, as long as we set aside all special objects of engagements, we may look upon the complete or partial destruction of the enemy as the sole object of all engagements.

Now we maintain that in the majority of cases, and especially in great engagements, the special object by which the engagement is individualized and connected with the great whole is only a weak modification of that general object. Or it is a secondary object connected with it, important enough to individualize the engagement, but always insignificant in comparison with that general object; so that if the secondary object alone should be obtained, only an unimportant part of its purpose is fulfilled. If this assertion is correct, then we see that the idea, according to which the destruction of the enemy's force is only the means, and something else always the object, is true only in form, and that it would lead to false conclusions, if we did not recollect that this very destruction of the enemy's force is also comprised in that object, and that this object is only a weak modification of it.

The fact that this had been forgotten led to completely false views before the wars of the last period, and created tendencies as well as fragments of systems, by which theory thought it raised itself so much the more above handicraft, the less it supposed itself to be in need of its proper instrument, that is, the destruction of the enemy's force.

Such a system could not, of course, have arisen unless supported by other false suppositions, and unless in place of the destruction of the enemy's forces other things had been substituted to which an efficacy was ascribed which did not rightly belong to them. We shall combat these things whenever the subject gives us occasion to, but we cannot treat of the engagement without insisting on the real importance and value which belong to it, and without giving warning against the errors to which merely formal truth might lead.

But how shall we prove that in most cases, and in those of most importance, the destruction of the enemy's military forces is the chief thing? How shall we meet that extremely subtle idea, which conceives it possible, by means of particularly artful methods, to effect by a small direct destruction of the enemy's forces a much greater destruction indirectly, or by means of small but very cleverly directed blows to produce such paralysis of the enemy's forces, such a bending of the enemy's will, that this mode of procedure should be viewed as a great shortening of the road? Undoubtedly an engagement at one point may be of more value than at another. Undoubtedly there is also a skilful arrangement of engagements in strategy, which is in fact nothing more than the art of this arrangement. To deny that is not our intention, but we maintain that the direct destruction of the enemy's forces is everywhere the predominating thing; we contend here for the predominating importance of the destructive principle and nothing else.

We must, however, call to mind that we are now concerned with strategy, not with tactics, and therefore we do not speak of the means which the former may have of destroying at a small expense a large body of the enemy's forces. If we speak of direct destruction, we mean tactical successes, and therefore our assertion is that only great tactical successes can lead to great strategical ones, or, as we have once before more distinctly expressed it, tactical successes are of paramount importance in the conduct of war.

The proof of this assertion seems to us simple enough; it lies in the time which every complicated (skilful) combination requires. The question whether a simple attack or a more complicated, skilful one will produce greater effects may undoubtedly be decided in favor of the latter as long as the enemy is assumed to be quite passive. But every complicated attack demands more time, and this time must be granted it without the whole being disturbed, while engaged in preparation for its effect, by a counterattack on one of its parts. Now if the enemy decides upon a simpler attack, which can be executed in a short time, he gets the start of us and disturbs the effect of the great plan. Therefore, in considering the value of a complicated attack we must take into account all the dangers which we run during its preparation, and can only adopt it if there is no reason to fear that the enemy will disturb us by means of a shorter one. Whenever this is the case we must ourselves choose the shorter one, and shorten it still further according as the character and conditions of the enemy and other circumstances may render necessary. If we abandon the weak impressions of abstract concepts, and descend to the region of practical life, a bold, courageous, resolute enemy will not give us time for far-reaching skilful combinations, and it is just against such an enemy that we would have most need of skill. By this is shown, it appears to us, the superiority of simple and direct successes over those that are complicated.

Our opinion, therefore, is not that the simple attack is the best, but that we must not aim further than our scope permits, and that this principle will lead more and more to direct combat the more warlike our opponent is. Therefore, far from trying to surpass the enemy in making more complicated plans, we must rather always seek to get ahead of him in making them more simple.

If we look for the ultimate foundation stones of these opposing principles, we find that in the one it is sagacity, in the other, courage. Now there is something very attractive in the notion that a moderate degree of courage joined to a great deal of sagacity will produce greater effects than moderate sagacity with great courage. But unless we conceive of these elements in an illogical disproportion, we have no right to assign to sagacity this advantage over courage in a field whose name is danger, and which must be regarded as the true domain of courage.

After this abstract investigation we shall only add that experience, very far from leading to a different conclusion, is rather the sole cause which has forced us in this direction and given rise to such investigations.

Whoever reads history without prejudice cannot fail to arrive at the conviction that, of all military virtues, energy in the conduct of war has always contributed the most to the glory and success of arms.

How we carry out our fundamental principle of regarding the destruction of the enemy's force as the principal object, not only in war as a whole but also in each separate engagement, and how we are going to adapt that principle to the forms and conditions necessarily demanded by the circumstances out of which war springs, the sequel will show. For the present, all we desire is to contend for its general importance, and with this result we return again to the engagement.

CHAPTER 4

THE ENGAGEMENT IN GENERAL
(CONTINUED)

In the last chapter we insisted on the destruction of the enemy as being the object of the engagement, and we sought to prove, through a special investigation, that this is true in the majority of cases, and in respect of great engagements, because the destruction of the enemy's forces is always the predominating object in war. The other objects which may be mixed up with this destruction of the enemy's forces, and may have more or less influence, we shall describe generally in the next chapter, and gradually we shall become better acquainted with them. Here we are divesting the engagement of them entirely, and considering the destruction of the enemy as the completely sufficient object of the individual engagement.

What are we to understand by the destruction of the enemy's force? A diminution of it relatively greater than that of our own force. If we have a great superiority in numbers over the enemy, then naturally the same absolute amount of loss on both sides will be for us a smaller one than for him, and consequently may be regarded in itself as an advantage. Since we are here considering the engagement as divested of all objects, we must also exclude from our consideration the case in which the engagement is used only indirectly for a greater destruction of the enemy's force. Consequently only that direct gain which we have made in the mutual process of destruction is to be regarded as the object, for this is an absolute gain, which runs through the reckoning of the whole campaign, and which always at the end turns out to be a pure gain. Every other kind of victory over our opponent would either have its motive in other objects, which we have completely excluded here, or would only yield a temporary relative advantage. An example will make clear what we mean.

If by a skilful disposition we have placed our opponent in such a disadvantageous position that he cannot continue the engagement without danger, and after some resistance he retreats, we may say that we have overcome him at that point; but if in thus overcoming him we have lost just as many forces as the enemy, then in closing the account of the campaign, there will be nothing left of this victory, if such a result can be called a victory. Therefore the overcoming of the enemy, that is, placing him in such a position that he must give up the engagement, counts for nothing in itself, and for that reason cannot belong to the definition of the object. Therefore there is left over, as we have said, nothing except the direct gain which we have made in the process of destruction. But to this belong not only the losses which have occurred in the course of the engagement, but also those which, after the withdrawal of the conquered party, occur as direct consequences of it.

Now it is known from experience that the losses in physical forces in the course of an engagement seldom show a great difference between victor and vanquished, often none at all, and that the most decisive losses on the side of the vanquished only begin with the retreat, that is, those which the victor does not share with him. The weak remnants of battalions already in disorder are cut to pieces by cavalry, exhausted men are left lying, disabled guns and broken caissons are abandoned, others cannot be moved quickly enough along bad roads and are captured by enemy cavalry, while during the night individual groups lose their way and fall defenseless into the enemy's hands. Thus the victory usually gains substance only after it is already decided. This would be a paradox, if it were not explained in the following manner.

The loss in physical forces is not the only one which both sides suffer in the course of the engagement; moral forces also become shattered, broken and destroyed. It is not only the loss in men, horses and guns, but in order, courage, confidence, cohesion and plan which come into consideration when it is a question whether the engagement can still be continued or not. It is principally the moral forces which decide here, and in all cases in which the victor has lost as heavily as the vanquished, it is these alone.

The comparative relation of the physical losses is anyhow difficult to estimate in the course of the engagement, but not so the relation of the moral ones. Two things principally make it manifest. The one is the loss of the ground on which the engagement has taken place, the other the superiority of the enemy's reserves. The more our reserves diminish as compared with those of the enemy, so much the more forces we have used to maintain the equilibrium; by this alone noticeable evidence of the moral superiority of the enemy is given, which seldom fails to stir up in the soul of the commander a certain bitterness of feeling and contempt for his own troops. But the principal thing is that troops who have been fighting for a long period are more or less like burnt-out cinders; their ammunition is consumed; they have melted away to a certain extent; their physical and moral strength is exhausted, and perhaps their courage broken as well. Such a force, irrespective of the diminution in its number, regarded as an organic whole, is far from being what it was before the engagement; and thus it is that the loss in moral force is shown by the measure of the reserves that have been used up, as by a yardstick.

Lost ground and lack of fresh reserves are, therefore, usually the principal causes which determine a retreat; but at the same time we by no means exclude or desire to underestimate other reasons, which may lie in the connection of the parts, in the plan of the whole, and so forth.

Every engagement is, therefore, the bloody and destructive measuring of the strength of forces, physical and moral; whoever has the greatest sum total of both left at the end is the conqueror.

In the engagement the loss of moral force was the chief cause of the decision; after the decision this loss continues to increase until it reaches its culminating point at the end of the whole action; it becomes, therefore, the means of making that gain in the destruction of the physical forces which was the real object of the engagement. The loss of all order and unity often makes even the resistance of individual units fatal to them. The courage of the whole is broken; the original tension over loss and gain, in which danger was forgotten, is gone, and to the majority danger now appears no longer an appeal to their courage, but rather the endurance of a cruel punishment. Thus, in the first moment of the enemy's victory, the instrument is weakened and blunted, and therefore no longer fit to repay danger by danger.

The victor must use this time to make his real gain in the destruction of the enemy's physical forces; only that which he attains in this respect is he sure of; the moral forces of the opponent gradually recover, order is restored, courage is revived, and, in the majority of cases, there remains only a very small part of the superiority obtained, often none at all. In some cases even, although rarely, the spirit of revenge and intensified hostility may bring about an opposite result. On the other hand, whatever is gained in killed, wounded, prisoners and captured guns can never disappear from the account.

The losses in the battle consist more in killed and wounded; those after the battle more in lost guns and in prisoners. The first the victor shares with the vanquished, more or less, but not the second; and for that reason they occur usually only on the one side of the combat; at least, they are considerably in excess on that side.

Guns and prisoners have therefore at all times been regarded as the true trophies of victory, and at the same time as its measure, because through these things its extent is made manifest beyond all doubt. Even the degree of moral superiority may be better judged by them than by any other circumstance, especially if the number of killed and wounded is compared to them, and hereby the moral effects are raised to a higher power.

We have said that the moral forces destroyed in the engagement and its immediate consequences recover gradually, and often bear no trace of their destruction; this is the case with small divisions of the whole, less frequently with large divisions. It may, however, be the case even with these in the army, but seldom or never in the state or government to which the army belongs. Here the situation is estimated more impartially, and from a higher point of view. In the number of trophies taken by the enemy and their relation to the number of killed and wounded, the extent of their own weakness and inefficiency is recognized only too easily and too well.

On the whole, the lost balance of the moral forces must not be treated lightly because it has no absolute value and because it does not necessarily appear in the sum total of success; it may become of such excessive weight as to overthrow everything with an irresistible force. On that account it may often become one of the great aims of action. Of this we shall speak elsewhere; here we must examine more of its fundamental relations.

The moral effect of a victory increases, not merely in proportion to the measure of the forces engaged, but in a progressive ratio—that is to say, not only in extent, but also in intensive strength. In a beaten division order is easily restored. As a single frozen limb is easily warmed by the rest of the body, so the courage of a defeated division is easily raised again by the courage of the army as soon as it rejoins it. If, therefore, the effects of a small victory do not entirely disappear, still they are partly lost to the enemy. This is not the case if the army itself has sustained a great defeat; then all parts collapse together. A great fire attains quite a different degree of heat from several small ones.

Another relation which should determine the moral value of a victory is the numerical relation of the forces which have been in conflict with each other. To defeat many with few is not only a double gain, but shows also a greater, especially a more general, superiority, which the conquered must always be fearful of encountering again. Nevertheless, this influence is in reality scarcely observable in such a case. In the moment of action, the notions of the actual strength of the enemy are generally so uncertain, the estimate of our own usually so incorrect, that the party superior in numbers either does not admit the disproportion at all, or at all events not nearly in its full truth, whereby he evades almost entirely the moral disadvantage which would arise from it for him. Only later, in history, does that strength, long suppressed through ignorance, vanity or a wise discretion, emerge, and then it may glorify the army and its leader, but it can then no longer by its moral weight do anything for events long past.

If prisoners and captured guns are those things by which the victory principally gains substance, its true crystallizations, then the plan of the engagement will also have these things especially in view; the destruction of the enemy by death and wounds appears here merely as a means to an end.

How far this may influence the arrangements in the engagement is not an affair of strategy, but the disposition of the engagement itself is closely connected with it, and that on account of the security of our own rear and the endangering of the enemy's. The number of prisoners and captured guns depends very much on this point, and it is a point with which, in many cases, tactics alone is not able to deal, namely, when the strategic circumstances are too much opposed to the tactical.

The risks of having to fight on two sides, and the still greater danger of having no line of retreat left open, paralyze the movements and the power of resistance and affect the alternatives of victory and defeat; further, in case of defeat, they increase the loss, often raising it to its extreme limit, that is, to destruction. Therefore, the threat to the rear makes defeat more probable, and, at the same time, more decisive.

From this arises a true instinct for the whole conduct of war, and especially for great and small engagements: to secure our own line of retreat and to seize that of the enemy. This follows from the concept of victory, which, as we have seen, is something more than mere slaughter.

In this effort we see the first more immediate purpose of the combat, and one which is quite general. No engagement is conceivable in which this effort, either in its double or single form, should not go hand in hand with the mere impact of force. Even the smallest detachment will not throw itself upon its enemy without thinking of its line of retreat, and, in most cases, it will have an eye upon that of the enemy also.

We would have to digress to show how often, in complicated cases, this instinct is prevented from going the direct road, how often it must yield to the difficulties arising from more important considerations; we shall, therefore, rest content with affirming it to be a general natural law of the engagement.

It is, therefore, everywhere active, presses everywhere with its natural weight, and so becomes the pivot on which almost all tactical and strategic maneuvers turn.

If we now take a look at the general concept of victory, we find in it three elements:

1. The greater loss of the enemy in physical forces.
2. The greater loss of the enemy in moral forces.
3. His open admission of this by his renunciation of his intention.

The reports of each side of losses in killed and wounded are never exact, seldom truthful and, in most cases, full of intentional misrepresentations. Even the number of trophies is seldom reliably stated; when it is not very considerable, it may leave the victory still a matter of doubt. Of the loss in moral forces there is no reliable measure, except in the trophies; therefore, in many cases, giving up the combat is the only real evidence of victory left. It is to be regarded as a confession of inferiority—as the lowering of the flag, by which, in this particular instance, right and superiority are conceded to the enemy, and this element of humiliation and shame, which further remains to be distinguished from all the other moral consequences of being no longer a match for the enemy, is an essential part of the victory. It is this part alone which acts upon public opinion outside the army, upon the people and the government in both belligerent states, and upon all others who are involved.

The renouncement of the intention is not quite identical with quitting the field of battle, even when the combat has been very obstinate and long kept up. No one will say of advanced posts, when they retire after an obstinate resistance, that they have given up their intention. Even in engagements aimed at the destruction of the enemy's army, the retreat from the battlefield is not always to be regarded as a renunciation of this intention, as, for instance, in retreats planned beforehand, in which the ground is disputed foot by foot. All this belongs to our discussion of the special object of engagements. Here we wish only to draw attention to the fact that in most cases the relinquishment of the intention is very difficult to distinguish from the withdrawal from the battlefield, and that the impression produced by the latter, both within and outside of the army, is not to be treated lightly.

For generals and armies whose reputation is not established, this is a peculiarly difficult side of many operations otherwise justified by circumstances, where a series of engagements, each ending in retreat, may appear as a series of defeats, without being so in reality, and where that appearance may exercise a very disadvantageous influence. It is impossible for the retreating general in this case, by making known his real intentions, to counteract the moral effect, for to do that effectively he would have to disclose his plans completely, which, of course, would run too much counter to his principal interests.

In order to draw attention to the special importance of this conception of victory we shall only refer to the battle of Soor, the trophies of which were not important (a few thousand prisoners and twenty guns), and where Frederick proclaimed his victory by remaining five days on the field of battle, although his retreat into Silesia had been previously determined, and was a measure natural to his whole situation. According to his own account, he thought he would get nearer to a peace by the moral weight of this victory. Now although other successes were necessary before this peace came to pass—the engagement at Katholisch-Hennersdorf in Lusatia and the battle of Kesseldorf—still we cannot say that the moral effect of the battle of Soor was nil.

If it is chiefly the moral force which is shaken by defeat, and if thereby the number of trophies mounts up to an unusual height, the lost engagement becomes a rout, which is not the necessary counterpart to every victory. Since in the case of such a rout the moral force of the vanquished is much more seriously shaken, there often ensues a complete incapability of further resistance, and the whole action consists of giving way, that is, of flight.

Jena and Waterloo were routs, but not so Borodino.

Although without pedantry we can give here no single line of separation, because the difference between the things is one of degree, yet adherence to concepts is essential as a central point to give clearness to our theoretical ideas, and it is a lack in our terminology that there is only one word for a victory over the enemy tantamount to a rout, and a conquest of the enemy only tantamount to a simple victory.

CHAPTER 5

ON THE SIGNIFICANCE OF
THE ENGAGEMENT

Having in the preceding chapter examined the engagement in its absolute form, so to speak, as the miniature picture of the whole war, we now turn to the relations which it bears as one part to the other parts of a greater whole. First we inquire what precise significance an engagement can have.

Since war is nothing but a process of mutual destruction, then the most natural answer in theory, and perhaps also in reality, appears to be that all the powers of each party unite in one great mass and all results in one great collision of these masses. There is certainly much truth in this conception, and on the whole it seems to be very advisable that we should adhere to it and should on that account look upon minor engagements in the beginning only as necessary waste, like the shavings from a carpenter's plane. However, the thing can never be settled so simply.

That the multiplication of engagements arises from a division of forces is a matter of course, and the more immediate objects of separate engagements will therefore be discussed under the subject of the division of forces. But these objects and, together with them, the whole mass of engagements may in a general way be divided into certain classes, and it will contribute to the clarity of our thought if we acquaint ourselves with these now.

Destruction of the enemy's military forces is certainly the object of all engagements; but other objects may be connected with it, and these other objects may even become predominant. We must therefore distinguish the case where the destruction of the enemy's force is the principal object, and that where it is more a means. Apart from the destruction of the enemy's force, the possession of a place or the possession of some object may also be the general motive for an engagement, and it may be either one of these alone or several together, in which cases, however, one usually remains the principal motive. Now the two principal forms of war, the offensive and the defensive, of which we shall soon speak, do not modify the first of these motives, but they certainly do modify the other two, and if we wanted to arrange them in a scheme, it would appear thus:

OFFENSIVE ENGAGEMENT	DEFENSIVE ENGAGEMENT
1. Destruction of enemy's forces.	1. Destruction of enemy's forces.
2. Conquest of a place.	2. Defense of a place.
3. Conquest of some object.	3. Defense of some object.

These motives, however, do not seem to embrace the whole of the subject, if we recollect that there are reconnaissances and demonstrations, in which obviously none of these three points is the object of the engagement. Actually we must, therefore, on this account introduce a fourth class. Strictly speaking, in reconnaissances in which we wish the enemy to show himself, in alarms by which we wish to wear him out, in demonstrations by which we wish to prevent his leaving some point or to draw him off to another, all the objects are such as can only be attained indirectly and *under the pretext of one of the three objects specified above,* usually of the second; for the enemy whose aim is to reconnoiter must draw up his force as if he really intended to attack and defeat us, or drive us off, and so forth. But this pretended object is not the real one, and our present question only concerns the latter; therefore, we must add a fourth to the three objects of the offensive mentioned above, namely, to induce the enemy to make a false step or, in other words, to make a feint of engagement. It is natural that only offensive means are conceivable in connection with this object.

On the other hand, we must observe that the defense of a place may be of two kinds: either absolute, if the point may not be given up at all, or relative, if it is only required for a certain time. The latter happens constantly in the engagements of advance posts and rear guards.

That the nature of these different intentions of an engagement must have an essential influence on the dispositions of the engagement is evident. We act differently if our object is merely to drive an enemy's post out of its place from what we do if our object is to defeat him completely; differently, if we mean to defend a place to the bitter end from what we do if our design is only to detain the enemy for a certain time. In the first case we concern ourselves little with the line of retreat, in the latter it is the principal point, and so forth.

But these reflections belong properly to tactics, and are only introduced here as an example for greater clarity. What strategy has to say on the different objects of the engagement will appear in the chapters which touch on these objects. Here we have only a few general observations to make.

First, that the importance of the objects decreases approximately in the order in which they stand above; secondly, that the first of these objects must always predominate in the main battle; finally, that the last two in a defensive engagement are in reality such as yield no fruit; that is to say, they are purely negative, and can only be useful indirectly, by facilitating something else which is positive. It is, therefore, *a sign of something wrong in the strategic situation if engagements of this kind become too frequent.*

CHAPTER 6

DURATION OF THE ENGAGEMENT

If we consider the engagement no longer in itself but in relation to the other military forces, then its duration acquires a special importance.

This duration is to be regarded to a certain extent as a second subordinate success. For the victor the engagement can never be decided too quickly, for the vanquished it can never last too long. The speedy victory is a higher degree of victory; a late decision is, on the side of the defeated, some compensation for the loss.

This is in general true, but it acquires a practical importance in its application to those engagements, the object of which is a relative defense.

Here the whole success often lies in the mere duration. This is the reason why we include duration among the strategic elements.

The duration of an engagement is necessarily bound up with its essential conditions. These conditions are: absolute amount of strength, relation in strength and arms between the two sides, and the nature of the country. Twenty thousand men do not exhaust themselves against one another as quickly as two thousand; we cannot resist an enemy double or three times our strength as long as one of the same strength. A cavalry engagement is decided sooner than an infantry engagement; and an engagement entirely between infantry more quickly than if artillery is present; in mountains and forests we do not advance as quickly as on level ground. All this is clear enough.

From this it follows, therefore, that strength, relation of the three arms, and position must be considered, if the engagement is to fulfil an object by its duration; but this rule was less important to us in this special discussion than to connect with it at once the chief results which experience offers on the subject.

The resistance of an ordinary division of 8,000 to 10,000 men of all arms, even opposed to an enemy considerably superior in numbers and in not very advantageous country, lasts several hours, and if the enemy is only slightly or not at all superior, even half a day. A corps of three or four divisions gains twice that time; an army of 80,000 to 100,000, three or four times as much. Therefore the masses may be left to themselves for that length of time, and no separate engagement will take place if within that time other forces, whose activity quickly fuses with the result of the engagement which has already taken place, can be brought up to form a whole.

These figures are taken from experience; but it is important to us at the same time to characterize more particularly the moment of the decision, and consequently of the termination.

CHAPTER 7

DECISION OF THE ENGAGEMENT

No engagement is decided in a single moment, although in every engagement there are moments of great importance, which principally bring about the decision. The loss of an engagement is, therefore, a gradual sinking of the scale. But there is in every engagement a point when it may be regarded as decided, in such a way that the renewal of the fight would be a new engagement, not a continuation of the old one. To have a clear conception of this point is very important, in order to be able to decide whether, with the prompt assistance of reinforcements, the engagement can again be resumed with advantage.

Often in engagements which cannot be resumed, new forces are sacrificed in vain; often the opportunity is lost of reversing the decision, where this might still easily be done. Here there are two examples, which could not be more to the point:

When the Prince of Hohenlohe, in 1806, at Jena, with 35,000 men opposed to from 60,000 to 70,000 under Bonaparte, had accepted battle and lost it—and lost it in such a way that the 35,000 could be regarded as annihilated—General Rüchel undertook to renew the battle with about 12,000; the result was that in a moment his force was likewise annihilated.

On the other hand, on the same day at Auerstädt, the Prussians with 25,000 men fought Davoust, who had 28,000, until mid-day—without success, it is true, but still without the force being reduced to a state of dissolution, even without greater loss than the enemy, who was absolutely without cavalry. But they neglected the opportunity of using the reserve of 18,000, under General Kalkreuth, to reverse the battle, which, under those circumstances, it would have been impossible to lose.

Each engagement is a whole in which the partial engagements combine into one total result. In this total result lies the decision of the engagement. This success need not be exactly a victory such as we have depicted in Chapter 4, for often the preparations for that have not been made, often there is no opportunity for that if the enemy gives way too soon, and in most cases the decision, even when the resistance has been obstinate, takes place before such a degree of success is attained as really satisfies the idea of a victory.

We therefore ask: Which is usually the moment of decision, that is to say, the moment when a new and presumably not disproportionate force is no longer able to reverse a disadvantageous engagement?

If we omit feint engagements, which in accordance with their nature are properly without decision, then:

1. If the possession of a movable object was the aim of the engagement, the loss of it is always the decision.
2. If the possession of a tract of country was the object of the engagement, then the decision usually lies in its loss. Yet not always, and only if this tract of country is of peculiar strength. Country easily accessible, however important it may be in other respects, can be retaken without much danger.
3. But in all other cases, when these two circumstances have not already decided the engagement, particularly therefore in a case where the destruction of the enemy's force is the principal object, the decision lies in that moment when the victor ceases to feel himself in a state of disintegration, that is, of a certain inefficiency, and when consequently, there is no further advantage in using the successive efforts spoken of in Book III, Chapter 12. For this reason we have given the strategic unity of the engagement its place here.

An engagement, therefore, in which our assailant has not lost his condition of order and efficiency at all, or at least only with a small part of his force, while our own forces are, more or less, disorganized, is past recovery, and it is just as much so if the enemy has already recovered his efficiency.

The smaller, therefore, that part of the force which has really been engaged and the greater that portion which as reserve has contributed to the result by its mere presence, so much the less will any new force of the enemy again wrest the victory from our hands. That commander, as well as that army which has succeeded to the greatest extent in conducting the engagement with the greatest economy of forces, and making the most of the moral effect of strong reserves, goes the surest way to victory. In this respect we must concede to the French, in modern times, especially when led by Bonaparte, a great mastery.

Further, the moment when the state of crisis in the engagement ceases for the victor, and his old efficiency returns, will come the sooner the smaller the unit he commands. A picket of cavalry pursuing an enemy at full speed will in a few minutes resume its old order, and also the crisis lasts no longer; a whole regiment of cavalry requires a longer time; it lasts still longer with the infantry, if extended in single lines of skirmishers, and longer again with divisions of all arms, when one part has happened to take one direction, and another, another, and when the engagement has thus caused a disorder, which usually becomes still worse, owing to the fact that no part knows exactly where the other is. Thus, the point of time when the conqueror has found again the instruments which he has been using and which have become mixed up and are partly in disorder; when he has rearranged them a little and put them in their proper places, and thus put the battle-workshop in order again—this moment occurs later and later, the greater the total force.

Again, this moment occurs later if night overtakes the conqueror in the crisis, and, lastly, it occurs later still if the country is broken and covered. But in regard to these two points we must observe that night is also a great means of protection, because circumstances are only seldom such that one may expect good success from night attacks, as on March 10, 1814, at Laon, where York against Marmont gives us a very apposite example of this. In the same way a covered and broken country will at the same time afford protection against a reaction to him who has been engaged in the prolonged crisis of victory. Both, therefore—night as well as covered and broken country—are obstacles which make the renewal of the same engagement more difficult, instead of facilitating it.

Hitherto, we have considered assistance arriving for the losing side as a mere increase of military force, therefore, as a reinforcement coming up directly from the rear, which is the most usual case. But the situation becomes quite different if these fresh forces come up on the enemy's flank or rear.

On the effect of flank or rear attacks so far as they belong to strategy, we shall speak in another place. Such a one as we have here in view, intended for the restoration of the engagement, belongs chiefly to tactics, and is only mentioned because we are here speaking of tactical results, and our conceptions, therefore, must intrude upon the field of tactics.

By directing a force against the enemy's flank and rear, its efficacy may be much intensified; this, however, is not necessarily so; the efficacy may thereby be just as much weakened. The circumstances under which the engagement takes place decide upon this part of the plan as well as upon every other, without our being able to consider them here. In this respect two things are important for our subject, of which the first is that *flank and rear attacks have, as a rule, a more favorable effect on the success after the decision than upon the decision itself.* Now in regard to the resumption of a battle, our first object is a favorable decision and not the magnitude of the success. From this point of view one would therefore think that a force which comes up to re-establish our engagement is of less assistance if it falls upon the enemy in flank and rear, therefore separated from us, than if it joined us directly; certainly cases are not wanting where it is so, but we must say that most cases are on the other side, and they are so because of the second point which is of importance to us here.

This second point is *the moral effect of the surprise, which a reinforcement coming up to re-establish an engagement usually has in its favor.* Now the effect of a surprise is always increased if it takes place in the flank or rear, and an enemy involved in the crisis of victory in his extended and scattered order is less in a state to counteract it. Who does not feel that an attack in flank or rear, which at the beginning of the engagement, when the forces are concentrated and prepared for such an event would be of little importance, acquires quite another importance in the last moment of the engagement?

We must, therefore, admit without hesitation that in most cases a reinforcement coming up on the flank or rear of the enemy will be much more effective, will be like the same weight at the end of a longer lever. Under these circumstances we may undertake to restore the engagement with the same force which, employed in a direct attack, would have been insufficient. Here, where results almost defy calculation, because the moral forces gain complete superiority, is the right field for boldness and daring.

Therefore the eye must be directed on all these objects and all these elements of co-operating forces must be taken into consideration, when we have to decide in doubtful cases whether or not it is still possible to restore an engagement which has taken an unfavorable turn.

If the engagement is to be regarded as not yet ended, then the new engagement which is opened by the arrival of assistance fuses with the old into one, that is, into a joint result, and the first disadvantage then disappears altogether from the account. But this is not the case if the engagement was already decided; then there are two results separate from one another. Now if the assistance which arrives is only of comparative strength, that is, if it is not in itself alone a match for the enemy, then a favorable result is hardly to be expected from this second engagement; but if it is so strong that it can undertake the second engagement without regard to the first, then it may be able by a favorable issue to compensate for and outweigh the first engagement, but never to make it disappear altogether from the account.

At the battle of Kunersdorf, Frederick the Great at the first onset captured the left wing of the Russian position and took seventy pieces of artillery; at the end of the battle both were lost again, and the whole result of this first engagement had disappeared from the account. Had it been possible to stop at the first success, and to put off the second part of the battle until the next day, then, even if the king had lost it, the advantages of the first success could always counterbalance this loss.

But when a disadvantage is restored and turned to our own advantage before its conclusion, its minus result on our side not only disappears from the account, but also becomes the foundation of a greater victory. That is, if we imagine exactly the tactical course of an engagement, we may easily see that, until it is concluded, all successes in partial engagements are only suspended decisions, which may not only be destroyed by the main decision, but changed into the opposite. The more our forces have suffered, the more the enemy will have exhausted himself; the greater, therefore, will be the crisis for the enemy, too, and the more the superiority of our fresh troops will tell. If now the total result turns in our favor, if we wrest from the enemy the field of battle and recover all the trophies, then all the forces which he has sacrificed in obtaining them become sheer gain for us, and our former defeat becomes the stepping-stone to greater triumph. The most brilliant feats of arms which, in the case of victory, the enemy would have prized so highly that the loss of forces which they cost would have been disregarded leave behind now nothing but regret for these sacrificed forces. Such is the alteration which the magic of victory and the curse of defeat produce in the specific gravity of the elements.

Therefore, even if we are decidedly superior in strength, and are able to repay the enemy his victory by a still greater one, it is always better to forestall the conclusion of a losing engagement, if it is of comparative importance, so as to reverse its course rather than to deliver a second engagement.

Field-Marshal Daun attempted in 1760 to come to the assistance of General Laudon at Liegnitz while the latter's engagement was still going on; but when that engagement was lost, he did not attack the king next day, although he did not lack strength.

For these reasons bloody engagements of advance guards which precede a battle are to be looked upon only as necessary evils, and when not necessary they are to be avoided.

We will also have another conclusion to consider.

If in a closed engagement the decision has gone against us, this does not constitute a motive for deciding on a new one. The determination for this new one must proceed from other conditions. This conclusion, however, comes into conflict with a moral force, which we must take into account; it is the feeling of revenge and retaliation. From the highest commander to the lowest drummer-boy this feeling is nowhere lacking, and therefore troops are never in better fighting spirits than when it is a question of squaring an account. This is, however, based on the supposition that the defeated part is not too great in proportion to the whole, because otherwise the above feeling would lose itself in that of impotence.

There is therefore a very natural tendency to use this moral force to repair the disaster on the spot, and chiefly on that account to seek another engagement if other circumstances permit. Naturally, this second engagement must usually be an offensive one.

In the catalogue of minor engagements there are to be found many examples of such acts of retaliation; but great battles usually have too many other determining causes to be occasioned by this weaker force.

It was such a feeling which led the noble Blücher with his third corps to the field of battle on February 14, 1814, when the other two had been beaten three days before at Montmirail. Had he known that he still would meet Bonaparte himself, then, naturally, overwhelming reasons would necessarily have induced him to postpone his revenge. But he hoped to revenge himself on Marmont, and instead of gaining the reward of his desire for honorable satisfaction, he suffered the penalty of his erroneous calculation.

On the duration of engagements and the moment of their decision depend the distances at which those masses should be disposed which are intended to fight *in conjunction*. This disposition would be a tactical arrangement in so far as it has in view one and the same engagement; it can, however, only be regarded as such, provided the position of the troops is so compact that two separate engagements cannot be imagined, and consequently that the space which the whole occupies can be considered strategically as a mere point. But in war, cases frequently occur where even those forces intended to fight *in conjunction* must be so far separated from each other that, while their union for a conjoint engagement certainly remains the principal object, still the occurrence of separate engagements remains possible. Such a disposition is therefore strategic.

Arrangements of this kind are: marches in separate columns and masses, advance guards and flanking corps, reserves intended to serve as supports for more than one strategic point, the concentration of separate corps from widely extended quarters, etc. We can see that they constantly occur, and constitute, so to speak, the small change in the strategic economy, while the main battles and everything on their level are the gold and silver pieces.

CHAPTER 8

MUTUAL UNDERSTANDING
AS TO AN ENGAGEMENT

No engagement can originate without mutual consent thereto; and in this notion, which constitutes the whole basis of a duel, is the root of a certain phraseology used by historical writers, which leads to many vague and false conceptions.

The treatment of the subject by writers frequently hinges on the idea that one commander has offered battle to the other, and the latter has not accepted it.

But the engagement is a very modified duel, and its foundation consists not merely in the mutual wish to fight, that is, in consent, but in the objects which are connected with the engagement; these belong always to the greater whole, and that so much the more, as even the whole war, considered as a "combat-unit," has political objects and conditions which belong to a greater whole. The mere desire to conquer each other becomes therefore quite a subordinate matter, or rather it ceases completely to be anything in itself and is only the nerve which conveys the impulse of action from the higher will.

Amongst the ancients, and then again during the early period of standing armies, the expression, "to offer battle to the enemy in vain," had more meaning than it has now. Among the ancients everything was arranged with a view to measuring each other's strength in the open field, free from anything in the nature of a hindrance, and the whole art of war consisted in the organization and composition of the army, that is, in the order of battle.

Now since their armies regularly entrenched themselves in their camps, the position in a camp was regarded as something unassailable, and a battle did not become possible until the enemy left his camp, and placed himself in the lists, so to speak, in accessible country.

If, therefore, we hear about Hannibal having offered battle to Fabius in vain, that says nothing with respect to the latter except that a battle was not part of his plan, and in itself proves neither the physical nor the moral superiority of Hannibal; but with respect to Hannibal, the expression is still correct enough in the sense that he really wished a battle.

In the early period of modern armies, the conditions were similar in the case of great engagements and battles. That is to say, great masses were brought into action, and led throughout it by means of an order of battle, which as a great unwieldy whole required a more or less level plain and was neither suited to attack nor yet to defense in a broken, covered or even mountainous country. Therefore, the defender possessed here also, to some extent, the means of avoiding battle. These conditions, although gradually becoming less stringent, continued until the first Silesian War, and it was not until the Seven Years' War that attacks on an enemy even in difficult country gradually became feasible and customary; ground certainly did not cease to be a source of strength to those making use of its aid, but it was no longer a charmed circle which shut out the natural forces of war.

During the past thirty years war has developed much more in this respect, and there is no longer anything which stands in the way of a general seriously desirous of a decision by battle. He can seek out his enemy and attack him; if he does not do so, then it cannot be said of him that he wanted the engagement, and the expression "he offered a battle which his opponent did not accept" now means nothing more than that he himself did not find circumstances advantageous enough for an engagement, an admission which the above expression does not suit and which it only strives to cloak.

It is true the defender can now no longer refuse an engagement, yet he may still avoid it by giving up his position and the role connected with it; that is, however, half a victory for the assailant, and an acknowledgment of his temporary superiority.

This idea relating to a challenge can, therefore, no longer be used to excuse with such rodomontade the inactivity of him whose part it is to advance, that is, to take the offensive. The defender, who, as long as he does not give way, must be credited with wanting battle, can, no doubt, if he is not attacked, say that he had offered it, if this were not already self-evident.

On the other hand, he who wishes and is able to avoid it, cannot now be forced into an engagement. Since the advantages which the aggressor gains by this avoidance are often not sufficient, and an actual victory becomes for him a pressing necessity, sometimes the few means which are available to force even such an opponent to an engagement are sought for and employed with particular skill.

The principal means for this are: first, *surrounding* the enemy so as to make his retreat impossible, or at least so difficult that he prefers to accept the engagement; and, second, *surprising him*. This latter way, which formerly had its basis in the awkwardness of all movements, has become in modern times very ineffective. Because of the flexibility and mobility of modern armies they do not hesitate to retreat even in sight of the enemy, and only some special obstacles in the nature of the country can cause serious difficulties in the operation.

The battle of Neresheim may be cited as an example of this kind, fought by the Archduke Charles with Moreau in the Rauhe Alp, August 11, 1796, merely with a view to facilitating his retreat, although we freely confess that we have never been quite able to understand the argument of the renowned general and author in this case.

The battle of Rossbach is another example, if we suppose that the commander of the allied army did not really have the intention of attacking Frederick the Great.

Of the battle of Soor the king himself says that it was only accepted because a retreat in the presence of the enemy appeared to him a critical operation; however the king has also given other reasons for the battle.

On the whole, with the exception of actual night attacks, such cases will always be of rare occurrence, and those in which an enemy is compelled to fight by being surrounded will really occur only in the case of single corps, like Fink's corps at Maxen.

CHAPTER 9

THE BATTLE

ITS DECISION

What is a battle? A conflict of the main body, but not an unimportant one for a secondary object, not a mere attempt which is given up when we see at an early stage that our object will be difficult to attain, but a conflict waged with all our efforts for the attainment of a real victory.

Even in such a battle, minor objects may be mixed up with the principal object, and it will take on many special shades of color from the circumstances from which it arises, for even a battle belongs to a greater whole of which it is only a part. But because the essence of war is combat, and the battle is the combat of the main armies, it is always to be regarded as the real center of gravity in war, and therefore on the whole its distinguishing character is that, more than any other engagement, it exists for its own sake.

This has an influence on the *manner of its decision,* on the *effect of the victory gained in it,* and determines the *value which theory must assign to it as a means to the end.* On that account we make it the subject of our special consideration, and that too at this stage, before we consider the special objects which may be bound up with it, but which do not essentially change its character if it really deserves to be called a battle.

If a battle is essentially an end in itself, the elements of its decision must be contained in itself; in other words, victory must be striven for as long as a possibility of it is present. It must not, therefore, be given up on account of particular circumstances, but only and solely if the forces appear completely insufficient.

Now how can this moment be more precisely indicated?

If a certain skilfully devised order and co-ordination of the army is the principal condition under which the bravery of the troops can gain a victory, as has been the case for some time in the modern art of war, then *the destruction of this order is the decision*. A defeated wing which gets out of line also influences decisively the fate of that which still stands its ground. If, as was the case at another time, the essence of the defense consists in an intimate alliance of the army with the ground on which it fights and its obstacles, so that the army and the position are a single whole, then *the conquest of an essential point in this position* is the decision. The key of the position is lost, we say; it cannot therefore be defended any further; the battle cannot be continued. In both cases the beaten armies are more or less like the broken strings of an instrument which refuse to function.

The principles, geometrical on the one hand and geographical on the other, which had the tendency to put combating armies in a state of crystallizing tension which did not permit the available powers to be used to the last man, have in our days at least, lost so much of their influence that they no longer predominate. Armies are still led into battle in a certain order, but the order is no longer of decisive importance; obstacles of ground are also still used to strengthen resistance, but they are no longer the only support.

We have tried in the second chapter of this book to take a general view of the nature of the modern battle. According to our conception of it, the order of battle is only a disposition of forces for their convenient use, and the course of the battle a mutual slow wearing away of these forces against one another, to see which will have exhausted his adversary soonest.

The resolution, therefore, to give up the engagement arises, in a battle more than in any other engagement, from the condition of the fresh reserves remaining available; for only these still possess all their moral forces, and the cinders of the battered and shattered battalions, already burnt out by the destructive element, cannot be placed on a level with them. Lost ground, as we have said elsewhere, is also a measure of lost moral force; therefore it must also be taken into account, but more as a sign of loss suffered than for the loss itself, and the number of fresh reserves remains always the main thing to be considered by both commanders.

Usually a battle inclines in one direction from the very beginning, but in a manner hardly noticeable. This direction is frequently fixed in a very decided manner by the arrangements which have been made previously, and then it shows a lack of insight in the general who begins battle under these unfavorable circumstances without being aware of them. Even when this does not occur the course of a battle naturally resembles a slow disturbance of equilibrium, which begins soon, but, as we have said, is almost imperceptible at first and with each fresh moment becomes stronger and more visible, rather than a wavering oscillation from one side to the other, as those who are misled by untrue descriptions of battle usually picture it to themselves.

But whether it happens that the balance is for a long time little disturbed, or that even after it has been lost on the one side it is regained, and is then lost on the other side, it is certain at all events that in most instances the defeated general foresees his fate long before he retreats, and that cases in which some event. acts with unexpected force upon the course of the whole usually exist only in the extenuating imagination with which every one tells of his lost battle.

Here we can only appeal to the judgment of unprejudiced men of experience, who will, we are sure, assent to what we have said, and answer for us to such of our readers as do not know war from their own experience. To develop the necessity of this course from the nature of the thing would lead us too far into the province of tactics, to which this subject belongs; here we are concerned only with its results.

If we say that the defeated general foresees the unfavorable result usually some time before he decides to give up the battle, we admit that there are also instances to the contrary, because otherwise we should be maintaining a proposition intrinsically absurd. If with each decisive turn of a battle, the battle were to be regarded as lost, then necessarily no more forces would be used to change its course, and consequently this decisive turn could not precede the retreat by any length of time. There are cases, certainly, of battles which, after having taken a decided turn to one side, have still ended in favor of the other; but these are rare, not usual. These exceptional cases, however, are reckoned upon by every general against whom fortune declares itself, and he must reckon upon them as long as there remains any possibility of reversing the battle. He hopes by stronger efforts, by stimulating what moral forces still survive, by surpassing himself, or by some fortunate chance, still to see things change in a moment and pursues this hope as far as his courage and his judgment agree. We wish to say something more about that, but first we will mention the signs of the scales turning.

The success of the whole engagement consists in the sum total of the successes of all partial engagements; but these successes of separate engagements can be distinguished by three different things.

First, by the mere moral force in the mind of the leading officers. If a general of division has seen how his battalions have been defeated, it will have an influence on his demeanor and his reports, and these again will have an influence on the measures of the commander-in-chief; therefore even those unsuccessful partial engagements, which to all appearance are retrieved, are not lost in their results, and their impressions add up in the mind of the commander without much trouble and even against his will.

Second, by the quicker melting away of our troops, which can easily be estimated in the slow and less tumultuous course of our battles.

Third, by lost ground.

All these things serve for the eye of the general as a compass by which to recognize the course of the battle on which he has embarked. If whole batteries have been lost and none of the enemy's taken; if battalions have been overthrown by the enemy's cavalry, while those of the enemy everywhere form impenetrable masses; if the line of fire of his order of battle withdraws involuntarily from one point to the other; if fruitless efforts have been made to gain certain points, and the assaulting battalions have each time been scattered by well-directed volleys of grape and case; if our artillery's reply to that of the enemy grows feeble; if the battalions under fire melt away unusually fast, because, with the wounded, crowds of unwounded men withdraw; if single divisions have been cut off and made prisoners through the disruption of the plan of battle; if the line of retreat begins to be endangered—then the commander must in all these things recognize very well in which direction his battle is going. The longer this direction continues, the more decided it becomes, so much the more difficult becomes a change for the better, so much the nearer the moment when he must give up the battle. We shall now make some observations concerning this moment.

We have already said more than once that the relative number of the fresh reserves remaining at the end is usually the main reason for the final decision. That commander who sees his adversary decidedly superior to him in this respect determines upon a retreat. It is quite a characteristic of modern battles that all misfortunes and losses which take place in the course of them can be retrieved by fresh forces, because the arrangement of the modern battle order and the way in which troops are brought into action permit their use in almost any place and in any situation. So long, therefore, as that commander, against whom the result seems to be going, still retains a superiority in reserve force, he will not give up. But from the moment that his reserves begin to become weaker than his enemy's, the decision may be regarded as settled. From now onward he depends partly on special circumstances, partly on the degree of courage and perseverance which he personally possesses, and which may perhaps degenerate into foolish obstinacy. How a commander can succeed in estimating correctly the reserves still remaining on both sides is a matter of technical skill in execution, which does not in any way belong here; we keep to the result as it forms itself in his mind. But even this result is not yet the real moment of decision, for a motive which arises only gradually is not suited to that, but is only a general determination of his resolution, and this resolution itself still requires special causes for it. Of these there are chiefly two, which constantly recur, that is: danger to his retreat and the arrival of the night.

If the retreat becomes more and more endangered with every new step which the battle takes in its course, and if the reserves are so much diminished that they are no longer adequate to provide fresh breathing space, then there is nothing left but to submit to fate, and by an orderly retreat to save what, by a longer delay ending in flight and disaster, would be lost.

However, as a rule, night puts an end to all engagements, because a night engagement offers no hope of advantage except under particular circumstances, and since night is better suited for a retreat than day, the commander who has to consider the retreat as inevitable, or extremely probable, will prefer to make use of the night for this purpose.

That there are, besides these two usual and chief causes many others also, which are smaller, more individual and not to be overlooked, is a matter of course; for the more a battle tends toward a complete upset of equilibrium, the more noticeable is the influence of each partial result on it. Thus the loss of a battery, the successful charge of some regiments of cavalry, may call into life the resolution to retreat already ripening.

In concluding this subject we must dwell for a moment on the point at which the courage of the commander engages in a sort of conflict with his reason.

If, on the one hand, the dictatorial pride of a victorious commander, if the inflexible will of a naturally obstinate spirit, if the convulsive resistance of a noble enthusiasm will not yield the battlefield, where they are to leave their honor, on the other hand, reason advises not to give away everything, not to stake all he has left on the game, but to retain as much as is necessary for an orderly retreat. However highly we must value courage and steadfastness in war, and however little prospect of victory there is for him who cannot resolve to seek it by the exertion of all his strength, still there is a point beyond which perseverance can only be called desperate folly, and therefore cannot be approved by any critic. In the most celebrated of all battles, that of Waterloo, Bonaparte used his last forces in an effort to retrieve a battle which was past retrieving. He spent his last penny, and then like a beggar, fled from the battlefield and from his empire.

CHAPTER 10

THE BATTLE (CONTINUED)

EFFECTS OF VICTORY

According to the point of view taken, one may feel as much astonished at the extraordinary results of many great battles as at the lack of results in others. We shall now dwell for a moment on the nature of the effect of a great victory.

Three things may easily be distinguished here: the effect upon the instruments themselves, that is, upon the generals and their armies; the effect upon the states interested in the war; and the particular result which these effects manifest in the subsequent course of the campaign.

If we only think of the insignificant difference which ordinarily exists between victor and vanquished in the way of killed, wounded, prisoners and lost artillery on the field of battle itself, the consequences which develop out of this insignificant point seem often quite incomprehensible, and yet, usually, everything happens only too naturally.

We have already said in Chapter 7 that the magnitude of a victory increases not merely in the same measure as the number of vanquished forces increases, but in a higher ratio. The moral effects resulting from the issue of a great battle are greater on the side of the conquered than on that of the conqueror: they lead to greater losses in physical force, which then in turn react on the moral element, and so they go on mutually supporting and increasing each other. We must therefore lay special stress upon this moral effect. It occurs on both sides but in opposite directions; as it undermines the strength of the vanquished so it increases the strength and activity of the victor. But its chief effect is upon the vanquished, because here it is the direct cause of fresh losses, and furthermore it is homogeneous in nature with danger, fatigue and hardship—in-fact, with all distressing circumstances among which war moves, and therefore allies itself with them and increases by their assistance, while with the victor all these things act like weights on the upward sweep of his spirits. We find, therefore, that the vanquished sinks much further below the original line of equilibrium than the victor rises above it. On this account, if we speak of the effects of victory we have in mind mainly those which manifest themselves in the vanquished army. If this effect is more powerful in an engagement on a large scale than in one on a small scale, then again it is much more powerful in the main battle than in a subordinate engagement. The main battle takes place for its own sake, for the sake of the victory which it is to give, and which is sought for in it with the utmost effort. Here, on this spot, in this very hour, to conquer the enemy is the purpose in which all the threads of the plan of the war converge, in which all distant hopes, all vague conceptions of the future meet; fate steps in before us to give an answer to our bold question. This is the state of mental tension not only of the commander but of his whole army down to the lowest camp follower, in decreasing strength no doubt, but also in decreasing importance. A great battle has, naturally, never at any time been an unprepared, unexpected, blind routine service, but a grand act, which, partly of itself and partly according to the purpose of the commander, stands out from among the mass of ordinary activities sufficiently to raise the tension of all minds to a higher degree. But the higher this tension with respect to the issue, the more powerful must be the effect of that issue.

Again, the moral effect of victory in our battles is greater than it was in the earlier ones of modern military history. If our battles are as we have depicted them, a real struggle of forces to the utmost, then the sum total of all these forces, of the physical as well as the moral, has more to do with the decision than certain special dispositions or even chance.

A mistake that we make may be repaired next time; from good fortune and chance we can hope for more favor on another occasion; but the sum total of moral and physical powers generally does not alter so quickly, and, therefore, what the verdict of a victory has decided concerning it appears of much greater importance for the entire future. It is, indeed, very probable that of all concerned in a battle, whether belonging to the army or not, very few have given a thought to this difference, but the course of the battle itself impresses on the minds of all those present in it such a conviction. The account of this course in public documents, however much it may be glossed over by the dragging in of irrelevant circumstances, shows also, more or less, to the rest of the world, that the causes were more of a general than of a particular nature.

He who has never been present at the loss of a great battle will have difficulty in forming a living and consequently entirely true conception of it, and the abstract notions of this or that small loss will never come up to the real conception of a lost battle. Let us stop for a moment to picture it.

The first thing which perhaps overpowers the imagination—and we may say, also the intellect—in a lost battle is the dwindling of the masses; then the loss of ground, which takes place more or less always, and therefore on the side of the assailant also, if he is not fortunate; then the disorganized original formation, the confusion of the troops, the dangers of the retreat, which, with few exceptions are always present in a greater or less degree; and finally the retreat itself which mostly takes place at night or, at least, is continued throughout the night. On this very first march we must leave behind a great number of men exhausted and scattered about, often the bravest, who have ventured forth the farthest and held out the longest. The feeling of being defeated, which on the battlefield only seized the superior officers now spreads through all ranks, down to the common soldier. It is aggravated by the horrible idea of being forced to leave in the enemy's hands so many brave comrades, whose worth we never rightly appreciated till this very battle; aggravated also by a rising distrust of the commander, whom every subordinate more or less blames for the fruitless efforts he has made. And this feeling of being defeated is no mere imagination which may be overcome. It is the evident truth that the enemy is superior to us, a truth which might originally have been so hidden that it could not before be perceived, but which always comes out clear and conclusive in the issue. We have, perhaps, suspected it before, but for lack of anything more real have had to set against it hope in chance, reliance on fortune and providence or bold daring. Now, all this has proved inadequate, and the stern truth faces us harshly and imperiously.

All these impressions are still widely different from a panic terror, which in an army fortified by military virtue is never, and in any other, only by exception, the result of lost battles. They must arise even in the best of armies, and although long habituation to war and victory, together with great confidence in the commander, may modify them a little here and there, they are never entirely lacking in the first moment. Nor are they the mere consequence of lost trophies—these are usually lost at a later period, and the loss of them does not become generally known so quickly—and they will, therefore, not fail to appear even when the scale turns in the slowest and most gradual manner, and they constitute that effect of a victory upon which we can count in every case.

We have already said that the extent of the trophies intensifies this effect.

How very much an army in this condition, regarded as an instrument, is weakened! In this weakened condition in which, as we said before, it finds new enemies in all the ordinary difficulties of warfare, how little can it be expected to be in a position to recover by fresh efforts what has been lost! Before the battle there was a real or imagined equilibrium between the two sides; this is lost, and therefore some external cause is required to restore it; every new effort without such external support will only lead to fresh losses.

Thus, therefore, the most moderate victory of the main army must tend to cause a constant sinking of the scale on the opponent's side, until new external circumstances bring about a change. If these are not near, if the victor is a restless opponent, who, thirsting for glory, pursues great aims, then a first-rate commander and a true military spirit in the army, hardened by many campaigns, are required, in order to prevent the swollen tide of preponderance from breaking through entirely, and to moderate·its course by small but reiterated acts of resistance, until the force of victory at the end of a certain period has spent itself.

And now as to the effect of defeat beyond the army, upon the nation and the government! It is the sudden collapse of hopes stretched to the utmost, the downfall of all self-reliance. In place of these extinct forces, fear, with its destructive properties of expansion, rushes into the vacuum left, and completes the prostration. It is a real apoplectic stroke, which one of the two athletes gets from the electric spark of the main battle. And this effect, too, however different in its degrees, is never completely lacking. Instead of every one hastening with a spirit of determination to aid in repairing the disaster, every one fears that his efforts will be in vain, and stops and hesitates, when he should rush forward; or in despondency he lets his weapons fall leaving everything to fate.

The consequences which this effect of victory brings forth in the course of the war itself depend in part on the character and talent of the victorious general, but more on the circumstances from which the victory proceeds, and to which it leads. Without boldness and enterprising spirit on the part of the commander, the most brilliant victory will lead to no great result, and its force exhausts itself much more quickly still on circumstances, if these offer a strong and stubborn opposition to it. How very differently from Daun, Frederick the Great would have used the victory at Kollin; and what different results France, in place of Prussia, might have given a battle of Leuthen!

The conditions which allow us to expect great results from a great victory we shall learn when we come to the subjects with which they are connected; only then will it be possible to explain the disproportion which may appear at first sight between the magnitude of a victory and its results, and which is only too readily attributed to a lack of energy on the part of the victor. Here, where we are concerned with the main battle in itself, we shall merely say that the effects now described never fail to attend a victory, that they increase with the intensive strength of the victory—increase in proportion as the battle becomes a main battle, that is, the more the entire military force is concentrated in it, the more the whole military power of the nation is contained in that force, and the whole state in that military power.

But can theory accept this effect of victory as absolutely necessary? Must it not rather endeavor to discover a counteracting means capable of neutralizing this effect? It seems so natural to answer this question in the affirmative; but heaven protect us from taking, as most theories do, this wrong course, which leads to a series of mutually destructive pros and cons.

The effect is, no doubt, absolutely necessary, for it lies in the nature of things, and it exists, even if we find means to counteract it; just as the motion of a cannon ball in the direction of the earth's rotation always persists, although when fired from east and west, part of the general velocity is destroyed by this opposite motion.

All war presupposes human weakness, and against that it is directed.

Therefore, if later, in another place, we consider what is to be done after the loss of a great battle, if we take into account the resources which still remain, even in the most desperate cases, if we express a belief in the possibility of retrieving all, even in such a case, we do not mean thereby that the effects of such a defeat are by degrees completely wiped out. For the forces and means used to repair the disaster might have been applied to the realization of some positive object, and this applies both to the moral and physical forces.

Another question is, whether, through the loss of a great battle, forces are not roused, which otherwise would never have come to life. This case is certainly conceivable, and it is what has actually occurred with many nations. But to produce this intensified reaction is beyond the province of the art of war, which can only take account of it where in any case it is to be presupposed.

If there are cases in which the results of a victory may appear rather of a destructive nature due to the reaction of the forces aroused by it—cases which certainly are very exceptional—then it must the more surely be granted that there is a difference in the effects which one and the same victory may produce according to the character of the people or state which has been conquered.

CHAPTER 11

THE BATTLE (CONTINUED)

THE USE OF THE BATTLE

Whatever form the conduct of war may take in particular cases, and whatever we may eventually have to recognize as necessary respecting it, we have only to refer to the conception of war in order to be convinced of the following statements:

1. The destruction of the enemy's military force is the leading principle of war, and for all positive action the main way to the object.
2. This destruction of the enemy's force is principally effected only by means of the engagement.
3. Only great and general engagements produce great results.
4. The results will be greatest when the engagements are united in one great battle.
5. It is only in a great battle that the general-in-chief commands in person, and he naturally prefers to entrust the direction of it to himself.

From these truths a double law follows, the parts of which mutually support each other; namely, that the destruction of the enemy's military force is to be sought principally by great battles and their results and that the chief object of great battles must be the destruction of the enemy's military force.

No doubt the principle of destruction is to be found more or less also in other means. No doubt there are instances in which, through favorable circumstances in a minor engagement, a disproportionately large number of enemy forces can be destroyed, and on the other hand in a great battle, the taking or holding of a single post may often predominate as a very important object. But as a general rule it remains a paramount truth that great battles are only fought with a view to the destruction of the enemy's force, and that this destruction can only be effected by means of the great battle.

The great battle must therefore be regarded as war concentrated—the center of gravity of the whole war or campaign. As the sun's rays unite in the focus of the concave mirror in a perfect image and in maximum heat, so the forces and circumstances of war are focused in the great battle for one concentrated utmost effect.

The very assemblage of forces in one great whole, which takes place more or less in all wars, indicates in itself the intention to strike a decisive blow with this whole, either voluntarily, in the case of the assailant, or at the instance of the enemy, in the case of the defender. Where this great blow does not take place, some modifying, and retarding, motives have attached themselves to the original motive of hostility, and have weakened, altered or completely checked the movement. But even in this state of inaction on both sides which has been the keynote in so many wars, the idea of a possible great battle serves always for both parties as a point of direction, a distant focus for the construction of their plans. The more war is a real war, the more it becomes a venting of hatred and hostility, a mutual struggle to overthrow, so much the more will all activities be concentrated in deadly combat, and also the more prominent in importance becomes the great battle.

Everywhere, when the object aimed at is of a great and positive nature, one, therefore, in which the interests of the enemy are deeply affected, the great battle offers itself as the most natural means; it is, therefore, also the best means, as we shall show in greater detail later, and, as a rule, when it is evaded from aversion to the great decision, punishment follows.

The positive object belongs to the assailant, and therefore the great battle is also more particularly his means. But without defining the concepts of offense and defense more closely here, we must observe that, even for the defender in most cases, there is no other effectual means with which to meet sooner or later the needs of his situation and to solve the problem presented to him.

The battle is the bloodiest way of solution. True, it is not merely reciprocal slaughter, and its effect is more a killing of the enemy's courage than of the enemy's soldiers, as we shall see more plainly in the next chapter—but still blood is always its price, and slaughter[1] its character as well as its name; from this the human side of the general recoils.

But the spirit of man trembles still more at the thought of the decision given with one single blow. All action is here compressed *into one point* of space and time, and at such a moment there is stirred up within us a dim feeling as if in this narrow space all our forces could not develop and come into activity, as if by mere time we had already gained much, although this time owes us nothing at all. This is all mere illusion, but even as illusion it is something, and this very weakness which comes upon man in every other momentous decision may well be felt more powerfully by the general, when he must stake interests of such enormous weight upon one venture.

Thus governments and generals have at all times endeavored to avoid the decisive battle, seeking either to attain their aim without it, or dropping that aim unobserved. Writers on history and theory have exhausted themselves to discover in some other feature in these campaigns not only an equivalent for the decision by battle which has been avoided, but even a higher art. In this way, in the present age, we almost look upon a great battle in the economy of war as an evil, rendered necessary through some error that has been committed, as a morbid eruption to which a regular prudent system of war would never lead. Only those generals were to deserve laurels who knew how to carry on war without spilling blood, and the theory of war—a real pundit's business—was to be expressly directed to teach this.

Contemporary history has destroyed this illusion, but no one can guarantee that it will not return here and there for a shorter or longer period of time, and lead those at the head of affairs to perversities which please man's weakness, and therefore have the greater affinity for his nature. Perhaps, by and by, Bonaparte's campaigns and battles will be looked upon as mere acts of barbarism and partial stupidity, and we shall once more turn with satisfaction and confidence to the dress-sword of obsolete and desiccated institutions and manners. If theory can warn us against this, it renders a real service to those who listen to its warning. May we succeed in lending a hand to those who in our dear native land are called upon to speak with authority on these matters, that we may be their guide in this field of inquiry and invite them to make a candid examination of the subject.

[1] *Schlacht* from *schlachten*, to slaughter.—Ed.

Not only the conception of war but experience also leads us to look for a great decision only in a great battle. From time immemorial, only great victories have led to great successes, on the offensive without exception, on the defensive side more or less so. Even Bonaparte would not have seen the day of Ulm, unique in its kind, if he had shrunk from shedding blood; it is rather to be regarded as only an aftermath from the victorious events in his preceding campaigns. It is not only the bold, daring and defiant generals who have sought to complete their work by the great venture of decisive battles, but all fortunate ones; and we may rest satisfied with the answer which they have given to this vast question.

Let us not hear of generals who conquer without bloodshed. If bloody slaughter is a horrible spectacle, then it should only be a reason for treating war with more respect, but not for making the sword we bear blunter and blunter by degrees from feelings of humanity, until once again someone steps in with a sword that is sharp and hews away the arms from our body.

We regard a great battle as a principal decision, but certainly not as the only one necessary for a war or a campaign. Instances of a great battle deciding a whole campaign have been frequent only in modern times; those which have decided a whole war are among the rarest exceptions.

A decision which is brought about by a great battle depends naturally not only on the battle itself, that is, on the mass of military forces engaged in it and on the intensity of the victory, but also on a great number of other relations between the war strengths of both sides and between the states to which these belong. But by the principal mass of the force available being brought to the great duel, a great decision is also ushered in, the extent of which may perhaps be foreseen in many respects, though not in all. And although it is not the only one, it is nevertheless the *first* decision, and, as such, has an influence on those which succeed. Therefore a deliberately planned great battle, according to its relations, is more or less, but always in some degree, to be regarded as the provisional middle point and center of gravity of the whole system. The more a general takes the field in what is the true spirit of war, as it is of every combat, with the feeling and the idea—that is, the conviction—that he must and will conquer, the more he will strive to throw every weight into the scale of the first battle, hoping and striving to win everything by it. Bonaparte no sooner entered on one of his wars than he was thinking of conquering his enemy at once in the first battle, and Frederick the Great, although in a more limited sphere, and confronted with more limited crises, thought the same when, at the head of a small army, he sought to get elbow room for himself in the rear of the Russians or the Federal Imperial Army.

The decision which is offered by the great battle, depends, we have said, partly on the battle itself, that is, on the number of troops engaged and on the magnitude of the success.

How the general may increase its importance in respect to the first point is self-evident and we shall merely observe that according to the scope of the great battle, the number of cases which are decided along with it increases, and that therefore generals, confident in themselves and inclined to great decisions, have always managed to make use of the greater part of their troops in it without neglecting on that account essential points elsewhere.

As far as the success, or more strictly speaking, the intensive strength of the victory is concerned, that depends chiefly on four points:

1. On the tactical form in which the battle is fought.
2. On the nature of the country.
3. On the relative proportions of the three arms.
4. On the relative strength of the two armies.

A battle with parallel fronts and without any enveloping action will seldom yield as great success as one in which the defeated army has been turned, or compelled to engage more or less with a change of front. In a broken or hilly country the successes are likewise smaller, because the power of the blow is everywhere weakened.

If the cavalry of the vanquished is equal or superior to that of the victor, then the effects of the pursuit are diminished, and by that a great part of the results of victory is lost.

Finally, it is easy to understand that if superior numbers are on the side of the victor, and he uses this advantage to turn the flank of his adversary, or to compel him to change front, greater results will follow than if the victor had been weaker in numbers than the vanquished. The battle of Leuthen might lead one to doubt the correctness in practice of this principle; but we beg permission for once to say what otherwise we do not like: no rule without exception.

In all these ways, therefore, the commander has the means of giving his battle a decisive character; it is true he thus exposes himself to an increased amount of danger, but all his actions are subject to that dynamic law of the moral world.

There is then nothing in war which can be compared with the great battle in the way of importance, and the acme of strategic ability is displayed in the provision of means for this, in the skilful determination of place and time and direction of troops, and in the good use of its success.

But it does not follow from the importance of these things that they must be of a very complicated and obscure nature; on the contrary, everything is very simple; the art required for making the plan is very slight. But there is a great need of quickness in judging of circumstances, need of energy, steady consistency, a youthful spirit of enterprise—heroic qualities, to which we shall have to refer often. There is, therefore, but little needed here of what can be taught by books, and much of what, if it can be taught at all, must come to the general through some other medium than printer's type.

The impulse toward a great battle, the voluntary, sure progress toward it, must proceed from a feeling of innate power and a clear sense of its necessity; in other words, it must proceed from inborn courage and from insight sharpened by great experiences in life.

Great examples are the best teachers, but it is certainly unfortunate if a cloud of theoretical prejudices intervenes, for even sunlight is refracted and tinted by the clouds. To destroy such prejudices, which many a time rise and spread like a miasma, is an imperative duty of theory, for the misbegotten offspring of human reason, mere reason can in turn destroy.

CHAPTER 12

STRATEGIC MEANS

OF UTILIZING VICTORY

The more difficult part, viz., that of preparing the victory as well as possible, is an unobtrusive service which strategy renders, but for which she hardly gets any praise. Her brilliance and renown she wins by turning to good account a victory gained.

What may be the special object of a battle, how it fits into the whole system of a war, up to what point the course of victory may lead according to the nature of circumstances, where its culminating point lies—all these are things which we shall not consider until later. But under any conceivable circumstances the fact holds good that, unless pursued, no victory can have a great effect, and that, however short the course of victory may be, it must always lead beyond the first steps of this pursuit. To avoid frequent repetition of this, we shall now dwell for a moment on this necessary supplement of victory in general.

The pursuit of a beaten opponent begins at the moment when, giving up the engagement, he leaves his position. All previous movements in one direction and another do not belong to this but to the development of the battle itself. Usually the victory at the moment here indicated, even if it is certain, is still very small and weak, and in the series of events it would not yield any very positive advantages if not completed by a pursuit on the first day. It is generally only then, as we have said before, that the trophies which give substance to the victory begin to be gathered. Of this pursuit we shall speak presently.

Usually both sides enter the battle with their physical powers considerably weakened, for the movements immediately preceding are usually of a very trying character. The exertion which the fighting out of a great combat costs completes the exhaustion. In addition to this, the victorious party is very little less disorganized and out of its original formation than the vanquished, and therefore feels the need to reform, to collect stragglers and issue fresh ammunition to those who are without. All these circumstances put the victor himself into a state of crisis of which we have already spoken. Now if the defeated force is only a subordinate portion of the enemy's army which can be retired, or if it can otherwise expect a considerable reinforcement, the victor may easily run into the obvious danger of forfeiting his victory, and this consideration, in such a case, very soon puts an end to pursuit, or at least restricts it materially. Even when a strong reinforcement of the enemy is not to be feared, the victor finds in the above circumstances a powerful check to the rapidity of his pursuit. There is no reason, it is true, to fear that the victory will be snatched away, but adverse engagements are still possible, and may diminish the advantages which up to the present have been gained. Moreover, at this moment the whole weight of physical humanity, with its needs and weaknesses, hangs heavy on the will of the commander. All the thousands under his command need rest and refreshment, and long to see a stop put to toil and danger for the present; only a few, forming an exception, can see and feel beyond the present moment. Only in these few is there sufficient mental vigor left still to think, after all that is necessary has been done, of those successes which at such a moment appear as mere embellishments of victory—as a luxury of triumph. But all these thousands have a voice in the council of the general, for through the various ranks of the military hierarchy these interests of physical humanity have their sure conductor into the heart of the commander. His own energies, through mental and bodily fatigue, are more or less weakened, and thus it happens that, mostly from these purely human causes, less is done than might have been done, and that generally what is done is to be ascribed entirely to the *thirst for glory*, the *energy*, perhaps also to the *hardheartedness* of the general-in-chief. It is only thus that we can explain the hesitating manner in which many generals pursue a victory which superior numbers have given them. The first pursuit of the victory we limit in general to the first day, including at most the following night. At the end of that period the necessity of our own recovery will in any case demand a halt.

This first pursuit has different natural degrees.

The first is, if cavalry alone is employed; in that case it usually amounts more to alarming and watching than to pressing the enemy in reality, because the smallest obstacle of ground is generally sufficient to check the pursuer. Much as cavalry can achieve against single bodies of weakened and demoralized troops, still when opposed to the bulk of the beaten army it is only the auxiliary arm, because the retreating enemy can employ his fresh reserves to cover his withdrawal, and, therefore, at the next trifling obstacle of ground, by combining all arms he can make a stand with success. The only exception to this is in the case of an army in actual flight and complete dissolution.

The second degree is when the pursuit is made by a strong advance-guard composed of all arms, containing naturally the greater part of the cavalry. Such a pursuit drives the enemy as far as the nearest strong position for his rear-guard, or the next position of his army. Neither can usually be found at once, and, the pursuit is thus carried farther; generally, however, it does not extend beyond the distance of from three to at most a half dozen miles, because otherwise the advance-guard would not feel itself sufficiently supported.

The third and most vigorous degree is when the victorious army itself continues to advance as far as its physical powers admit. In this case the beaten army will quit most of the positions which a country offers on the mere preparations for an attack, or a movement to turn its flank, and the rear-guard will be still less likely to get involved in an obstinate resistance.

In all three cases the night, if it sets in before the completion of the whole action, usually puts an end to it, and the few instances in which this does not take place, and the pursuit is continued throughout the night, must be regarded as pursuits in an exceptionally vigorous form.

If we reflect that in fighting by night everything must be, more or less, left to chance, and that at the conclusion of a battle the regular organization and routine of an army must in any case be greatly upset, we may easily conceive the reluctance of both generals to carry on their business in the darkness of the night. Unless a complete dissolution of the vanquished army, or a rare superiority of the victorious army in military virtue, ensures success, everything would be more or less left to fate, which can never be in the interest of any general, even the most foolhardy. As a rule, therefore, night puts an end to the pursuit, even when the battle has only been decided shortly before darkness sets in. This allows the defeated either time to rest and to rally immediately, or, if he continues to retreat during the night, it gives him a march in advance. After this break the defeated is decidedly in a better condition; much of what had been scattered and thrown into confusion has been restored, fresh ammunition has been issued, the whole put into a fresh formation. Whatever further encounter now takes place with the victor is a new engagement, not a continuation of the old, and although it may be far from promising absolute success, still it is a new combat, and not merely a gathering up of the crumbled ruins by the victor.

When, therefore, the victor can continue the pursuit even throughout the night, if only with a strong advance-guard composed of all arms, the effect of the victory will be immensely increased. Of this, the battles of Leuthen and Waterloo are examples.

The whole action of the pursuit is fundamentally tactical and we only dwell upon it here in order to make plain the difference which it produces in the effect of a victory.

This first pursuit, as far as the enemy's nearest point of resistance, is a right of every victor, and hardly depends in any way on his further plans and conditions. These may considerably diminish the positive results of a victory gained with the main body of the army, but they cannot make this first use of it impossible; at least cases of that kind, if conceivable at all, would be so uncommon that they could have no appreciable influence on theory. And here certainly we must say that the example afforded by modern wars opens up quite a new field for energy. In preceding wars, resting on a narrower basis, and more circumscribed in their scope, there had grown up unnecessary conventional restrictions in many other matters, but particularly in this one. *The conception, honor of victory,* seemed to generals so much the main point that in consequence they thought less of the real destruction of the enemy's military force, as in point of fact that destruction appeared to them only as one of the many means in war, not in any regard as the principal, much less as the only, one. All the more readily they put the sword in its sheath the moment the enemy had lowered his. Nothing seemed more natural to them than to stop the combat as soon as the decision was obtained, and to regard all further carnage as unnecessary cruelty. Even if this false philosophy did not determine their resolutions entirely, still it introduced a way of looking at things in which ideas of the exhaustion of all powers and of the physical impossibility of continuing the struggle, obtained more immediate consideration and carried great weight. Certainly the sparing of one's own instrument of victory is a vital question if this is the only one we possess, and we foresee that soon a moment may arrive when it will not be sufficient in any case for all that has to be done, as in fact is usually the result of every continuation of the offensive. But this calculation was still false in so far as the further loss of forces, which we might suffer by a continuance of the pursuit, was out of all proportion to that of the enemy. That view again could, therefore, only exist so long as the destruction of the military forces was not regarded as the main point. And so we find that in former wars real heroes only—such as Charles XII, Marlborough, Eugene, Frederick the Great—added a vigorous pursuit to their victories when they were decisive enough, and that other generals usually contented themselves with the possession of the field of battle. In modern times the greater energy infused into the conduct of wars, through the greater importance of the circumstances from which they have proceeded, has destroyed these conventional limitations; the pursuit has become a chief business for the victor; trophies have on that account multiplied in extent, and although there are cases also in modern warfare in which this has not been so, still they belong to the exceptions, and are to be accounted for by peculiar circumstances.

At Görschen and Bautzen nothing but the superiority of the allied cavalry prevented a complete rout; at Gross Beeren and Dennewitz the ill-will of the Crown Prince of Sweden; at Laon the enfeebled personal condition of old Blücher.

But Borodino is also a pertinent example here, and we cannot resist saying a word or two more about it, partly because we do not think that the affair is explained simply by blaming Bonaparte, partly because it might appear as if this, and with it a great number of similar cases, belonged to that class which we have considered as so extremely rare, cases in which the general conditions seize and fetter the general at the very beginning of the battle. French authors in particular, and great admirers of Bonaparte (Vaudancourt, Chambray, Ségur), have blamed him decidedly because he did not drive the Russian army completely off the battlefield, and did not use his last reserves to destroy it, because in that case what was then only a lost battle would have become a complete defeat. It would carry us too far to describe circumstantially the mutual situation of the two armies; but this much is evident: when Bonaparte crossed the Niemen, he had 300,000 men in the corps which afterward fought at Borodino, and after this battle only 120,000 remained. He might well therefore have been apprehensive that he would not have enough left to march upon Moscow, the point on which everything seemed to depend. A victory such as he had just gained gave him almost a certainty of taking that capital for it seemed in the highest degree improbable that the Russians would be in a condition to fight a second battle within a week; and in Moscow he hoped to dictate the peace. No doubt the complete destruction of the Russian army would have made this peace much more certain; but still the first consideration was to get to Moscow, that is, to get there with a force with which he would appear master of the capital and through that of the empire and the government. The force which he brought with him to Moscow was no longer sufficient for that, as the result has shown, but it would have been still less so if, in destroying the Russian army, he had destroyed his own at the same time. Bonaparte was thoroughly aware of all this, and in our eyes he stands completely justified. But on that account this case is still not to be counted among those in which, through the general circumstances, the general is prevented from first following up his victory with a pursuit, for it was not yet a question of mere pursuit. The victory was decided at four o'clock in the afternoon, but the Russians still occupied the greater part of the battlefield; they were not yet disposed to give up the ground, and if the attack had been renewed, they would still have offered a most determined resistance, which, it is true, would have certainly ended in their complete defeat, but would have cost the victor much further bloodshed. We must, therefore, reckon the battle of Borodino as among battles, like Bautzen, left unfinished. At Bautzen it was the vanquished who preferred first to quit the field of battle. At Borodino, it was the victor who preferred to be content with a partial victory, not because the decision appeared doubtful, but because he could not afford to pay for the whole.

Returning now to our subject, the conclusion we draw from our reflections with reference to the pursuit is that the energy thrown into it chiefly determines the value of the victory; that the pursuit is a second act of the victory, in many cases even more important than the first, and that strategy, in here approaching tactics to receive from her the finished work, lets the first act of her authority be to demand this perfecting of the victory.

But the effect of victory is seldom found to come to an end with the pursuit; only now begins the real course to which victory lends speed. This course is conditioned, as we have said, by other circumstances, of which it is not yet time to speak. But we may here mention what there is of a general character in the following up of a victory, in order to avoid repetition when the subject occurs again.

In the further following up of a victory we can again distinguish three degrees: merely following the enemy, actually pressing on him and a parallel march to intercept him.

Simply *to follow* the enemy causes him to continue his retreat, until he thinks he can risk another engagement. It would therefore suffice to give us all the effect of the advantage gained, and, besides that, all that the beaten enemy cannot carry away with him; sick, wounded and disabled from fatigue, quantities of baggage and carriages of all kinds, will fall into our hands. But this mere following does not tend to heighten the state of disintegration in the enemy's army, an effect which is produced by the two next degrees.

If, for instance, instead of contenting ourselves with following the enemy into the camp which he has just vacated and occupying just as much of the country as he chooses to abandon, we make our arrangements so as to demand something more from him every day, and accordingly with our advance-guard organized for the purpose, attack his rear-guard every time it attempts to halt, then this will hasten his retreat, and consequently tend to increase his disorganization. This it will principally effect by the character of unremitting flight which his retreat will thus assume. Nothing makes such a depressing impression on the soldier as hearing the enemy's cannon again at the very moment when, after a strenuous march, he seeks some rest. If this impression is repeated day after day for some time, it may lead to panic. There lies in it a constant admission of being obliged to obey the decree of the enemy, and of being incapable of any resistance, and the consciousness of this cannot but weaken the morale of an army in a high degree. The effect of pressing the enemy in this way attains a maximum when it forces him to make night marches. If the victor scares away his defeated opponent at sunset from a camp which has just been taken up either for the main body of the army, or for the rear-guard, the vanquished must either make an actual night march, or at least alter his position in the night, moving it farther to the rear, which is much the same thing. The victor can, on the other hand, pass the night in quiet.

The arrangement of marches and the choice of positions depend in this case also upon so many other things, especially on the subsistence of the army, on strong natural obstacles in the ground, on large cities, and so forth, that it would be ridiculous pedantry to show by a geometrical analysis how the pursuer, being able to force the hand of the retreating enemy, can compel him to march every night while he himself takes his rest. But nevertheless it remains true and practicable that marches in pursuit may be so planned as to have this tendency, and that the efficacy of the pursuit is very much increased thereby. If this is seldom considered in the execution, it is because such a procedure is also more difficult for the pursuing army than a regular adherence to stations and hours of the day. To start in good time in the morning, to encamp at mid-day, to occupy the rest of the day in providing for the needs of the army, and to use the night for repose, is a much more comfortable procedure than regulating one's movements exactly according to those of the enemy, therefore determining nothing till the last moment, starting on the march, sometimes in the morning, sometimes in the evening, always being in the presence of the enemy for several hours, exchanging cannon shots with him and keeping up skirmishing fire, planning maneuvers to turn him—in short, using every tactical means which such a course renders necessary. All this naturally bears with a heavy weight on the pursuing army, and in war, where there are so many burdens to be borne, men are always inclined to get rid of those which do not seem absolutely necessary. These observations remain true, whether applied to a whole army or, as in the more usual case, to a strong advance-guard. For the reasons just mentioned, this second method of pursuit, this continued pressing of the defeated enemy is rather a rare occurrence. Even Bonaparte in his Russian campaign, in 1812, practiced it but little, for the very obvious reason that the difficulties and hardships of this campaign threatened his army in any case with complete destruction before it had reached its object. On the other hand, the French in their other campaigns have distinguished themselves by their energy in this point also.

Lastly, the third and most effective form of pursuit is the parallel march to the immediate goal of the retreat.

Every defeated army will naturally have behind it, at a greater or less distance, some point, the attainment of which is the first purpose in view, be it that the army's further retreat can be endangered thereby, as in the case of a defile, or that it is important for the point itself that it should be reached before the enemy arrives, as in the case of capital cities, magazines, etc., or, lastly, that the enemy at this point can gain new powers of defense, as in the case of a fortified position, or of a junction with other corps.

Now if the victor directs his march on this point by a lateral road, it is evident how this may quicken the retreat of the beaten army in a destructive manner, convert it into hurry, perhaps finally into flight. The vanquished has only three ways to counteract this. The first would be to throw himself in front of the enemy, in order by an unexpected attack to gain that probability of success which is lost to him in general from his position; this plainly supposes an enterprising bold general, and an excellent army, beaten but not utterly defeated; therefore, it can only be employed by a beaten army in very few cases.

The second way is hastening the retreat; but this is just what the victor wants, and it easily leads to excessive efforts on the part of the troops, by which enormous losses are sustained, in stragglers, broken guns and carriages of all kinds.

The third way is to make a detour, in order to get round the nearest point of interception, to march with more ease at a greater distance from the enemy, and thus to render the haste less damaging. This last way is the worst of all, since it is to be regarded as a new debt contracted by an insolvent debtor, and leads to still greater embarrassment. There are cases in which this course is advisable; others where there is nothing else left; also instances in which it has been successful. But on the whole it is certainly true that its adoption is usually decided less by a clear conviction of its being the surest way of attaining the aim than by another, inadmissible, reason. This reason is the dread of encountering the enemy. Woe to the commander who gives in to this! However much the morale of his army may have deteriorated, and however well-founded may be his apprehensions of being on account of this at a disadvantage in any conflict with the enemy, the evil will only be made worse by too anxiously avoiding every possible risk of collision. Bonaparte in 1813 would not even have brought the 30,000 or 40,000 men who were left to him after the battle of Hanau over the Rhine, if he had tried to avoid that battle and to pass the Rhine at Mannheim or Coblenz. It is just by means of small engagements, carefully prepared and executed, in which the defeated army, being on the defensive, has always the assistance of the ground—it is just by these that the moral strength of the army can most easily be encouraged.

The beneficial effect of the smallest successes is incredible; but with most generals to make such an attempt demands great self-command. The other way, that of evading all encounters, appears at first so much easier, that it is most generally preferred. It is therefore usually just this system of evasion which best promotes the intention of the pursuer, and often ends with the complete downfall of the pursued. We must, however, remember here that we are speaking of a whole army, not of a single division, which, having been cut off, is seeking to rejoin the main army by making a detour. In such a case circumstances are different, and success is not uncommon. But one condition in this race for the goal is that a division of the pursuing army should follow by the same road which the pursued has taken, in order to pick up what has been left behind, and keep up the impression which the presence of the enemy never fails to make. Blücher neglected this in his, in other respects, model pursuit from Waterloo to Paris.

Such marches, it is true, tell upon the pursuer as well as the pursued, and they would not be advisable if the enemy's army falls back upon another army of considerable strength, if it has a distinguished general at its head, and if its destruction is not already well prepared. But when this means can be adopted, it acts like a mighty engine. The losses of the beaten army from sickness and fatigue are on such a disproportionate scale, the spirit of the army is so weakened and lowered by constant anxiety about impending ruin, that at last anything like a well-organized resistance is almost out of the question; every day thousands of prisoners fall into the enemy's hands without striking a blow. In such a time of abundant good fortune, the victor need not hesitate about dividing his forces in order to draw into the vortex of destruction everything within reach of his army, to cut off detachments, to take fortresses unprepared for defense, to occupy large cities, and so forth. He may do anything until a new state of things arises, and the more he ventures in this way, the longer will it be before that change will take place.

There is no lack of examples of such brilliant results from great victories and of magnificent pursuits in the wars of Bonaparte. We need only recall Jena, Regensburg, Leipzig and Waterloo.

CHAPTER 13

RETREAT AFTER A LOST BATTLE

In a lost battle the power of an army is broken, the moral to a greater degree than the physical. A second battle, unless new favorable circumstances come into play, would lead to a complete defeat, perhaps to destruction. This is a military axiom. Naturally, the retreat is continued up to that point where the equilibrium of forces is restored, either by reinforcements, or by the protection of strong fortresses, or by great obstacles of the ground, or by a dispersion of the enemy's force. The magnitude of the losses sustained, the extent of the defeat, but still more the character of the enemy, will bring nearer or put off this moment of equilibrium. How many instances there are of a beaten army which has rallied again at a short distance, without its circumstances having altered in any way since the battle! The cause of this may be traced to the moral weakness of the adversary, or to the preponderance gained in the battle not being great enough to make a deep impression.

In order to profit by this weakness or mistake of the enemy, not to yield one inch breadth more than the pressure of circumstances demands, but above all, to keep the moral forces up to as advantageous a point as possible, a slow retreat, offering incessant resistance is absolutely necessary together with a bold and spirited counterstroke, whenever the pursuing enemy seeks to push his advantage too far. Retreats of great generals and of armies accustomed to war have always resembled that of a wounded lion, and such is, undoubtedly, also the best theory.

It is true that at the moment of quitting a dangerous position we have often seen trifling formalities observed which caused a waste of time and therefore became dangerous, while in such cases everything depends on getting out of the place quickly. Practiced generals consider this principle a very important one. But such cases must not be confounded with a general retreat after a lost battle. Whoever thinks here by a few rapid marches to gain a start, and to recover more easily a firm standing, commits a great error. The first movements must be as small as possible, and it must be a general principle not to let our hand be forced by the enemy. This principle cannot be followed without bloody engagements with the enemy at our heels, but the principle is worth the sacrifice; without it we get into a hurried movement which soon turns into a headlong rush, and costs in stragglers alone more men than rear-guard actions would have done, and, besides that, extinguishes the last remnants of courage.

A strong rear-guard composed of our best troops, commanded by our bravest general, and supported by the whole army at critical moments; a careful utilization of ground, strong ambuscades wherever the boldness of the enemy's advance-guard, and the ground, afford opportunity; in short, the preparing and the planning of regular small battles—these are the means of following this principle.

The difficulties of a retreat are naturally greater or less according as the battle has been fought under more or less favorable circumstances, and according as it has been more or less obstinately contested. If we hold out to the last man against a superior enemy, the battles of Jena and Waterloo show how impossible anything like a regular retreat may become.

Now and again it has been suggested (Lloyd, Bülow) to divide for the purpose of retreating, therefore to retreat in separate divisions or even on divergent lines. Such a separation as is made merely for its great convenience, and where concentrated action continues possible and is intended, is not what we now refer to. Any other kind is extremely dangerous, contrary to the nature of the thing, and therefore a great error. Every lost battle is a weakening and disintegrating influence; the immediate need is to concentrate and, in concentration, to recover order, courage and confidence. The idea of harassing the enemy by separate corps on both flanks, at the moment when he is following up his victory, is a perfect anomaly; a faint-hearted weakling of an enemy might be impressed in that manner, and for such a case it may answer; but where we are not sure of this weakness in our opponent we had better let it alone. If the strategic conditions after a battle require that we should cover ourselves right and left by separate detachments, so much must be done as in the circumstances is unavoidable. But this separation must always be regarded as an evil, and we are seldom in a state to begin it the day after the battle itself.

If Frederick the Great after the battle of Kollin, and the raising of the siege of Prague retreated in three columns, that was done not out of free choice, but because the position of his forces, and the necessity of covering Saxony, left him no alternative. Bonaparte, after the battle of Brienne, sent Marmont back to the Aube, while he himself passed the Seine and turned toward Troyes. That this did not end in disaster was solely owing to the fact that the Allies, instead of pursuing, divided their forces in like manner, turned with the one part (Blücher) toward the Marne, while with the other (Schwarzenberg), from fear of being too weak, they advanced very slowly.

CHAPTER 14

NIGHT FIGHTING

The manner of conducting an engagement at night, and the details of its course, is a subject of tactics. We only examine it here in so far as the whole appears as a special strategic means.

Fundamentally, every night attack is only a more intense form of surprise. Now at first sight such an attack appears quite pre-eminently advantageous, for we suppose the enemy to be taken by surprise, the assailant naturally to be prepared for what is to happen. What an inequality! Imagination paints a picture of the most complete confusion on the one side, and on the other the assailant only occupied in reaping the fruits of it. Hence the frequent schemes for night attacks by those who have nothing to lead, and no responsibility, while these attacks so seldom take place in reality.

These ideas are all based on the hypothesis that the assailant knows the measures of the defender because they have been made and announced beforehand, and cannot escape his reconnaissances and inquiries; that on the other hand, the measures of the assailant, being only made at the moment of execution, cannot be known to the enemy. But even the latter supposition is not always quite correct, and still less is the first. If we are not so near the enemy as to have him right before our eyes, as the Austrians had Frederick the Great before the battle of Hochkirch, all that we know of his position must always be imperfect, as it is obtained by reconnaissances, patrols, information from prisoners and spies, sources on which no firm reliance can be placed if only because intelligence thus obtained is always more or less of an old date, and the position of the enemy may have altered in the meantime. Moreover, with the tactics and mode of encampment of former times it was much easier than it is now to investigate the position of the enemy. A line of tents is much easier to distinguish than a line of huts or even a bivouac, and an encampment on regularly extended front lines also easier than one of divisions formed in columns, the mode often used at present. We may have the ground on which a division has pitched camp in that manner completely under our eye, and yet not be able to arrive at any accurate idea of it.

But the position again is not all that we must know; the measures which the defender may take in the course of the engagement are just as important, and do not after all consist in mere random shots. These measures, too, make night attacks more difficult in modern wars than in former wars, because in modern wars they are of greater importance than those taken at an earlier stage. In our engagements the position of the defender is more temporary than definitive, and on that account the defender is better able to surprise his adversary with unexpected blows than he could formerly.

Therefore what the assailant knows of the defender in the case of a night attack is seldom or never sufficient to compensate for the lack of direct observation.

But the defender has on his side another small advantage as well, which is that he is more at home than the assailant on the ground which forms his position, just as the inhabitant of a room even in darkness is able to find his way around in it with greater ease than a stranger. He is able to find each part of his force more quickly and therefore can more readily get at it than is the case with his adversary.

From this it follows that the assailant in an engagement at night needs his eyes just as much as does the defender and that, therefore, only particular reasons can make a night attack advisable.

Now these reasons arise mostly in connection with subordinate parts of an army, rarely with the army itself. Hence it follows that a night attack also as a rule can only take place in subordinate engagements and seldom in great battles.

We may attack a subordinate portion of the enemy's army with a very superior force, consequently enveloping it with a view either to take the whole, or to inflict very severe loss on it by an engagement in which he is at a disadvantage, provided that other circumstances are in our favor. But such a scheme can never succeed except by a great surprise, because no subordinate part of the enemy's army would enter such a disadvantageous engagement, but would refuse it. But a high degree of surprise, except in rare instances in a very close country, can only be attained at night. If therefore we wish to gain such an advantage from the faulty disposition of a subordinate military force of the enemy, then we must make use of the night, to complete at least the preliminary arrangements, even if the engagement itself is not to open until toward daybreak. This is therefore the reason for all the little enterprises by night against outposts, and other small bodies, the main point being invariably through superior numbers, and by getting round his position, to entangle him unexpectedly in an engagement at such a disadvantage that he cannot disengage himself without great loss.

The larger the body attacked, the more difficult the undertaking, because a stronger force has greater resources within itself to maintain the fight for some time, till help arrives.

On that account the whole of the enemy's army can never in ordinary cases be the object of such an attack, for although it has no assistance to expect from outside, still, it contains within itself sufficient means of repelling attacks from several sides, particularly in our day, when everyone from the outset is prepared for this very common form of attack. Whether the enemy can attack us successfully on several sides depends generally on conditions quite different from that of its being done unexpectedly. Without entering here into the nature of these conditions, we confine ourselves to observing that with turning an enemy, great results, but also great dangers, are connected; that, therefore, if we set aside special circumstances, nothing justifies it but a great superiority, just such as we can use against a subordinate part of the enemy's army.

But the turning and surrounding of a small enemy corps, and particularly in the darkness of night, is also more practicable for this reason, that whatever we stake upon it, and however superior the force used may be, still probably it constitutes only a limited portion of our army, and we can sooner stake that than the whole on the hazard of a great venture. Besides, a greater part, or even the whole, serves as a support and rallying point for the part taking the risk, which again diminishes the danger of the enterprise.

Not only the risk, but the difficulty of execution as well confines night enterprises to somewhat small detachments. Since surprise is the real essence of it, it follows that stealthy approach is also the chief condition of its execution. But this is more easily done with small bodies than with large ones, and is seldom practicable for the columns of a whole army. For this reason such enterprises are in general only directed against single outposts, and can only be used against larger bodies if they are without sufficient outposts, like Frederick the Great at Hochkirch. Again this will happen more seldom with the army itself than with minor divisions of it.

In recent times, when war has been carried on with so much greater rapidity and vigor, it was bound to happen more frequently that armies encamped very close to each other and without having a very strong system of outposts, because both things always occur at the crisis which shortly precedes a great decision. But then at such times the readiness for battle on both sides is also greater. On the other hand, in former wars it was a frequent practice for armies to take up camps in sight of each other, even when they had no other object but that of mutually holding each other in check, and consequently for a longer period. How often Frederick the Great stood for weeks so near to the Austrians that the two might have exchanged cannon shots with each other!

But these practices, certainly more favorable to night attacks, have been discontinued in more recent wars; and armies, being now no longer, in regard to subsistence and requirements for encampment, such independent bodies complete in themselves, find it necessary to leave usually a day's march between themselves and the enemy. If we now once more give our particular attention to the night attack of an army, we see that adequate motives for it can but seldom occur, and that they can be reduced to one or other of the following cases:

1. A quite unusual degree of carelessness or audacity on the part of the enemy, which very rarely occurs, and when it does, is usually compensated for by a great superiority in moral force.
2. A panic in the enemy's army, or generally such a degree of superiority in moral force on our side that this alone is sufficient to supply the place of guidance in action.
3. Cutting through an enemy's army of superior force which keeps us enveloped, because in this everything depends on surprise, and the object of a mere escape allows a much greater concentration of forces.
4. Finally, in desperate cases, when our forces have such a disproportion to the enemy's that we see no possibility of success, except through extraordinary daring.

But in all these cases there still remains the condition that the enemy's army is under our eyes, and protected by no advance-guard.

As for the rest, most night engagements are so conducted as to end with daybreak, so that only the approach and the first attack are made under cover of darkness, because the assailant in that manner can make better use of the results of the confusion into which he plunges his adversary. On the other hand, engagements which do not begin until daybreak, in which the night therefore is merely used for the approach, are not to be counted as night fights.

Book V

MILITARY FORCES

CHAPTER 1

GENERAL SCHEME

We shall consider the military forces

1. As regards their numerical strength and organization
2. As regards their condition apart from fighting
3. In respect of their maintenance; and, lastly,
4. In their general relations to country and terrain.

Thus we shall concern ourselves in this book with those aspects of the military forces which must be regarded only as *necessary conditions of combat* and not as the combat itself. They stand in more or less close connection and reciprocal relation; therefore they will often be mentioned in dealing with the employment of the combat; but first we must consider each by itself, as a whole, in its essence and peculiarities.

CHAPTER 2

Theater of War, Army, Campaign

The nature of the subject does not allow of an exact definition of these three factors, denoting, respectively, space, mass and time in war; but in order that we may not at times be entirely misunderstood, we must seek to make somewhat clearer the common usage of these terms, to which we prefer in most cases to adhere.

1. THEATER OF WAR

This term denotes properly such a portion of the whole sphere of war as has its boundaries protected and thus possesses a kind of independence. This protection may consist of fortresses or important natural obstacles presented by the country or in its being separated by a considerable distance from the rest of the sphere of war. Such a portion is not a mere part of the whole, but a small whole complete in itself. Consequently it is more or less in such a condition that changes which take place at other points of the area embraced in operations have only an indirect and no direct influence upon it. If we wanted an exact distinguishing mark, it could only be the possibility of our conceiving an advance in the one while a retreat was taking place in the other, or a defense in the one while an attack was going on in the other. Such a clearly defined concept as this cannot be universally applied; it is used here merely to indicate the most essential point.

2. ARMY

With the assistance of the concept of a theater of war, it is very easy to say what an army is; it is, in point of fact, the mass of troops in the same theater of war. But this plainly does not include all that is meant by the term in its common usage. Blücher and Wellington each commanded a separate army in 1815, although the two were in the same theater of war. The chief command is, therefore, another distinguishing mark of an army. This distinguishing mark, however, is closely related to the preceding, for where things are well organized, there should exist only one chief command in a theater of war, and the commander-in-chief in a separate theater of war should always have an adequate degree of independence.

The mere absolute strength of a body of troops plays a less decisive part in this designation than might at first appear. For where several armies are acting under one command, and in one and the same theater of war, they are called armies, not by reason of their strength, but from conditions antecedent to the campaign (1813, the Silesian Army, the Army of the North, etc.), and although we could divide a great mass of troops intended to remain in the same theater into corps, we would never divide them into different armies; at least, such a division would be contrary to the usage which seems to have attached itself firmly to this term. On the other hand, it would be pedantry to claim the term army for each band of irregular troops acting independently in a remote province. Still we must not leave unnoticed the fact that it surprises no one when we speak of the "army" of the Vendeans in the Revolutionary War, and yet it was not much stronger.

The concepts of army and theater of war, therefore, as a rule go together and are correlative.

3. CAMPAIGN

Although the sum of all military events which happen in all the theaters of war in one year is often called a *campaign*, it is more usual and more definite to understand by the term the events in *one single* theater of war. A worse mistake is to connect the concept of a campaign with a period of one year, for wars no longer divide themselves naturally into campaigns of a year's duration by definite and long periods in winter quarters. Since, however, the events in a theater of war fall naturally into certain larger units—when, for instance, the direct effects of some more or less great catastrophe cease and new complications begin to develop—these natural divisions must be taken into consideration in order to allot to each year (campaign) its complete share of events. No one would maintain that the campaign of 1812 terminates at Memel, where the armies were on the 1st of January, and count the further retreat of the French beyond the Elbe as belonging to the campaign of 1813, since this was plainly only a part of the whole retreat from Moscow.

That we cannot give to these concepts any greater degree of distinctness is not at all a disadvantage, since they cannot be used, as in the case of philosophical definitions, to form any sort of basis for decisions. They only serve to give more clarity and precision to the language we use.

CHAPTER 3

Relative Strength

In Book III, Chapter 8, we have determined the value of superior numbers in engagements, and consequently the value of general superiority in strategy, from which proceeds the importance of relative strength, concerning which we must now make a few more detailed observations.

If we examine modern military history without bias, we must admit that superiority in numbers becomes every day more decisive; we must therefore value somewhat more highly than perhaps has been done before, the principle of being as strong as possible in a decisive engagement.

Courage and morale have, in all ages, increased an army's physical powers, and will continue to do so in the future; but we find that at certain periods in history a superiority in the organization and equipment of an army has given it a great moral preponderance, while at others a great superiority in mobility has had a like effect. At one time it was new tactical systems; at another, the art of war involved itself in an effort to make a skilful use of ground on great general principles; and by such means here and there we find one general gaining great advantages over another. But this tendency has itself disappeared, and has given place to a more natural, and simpler method of procedure. If we look at the experiences of recent campaigns without any preconceived opinion, we must admit that there are but few traces of any of the above-mentioned phenomena, either throughout any whole campaign or in engagements of a decisive character—especially the main battle, in regard to which we refer to Book IV, Chapter 2.

Armies are in our days so much on a par in regard to arms, equipment and training that there is no very notable difference between the best and the worst in these matters. Between the various technical corps scientific knowledge still makes a noticeable difference, but in general it amounts merely to some being the inventors and introducers of improved arrangements, which the others immediately imitate. Even the subordinate generals, leaders of corps and divisions, have everywhere adopted, so far as their own sphere of activity is concerned, somewhat the same views and methods, so that, apart from the talent of the commander-in-chief, a factor dependent on chance and not bearing a constant relation to the standard of education among the people and in the army, there is nothing now but habituation to war which can give one army a decided superiority over another. The nearer we approach to a state of equality in all these things, the more decisive becomes relative numerical strength.

The character of modern battles is the result of this equality. One need only read without prejudice of the battle of Borodino, in which the best army in the world, the French, measured its strength with the Russian, which, in many parts of its organization and in the training of its individual members might be considered the most backward. In the whole battle there is not a single trace of superior skill or intelligence; it is a simple trial of strength between the respective armies throughout; and as they were nearly equal, the result could be nothing but a gentle turning of the scale in favor of that side where there was the greatest energy on the part of the commander and the greater habituation to war on the part of the troops. We have taken this battle as an illustration, because in it there was an equality of numbers on each side such as is rarely to be found.

We do not maintain that all battles are like this, but such is the keynote of most of them.

In a battle in which the forces try their strength on each other in so slow and methodical a manner, superior numbers on one side must make the result in its favor much more certain. And it is a fact that we may search modern military history in vain for a battle in which an army has beaten another double its own strength, an occurrence by no means uncommon in former times. Bonaparte, the greatest general of modern times, had managed to assemble in all his victorious main battles—with one exception, that of Dresden, 1813—a superior army, or at least one very little inferior to that of his opponent, and when it was impossible for him to do so, as at Leipzig, Brienne, Laon and Waterloo, he was beaten.

The absolute strength is in strategy generally a given quantity, which the commander can no longer alter. But from this it by no means follows that it is impossible to carry on a war with a decidedly inferior force. War is not always a voluntary decision of state policy, and least of all is it so when the forces are very unequal. Consequently, any state of relative strength is imaginable in war, and it would be a strange theory of war which would resign completely just where it is most needed.

However desirable theory may consider an adequate force, still it cannot say that no use can be made of the most inadequate. No limits can be prescribed in this respect.

The weaker the force, the smaller must be its objects; and the weaker the force, the shorter time it will last. In these two directions weakness has room to deviate, if we may use the expression. Concerning the changes which the degree of strength produces in the conduct of war we shall only be able to speak later on, when they present themselves; at present it is sufficient to have indicated the general point of view, but to complete that we shall add one more observation.

The more one of the parties in an unequal combat is lacking in amount of forces, so much the greater under the pressure of danger must their inner tension and energy become. Where the reverse takes place, where instead of a heroic, a cowardly desperation appears, there, of course, all art of war ceases.

If with this energy of forces there is combined a wise moderation in the objects proposed, there arises that play of brilliant blows and prudent forbearance which we admire in the wars of Frederick the Great.

But the less this moderation and caution are able to do, the more predominant the tension and energy of the forces must become. When the disproportion of power is so great that no limitation of our own object can ensure us safety from a catastrophe, or where the probable duration of the danger is such that the greatest economy of forces can no longer bring us to our object, then the tension of forces will, or should, be concentrated in one desperate blow. He who is hard pressed, expecting little help from things which promise none, will place his whole and last trust in the moral superiority which despair always gives the brave. He will regard the greatest daring as the greatest wisdom—at most, perhaps, employing the assistance of subtle stratagem—and if he does not succeed, will find in an honorable downfall the right to rise again in the future.

CHAPTER 4

RELATION OF THE THREE ARMS

We shall only speak of the three principal arms: infantry, cavalry and artillery.

We must be excused for making the following analysis, which belongs more properly to tactics, but is necessary in order to clarify our thought.

The engagement employs two essentially different means: the use of firearms, and the hand-to-hand or personal combat. This again is either attack or defense. (Since we are speaking here of elements, attack and defense are to be understood in an absolute sense.) Artillery operates, obviously, only as a firearm, cavalry only through personal combat, infantry through both.

In personal combat the essence of defense consists in standing firm, as if rooted to the ground; the essence of attack is movement. Cavalry is entirely deficient in the first quality but enjoys the latter pre-eminently. It is therefore only suited to attack. Infantry has pre-eminently the property of standing firm, but is not altogether without mobility.

From this distribution of the elementary forces of war into different arms, we have the superiority and general utility of infantry as compared with the other two arms, since it is the only arm which unites in itself all three elementary forces. Hence it is evident how the combination of the three arms in war leads to a more perfect use of the forces, by affording the means of strengthening at will either the one or the other of the principles which are united in an unalterable proportion in the infantry.

In the wars of the present time, firearms are by far the most effective means. Nevertheless, the personal combat, man to man, is just as clearly to be regarded as the real basis of the engagement. For that reason an army of artillery alone would be an absurdity in war. An army of nothing but cavalry is conceivable, only it would possess very slight intensive strength. An army of infantry alone is not only conceivable but also much stronger. The three arms, therefore, stand in this order in respect to independence: infantry, cavalry, artillery.

But this order does not hold good if applied to the relative importance of each arm when they are all three acting in conjunction. As destructive capacity is much more effective than that of motion, the complete lack of cavalry would weaken an army less than the total lack of artillery. An army consisting of infantry and artillery alone would certainly find itself in an awkward position if opposed to one composed of all three arms; but if what it lacked in cavalry were compensated for by a relatively large number of infantry, it would still, by a somewhat differently organized mode of procedure, be able to manage very well tactically. It would find itself considerably embarrassed in the matter of outposts; it would never be able to pursue a defeated enemy with great vivacity, and in making a retreat would be exposed to greater hardships and exertions, but these inconveniences would still hardly be sufficient in themselves to drive it completely out of the field. On the other hand, such an army opposed to one composed of infantry and cavalry only would be able to play a very good part, while it is hardly conceivable that the latter could keep the field at all against an army made up of all three arms.

Let it be understood that these reflections on the relative importance of each single arm are deduced only from the generality of all wars, where one case is analogous to another; and therefore it cannot be our intention to apply the truth thus ascertained to each individual case of a particular engagement. A battalion on outpost service or in retreat might prefer to have with it a squadron of cavalry rather than a few cannon. A body of cavalry and mounted artillery, sent in rapid pursuit of, or to cut off, a fleeing enemy, cannot use infantry, and so forth.

If we summarize once more the result of these considerations, they amount to this:

1. Infantry is the most independent of the three arms.
2. Artillery is entirely dependent.
3. Infantry is the most important in a combination of the three arms.
4. Cavalry can most easily be dispensed with.
5. A combination of the three arms gives the greatest strength.

Now, if the combination of the three gives the greatest strength, it is natural to inquire concerning the absolutely best combination, and this question is almost impossible to answer.

If we could form a comparative estimate of the cost of organizing, provisioning and maintaining each of the three arms, and furthermore an estimate of the service rendered by each in war, we should obtain a definite result which would express abstractly the best combination. But this is little more than a play of the imagination. To begin with, the first of the two estimates thus compared is difficult to determine; one of the factors, the cost in money, it is true, is not difficult to find; but another, the value of men's lives, is something which no one will wish to estimate in figures.

Also the circumstance that each of the three arms chiefly depends on a different kind of national resources—infantry on the number of population, cavalry on the number of horses, artillery on available financial means—introduces a foreign determining element, the predominant influence of which may be plainly observed in the great outlines of the history of different peoples at various periods.

Since, however, for other reasons we cannot altogether dispense with some standard of comparison, we must take, in place of the whole of the first of the two things to be compared, only that one of its factors which can be ascertained, namely, the cost in money. Now on this point it is sufficient for our purpose to state that, in general, a squadron of 150 horses, a battalion of 800 men, a battery consisting of eight six-pounders, cost nearly the same, both in respect to the expense of equipment and of maintenance.

With regard to the other member of the comparison, that is, how much service one arm is capable of rendering as compared with the others, it is much more difficult to fix a definite quantity. An estimate might perhaps be possible if it depended merely on destructive capacity; but each arm has its own particular purpose, therefore has its own particular sphere of effectiveness. This again is not so definite that it might not be greater or smaller—which would merely cause modifications in the conduct of war, but no decisive disadvantages.

We often speak of what experience teaches on this subject, and we suppose that military history affords sufficient information for a settlement of the question, but all this must be regarded as nothing more than a way of talking, which, since it is not derived from anything of a fundamental and necessary nature, deserves no consideration in a critical examination.

Now although the best ratio between the arms is, certainly, conceivable as a fixed quantity, this quantity is impossible to find; it is a mere sport of fancy. Nevertheless it will still be possible to state the effects of having a great superiority or a great inferiority in one particular arm as compared with the same arm in the enemy's army.

Artillery increases the destructive capacity of fire; it is the most formidable of the arms, and its lack, therefore, diminishes very considerably the intensive force of an army. On the other hand, it is the least mobile arm, and consequently makes an army more unwieldy; further, it always requires a force for its protection, because it is incapable of personal engagement. If it is too numerous, so that the troops which can be appointed for its protection are not able to resist the attacks of the enemy at every point, it is often lost. Here a new disadvantage shows itself, namely, that it is the only one of the three arms, the principle parts of which—that is, guns and carriages—the enemy can very soon use against us.

Cavalry increases mobility in an army. If it is too few in number, the brisk flame of the element of war is thereby weakened, because everything must be done more slowly (on foot); everything must be organized with more care. The rich harvest of victory, instead of being cut with a scythe, can only be reaped with a sickle.

An excess of cavalry can certainly never be regarded as a direct weakening of the military force, as an intrinsic disproportion, but it may certainly be so indirectly, owing to the difficulty of maintenance, and if we reflect that instead of a surplus of 10,000 horsemen not required, we might have 50,000 infantry.

These peculiarities arising from the preponderance of one arm are the more important to the art of war in its limited sense, as that art teaches the use of whatever forces are available; and when forces are placed under the command of a general, the proportion of the three arms is usually already settled without his having had much voice in the matter.

If we wish to conceive of the character of a type of war as modified by the preponderance of one of the three arms, it is to be done in the following manner:—

An excess of artillery must lead to a more defensive and passive character in our undertakings; we shall seek security in strong positions, in very broken ground, even in mountain positions, in order that the natural obstacles of the ground may serve as defense and protection for our numerous artillery, and that the enemy's forces may come themselves to seek their destruction. The whole war will be carried on in a serious, formal minuet step.

On the other hand, a lack of artillery will cause us to permit the offensive, the active, the mobile principle to predominate; marching, fatigue, effort become our special weapons, and the war will thus become more diversified, more lively, rougher; for great events will be substituted a number of small ones.

With a very numerous cavalry we shall seek wide plains and have a preference for great movements. At a greater distance from the enemy we shall enjoy more rest and greater conveniences without conferring the same advantages on our adversary. We shall venture on bolder measures to outflank him, and on more daring movements generally, since we have command over space. In so far as diversions and invasions are true auxiliary means of war we shall be able easily to avail ourselves of them.

A decided lack of cavalry diminishes the mobility of an army without increasing its destructive power as an excess of artillery does. A prudent and methodical procedure becomes then the leading characteristic of war. Always to remain near the enemy in order to keep him constantly in view—no rapid, still less, hasty movements, everywhere a slow pushing on of well-concentrated masses, a preference for the defensive and for broken country, and, when the offensive must be resorted to, the shortest road to the center of force in the enemy's army—these are the natural tendencies in this case.

These different forms which warfare assumes, according to the predominance of one or another of the three arms, will seldom be so extensive and decisive as alone or primarily to determine the direction of an entire undertaking. Whether we should choose the strategic attack or the defensive, the choice of the theater of war, the determination to fight a great battle, or the adoption of some other means of destruction, are points which will probably be determined by other and more essential considerations; at least, if this is not the case, it is much to be feared that we have mistaken a minor point for the main one. But even if that is the case, if the great questions have been already decided, on other grounds, there always remains a certain margin for the influence of the preponderating arm, for in the offensive we can always be prudent and methodical, in the defensive bold and enterprising, and so forth, through all the different stages and gradations of military life.

On the other hand, the nature of a war may have a notable influence on the proportions of the three arms.

First, a national war, supported by militia and a general levy, must naturally bring into the field a very numerous infantry; for in such wars there is a greater lack of means of equipment than of men, and as the equipment in such a case is anyhow confined to what is absolutely necessary, we may easily imagine that for every battery of eight pieces, not one, but two or three battalions of infantry might be raised.

Second, if a weak state opposed to a powerful one cannot take refuge in a general levy, or in a militia system resembling it, then the increase of its artillery is, of course, the shortest way of bringing up its weak army nearer to an equality with that of the enemy, for it saves men, and intensifies the most essential principle of military force, that is, its destructive capacity. Anyway, such a state will be, for the most part, confined to a limited theater of war, and therefore this arm will be better suited to it. Frederick the Great adopted this means in the later period of the Seven Years' War.

Third, cavalry is the arm for movement and great decisions; its increase beyond the ordinary proportions is therefore important if the war extends over a great space, if expeditions are to be made in various directions, and great and decisive blows are intended. Bonaparte provides an example of this.

That the offensive and defensive cannot really in themselves exercise an influence on the proportion of cavalry will not become clear until we come to speak of these two forms of action in war. In the meantime, we shall only remark that both assailant and defender as a rule traverse the same spaces in war, and may have also, at least in many cases, the same decisive intentions. We remind our readers of the campaign of 1812.

It is commonly believed that, in the Middle Ages, cavalry was much more numerous than infantry and has gradually decreased down to the present day. Yet this is a mistake, at least, in part. The proportion of cavalry was, according to numbers, on the average perhaps, not much greater. Of this we may convince ourselves by tracing, through the history of the Middle Ages, the detailed statements of the armed forces then employed. We have only to think of the masses of infantry which composed the armies of the Crusaders, or the masses which followed the Emperors of Germany on their Roman expeditions. It was the importance of the cavalry which was so much greater in those days; it was the stronger arm, composed of the flower of the people, so much so that, although always very much weaker actually in numbers, it was nevertheless always looked upon as the important force; infantry was little valued, hardly spoken of. Hence has arisen the belief that its numbers were few.

No doubt it happened oftener than at present that in military attacks of lesser importance in the interior of France, Germany and Italy, a small army was composed entirely of cavalry. Since it was the chief arm, there is nothing inconsistent in that; but these cases cannot decide, if we consider the majority of cases, in which they are greatly outnumbered by larger armies. It was only when the obligations to military service imposed by the feudal laws had ceased, and wars were carried on by soldiers recruited, hired and paid—when, therefore, wars depended on money and enlistment, that is, at the time of the Thirty Years' War and the wars of Louis XIV—that this employment of a great mass of not particularly useful infantry was checked. Perhaps in those days there might have been a return to the exclusive use of cavalry, if infantry had not already risen in importance through the improvements in firearms, by which it maintained, to some extent, its numerical superiority in proportion to cavalry. At this period, if infantry was weak, the proportion was as one to one, if numerous, as three to one.

Since then, cavalry has steadily decreased in importance, according as improvements in the use of firearms have advanced. This is intelligible enough in itself, but the improvement we speak of does not relate solely to the weapon itself and the skill in handling it, but also to the greater ability in using troops armed with this weapon. At the battle of Mollwitz the Prussian army had brought the fire of their infantry to so high a state of perfection that even since then no improvement in that respect has been possible. On the other hand, the use of infantry on broken ground and of firearms in an engagement of skirmishers has only developed since that time, and is to be regarded as a great advance in destructive action.

It is our opinion, therefore, that the proportion of cavalry has changed little, in regard to numbers, but much in respect to its importance. This seems to be a contradiction, but in reality it is not so. The infantry of the Middle Ages, although numerous in an army, had not reached that proportion by reason of its intrinsic value as compared with cavalry, but because all that could not be placed in the more costly cavalry was handed over to the infantry; this infantry was, therefore, merely an improvisation; and if the number of cavalry had depended merely on the value set on that arm, it could never have been too great. Thus we can understand how cavalry, in spite of its constantly decreasing importance, still, perhaps, is important enough to keep its numerical relation at that point which it has hitherto so constantly maintained.

It is a remarkable fact that, at least since the War of the Austrian Succession, the proportion of cavalry to infantry has not changed, being constantly between a fourth, a fifth or a sixth of the infantry. This seems to indicate that these proportions exactly meet the natural requirements of an army, and give us the figures which it is impossible to find in a direct manner. We doubt, however, that this is the case, and find that the other reasons for a numerous cavalry in the principle cases are perfectly evident.

Austria and Russia are states which have kept up a numerous cavalry, because they have in their dominions the fragments of a Tartar organization. Bonaparte for his purposes could never be strong enough; after having made all possible use of conscription, he had no other way of strengthening his armies except by increasing the auxiliary arms, which depend more on money than on the use of men. Moreover, it stands to reason that in military enterprises of such enormous extent as his, cavalry was bound to have a greater value than in ordinary cases.

Frederick the Great, as is well known, calculated carefully every recruit that could be saved to his country; it was his main concern to keep up the strength of his army, as far as possible at the expense of other countries. His reasons for this are easy to understand, if we remember that his small dominions did not at that time include East Prussia and the Westphalian provinces. Cavalry was kept up to strength more easily by recruiting than infantry, irrespective of fewer men being required; in addition to which, his system of war was completely founded on the mobility of his army, and so it happened that while his infantry diminished in number, his cavalry was always increasing till the end of the Seven Years' War. Still even at the end of that war it was hardly more than one-fourth of the number of infantry that he had in the field.

At the period referred to there is also no lack of instances of armies entering the field unusually weak in cavalry, and yet carrying off the victory. The most remarkable is the battle of Grossgörschen. If we only count the divisions which took part in the battle, Bonaparte was 100,000 strong, of which 5,000 were cavalry, 90,000 infantry. The Allies had 70,000, of which 25,000 were cavalry and 40,000 infantry. Thus, in place of the 20,000 cavalry on the side of the Allies in excess of the total of the French cavalry, Bonaparte had only 50,000 additional infantry, when he ought to have had 100,000. Since he gained the battle with that superiority in infantry, we may ask whether it was at all possible that he would have lost it if the proportions had been 140,000 to 40,000.

Certainly the great advantage of our superiority in cavalry was shown immediately after the battle, for Bonaparte gained hardly any trophies by his victory. The winning of a battle is therefore not everything; but is it not always the important thing?

In view of these considerations, we can hardly believe that the numerical proportion between cavalry and infantry which has been established and maintained for eighty years is the natural one, founded solely on their absolute value. We are rather inclined to think that, after some fluctuations, the relative proportions of these arms will change further in the same direction as hitherto, and that the normal number of cavalry at last will be considerably less.

With respect to artillery, the number of guns has naturally increased since its first invention, and according as they have become lighter in weight and otherwise improved. Still, since the time of Frederick the Great, it has also kept very much to the same proportion of two or three guns per 1,000 men, that is, at the beginning of a campaign; for during its course artillery does not melt away as fast as infantry. Therefore at the end of a campaign the proportion is generally notably greater, and can be estimated at three, four or five guns per 1,000 men. Whether this is the natural proportion, or whether the increase of artillery can be carried still further, without disadvantage to the whole conduct of war, must be left for experience to decide.

The chief result of these considerations is as follows:

1. That infantry is the chief arm, to which the other two are subordinate.
2. That by the exercise of great skill and activity in the conduct of war the lack of the two subordinate arms may in some measure be compensated for, provided that we are so much stronger in infantry, and the better the infantry, the more easily this may be done.
3. That it is more difficult to dispense with artillery than with cavalry because it has the chief capacity for destruction, and its use in action is more closely associated with that of the infantry.
4. That artillery being the strongest arm, as regards destructive action, and cavalry the weakest in that respect, we must always ask: How much artillery can we have without inconvenience, and what is the least proportion of cavalry with which we can manage?

CHAPTER 5

BATTLE ORDER OF THE ARMY

The order of battle is that division and combination of the different arms into separate parts of the whole army, and that form of disposition of those parts which is to be the rule throughout the whole campaign or war.

It consists, therefore, in a certain sense, of an arithmetical and of a geometrical element, *the division* and *the disposition*. The first proceeds from the permanent peace organization of the army, adopts as units certain parts, such as battalions, squadrons, regiments, and batteries, and with them forms units of a higher order up to the highest of all, the whole army, according to the requirements of given circumstances. In like manner, disposition proceeds from the elementary tactics, in which the army is instructed and trained in time of peace, and which must be looked upon as a property in the troops that is not to be essentially modified once war has broken out. The disposition connects these tactics with the conditions which the use of the troops in war and in large masses demands, and thus it determines in a general way the form in which the troops are to be drawn up for engagement.

This has invariably been the case when great armies have taken the field, and there have even been times when this form was considered as the most essential part of the engagement.

In the seventeenth and eighteenth centuries, when the improvements in firearms occasioned a great increase of infantry and permitted it to be deployed in long, thin lines, the order of battle was thereby simplified, but at the same time it became more difficult and required more skill in execution. Since no other way of disposing of cavalry was known, except that of posting them on the wings, where they were out of the fire and had room to move about, therefore in the order of battle the army always became a closed, inseparable whole. If such an army was cut in the middle, it was like an earthworm cut in two; the wings still had life and mobility, but they had lost their natural functions. The army lay, therefore, under a kind of spell of unity, and whenever any parts of it had to be placed in a separate position, each time a small reorganization and disorganization was necessary. The marches which the whole army had to make were a state in which it found itself, to a certain extent, outside the rules of the game. If the enemy was at hand, the march had to be arranged in the most elaborate manner, and in order that one line or one wing might be always at a reasonable distance from the other, the troops had to scramble along over every obstacle. Marches had to be constantly stolen on the enemy, and this constant theft only escaped severe punishment through the circumstance that the enemy lay under the same spell.

Hence, when, in the latter half of the eighteenth century, it was discovered that cavalry would serve just as well to protect a wing if it stood in rear of the army as if it were placed on the prolongation of the line, and that, besides this, it might be applied to other purposes than merely fighting a duel with the enemy's cavalry, a great advance was made. The army in its principal extension or front, which is always the breadth of its battle order, now consisted entirely of homogeneous members, so that it could be divided into any number of parts at will, each part like another and like the whole. In this way it ceased to be one single piece and became a many-membered whole, consequently flexible and articulated. The parts could be separated from the whole and then joined on again without difficulty; the order of battle always remained the same. Thus arose the corps of all arms; that is, such an organization became possible. The need for it had probably been felt much earlier.

That this all arises out of the battle is very natural. The battle was formerly the whole war, and will always continue to be the principal part of it; but the order of battle belongs far more to tactics than to strategy, and in introducing this deduction here our only object has been to show how tactics, in organizing the whole into smaller wholes, made preparation for strategy.

The greater armies become, the more they are distributed over wide spaces, and the more diversified the action and reaction of the different parts among themselves, the wider becomes the field of strategy. Therefore, the order of battle, in the sense of our definition, has also had to come into a kind of reciprocal action with strategy, which manifests itself chiefly at the points where tactics and strategy meet, that is, at those moments where the general distribution of the military forces passes into the special dispositions for the engagement.

We now turn to the three points of the *division, combination of arms* and *disposition* from the strategic point of view.

1. DIVISION

In strategy we should never ask what must be the strength of a division or corps, but how many corps or divisions an army must have. There is nothing more unmanageable than an army divided into three parts, except it be one divided into only two, in which case the supreme command must be almost neutralized.

To fix the strength of great and small corps, either on the grounds of elementary tactics or on higher grounds, leaves an incredibly wide field for arbitrary judgment, and heaven knows what modes of reasoning have disported themselves in it. On the other hand, the necessity of dividing an independent whole into a certain number of parts is a thing as obvious as it is definite, and this idea furnishes real strategic reasons for determining the number of the greater members—such as corps and divisions—of an army, and consequently their strength, while the strength of the smaller subdivisions, such as companies, battalions, etc., is left to be determined by tactics.

We can hardly conceive the smallest isolated body in which there are not at least three parts to be distinguished, so that one part may be sent ahead and another part left in the rear. That four are still more convenient is obvious, if we reflect that the middle part, being the main force, ought to be stronger than either of the others. In this way, we can go on to eight parts, which appears to us to be the most suitable number for an army, if we take as a constant necessity one part for an advance guard, three for the main body, that is, a right wing, center and left wing, two parts for reserve, and one to dispatch to the right, one to the left. Without pedantically ascribing a great importance to these numbers and figures, we believe that they represent the most usual and frequently recurring strategic disposition, and therefore one that is convenient.

Certainly it seems to make the supreme command of an army (and the command of every whole) immensely easier if there are only three or four subordinates to command, but the commander-in-chief must pay dearly for this convenience and in two ways. In the first place, an order loses more in rapidity, force and precision, the longer the stepladder down which it must descend, and this must be the case if there are corps commanders between the division leaders and the chief. Second, the chief generally loses his own real power and efficiency the wider the spheres of action of his immediate subordinates become. A general commanding 100,000 men in eight divisions exercises a power intensively greater than if the 100,000 men were divided into only three corps. There are many reasons for this, but the most important is that each commander looks upon himself as having a kind of proprietary right in all parts of his corps, and almost always opposes the withdrawal from him of any portion of it for a longer or shorter time. A little experience of war will make this evident to any one.

But on the other hand, the number of parts must not be too great, or disorder will ensue. It is difficult enough to manage eight divisions from one headquarters, and the number should probably not be allowed to exceed ten. While in a division, in which the means of putting orders into effect are much less, the smaller and usual number of four or, at the most, five brigades must be regarded as the more suitable.

If ten divisions to an army and five brigades to a division are not sufficient, that is, if the brigades would become too strong, then corps commands would have to be introduced; but we must remember that thereby a new factor is created, which at once very much reduces the number of all other subdivisions.

But what is too strong a brigade? The custom is to make brigades from 2,000 to 5,000 men strong, and there appear to be two reasons for making the latter number the limit; the first is that a brigade is supposed to be a subdivision which can be commanded by one man directly and must therefore be all within range of his voice; the second is that any larger body of infantry should not be left without artillery, and through this first combination of arms a special subdivision naturally arises.

We do not wish to involve ourselves in these tactical subtleties; neither shall we enter upon the disputed point: when and in what proportions the combination of all three arms should take place, whether with divisions of 8,000 to 12,000 men, or with corps from 20,000 to 30,000 men strong. The most decided opponent of these combinations will scarcely take exception to the assertion that nothing but this combination of the three arms can make a member of the army independent, and that, therefore, for such members as are intended to be frequently isolated in war, it is at least very desirable.

An army of 200,000 men in ten divisions, the divisions composed of five brigades each, would give brigades 4,000 strong. We see here no disproportion. Certainly this army might also be divided into five corps, the corps into four divisions, and the division into four brigades, which makes the brigade 2,500 men strong; but the first distribution, looked at in the abstract, appears to us preferable, for besides that, in the other, there is one more degree of rank, five parts are too few to make an army manageable; four divisions, in like manner, are too few for a corps, and 2,500 men is a weak brigade, while, in this manner, there are eighty brigades, whereas the first distribution has only fifty, and is therefore simpler. All these advantages are given up merely for the sake of having to give orders only to half as many generals. For smaller armies of course, the distribution into corps is still more unsuitable.

This is the abstract view of the matter. The particular case may present good reasons for deciding otherwise. We must admit that, whereas eight or ten divisions may still be controlled when united in a level country, in widely extended mountain positions this might perhaps be impossible. A great river which divides an army into halves, makes a commander for each half indispensable. In short, there are hundreds of local and individual circumstances of a decisive character, before which abstract rules must give way.

But still experience teaches us that these abstract reasons come most frequently into use and are more seldom overruled by others than we should perhaps suppose.

We permit ourselves to make clear the scope of the foregoing considerations by a simple outline, for which purpose we now place the different points of importance beside one another.

Inasmuch as we understand by the term *"members of a whole"* only those which result from the first division, we say:

1. If a whole has too few members, it is unwieldy.
2. If the parts of a whole body are too large, the power of the commander-in-chief is thereby weakened.
3. With every additional step through which an order has to pass, it is weakened in two ways; first, by the loss of force, which it suffers in its passage through an additional step; second, by the longer time it needs for transmission.

The tendency of all this is to show that the number of co-ordinate members should be as great, and the series of gradations as small, as possible, and the only limitation to this conclusion is that in an army no more than from eight to ten members, and in smaller units no more than from four to six subdivisions, can be conveniently controlled.

2. COMBINATION OF ARMS

For strategy the combination of the three arms in the order of battle is only important in regard to those parts of the army which, according to the usual order of things, are likely to be frequently employed in a detached position, where they may be forced to fight an independent engagement. Now it is natural that the members of the first order, i.e., corps or divisions, and for the most part only these, are intended for detached positions, because, as we shall see elsewhere, a detached position generally involves the idea and the necessity of an independent whole.

In a strict sense strategy would therefore only require a permanent combination of arms in corps, or where these do not exist, in divisions, leaving it to circumstances to determine when a provisional combination of the three arms shall be made in subdivisions of an inferior order.

But it is easy to see that, when corps are of considerable size, such as 30,000 or 40,000 men, they will seldom be in a situation to take up a position in one mass. With corps of such strength a combination of the arms in the divisions is therefore necessary. No one who has had any experience in war will treat lightly the delay which occurs when troops have to be hurriedly detailed and a detachment of cavalry has to be brought to the support of the infantry from some other perhaps distant point—to say nothing of the confusion which takes place.

The details of the combination of the three arms, how far it should go, how thorough it should be, what proportions should be observed, the strength of the reserves of each to be set apart—these are all purely tactical considerations.

3. DISPOSITION

The determination as to the distribution in space according to which the parts of an army should be drawn up in battle order is likewise completely a tactical matter, referring solely to the battle. No doubt there is also a strategic disposition of the parts; but it depends almost entirely upon the decisions and needs of the moment, and the theoretical element contained in it does not come within the meaning of the term "order of battle." We shall therefore treat of it in the following chapter under the heading: *General Disposition of an Army.*

The battle order of an army is therefore the division and disposition of it in mass, ready for battle. Its parts are so combined that both the tactical and strategical requirements of the moment can be easily satisfied by the employment of separate parts drawn from this mass. When such momentary need ceases, these parts resume their original places, and thus the battle order becomes the first step to, and principal foundation of, that wholesome "methodism" which, like the beat of a pendulum, regulates the work in war, and of which we have already spoken in Book II, Chapter 4.

CHAPTER 6

GENERAL DISPOSITION OF AN ARMY

From the moment of the first assembling of military forces to that of mature decision, where strategy has brought the army to the decisive point, and each particular part has had its position and role pointed out by tactics, there is in most cases a long interval. It is the same between one decisive catastrophe and another.

Formerly these intervals, to a certain extent, did not belong to war at all. Take for example the manner in which Luxemburg encamped and marched. We single out this general because he is celebrated for his camps and marches, and therefore may be considered a representative general of his period, and from the *Histoire de la Flandre militaire* we know more about him than about other generals of the time.

The camp was regularly pitched with its rear close to a river, or morass, or a deep valley, which in the present day would be considered madness. The direction in which the enemy lay had so little to do with determining the front of the camp, that cases are very common in which the rear was toward the enemy and the front toward their own country. This now unheard-of method of procedure is only to be understood when we regard the convenience of the troops as the chief, indeed almost as the only, consideration in the choice of camps, and therefore look upon the state of being in camp as a state outside of the action of war—behind the scenes, so to speak, where one is not molested. The practice of always resting the rear upon some obstacle may be considered the only measure of security which was taken—security, of course, as understood in the mode of conducting war in that day, for such a measure was quite inconsistent with the possibility of being compelled to fight in such a camp. There was, however, little reason to fear this, because engagements generally depended upon a kind of mutual understanding, like a duel, in which the parties repair to a convenient rendezvous. Since armies could not fight in every region—partly on account of their numerous cavalry, which in the decline of its splendor was still regarded, particularly by the French, as the principal arm, partly on account of the unwieldy organization of their battle order—an army in a broken country was as it were under the protection of a neutral territory. Since the army could make but little use of broken ground, it was therefore deemed preferable to go to meet an enemy who was seeking battle. We know, indeed, that Luxemburg's battles at Fleurus, Steenkerke and Neerwinden were conceived in a different spirit; but this spirit had only just then freed itself under this great general from the old method, and it had not yet reacted upon the method of encampment. Alterations in the art of war originate always in decisive actions, and then lead by degrees to modifications in other things. The expression *il va à la guerre*—"He is going to the war"—which was used in reference to an irregular setting out to watch the enemy, shows how little the state of an army in camp was considered to be a state of real warfare.

It was not much different with the marches, when the artillery separated itself completely from the rest of the army, in order to take advantage of better and more secure roads, and the cavalry detachments generally took the right wing alternately, so that each might have in turn its share of the glory of marching on the right.

At present (that is, chiefly since the Silesian Wars) the situation outside of the engagement is so thoroughly influenced by its connection with the engagement that the two states are in intimate correlation, and the one can no longer be conceived completely without the other. Formerly in a campaign the engagement was the real weapon, the state outside of the engagement only the handle—the former the steel blade, the latter the wooden haft glued to it, the whole therefore composed of heterogeneous parts. Now the engagement is the edge, the situation outside of the engagement the back of the blade, the whole to be regarded as metal completely welded together, in which it is impossible any longer to distinguish where the steel ends and the iron begins.

This state in war outside of the engagement is now regulated partly by the institutions and regulations with which the army comes prepared from a state of peace, partly by the tactical and strategic arrangements of the moment. The three situations in which an army may be are: in quarters, on the march or in camp. All three belong as much to tactics as to strategy, and both tactics and strategy which here, in many ways, border on each other, often seem to, or actually do, interlock, so that many dispositions may be looked upon at the same time as both tactical and strategic.

We shall treat of these three situations of an army outside of the engagement in a general way, before any special ends they serve are attached to them; for this purpose, however, we must first of all consider the general disposition of the forces, because it exercises a superior and more comprehensive influence on camps, quarters and marches.

If we consider the disposition of forces in a general way, that is without special objects, we can only conceive it as a unity, that is, only as a whole intended to strike a joint blow, for any deviation from this simplest form would already imply a special object. Thus arises the concept of an army, be it large or small.

Further, where all special objects are lacking, there only remains as the sole object the preservation of the army itself, and consequently also its security. That the army should exist without particular disadvantage, and that it should be capable of a united blow without particular disadvantage, are, therefore, the two conditions. From these result the following considerations more immediately applying to what concerns the existence and security of the army.

1. Facility of subsistence.
2. Facility of providing shelter for the troops.
3. Security of the rear.
4. An open country in front.
5. The position itself in a broken country.
6. Strategic points of support.
7. A suitable distribution of the troops.

Our comments on these individual points are as follows:—

The first two lead us to seek out cultivated districts and great towns and roads. They determine measures in general rather than in particular.

In the chapter on lines of communication (Book V, Chapter 16) will be found what we mean by security of the rear. The first and most important point in this respect is that the position should be at a right angle with the principal line of retreat adjoining the position.

Respecting the fourth point, an army, of course, cannot have a view over a stretch of country as it has over its front when in a tactical position for battle. But the advance guard, scouts and patrols, etc., sent forward, serve as strategic eyes, and the observation will naturally be easier for these in an open than in a broken country. The fifth point is merely the reverse of the fourth.

Strategical points of support differ from tactical in two respects: the army need not be in immediate contact with them and they must be of far greater extent. The cause of this is that, naturally, the circumstances of time and space in which strategy moves are generally on a larger scale than those of tactics. If, therefore, an army takes up a position a few miles from the sea coast or the banks of a great river, it rests strategically on these objects, for the enemy cannot make use of such a space as this to effect a strategic turning movement. He will not venture into this district on marches miles in length, occupying days and weeks. On the other hand, in strategy, a lake some miles in circumference is hardly to be looked upon as an obstacle; in its method of procedure a few miles to the right or left are not of much consequence. Fortresses will become strategic points of support, in proportion to their size and the extent of their sphere of action for offensive undertakings.

The disposition of the army in separate parts is determined either by special objects and requirements, or by those of a general nature. Here we can only speak of the latter.

The first general requirement is to push forward the advance guard with the other troops required to observe the enemy.

The second is that, with very large armies, the reserves should usually be placed several miles in the rear, thus leading to a separated disposition.

Lastly, the covering of both wings of an army usually requires a separate disposition of particular corps.

By this covering it is not meant that a portion of the army is to be detached to defend the space round its wings, in order to prevent the enemy from approaching these so-called weak points. Who would then defend the wings of these flanking corps? This kind of conception, which is so common, is complete nonsense. The wings of an army are not in themselves weak points, for the enemy also has wings, and cannot menace ours without placing his own in jeopardy. It is only when circumstances are unequal, when the enemy's force is larger than ours, when his lines of communication are more secure (see "Lines of Communication"), it is only then that the wings become weaker parts. But of these special cases we are not now speaking, nor consequently of the case in which a flanking corps is appointed in connection with other designs to defend the space on our wing, for that no longer belongs to the category of general dispositions.

But although the wings are not particularly weak parts, still they are particularly important, because here, on account of flanking movements, the defense is not so simple as in the front; therefore, measures become more complicated and require more time and preparation. For this reason it is necessary in the majority of cases to give the wings special protection against unforeseen enterprises on the part of the enemy. This is done by placing stronger masses on the wings than would be required for mere purposes of observation. The greater these masses are, even if they offer no very formidable resistance, the more time is required to dislodge them and the more the enemy's forces and intentions are unfolded. By that means the object of the measure is attained. What is to be done further depends on the particular plans of the moment. We may therefore regard corps placed on the wings as lateral advance guards, intended to retard the advance of the enemy through the space beyond our wings and give us time to make dispositions to counteract his movement.

If these corps are to fall back on the main army and the latter is not to make a backward movement at the same time, it obviously follows that they must not be placed in the same line with the front of the main army, but thrown out somewhat forward, because a retreat, in order not to become involved in a serious engagement at its starting point, must not be made directly to the side of the position.

There arises, therefore, out of these inherent reasons for a divided disposition, a natural system consisting of four or five segregated parts, according as the reserve remains with the main body or not.

As the subsistence and shelter of the troops as a rule help to decide the general disposition, they also contribute to a disposition in separate sections. The attention that both demand adds another to the reasons above discussed, and we seek to satisfy the one without prejudice to the other. In most cases, by the division of an army into five separate corps, the difficulties of accommodation and subsistence will be overcome, and no great alteration will be required on their account.

Now we must turn our attention to the distances at which these separated corps may be placed, if the intention of mutual support, and consequently of joint attack, is to be carried out. At this point we remind our readers of what was said in the chapters on the duration and decision of the engagement, according to which no absolute determination, but only the most general, as it were, an average one, can be made, because absolute and relative strength, weapons and region have a very great influence.

The distance of the advance guard is the easiest to fix. Since in retreating it falls back on the main body of the army, it might be approximately at a distance of a long day's march without incurring the risk of being obliged to fight an independent battle. But it will not be sent farther in advance than the security of the army requires, because the farther it has to fall back, the more it suffers.

Respecting the detachments on the flanks, as we have said above, the engagement of an ordinary division of 8,000 to 10,000 men usually lasts for several hours, even for a half day before it is decided; on that account, there need be no hesitation in placing such a division at a distance of some hours, that is, five to ten miles, and for the same reason, corps of three or four divisions may be placed a day's march away, that is, at a distance of fifteen or twenty miles.

From this natural general disposition of the main army into four or five divisions at particular distances there will arise a certain system, which divides the army mechanically so long as no special objects intervene.

But although we assume that each of these distinct parts of an army is suited to fight an independent engagement, and that this may be necessary, it does not by any means follow that the real object of a disposition in separate sections is that the parts should fight separately; the necessity for the disposition in separate sections is mostly only a condition of existence imposed by time. If the enemy approaches in order to come to a decision by means of a general engagement, the strategic period is over, and everything concentrates itself into the one issue, that of the battle, and the objects of disposition in separate sections come to an end and disappear. As soon as the battle begins, consideration of quarters and subsistence is suspended; the observation of the enemy before our front and on our flanks, and the checking of his impetus by means of moderate resistance has fulfilled its purpose, and now all resolves itself into the great unity of the main battle. The best criterion of the value of the distribution into separate sections is whether this is actually the case, in other words whether it has been merely regarded as a condition of the disposition, a necessary evil, and having united action in battle as its object.

CHAPTER 7

ADVANCE GUARD AND OUTPOSTS

These two things are among those in which the tactical and strategic elements overlap. On the one hand we must reckon them amongst those provisions which give the engagement its form and secure the execution of our tactical plans; on the other hand, they frequently cause independent engagements, and on account of their position, more or less distant from the main army, they are to be regarded as links in the strategic chain. It is this very position which obliges us to supplement the preceding chapter by devoting a few moments to their consideration.

Every body of troops, when not completely ready for battle, requires an advance guard to learn the approach of the enemy and to reconnoiter before he comes into sight, for its range of vision, as a rule, does not extend much farther than the range of its weapons. But what sort of man would that be who could not see farther than his arms can reach! The outposts are the eyes of the army, as has been said before. The need of them, however, is not always equally great; it has its degrees. The strength of armies and the extent of ground they cover, time, place, circumstances, the type of war, even chance, are all points which have an influence upon the matter. Therefore we cannot be surprised that military history, instead of furnishing any definite and simple outlines of the method of using advance guards and outposts, only presents the subject in a kind of chaos of examples of the most diversified nature.

Sometimes we see the security of an army entrusted to a definite corps of advance guard; at another time to a long line of separate outposts; sometimes both these arrangements co-exist, sometimes there is no question of one or the other; at one time there is one advance guard in common for all the advancing columns; at another time, each column has its own advance guard. We shall endeavor to get a clear idea of the subject, and then see whether we can reduce it to a few principles for application.

If the troops are on the march, a more or less strong detachment forms its vanguard, or advance guard, and in the case of the movement of the army being reversed, this same detachment will form the rear guard. If the troops are in quarters or camp, an extended line of weak posts forms its vanguard, *the outposts.* Naturally, when the army is halted, a greater extent of space can and must be watched than when the army is in motion, and therefore in the one case the conception of a chain of posts, in the other, that of a concentrated corps, arises spontaneously.

The actual strength of an advance guard, as well as of outposts, ranges from a considerable corps, composed of all three arms, to a regiment of hussars, and from a strongly entrenched defensive line, occupied by portions of troops from each arm, to mere outlying pickets and infantry scouts detached from the camp. The services assigned to such vanguards range, therefore, from mere observations to resistance, and this resistance is not only capable of giving the main army the time which it requires to prepare for battle, but also of forcing the enemy to an earlier unfolding of his plans and intentions, thereby considerably raising the value of the observation.

According as the main army requires more or less time and its resistance is more or less intended and arranged to meet the special measures of the enemy, so much the more or less does it need a strong advance guard and outposts.

Frederick the Great, who, of all generals, may be considered the most ready for battle, and who almost led his army into battle by a mere word of command, did not require strong outposts. We see him, therefore, always encamping close under the eyes of the enemy, without any great apparatus of outposts, relying for his security at one place on a hussar regiment, at another on a light battalion, or perhaps on pickets and scouts detached from the camp. On the march a few thousand horse, generally belonging to the cavalry on the flanks of the first line, formed his advance guard, and at the end of the march rejoined the main army. He seldom had any corps permanently employed as advance guard.

When it is the intention of a small army, by using the whole weight of its mass with great vigor and activity, to make the enemy feel the effect of its superior discipline and the greater resolution of its commander, then almost everything must be done *sous la barbe de l'ennemi,* in the same way as

Frederick the Great did when opposed to Daun. A cautious disposition of troops, an elaborate system of outposts, would wholly neutralize its superiority. The fact that errors and an exaggeration of this system may occasionally lead to a battle of Hochkirch is no argument against such a procedure; we should rather say that as there was only one battle of Hochkirch in all the Silesian Wars, we ought to recognize in this system a proof of the king's mastery.

We see that Bonaparte, however, whose army certainly did not lack steadiness nor himself determination, almost always moved with a strong advance guard. There were two reasons for this.

The first is to be found in the alteration in tactics. A whole army is no longer led into battle by mere word of command, to settle the affair like a great duel by more or less skill and bravery. Military forces are now arranged more to suit the peculiarities of ground and the circumstances. The battle order, and consequently the battle itself, is a whole made up of many parts, from which it follows that the simple decision to fight becomes a complicated plan, and the word of command a more or less lengthy preparation. For this, time and data are required.

The second cause lies in the great size of modern armies. Frederick led thirty to forty thousand men into battle; Napoleon from one to two hundred thousand.

We have selected these examples because we can assume that generals like these would not have adopted an elaborate method of procedure without some good reason. On the whole, the use of advance guards and outposts in modern wars has become more prominent; not that in the Silesian Wars every one acted as Frederick did, for the Austrians had a much stronger system of outposts and much more frequently sent forward a corps of advance guard, for which they had sufficient reason because of their situation and circumstances. Likewise there are in most modern wars plenty of variations. Even the French marshals, Macdonald in Silesia, Oudinot and Ney in the Mark (Brandenburg), advanced with armies of sixty to seventy thousand men, without our learning of a corps of advance guard.

We have hitherto been discussing advance guards and outposts with regard to their numerical strength; but there is another difference which we must make clear. It is that, when an army advances or retires on a certain breadth of ground, it may have a van and rear guard in common for all the columns which are marching side by side, or each column may have one for itself. In order to form a clear idea on this subject, we must look at it in the following way:

Fundamentally the advance guard, when there is a corps which specially bears this name, is only intended for the security of the main army proceeding in the middle. If this main army is marching upon several roads which lie close together and can easily serve also for the advance guard, and therefore be covered by it, then the flank columns naturally require no special covering.

Those corps, however, which go ahead at greater distances, being really detached corps, must provide their own vanguards. The same applies also to any of those corps which belong to the central mass, and owing to the direction that the roads may happen to take, are too far from the center. Therefore there will be as many advance guards as there are separate, parallel columns in which the army advances. If each of these advance guards is much weaker than a single general one would be, then they fall more into the class of other tactical dispositions, and the advance guard will entirely disappear from the strategic picture. But if the main army in the center has a much larger corps for its advance guard, then that corps will appear as the advance guard of the whole, and will be so in many respects.

But what can be the reasons for giving to the center so much stronger a vanguard than to the wings? The following three:

1. Because the mass of troops composing the center is usually more considerable.
2. Because it is evident that the central point of the tract of country in which the front of an army is extended remains always the most important point, since all plans relate mostly to it, and therefore the field of battle is also usually nearer to it than to the wings.
3. Because, although a corps thrown forward in front of the center does not directly protect the wings as a real vanguard, it still contributes greatly to their security indirectly. The enemy cannot in ordinary cases pass by such a corps within a certain distance in order to effect any enterprise of importance against one of the wings, because he would have to fear an attack in flank and rear. Even if this check which a corps thrown forward in the center imposes on the enemy is not sufficient to constitute complete security for the wings, it is at all events sufficient to relieve the flanks from fear in a great many cases.

The vanguard of the center, if much stronger than that of the wings, that is to say, if it consists of a special corps of advance guard, has no longer the mere mission of a vanguard to protect the troops in its rear from sudden surprise; it also operates in a more general strategic sense as a corps thrown forward in advance.

The following are the purposes for which such a corps may be used, and therefore those which determine its employment:

1. To ensure a stouter resistance, and make the enemy advance with more caution; consequently to increase the effects of an ordinary vanguard, whenever our arrangements are such as to require much time.
2. If the central mass of the army is very large, to be able to keep this unwieldy body at some distance from the enemy, while we still remain close to him with a mobile body of troops.
3. That we may have a corps of observation close to the enemy, if there are any other reasons which require us to keep the principal mass of the army at a considerable distance.

The idea that weak look-out posts, mere irregulars, might answer just as well for this observation is set aside at once if we reflect how easily such might be dispersed, and how very limited also are their means of observation as compared with those of a large corps.

4. In pursuit of the enemy. A single corps of advance guard, to which the greater part of the cavalry is to be attached, can move more quickly than the whole army, arrive later at night, and be ready earlier in the morning.
5. Lastly, on a retreat, as a rear guard, to be used in defending the principal natural obstacles of ground. In this respect also the center is specially important. At first sight it certainly appears as if such a rear guard would be constantly in danger of having its flanks turned. But we must remember that, even if the enemy should already have pressed forward somewhat farther upon the flanks, he has yet to march the whole way from there to the center, if he wants seriously to threaten the latter, and therefore the rear guard of the center can resist somewhat longer and move later than the rest. On the other hand, the situation becomes at once critical if the center falls back quicker than the wings; it at once looks as if the line had been broken through, and even the very look of that is to be dreaded. At no time is there a greater necessity for concentration and holding together, and at no time is this more acutely felt by every one than on a retreat. The intention always is that the wings, in the last instance, should unite again with the center; and if, on account of subsistence and roads, the retreat has to be made on a considerable breadth of

front, still the movement generally ends in a concentration on the center. If we add to these considerations the fact that the enemy usually advances with his principal force and with the greatest pressure in the center, we must realize that the rear guard of the center is of special importance.

Accordingly, the throwing forward of a special corps of the advance guard will be appropriate in all those cases where one of the above situations occurs. Hardly any of them arise if the center is not stronger than the wings, as, for example, when Macdonald advanced against Blücher, in Silesia in 1813, and when the latter made his movement toward the Elbe. Both of them had three corps, which usually moved in three columns by different roads, the heads of the columns in line. On this account no mention is made of their having had advance guards.

But this disposition in three columns of equal strength is partly for that very reason anything but recommendable, just as the division of a whole army into three parts is very unmanageable, as stated in Chapter 5 of this book.

When the whole is formed into a center, with two wings separate from it, which we have represented in the preceding chapter as the most natural formation as long as there are no special objects in view, the corps forming the advance guard, according to the simplest idea, will have its place in front of the center, and also therefore before the line of the wings. However, since the corps thrown out on the flanks fundamentally perform a similar office for the flanks as the advance guard for the front, it will very often happen that these corps will be in line with the advance guard, or even thrown forward still farther, according to circumstances.

With respect to the strength of an advance guard we have little to say, as now, very properly, it is the general custom to detail for that duty one or more of the members of the first order into which the army is divided, and to reinforce them with a part of the cavalry; i.e., a corps, if the army is divided into corps, one or more divisions, if the army is divided into divisions.

It is easy to perceive that in this respect also the greater number of divisions is an advantage.

How far the advance guard should be pushed to the front must depend entirely on circumstances; there are cases in which it may be more than a day's march in advance, and others in which it stands immediately before the front of the army. If we find it in most cases at a distance of from five to fifteen miles, this proves, certainly, that this distance is most frequently necessary; but we cannot make of it a rule by which we are always to be guided.

In the foregoing consideration we have lost sight of *outposts* altogether, and therefore we must now return to them again.

In saying at the beginning that outposts belong to stationary troops and advance guards to troops in motion, our object was to refer the conceptions back to their origin and keep them distinct for the present; but it is evident that if we wanted to adhere strictly to the words, we should get little more than a pedantic distinction.

If an army on the march halts at night to resume the march next morning, the advance guard must naturally do the same, and so must always organize the outpost duty, required both for its own security and that of the main body, without on that account being changed from an advance guard into mere outposts. If the latter are to be regarded as something opposed to the idea of an advance guard, it can only happen in a case where the main mass of troops intended as advance guard breaks up into individual posts, and of the united corps little or nothing remains—where, therefore, the idea of a long line of posts predominates over that of a united corps.

The shorter the time of rest, the less complete is the covering of the army required to be, for the enemy has no opportunity at all to learn from one day to another what is covered and what is not. The longer the halt, the more complete must be the observation and covering of all points of approach. As a rule, therefore, when the halt is long, the vanguard becomes more and more extended into a line of posts. Whether the change becomes complete, or whether the idea of a concentrated corps remains predominant, depends chiefly on two circumstances. The first is the proximity of the contending armies; the second is the nature of the country.

If the armies are very close in comparison to the width of their front, then it will often be impossible to post a corps of advance guard between them, and the armies will be able to maintain their security only through a series of small posts.

A concentrated corps, as it covers the approaches of the army less directly, generally requires more time and space to become effective. Therefore, if the army covers a great extent of front, as in quarters, and a concentrated corps is to cover all approaches, it is necessary that it should be at a considerable distance from the enemy. On this account winter quarters, for instance, have generally been covered by a cordon of outposts.

The second circumstance is the nature of the country. Where, for example, a great obstacle of the ground affords the means of forming a strong line of posts with but few troops, it will not be left unused.

Lastly, in winter quarters, the severity of the season may also be a reason for breaking up the corps of advance guard into a line of posts, because it is easier to find shelter for it in that way.

The use of a reinforced line of outposts was brought to the greatest perfection by the Anglo-Dutch army, during the winter campaign of 1794 and 1795 in the Netherlands, when the line of defense was formed by brigades composed of all arms, in single posts, and supported by a reserve. Scharnhorst, who was with that army, introduced this system into the Prussian army in East Prussia on the Passarge in 1807. Elsewhere in modern times it has seldom occurred, chiefly because the wars have been too full of movement. But even when there has been occasion for its use, it has been neglected, as, for example, by Murat at Tarutino. A wider extension of his defensive line would have spared him the loss of some thirty pieces of artillery in an engagement of outposts.

It cannot be disputed that, where circumstances permit, great advantages may be derived from this system. We propose to return to the subject on other occasions.

CHAPTER 8

MODE OF ACTION

OF ADVANCED CORPS

We have just seen how the security of the army is supposed to be assured by the effect which an advance guard and flank corps produce on an advancing army. Such corps are always to be considered as very weak whenever we think of them as in conflict with the main army of the enemy, and therefore a peculiar mode of using them is required, that they may fulfill the purpose for which they are intended, without having to fear serious losses from this disproportion in strength.

The object of these corps is to observe the enemy and delay his progress.

Even for the first of these purposes a smaller body would never accomplish as much as a larger, partly because it is easier to drive back, partly because its means of observation—that is, its eyes—do not reach as far.

But observation is also to have a higher task; the enemy is to be made to unfold his whole strength before such a corps, and thereby to reveal, to a certain extent, not only his strength but also his plans.

For this the mere presence of such a corps would be sufficient, and it would have merely to wait and see the measures by which the enemy seeks to drive it back, and then begin its retreat.

But further, it must also delay the advance of the enemy, and that implies actual resistance.

Now how can we conceive this waiting until the last moment, as well as this resistance, without such a corps being in constant danger of serious loss? Chiefly in this way, that the enemy, too, is preceded by an advance guard, and therefore does not advance at once with all the outflanking and overpowering weight of his whole force. Now even if this advance guard is from the beginning superior to ours, as it is naturally arranged to be, even if the enemy's main army is nearer to his advance guard than we are to ours, and being already on the march, will soon be on the spot to support the attack of his advance guard with all his strength—still this first stage, in which our advanced corps has to contend with the enemy's, that is, with approximately its equal, gives it a certain gain in time and an opportunity to watch the approach of the enemy for some time without endangering its own retreat.

But even a certain amount of resistance, which such a corps offers in a suitable position, does not bring such disadvantage as we might expect in other cases, considering the disproportion in the strength of the forces engaged. The chief danger in resisting a superior enemy consists always in the possibility of being turned and placed at a great disadvantage by an enveloping attack. In the case to which our attention is directed, however, such a risk is greatly lessened, owing to the advancing enemy never knowing exactly how near at hand support from his opponent's main army may be, which might place his advanced columns, in their turn, between two fires. The consequence is that the enemy in advancing keeps the heads of his single columns as nearly as possible in line, and, only after he has carefully reconnoitered the situation of his opponent, begins cautiously and warily to turn one or the other wing. While he is thus groping about and being so cautious, the corps thrown forward is enabled to fall back before it is in any serious danger.

The length of the resistance, however, which such a corps may offer against frontal attack or against the beginning of a turning movement depends chiefly on the nature of the ground and the proximity of its own supports. If this resistance is continued beyond its natural measure, either from lack of judgment or from a sacrifice being necessary in order to give the main army the time it requires, the consequence must always be a very considerable loss.

It is only in the rarest instances, when some considerable obstacle of the ground gives an opportunity for it, that the resistance actually made in such an engagement can be of importance, and the duration of the little battle which such a corps could offer would in itself hardly gain the time required. That time is afforded in three very natural ways:

1. By the more cautious, and consequently slower, advance of the enemy.
2. By the duration of the actual resistance offered.
3. By the retreat itself.

This retreat must be made as slowly as is consistent with safety. Wherever the country affords good new positions, they should be made use of, as that obliges the enemy to prepare fresh attacks and turning movements, and by that means more time is gained. Perhaps in the new position a real engagement may even be fought.

We see that the resistance by an engagement and the retreat are closely bound up with one another, and that the shortness of the duration of the engagements must be made up for by their multiplication.

This is the kind of resistance which an advanced force should offer. The effect depends chiefly on the strength of the corps and the configuration of the country; next, on how far the corps has to go, and on how it is reinforced and retired.

A small detachment, even when the forces on both sides are equal, can never make as long a stand as a considerable corps, for the larger the masses, the more time they require to complete their action, of whatever kind it may be. In a mountainous country the mere marching is of itself slower, the resistance in the different positions longer, and attended with less danger, and at every step favorable positions may be found.

As the distance to which a corps is pushed forward increases, so too will the length of its retreat, and therefore also the absolute gain of time by its resistance; but since such a corps by its position has even less power of resistance and less reinforcement, its retreat must be made comparatively more rapidly than if it had been nearer the main army and had a shorter distance to traverse.

The reinforcement and retirement of an advanced corps must naturally have an influence upon the duration of the resistance, as all the time that prudence and security require for the retreat is so much taken from the resistance, and therefore diminishes its amount.

There is a marked difference in the time gained by the resistance of an advance corps when the enemy makes his first appearance only after midday; in such a case the length of the night is so much additional time gained, as the advance is seldom continued throughout the night. Thus it was that, in 1815, on the short distance from Charleroi to Ligny, not quite ten miles, the first Prussian corps under General Ziethen, about 30,000 strong, against Bonaparte at the head of 120,000 men, was enabled to gain more than twenty-four hours for the Prussian army to effect its concentration. The first attack was made on General Ziethen about nine o'clock in the morning of the 15th of June, and the battle of Ligny did not begin until about two in the afternoon of the 16th. General Ziethen suffered, it is true, very considerable loss, amounting to five or six thousand men killed, wounded or prisoners.

If we refer to experience, the following are the results, which may serve as a basis in any calculations of this kind:

A division of ten or twelve thousand men supported by cavalry, a day's march of fifteen to twenty miles in advance in an ordinary country, not particularly strong, will be able to detain the enemy (including time occupied in retreat) about half as long again as he would otherwise require to march over the same ground, but if the division is only five miles in advance, then the enemy ought to be detained about twice or three times as long as his unopposed march would take.

Therefore supposing the distance to be twenty miles, for which usually ten marching hours are required, then from the moment that the enemy appears in force in front of the advanced division, we may reckon upon fifteen hours before he is in condition to attack our main army. On the other hand, if the advanced guard is posted only five miles in advance, then the time which will elapse before our army can be attacked will be more than three or four hours, and may very easily come up to double that, for the enemy still requires just as much time to develop his first measures against our advance guard, and the time of resistance offered by that guard in its original position will be even longer than it would be in a position farther forward.

The consequence is that in the first of these supposed cases the enemy cannot easily make an attack on our main body on the same day that he dislodges our advance corps, and this exactly coincides with experience. Even in the second case the enemy must succeed in dislodging our advance guard in the first half of the day to have the requisite time for a battle.

Since the night comes to our help in the first of these supposed cases, we can see how much time may be gained by an advance guard thrown farther forward.

With reference to troops placed on the flanks, the object of which we have explained before, the mode of action is in most cases more or less connected with circumstances which belong to the province of more detailed application. The simplest way is to look upon them as advance guards which are placed on the flanks, and, being at the same time thrown out somewhat in advance, retreat in an oblique direction upon the army.

Since these bodies are not immediately in front of the army, and thus cannot be so easily retired on both flanks as a regular advance guard, they would be exposed to greater danger if it was not that the enemy's offensive power also in most cases is somewhat less at the outer extremities of his line, and that, if it comes to the worst, such detachments would have sufficient room to give way without exposing the army so directly to danger as a fleeing advance guard might do.

The most usual and best means of retiring an advanced corps is by a considerable body of cavalry, for which reason, when necessary because of the distance which the corps is advanced, the reserve cavalry is posted between the main body and the advanced corps.

The final conclusion is that an advanced corps is effective less by its actual expenditure of strength than by its mere presence, less by the engagements in which it takes part than by the possibility of taking part in such engagements. It should never attempt to stop the enemy's movement, but only serve like the weight of a pendulum to moderate and regulate them, so as to make it possible to calculate them.

CHAPTER 9

CAMPS

The three situations of an army when not involved in an engagement are here considered only strategically, that is, only in so far as they are conditioned by place and time and the number of the military forces. All those subjects which relate to the internal arrangement of the engagement and the transition into the state of engagement belong to tactics.

The disposition in camps, by which we mean every disposition of an army except in quarters, whether it be in tents, huts or bivouac, is strategically completely identical with the engagement which is conditioned by it. Tactically, it is not always so, for we can, for many reasons, choose a site for encamping which is different from the proposed battlefield. Having already said all that is necessary on the disposition of an army, that is, concerning the place which the different parts will occupy, the subject of camps gives us occasion only for historical treatment.

In former times, that is, before armies grew once more to considerable dimensions, before wars became of greater duration and more connected in their individual parts, and up to the time of the French Revolution, armies always used tents. This was their normal state. With the beginning of the mild season of the year they left their quarters, and did not again take them up until winter set in. Winter quarters at that time must be looked upon as, so to speak, a state of warlessness, for in them the forces were neutralized and the whole machine stopped. Quarters to refresh an army which preceded the real winter quarters, and other temporary and limited cantonments, were transitional and exceptional conditions.

This is not the place to inquire how such a periodical voluntary neutralization of forces was and still is consistent with the object and nature of war; we shall come to that subject later. Enough that it was so.

Since the war of the French Revolution armies have completely done away with tents because of the large baggage train which they cause. It is found better for an army of 100,000 men to have, in place of 6,000 tent horses, 5,000 additional cavalry or several hundred extra guns, and in great and rapid operations such a baggage train is an obvious hindrance and of little use.

But this change is attended by two drawbacks: an increase of casualties in the forces and a greater devastation of the country.

However slight the protection afforded by a roof of inferior canvas, it cannot be denied that in the long run it is great relief to the troops. For a single day the difference is small, because a tent is little protection against wind and cold, and does not completely exclude dampness; but this small difference, if repeated two or three hundred times a year, becomes important. A greater loss through sickness is the natural result.

How the devastation of the country is increased through the lack of tents for the troops requires no explanation.

One would suppose therefore that on account of these two drawbacks the doing away with tents must have diminished again the energy of war in another way, viz., that troops must remain longer and more often in quarters, and from want of the requisites for encampment must forgo many a position which would have been possible by means of tents.

This would indeed have been the case had there not been, in the same epoch of time, an enormous revolution in war generally, which swallowed up all these small, subordinate effects.

Its elemental fire has become so overpowering, its energy so extraordinary, that these regular periods of rest have disappeared, and every power presses forward with persistent energy toward the great decision, which will be treated more fully in Book IX.[1] Under these circumstances, therefore, there is no question of a change which the lack of tents caused in the use of the military forces. Troops now occupy huts or bivouac in the open air, without any regard to season of the year, weather or locality, according as the general plan and object of the campaign require.

Whether war will continue to maintain this energy, under all circumstances and at all times, is a question we shall consider later. Where this

---

[1] This book was never written.—Ed.

energy is wanting, the lack of tents may indeed exercise some influence on the conduct of war; but that this reaction will ever be strong enough to bring back the use of tents is very doubtful, because now that much wider limits have been opened for the elemental force of war, it will not return within its old narrow bounds, except occasionally for a certain time and under certain circumstances, only to break out again with the overpowering violence of its nature. Permanent arrangements for an army can, therefore, be based only upon that nature.

CHAPTER 10

MARCHES

Marches are a mere passing from one position to another, and this makes them subject to two primary conditions.

The first is the mobility of the troops, so that forces shall not be squandered uselessly when they might be usefully employed; the second is precision in the movements, so that they are executed correctly. If we marched 100,000 men in one single column, that is, upon *one* road, without intervals, the rear of the column would never arrive at the proposed destination on the same day as the head of the column; we would either have to advance at an unusually slow pace, or the mass would scatter as a falling stream of water scatters in drops; and this scattering, together with the excessive exertion laid upon those in the rear owing to the length of the column, would soon throw everything into confusion.

In contrast to this extreme, the smaller the mass of troops in one column, the greater will be the ease and precision with which the march can be performed. The result of this is the need of a *division* quite distinct from the division due to disposition. Therefore, although the division into columns of march generally originates from the disposition, it does not do so in every particular case. A large mass which is to be concentrated on one point must necessarily be divided for the march. But even if a disposition of the army in separate parts causes a march in separate divisions, sometimes the requirements of the disposition, sometimes those of the march, are predominant. For instance, if the disposition of the troops is one made merely for rest, one in which an engagement is not expected, then the requirements of the march predominate, and these requirements are chiefly the choice of good paved roads. Keeping in view this difference, we choose a road in the one case on account of the quarters and camping ground, in the other the quarters and camps on account of the road. Where a battle is expected, and everything depends on our reaching that particular point with a mass of troops, then we have no objections to reaching that point even by the most difficult by-roads, if necessary. If, on the other hand, we are only setting out, so to speak, with the army for the theater of war, then the nearest main roads are selected for the columns, and we look for as good quarters and camps as can be obtained near them.

Whether the march is of the one kind or the other, it is a fundamental principle of the modern art of war, in all cases in which even the possibility of an engagement is conceivable, that is, in the whole realm of real war, so to organize the columns that the mass of troops composing each column is capable of independent engagement. This requirement is met by the combination of the three arms, by an organic division of the whole, and by the appointment of a suitable commander. Marches, therefore, have been the chief cause of the new battle order, and they profit most by it.

When in the middle of the eighteenth century, especially in the theater of war of Frederick the Great, movement began to be regarded as a specific principle of fighting, and victory to be wrested from the enemy by means of unexpected movements, the lack of an organic battle order made the most complicated and clumsy arrangements in marches necessary. In order to carry out a movement near the enemy, an army always had to be ready to fight; but they were not ready to fight unless the whole army was concentrated, because nothing less than the army constituted a complete whole. In a march to a flank, the second line, in order to be always at the required distance, that is, not over a mile from the first, had to march up hill and down dale, which demanded immense exertion, as well as considerable knowledge of the locality; for where can one find two good roads running parallel at a distance of a mile from each other? The cavalry on the wings had to encounter the same difficulties when the march was directly to the front. There was further difficulty with the artillery, which required a road for itself, protected by the infantry; for the files of the infantry should form unbroken lines, and the artillery would have increased the length of their already long, trailing columns still more, and thrown all their regulated distances into disorder. It is only necessary to read the dispositions of marches in Tempelhoff's *History of the Seven Years' War* to be convinced of all these circumstances and of the restraints thus imposed on the action of war.

But the modern art of war has divided armies in a natural or organic way, so that each of the principal parts forms in itself a complete whole, of small proportions, capable in battle of all the actions of the large whole, with the one difference, that the duration of its action is shorter. It is, therefore, no longer necessary, even where united action is intended, to have all columns in a mass near each other, that they may all be able to unite before the engagement. It is sufficient if this unification takes place during the course of the engagement.

The smaller a body of troops the more easily it can be moved, and therefore the less it requires that division which is not a result of the disposition in separate parts, but of the unwieldiness of the mass. A small body can thus march upon one road, and if it is to advance on several lines it easily finds roads near each other good enough for its needs. The greater the mass the greater becomes the necessity for dividing, the greater becomes the number of columns, and the need of paved roads, or even great high roads, consequently also the distance of the columns from each other. Now the danger of this division is—arithmetically expressed—in an inverse ratio to the necessity for it. The smaller the parts are, the more readily must they be able to render assistance to each other; the larger they are, the longer they can be left to depend on themselves. If we only call to mind what has been said in the preceding book on this subject, and also consider that in cultivated districts at a few miles distance from the main road there are always fairly well-paved roads running in a parallel direction, it is easy to see that, in planning a march, there are no great difficulties which make speed and precision incompatible with the proper concentration of forces. In a mountainous country parallel roads are scarcest and the difficulties of communication between them greatest; but on the other hand, the defensive powers of a single column are very much greater.

In order to make this idea clearer, let us look at it for a moment in concrete form.

A division of 8,000 men, with its artillery and other vehicles, takes up, as we know by experience in ordinary cases, a space of about three miles. If, therefore, two divisions march one after the other on the same road, the second arrives one hour after the first. But as said in Book IV, Chapter 6, a division of this strength is quite capable of maintaining an engagement for several hours, even against a superior force. Therefore, supposing the worst, that is, supposing the first division had to go into action at once, still the second division would not arrive too late to support it. Further, within three miles, right and left of the road on which we march, in the cultivated countries of Central Europe, there are, generally, lateral roads which can be used for a march, so that there is no necessity to go across country, as was so often done in the Seven Years' War.

Further, it is known by experience that the head of a column composed of four divisions and a reserve of cavalry, even on poor roads, generally does a march of fifteen miles in eight hours. Now if we reckon for each division three miles in depth, and the same for the reserve cavalry and artillery, the whole march will last thirteen hours. This is no great length of time, and yet in this case about forty thousand men would have marched over the same road. But with such a mass as this we can make use of lateral roads, which are to be found at a slightly greater distance, and therefore easily shorten the march. If the mass of troops marching on the same road is still greater than the above supposed, then it is a case in which the arrival of the whole on the same day is no longer indispensable, for now such masses never give battle in the first hour of their meeting, but usually not until the next day.

We have introduced these concrete cases, not in order to exhaust the list of circumstances of the kind, but to make ourselves more intelligible, and by means of this glance at the results of experience to show that in the present mode of conducting war the organization of marches no longer offers such great difficulties; that the most rapid marches, executed with the greatest precision, no longer require either that particular skill or that exact knowledge of the country which was needed for Frederick's rapid and exact marches in the Seven Years' War. Through the natural or organic way of dividing armies, they take place almost of themselves, at least without any great preparatory plans. In times past, battles were conducted by mere word of command, but marches required elaborate plans, whereas now it is the battle order that requires the elaborate plans, and for a march the mere word of command almost suffices.

As is well known, all marches are either perpendicular to the front or parallel. The latter, also called flank marches, alter the geometrical position of the parts; those parts which, in disposition, were in line, will on the march follow one another, and vice versa. Now, although all of the degrees lying within the right angle can just as well be taken as direction of the march, yet its order must definitely be of one kind or the other.

This geometrical alteration could only be completely carried out by tactics, and by it only through the file march, as it is called, which, with great masses, is impossible. Far less is it possible for strategy to do it. The parts which change their geometrical relation, are in the old order of battle only the center and the wings; in the new, they are the members of the first order—corps, divisions or even brigades, according to the organization of the army. Now, the consequences above deduced from the new battle order have an influence here also, for as it is no longer so necessary, as formerly, that the whole army should be assembled before action starts, the greater care is taken that those troops which march together form one whole. If two divisions were so placed that the one behind formed the reserve for the other, and that they were to advance against the enemy upon two roads, no one would think of sending a portion of each division by each of the roads, but a road would without hesitation be assigned to each division; they would therefore march side by side, and each general of division would be left to provide a reserve for himself in case of an engagement. Unity of command is much more important than the original geometrical relation. If the divisions reach their new position without an engagement, they can resume their previous relations. Much less, if two divisions, standing next to each other, are to make a *parallel* march upon two roads, should we think of placing the second line or reserve of each division on the rear road. Instead of that, we should allot to each of the divisions one of the roads, and therefore during the march consider one division as forming the reserve to the other. If an army of four divisions, of which three form the front line and the fourth the reserve, is to march against the enemy in that order, it is natural to assign a separate road to each of the three divisions in front, and make the reserve follow the center. If there are not three roads at a suitable distance apart, then we need not hesitate to march upon two roads, without any serious inconvenience arising from so doing.

It is the same in the opposite case of the flank march.

Another point is the marching off of columns from the right flank or left. In parallel marches this is the natural thing. No one would march off from the right to make a movement to the left flank. In a march to the front or rear, the order of march should properly be chosen according to the direction of roads in respect to the future line of deployment. This may, indeed, be done frequently in tactics, since its space is smaller and its geometrical relations easier to survey. In strategy it is quite impossible, and if, nevertheless, we have seen here and there in strategy a certain analogy to tactics, it was mere pedantry. Formerly the whole order of march was a purely tactical affair, because the army even on the march remained an indivisible whole and contemplated nothing but *one* total engagement. Nevertheless Schwerin, for example, when he marched off from the district of Brandeis, on the 5th of May, could not tell whether his future field of battle would be on his right or left, and on this account he was obliged to make his famous countermarch.

If an army in the old battle order advanced against the enemy in four columns, the cavalry wings of the first and second lines formed the two exterior columns, the infantry wings of both lines the two central columns. Now these columns could march off, all from the right or all from the left, or the right wing from the right, the left wing from the left, or the left from the right, and the right from the left. In the latter case it would have been called "double column from the center." But all these forms, although they were supposed to have had a direct relation to the future deployment, were really all quite indifferent in that respect. When Frederick the Great entered on the battle of Leuthen, his army had been marched off by wings from the right in four columns. This made the transition to that marching off in lines, so much admired by all historians, very easy, because it happened to be the left wing of the Austrians that the king wanted to attack. Had he wanted to turn their right wing, he would have had to make a countermarch, as he did at Prague.

If these forms did not meet their object even in those days, they would be mere trifling in relation to it now. We know now just as little as formerly the situation of the future battlefield in relation to the road we take; and the little loss of time occasioned by marching off in the wrong order is now infinitely less important than formerly. The new order of battle has a beneficial influence also in this respect, that it is now quite immaterial which division arrives first or which brigade is brought under fire first.

Under these circumstances the march off from the right or the left is of no consequence now, except that when it is done alternately it tends to equalize the fatigue which the troops undergo. And this is the only reason, though certainly an important one, for generally retaining both modes of marching off.

The marching off from the center as a definite formation disappears of itself under these circumstances, and can take place only accidentally. A marching off from the center by one and the same column is in strategy, in point of fact, nonsense, for it supposes a double road.

The order of march belongs, moreover, more to the province of tactics than to that of strategy, for it is the division of a whole into parts, which, after the march, are once more to resume the state of a whole. Since, however, in modern warfare the exact keeping together of the parts is no longer necessary, but the parts during a march are more separated from each other and therefore left to themselves, it is much easier now for independent engagements to take place, which the parts fight independently, and which, therefore, must be considered as complete engagements in themselves. On that account we have thought it necessary to say so much on the subject.

Furthermore, since a battle order in three parts in juxtaposition is, as we have seen in Chapter 5 of this Book, the most natural where no special object predominates, it follows that the order of march in three large columns is also the most natural.

It remains only to observe that the notion of a column does not originate only from the road which a body of troops takes, but that in strategy also one must so designate masses of troops which take the same road on different days. For the division into columns is made chiefly to shorten and facilitate the march, as a small number always marches more quickly and conveniently than large bodies. But this end may also be attained by marching troops on different days, as well as by marching them on different roads.

CHAPTER 11

MARCHES (CONTINUED)

Respecting the length of a march and the time it requires, it is natural for us to adhere to the general results of experience.

For our modern armies it has long been settled that a march of fifteen miles should be the usual day's work, which, on long distances, must even be reduced to ten miles per day, allowing for the necessary rest days, to make such repairs of all kinds as may be required.

Such a march in level country, and on average roads, will occupy a division of 8,000 men from eight to ten hours; in a mountainous country from ten to twelve hours. If several divisions are united in one column, the march will occupy several hours longer, even after deducting the intervals which must elapse between the departure of the first and succeeding divisions.

We see, therefore, that the day is pretty well occupied with such a march; that the fatigue endured by a soldier loaded with his pack for ten or twelve hours is not to be judged by that of an ordinary journey of fifteen miles on foot, which a single person, on tolerable roads, might easily make in five hours.

The longest marches on single occasions are of twenty-five or, at most, thirty miles a day, or, if repeated on successive days, of twenty.

A march of twenty-five miles requires a halt of several hours, and a division of 8,000 men will not do it, even on good roads, in less than sixteen hours. If the march is one of thirty miles, and there are several divisions in the column, we may reckon upon at least twenty hours.

We mean here the march of a number of whole divisions from one camp to another, for that is the usual form of marches made on a theater of war. When several divisions are to march in one column, the first division to move is assembled and marched off somewhat earlier than the rest, and therefore arrives at its camping ground so much the sooner. But this difference can never amount to the whole time, which corresponds to the depth of a division on the line of march, and which is so well expressed in French as the time it requires for its *découlement* (trickling down). The soldier is, therefore, saved very little fatigue in this way, and every march is very much lengthened in duration in proportion as the number of troops to be moved increases. To assemble and march off the different brigades of a division, in like manner at different times, is seldom practicable, and for that reason we have taken the division itself as the unit.

In long distances, when troops march from one quarters to another, and go over the roads in small bodies, and without points of assembly, the distance in itself can, no doubt, be greater, but in point of fact it is already so because of the detours which quarters necessitate.

But those marches, on which troops have to assemble daily in divisions, or perhaps even in corps, and must nevertheless withdraw into quarters, take up the most time, and are only advisable in rich countries, and where the masses of troops are not too large, because in such cases the better facility of subsistence and shelter compensates sufficiently for the fatigue of a longer march. The Prussian army in its retreat in 1806 undoubtedly followed an erroneous system in taking up quarters for the troops every night on account of subsistence. They could have procured subsistence in bivouacs, and the army would not have been obliged to spend fourteen days and tremendous exertion on about 250 miles.

If a bad road or hilly country has to be marched over, all these calculations as to time and distance undergo such modifications that it is difficult to estimate, with any certainty, in a particular case the time required for a march; much less, can any general theory be established. All that theory can do, therefore, is to direct attention to the dangerous errors which one is liable to make in this case. To avoid them, the most careful calculation is necessary, and a large margin for unforeseen delays. The influence of weather and the condition of the troops also come into consideration.

Since the doing away with tents and the introduction of the system of subsisting troops by forcible requisition of provisions on the spot, the baggage of an army has been noticeably diminished, and naturally we look first of all for the most important consequence, an acceleration in the movements of an army, and, consequently, for an increase in the length of the day's march. This, however, is only realized under certain circumstances.

Marches within the theater of war have been very little accelerated by this means, for it is well known that whenever the object required marches of unusual length, it had commonly been the practice to leave the baggage behind or send it on beforehand, and, generally, to keep it separate from the troops during the continuance of such movements. Thus it had in general no influence on the movement, and as soon as it ceased to be a direct impediment, no further trouble was taken about it, whatever damage it might suffer. Marches, therefore, took place in the Seven Years' War, which even now cannot be surpassed; as an instance we cite Lascy's march in 1760, when he had to support the diversion of the Russians on Berlin. He got over the road from Schweidnitz to Berlin through Lusatia, a distance of 225 miles, in ten days, averaging, therefore, twenty-two miles a day, which, for a corps of 15,000 would be an extraordinary march even in these days.

On the other hand, just on account of the change of method of subsistence the movements of armies have been affected by a new *retarding* influence. If troops have partly to procure supplies for themselves, which often happens, they require more time than would be necessary merely to receive rations from provision wagons. Besides this, on marches of considerable duration troops cannot be encamped in such large numbers at any one point; the divisions must be separated from one another, in order the more easily to provide for them. Lastly, it happens frequently that it is necessary to place part of the army, particularly the cavalry, in quarters. All this occasions on the whole a noticeable delay. We find, therefore, that Bonaparte in pursuit of the Prussians in 1806, with a view to cutting off their retreat, and Blücher in 1815, in pursuit of the French, with a like object, only accomplished about 150 miles in ten days, a rate which Frederick the Great was also able to attain in his marches from Saxony to Silesia and back, notwithstanding the baggage that he had to carry with him.

Nevertheless, the mobility and manageableness, if we may use the expression, of the parts of an army, both great and small, on the theater of war have very perceptibly gained by the diminution of baggage. Partly because, with the same number of cavalry and guns, there are fewer horses, and therefore less often trouble about forage, partly because armies are no longer so tied to their positions, and there is no need to consider constantly a long train of baggage trailing in the rear.

Marches such as that which, after raising the siege of Olmütz, in 1758, Frederick the Great made with 4,000 vehicles, the escort of which employed half his army, broken up into single battalions and companies, could hardly be effected now in presence of even the most timid adversary.

On long marches, such as from the Tagus to the Niemen, that alleviation is more sensibly felt, for although the usual measure of the day's march remains the same on account of the vehicles still retained, yet in cases of great urgency, we can deviate from that usual measure with smaller sacrifices.

Generally the diminution of baggage tends more to a saving of energy than to the acceleration of movement.

CHAPTER 12

MARCHES (CONTINUED)

We have now to consider the destructive influence which marches exercise upon an army. It is so great that as a specific agency of destruction it may be ranked with the engagement.

One single moderate march does not wear down the army, but a succession of even moderate marches is certain to tell upon it, and a succession of severe ones will, of course, do so much more.

At the actual scene of war, want of food and shelter, bad, worn-out roads, and the necessity of being in a perpetual state of readiness for battle are causes of excessive exertions, by which men, beasts and vehicles, as well as clothing, are ruined.

It is commonly said that a long rest does not suit the physical health of an army; that at such time there is more sickness than during moderate activity. No doubt sickness will and does occur if soldiers are packed too close in confined quarters; but the same thing will occur in quarters taken up on the march, and the want of air and exercise can never be the cause of such sickness, as it is so easy to give the soldier both by means of exercises.

Let us consider merely what a difference it makes to the disturbed and faltering organism of a human being whether he falls sick in a house or is seized on an open highway, up to his knees in mud, under torrents of rain and loaded with a knapsack on his back. Even if he is in a camp he can soon be sent to the next village, and will not be entirely without medical assistance, while on a march he remains lying for hours without any sort of aid and then drags himself along for miles as a straggler. How many trifling illnesses become serious because of this, how many serious ones become mortal! Let us consider how even an ordinary march in the dust, and under the burning rays of a summer sun, may cause the most excessive heat, in which state, suffering from intolerable thirst, the soldier then rushes to the first fresh spring, to draw from it sickness and death.

It is not our object by these reflections to recommend less activity in war; the instrument is there for use, and if the use wears the instrument away that is only natural. We only wish to see everything put in its right place, and to oppose that theoretical bombast according to which the most overwhelming surprise, the most rapid movement, the most incessant activity cost nothing, and are represented as rich mines which the indolence of the general leaves unworked. It is the same with the exploitation of these mines as with that of those from which gold and silver are obtained; nothing is seen but the product, and no one asks about the value of the work which has brought this product to light.

On long marches outside a theater of war, the conditions under which the march is made are no doubt usually easier, and the daily losses smaller, but on the other hand men with the slightest sickness are generally lost to the army for some time, as it is difficult for convalescents to catch up with an army constantly advancing.

Amongst the cavalry the number of lame horses and horses with sore backs rises in an increasing ratio, and many vehicles break down or require repair. At the end of a march of 500 miles or more, an army invariably arrives much weakened, particularly as regards its cavalry and vehicles.

If such marches are necessary on the theater of war itself, that is, under the eyes of the enemy, then that disadvantage is added to the other, and from the two combined the losses with large masses of troops and under conditions otherwise unfavorable may amount to something incredible.

Just one or two examples to illustrate what we have been saying.

When Bonaparte crossed the Niemen on the 24th of June, 1812, the enormous center of his army with which he subsequently marched against Moscow numbered 301,000 men. At Smolensk, on the 15th of August, he detached 13,500, leaving, it is to be supposed, 287,500. The actual state of his army, however, at that date was only 182,000; he had therefore lost 105,500.[1] Bearing in mind that up to that time only two engagements worthy of consideration had taken place, one between Davoust and Bagration, the other between Murat and Tolstoy-Ostermann, we may put down the losses of the French army in action at 10,000 men at most, and therefore

· [1] All these figures are taken from Chambray.

the losses in sick and stragglers, within fifty-two days on a march of about 350 miles direct to his front, amounted to 95,000, that is, one-third of the whole force.

Three weeks later, at the time of the battle of Borodino, the loss amounted already to 144,000 (including the casualties in battle) and eight days after that again, at Moscow, to 198,000. The losses of this army in general were at the beginning of the campaign at the rate of a hundred-and-fiftieth daily, subsequently they rose to a hundred-and-twentieth, and in the last period they increased to one-nineteenth of the original strength.

The movement of Napoleon from the crossing of the Niemen up to Moscow may, it is true, be called an impetuous one; still, we must not forget that it lasted eighty-two days, in which time he covered only about 600 miles, and that the French army upon two occasions made regular halts, once at Vilna for about fourteen days, and again at Witebsk for about eleven days, during which time many stragglers had time to rejoin. This fourteen weeks' advance was not made at the worst season of the year, nor over the worst of roads, for it was summer, and the roads along which they marched were mostly sand. It was the immense mass of troops collected on one road, the lack of adequate subsistence, and an enemy who was retreating but not in flight, which were the adverse conditions.

Of the retreat of the French from Moscow to the Niemen we shall say nothing, but this we may mention: that the Russian army following them left the region of Kaluga 120,000 strong, and reached Vilna with 30,000. Everyone knows how few it lost in engagements during the period.

One more example from Blücher's campaign of 1813 in Silesia and Saxony, a campaign distinguished not by a long march but by many movements to and fro. York's corps of Blücher's army began this campaign on the 16th of August about 40,000 strong, and was reduced to 12,000 at the battle of Leipzig, the 19th of October. The principal engagements which this corps fought at Goldberg, Lowenberg, on the Katzbach, at Wartenburg and Mockern (Leipzig) cost it, on the authority of the best writers, 12,000 men. According to that their losses from other causes in eight weeks amounted to 16,000, or two-fifths of the whole.

We must therefore expect great wear and tear on our own forces, if we are to carry on a war full of movements; we must arrange the rest of our plan accordingly, and, above all things, the reinforcements which are to follow.

CHAPTER 13

QUARTERS

In the modern system of war, quarters have again become indispensable, because tents and a complete military train no longer make an army independent of them. Huts and open-air camps (bivouacs as they are called), however extensive and well arranged, cannot be regularly used to shelter troops without sickness sooner or later, according to the climate, gaining the upper hand, and prematurely exhausting their strength. The campaign in Russia in 1812 is one of the few in which, in a very severe climate, the troops, during the whole six months that it lasted, hardly ever lay in quarters. But what was the consequence of this extreme effort, which should be called an extravaganza, if that term was not even more applicable to the political conception of the enterprise!

Two things interfere with the occupation of quarters—proximity of the enemy and rapidity of movement. For these reasons they are left as soon as the decision approaches, and cannot be taken up again until the decision is over.

In modern wars, that is, in all the campaigns we have seen during the last twenty-five years, the elemental force of war has acted with all its energy. Nearly all that was possible has generally been done in them, as far as activity and effort are concerned; but all these campaigns have been of short duration, they have seldom exceeded half a year; in most of them a few months have sufficed to lead to the goal, that is, to a point where the vanquished enemy saw himself compelled to sue for an armistice or even for peace, or to a point where, on the conqueror's part, the impetus of victory had exhausted itself. During the period of extreme effort there could be little question of quarters, for even in the victorious march of the pursuer, when there was no longer any danger, rapidity of movement made that kind of relief impossible.

But when for some reason the course of events is less impetuous, when an even poising and balancing of forces is more the rule, then the housing of troops becomes a foremost subject for attention. The need of this has some influence on the conduct of the war itself, partly by making us seek to gain more time and security by a stronger system of outposts, by a more considerable advance guard thrown forward; and partly by causing our measures to be governed more by the richness and fertility of the country than by the tactical advantages of the ground and the geometrical relations of lines and points. A commercial city of twenty or thirty thousand inhabitants, a road thickly studded with large villages or flourishing cities, give such facilities for the concentration in one position of large bodies of troops, and this concentration gives such an ease and freedom of action that these advantages abundantly compensate for those which a better situation of the point could give.

Concerning the form of the arrangement of quarters we have only a few observations to make, since this subject belongs for the most part to tactics.

The housing of troops is of two kinds, inasmuch as it can be either the main point or only a secondary consideration. If the disposition of the troops in the course of a campaign is regulated by reasons purely tactical and strategical, and if, as is done more especially with cavalry, the troops are directed for their comfort to occupy quarters in the vicinity of the point of concentration, then the quarters are subordinate considerations and take the place of a camp. They must, therefore, be chosen within such a radius that the troops can reach the point of assembly in good time. But if an army takes up quarters for rest, then the housing of the troops is the main point, and other measures, consequently also the specific selection of the point of assembly, will be influenced by that object.

The first question for examination here concerns the form of the quarters area as a whole. The usual form is that of a very long oval, a mere widening as it were of the tactical battle order. The point of assembly for the army is in front, the headquarters in rear. Now these three arrangements are, in point of fact, adverse, indeed almost opposed, to the safe assembly of the army before the arrival of the enemy.

The more the quarters form a square, or rather a circle, the quicker the troops can concentrate at one point, that is, the center. The further the place of assembly is situated in the rear, the longer the enemy will be in reaching it, and, therefore, the more time is left us to assemble. A point of assembly in rear of the quarters can never be in danger. And, on the other hand, the farther the headquarters are to the front, so much the sooner reports arrive, and so much the better is the commander informed of everything. At the same time the first-mentioned arrangements are not without reasons which deserve some attention.

By the extension of quarters in width we have in view the protection of the country which the enemy would otherwise use for supplies. But this reason is neither entirely true nor very important. It is only true as far as regards the outermost wings, but does not apply at all to intermediate spaces existing between separate groups of the army, if the quarters of those groups are drawn closer around their point of assembly, for no enemy will venture into those intervals of space. And it is not very important, because there are simpler means of shielding the areas in our vicinity from the enemy's requisitions than by scattering the army itself.

The placing of the point of assembly in front is intended to protect the quarters, for the following reasons: In the first place, a body of troops, suddenly called to arms, always leaves behind it in quarters a tail of stragglers—sick, baggage, provisions, etc.—which may easily fall into the enemy's hands if the point of assembly is placed in rear. In the second place, we have to apprehend that if the enemy with some cavalry detachments passes by the advance guard, or if the latter should have been really scattered, he would fall upon isolated regiments or battalions. If he encounters a force drawn up in formation, although it is weak, and in the end must be overpowered, still he is brought to a stop, and in that way time is gained.

In regard to the position of the headquarters, it has been generally supposed that it cannot be made too secure.

Taking these different points into consideration, we may conclude that the best arrangement for the areas of quarters is where they take an oblong form, approaching the square or circle, have the point of assembly in the center, and the headquarters placed on the front line, well protected by somewhat considerable masses of troops.

What we have said as to covering of the wings, in treating of the disposition of the army in general, applies here also; therefore corps detached from the main army, right and left, even if intended to fight in conjunction with the rest, will have points of assembly of their own in the same line with the main army.

If, further, we reflect that the nature of a country, on the one hand, by favorable features of the ground determines the natural point of assembly, and on the other hand, by the position of cities and villages determines the most suitable situation for quarters, we will understand how very rarely any geometrical form can be decisive in our present subject. But yet it was necessary to direct attention to it, because, like all general laws, it affects the majority of cases in a greater or less degree.

What now remains to be said as to an advantageous position for quarters is that they should be taken up behind some natural obstacle of ground affording cover, while the enemy can be watched by small but numerous parties; or they may be taken up behind fortresses, which, when circumstances prevent any estimate of the strength of their garrisons, impose upon the enemy much greater respect and caution.

We reserve the subject of winter quarters, covered by defensive works, for separate treatment.

The quarters taken up by troops on a march differ from those of stationary troops in that in order to avoid detours they are spread out very little, but extend lengthwise along the road; if this extension does not exceed a short day's march, the arrangement is not at all unfavorable to quick assembly.

In all cases where we are in face of the enemy, according to the technical phrase in use, that is, in all cases where there is no considerable space between the advance guards of the two armies, the extent of the quarters and the time required to assemble the army determine the strength and position of the advance guard and outposts; or when these are conditioned by the enemy and circumstances, then, on the contrary, the extent of the quarters must depend on the time which the resistance of the advance guard can give us.

In Chapter 8 of this Book we have stated how this resistance, in the case of an advanced body, may be conceived. From the time of that resistance we must deduct the time required for transmission of reports and getting the men under arms, and the remainder only is the time available for reaching the point of assembly.

In order here finally to embody our ideas in a conclusion applicable under ordinary circumstances, we wish to observe that if the quarters area were a circle with the distance of the advance guard as radius, and the point of assembly lay pretty much in the center, the time which is gained by checking the enemy's advance would be available for the transmission of reports and getting under arms. This should in most cases be sufficient, even if the transmission is not made by means of signals, cannon-shots, etc., but simply by relays of orderlies, which alone gives security.

With an advance guard pushed forward fifteen miles in front, our quarters might therefore cover a space of about 700 square miles. In a moderately populated country there would be about 10,000 houses in this area, which for an army of 50,000, after deducting the advance guard, would give about four men to a billet, therefore very comfortable quarters, and for an army of twice the strength nine men to a billet, still therefore not very cramped quarters. On the other hand, if the advance guard could not be pushed forward more than five miles, we shall get only a space of eighty square miles, for although the time gained does not diminish exactly in proportion to the distance of the advance guard, and even with a distance of five miles we may still calculate on a gain of six hours, yet the necessity for caution increases when the enemy is so close. But in such a space an army of 50,000 men could find some measure of accommodation only in a very thickly populated region.

From all this we see what an important role is played by large or at least good-sized cities, which afford convenience for sheltering 10,000 to 20,000 men almost at one point.

From this result it would follow that, if we are not very close to the enemy, and have a suitable advance guard, we might remain in quarters even if the enemy is concentrated, as Frederick the Great did at Breslau in the beginning of the year 1762, and Bonaparte at Witebsk in 1812. But although by preserving a proper distance and by suitable arrangements we have no reason to fear for the security of assembling, even opposite an enemy who is concentrated, yet we must not forget that an army engaged in assembling in haste can do nothing else in that time; that it is, therefore, at that moment not in a condition to avail itself of developing circumstances, and is thus deprived of the greater part of its efficiency. The consequence of this is that an army should only go entirely into quarters in the three following cases:

1. If the enemy does the same.
2. If the condition of the troops makes it absolutely unavoidable.
3. If the immediate task of the army is entirely confined to the defense of a strong position, and nothing matters but to assemble the troops at that point in good time.

The campaign of 1815 gives a very remarkable example of the assembly of an army from quarters. General Ziethen, with Blücher's advance guard, 30,000 men, was posted at Charleroi, only ten miles from Sombreff, the place appointed for the assembly of the army. The farthest quarters of the army were about forty miles from Sombreff, that is, on the one side beyond Ciney, and on the other as far as Liége. Notwithstanding this, the troops quartered beyond Ciney were assembled at Ligny several hours before the battle began, and those near Liége (Bülow's corps) would have been also, had it not been for chance and faulty arrangements in the communication.

Unquestionably, proper care for the security of the Prussian army was not taken; but in explanation we must say that the arrangements were made at a time when the French army itself was still dispersed over widely extended quarters, and that the fault consisted simply in not changing them the moment the first news was received that the enemy's troops were in movement, and that Bonaparte had joined his army.

Still it is noteworthy that the Prussian army could have concentrated at Sombreff before the attack of the enemy. It is true that on the night of the 14th, that is, twelve hours before General Ziethen was actually attacked, Blücher received information of the advance of the enemy, and began to assemble his army; but on the 15th at nine in the morning, Ziethen was already fully engaged, and it was not until that same moment that General Thielemann at Ciney first received orders to march to Namur. He had therefore first to assemble his troops in divisions and then to march thirty-two and a half miles to Sombreff, which he did in twenty-four hours. General Bülow would also have been able to arrive about the same time, if the order had reached him properly.

But Bonaparte was not able to make his attack on Ligny until two o'clock in the afternoon on the 16th. The apprehension of having Wellington on the one side of him, and Blücher on the other, in other words, the disproportion in strength, contributed to this slowness. We see, however, how the most resolute commander is detained by having cautiously to feel his way, as he always must in cases which are in some degree complicated.

Some of the considerations here raised are obviously more tactical than strategic in their nature; but we have preferred rather to encroach a little than to run the risk of not being clear.

CHAPTER 14

SUBSISTENCE

In modern warfare this subject has acquired a much greater importance and that for two reasons. The first is that armies in general are now much larger than those of the Middle Ages, and even those of the ancient world. Here and there in earlier times we do find armies which equal or even greatly surpass the more modern ones in size, but nevertheless these are rare and transient phenomena. In modern military history, however, since the time of Louis XIV, armies have always been very large. But the second reason is still much more important and more peculiar to modern times. It consists in the closer internal connection of our wars, in the constant state of readiness for battle of the military forces with which they are waged. Most of the old wars consist of single unconnected enterprises, separated by intervals during which the war either actually ceased altogether and only continued to exist politically, or the military forces had, at all events, got so far removed from each other that each occupied itself solely with its own wants, without paying any attention to the army opposing it.

The more modern wars, that is, the wars since the Peace of Westphalia, have through the efforts of the governments assumed a more systematic and coherent form. The military object everywhere predominates, and demands such arrangements with regard to subsistence that it can everywhere be adequately supplied. There were also, it is true, long periods of inaction in the wars of the seventeenth and eighteenth centuries, almost amounting to a cessation of war, namely, the periods regularly passed in winter quarters, but even these were still subordinate to the military object. They were caused by the bad season and not by lack of maintenance for the troops, and as they regularly ended with the return of summer, uninterrupted military action, at all events during the good season, is the established rule.

As the transitions from one condition or mode of action to another always have taken place gradually, so it was in this case. In the wars with Louis XIV the Allies still used to send their troops into winter quarters in distant provinces in order to subsist them more easily. In the Silesian Wars this was no longer done.

This systematic and coherent form of carrying on war did not actually become possible for states until they substituted professional soldiers for the feudal armies. Feudal duty was commuted for a contribution, and personal service either entirely ceased—recruiting taking its place—or survived only in the lowest class, the nobility regarding the furnishing of a quota of men (as is still done in Russia and Hungary) as a kind of contribution, a tax in men. In every case, as we have elsewhere observed, armies became henceforward an instrument of the Cabinet, primarily supported by the Treasury or the revenue of the government.

The same kind of change which took place in the mode of raising and keeping up the military forces necessarily took place in the mode of subsisting them. The upper classes, having been released from the former service in return for a money payment, could not have the burden of the latter reimposed upon them quite so easily. The Cabinet and the Treasury had therefore to provide for the subsistence of the army, and could not allow it to be maintained in its own country at the expense of the people. Governments were therefore obliged to look upon the subsistence of the army entirely as their own affair. The subsistence thus became more difficult in two ways: first, by becoming an affair of the government, and, next, because the military forces were supposed to be permanently kept up to confront those of the enemy.

Thus not only an independent military class, but also an independent organization for its subsistence, was created and developed to the highest degree possible.

Not only were stores of provisions collected, either by purchase or by delivery from the state demesne, and so from different points, and lodged in magazines, but they were also forwarded from these by a special transport system to the troops, baked in their neighborhood in a special bakery, and then by means of another transport system—this time finally attached to the troops themselves—carried away by them. We take a glance at this system not merely because it throws light on the character of the wars in which it arose, but because it can never entirely cease to exist. Some parts of it will always reappear.

Thus military organization strove to become more and more independent of the people and of the country.

The consequence was that in this manner war certainly became more systematic, more coherent and more subordinated to its object, that is, to the political object sought, but also at the same time much more limited and impeded in its movements and infinitely more weakened in energy. For now an army was tied to its magazines, limited to the effective range of its transport service, and very naturally everything tended to organizing the subsistence of the army as economically as possible. Fed on a wretched scrap of bread, the soldier often tottered about like a shadow, and no prospect of a change in his luck comforted him for his present privations.

Whoever treats this wretched way of feeding soldiers as a matter of indifference, and only thinks of what Frederick the Great achieved with soldiers subsisted in this manner, does not take a wholly impartial view of the matter. The power of enduring privations is one of the finest virtues in a soldier, and without it no army is animated with the true military spirit; but such privation must be a temporary one, due to the force of circumstances and not the consequence of a miserable system or of a parsimonious, abstract calculation of the minimum necessary to support life. When such is the case, the strength of the individual soldier will always deteriorate physically and morally. What Frederick the Great managed to do with his soldiery cannot be taken as a standard for us. For one thing, the same system obtained among his opponents; for another, we do not know how much more he might have undertaken if he had been able to let his troops live as Bonaparte allowed his to do whenever circumstances permitted.

To the feeding of horses no one had dared to extend the artificial system of supply, because forage is much more difficult to provide on account of its bulk. A ration for a horse weighs about ten times as much as one for a man. The number of horses with an army is, however, more than one-tenth of the number of men, but even now from one-fourth to one-third, and formerly it was one-third to one-half; therefore the weight of the forage required is three, four or five times as much as that of the soldier's rations. On this account the shortest and most direct means were taken to meet the wants of an army in this respect, that is, by foraging expeditions. Now these expeditions in another way interfered greatly with the conduct of the war; first, by making it a principal object to keep the war in the enemy's country, and, next, because they made it impossible to remain very long in one part of the country. However, at the time of the Silesian Wars foraging expeditions had become much less common. They were found to waste a country and strain its resources much more than if the requirements were met by means of requisitions and forced contributions from the district.

When the French Revolution suddenly brought a national army once more upon the stage of war, the means which governments could provide were found insufficient, and the whole system of war, which had its origin in the limited extent of those means and likewise found its security in this limitation, fell to pieces. Of course in the downfall of the whole was included that of the part of which we are here treating, the system of subsistence. Troubling themselves little about magazines, and thinking still less about organizing that artificial clockwork, which made the different sections of the transport service go round and round like wheels in a machine, the revolutionary leaders sent their soldiers into the field, forced their generals into battle, subsisted and reinforced their armies, and filled everything with life and enthusiasm by taking whatever they wanted by means of exactions, robbery and plunder.

Between these two extremes the war under Bonaparte and against him kept a sort of middle way; that is to say, among all the means available it used those which suited it best. And so it will also probably be in the future.

The modern method of subsisting troops, that is, seizing everything which is to be found in the country without regard to *meum* and *tuum*, may also be carried out in four different ways, namely, by subsisting on the individual inhabitant, exactions which the troops themselves look after, general requisitions and magazines. All four are usually applied together, one generally prevailing over the others. Still it sometimes happens that only one is applied and no other.

1. SUBSISTING ON THE INDIVIDUAL INHABITANT, OR ON THE COMMUNITY, WHICH IS THE SAME THING

If we bear in mind that in a community consisting even, as it does in great cities, of consumers only, there must always be provisions enough to last for several days, we may easily see that even the most densely populated place can furnish food and quarters for one day for about as many troops as there are inhabitants, and for a less number of troops for several days without the necessity of any particular previous preparation. In cities of considerable size this gives a very satisfactory result, because it enables us to subsist a large force at one point. But in smaller cities, and still more in villages, the result would be very unsatisfactory. A population of three or four thousand to twenty-five square miles, which is very considerable, would only suffice to feed from 3,000 to 4,000 soldiers, and if the forces were large, that would demand the scattering of the troops over so much ground that all other conditions might thereby be seriously affected. But in open country and even in small towns the quantity of those kinds of provisions which are essential in war is generally much greater. The bread supply of a farmer is generally enough for the consumption of his family for several days, perhaps from one to two weeks; meat can be obtained daily, and vegetables are generally forthcoming in sufficient quantity to last till the next crop. Therefore in quarters which have not yet been occupied, there is no difficulty in subsisting troops three or four times the number of the inhabitants for several days, which again is a very satisfactory result. According to this, where the population is about 2,000 or 3,000 to twenty-five square miles, and if no large city is included, a column of 30,000 would require about a hundred square miles, a square with sides ten miles long. Therefore, for an army of 90,000, which we may reckon at about 75,000 combatants, if marching in three columns contiguous to one another, we should require to take up a front of only thirty miles in breadth in the event of there being three roads within that breadth.

If several columns follow one another in these quarters, special measures must be adopted by the civil authorities, which for the needs of a day or two longer is not difficult. Thus if the 90,000 are followed a day later by a like number, even these last would suffer no want. This makes up the large number of 150,000 combatants.

Forage for the horses raises still fewer difficulties, as it requires neither grinding nor baking, and as there must be forage forthcoming in sufficient quantity to last the horses of the country until next harvest, even when there is little stall-feeding, there will hardly be want. Only the deliveries of forage should, of course, be demanded from the community at large and not from the individual inhabitants. Besides, it is of course assumed that in making arrangements for the march some attention is paid to the nature of the district so as not to send cavalry straight into commercial and industrial centers and into districts where there is no forage.

The conclusion to be drawn from this hasty glance is, therefore, that in a moderately populated country, that is, a country with from 2,000 to 3,000 souls to the twenty-five square miles, an army of 150,000 combatants will be subsisted by the inhabitants, taken individually and by communities, for from one to two days within a very narrow space which admits of all taking part in a battle. This means that such an army can be subsisted on an uninterrupted march without magazines or other previous preparations.

On this result were based the enterprises of the French army in the Revolutionary War and under Bonaparte. They marched from the Adige to the lower Danube and from the Rhine to the Vistula with little means of subsistence except upon the individual inhabitants and without ever suffering any want. As their undertakings were aided by moral and physical superiority and attended by unquestionable successes, and as at all events they were never delayed in any case by hesitation or caution, their progress on their career was that of an uninterrupted march.

If circumstances are less favorable, if the population is not so great, or if it consists more of tradespeople than of farmers, if the soil is bad, if the country has already been several times overrun—then, of course, the results will fall short of what we have supposed. But if we reflect that, with the front of the column extended from ten to fifteen miles, it at once covered more than twice the previous ground—225 square miles instead of 100—and that this is still an extent which ordinarily admits of all taking part in a battle, we see that even under unfavorable circumstances this method of subsistence will still make possible an uninterrupted progress.

But as soon as a halt of several days takes place, great distress would necessarily ensue if other arrangements were not made. These arrangements consist of two kinds, and without them even now no considerable army can continue to exist. The first is that of providing the troops with a transport service, by means of which enough bread or flour, the most necessary part of their subsistence, can be carried along with them for some three or four days. If to this we add three or four days' rations which the soldier himself can carry, the indispensable minimum of subsistence is always assured for eight days.

The second arrangement is that of a regular commissariat, which spends every moment of the halt in collecting provisions from distant districts so that at any moment we can change over from the system of quartering on the individual inhabitants to another system.

Subsistence in quarters has the immense advantage that hardly any transport is required and that it is done in the shortest time, but, of course, it assumes that quarters can, as a rule, be provided for all the troops.

2. SUBSISTENCE THROUGH EXACTIONS ENFORCED BY THE TROOPS THEMSELVES

If a single battalion occupies a camp, this camp may in any case be placed in the vicinity of some villages, and these can be ordered to furnish subsistence. The method of subsistence would then not differ essentially from that just described. If, however, as is usual, the body of troops which is to encamp at one point is much greater, the only thing is for a larger unit, a brigade, for instance, or a division, to levy on definite districts a joint requisition, and then divide the proceeds.

The first glance shows that by this procedure the subsistence of a large army can never be provided. The yield from the stores of the country will be much less than if the troops had taken up their quarters in the same district. When thirty or forty men take possession of a farmer's house, they will be able, if necessary, to exact the last mouthful. But an officer sent with one or two men to exact food has neither time nor means to hunt out all stores, and often, too, he lacks the means of transport. So he will only be able to collect a small fraction of what is there. Besides, in camps the troops are crowded together in such a way at one point that the range of country from which provisions can be collected in a hurry is not of sufficient extent to furnish all that is required. What could be done in the way of supplying 30,000 men within a circle of five miles in diameter, that is, from an area of fifteen or twenty square miles? Moreover it would seldom be possible to collect even what there is, for most of the nearest adjacent villages would be occupied by separate bodies of troops, who refuse to let anything go. Lastly, by such a measure there would be the greatest waste, because some men would get more than their share and a great deal would be lost without benefit to anyone.

The result is, therefore, that the subsistence of troops by exactions of this kind can only be successfully accomplished when the numbers are not too large, not exceeding a division of 8,000 or 10,000 men, and even then it will only be accepted as a necessary evil.

It cannot usually be avoided in the case of troops directly in front of the enemy, such as advance guards and outposts, when the army is advancing, because these bodies must arrive at points where no preparations whatever could have been made beforehand, and they are usually too far from the stores collected for the rest of the army; further, in the case of mobile columns which are left to themselves, and lastly, in all cases where by chance there was neither time nor means to procure subsistence in any other way.

The more the troops are organized for regular requisitions, and the greater the extent to which time and circumstances permit the adoption of this way of obtaining subsistence, the better will be the result. But usually time is lacking, for what the troops get for themselves directly comes to them much more quickly.

3. BY REGULAR REQUISITIONS

This is unquestionably the s^implest and most effective means of subsisting troops and it has also been the basis of all modern wars.

It differs from the preceding way chiefly by the co-operation of the local authorities. The supply in this case must not be carried off forcibly just from the spot where it is found, but must be regularly delivered in accordance with a reasonable apportionment. This apportionment can only be made by the local authorities.

In this, all depends on time. The more time there is, the more general can the apportionment be made, the less oppressive it will be, and the more regular will be the result. Even purchases may be made for cash to assist, and in that way this method of providing subsistence approaches the next to be considered. In all assemblages of troops in their own country there is no difficulty in subsisting by regular requisitions; neither, as a rule, is there any in movements to the rear. On the other hand, in all movements into a country of which we are not yet in possession, there is very little time for such arrangements, seldom more than the one day by which the advance guard usually precedes the army. By means of the advance guard, orders are then sent to the local authorities, specifying how many portions and rations, in this place and that, they are to have ready. As these can only be furnished from the immediate neighborhood, that is, within a circuit of a few miles round each point, the collections so made in haste would never be nearly sufficient for a considerable army if the army did not bring with it a sufficient supply for several days. It is therefore the duty of the commissariat to economize what is received, and only to issue provisions to those troops which have nothing. With each succeeding day, however, the embarrassment will diminish; that is to say, if the distances from which provisions can be procured increase in proportion to the number of days, the area to be drawn upon, and consequently the result, increases as the squares of these distances. If on the first day only twenty square miles have been able to deliver provisions, on the next, we shall have eighty, on the third, one hundred and eighty; on the second day, therefore, twelve more than on the first, on the third, twenty more than on the second.

This, of course, is only an indication of what may take place, subject to many restricting circumstances which may intervene, of which the principal is that the district out of which the army has just come cannot contribute to the same extent as the others. But on the other hand, we must also remember that the radius within which we can levy may increase more than ten miles a day in width, perhaps fifteen or twenty, or in many places still more.

The actual delivery of these requisitions is enforced, at all events for the most part, by the executive power of individual detachments assigned to the local officials, but still more by the fear of being held responsible and punished and ill-treated, which in such cases usually weighs heavily on the whole population like a general oppression.

We cannot, however, propose to enter into details—into the whole machinery of commissariat and army subsistence. We have only the result in view.

This result, derived from a common-sense view of the general circumstances and confirmed by the experiences of the wars since the French Revolution, is that even the largest army, if it carries with it provisions for a few days, may without hesitation be subsisted by such levies, which only begin at the moment of its arrival and affect at first only the districts in the immediate vicinity, but afterwards, as time goes on, are extended over an ever-widening circle and administered by ever higher authorities.

This resource has no limits except those of the exhaustion, impoverishment and devastation of the country. When the stay of an invading army is of some duration, the administration of this system is at last handed over to the highest authorities. These, naturally, do all they can to equalize its pressure as much as possible and to lighten the burden of the delivery demanded. At the same time, even the invader, when his stay is prolonged in the enemy's country, is usually not so barbarous and reckless as to lay upon that country the entire burden of his support. Thus the system of requisitions of itself gradually approaches that of magazines, at the same time without ever entirely ceasing or noticeably modifying the influence it has upon the operations of war. For there is a wide difference between a case in which some of the resources which have been drawn from a district are replaced by supplies brought from more distant parts (the district, however, still remaining substantially the source on which the army depends for its supplies), and the case of an army which—as in the eighteenth century—provides for all its wants from its own resources, the country in which it is operating contributing, as a rule, nothing toward its support.

The chief difference consists in two things, the employment of the transport system of the country and the employment of its bakeries. In this way that enormous army baggage train, which is almost always destroying its own work, is got rid of.

It is true that even now no army can do entirely without a commissariat train, but it is immensely reduced and serves only to carry over, so to speak, the surplus of one day to the next. Peculiar circumstances, as in Russia in 1812, have even in more recent times compelled an army to take with it an enormous baggage train, and also field bakeries. But, in the first place, these are exceptional cases, for how seldom will it happen that 300,000 men advance six hundred and fifty miles upon almost a single road, and that through countries such as Poland and Russia, shortly before the season of harvest! In the second place, even in such cases any means of supply attached to an army will be regarded as only an assistance in case of need, the contributions of the country being consequently always looked upon as the basis of the whole system of supply.

Since the first campaigns of the war of the French Revolution the requisition system has constantly formed such a basis. The armies opposed to them were also obliged to adopt the same system, and it is not at all likely that it will ever be abandoned. No other yields such results, alike in energy for the prosecution of the war and in ease and flexibility of working. Whatever direction an army takes, it usually for the first three or four weeks has no difficulty about subsistence, and later on it can be helped out with magazines, so that we may very well say that by this method war has acquired the most perfect freedom of movement. Difficulties, certainly, may be greater in one direction than in another and that may carry some weight in preliminary deliberation, but we shall never come up against an absolute impossibility nor will the attention given to subsistence ever dictate our decision.

To this there is only one exception, which is a retreat through the enemy's country. In such a case we encounter a combination of many conditions unfavorable to subsistence. We have to keep continuously moving, generally, indeed, without any halt worth speaking of, so there is no time to collect provisions. The circumstances under which the retreat is begun are in most cases highly unfavorable, so that we have to keep together, concentrated in one mass. There can, usually, therefore, be no question of dispersing into quarters or of any considerable extension in the width of the columns, and the hostile attitude of the country does not permit collection of provisions by mere requisition not backed by executive force. And, lastly, the moment is further in itself especially suited to bring out the resistance and ill-will of the inhabitants. On account of all this, an army in such cases is as a rule confined to its previously prepared lines of communication and retreat.

When Bonaparte had to retreat in 1812, it was impossible for him to do so by any other line but the one by which he had advanced, on account of the subsistence of his army. If he had attempted any other he would only have plunged into more speedy and certain destruction. All the censure, therefore, passed on him even by French writers with regard to this point is foolish to the last degree.

4. SUBSISTENCE FROM MAGAZINES

If we were to make a generic distinction between this method of subsisting troops and the preceding, we could only do so in the case of an organization such as existed during the last thirty years of the seventeenth century and during the eighteenth. Can this organization ever return?

It is certainly hardly conceivable how it could have been otherwise when we think of war with great armies condemned to remain on one spot for seven, ten or twelve years, as happened in the Netherlands, on the Rhine, in Upper Italy, Silesia and Saxony. For what country could continue so long to be the main source of supply of the armies of both sides, without being utterly ruined and thus gradually failing to do its service?

But here the question naturally arises: Shall the war determine the nature of the system of subsistence, or the system of subsistence determine the nature of the war? To this we answer: At first the system of subsistence will determine the war in so far as the other conditions on which war depends permit. But when these other conditions begin to offer too much resistance, the war will react upon the system of subsistence and, in this case, therefore, determine it.

A war carried on by means of requisitions and local supplies has such an advantage over one that depends on subsistence from magazines that the latter no longer seems to be at all the same thing. No state, therefore, will venture to encounter the former with the latter. If anywhere there were a war minister, so narrow-minded and so ignorant as to fail to recognize the universal necessity of these circumstances, and to send out an army at the opening of the war to be subsisted in the old way, the force of circumstances would soon carry the commander away with it in its course and the system of requisitions would break out spontaneously. If we further reflect that the great expense which a system of magazines entails must necessarily reduce the expenditure on armament and the strength of the armed forces—no state being too well off for money—almost the only remaining possibility of re-establishing such a system lies in a diplomatic agreement for the purpose between the belligerent parties, a case which must be regarded as a mere play of fancy.

Wars, therefore, may be expected henceforward always to begin with the requisition system. How much one or another government will do to supplement this by an artificial organization to spare their own country, etc., remains to be seen. That it will not be overmuch we may be certain, for at such moments the tendency is to look to the most urgent wants, and an artificial system of subsisting troops is no longer included in that category.

But now, if a war is not so decisive in its results and its operations are not so extensive as its nature properly demands, the requisition system will begin to exhaust the district subject to it to such a degree that either peace must be concluded or means found to lighten the burden on the country and to become more independent of it for the subsistence of the army. The latter was the case of the French army under Bonaparte in Spain, but the first happens much more frequently. In most wars the exhaustion of the states increases to such a degree that, instead of thinking of prosecuting the war at a still greater expense, the necessity for peace becomes so urgent as to be imperative. Thus from this point of view too the modern method of carrying on war will make wars shorter.

We do not want, however, positively to deny the possibility of wars conducted on the old system of subsistence. When the relations between the two sides are such as to urge them to it and other circumstances favorable to it arise, it will perhaps once more show itself. Only we can never find in this type of system a natural organism. It is rather a mere abnormality, permitted by circumstances but which can never spring from the essential nature of war. Still less can we consider it as an improvement in war on the ground of its being more humane, for war itself is not a humane thing.

Whatever method of providing subsistence may be chosen, it will naturally be more easily carried out in rich and populous districts than in poor and thinly populated ones. That the population also must be taken into account in this connection is explained by the double relation it bears to the amount of the stores available in the country. In the first place, where much is consumed, much must also be in store, and in the second place, where the poulation is greater, the greater also is the production. To this, it is true, such districts as are inhabited by industrial workers form an exception, especially when, as is often the case, they consist of mountain valleys surrounded by barren land. But generally speaking it is always much easier to provide for the needs of an army in a well-populated, than in a poorly populated, country. An army of 100,000 men cannot be as well supported on 10,000 square miles inhabited by 400,000 people, however fertile the soil may be, as it would be on 10,000 square miles inhabited by

2,000,000 people. Besides, the roads and waterways are much better and more numerous in rich countries, the means of transport more abundant, the trade communications easier and safer. In a word, there is infinitely less difficulty in supporting an army in Flanders than in Poland.

The consequence is that war with its manifold tentacles settles by preference on high roads, on populous cities, on the fertile valleys of great rivers or along the coasts of well-frequented seas.

This shows clearly how the subsistence of troops may have a general influence upon the direction and form of military undertakings, and upon the choice of a theater of war and lines of communication.

How far this influence may extend and what weight in our calculations may be attached to the difficulty or ease of subsistence, depends, of course, very much on the manner in which the war is to be carried on. If war is carried on in its own proper spirit, that is with the unbridled force which is its essence, with urgent striving and craving for combat and decision, the subsistence of the troops is an important, but a subordinate, problem. But if a state of equilibrium ensues, in which the armies go marching up and down for many years in the same province, then subsistence often becomes the principal thing. The quartermaster becomes the commander-in-chief, and the conduct of the war becomes a management of transport wagons.

There are numberless campaigns in which nothing happened. The object was not achieved, the forces were used to no purpose, and the excuse given for it all is lack of subsistence. On the other hand Bonaparte used to say: *"Qu'on ne me parle pas des vivres!"* ("Don't talk to me of subsistence!")

In the Russian campaign that general certainly made it clear that such recklessness may be carried too far. This is not to say that perhaps his whole campaign came to grief through that cause alone, which after all would still remain conjectural, yet it is beyond doubt that to his lack of regard for the subsistence of his troops was due the unprecedented melting away of his army on his advance and its utter ruin on his retreat.

But while fully recognizing in Bonaparte the eager gambler who ventures on many a mad extreme, we may justly say that he and the Revolutionary generals who preceded him dispelled a powerful prejudice with regard to the subsistence of troops and showed that this must never be looked upon in any other light than as a means of war, never as an end.

Besides, it is with privation in war just as with physical effort and danger. The demands which the general can make on his army are without any defined limit. A man of strong character demands more than does a feeble sentimentalist. The behavior of the army varies also according as the soldier's will and energies are sustained by habit, military spirit, confidence in the commander and affection for him, or by enthusiasm for the national cause. But we should be able to make it an established principle that privation and want, to whatever pitch they may be carried, are always regarded as transitory states and that they must lead to an abundance of subsistence and indeed even, on occasion, to a superfluity.

Can there be anything more touching than the thought of so many thousand soldiers, badly clothed, with packs on their backs weighing thirty or forty pounds, marching wearily all day long in every kind of weather and on every kind of road, their life and health continually at stake, and for all this never once being able to get so much as a full meal of dry bread. Anyone who knows how often this happens in war is at a loss to understand how it does not oftener lead to a failure of their will and strength, and how a mere direction given to men's thoughts is able by its lasting effect to call forth and maintain such efforts.

Whoever, therefore, imposes great privations on the soldier because great objects demand it, will also, whether it be from sympathy or from prudence, bear in mind the compensation which he owes him for them in days to come.

We have now still to consider the difference between subsistence in attack and subsistence in defense.

The defender is in a position, while his defense continues, to make uninterrupted use of the subsistence which he has been able to lay in beforehand. He cannot, therefore, well be in want of the necessaries of life, especially in his own country; but even in the enemy's country this still holds good. The assailant, on the other hand, is moving away from his sources of supply, and as long as he is advancing, and even during the first weeks after he stops, must procure from day to day what he requires. This can rarely be done without want and inconvenience.

This difficulty usually reaches a maximum at two periods. The first is in the advance, before the decision has taken place. At that time, while the defender still has all his own stores in his hands, the assailant has been obliged to leave his behind him. He is obliged to keep his masses concentrated and cannot, therefore, spread his army over any considerable space. Not even a transport train has been able any longer to follow him as soon as the movements of the battle have begun. If his preparations have not been very well made, it may easily happen at this moment that his troops are in need and want of supplies for some days before the decisive battle, which is certainly not the way to bring them into the battle in good condition.

The second period at which want is mainly felt is at the end of a victorious campaign, when the lines of communication begin to become too long, especially if the war is being carried on in a poor, thinly populated country, where perhaps the feeling too is hostile. What an enormous difference between a line of communication from Vilna to Moscow, where every vehicle has to be procured by force, and that from Cologne by Liége, Louvain, Brussels, Mons and Valenciennes to Paris, on which a merchant's order or a bill of exchange suffices to procure millions of rations!

This difficulty has often resulted in obscuring the splendor of the most brilliant victories, in wasting away the forces of the victor and in making necessary a retreat which then by degrees takes on all the symptoms of a real defeat.

Forage, of which, as we have said before, there is usually at first the least deficiency, will run short soonest if a country begins to be exhausted, for on account of its bulk it is most difficult to procure from a distance, and a horse feels the effect of low feeding much sooner than does a man. For this reason an over-numerous cavalry and artillery may become a real burden, and an actual source of weakness to an army.

CHAPTER 15

BASE OF OPERATIONS

If an army sets out from its point of origin on any expedition, whether it be to attack the enemy and his theater of war or to take post on the frontiers of its own, it remains in a state of necessary dependence on its sources of subsistence and material and must maintain its communications with them, as they are the conditions of its existence and maintenance. This dependence increases in intensity and extent with the size of the army. But it is not always either possible or necessary that the army should remain in direct communication with the whole of its own country, but only with that portion of it which is directly in its rear and consequently covered by its position. In this portion of the country, as far as is necessary, special depots of provisions are established and arrangements made for regularly forwarding reinforcements and supplies. This piece of the country is therefore the base of the army and all its undertakings, and must be regarded as forming with the army a single whole. If the supplies, for their greater security, are stored in fortified places, the idea of a base becomes more distinct; but it does not originate from this, and, in a number of cases, no such arrangement is made.

A portion of the enemy's country may also become a base for our army, or, at least, form part of it, for when an army penetrates into an enemy's country, a number of its needs are supplied from the part occupied. But it is then a necessary condition that we are really masters of this tract of country, certain, that is to say, of our orders being obeyed there. This certainty, however, seldom extends beyond the reach of our ability to keep the inhabitants in awe by small garrisons and by detachments moving about from place to place, and that usually does not mean very much. The consequence is that in the enemy's country, the district from which we draw supplies of all kinds is very limited compared with the army's needs and in most cases inadequate. Our own country, therefore, must give us much, and consequently that part of it which lies in the rear of the army must once more be regarded as a necessary part of our base.

The wants of an army must be divided into two classes: those which every cultivated country can furnish, and those which can only be obtained from the places where they are produced. The first are chiefly means of subsistence, the second, means of outfit and upkeep. The former can be obtained from the enemy's country as well as from our own; the latter—men, for instance, arms, and usually also munitions—as a rule, only from our own. Although there are exceptions to this classification in certain cases, still they are few and trifling, and the distinction we have drawn is of great and permanent importance and proves again that the communication with our own country is indispensable.

Depots of provisions and forage are generally formed in open cities, both in the enemy's country and in our own. There are not as many fortresses as would be required to contain these much bulkier stores, which are quickly consumed and wanted now here and now there, and the loss of them is easier to replace. Stores of outfit and upkeep, on the other hand—of such things as arms, munitions and equipment—in the neighborhood of the theater of war are hardly ever stored in open cities, but are rather brought in from greater distances, while in the enemy's country they are never stored anywhere but in fortresses. From this point again it may be inferred that the base is of more importance with regard to the means of outfit and upkeep than with regard to the means of subsistence.

Now, the more that means of both kinds are collected together in great magazines before being brought into use, and the more, therefore, that all separate streams unite in great reservoirs, so much the more may these magazines be regarded as taking the place of the whole country, and so much the more will the conception of a base connect itself with these great depots of supply. But it can never go so far that any such places could in themselves be taken to constitute a base.

If these sources of supply of both kinds are abundant, that is, if the tracts of territory are wide and rich, if the stores are collected in larger depots so as to be more quickly available, if these depots are protected in one way or another, if they are in close proximity to the army and accessible by good roads, if they extend along a considerable width in the rear of the army or even surround it in part as well—the result is a greater vitality for the army and also a greater freedom of movement. Attempts have been made to sum up all the advantages which an army derives from being so situated in one single conception—the magnitude of the base of operations. By the relation which this base bears to the object of the undertakings, by the angle which its extremities make with this object (supposed as a point), it has been attempted to express the whole sum of the advantages and disadvantages which accrue to an army from the position and nature of its sources of subsistence and material. But it is obvious that this pretty piece of geometry cannot be taken seriously, as it rests on a series of substitutions which have all had to be made at the expense of truth. The base of an army, as we have seen, forms a structure in which three elements are blended, and in which the army is placed. The three elements are (1) the resources of the district, (2) the depots for stores established at various points, and (3) the *province* out of which these stores are collected. These three things are separated in space, they do not admit of being reduced to one whole and, least of all, of being represented by a line which is to indicate the breadth of the base and which is usually, quite arbitrarily, drawn either from one fortress to another or from one provincial capital to another, or along the political frontier of the country. Furthermore a definite relation cannot be established between these three elements, for in reality they always more or less blend into one another. In one case the surrounding district affords many kinds of military equipment which otherwise are usually only imported from a great distance; in another, even the means of subsistence have to be brought from afar. Sometimes, the nearest fortresses are great arsenals, ports or commercial cities in which the military resources of a whole state are all to be found together; sometimes, they are nothing but wretched ramparts, hardly sufficient for their own defense.

The consequence is that all deductions from the dimensions of the base of operations and of the angle of operations, and the whole theory of war founded on these data, so far as it was geometrical in character, has never had the slightest attention paid to it in actual war, and in the world of ideas has only given rise to wrong tendencies. But as the basis of this train of reasoning is a truth, and only the conclusions are false, the same view will easily and frequently force its way to the front again.

We think, therefore, that we cannot go beyond acknowledging generally the influence of a base on military enterprises, that there is, however, no means of reducing this to one or two simple ideas to serve as rules for practice, but that in every individual case all the things we have specified must be kept in view *together*.

Once the arrangements for the subsisting and refitting of the army have been made in a certain district and for operating in a certain direction, then even in our own country only this district is to be regarded as the base of the army. As any change of it requires time and labor, an army cannot change its base from one day to another, even in its own country, and therefore it is also always more or less limited in the direction of its operations. If then, when operating in the enemy's country, we wished to regard the whole line of our own frontier on that side as the basis of our army, we might very well do so in a general sense, in so far as potentially arrangements might be made anywhere, but not at any particular moment, since arrangements have not been actually made everywhere. When the Russian army retreated before the French in 1812, at the beginning of the campaign, the whole of Russia might certainly have been considered as its base, the more so because the vast extent of the country offered the army abundance of space in any direction in which it chose to move. This is no illusory idea, as it was actually realized later, when other Russian armies moved against the French from several different sides. But still, at any given period of the campaign the base of the Russian army was not so extensive. It was chiefly confined to the roads on which the whole train of transport to and from the army had been organized. This limitation prevented the Russian army, for instance, from making the further retreat which became necessary after the three days' fighting at Smolensk in any direction but that of Moscow and so hindered their turning suddenly in the direction of Kaluga, as had been proposed, in order to draw the enemy away from Moscow. Such a change of direction would only have been possible if preparations for it had long before been made.

We have said that the dependence on the base increases in intensity and extent with the size of the army, a thing which is obvious. An army is like a tree. From the ground out of which it grows it draws its vital forces; if it is small, it can easily be transplanted, but this becomes more and more difficult as it increases in size. A small body of troops has also its channels from which it draws its life, but it strikes root easily where it happens to be; not so a large army. When, therefore, we talk of the influence of the base on the operations of an army, the dimensions of the army must always furnish the standard on which all our ideas are founded.

Further, it is natural that for the immediate wants of the present hour *subsistence* is the main point, but for the general condition of the army through a longer period of time, *refitment and recruitment* are more important, because the latter can only be provided from particular sources, while the former can be obtained in many ways. This again more distinctly defines the influence of the base on the operations of the army.

However great that influence may be, we must never forget that it belongs to those things which can only show a decisive effect after some considerable time, and that therefore the question always remains what may happen in that time. The value of a base of operations will, therefore, seldom determine the choice of an undertaking in the first instance. Mere difficulties which may arise in this respect must be set aside and compared with the other effective means. Obstacles of this kind often vanish before the force of decisive victories.

CHAPTER 16

LINES OF COMMUNICATION

The roads which lead from the position of an army to those points in its rear at which its sources of subsistence and refitment are chiefly concentrated, and which in all ordinary cases it chooses for its retreat, have a double purpose. In the first place they are its *lines of communication* for the constant sustenance of the forces and, next, they are *lines of retreat*.

We have said in the preceding chapter that, although according to the present system of subsistence, an army is chiefly fed from the district in which it is operating, it must still be looked upon as forming a whole with its base. The lines of communication belong to this whole; they form the connection between the army and its base, and are to be considered as so many great vital arteries. Supplies of every kind, munition transports, detachments moving backward and forward, posts, orderlies, hospitals, depots, reserves of munitions, administrative officials—all these are constantly making use of these roads, and the total value of these services is of the utmost importance to the army.

These life channels must therefore neither be permanently severed, nor must they be too long and difficult, because on a long road some strength is always lost and the result is a weakening of the army.

By their second purpose, that is, as lines of retreat they constitute in a real sense the strategic rear of the army.

For both purposes the value of these roads depends on their *length*, their *number*, their *situation*, that is, their general direction and their direction close to the army, their *quality* as roads, *difficulties of ground*, the *politi-*

cal relations and feeling of the local population, and, lastly, on the *protection* they derive from fortresses or natural obstacles of the country.

But not all the roads that lead from the point occupied by the army to the sources of its life and strength are necessarily its proper lines of communication. They may no doubt be used for that purpose, and may be considered as auxiliary to the communication system, but this system confines itself to the roads prepared for it. Only those roads on which magazines, hospitals, stations, posts for despatches and letters are established, commandants appointed and police and garrisons distributed, can be regarded as the real lines of communication. But here arises a very important difference, though one often overlooked, between an army in its own country and one in the enemy's. The army in its own country will, it is true, also have its prepared lines of communication, but it is not entirely confined to them and can in case of need abandon them and choose any other road, which is still at all available. For it is everywhere at home, has everywhere its own officials, and finds good will everywhere. Although, therefore, other roads are not as good and as suitable for its purposes, yet the choice of them is not impossible, and the army, if it should see itself turned and forced to change front, will not regard these roads as *impossible*.

An army in the enemy's country, on the other hand, can as a rule regard as lines of communication only those roads on which it has itself advanced, and here, in effect, a great difference arises from small and insignificant causes. An army advancing in the enemy's country makes the arrangements which are essential for its lines of communication as it advances and at the same time protects them. Owing to the fear and terror the presence of the army inspires, it can give these measures, in the eyes of the inhabitants, the stamp of unalterable necessity, and even cause them to be regarded as a mitigation of the general evils of war. Small garrisons left behind here and there support and maintain the whole system. But if, on the other hand, the commissaries, commandants of stations, police, field posts and other administrative services were to be sent on some remote road, on which the army had not passed, the inhabitants would look upon it all as a burden, from which they could very conveniently keep free. Unless the most decisive defeats and disasters have thrown the enemy country into a panic terror, these functionaries are treated in a hostile manner and are driven away with broken heads. Above everything else garrisons are therefore required to subjugate the new road, garrisons in this case no doubt stronger than usual. Nevertheless the danger will always remain of the inhabitants trying to resist them. In short, an army advancing in a hostile country lacks all the instruments of enforcing obedience. It must first set up its administrative officials, and that, too, by the authority of arms; and this cannot be done everywhere—not without sacrifices and difficulties, and not in a moment.

It follows from this that an army in an enemy country is much less able by changing its system of communications to switch over from one base to another than it is in its own country, where that is, at all events, possible. It follows, too, that in general it becomes more hampered in its movements and more sensitive about its lines of communication.

But the choice and organization of these lines is also from the very beginning subject to many conditions by which it is restricted. They must not only generally follow major roads, but they will in many respects be all the better, the wider these roads are, the more populous and flourishing are the cities they touch, and the more protected they are by strong places. Rivers also, as waterways, and bridges, as points for crossing, count for much in the choice. For these reasons the situation of the lines and, consequently, the route which an army takes in the attack are only up to a certain point the subject of a free choice, while the more precise situation depends on geographical circumstances.

All the above-mentioned things taken together determine the strength or weakness of an army's communication with its base, and the result, compared with that obtained in the same way for the army of the enemy, decides which of the two opponents is in a better position to sever the communications of the other or even his line of retreat, that is, to *turn* him. Apart from moral or physical superiority, only that party can accomplish this effectually whose lines are superior to those of his enemy, for otherwise his adversary saves himself in the shortest way by retaliating in kind.

Now this turning can, by reason of the double purpose of the roads, also have a double object. Either the communications may be interfered with or broken, in order that the army may begin to grow feeble and die away and thus be compelled to retreat, or the object may be to cut off the retreat itself.

With regard to the first object, it is to be observed that a momentary interruption, when armies are subsisted as they now are, seldom has any effect. For this, on the contrary, a certain time is necessary in order that the number of losses may offset what individually they lack in importance. A single flank attack which might have accomplished a decisive stroke at one time, when in the artificial subsistence system thousands of wagons were going back and forth, will now effect absolutely nothing, however successful it may be. For it could at most carry off one transport train, which would cause a partial weakness, but would not make any retreat necessary.

The consequence is that flank attacks, which have always been more the fashion in books than in real life, now seem more than ever unpractical, and it may be said that only very long communication lines in unfavorable circumstances, and above all their being exposed everywhere and at any moment to the attacks of an *insurgent population*, make them dangerous.

As regards the cutting off of the retreat, we must not even, in this respect, exaggerate the danger of hampered or threatened lines. Recent experience makes us aware that with good troops and resolute leaders it is *more difficult* to encircle them than it is for them to cut their way out.

The means of shortening and protecting long lines of communication are very limited. The seizure of some fortresses adjacent to the position taken up or on the roads leading to the rear—or in the event of there being no fortresses in the country, the strengthening of suitable points—kind treatment of the inhabitants, strict discipline on the military road, good police in the country, assiduous repair of the roads—these are the only means whereby the evil may be diminished, but certainly never quite removed.

Furthermore, what has been said in treating of the upkeep of the routes which armies preferably take must also be considered as applying particularly to lines of communication. The widest roads through the richest cities and the best cultivated provinces are the best lines of communication. Even when they make considerable detours they deserve the preference and in most cases enable us to fix in greater detail the disposition of the army.

CHAPTER 17

COUNTRY AND GROUND

Quite apart from their influence upon the means of subsistence, which form another aspect of the subject, country and ground bear a very close and ever-present relation to military action. They have a decisive influence upon the engagement both as regards its actual course and as regards the preparations for it and the use made of it. We now proceed to consider country and ground from this point of view, that is, in the full meaning of the French term *terrain*.

The ways in which their influence is exerted lie for the most part within the province of tactics, but the effects appear in strategy. An engagement in mountains is, in its consequences as well as in itself, quite a different thing from an engagement on a level plain.

But until we have distinguished attack from defense and given a closer consideration to both, we cannot consider the leading features of the terrain with reference to their effects, and must therefore confine ourselves here to their general character. There are three properties through which country and ground influence military activity, namely, as an obstacle to approach, as an obstacle to an extensive view and as protection against the effect of firearms. All other properties may be traced back to these three.

Unquestionably this threefold influence of ground has a tendency to make warfare more diversified, more complicated and more scientific, for they are obviously three additional quantities to be taken into account in our calculations.

A completely level, and completely open, plain, a ground which has thus no influence at all, does not exist in reality except in relation to very small bodies of troops and in relation to them only for the duration of a given moment. When larger bodies are concerned and a longer duration of time, the features of the ground begin to influence the action, and in the case of a whole army it is almost inconceivable that even for a single moment, such for instance as that of a battle, the ground should not have an influence upon it.

This influence, therefore, is practically always present, but it is certainly stronger or weaker according to the nature of the country.

If we keep in view the great mass of topographical phenomena, we shall find that it is chiefly in three ways that a tract of country departs from the idea of an open, unencumbered plain: first, by the formation of the ground, that is, hills and valleys; second, by the existence of woods, marshes and lakes as natural features; and last, by such changes as have been produced by the hand of man. In all three directions the influence of the ground on military operations increases. If we follow these directions a certain distance, we get mountain country, country that is little cultivated and covered with woods and marshes, and highly cultivated country. In all three cases, therefore, war thus becomes more complicated and more of an art.

The influence of cultivation is of course greater or less according to the kind of cultivation. It is strongest with the kind usual in Flanders, Holstein and other countries in which the land is intersected by many ditches, fences, hedges and dykes, interspersed with many single dwellings and small copses.

The conduct of war will thus be easiest in a country which is flat and moderately cultivated. This, however, only holds good in quite a general sense, and if we leave entirely out of consideration the use which the defensive makes of obstacles of ground.

Each of these three kinds of ground has its own effect on accessibility, facility of observation and cover.

In a thickly wooded country the obstacle to sight predominates; in a mountainous country, the obstacle to approach, and in highly cultivated districts both obstacles exist in a moderate degree.

A thickly wooded country renders a great part of the ground to a certain extent impracticable for military movements, because, apart from difficulties of approach, the entire lack of facilities for observation does not allow use to be made of all the ways of passing through it. This, on the one hand, again simplifies the conduct of operations, but, on the other, makes it much more difficult. If in such a country it is hardly practicable entirely to concentrate our forces for an engagement, yet on the other hand they have not to be split up into so many sections as are usual in mountains and in districts that are very much intersected. In other words, in such a country, the splitting up of our forces is less avoidable, but also less considerable.

In mountains the obstacles to approach predominate and take effect in two ways, either by making it impossible for us to get through except at certain points, or by forcing us where we can get through to move more slowly and with greater effort. On this account the elasticity of all movements is much diminished in mountains, and all operations take much more time. But a mountain ground, as opposed to all others, has the further peculiarity that one point commands another. We shall devote the following chapter to the discussion of higher ground generally, and shall here only remark that it is this peculiarity which causes the great splitting up of forces in mountainous country, for the points there are not only important in themselves, but also for the influence they have upon other points.

As we have elsewhere observed, all three kinds of country and ground, as they tend to an extreme, have the effect of weakening the influence of the supreme command upon the result in proportion as the energies of the subordinates, down to the common soldier, come out more strongly. The greater the splitting up of the forces, and the less possible it becomes to keep them under observation, so much the more, obviously, each leader is left to himself. Certainly as the action becomes more divided, more various, and more many-sided, the influence of intelligence will necessarily, on the whole, increase, and even the commander-in-chief will be able in such circumstances to show a greater ability than usual. But here again we must come back to what we have said before: that in war the sum of the single successes is more decisive than the form in which they are connected. If we push our present examination to the extreme limit and figure to ourselves a whole army extended in a long line of skirmishers, where each soldier is waging his own little battle, more depends upon the sum of the single victories than of the form of their connection. For the effectiveness of sound plans can only proceed from positive results, not from negative. It will therefore in this case be the courage, the skill and the spirit of the individual that more than anything else will prove decisive. It is only when two armies are equally matched, or the qualities of both are equally balanced, that the talent and judgment of the commander again become decisive. The consequence is that national armies, and insurgent levies, etc., etc., in which at all events the warlike spirit of the individual is usually high, though his skill and valor may not be exactly superior, are still able to maintain a superiority, when their forces are very widely dispersed and when they have the advantage of a ground that is very much intersected. But it is only on such ground that they hold out for any length of time, because forces of this kind are usually devoid of all the qualities and virtues which are indispensable when tolerably large numbers have to act as one united body.

Also in the nature of armed forces there are many gradations between one of these extremes and the other, for the very circumstance of being engaged in the defense of its own country gives to even a regular standing army something of the character of a national army, and makes it more suited for acting in small detachments.

Now the more these qualities and circumstances are wanting in an army and the greater they are on the side of its opponent, so much the more will it dread being split up into small detachments and the more it will avoid broken country. But to avoid such country is seldom a matter of choice. We cannot choose a theater of war like a piece of merchandise from among several patterns, and thus we find generally that troops which naturally fight with advantage in concentrated masses exhaust all their ingenuity in trying to carry out this system, as far as possible, *in direct opposition to the nature of the country.* They must in consequence submit to other disadvantages, such as scanty and difficult subsistence, bad quarters, and numerous attacks from all sides in an engagement. But the penalty of giving up their own special advantage would be much greater.

These two opposite tendencies, the one to concentration, the other to dispersion of forces, prevail more or less according as the nature of the troops engaged is inclined more to the one side or the other. But however decided the tendency, the one side cannot always remain with its forces concentrated, neither can the other expect success solely from the efficiency it gets from dispersion. Even the French were obliged in Spain to divide their forces, and even the Spaniards in the defense of their country by means of a national rising were obliged to hazard part of their forces on great battlefields.

Next to the connection which country and ground have with the general, and especially with the political, composition of the armed forces, the most important is that which they have with the relative proportions of the three arms in the forces.

In all countries which are difficult to traverse, whether the obstacles are mountains, forests or cultivation, a numerous cavalry is useless. That is obvious. It is just the same with artillery in a wooded country; there may easily be no room to use it to full advantage, no roads to bring it up and no forage for the horses. For this arm, highly cultivated countries are less disadvantageous, and least of all mountains. Both, it is true, afford cover against its fire and in that respect they are unfavorable to an arm which is pre-eminently effective through its fire. Both also furnish means for the enemy's infantry frequently to put the unwieldy artillery in jeopardy, as infantry can pass anywhere. In neither of them, however, is there in general any want of space for the use of an numerous artillery, and in mountainous countries it has the great advantage that the slower movements of the enemy double its efficiency.

But it is undeniable that on any difficult ground infantry has a decided advantage over all other arms, and that on such ground its number may considerably exceed the usual proportion.

CHAPTER 18

HIGHER GROUND

The word "command" in the art of war has a peculiar charm, and, in fact, to this element belongs a very large part, perhaps the larger half, of the influences which ground exercises on the use of armed forces. Here many of the sacred relics of military erudition have their root, as for instance, commanding positions, key positions, strategic maneuvers, and so forth. We shall examine the subject as searchingly as we can without prolixity, and pass in review the true and the false, reality and exaggeration.

Every exertion of physical force if made upward is more difficult than if made in the opposite direction; consequently it must be so in an engagement, and there are three obvious reasons why it is so. First, every height may be regarded as an obstacle to approach; second, although the range is not perceptibly greater in shooting down from a height, yet, all geometrical relations being taken into account, we have a distinctly *better* chance of hitting the mark than in the opposite case; and, third, we have the advantage of a better general view. How all these advantages unite in an engagement does not concern us here. We combine into one whole the sum of the advantages which tactics derives from an elevated position and regard it as the first strategical advantage.

But the first and last of the advantages just enumerated must appear once more as advantages in strategy itself, for we march and reconnoiter in strategy as well as in tactics. If, therefore, a higher position is an obstacle to the approach of those on the lower ground, this is the second strategic advantage; and the better command of view which this higher position affords is the third advantage which strategy may derive from it.

Of these elements is composed the power of dominating, overlooking, commanding. From these sources springs the feeling of superiority and security in him who stands on a mountain ridge and looks down on his enemy below, and the feeling of weakness and anxiety in him who is below. Perhaps the total impression made is even stronger than it ought to be, because the advantage of higher ground strikes the senses more than do the circumstances which modify that advantage. Perhaps the impression made surpasses that which the truth warrants, in which case this effect of imagination must be regarded as a new element which strengthens the effect produced by being on higher ground.

The advantage of great facility of movement, of course, is not absolute, and not always in favor of the side occupying the higher position; it is only so when his opponent wishes to attack him; it is not so if the two opponents are separated by a great valley, and it is actually in favor of the army on the lower ground if both wish to fight in the plain (battle of Hohenfriedberg). The command of view, also, has likewise great limitations. A wooded country in the valley below, and often the very masses of the mountains themselves on which we stand obstruct the vision. Countless are the cases in which we might seek in vain on the spot for those advantages of a position on higher ground which we have chosen on a map, and we might often be led to think we had only involved ourselves in all the disadvantages which are their opposites. But these limitations and conditions do not destroy the superiority which the occupant of the higher position possesses alike for defense and for attack. We shall point out in a few words how this is the case with each.

Of the three strategic advantages of higher ground, *the greater tactical strength, the more difficult approach* and *the better view,* the first two are of such a nature that they belong really to the defensive only. For it is only in holding firmly to a position that we can make use of them, while the assailant in moving cannot take them with him. But the third advantage can be made use of by the offensive just as well as by the defensive.

From this results the importance of higher ground to the defensive, and as it can only be gained in a decisive way in mountainous countries, it would seem to follow, as a consequence, that the defensive has an important advantage in mountain positions. How it is that, through other circumstances, this is not so in reality, we shall show in the chapters (Book VI, Chapters 15, 16, 17) on the defense of mountains.

We must first of all make a distinction. If it is merely a question of higher ground at a single point, a position, for instance, the strategic advantages rather merge in the single tactical one of an advantageous position. But if, on the other hand, we are thinking of an important tract of country, a whole province, for instance, as a regular slope, like the declivity of a general watershed, so that we can march for several days and still remain on higher ground than the country before us, the strategic advantages become greater. We can now use the advantage of the higher ground not only in the combination of our forces in a single engagement, but also in the combination of several engagements with one another. Thus it is with the defensive.

As regards the offensive, it enjoys to a certain extent the same advantages from higher ground as the defensive, because the strategic attack does not, like the tactical, consist of a single act. The advance of the strategic attack is not the continuous movement of a machine; it is made in single marches with longer or shorter pauses between, and at each halting point the assailant is just as much on the defensive as his opponent.

Through the advantage of a better view, a position on higher ground confers on the offensive and the defensive alike a certain kind of efficiency for action which we must not omit to notice. It is the facility for acting with separate masses. For each portion of a force separately derives from higher ground the same advantages as the whole. Consequently, a separate corps, be it strong or weak in numbers, is stronger than it would otherwise be, and we can venture to take up a position with less danger than we could if we were not in a commanding position. The advantages which are to be derived from such separate bodies of troops is a subject for another place.

If higher ground is combined with other geographical advantages in our favor, if the enemy finds himself cramped in his movements from other causes, as, for instance, by the proximity of a large river, such disadvantages of his position may become quite decisive, so that he cannot soon enough extricate himself from them. No army can maintain itself in the valley of a great river if it is not in possession of the heights by which the valley is formed.

The possession of higher ground may thus become actual command, and we can by no means deny that this idea represents a reality. But nevertheless the expressions, "commanding ground," "covering position," "key of the country," and so forth, in so far as they are founded on the nature of higher ground and lower ground, are usually hollow shells without any sound kernel. These imposing elements of theory have been chiefly resorted to in order to give a flavor to the seeming commonplaceness of military combinations. They have become the darling themes of learned soldiers, the magic wands of strategical adepts, and neither the emptiness of these fanciful conceits nor their contradiction by experience has sufficed to convince authors and readers that they were here drawing water in the leaky vessel of the Danaids. The conditions have been mistaken for the thing itself, the instrument for the hand. The occupation of such and such a position or tract of country has been looked upon as an exercise of power like a thrust or a blow, and the position or tract of country as in itself a real quantity. Whereas the former, like the lifting of an arm, is nothing, and the latter nothing but a dead instrument, a mere property which can only attain reality in an object, a mere sign of plus or minus prefixed to quantities that are still lacking. This thrust and blow, this object, this quantity, is a *victorious engagement*. That alone really counts, that alone can we reckon with and that we must keep constantly in view, alike in passing judgment in books and in taking action in the field.

Consequently, if the number and importance of victorious engagements alone are what decide, it is clear that the relation of the two armies and their leaders to one another again becomes the first object for consideration and that the part played by the influence of the ground can only be a subordinate one.

Book VI

DEFENSE

CHAPTER 1

OFFENSE AND DEFENSE

1. CONCEPTION OF DEFENSE

What is the conception of defense? The warding off of a blow. What then is its characteristic sign? The awaiting of this blow. This is the sign which makes any act a defensive one, and by this sign alone can defense be distinguished from attack in war. But inasmuch as an absolute defense completely contradicts the conception of war, because there would then be war carried on by one side only, it follows that defense in war can only be relative, and the above characteristic sign must therefore only be applied to the conception as a whole; it must not be extended to all parts of it. A partial engagement is defensive if we await the onset, the charge of the enemy; a battle is so if we await the attack, that is, the appearance of the enemy before our position and within range of our fire; a campaign is defensive if we await the entry of the enemy into our theater of war. In all these cases the sign of awaiting and warding off belongs to the general conception, without any contradiction arising with the conception of war, for it may be to our advantage to await the charge against our bayonets, or the attack on our position and our theater of war. But as we must return the enemy's blows if we are really to carry on war on our side, therefore this offensive action in defensive war falls in a certain sense under the heading of defense—that is to say, the offensive of which we make use falls under the conception of position, or theater of war. We can, therefore, in a defensive campaign fight offensively, in a defensive battle we may use some divisions for offensive purposes, and lastly, while simply remaining in position awaiting the enemy's onslaught, we still send offensive bullets into his ranks to meet him. The defensive form in war is therefore not a mere shield, but a shield formed of skilfully delivered blows.

2. ADVANTAGES OF THE DEFENSIVE

What is the object of defense? *To preserve.* To preserve is easier than to gain; from which it follows at once that the means on both sides being supposed equal, defense is easier than attack. But wherein does the greater ease of preservation and protection lie? In this, that all time which elapses unused falls into the scale in favor of the defender. He reaps where he has not sown. Every intermission of the attack, either from erroneous views, from fear or from indolence, is in favor of the defender. This advantage saved the state of Prussia from ruin more than once in the Seven Years' War. This advantage, which is derived from the conception and object of the defensive, lies in the nature of all defense, and in other spheres of life, too, particularly in legal business, which bears so much resemblance to war, it is expressed by the Latin proverb, *beati sunt possidentes.* Another advantage arising only from the nature of war is the assistance afforded by the lie of the land, and of this the defense has the preferential use.

Having established these general conceptions we now turn more directly to the subject.

In tactics every engagement, great or small, is a *defensive one* if we leave the initiative to the enemy, and await his appearance on our front. From that moment on we can make use of all offensive means without losing the two advantages of defense mentioned above, namely, that of awaiting and that of the terrain. In strategy, first of all, the campaign takes the place of the battle, and the theater of war that of the position; then, later, the whole war takes the place of the campaign, and the whole country that of the theater of war, and in both cases the defensive remains what it was in tactics.

It has been observed before in a general way that defense is easier than attack. But since the defensive has a negative object, that of *preserving,* and the offensive a positive object, that of *conquering,* and since the latter increases our own war resources, but the preserving does not, we must, in order to express ourselves distinctly, say that the *defensive form of war is in the abstract stronger than the offensive.* This is the result we have been aiming at; for although it is absolutely natural and has been confirmed by experience a thousand times, it is still entirely contrary to prevalent opinion—a proof of how ideas may be confused by superficial writers.

If the defensive is the stronger form of conducting war, but has a negative object, it is self-evident that we must make use of it only as long as our weakness compels us to do so, and that we must give up that form as soon as we feel strong enough to aim at the positive object. Now as our relative strength is usually improved if we gain a victory through the assistance of the defensive, it is therefore, also, the natural course in war to begin with the defensive, and to end with the offensive. It is therefore just as much in contradiction with the conception of war to suppose the defensive the ultimate object of the war as it was a contradiction to understand passivity to belong not only to the defensive as a whole, but also to all the parts of the defensive. In other words: a war in which victories are merely used to ward off blows, and where there is no attempt to return the blows, would be just as absurd as a battle in which the most absolute defense (passivity) should prevail in all measures.

Against the correctness of this general view many examples might be cited of wars in which the defensive continued to be defensive to the last, and an offensive reaction was never contemplated; but such an objection could only be made if we lost sight of the fact that here it is a question of a general conception, and that the examples which one might oppose to this general conception are all to be regarded as cases in which the possibility of offensive reaction had not yet arrived.

For example, in the Seven Years' War, at least in the last three years of it, Frederick the Great did not think of an offensive. Indeed we even believe that he really regarded the offensive in this war only as a better means of defending himself; his whole situation compelled him to this course, and it is natural that a general should have only that in sight which is immediately connected with his situation. Nevertheless, we cannot look at this example of a defense on a great scale without supposing that the idea of a possible offensive reaction against Austria lay at the bottom of it all, and without saying to ourselves that the moment for that offensive reaction had merely not yet arrived. The conclusion of peace shows that this idea is not without foundation even in this instance; for what could have induced the Austrians to make peace except the thought that they were not in a condition to make head against the talent of the king with their own forces alone; that in any case their efforts had to be even greater than heretofore and that the slightest relaxation of them would probably lead to fresh losses of territory. And, in fact, who can doubt that Frederick the Great would have tried to conquer the Austrians again in Bohemia and Moravia, if Russia, Sweden and the army of the Holy Roman Empire had not diverted his forces?

Having thus defined the conception of the defensive in its true meaning and laid down the limits of defense, we return once more to the assertion that the defensive *is the stronger form of making war.*

Upon a closer examination and comparison of the offensive and defensive, this will appear perfectly plain. But for the present we shall confine ourselves to noticing the contradiction in which the opposite view would stand with itself and with the results of experience. If the offensive form were the stronger, there would be no occasion ever to use the defensive. As it has in any case merely a negative object, everyone would necessarily want to attack, and the defensive would be an absurdity. On the other hand, it is very natural that the higher object should be purchased by greater sacrifices. Whoever feels himself strong enough to make use of the weaker form may aim at the greater object; whoever sets before himself the lesser object can only do so in order to have the benefit of the stronger form. If we look to experience, it would probably be something unheard of if in the case of two theaters of war the offensive were taken with the weaker army, and the stronger army were left for the defensive. But if everywhere and at all times the reverse of this has taken place, it indicates plainly that generals, although their own inclination prompts them to the offensive, still hold the defensive to be the stronger form. In the next chapters we must explain some further preliminary points.

CHAPTER 2

THE RELATIONS OF THE OFFENSIVE AND DEFENSIVE TO EACH OTHER IN TACTICS

First of all we must inquire into the circumstances which lead to victory in an engagement.

Of superiority in numbers, and bravery, discipline or other qualities of an army, we say nothing here, because, as a rule, they depend on things which lie outside the province of the art of war in the sense in which we are now considering it. Besides, they would exercise the same effect in the offensive and in the defensive. Even the superiority *in numbers in general* cannot come into consideration here, as the number of troops is likewise a given quantity, and does not depend on the will of the general. These things have, furthermore, no particular connection with attack and defense. But, apart from these, there are only three other things which appear to us of decisive importance, namely, *surprise, advantages of ground* and *attack from several sides*. The surprise produces an effect by opposing to the enemy at some particular point a great many more troops than he expected. The superiority in numbers in this case is very different from the general superiority of numbers; it is the most powerful agent in the art of war. The way in which the advantage of ground contributes to the victory is in itself quite understandable, and we have only to observe that it is not merely a question of obstacles which obstruct the advance of an enemy, such as steep grounds, high mountains, marshy streams, hedges, etc., but that it is also an advantage of ground if it affords us the opportunity of lining up troops on it without their being seen. Indeed, we may say that even from ground which is quite without special features the person who knows it derives assistance. The attack from several sides includes all tactical turning movements, great and small, and its effects are derived partly from the doubled efficiency of fire and partly from the enemy's fear of being cut off.

Now how are the offensive and defensive related to one another with respect to these things?

Having in view the three principles of victory just described, the answer to this question is that only a small portion of the first and last of these principles is in favor of the offensive, while the greater part of them, and the second principle exclusively, are at the disposal of the defensive.

The assailant has only the advantage of the actual surprise of the whole mass with the whole, while the defender is in a condition to surprise incessantly, throughout the course of the engagement, by the force and form which he gives to his attacks.

The assailant has greater facilities than the defensive for surrounding and cutting off the whole, as the latter is in a fixed position while the former is in a state of movement with reference to that position. But then again this enveloping movement applies only to the whole, for in the course of the engagement and for the separate sections an attack from several sides is easier for the defensive than for the offensive, because, as was said above, *the former is in a better condition to surprise by the force and form of his attacks.*

That the defender enjoys to a higher degree the assistance of ground is self-evident; while his superiority in surprise by means of the force and form of his attacks results from the fact that the assailant is obliged to approach by roads and paths where he becomes easy to observe, while the defender conceals his position, and, until the decisive moment, remains almost invisible to his assailant. Since the right method of defense has become general, reconnaissances have gone quite out of fashion, that is to say, they have become impossible. Reconnaissances, it is true, are still made at times, but they seldom bring back much information. Immense as is the advantage of being able to select the ground for the disposition of troops, and become perfectly acquainted with it before a battle, plain as it is that he (the defender) who lies in wait in such a chosen position can much more easily surprise his adversary than can the assailant, yet still to this very hour the old notion has not been discarded that a battle which is accepted is half lost. This is due to the old kind of defensive practiced twenty years ago, and partly also during the Seven Years' War, when the only assistance expected from the terrain was that it should form a front which could be penetrated only with difficulty (steep mountain slopes, etc.), where the lack of depth in the disposition and the difficulty of moving the flanks produced such weakness that the armies dodged one another from one mountain to another, thereby making things worse and worse. If some kind of support had been found on which to rest the wings, then everything depended on preventing the army, stretched out between these points, like a piece of work on an embroidery frame, from having a hole broken in it. The ground occupied possessed a direct value at every point, and therefore a direct defense was required everywhere. Under such circumstances, a movement or a surprise during the battle was out of the question; it was the exact opposite of what constitutes a good defense and of what the defense has actually become in modern warfare.

In reality, contempt for the defensive has always been the result of an epoch in which a certain style of defense has outlived its day; and this was also the case with the method mentioned above, for in times previous to the period we refer to that method was actually superior to the offensive.

If we follow the development of the modern art of war, we find that in the beginning, that is, the Thirty Years' War and the War of the Spanish Succession, the deployment and disposition of the army was one of the most important points in the battle. It was the most important part of the plan of the battle. This gave the defender, as a rule, a great advantage, as he was already in his position and deployed before the attack could begin. As soon as the troops acquired greater capability of maneuvering, this advantage ceased, and for a time the superiority passed over to the side of the offensive. Then the defensive sought protection behind rivers or deep valleys, or on mountains. It thus recovered a decisive advantage and continued to maintain it until the assailant acquired such increased mobility and expertness in maneuvering that he could himself venture into broken ground and attack in separate columns, and therefore was able to *turn* his opponent. This led to a continually increased extension, as a result of which, it naturally occurred to the offensive to concentrate at a few points, and break through the thin line of the enemy. Thus for a third time the offensive gained the superiority, and the defense was again obliged to alter its system. This it has done in the most recent wars by keeping its forces concentrated in large masses, the greater part not deployed, and, where possible, concealed, thus merely taking up a position in readiness to act according to the measures of the enemy as soon as they were sufficiently revealed.

This does not exclude entirely a partially passive defense of the ground; its advantage is too great to prevent its being used a hundred times in a campaign. But this passive defense of the ground is usually no longer the main point, the point with which we are here concerned.

If the offensive should discover some new and powerful method which it can bring to its assistance—an event not very probable, considering the simplicity and inner necessity to which everything has now progressed—then the defense will have again to alter its method. But it will always have the assistance of the ground, which will ensure to it in general its natural superiority, as the special properties of country and ground now exercise a greater influence than ever on warfare.

CHAPTER 3

THE RELATIONS OF THE OFFENSIVE AND DEFENSIVE TO EACH OTHER IN STRATEGY

First of all, let us ask the following question again: What are the circumstances which ensure a successful result in strategy?

In strategy, as we have said before, there is no victory. On the one hand, strategic success is the advantageous preparation of the tactical victory; the greater this strategic success, so much the less doubtful the victory in the engagement. On the other hand, strategic success lies in making use of the victory gained. The more events that, *after* a battle is won, strategy, by means of its combinations, can include in the results of it, the more it can carry off from the tumbling ruins whose foundations have been shattered by the battle; the more it sweeps up in great masses what in the battle itself had to be painfully won piece by piece, the grander will be its success. Those things which chiefly lead to this success or facilitate it—the leading principles, therefore, of strategic efficiency—are as follows:

1. The advantage of ground.
2. Surprise, let it be either in the form of an actual surprise assault or by the unexpected disposition of superior forces at certain points.
3. The attack from several quarters (all these three, as in tactics).
4. The assistance of the theater of war by fortresses, and everything belonging to them.
5. The support of the people.
6. The utilization of great moral forces.

Now, what are the relations of offensive and defensive with respect to these things?

The defender has the advantage of ground, the assailant that of the surprise attack. This is the case in strategy as well as in tactics. But concerning the surprise, we must observe that it is an infinitely more effective and important means in strategy than in tactics. In tactics, a surprise seldom reaches the level of a great victory, while in strategy it has often finished the whole war at one stroke. But we must observe again that the advantageous use of this means presupposes some *great* and *rare*, as well as *decisive*, errors committed by the adversary. Therefore, it does not alter the balance much in favor of the offensive.

The surprise of the enemy, by placing superior forces in position at certain points, has again a great resemblance to the analogous case in tactics. If the defender were compelled to distribute his forces upon several points of approach to his theater of war, then the assailant would have plainly the advantage of being able to fall upon one point with all his weight. But here also the new art of defense has imperceptibly brought about new principles by means of a different procedure. If the defender does not apprehend that the enemy, by making use of an undefended road, will throw himself upon some important magazine or depot, or on some unprepared fortress, or on the capital, and if he is not on that account obliged to oppose the enemy on the very road he has chosen, because otherwise he would have his retreat cut off, then there are no grounds for dividing his forces. For if the assailant chooses a different road from that on which the defender happens to be, then some days later, the latter can still with his whole force seek him out on that road; indeed, he may in most cases even be sure that he will himself have the honor of being sought out by his opponent. If the latter is obliged himself to advance with his forces in separate columns, which is often almost unavoidable on account of subsistence, then plainly the defensive has the advantage of being able to fall with his entire force upon one part of the enemy.

In strategy, attacks in flank and rear, which there have reference to the sides and reverse of the theater of war, greatly change their character.

1. There is no bringing the enemy under two fires, because we cannot fire from one end of a theater of war to the other.
2. The apprehension of losing the line of retreat is very much weaker, for the spaces in strategy are so great that they cannot be barred as in tactics.
3. In strategy, on account of the greater space involved, the effectiveness of interior, that is of shorter, lines is much more considerable, and this forms a great countercheck against attacks from several directions.

4. A new principle makes its appearance in the sensitiveness of the lines of communication, that is, in the effect which is produced by merely interrupting them.

Now it is natural no doubt, that on account of the greater spaces in strategy, the enveloping attack, or the attack from several sides, is only possible as a rule for the side which has the initiative, that is, the offensive, and that the defender is not in a condition, as he is in tactics, to turn the tables on the enemy, in the course of the action, by surrounding him. He cannot do this because he is neither able to draw up his forces in such comparative depth nor to conceal them so well. But, then, of what use to the assailant is the facility of enveloping, if its advantages are not forthcoming? Consequently, in strategy we could not bring forward the enveloping attack as likely to have any success at all, if its influence on the lines of communication did not come into consideration. But this factor is seldom great at the first moment, when attack and defense meet each other and are still opposed to each other in their original position. It only becomes great as a campaign advances, if the offensive in the enemy's country gradually becomes a defensive. Then the lines of communication of this new defender become weak, and the party originally on the defensive, in assuming the offensive, can derive advantage from this weakness. But who does not see that this superiority of the offensive is not to be attributed to it as a general thing? In reality it has been created by the superior situation of the defensive.

The fourth principle, the *assistance of the theater of war*, is naturally an advantage on the side of the defensive. If the attacking army opens the campaign, it breaks away from its own theater, and is thus weakened, that is, it leaves fortresses and depots of all kinds behind it. The greater the sphere of operations which it has to traverse, the more the attacking army will be weakened (by marches and garrisons); the defending army continues to keep up its connection with everything; that is, it enjoys the support of its fortresses, is not weakened in any way and is near to its sources of supply.

The support of the people, as a fifth principle, is not, it is true, to be had in every defense, for a defensive campaign may be carried on in the enemy's country, but still this principle is only derived from the idea of the defensive, and finds its application in the great majority of cases. Moreover, by this is meant chiefly, although not exclusively, the operation of a general levy, and even of a national rising. Another advantage of this is that all friction is diminished, and that all sources of supply are nearer and flow more abundantly.

The campaign of 1812 gives as in a magnifying glass a very clear illustration of the effectiveness of the means specified under principles three and four. A half million men passed the Niemen, 120,000 fought at Borodino, and still fewer reached Moscow.

We may say that the effect itself of this stupendous attempt was so great that even if the Russians had not followed up with any offensive at all, they would still have been secure from any fresh attempt at invasion for a considerable time. It is true that with the exception of Sweden there is no country in Europe which is in a position similar to that of Russia, but the efficient principle is always the same, the only distinction being in the greater or less degree of its strength.

If we add to the fourth and fifth principles, the consideration that these forces of the defensive belong to the original defensive, that is, the defensive carried on in our own soil, and that they are much weaker if the defense takes place in an enemy's country and is mixed up with offensive undertakings, then from that a new disadvantage for the offensive is derived, much the same as above, in respect to the third principle. For the offensive is just as little composed entirely of active elements as the defensive of mere warding off blows. Indeed every attack which does not lead directly to peace must inevitably end in a defensive.

Now, if all defensive elements which are brought into use in the attack are weakened by its nature, that is, by belonging to the attack, then it must also be considered as a general disadvantage of the offensive.

This is so far from being an idle subtlety that, on the contrary, we should rather say that in it lies the chief disadvantage of the offensive in general. Therefore, in every plan for a strategic attack, most particular attention ought to be directed from the very beginning to this point, that is, to the defensive, which will follow. This we shall see more plainly when we come to the book on the "Plan of War."

The great moral forces which at times saturate the elemental violence of war, like a spontaneous ferment, which, therefore, the commander can use in certain cases to assist the other means at his disposal, may be assumed to exist as much on the side of the defensive as of the offensive; at least those which are more especially in favor of the attack—such as confusion and disorder in the enemy's ranks—do not generally appear until after the decisive stroke is delivered, and consequently seldom help to influence it.

We think we have now sufficiently established our proposition, that *the defensive is a stronger form of war than the offensive;* but there still remains to be mentioned one small factor hitherto unnoticed. It is the courage—the feeling of superiority in an army—which springs from a consciousness of belonging to the attacking party. The thing is in itself a fact, but the feeling very soon merges into the more general and more powerful one which is imparted to an army by victory or defeat, by the talent or incapacity of its general.

CHAPTER 4

CONVERGENCE OF ATTACK AND
DIVERGENCE OF DEFENSE

These two conceptions, these forms in the use of offensive and defensive, appear so frequently in theory and in practice that the imagination is involuntarily disposed to look upon them as forms inherent in attack and defense. This, however, as the slightest reflection shows, is not really the case. We take, therefore, the earliest opportunity of examining them, in order to obtain, once for all, clear ideas respecting them, and, in proceeding with our consideration of the relations of attack and defense, to set these conceptions aside altogether and not have our attention forever distracted by the appearance of advantage or disadvantage which they cast upon things. We treat them here as pure abstractions, extract the conception of them like an essence and reserve till later our remarks on the part this plays in practice.

The defender, both in tactics and in strategy, is conceived as awaiting the enemy, and so, as standing still; the assailant, as being in movement, and in movement having relation to that standing still. It necessarily follows from this that turning and enveloping lie wholly in the discretion of the assailant, so long, that is, as he continues to move and the defender to stand still. This freedom in the choice of the mode of attack, whether it shall be convergent or not, according as it is advantageous or otherwise, would have to be reckoned as a general advantage to the offensive. But this choice is free only in tactics, and not always in strategy. In the former, the points on which the wings rest hardly ever afford absolute security, but they very frequently do in strategy, if the front to be defended stretches in a straight line from one sea to another, or from one neutral territory to another. In such cases the attack cannot be made in a convergent form, and the freedom of choice is limited. It is limited in a still more embarrassing manner if the assailant is *obliged* to operate on converging lines. France and Russia cannot attack Germany in any other way than on converging lines; therefore they cannot attack with their forces united. Now if we might assume that the convergent form in the operation of forces is in most cases the weaker, the advantage which the assailant possesses in his greater freedom of choice would probably be completely outweighed by the fact that in other cases he is compelled to use the weaker form.

We shall now examine more closely the operation of these forms both in strategy and in tactics.

It has been considered one of the chief advantages of giving a converging direction to forces, that is, in operating from the circumference of a circle toward its center, that the farther the forces advance, the more they concentrate. The fact is true but not the supposed advantage, for the concentration occurs on both sides, and the equilibrium, therefore, is maintained. The same is the case with the dispersion in a diverging operation.

But another and real advantage is that forces moving on converging lines operate toward a *common* point, while those moving on diverging lines do not. Now what are the advantages of these two ways of operating? To answer this we must separate tactics from strategy.

We do not want to push the analysis too far and we therefore give the following points as the advantages of these modes of operation in tactics:

1. A doubled or, at all events, an increased, effect of fire, as soon, that is, as the concentration has reached a certain point.
2. Attack on one and the same section from several sides.
3. The cutting off of the retreat.

The cutting off of the retreat may also be conceived strategically, but it is obviously much more difficult because great spaces are not easily blocked. The attack on one and the same section from several sides in general becomes the more effective and decisive, the smaller this section is and the more nearly it is conceived as approaching the extreme limit, that is, the single combatant. An army can easily give battle on several sides at the same time, a division less easily, a battalion only if formed in mass, and a single man not at all. Now strategy has for its province large masses, spaces and periods of time, while with tactics it is just the reverse. From this it follows that the attack from several sides in strategy cannot have the same results as in tactics.

The effect of fire is something quite outside the scope of strategy, but its place is there taken by something else. It is that insecurity of its base which every army more or less feels if there is a victorious enemy in its rear, whether near or far off.

It is, therefore, certain that the convergent mode of operating has an advantage in the fact that in operating against *a* the forces are at the same time operating against *b*, without on that account operating less strongly against *a*, and that in operating against *b* they are also operating at the same time against *a*. The total effect, therefore, is greater than the sum of what the effects on *a* and *b* would have been had the operation not been convergent. This advantage arises both in tactics and in strategy, though in a somewhat different way in each.

Now in the divergent mode of operating, what is there to set against this advantage? Obviously the fact of having the forces in closer proximity to one another and of moving on interior lines. It is not necessary to demonstrate how this can increase the effectiveness of the forces to such an extent that without a great superiority of force the assailant dare not expose himself to this disadvantage.

Once the defender has decided to move—which he does later, it is true, than the assailant but always in time to break the fetters of paralyzing inaction—this advantage of greater concentration and interior lines becomes a very decisive one and one usually more effective toward gaining the victory than the convergent form of the attack. Victory, however, must come before its result can be achieved; we must conquer before we can think of cutting off the enemy's retreat. In short, we see that there is here a relation similar to that which exists between attack and defense generally. The convergent form leads to brilliant results, but those of the divergent form are surer. The former is the weaker form but has a more positive, that is, a more effective, object; the latter is a stronger form but has a negative object. In this way the two forms, it seems to us, have been brought into a sort of fluctuating equilibrium. If we now add to this the fact that the defense, because it is not everywhere an absolute defense, that is, defense and nothing else, can also itself on occasion avail itself of convergently directed forces, we shall at least no longer have the right to believe that this mode of operating is in itself sufficient to assure to the attack a quite universal advantage over the defense, and we shall set ourselves free from the influence that this sort of idea at every opportunity is wont to exercise upon our judgment.

What we have said so far applied both to tactics and strategy. We must now raise an extremely important point which concerns strategy alone.

The advantage of interior lines increases with the spaces to which those lines relate. With distances of a few thousand paces or a mile or two, the time gained can naturally not be so great as with distances of several days' march, or, indeed, of a hundred or a hundred and fifty miles; the former, that is, the smaller, distances concern tactics, the greater ones, strategy. Now though we certainly in strategy need more time to reach our object than we do in tactics, and an army is not so quickly defeated as a battalion, still these periods of time in strategy only increase up to a certain point, namely, up to the duration of a battle or at most the couple of days or so for which a battle may be avoided without serious loss. Further, there is still a much greater difference in the actual start we get of the enemy in the one case and in the other. With the small distances of tactics, the movements of the one side in a battle take place almost under the eyes of the other, and the army standing on the exterior line will therefore usually be quickly aware of the movements of its adversary. With the long distances of strategy it very rarely indeed happens that a movement of one army should not remain concealed from the other for at least a day, and there are cases enough in which, if the movement involved only a part of the army and consisted in the despatch of a considerable detachment, this has remained concealed for weeks. It is easy to see how great is the advantage of concealment to that side which, from the nature of its position, is most capable of making use of it.

We here close our discussion of the convergent and divergent modes of operating with our forces and their relation to attack and defense, proposing in both cases to come back again to the subject.

CHAPTER 5

Character of the Strategic Defensive

It has already been stated what defense is. It is nothing but a stronger form of the conduct of war, by means of which we intend to gain the victory in order, after superiority has been gained, to pass over to attack, that is to say, to the positive object of war.

Even if the intention of war is only the maintenance of the *status quo*, still the mere parrying of a blow is something contrary to the idea of war, because warfare is unquestionably no passive endurance. If the defender has gained an important advantage, defense has done its part, and under the protection of this advantage he must return the blow if he does not want to expose himself to certain ruin. Common sense demands that we should strike while the iron is hot, that we should use the advantage we have gained to prevent a second attack. How, when and where this reaction is to begin is, of course, subject to many other conditions which can only later be fully explained. Here we confine ourselves to saying that this transition to the counter-blow must be considered as a natural tendency of the defensive, therefore an essential element of it, and that in all cases in which a victory won by the defensive form has not been turned to account in some way in the military economy, but allowed, so to speak, to wither away unused, a great blunder is being committed.

A swift and vigorous transition to attack—the flashing sword of vengeance—is the most brilliant point of the defensive. He who does not bear this in mind from the first, who does not from the first include it in his conception of defense, will never understand the superiority of the defensive. He will be forever thinking only of the means which are being lost to the enemy by the attack and gained by himself, which means, however, depend not on how we start a battle but how we finish it. Further, it is a gross confusion of ideas if by attack a sudden onset is always to be understood, and by defense nothing is suggested but embarrassment and confusion.

It is true that a conqueror makes his decision to go to war sooner than does the innocent defender, and if he knows how to keep his measures properly secret, he may also perhaps take the defender unawares; but that is something quite foreign to war itself, for it ought not to be so. War comes into being more for the benefit of the defender than for that of the conqueror for not till the invasion has called forth the defense does war begin. A conqueror is always a lover of peace (as Bonaparte constantly asserted of himself); he would like to make his entry into our state unopposed. In order to prevent this, we must choose war, and thus also make our preparations beforehand. In other words, it is just the weak, the side that must defend itself, which should always be armed in order not to be taken by surprise. So the art of war will have it.

The appearance of one side sooner than the other in the theater of war depends, moreover, in most cases on things quite different from offensive or defensive intentions. These intentions, therefore, are not the cause but often the result of it. Whoever is ready first, if the advantage of surprise is sufficiently great, goes to work offensively for *this* reason, and he who is ready last can only in some measure, by the advantages of the defensive, make up for the disadvantage which threatens him.

At the same time it must be looked upon in general as an advantage for the offensive that it can make that good use of being ready first which has already been recognized in Book III; only this general advantage is not a necessary element in every individual case.

If, therefore, we picture to ourselves a defensive as it should be, it includes the greatest possible preparation of all means, an army inured to war, a general who awaits his adversary not in anxiety from a feeling of uncertainty, but from free choice, a cool presence of mind, fortresses which do not dread a siege and, lastly, a healthy people who fear the enemy as little as he fears them. So provided, defense, confronted with attack, will no longer play so poor a part, and attack will no longer seem so easy and infallible as it does to the vague imagination of those who can only see in the offensive courage, strength of will and movement; in the defensive helplessness and apathy.

CHAPTER 6

EXTENT OF THE MEANS OF DEFENSE

We have shown in the second and third chapters of this book how the defense has a natural advantage in the employment of those things which—apart from the absolute strength and the quality of the military forces—determine tactical as well as strategic success, namely, advantage of ground, surprise, attack from several sides, assistance of the theater of war, support of the people and utilization of great moral forces. We think it may be useful if here we glance once more at the extent of the means which are more especially at the disposal of the defender and which are to be regarded as the various orders of columns in his edifice.

1. LANDWEHR (MILITIA)

This force has also been used in modern times abroad for attack on an enemy country, and it is not to be denied that its organization in many states, for instance in Prussia, is of such a kind that it must almost be regarded as part of the standing army, and that, therefore it does not belong to the defensive exclusively. At the same time we must not overlook the fact that the very vigorous use made of it in 1813, 1814 and 1815 was the result of a defensive war; that it is organized in very few places to the same degree as in Prussia, and in so far as its organization falls below the level of complete efficiency, it is necessarily better suited for the defensive than for the offensive. But, besides that, there always lies in the idea of a *Landwehr* the notion of a very extensive more or less voluntary co-operation of the whole mass of the people in support of the war, with their physical powers, their possessions and their convictions. The more its organization deviates from this type, so much the more will the force thus created be a standing army under another name, and the more it will have the advantages of such a force; but it will also lose in proportion the advantages of a real *Landwehr*, which is a store of strength much wider in its extent, much less narrowly limited in its scope and much more easily increased by appeals to national spirit and patriotism. It is in these things that the essence of a *Landwehr* consists. In the lines of its organization room must be left for the co-operation of the whole people, or in expecting any notable service from it, we are following a mere phantom.

But the close relation between the essence of a *Landwehr* and the idea of defense is unmistakable, and equally unmistakable is the fact that such a *Landwehr* will always belong more to defense than to attack and will manifest chiefly in defense those effects through which it is superior to the attack.

2. FORTRESSES

The assistance derived by the assailant from fortresses is confined to those lying close to the frontier and is but slight. That derived by the defense strikes deeper into the country, thus bringing more fortresses into operation, and this operation is of an incomparably greater intensive strength. A fortress which is the object of a siege and can hold out against it naturally throws a much greater weight into the scales of war than one which by the strength of its works merely forbids the idea of its capture and which, therefore, neither occupies nor destroys any of the enemy's forces.

3. THE PEOPLE

Although the influence of a single inhabitant of the theater of war on the course of the war in most cases is not more perceptible than the co-operation of a drop of water in a whole river, still even in cases where there is no such thing as a general rising of the people, the *total influence* which the inhabitants of a country have on a war is anything but insignificant. Everything works more easily in our own country, provided there is nothing in the feeling of the subjects to prevent this. No contributions great or small are made to the enemy except under the compulsion of open violence, which must be exerted by the troops and costs them dear in force and effort. The defender gets all this, even if not always freely offered, as in cases of enthusiastic devotion, still through the long practiced habits of a civil obedience which has become a second nature to the inhabitants and, further, is kept alive by other means of inspiring fear and exerting compulsion which do not proceed from the army and have nothing to do with it. But the voluntary co-operation which proceeds from genuine loyalty is also in all cases very important, since it never fails in any point that demands no sacrifices. We shall only single out one such point, which is of great importance for the conduct of war: that is *intelligence*, not so much the special, great, important pieces of intelligence that espionage gives, as that concerning the innumerable little uncertainties which attend the daily service of an army and with regard to which an understanding with the inhabitants gives the defenders a general advantage.

Now if we ascend from these quite general and never failing relations to the exceptional cases in which the population begins to take part in the struggle, and then further, up to the highest degree, where, as in Spain, the war, as regards its leading events, is chiefly a war carried on by the people themselves, we realize that here we have not merely an intensified form of popular co-operation but really new force, and that, therefore

4. THE NATIONAL RISING

or national call to arms may be cited as a special means of defense.

5. ALLIES

Finally, we may further reckon *allies* as the last support of the defensive. Naturally we cannot mean thereby ordinary allies, which the assailant may likewise have; we speak of those *essentially interested in maintaining* a country's integrity. If, for instance, we look at the various states composing Europe at the present time, we find—without speaking of a systematically regulated balance of power and interests such as does not exist and therefore is often with justice disputed—that still unquestionably the interests, great and small, of states and nations are interwoven with one another in a most complicated and changeable manner. Each point at which they cross forms a strengthening knot, for in it the tendency of the one counterbalances the tendency of the other. By means of all these knots, therefore, a more or less close inter-connection of the whole is created and for any change to take place this inter-connection must be partially overcome. In this way the sum total of the relations of all the states to one another serves rather to maintain the *status quo* of the whole than to introduce changes in it, that is to say, that in general the course of events tends to *the maintenance of the status quo*.

Thus, we believe, must the idea of a balance of power be conceived, and in this sense such a balance will always spontaneously arise wherever several civilized states have many points of contact.

How effective the tendency of these collective interests toward the maintenance of the existing condition may be is another question. We can, indeed, conceive changes in the relations of individual states to one another which promote this effectiveness of the whole, and others which obstruct it. In the former case they are efforts to strengthen the political balance, and as these have the same tendency as the collective interests, they will also have the majority of these interests on their side. In the latter case, however, they are abnormalities, excessive activity of individual parts, real diseases. That these should occur in a whole so feebly bound together as the multitude of greater and smaller states is not to be wondered at. After all, they occur in the marvelously ordered organic whole of all living nature.

If, therefore, we are reminded of the cases in history in which single states have been able to effect important changes solely for their own advantage, without even so much as an attempt having been made by the whole to prevent them, and, indeed, of cases in which a single state has been in the position to raise itself so much above the rest as to become the almost absolute arbiter of the whole, our answer is that these cases in no way prove that the tendency of the collective interests to the maintenance of the *status quo* does not exist, but only that their effectiveness at the moment was inadequate. Effort toward an object is not the same thing as motion toward it. But it is by no means a nullity on that account, a truth of which the dynamics of the heavens afford the best illustration.

When we say that the tendency of equilibrium is the maintenance of the existing condition, we certainly assume that in this condition rest, that is, equilibrium, existed. For where this has already been disturbed and a tension already been introduced, the tendency of equilibrium may also, certainly, be directed toward a change. But if we regard the nature of the thing, this change can never affect more than a few single states, and never, therefore, the majority of them. It is certain then that this majority sees its maintenance always represented and assured by the collective interests of all, certain also that each single state, which is not in the position of finding itself already in tension against the whole, will have more interests in its favor than against it in defending itself.

Whoever laughs at these reflections as utopian dreams does so at the expense of philosophical truth. Although the latter teaches us to recognize the relations in which the essential component parts of things stand to one another, it would certainly be rash to expect to deduce from them laws by which each individual case could be governed, without regard to any accidental disturbing influences. But when a person, in the words of a great writer, *"never rises above anecdote,"* builds all history on it, begins always with the most individual points, with the climaxes of events; when he never goes deeper than just so far as he has cause, and thus never reaches the deepest foundation of existing general relations—such a person's opinion will never have value beyond a single case, and for him, certainly, what philosophy settles for the generality of cases will only appear a dream.

Without that general tendency toward rest and the maintenance of the existing state of things, a number of civilized states would never be able to exist for a long time side by side; they would necessarily have to fuse into one. Therefore, as the Europe of today has existed in this form for over a thousand years, we can only ascribe this result to that power of the collective interests which makes for stability. If the protection of the whole has not always been adequate to the maintenance of each single state, such exceptions are irregularities in the life of the whole, which, however, have not destroyed that life but have been overmastered by it.

It would be very superfluous to run through the mass of events in which changes which disturbed too much the balance of power have been either prevented or reversed, by the more or less open opposition of the other states. The most cursory glance at history reveals them. Of one case only shall we speak because it is always on the lips of those who ridicule the idea of political balance and because it seems to be quite peculiarly in place here, as an instance in which an unoffending state acting on the defensive perished without receiving any foreign aid. We allude to Poland. That a state of eight million inhabitants should disappear, should be divided among three others without a sword being drawn by any of the rest of the European states, appears at first sight a case which would either sufficiently prove the general ineffectiveness of the political balance or at least show how little it could do in individual instances. That a state of such extent should disappear and become the prey of others, which were already among the most powerful (Russia and Austria), seemed to be a case of the most extreme kind. And if such a case could not rouse the collective interests of the whole commonwealth of states, then the efficacy which these collective interests should have for the maintenance of individual states is to be regarded as imaginary. But we still maintain that a single instance, however striking, proves nothing against the general truth, and we furthermore assert that the downfall of Poland is also not so unintelligible as it seems. For was Poland really to be regarded as a European state, as a homogeneous member of the European commonwealth of states? No! It was a Tartar state, which, instead of being located on the borders of the European political world, like the Tartars of the Crimea on the Black Sea, lay in their midst on the Vistula. We neither desire by this to speak disrespectfully of the Polish people nor to justify the partition of their country, but only to see things as they are. For a hundred years this state had fundamentally ceased to play any independent political part; it had merely become the apple of discord for other states. In its condition and with its constitution it could not in the long run maintain its existence among them. But any essential change in this Tartar condition could not have been the work of less than a half, or even of a whole, century, even if the leaders of this people had been willing to attempt it. But these men were themselves much too thorough Tartars to desire it. Their turbulent politics and their immeasurable frivolity went hand in hand, and so they staggered into the abyss. Long before the partition of Poland, the Russians had become quite at home there. The idea of an independent state with boundaries of its own had absolutely ceased to exist, and nothing is more certain than that if it had not been partitioned, Poland must have become a Russian province. If this had not been so, and if Poland had been a state capable of putting up a defense, the three powers would not so readily have proceeded to its partition, and those powers most interested in maintaining its integrity, like France, Sweden and Turkey, would have been able to co-operate in a very different manner toward its preservation. But if the preservation of a state is entirely dependent on external support, that certainly is asking too much.

Over a hundred years before, the partition of Poland had several times been talked of, and, ever since, the country had been not so much like a private house as like a public road on which foreign armies were continually jostling one another. Was it the duty of the other states to prevent this? Were they constantly to keep the sword drawn to protect the political sanctity of the Polish frontiers? That meant to ask a moral impossibility. Poland was at that time politically not much more than an uninhabited steppe, and impossible as it would have been to be always protecting this defenseless steppe from the attacks of the other states among which it lay, the integrity of this so-called state could not be guaranteed. On all these grounds the noiseless downfall of Poland is as little to be wondered at as the silent downfall of the Crimean Tartars. The Turks had a greater interest in preserving the latter than had any European state in preserving the indepéndence of Poland, but they saw that it would be a vain effort to try to protect a defenseless steppe.

We return to our subject, and think we have proved that the defender in general can count more on foreign aid than can the assailant. He may reckon the more certainly on it in proportion as his existence is of importance to others, that is to say, the sounder and more powerful is his political and military condition.

Of course the subjects which have been here enumerated as means properly belonging to the defensive will not be at the command of each particular defensive. Sometimes one, sometimes another, may be wanting, but they all belong to the idea of the defensive as a whole.

CHAPTER 7

RECIPROCAL ACTION AND REACTION OF ATTACK AND DEFENSE

We shall now consider attack and defense separately, as far as they can be separated from each other. We begin with the defensive for the following reasons: It is certainly very natural and necessary to base the rules for the defense upon those of the offensive, and *vice versa*; but one of the two must still have a third point of departure, if the whole chain of ideas is to have a beginning, that is, to become possible. The first question concerns this point.

If we reflect philosophically how war arises, the conception of war does not properly arise with the *offensive*, because this has for its absolute object not so much *combat* as the *taking possession of something*. It first arises with the *defensive*, for this has combat for its direct object, warding off and combat being obviously one and the same thing. The warding off is directed entirely against the attack, and therefore necessarily presupposes it; but the attack is not directed against the warding off, but upon something else—the *taking possession of something*, and consequently does not presuppose the warding off. It is therefore natural that he who first brings the element of war into the action, he from whose point of view two opposite parties are first conceived, also establishes the first laws for the war, and that he is the *defender*. We are not speaking of any individual case; we are only dealing with a general, an abstract, case, which theory postulates in order to determine the course it is to take.

We now know where to look for the fixed point, outside of the reciprocal action of attack and defense, and we find that it lies in the defense.

If this deduction is correct, it must afford determining reasons for the defender even when as yet he knows nothing of what the assailant means to do, and further, these determining reasons must decide the disposition of his means of combat. On the other hand, for the assailant, so long as he knew nothing of his opponent, it would afford no determining reasons for his procedure and the employment of his means of combat. He would necessarily be unable to do anything but take these means with him, that is, take possession by means of his army. And so it is in fact. For to provide the means of combat is one thing, to use them is another, and the assailant who takes them with him on the perfectly general assumption that he will need them, and who, instead of taking possession of the country with commissaries and proclamations, does so with armies, has not yet, properly speaking, committed any positive act of war. It is the defender, who not merely assembles his means of combat but also disposes them for combat as he proposes to conduct the war, who is the first to practice an activity which the conception of war really fits.

The second question now is: Of what kind, theoretically, can the determining reasons be which arise in the mind of the defender first, before the attack itself has been thought of? Obviously, it is the advance to take possession, which is theoretically extraneous to the war but which provides the foundation for the first rules for military action. The defense has to oppose this advance, which must therefore be thought of in relation to the country, and thus arise the first and most general determining factors for the defense. Once these are established, the attack is framed to meet them, and from the consideration of the means which it employs, new principles for the defense are derived. We now get the reciprocal action which theory in its investigation can continue to follow as long as it finds the new results accruing worth its consideration.

This short analysis was necessary in order to give more clarity and solidity to all our future discussions. This kind of thing is not made for the field of battle nor for future generals, but for the host of theorists who have so far treated the subject altogether too lightly.

CHAPTER 8

METHODS OF RESISTANCE

The conception of defense is warding off; in this warding off lies an awaiting, and this awaiting we have taken as the chief characteristic of the defense, and at the same time as its principal advantage.

But as defense in war cannot be a mere endurance, so the awaiting cannot be an absolute state but only a relative one. The thing to which it is relative is, as regards space, the country or the theater of war, or the position, and as regards time, the war, the campaign or the battle. That these things are not immutable units, but only the centers of regions which overlap and get blended together, we know; but in practical life we must often be content to divide things into groups without rigidly separating them, and these conceptions have in practical life itself acquired sufficient definiteness to enable us conveniently to group our other ideas around them.

A defense of a country, therefore, only awaits an attack on the country; a defense of a theater of war awaits an attack on the position. Every positive activity, and consequently every activity more or less in the nature of an attack, which defense uses after this moment of awaiting has passed, will not invalidate our conception of defense, for its chief distinguishing mark and its chief advantage, namely, the awaiting has been present.

The conceptions of war, campaign and battle with regard to time are coupled, respectively, with the conceptions of country, theater of war and position, and on that account they have the same relation to our subject.

Defense, therefore, consists of two heterogeneous parts, that of awaiting and that of acting. By relating the former to a definite object and therefore giving it precedence of action, we have made it possible to unite them both in one whole. An act of defense, especially a long one, such as a campaign or a whole war, will not, as regards time, consist of two great halves, the first of nothing but awaiting, the second of nothing but acting, but of an alternation of these two states, in which awaiting may run like a continuous thread through the whole act of defense.

We attach so much importance to this awaiting simply because the nature of the subject demands it. In previous theories it has certainly never been put forward as an independent conception, but in the practical world it has constantly, though often unconsciously, served as a guiding thread. It is such a fundamental part of the whole act of war that the one without the other seems scarcely possible; and consequently in what follows we shall often come back to it by calling attention to its effects in the dynamic play of forces.

For the present we shall occupy ourselves in making it clear how the principle of awaiting permeates the act of defense and what successive stages in the defense itself spring from it.

In order to fix our ideas on the simpler subject, we shall defer the defense of a country, a subject in which political circumstances are more diversified and have a stronger influence, until Book VIII, "Plan of a War." On the other hand, the act of defense in a position or a battle is a matter of tactics and only *as a whole* forms a starting point for strategical activity. The defense of a *theater of war* will consequently be the subject on which we shall best be able to exhibit the conditions of defense.

We have said that awaiting and action—which last is always a counterstroke and therefore a reaction—are both essential parts of defense, for without the former there would be no defense, and without the latter no war. This view has earlier led us to the idea of the defensive being nothing but *the stronger form of the conduct of war enabling us to conquer the adversary the more surely.* This idea we must adhere to firmly, partly because in the last instance it alone saves us from absurdity, partly because the more vividly and closely it is present to us, the greater is the energy it imparts to the whole act of defense.

We must not, therefore, make a distinction between the counter-attack and the defense, of which the counter-attack is the second necessary element. We must not consider the element which consists in the actual warding off of the enemy from the country, the theater of war or the position as alone the *necessary* part, which would only go so far as is required to secure those objects. Nor, on the other hand, must we regard the possibility of a reaction pushed farther and *passing over into the province of the real strategic attack* as something foreign to the defense and having nothing to do with it. Such ideas would be *opposed* to the idea set forth above, and consequently we cannot consider such a distinction as an essential one, but must adhere to our contention that at the bottom of all defense must lie the idea of a *retaliation*. Otherwise, however much damage in case of success might have been inflicted on the enemy in that first counter-attack, the necessary balance in the dynamic relation between attack and defense would be lacking.

We say, therefore, that defense is the stronger form of warfare, enabling us to overcome the enemy more easily, and that we leave it to circumstances to decide whether this victory does or does not go beyond the object to which the defense related.

But as defense is inseparable from the idea of awaiting, the object of *defeating the enemy* can only exist conditionally, that is, only if attack follows. It is, therefore, obvious that if this does not take place, the defensive is content with maintaining what it possesses. This, then, is its object in the state of awaiting, that is to say, its immediate object, and only while it contents itself with this more modest end can it enjoy the advantages of the stronger form of warfare.

If we suppose an army with its theater of war intended for defense, the defense may be made as follows:

1. By attacking the enemy's army as soon as it enters the theater of war (Mollwitz, Hohenfriedberg).
2. By taking up a position close to the frontier and waiting till the enemy appears with the intention of attacking it, in order then to attack him (Czaslau, Soor, Rossbach). Obviously in this case our procedure is of a more passive type, our state of awaiting lasts longer, and though the *time* gained by this second mode of procedure will be little or nothing in addition to that gained in the first, in the event of the enemy's attack really taking place, still the battle, which in the first case was certain, is now less so. It may be that the enemy's resolution will not be equal to attacking at all. The advantage of the state of awaiting is already, therefore, greater.
3. By the army in such a position waiting not merely for the enemy to make up his mind to give battle, that is, till he appears in front of our position, but till he actually attacks (viz., to keep to the same general, Bunzelwitz). In this case we shall, therefore, fight a true defensive battle, which, however, as has already been said, may include an offensive movement with one or more parts of the army. Here, too, as before, the gain in time will not amount to much, but the resolution of the

enemy will be put to a fresh proof. Many a one after advancing to the attack has yet at the last moment, or on the first attempt, given it up, because the enemy's position was too strong.

4. By the army transferring its resistance to the heart of the country. The object of this retreat is to cause and to await such a weakening of the enemy that he must either of himself stop in his advance or can, at least, no longer overcome the resistance which we finally offer to his progress.

This case is exhibited in the simplest and clearest manner if the defender can leave one or more of his fortresses behind him, which the assailant is obliged to besiege or blockade. How greatly his forces are weakened by this and what an opportunity is given to the defender of attacking them at some point with greatly superior numbers is obvious.

But even where there are no fortresses, such a retirement into the interior can gradually create for the defender the equality or superiority which he needs and which at the frontier he did not possess. For every advance in the strategic attack weakens him who makes it, partly of itself, partly by reason of the division of forces that becomes necessary, a point on which we shall have more to say in discussing the attack. We assume here the truth of this assertion, as we regard it as sufficiently proved by every war to be a fact.

Now in this fourth case the time gained is to be regarded before all else as an important advantage. If the assailant lays siege to our fortresses we have time till their probable fall, which may, indeed, be several weeks, or in some cases, several months. But if the weakening, that is, the exhaustion of his attacking force is brought about merely by his advance and his having to garrison the necessary points, merely therefore by the length of his march, our gain in time becomes in most cases still greater and our action will not be so tied to a definite point of time.

Apart from the alteration in relative strength between defender and assailant which is brought about at the end of this march we must for the former bring into account the *increased advantage* of awaiting the enemy. Even if the assailant had not in his advance been really so weakened that he could not still attack our main force where it is halted, yet resolution might perhaps fail him to do so, for this resolution will here always necessarily have to be stronger than it would have needed to be at the frontier. For one thing his forces are weakened and no longer fresh, and the danger is increased; for another, with an irresolute commander the possession of the country into which he has come is often enough to banish entirely from his mind the idea of a battle, either because he really believes, or pretends to believe, that a battle is no longer necessary. From this abandoned attack the defender can now obtain not, certainly, a satisfactory negative success, but still a great gain of time.

It is clear that in all four of the cases indicated the defender has the benefit of the ground, and likewise that on account of this he can bring into action the co-operation of his fortresses and of the people. These effective forces, indeed, increase with each fresh type of defense, and it is precisely these things which in the fourth type produce the weakening of the enemy's power. Now as the advantages of the state of awaiting increase in the same direction, it follows, of course, that those types are to be regarded as a real ascending scale of defense, and that this form of warfare gets stronger and stronger the further it is removed from the attack. We are not afraid of anyone accusing us on this account of thinking that the most passive of all forms of defense is the strongest. The action of resistance is not to be weakened with each fresh type, but only *delayed, postponed*. But that in a strong and suitably entrenched position a stronger defense is possible, and that if the forces of the enemy have half exhausted themselves on it, the counterstroke that follows is also more effective, is surely nothing unreasonable. Without his advantages of position at Kollin, Daun would not have gained the victory, and if, when Frederick the Great brought off not more than 18,000 men from the battlefield, Daun had pressed the pursuit more strongly his success might have been one of the most brilliant in military history.

We therefore maintain that with each new stage of defense the preponderance or, more correctly speaking, the counterpoise obtained by the defender increases, and, consequently, the strength of his counterstroke.

Now are these advantages of the increasing force of the defensive to be had absolutely for nothing? By no means. The sacrifices with which they are bought increase in the same proportion.

If we await the enemy within our own theater of war, however near the frontier the decision may take place, this theater of war will nevertheless always be entered by the enemy force, which must entail some sacrifice on its part, whereas had we made the attack, this disadvantage would have fallen on the enemy. If we do not march against the enemy at once to attack him, our sacrifice becomes somewhat greater. The extent of country that he overruns and the time that he requires to reach our position increase it. If we wish to fight a defensive battle and if we, therefore, leave its determination and the choice of time for it to the enemy, it may be that he remains for some time in occupation of the piece of territory of which he is master, and the time which through his lack of resolution he allows us to gain is in that way paid for by us. Still more noticeable do our sacrifices become if a retirement into the interior of the country takes place.

But all these sacrifices which the defender makes, usually cause him only a falling off in power which has merely an *indirect*, and therefore a later and not a direct, effect on his military forces. Often it is so indirect that its consequences are hardly noticeable. The defender, therefore, seeks to strengthen himself in the present moment at the expense of the future; he borrows—as everyone must who is too poor for his circumstances.

Now if we want to examine the result of these different types of resistance, we must look at the *object of the attack*. This is to obtain possession of our theater of war, or at least of a considerable portion of it, for, under the conception of the whole, at least the greater part of it must be understood, and the possession of a few miles of territory has in strategy, as a rule, no independent importance. As long, therefore, as the assailant is not yet in possession of this, that is, so long as, from fear of our power, he has not yet advanced to the attack of our theater of war at all, or has not yet sought us out in our position, or has refused the battle we wished to offer him—for so long the object of the defense has been attained, and the measures taken for the defense have worked successfully. This success is, certainly, a merely *negative* one which cannot directly provide the forces for a real counterstroke. It can, however, provide them *indirectly*, that is, it tends to do so. For the time which elapses is *lost to the assailant*, and every loss of time is a disadvantage and must weaken in some way the party that suffers it.

Therefore in the first three types of the defense, that is if it takes place on the frontier, *the non-decision is in itself a success for the defensive*.

But it is not so with the fourth type.

If the enemy besieges our fortresses, we must relieve them in time. Therefore it rests with us to bring about the decision by positive action.

This is likewise the case if the enemy has followed us into the interior of the country without besieging any of our places. Certainly in this case we have more time, we can wait for the moment when the enemy is at his weakest, always, however, on the assumption that finally we must pass over to action. The enemy, indeed, is now perhaps in possession of the whole of the piece of territory which constituted the object of his attack, but it is only loaned to him. The tension still continues, and the decision is still pending. So long as the defender is daily strengthened and the assailant weakened, the absence of a decision is to the interest of the former. As soon, however, as the culminating point is reached, as it must be, were it only by the final operation of the general losses to which the assailant has exposed himself, then it is for the defender to take action and bring about a decision, and the advantage of awaiting the enemy is to be regarded as wholly exhausted.

This point of time naturally cannot be fixed by any general standard. A multitude of circumstances and relations may determine it, but we must note that the approach of winter usually makes a very natural turning-point. If we cannot prevent the enemy from wintering in the piece of territory he has occupied, then as a rule it will have to be regarded as given up. We have only, however, to call to mind Torres Vedras to see that this rule is not universal.

What now is the decision generally?

We have always in our discussions conceived it in the form of a battle, but this is not, of course, necessary. It may be conceived as a number of combinations of engagements with separate corps, which lead to a change in affairs either by actually finding vent in bloodshed or making the retreat of the enemy necessary by their probable effects.

There can be no other decision on the theater of war itself. That is the necessary consequence of the view of war that we have set forth, for even if an enemy's army, from sheer lack of provisions, starts its retreat, yet this retreat is still caused only by the restraint in which our sword holds that army. If our forces were not there at all, it would no doubt be able to devise means for provisioning itself.

Therefore, even if at the end of his aggressive course, when the enemy falls a victim to the difficulties of his attack, when detachments and hunger and sickness have weakened and worn him out, it is always only the dread of our sword that can cause him to turn about and let everything go. But, nevertheless, there is certainly a great difference between such a decision and one brought about at the frontier.

In the latter case, it is only our arms that are opposed to his, only our arms that keep his in check or work destruction upon them. But in the former case, at the end of their course of aggression, the enemy's forces are already half destroyed by their own efforts. This gives quite another weight to our arms, and they are, therefore, although the last, yet no longer the only ground for the decision. This destruction of the enemy's forces in their advance prepares the decision, and may do so to such an extent that the mere possibility of our counter-attack may cause their retreat, and therefore the reversal of the situation. In this case, therefore, we can practically ascribe the decision to nothing else than the efforts made in the advance. Now, in point of fact, we shall find no case in which the sword of the defender has not co-operated; but for the practical view it is important to distinguish which of the two modes of operating has been the predominating one.

In this sense we think we may say that in defense there are two kinds of decision and consequently two kinds of reaction according as the assailant is to be brought low by *the sword of the defender or by his own efforts.*

That the first kind of decision predominates in the first three types of defense, the second kind of decision in the fourth, is obvious. The second kind will in most cases only occur, if the retirement deep into the interior of the country takes place, and nothing else can be a sufficient motive for such a retirement, considering the great sacrifices which it costs.

We have therefore become acquainted with two different fundamental methods of resistance. There are cases in military history in which they stand out as clear and distinct as in practical life it is ever possible for an abstract conception to do. When Frederick the Great in 1745 attacked the Austrians at Hohenfriedberg just as they were proposing to descend from the Silesian mountains, their strength could not have been weakened in any noticeable degree either by detachments or by fatigue; when, on the other hand, Wellington waited in the entrenched position of Torres Vedras till hunger and cold had reached such a pitch in Massena's army that it started of itself on its retreat; the sword of the defender had had no part in the actual weakening of the assailant. In other cases, in which the two kinds of decision are combined with each other in various proportions, still one of them definitely predominates. So it was in 1812. In that celebrated campaign such a number of bloody engagements took place that in other cases we might on that account have quoted it as a most perfect example of decision by the sword. Nevertheless, no other case, perhaps, has so clearly shown how the assailant may be brought to destruction by his own efforts. Of the 300,000 men composing the French center only about 90,000 reached Moscow. Not more than 13,000 had been detached, so there had been a loss of 197,000 men, and certainly not more than one-third of that loss can be put down to engagements.

All campaigns which have been distinguished by so-called temporizing, like those of the famous Fabius Cunctator, have chiefly calculated on the destruction of the adversary by his own efforts. In general there are numbers of campaigns in which this strategic holding back has been the principal feature without its ever being properly mentioned. Only if we shut our eyes to the fictitious reasons given by the historians and fix our gaze keenly on the events themselves shall we be led to this true cause of many a decision.

By this we think we have sufficiently explained the ideas which lie at the root of defense, the types of it and in these to have clearly indicated the two main kinds of resistance and made intelligible how the principle of awaiting the enemy permeates the whole group of ideas and is combined with positive action in such a way that the latter sometimes sooner or sometimes later comes to the front, and the advantage of awaiting the enemy seems then to be exhausted.

We think that in this way we have gone over and embraced the whole province of defense. Of course there still remain points in it of sufficient importance to form separate chapters, to become, in other words, centers of special groups of ideas, and we must therefore study these also. Such, for example are the nature and influence of fortresses, entrenched camps, defense of mountains and of rivers, operations against the flanks, and so forth. We shall treat of them in the following chapters, but we do not regard any of these points as lying outside the group of ideas above set forth. They are only more detailed applications of them to locality and circumstances. That group of ideas has been deduced from the conception of defense and its relation to attack. We have connected these simple ideas with reality and so pointed out the way in which we can from reality pass back again to those simple ideas and thus gain firm ground, so as not to be forced in our reasoning to take refuge on points of support which themselves float in air.

But there are many ways in which engagements may be combined, especially when they do not find vent in bloodshed but operate through their mere possibility. Resistance by the sword, therefore, acquires such a changed aspect, such a different character that we might be inclined to think there must be a possibility of discovering some other effective mode of combat. We might think that between bloody defeat in a simple battle and the effects of a fine-spun strategy that never lets things go that far, there is such a difference that we must necessarily assume the existence of some new force; just as astronomers from the great space between Mars and Jupiter have been led to infer the existence of other planets.

If the assailant finds the defender in a strong position, which he does not think he can take, or behind a considerable river, which he does not think he can cross, or even if he is afraid, on a further advance, of being unable properly to ensure his subsistence, then it is always the sword of the defender that produces these effects. For it is the fear of being conquered by this sword, either in main engagements or at some specially important points, which brings the assailant's action to a standstill; only he will either not admit this at all or at most in some roundabout way.

Now if it be granted that, even in the case of a bloodless decision, it is in the last instance the engagements which have not actually taken place but have *only been offered* that have brought the decision about, it will still be thought that in such a case the *strategic combination* of these engagements would have to be regarded as the most effective agency, and not their tactical decision, and that only this superiority of the strategic combination can be meant, if we are thinking of other means of defense than those of the sword. We admit this, and it brings us just to the point at which we wanted to arrive. What we say is that if the tactical result in the engagements must constitute the *foundation* of all strategic combinations, it is always possible and to be feared that the assailant may push through to this foundation and before all things direct his efforts to getting the upper hand in these tactical results in order by this means to frustrate the strategic combination. This combination, therefore, must *never be regarded as something independent*, but it only has its value if on account of the tactical results on this or that ground we can be without anxiety. To make ourselves intelligible here in a few words we shall only remind our readers that a general like Bonaparte strode ruthlessly through his opponent's whole strategical web to seek the actual combat *itself*, because in this combat he hardly ever had a doubt as to the issue. Wherever, therefore, strategy did not direct its whole effort to crush him in this combat with a superior strength, wherever it ventured on more subtle (and feebler) expedients, it was torn asunder like a spider's web. A general like Daun, however, could easily be checked by such expedients, but it would be folly to offer Bonaparte and his army what the Prussian army in the Seven Years' War dared to offer Daun and his army. Why? Because Bonaparte knew very well that everything depended on tactical successes and he was sure of them, while with Daun neither of these was the case. *On this account* we think it useful to show that every strategic scheme rests solely on tactical results, and that these, in all cases, whether obtained with bloodshed or without it, are the real, fundamental causes of the decision. Only if we have no reason to be afraid of this decision, whether on account of the character or the circumstances of our opponent, or the moral and physical equality of the two armies, or even the superiority of our own—only then can we, without an engagement, expect something from strategical combinations in themselves.

Now if in the whole compass of military history we find a great number of campaigns in which the assailant gives up his attack without a combat in which blood is shed, and in which, therefore, mere strategic combinations produce such an effect, we might be led to think that these combinations at least possess a great strength in themselves, and that, where in the tactical results a too decisive superiority on the part of the assailant is not to be presupposed, they could alone in most cases decide the affair. To this we must reply that if we are speaking about things which have their origin in the theater of war, and which, therefore, belong to war itself, this idea is mistaken, and that the ineffectiveness of most attacks has its cause in the higher, the political, circumstances of war.

The general circumstances out of which a war arises, and which naturally form its foundation, also determine its character, a point on which we shall have more to say later in discussing the "Plan of a War." These general circumstances, however, have made of most wars half-hearted affairs, in which real hostility has to force its way through such a mass of conflicting relations that it can only remain a very weak element in them. This effect must naturally show itself chiefly and most strongly on the side of the offensive, *the side of positive action.* So certainly we need not wonder if such a breathless, feverish attack could be brought to a standstill by the touch of a finger. Against a feeble resolution, crippled by a thousand considerations, and scarcely yet existing, a mere show of resistance is often enough.

It is not the number of unassailable positions everywhere to be found, not the formidable look of the dark mountain masses that overhang the theater of war, nor of the broad river that sweeps through it, nor the ease with which by certain combinations of engagements the arm that is to strike the blow against us can actually be paralyzed—it is none of these things that is the real cause of the frequent successes which the defender gains without shedding of blood. The cause lies in the weakness of the will with which the assailant puts forward his hesitating foot.

Those counteracting influences can and ought to be taken into account, but we should recognize them as being only what they are and not ascribe their effects to other things, to the things, that is, of which alone we are here speaking. We must not omit expressly to point out how military history in this respect can easily become a perpetual liar and deceiver if criticism has not been careful to take a corrective standpoint.

Let us now consider, in what we may call their common form, the many offensive campaigns without a bloody solution which have failed.

The assailant advances into the enemy's country, pushes his opponent back a little, but finds it too risky to let matters come to a decisive battle. He stops therefore in front of him and acts as though he had made a conquest and had nothing else to do but protect it; as if it was the enemy's business to seek a battle and as if he were offering it to him daily, and so on. All these are *pretenses* with which the commander deludes his army, his government, the world, even himself. The real reason, however, is that he finds the position of the enemy too strong for him. We are not now speaking of the case in which the assailant desists from his attack because he can make no use of his victory, because at the end of his advance he no longer has impulsive force enough to begin a new one. Such a case presupposes an attack that has already succeeded, a real conquest; but we have in view the case where in the middle of his intended conquest the assailant is brought to a standstill.

He is now waiting to take advantage of favorable circumstances, of which there is as a rule no prospect, for the intended attack already shows that no more could be expected from the immediate future than from the present; it is therefore another illusion. If now, as is commonly the case, the operation is in connection with other simultaneous operations, what we shrink from doing ourselves we put upon the shoulders of other armies, and the reasons for our own inactivity are sought in the lack of their support and co-operation. Insurmountable obstacles are talked of, and reasons discovered in the most complicated and subtle considerations. Thus the forces of the assailant waste away in inactivity or, rather, in an activity that is inadequate and therefore without result. The defender gains time, which is for him the most important thing, the season of bad weather draws near, and the attack ends in the assailant's retiring into winter quarters in his own theater of war.

That tissue of false representations now passes into history and suppresses the quite simple, real reason for the absence of success, namely, *the fear of the enemy's sword*. When criticism takes up such a campaign it is at pains to find reasons for and against, which yield no convincing result, because they all float in the air and we never descend into the real foundation of truth. The opposition through which the elemental energy of war, and therefore of the attack in particular, is weakened, lies for the most part in the political relations and intentions of the state, and these are always concealed from the world, the state's own people and the army, in many cases even from the commander-in-chief. No one will account for his own faint-heartedness by the confession that he fears he cannot attain the desired object with the force at his disposal or that he will rouse new enemies or that he does not wish to make his allies too powerful, and so on. Such things are hushed up for a long time or forever, but what has happened has to be put before the world in a coherent form, so the commander-in-chief is obliged, either on his own account or on that of the government to give currency to a tissue of fictitious reasons. These ever-recurring sham fights of military dialectic have ossified in theory into systems, which naturally contain just as little truth. Only by following the simple thread of internal connection, as we have tried to do, can theory return to the essential reality of things.

If we regard military history with this suspicion, a mighty pile of offensive and defensive lore, all mere idle verbiage, collapses, and the simple sort of idea we have given of the subject spontaneously emerges. We think, therefore, that it must be applied to the whole ground of defense, and that only by holding fast to it are we able with clear judgment to deal with the mass of events.

We have now still to consider the question of the employment of these different forms of defense.

They are all gradations of the same thing, purchased by continually increasing gradations of sacrifice, and on that account, if other circumstances did not enter in, the general's choice between them would already be sufficiently determined. He would choose that form which seemed to him just sufficient to give his force the degree of defensive capacity required, but to avoid useless sacrifices he would refuse to retire further. But we must not overlook the fact that the choice between these different forms is usually very limited, because the other essential points to be considered in the defense necessarily force him to one or another of these forms. For a retreat into the interior of the country the theater of war must have a considerable area, or there must be circumstances like those in Portugal in 1810, where *one* ally (England) gave support in the rear, and another (Spain), with its wide territory, considerably weakened the force of the enemy's attack. The situation of the fortresses, whether they are more on the frontier or more in the interior of the country, can likewise be decisive for or against such a plan, but still more the character, habits and sentiments of the inhabitants. The choice between an offensive and a defensive battle may be decided by the plan of the enemy, or by the peculiar qualities of both the armies and their generals. Finally, the possession of an exceptionally good position or line of defense, or the lack of it, may indicate one form or another. In short it is enough to mention these things to make us feel that it is more by them than by the mere relative strength of the armies that in many cases the choice of the defense will be determined. As we shall become better acquainted with the most important of the subjects here touched upon, the influence they have upon the choice can then be more clearly explained, and finally in Book VIII, "Plan of a War," all will be put together in one whole.

But this influence will mostly be decisive only if the two armies are more or less equal in strength; in the opposite case, and therefore in most cases, their relative strength will prevail. There is ample proof in military history that it has done so, without the chain of reasoning which has been here unfolded, and therefore vaguely *by mere instinctive judgment*, like most things that happen in war. It was the same general, the same army, that on the same theater of war, on one occasion fought the battle of Hohenfried-berg, and on another took up the camp at Bunzelwitz. Therefore even Frederick the Great, who, as regards battle, was the most inclined of all generals to the offensive, saw himself compelled at last, when greatly inferior in strength, to take up a real defensive position. And Bonaparte, who earlier would fall on his foe like a wild boar, do we not see him when the proportion of force turned against him, in August and September, 1813, turn this way and that, as if penned in a cage, instead of rushing forward ruthlessly upon some one of his opponents? In October of the same year, when the disproportion reached its climax, do we not see him at Leipzig seeking shelter in the angle formed by the Parthe, the Elster and the Pleisse, as if he was waiting for his enemy in the corner of a room with his back against the wall?

We cannot refrain from observing that from this chapter more than from any other in our book it becomes clear that our object is not to lay down new principles and methods of conducting war, but merely to investigate in its innermost relations what has long existed and to reduce it to its simplest elements.

CHAPTER 9

DEFENSIVE BATTLE

We have said, in the preceding chapter, that the defender, in the conduct of his operations, can make use of a battle which is tactically of a purely offensive character, if, at the moment the enemy invades his theater of war, he marches against him and attacks him; but that he may also await the appearance of the enemy before his front and then proceed to the attack. In this case the battle, tactically, will again be an offensive battle, although in modified form. Lastly, he may actually await the enemy's attack in his position, and then oppose him by local defense and by offensive action with portions of his force. In all this we can naturally conceive several different degrees and gradations, deviating more and more from the principle of a positive counterstroke and passing into that of local defense. We cannot here venture to say how far this may be carried, and which might be the most advantageous relation of the two elements of offensive and defensive for winning a decisive victory. But we maintain that when such a victory is desired, the offensive part of the battle should never be completely lacking, and we are convinced that all the effects of a decisive victory may and must be produced by this offensive part, just as in a purely tactical offensive battle.

As the field of battle is only a point in strategy, so the duration of a battle is, strategically, only an instant of time, and it is the end and result, not the course, of a battle that constitutes a strategic quantity.

Now if it were true that a complete victory may result from the offensive elements which lie in every defensive battle, there could be no fundamental difference between an offensive and a defensive battle, as far as regards strategic combinations. We are indeed convinced that this is so, but no doubt it appears otherwise. In order to examine the subject more closely, to make our view clear and thereby remove that appearance, we shall sketch, hastily, the picture of a defensive battle, such as we conceive it.

The defender waits for the attack in a position; for this he has selected and prepared suitable ground, that is, he has made himself well acquainted with the locality, constructed strong entrenchments at some of the most important points, opened and cleared paths of communication, posted batteries, fortified villages and chosen suitable places where he can draw up his masses under cover, etc. While the forces on both sides are consuming each other at the different points where they come into contact, a front more or less strong, the approach to which is made difficult by one or more parallel trenches or other obstacles, or by the influence of some strong commanding points, enables him with *a small part of his force to destroy great numbers of the enemy* at the various stages of the defense up to the heart of the position. The points of support which he has given his wings secure him from any sudden attack from several quarters; the covered ground which he has chosen for his position makes the enemy cautious, indeed timid, and affords the defensive the means of diminishing by partial and successful attacks the general backward movement which goes on as the engagement becomes gradually concentrated within narrower limits. Thus the defender regards with contentment the battle which rages with moderated violence in front of him—but he does not reckon that his resistance in front can last forever. He does not think his flanks impregnable; he does not expect that the whole course of the battle will be changed by the successful charge of a few battalions or squadrons. His position is *deep*, for each part in the graduated scale of the order of battle, from the division down to the battalion, has its reserve for unforeseen events and for renewal of the engagement. At the same time a considerable mass, one-quarter to one-third of the whole, is kept quite in the rear out of the battle, so far back that there can be no question of loss due to the enemy's fire, and if possible so far as to be beyond the encircling line by which the enemy might attempt to turn either flank. With this body he intends to cover his flanks from wider and greater turning movements and secure himself against unforeseen events. In the last third of the battle, when the assailant's plan is fully developed and most of his troops have been spent, the defender will throw this mass on a part of the enemy's army, and against this develop his own smaller offensive, using all the elements of attack, such as charges, surprise, turning movements, and by means of this pressure against the center of gravity of the battle, which is still in the balance, force the whole backward.

This is our normal conception of a defensive battle, based on our present-day tactics. In this battle the general turning movement by which the assailant intends to give his attack a better chance, and at the same time to make his success more complete, is answered by a subordinate turning movement on the part of the defensive against that part of the assailant's force used in his own turning movement. This subordinate turning movement may be regarded as sufficient to destroy the effect of the enemy's attempt, but it cannot give rise to a similar general enveloping of the assailant's army. The difference, therefore, between the forms which the victory takes will always be that in an offensive battle the enemy's army is encircled, and the action is directed toward its center, while in a defensive battle the action is more or less from the center to the circumference, in the direction of the radii.

On the field of battle itself, and in the first stage of the pursuit, the enveloping form must always be considered the more effective; not so much, generally, on account of its form, but rather only in the event of the envelopment being carried through to completion, to the point, that is to say, when in the battle itself it succeeds in essentially limiting the enemy's possibilities of retreat. But it is just against this extreme point that the defender's positive counterstroke is directed, and in many cases where this effort is not sufficient to obtain a victory, it will at least suffice to protect him from such an extreme as we allude to. But we must always admit that this danger, namely, of having the line of retreat seriously limited, is particularly great in defensive battles, and if it cannot be averted, the success in the battle itself, and in the pursuit, is thereby very much enhanced in favor of the enemy.

But it is usually only so in the first stage of pursuit, that is, until nightfall; on the following day the envelopment has reached its end, and both parties are again on an equality in this particular respect.

The defender may, no doubt, have lost his principal line of retreat, and therefore be placed strategically in a permanently disadvantageous position; but in most cases the turning movement itself will be at an end, because it was only planned for the field of battle, and therefore cannot reach much further. But what will take place, on the other hand, if the *defender* is victorious? A *division* of the defeated force. This facilitates the retreat at the first moment, but on the *next day a concentration of all parts* is the greatest necessity. Now if the victory has been a very decisive one, if the defender pursues with great energy, this concentration will often become impossible, and from this division of the beaten force the worst consequences may follow, which may gradually culminate in a complete dispersion. If Bonaparte had been victorious at Leipzig, the allied armies would have been completely divided, which would have reduced considerably the level of their strategic condition. At Dresden, although Bonaparte did not fight a regular defensive battle, the attack had the geometrical form of which we have been speaking, that is, from the center to the circumference. The embarrassment of the allies, in consequence of their separation, is well known, an embarrassment from which they were relieved only by the victory on the Katzbach, the tidings of which caused Bonaparte to return to Dresden with the Guard.

This battle on the Katzbach is itself a similar example. In it the defender at the last moment passed over to the offensive, and consequently operated on diverging lines; the French corps were thus forced apart, and several days after, as the fruits of the victory, Puthod's division fell into the hands of the Allies.

The conclusion we draw from this is that if the assailant, by converging lines, which are more natural to him, has a means of increasing his victory, on the other hand, the defender, by the divergent form which is natural to him, acquires also a means of giving greater results to his own victory than would be the case by a merely parallel position and frontal attack, and we think that the one means is at least as good as the other.

If, however, in military history we rarely find such great victories resulting from the defensive battle as from the offensive, that proves nothing against our assertion that in itself the one is as well suited to produce victory as the other; the real cause lies in the very different condition of the defender. The defender is generally the weaker of the two, not only in the amount of his forces, but also in every other respect; he either is, or thinks he is, not in a condition to follow up his victory with great results, and contents himself with merely fending off the danger and saving the honor of his arms. That the defender by inferiority of force and other circumstances may be tied down to that degree we do not dispute. But there is no doubt that this, which should be only the consequence of a necessity, has often been assumed to be the consequence of that part which every defender has to play. Thus in a truly absurd way it has become an axiom that defensive battles should really be confined to warding off attacks and not directed to the destruction of the enemy. We hold this to be one of the most harmful errors, a real confusion between the form and the thing it-self, and we maintain unreservedly that in the form of war which we call *defense*, the victory is not only more probable but may also attain the same magnitude and efficacy as in the attack, and that this may be the case not only in *the total result* of all the engagements which constitute a campaign, but also in any *particular* battle, if the necessary degree of force and energy is not wanting.

CHAPTER 10

FORTRESSES

Formerly, and up to the time of great standing armies, fortresses, that is, castles and fortified towns, were only built for the protection of the inhabitants. The baron, if he saw himself pressed on all sides, took refuge in his castle to gain time and wait for a more favorable moment, and towns sought by their walls to keep off the passing hurricane of war. This simplest and most natural object of fortresses did not continue to be the only one. The relations which such a place acquired to the whole country and to troops fighting here and there in it soon gave these fortified points a wider importance, a significance which made itself felt beyond their walls and contributed essentially to the conquest or retention of the country, to the successful or unsuccessful issue of the whole contest. In this manner they even became a means of making war more of a connected whole. Thus fortresses acquired that strategic significance which for a time was regarded as so important that it dictated the fundamental lines of plans of campaign, which were directed more to the taking of one or more fortresses than to the destruction of the enemy's army. People now turned to the reason for this significance, the relations, that is to say, of a fortified point to the country and the armies, and it was now thought that in defining the purpose of the points to be fortified it was impossible to be too careful, subtle and abstract. In this abstract purpose the original one was almost lost sight of, and the idea was conceived of fortresses apart from either towns or inhabitants.

On the other hand, the times are past in which the mere enclosure of a place with walls, without any other military preparations, could keep it perfectly dry when an inundation of war was sweeping over the whole country. Such a possibility rested partly on the former division of nations into small states, partly on the periodical character of the incursions then in vogue, which, almost like the seasons, had their fixed and very limited duration, since either the feudal forces hastened home, or the pay for the *condottieri* regularly used to run short. Now that large standing armies, with powerful trains of artillery mow down the opposition of walls or ramparts as it were with a machine, neither town nor other small corporation has any longer an inclination to put their forces to the hazard, only to be captured a few weeks or months later, and then to be treated all the more severely. Still less can it be to the interest of an army to break itself up by providing garrisons for a number of fortified places, which may for a time, no doubt, retard the progress of the enemy, but must in the end necessarily submit. We must always retain enough forces to be a match for the enemy in the open field, unless we can depend on the arrival of an ally who will relieve our strong places and set our army free. Consequently, the number of fortresses has necessarily much diminished, and this has again led to the abandonment of the idea of directly protecting the population and property in towns by fortifications, and promoted the other idea of regarding the fortresses as an *indirect* protection to the country, which they afford by their strategic importance as knots which hold together the strategic web.

Such has been the course of ideas, not only in books but also in practical life, though in books, as usually happens, it has certainly been much more spun out.

Natural as was this tendency of things, still these ideas were carried too far, and mere subtleties and fancies displaced the sound core of a natural and urgent need. We shall look only into these simple and important needs when we enumerate the objects and conditions of fortresses; we shall thereby proceed from the simple to the more complicated, and in the succeeding chapter we shall see what is to be deduced therefrom as to the determination of the position and number of fortresses.

The effectiveness of a fortress is plainly composed of two different elements, the passive and the active. By the first it protects the place, and all that it contains; by the other it exercises a certain influence over the adjacent country, even beyond the range of its guns.

This active element consists in the attacks which the garrison may undertake upon every enemy who approaches within a certain distance. The larger the garrison, so much the stronger numerically will be the detachments that may be employed for such objects, and the stronger the detachments, the further as a rule they can go. From which it follows that the sphere of the active influence of a great fortress is not only greater in intensity but also more extensive than that of a small one. But the active element itself is again, to a certain extent, of two kinds, consisting, namely, of operations of the garrison proper and of operations which other bodies of troops, great and small, not belonging to the garrison but in connection with it, may be able to carry out. For instance, bodies which independently would be too weak to face the enemy may through the shelter which, in case of necessity, the walls of a fortress afford them, be able to maintain themselves in the country and to a certain extent to command it.

The operations which the garrison of a fortress can venture to undertake are always rather restricted. Even in the case of large places and strong garrisons, the detachments which can be employed for them are mostly inconsiderable as compared with the forces in the field, and their sphere of influence seldom exceeds a couple of days' marches. If the fortress is small, the detachments it can send out are quite insignificant and the range of their effectiveness will generally be confined to the nearest villages. But bodies which do not belong to the garrison, and therefore are not under the necessity of returning to the place, are on that account much more at liberty in their movements, and by their means, if other circumstances are favorable, the active sphere of effectiveness of a fortress may be extraordinarily extended. Therefore if we speak of the active influence of fortresses in general, we must always keep this feature of them principally in view.

But even the smallest active influence of the weakest garrison is still essential for the different objects which fortresses are destined to fulfill, for, strictly speaking, even the most passive of all the functions of a fortress (defense against the attack) cannot be imagined apart from that active influence. At the same time it is evident that among the different purposes which a fortress may have to answer generally, or at this or that particular moment, the passive influence will be more required in one case, the active in another. Some of these purposes are simple, and the influence of the fortress will in such case be to a certain extent direct; others are complicated, and the influence then becomes more or less indirect. We shall examine these subjects separately, beginning with the first; but at the outset we must state that a fortress may, of course, be intended to answer several of these purposes, or even all of them, either simultaneously, or at all events at different stages of the war.

We say, therefore, that fortresses are great and most important supports of the defensive in the following ways:

1. *As secure depots of stores of all kinds.* The assailant during his aggression subsists his army from day to day; the defender usually must have made preparations long beforehand; he cannot therefore draw provisions exclusively from the district he occupies, and which he in any case desires to spare. Therefore storehouses are a great necessity for him. The provisions of all kinds which the aggressor possesses are in the rear as he advances, and are therefore withdrawn from the dangers of the theater of war, while those of the defensive remain exposed to them. If these provisions of all kinds are not in *fortified places,* a most injurious effect on the operations in the field must be the consequence, and the most extended and forced positions often become necessary in order to cover them.

   An army on the defensive without fortresses has a hundred vulnerable spots; it is a body without armor.

2. *As a protection to great and wealthy cities.* This purpose is very closely related to the first, for great and wealthy cities, especially commercial cities, are the natural storehouses of an army; as such, their possession and loss affects the army directly. Besides this, it is always worth while to preserve this portion of the national wealth, partly on account of the resources which it furnishes indirectly, partly because, in negotiations for peace, an important place is in itself a noticeable weight thrown into the scale.

This use of fortresses has been too little esteemed in modern times, and yet it is one of the most natural, and one which has the most powerful effect and is the least liable to mistakes. If there were a country in which not only all great and rich cities, but all populous places as well were fortified, and defended by the inhabitants and the farmers belonging to the adjacent districts, then by that means the rapidity of military operation would be so much reduced, and the people attacked would bear down the scale with so great a part of their whole weight that the talent as well as the force of will of the enemy's general would sink to nothing. We only mention this ideal of the fortification of a country to do justice to what we have just supposed to be the proper use of fortresses, and to prevent the importance of the *direct* protection which they afford being overlooked at any time. But in any other respect this idea is not to interrupt our investigation again; for among the whole number of fortresses there would always have to be some which, being more strongly fortified than others, would serve as the real supports of the military forces.

The purposes specified under 1 and 2 hardly call forth any other but the passive action of fortresses.

3. *As real barriers.* They close the roads, and in most cases also the rivers, on which they are situated.

   It is not as easy as is generally supposed to find a usable lateral road which passes round a fortress, for this detour must be made, not only out of reach of its guns, but also in more or less wide circuits, to avoid sorties of the garrison.

   If the country is in the least degree difficult, there are often delays connected with the slightest detour from the road which may cause the loss of a whole day's march, and, if the road is repeatedly used, may become of great importance.

   How they may interfere with enterprises by closing the navigation of a river is self-evident.

4. *As tactical* points d'appui. Since the diameter of the zone effectively covered by the fire of a not entirely insignificant fortress is usually some miles, and the offensive sphere of effectiveness in each case reaches somewhat farther, fortresses may be considered always as the best *points d'appui* for the flanks of a position. A lake several miles long can certainly be considered an excellent support for the wing of an army, and yet a fortress of moderate size achieves more. The flank does not require to rest close upon it, as the assailant, for the sake of his retreat, would not throw himself between it and our flank.

5. *As a station.* If fortresses are situated on the line of communication of the defensive, as is generally the case, they serve as convenient stations for all that passes up and down this line. The danger to lines of communication is mostly from raiding parties whose operation is always only intermittent. If a valuable convoy, on the unexpected approach of such a party, can reach a fortress by hastening the march or turning quickly, it is saved, and may wait there till the danger is past. Further, all troops marching to or from the army, after halting here for a few days, are better able to hasten the remainder of the march, halting days being just the time of greatest danger for them. In this way a fortress situated half way on a line of communication of one hundred and fifty miles shortens the line one half, so to speak.

6. *As places of refuge for weak or defeated corps.* Under the guns of a not too small fortress every corps is safe from the enemy's blows, even if no entrenched camp is specially prepared for it. No doubt such a corps must give up its further retreat, if it wants to stay there; but this is no great sacrifice in cases where a further retreat would only end in complete destruction.

In many cases a fortress can ensure a few days' halt without the retreat being altogether stopped. For the slightly wounded and stragglers, etc., who precede a beaten army, it is especially suited as a place of refuge, where they can wait to rejoin the army.

If Magdeburg had lain on the direct line of the Prussian retreat in 1806, and if that line had not been already lost at Auerstädt, the army could easily have halted for three or four days near that great fortress, and rallied and reorganized itself. But even as things were, it served as a rallying point for the remains of Hohenlohe's army, which only there resumed its visible existence.

It is only by actual experience in war itself that the beneficial influence of fortresses close at hand under unfavorable circumstances can be rightly understood. They contain powder and arms, forage and bread, give covering to the sick, security to the sound, and presence of mind to the frightened. They are like an hostelry in the desert.

In the four last-named purposes it is obvious that the active influence of fortresses is somewhat more called for.

> 7. *As a real shield against the enemy's aggression.* Fortresses which the defender leaves in his front break like blocks of ice the stream of the enemy's attack. The enemy must at least surround them, and requires for that, if the garrisons are brave and enterprising, perhaps double their strength. Moreover, these garrisons may and do mostly consist in part of troops, who, although competent in a garrison, are not fit for the field—half-trained militia, partial invalids, armed citizens, *landsturm*, etc. The enemy, therefore, is perhaps weakened four times more than we are in such a case.

This disproportionate weakening of the enemy's power is the first and most important but not the only advantage which a besieged fortress affords by its resistance. From the moment that the enemy breaks through our line of fortresses, all his movements become much more restricted; he is limited in his lines of retreat and must constantly attend to the direct covering of the sieges which he undertakes.

Here, therefore, fortresses co-operate with the defense in a splendid and very decisive manner, and of all the purposes they can have, this must be regarded as the most important.

If, nevertheless, this use of fortresses—far from being seen regularly repeating itself—occurs comparatively seldom in military history, the cause of this is to be found in the character of most wars, this means being in a certain sense too decisive and too thoroughly effective for them, which will only later be made clearer.

In this use of fortresses it is at bottom chiefly their offensive power that is called for; at least it is this by which their effect is chiefly produced. If a fortress were no more to an aggressor than a point which could not be occupied by him, it might become an obstacle to him, but not to such a degree as to be able to induce him to lay siege to it. But as he cannot leave six, eight or ten thousand men to do as they like in his rear, he is obliged to invest the place with a sufficient force, and in order not to have to go on doing this indefinitely, he must lay siege to it and take it. From the moment the siege begins it is then chiefly the passive efficacy of the fortress which comes into action.

All the purposes of fortresses which we have been hitherto considering are fulfilled in a comparatively simple and direct manner. On the other hand, in the next two objects the method of action is more complicated.

8. *As a protection to extended quarters.* That a moderate-sized fortress closes the approach to quarters lying behind it for a width of fifteen to twenty miles is a mere result of its existence; but how such a place comes to have the honor of covering a line of quarters seventy-five to one hundred miles in length, a feat so frequently mentioned in military history as a fact, is something that requires investigation, as far as it has really taken place, and refutation, as far as it may be mere illusion.

The following points have to be considered here:

(1) That the place in itself blocks one of the main roads, and really covers a breadth of fifteen to twenty miles of country.
(2) That it may be regarded as an exceptionally strong advanced post, or that it affords a more complete observation of the country, which is still further increased through secret information afforded by the relations of civil life which exist between a great city and the adjacent districts. It is natural that in a place of six, eight or ten thousand inhabitants, we should be able to learn more of what is going on in the neighborhood than in a mere village, the quarters of an ordinary outpost.
(3) That smaller bodies rest upon it, find in it protection and security, and can, from time to time, advance against the enemy, be it to bring in intelligence, or, in case he passes by the fortress, to undertake something against his rear; that, therefore, although it cannot leave its place, a fortress may still in a certain sense perform the functions of an advanced corps. (Book V, Chapter 8.)

(4) That the defender, after assembling his troops, can take up his position at a point directly behind this fortress, which the assailant cannot reach without becoming exposed to danger from leaving the fortress in his rear.

Every attack, no doubt, on a line of quarters as such is to be taken in the sense of a surprise, or, rather, we are only speaking here of that kind of attack. Now it is obvious that an attack by surprise accomplishes its effect in a much shorter space of time than the actual attack on a theater of war. Therefore, although in the latter case, a fortress which is to be passed by must necessarily be invested and kept in check, this will not be so necessary in the case of a mere sudden attack on a line of quarters, and therefore in the same proportion the fortress will be less an obstacle to the surprise of the quarters. That is true enough; also the quarters lying on the flanks at a distance of thirty to forty miles from the fortress cannot be directly protected by it; but the object of such a surprise does not consist merely in the attack of a few quarters. Until we reach the book on attack we cannot describe circumstantially the real object of such a surprise and what may be expected from it; but this much we may say at present: that its principal result is obtained, not by the actual attack on the separate quarters, but by the series of engagements which the pursuing aggressor forces on isolated detachments which are not in proper order, and more prepared for hurrying to certain points than for fighting. But this attack and pursuit will always have to be in a direction more or less toward the center of the enemy's quarters, and, therefore, an important fortress lying in front of this center will certainly prove a very great impediment to the assailant.

If we consider these four points in the whole of their joint effects, we see that an important fortress in a direct and in an indirect way certainly gives some security to a much greater extent of quarters than we should think at first sight. "Some security" we say, for all these indirect effects do not render the advance of the enemy *impossible;* they only make it *more difficult,* and, a more serious matter, consequently less probable and less of a danger to the defender. But that is all that is required, and all that is understood in this case under the term "covering." The real direct security must be attained by means of outposts and the arrangement of the quarters themselves.

There is, therefore, some truth in ascribing to a great fortress the capability of covering a wide extent of quarters lying in rear of it; but it is also not to be denied that often in actual plans for campaigns and still oftener in historical descriptions we meet with empty expressions or illusory views in connection with this subject. For if that covering is only realized by the co-operation of several circumstances, if even then it only produces a diminution of the danger, we can easily see that, in particular cases, through special circumstances, above all through the boldness of the enemy, this whole covering may prove an illusion. In actual war we must therefore not content ourselves with assuming offhand the efficacy of such and such a fortress, but must definitely study each single case.

9. *As covering a province not occupied.* If during the war a province is either not occupied at all, or only by an insufficient force, and likewise exposed more or less to incursions from enemy raiding parties, then a fortress, if not too insignificant in size, may be looked upon as a covering, or, if we prefer, as a security, for this province. It may, no doubt, be regarded as a security, for an enemy cannot become master of the province until he has taken the fortress, and that gives us time to hasten to its defense. But actual covering can certainly only be considered very indirect, or understood in a *figurative* sense. That is, the fortress can only by its active operation in some measure check the incursions of enemy raiders. If this operation is limited merely to what the garrison can effect, then the result will be small, for the garrisons of such places are generally weak and usually consist of infantry only, and that not of the best quality. The idea becomes a little more practical if small columns keep themselves in communication with the place, making it their base and place of retreat.

10. *As a focus of a general arming of the nation.* Provisions, arms and munitions cannot be supplied in a regular manner in a national war. On the other hand, it is just in such a war that each man naturally shifts for himself in these matters. In that way a thousand small sources furnishing means of resistance are opened which otherwise might have remained unused; and it is easy to see that a significant fortress, as a great magazine of these things, can well give to the whole resistance more force and solidity, more cohesion and greater results.

Besides, a fortress is a place of refuge for the wounded, the seat of the civil authorities, the treasury, the point of assembly for greater enterprises, and so forth. Lastly, it is a nucleus of resistance which, during the siege, places the enemy's forces in such a condition as facilitates and favors attacks of national levies upon them.

11. *For the defense of rivers and mountains.* Nowhere can a fortress answer so many purposes, undertake to play so many parts, as when it is situated

on a great river. It secures the passage at any time at that spot and hinders that of the enemy for several miles each way; it commands the use of the river for commercial purposes, gives shelter to all ships, blocks bridges and roads, and offers an opportunity for the indirect defense of the river, that is, the defense by means of a position on the enemy's side of it. It is evident that, by its influence in so many ways, it very greatly facilitates the defense of the river, and may be regarded as an essential part of that defense.

Fortresses in mountains are important in a similar manner. There they open or close whole systems of roads which have their junction at that spot; they thus command the whole country which is traversed by these mountain roads, and may be regarded as the true buttresses of the whole defensive system.

CHAPTER 11

FORTRESSES (CONTINUED)

We have discussed the purposes of fortresses; now for their situation. At first the subject seems very complicated, when we think of how many purposes there are, each of which may again be modified by the locality, but such apprehension has very little foundation if we keep to the essence of the subject and guard against unnecessary subtleties.

It is evident that all these demands are satisfied at the same time if, in those districts of country which are to be regarded as the theater of war, all the largest and richest cities on the great high roads connecting the two countries with each other are fortified, more particularly those adjacent to ports and inlets, or situated on large rivers and in mountains. Great cities and great high roads go hand in hand, and both have also a natural connection with great rivers and the sea coasts. All these four conditions, therefore, will agree very well with each other, and give rise to no contradiction. On the other hand, it is not the same with mountains, for large cities are seldom found there. If, therefore, the position and direction of a mountain range makes it suitable to a defensive line, it is necessary to close its roads and passes by small forts, built for this purpose alone and at the least possible cost, the great fortifications being reserved for the important cities in the level country.

We have not yet considered the frontiers of the state, nor said anything about the geometrical form of the whole system of fortresses, nor about the other geographical points connected with their situation, because we regard the conditions above mentioned as the most essential, and think that in many cases they alone are sufficient, particularly in small states.

But at the same time other conditions may be admitted, and even become necessary in countries of greater superficial extent, which either have a great many important cities and roads, or, on the contrary, are almost without any; in countries which are either very rich, and though already possessing many fortresses, still desire new ones, or, on the other hand, are very poor and must make a few do—in short, in cases where the number of fortresses does not fairly well correspond with the number of important cities and roads which present themselves, being either considerably greater or less. We shall now glance at these conditions.

The main questions which remain relate to:

1. The choice of the principal roads, if the two countries are connected by more roads than we wish to fortify.
2. Whether the fortresses are to be placed on the frontier only or spread over the country, or,
3. Whether they shall be distributed uniformly, or in groups.
4. The nature of the geographical features of the country to which it is necessary to pay attention.

A number of other questions, derived from the geometrical form of the line of fortifications, such as whether they should be placed in a single line or in several lines, that is, whether they do more service when placed one behind the other, or side by side in line with each other; whether they should be checker-wise, or in a straight line; or with salients and re-entering angles like the fortresses themselves—all these we regard as empty subtleties, that is, as considerations so insignificant, that the more important points will not allow them to be mentioned. We only touch on them here because they are not merely treated in many works, but also a great deal more is made of these pitiful trivialities than they are worth.

As regards the first question, in order to place it more clearly before our eyes, we shall only call to mind the relation of the south of Germany to France, that is, to the upper Rhine. If, without considering the number of separate states composing this district of country, we suppose it to be a whole which is to be fortified strategically, much doubt would necessarily arise, for a great number of the finest high roads lead from the Rhine into the interior of Franconia, Bavaria and Austria. Certainly, cities are not lacking which surpass others in size and importance, as Nuremburg, Würzburg, Ulm, Augsburg and Munich; but if we are not disposed to fortify all, there is no alternative but to make a selection. If, further, in accordance with our view, the fortification of the greatest and wealthiest is held to be the principal thing, still it is not to be denied that, owing to the distance between Nuremburg and Munich, the first has a very different strategic significance from the second. Therefore it always remains to be considered whether it would not be better, in place of Nuremburg, to fortify some other place in the neighborhood of Munich, even if the place is of less importance in itself.

As concerns the decision in such cases, that is, answering the first question, we must refer to what has been said in the chapters on the general plan of defense, and on the choice of the point of attack. Wherever the most natural point of attack is situated, there, above all, we shall place our defensive arrangement.

Therefore, amongst a number of great roads leading from the enemy's country into ours, we shall, above all, fortify that which leads most directly to the heart of our dominions, or that which, traversing fertile provinces or running alongside of navigable rivers, most facilitates the enemy's undertaking. We may then rest secure that the assailant either encounters there fortifications, or, should he resolve to pass them by, he will naturally offer a favorable opportunity for operations against his flank.

Vienna is the heart of South Germany, and plainly Munich or Augsburg, in relation to France alone (Switzerland and Italy being supposed neutral), would be more effective as a principal fortress than Nuremburg or Würzburg. But if, at the same time, we look at the roads leading into Germany from Switzerland through the Tyrol and from Italy, this will become still more evident, because, in relation to these, Munich and Augsburg will always be places of some effectiveness, whereas Würzburg and Nuremburg, in this respect, might just as well not exist at all.

We now turn to the second question: whether the fortresses should be placed on the frontier only, or distributed over the whole country. In the first place, we must observe that, as regards small states, this question is superfluous, for what in strategy can be called *frontiers* coincides, in their case, nearly with the whole country. The larger the state is supposed to be that we are considering in relation to this question, the more obvious appears the necessity of answering it.

The most natural answer is that fortresses belong to the frontiers, for they are to defend the state, and the state is defended as long as the frontiers are defended. This conclusion may be valid in general, but the following considerations will show how many limitations it may have.

Every defense which reckons chiefly on foreign assistance lays great value on gaining time. It is not a vigorous counterstroke, but a slow proceeding, in which the chief gain consists more in delay than in the weakening of the enemy. But it is natural that, all other circumstances being supposed alike, fortresses which are spread over the whole country, and include between them a very considerable area of territory, will take longer to be captured than those squeezed together in a close line on the frontier. Further, in all cases in which the object is to overcome the enemy through the length of his communications and the difficulty of maintaining himself in countries which are specially able to count on this kind of counterstroke, it would be a complete contradiction to make the defensive preparations only on the frontier. Lastly, let us also remember that, if circumstances will in any way permit it, the fortification of the capital is a main point; that according to our principles the chief cities and centers of trade demand it also; that rivers passing through the country, through mountains and other obstacles of ground afford advantages for new lines of defense; that many cities, through their strong natural situation, invite fortification; moreover, that certain accessories of war, such as munition factories, etc., are better placed in the interior of the country than on the frontier, and their value well entitles them to the protection of fortifications. Then we see that there is always more or less occasion for the construction of fortresses in the interior of a country. On this account we think that although states which possess a great number of fortresses are right in placing the greater number on the frontier, still it would be a great mistake if the interior of the country was left entirely destitute of them. We think, for instance, that this mistake has been made in a remarkable degree even in the case of France. A great doubt may in this respect justifiably arise, if the border provinces of a country contain no considerable cities, such cities only lying farther in the interior, as is the case in South Germany in particular, since Swabia is almost destitute of great cities, while Bavaria contains a large number. We do not hold it to be necessary to remove these doubts once for all on general grounds, believing that in such cases, in order to arrive at a solution, reasons derived from the particular situation must be taken into account. Still we must call attention to the closing remarks in this chapter.

The third question—whether fortresses should be kept together in groups or more equally distributed—will, if we reflect upon it, seldom arise. Still we must not, for that reason, count it among the useless subtleties, because, no doubt, a group of two, three or four fortresses, which are only a few days' march from a common center, give that point and the army placed there such strength, that, if the other conditions to some extent allow of it, there must be a great temptation to form such a strategic bastion.

The last point concerns the remaining geographical relations of the point to be chosen. That fortresses on the sea, on streams and great rivers and in mountains are doubly effective, we have already stated because it is one of the principal things to take into account, but there are a number of other points which must be considered.

If a fortress cannot lie on the river itself, it is better not to place it near, but at a distance of fifty to sixty miles from it; otherwise the river intersects and disturbs the sphere of action of the fortress in all those respects above mentioned.[1]

This is not the same in mountains, because there the movement of large or small masses is not restricted in the same degree to particular points as it is by a river. But fortresses on the enemy's side of a mountain range are not well placed, because they are difficult to relieve. If they are on our side, the difficulty of the enemy's laying siege to them is very great, as the mountains cut across his line of communication. We cite Olmütz, 1758, as an example.

It is easily seen that large impassable forests and marches have a similar effect to that of rivers.

The question has often been raised as to whether cities situated in a very inaccessible country are well or ill suited for fortresses. Since they can be fortified and defended at lesser expense, or be made much stronger, often impregnable, at an equal expenditure of forces, and since the services of a fortress are always more passive than active, it does not seem necessary to attach an excessive importance to the objection that they can easily be blockaded.

If we now, in conclusion, cast a retrospective glance over our simple system of fortification for a country, we may assert that it rests on large and lasting things and circumstances, directly connected with the foundations of the state itself, not on transient views on war, fashionable only for a day; not on imaginary strategic niceties, not on entirely individual needs of the moment—an error which might be attended with irreparable consequences if allowed to influence the construction of fortresses intended to last five hundred, perhaps a thousand, years. Silberberg in Silesia, built by Frederick II on one of the ridges of the Sudeten, has, from the complete alteration in circumstances which has since taken place, almost entirely lost its importance and object, while Breslau, if it had been, and continued to

---

[1] Philippsburg was the pattern of a badly placed fortress; it resembled a fool standing with his nose close to a wall.

remain, a strong fortress, would have always maintained its value against the French, as well as against the Russians, Poles and Austrians.

Our reader will not overlook the fact that these considerations have not been set forth for the case of a state providing itself with a brand-new set of fortifications. They would be useless, if such was their object, since such a case seldom, if ever, occurs; but they may all occur at the construction of any single fortress.

CHAPTER 12

DEFENSIVE POSITION

Any position in which we accept battle, by making use in it of the ground as a means of protection, is a *defensive position*, and it makes no difference whether our attitude in so doing is one of passivity or of offense. This follows from the general view we have taken of defense.

Now we might further apply the term to any position in which an army, marching to meet the enemy, would probably accept battle, if the enemy sought it there. In point of fact most battles come about in this way, and all through the Middle Ages there was no question of anything else. This, however, is not the sort of position of which we are speaking here; the great majority of all positions are of this kind, and the conception of a *position*, as opposed to a *camp taken up on the march*, will suffice for that. A position which is specially designated as a *defensive position* must, therefore, be something different.

In decisions which take place in an ordinary position, the conception of *time* obviously predominates. The armies march against each other in order to come to an engagement; the place is a subordinate point—all that is required of it is that it should not be unsuitable. But in a real defensive position the conception of *place* predominates. The decision is to be given at this place, or, rather, principally *through* this place. It is only of such a position as this that we are here speaking.

Now the place will have relation to two things, of which the first is that a force stationed at this point exerts a certain influence on the whole, and the second, that the locality serves the force as protection and as a means of strengthening. In a word, it is related both to strategy and to tactics.

Strictly speaking, the term *defensive position* originates from its relation to tactics. For its relation to strategy, the fact, namely, that a force stationed at this place serves by its presence as a defense of the country, also holds for a position taken up for purposes of offense.

The strategic effect of a position can only be shown in its full light later on when we treat of the defense of a theater of war. Here we shall only consider it so far as can now be done, and for this purpose we must examine more closely two ideas which resemble each other and which are often confused, namely, that of *turning* a position and that of *passing by it.*

The turning of a position relates to its front and is done either in order to attack it from the flank, or even from the rear, or to cut its line of retreat and communication.

The first of these—the attack from the flank or rear—is of a tactical nature. In our days, when the mobility of troops is so great and all plans of engagements are more or less directed to turning or enveloping the enemy, every position must be prepared for this, and if it is to deserve to be described as strong, must, along with a strong front, at least offer good facilities for engagements on the flanks and rear, if these are threatened. A position is, therefore, not made useless if the enemy turns it with the intention of attacking it on the flank or in the rear. The battle which takes place there was provided for in the choice of the position and must ensure to the defender the advantages which, as a rule, he could expect from this position.

If the position is *turned* by the assailant with a view to operating against the defender's line of retreat and communication, this relates to *strategy* and the question is how long the position can hold out in these circumstances, and whether we cannot beat the enemy at his own game. The answers to these questions depend on how the spot is situated, that is, chiefly on the relation between the lines of communication on the one side and the other. A good position should ensure to the army defending it a preponderance in this respect. In any case, the position is not even on this account made useless. Our opponent who is in this way kept employed by it is thereby at least neutralized.

But if the assailant, without troubling himself about the existence of the force awaiting him in the defensive position, presses on with his main force in another direction in pursuit of his object, he *passes by the position.* If he can do this with impunity and actually does it, he immediately forces us to abandon the position, which thus becomes useless.

There is hardly a position in the world which could not, in the merely verbal sense, be passed by, for cases, such as the isthmus of Perekop, on account of their rarity, scarcely deserve notice. The impossibility of passing by, therefore, must be relative to the disadvantages to which the assailant by passing by would subject himself. In what these disadvantages consist we shall have a better opportunity of stating in Chapter 28. They may be either great or small; in any case they are the equivalent of what would have been the tactical usefulness of the position, and, together with it, jointly constitute the object of the position.

From what has so far been said, two strategic properties of the defensive position result:

1. That it cannot be passed by.
2. That in the struggle for the lines of communication it gives the defender an advantage.

Here we have to add two other strategic properties:

3. That the relation between the lines of communication also advantageously affects the form of the engagement.
4. That the general influence of the ground is advantageous.

For the relation between the lines of communication has an influence not only on the possibility or impossibility of passing by a position or cutting off our opponent's means of subsistence, but also on the whole course of the battle. An oblique line of retreat facilitates a tactical turning movement on the part of the assailant and paralyzes our own tactical movements during the battle. An oblique disposition with regard to the line of communication is, however, often not the fault of tactics but a result of a mistake in the choice of the strategic point. It is, for instance, absolutely unavoidable if the road changes direction in the vicinity of the position (Borodino, 1812). The assailant is then heading in the direction to turn our line *without deviating from his own perpendicular disposition.*

Further, the assailant, if he has many ways for his retreat while we are confined to one, likewise has the advantage of a much greater tactical freedom. In all these cases the tactical skill of the defender is exerted in vain to overcome the disadvantageous influence which the strategic conditions exercise.

Lastly, as regards the fourth point, the ground may also in other respects create a situation so generally disadvantageous that even the most careful choice and most efficient employment of tactical aids can effect nothing to remedy it. In such circumstances the chief points are as follows:

1. The defender must above all try to get the advantage of overlooking his adversary, so as to be able swiftly to fall upon him within the limits of his position. It is only when the obstacles the ground offers to approach are combined with these two conditions that it is really favorable to the defense.

   Points which are under the influence of commanding ground are, therefore, disadvantageous to him; all or most positions in mountains (with which we shall deal more particularly in the chapters on mountain war); all positions which rest one flank on mountains, for such, though making it more difficult for the assailant to pass by, facilitate a turning movement; likewise, all positions which have a mountain range near them in front, and generally all ground which can create such situations as those above mentioned.

   As the reverse of these disadvantageous situations we shall only distinguish one case, that in which the position has a mountain range in its rear. This affords so many advantages that in general it may be accepted as one of the most favorable locations for defensive positions.

2. The ground may more or less correspond to the character of the army and its composition. A very numerous cavalry is a proper reason for seeking an open ground. Lack of this arm and perhaps also of artillery and the possession of a courageous infantry, inured to war and acquainted with the country, make it advisable to take advantage of a difficult, close ground.

The tactical relation which the location of a defensive position bears to the armed force we do not here have to discuss in detail, but only the total result, for this alone is a strategic quantity.

Undoubtedly a position in which an army proposes only to await the attack of the enemy should afford it such considerable advantages of ground as may be regarded as multiplying its forces. Where nature does much, but not so much as we want, the art of entrenchment comes to our aid. By this means single portions not unfrequently become *impregnable*, and not unfrequently so does the whole. Obviously, in the latter case the whole character of the measure is changed; it is then no longer a battle under favorable conditions that we are seeking, and in this battle the success of the campaign, but a success without a battle. By keeping our forces in an impregnable position, we are definitely refusing battle and forcing our opponent to adopt some other way of obtaining a decision.

We must, therefore, completely separate the two cases and shall deal with the latter of them in the following chapter under the title of "Fortified Positions and Entrenched Camps."

But the defensive position with which we have now to do is to be nothing more than a field of battle with advantages increased in our favor. In order, however, that it may become a field of battle, the advantages in our favor must not be *too great*. What degree of strength then may such a position have? Obviously, it must be in proportion to the degree of the enemy's determination to attack, and that is a matter for judgment in each individual case. Opposed to a Bonaparte, we may and must withdraw behind stronger ramparts than when faced by a Daun or a Schwarzenberg.

If single portions of a position are impregnable, for instance, the front, that is to be regarded as a separate factor in its total strength, for the forces not required at this point can be employed elsewhere. But we must not omit to observe that by the enemy being entirely beaten off from such impregnable points, the form of his attack acquires quite a different character, and we must ascertain beforehand whether this suits our situation.

For instance, to take up a position as has often been done, so close behind a great river that it is regarded as strengthening the front, is merely to make the river a point of support for the right or left flank; for the enemy is naturally forced to cross farther to the right or left and attack us with a changed front. The main question must therefore be: What advantages or disadvantages does that give us?

In our opinion a defensive position will the more closely approach the ideal, the better its strength is concealed and the more opportunity it affords us of so conducting our engagements as to effect a surprise. Just as, with regard to our forces, we endeavor to conceal from the enemy their real strength and the direction in which that strength is to be employed, so on the same principle, we should try to conceal from him the advantages we expect to derive from the formation of the ground. This, of course, can only be done up to a certain point and demands, perhaps, a special mode of procedure, as yet but little attempted.

The vicinity of an important fortress, in whatever direction it may be, gives any position a great advantage over the enemy in respect to the movement and use of its forces. An appropriate use of single field-works can make up for the lack of natural strength at single points and make it possible for the main features of the engagement to be determined in advance at our pleasure. With these artificial means of strengthening our position must be combined a judicious choice of those natural obstacles in the ground which will render the operation of the enemy's forces difficult without making it impossible. We must try to turn to good account every advantage of the situation—the fact that we know the battlefield and the enemy does not, that we can conceal our movements better than he can his, and the general superiority we have over him in the means of effecting a surprise in the course of the engagement. Then as a result of all these things combined, it is possible for the locality to exercise an overwhelming and decisive influence, to which the enemy succumbs without realizing the real cause of his overthrow. This is what we understand by a defensive position, and we consider it to be one of the greatest advantages of a defensive war.

Leaving out of consideration particular situations, we may assume that an undulating country, not too well, nor yet too little, cultivated, affords the most positions of this kind.

CHAPTER 13

FORTIFIED POSITIONS AND
ENTRENCHED CAMPS

We have said in the preceding chapter that a position made so strong by nature and art as to be impregnable lies quite outside the category of advantageous fields of battle and therefore constitutes one of its own. In this chapter we shall consider its peculiar character and, on account of its fortress-like nature, call it a *fortified position.*

Fortified positions cannot be produced by entrenchments alone, except as entrenched camps resting on fortresses, and still less by natural obstacles alone. Nature and art usually join hands in them, and on that account they are frequently called entrenched camps or positions. These terms, however, can really be applied to any position more or less provided with entrenchments, which has nothing in common with the nature of the positions here being discussed.

The object of a fortified position, therefore, is to make the military force stationed there practically unassailable, and by that means either directly to provide actual protection for a certain tract of country, or only for the *forces* stationed in that tract, in order then to use these forces for covering the country indirectly in another way. The former was the meaning of the lines of earlier wars, more particularly on the French frontier; the latter that of the entrenched camps showing a front in every direction and those laid out near fortresses.

If the front of a position through entrenchments and obstacles to approach is so strong that an attack becomes impossible, the enemy is compelled to turn it, in order to make his attack from a flank or from the rear. To prevent this being easily done, supporting points were sought for these lines, to give them a certain amount of support on the flanks, such as the Rhine and the Vosges give to the lines in Alsace. The longer the front of such a line, the more readily can it be protected from being turned, because every turning movement always involves some danger to the forces making it, and this danger increases in direct proportion to the deviation required from their original direction. Therefore a considerable length of front which could be made impregnable and good flank supports ensure the possibility of directly protecting a considerable tract of country against enemy invasion. This, at least, was the view from which works of this class originated, and this is the meaning of the lines in Alsace, with their right flank on the Rhine and their left on the Vosges, and of the lines in Flanders, seventy-five miles long, resting their right flank on the Scheldt and the fortress of Tournay, and their left on the sea.

But when we have not the advantages of so long and strong a front and of good flank supports, if the country is to be held at all by a well-entrenched force, this force must protect itself against being turned by being allowed—both itself and its position—to show a front in every direction. But then the conception of a *tract of country actually covered* vanishes, for such a position is strategically to be regarded as only a point that covers the force occupying it and thereby assures to it the possibility of maintaining its hold on the country, that is to say, of *maintaining itself in the country.* Such a camp can no longer be *turned;* that is, it cannot be attacked on the flank or in the rear by reason of those parts being weaker than the front, for it shows a front in all directions and is equally strong everywhere. But such a camp can be *passed by,* and, indeed, much more easily than an entrenched line, for it has practically no extension.

Entrenched camps connected with fortresses are fundamentally of this second kind, for their object is to protect the forces assembled in them. Their further strategical significance, namely, the employment of these protected forces, is, however, somewhat different from that of the other entrenched camps.

After this explanation of the origin of these three different means of defense we shall consider their value and distinguish between them under the names of *fortified lines, fortified positions* and *entrenched camps resting on fortresses.*

1. *Lines.* These are the most pernicious kind of cordon warfare. The obstacle they present to the assailant is absolutely of no value unless it is defended by a powerful fire; in itself it is no better than nothing at all. But now the extent of the field over which the fire of an army can still be thus effective is very small compared with the extent of the country. The lines

can only be very short and consequently cover very little country, or the army will not be able really to defend all points of them. The idea was then hit upon of not garrisoning all points of these lines, but only keeping them under observation and defending them by means of suitably posted reserves in the same way that a not too wide river can be defended; but this procedure is against the nature of the means employed. If the natural obstacles of the ground are so great that we could employ such a kind of defense, the entrenchments would be useless and dangerous, for that kind of defense is not local, and it is only for a local defense that entrenchments are suitable. But if the entrenchments themselves are to be regarded as the chief obstacle to approach, we can well understand how little an *undefended* entrenchment will mean. What is a trench ten or fifteen feet deep and a wall ten to twelve feet high against the united effort of many thousands, undisturbed by enemy fire? The consequence, therefore, is that such lines, if they were short and at the same time comparatively strongly manned, have been *turned*, or, if they were extensive and inadequately manned, have without great difficulty been attacked in front and taken.

Now as lines of this kind tie the force manning them down to local defense and take away all its mobility, they are a very ill-conceived means to employ against an enterprising enemy. If nevertheless they have been retained for a fairly long time in modern wars, the reason for that lies in the weakening of war's elemental violence, where a seeming difficulty produced all the effect of a real one. Furthermore, in most campaigns these lines were used merely for a subordinate defense against raiding parties. If for that purpose they have proved not wholly ineffective, we have only at the same time to remember how much more usefully the troops required for their defense might have been employed at other points. In the most recent wars there could be no question of them at all, nor do we find any trace of them; and it may be doubted if they will ever reappear.

2. *Positions.* The defense of a tract of country continues (as we shall show in greater detail in Chapter 27) as long as the force designated for it maintains itself there, and only ceases if that force quits the country and gives it up.

If a force is to maintain itself in a country which is attacked by an enemy of greatly superior strength, an impregnable position affords the means of defending this force against the violence of the sword.

Such positions, as we have said, must show a front on all sides. According to the ordinary extension of a tactical disposition, and if the force were not *too large* (and a large force would be out of keeping with the supposed case), it would therefore occupy a *very small space*. This, in the course of the engagement, would be exposed to so many disadvantages that, in spite of every possible strengthening by means of entrenchments, a successful resistance would be almost inconceivable. Such a camp, showing front in all directions, must necessarily, therefore, have a comparatively considerable extension of its flanks, and these flanks should at the same time be practically impregnable. But to give them this strength in spite of their great extension is more than the art of entrenchment can do, and it is therefore a fundamental condition for such a camp that it should be strengthened by natural obstacles of the ground such as to make many portions quite impossible to approach, and others difficult. In order, therefore, to employ this means of defense, a position of this kind needs to be found, and when it is lacking, it is impossible to attain our object by mere entrenchment. These considerations relate to the tactical results and are mentioned only in order that we may first properly establish the existence of this strategic means. For the sake of clearness we mention in this connection the examples of Pirna, Bunzelwitz, Colberg, Torres Vedras and Drissa. Let us now consider its strategic properties and effects.

The first condition is naturally that the force stationed in this camp should have its subsistence assured for some time, for so long, that is, as is thought necessary to keep the camp going. This will only be possible if the rear of the position rests on a port, as at Colberg and Torres Vedras, or is in close connection with a fortress as at Bunzelwitz and Pirna, or if stores have been accumulated inside it or in its vicinity, as at Drissa.

It is only in the first case that the provisioning can be assured more or less permanently; in the second and third case, however, only for a more or less limited time, so that on this point danger is ever threatening. From this it is evident how the difficulty of provisioning excludes a number of strong points, which otherwise would be suitable for an entrenched position, and makes those suitable *rare*.

In order to understand the effectiveness of such a position and its advantages and dangers, we must ask ourselves what the assailant can do against it.

a. The assailant can pass by the fortified position, pursue his enterprise and watch the position with more or less troops.

We must here distinguish between the case of a position occupied by the main body and one occupied only by a subordinate force.

In the first case, the passing by the position can benefit the assailant only if, besides the main force of the defender, there is also some other attainable and *decisive object for the attack*, as, for instance, the taking of a fortress or the capital, and so on. But even if there is such an object, he can only pursue it if the strength of his base and the lie of his communication line relieve him from fear of an operation against his strategic flank.

The conclusions we draw from this as to the admissibility and effectiveness of a fortified position for the main body of the defender's army is that they depend on two conditions. In the first place, its effectiveness against the strategic flank of the assailant must be so decisive that we may be sure in advance of keeping him at a point where he can do us no harm. In the second place there must be for the assailant absolutely no attainable object for which the defender might feel anxiety. If there is such an object and if the assailant's strategic flank cannot be sufficiently menaced, then the position either cannot be taken up at all, or only as a feint or to test the assailant's willingness to accept its importance as serious. But this is always attended with the danger that in case of failure we may be too late to reach the point threatened.

If the fortified position is only manned by a subordinate force, the assailant can never lack a further object to attack because this can itself be the main body of the enemy's army. In this case the importance of the position is therefore strictly limited to the means which it affords of operating against the enemy's strategic flank, and depends on this condition.

b. If the assailant does not dare to pass by a position, he can actually invest it and starve it into surrender. But this presupposes two conditions, first, that the position is not open in the rear, and second, that the assailant is strong enough for such an investment. If these two conditions are fulfilled, the assailant's army would certainly be neutralized for a time by the fortified camp, but the price the defender would have had to pay for this advantage would be the loss of the defensive force.

From this it is clear that *with the main body* of the army the expedient of such a fortified position will only be adopted:

aa) When the rear is perfectly safe (Torres Vedras).

bb) When we foresee that the enemy's superiority in numbers is not great enough actually to invest us in our camp. Should the enemy with an insufficient superiority still decide to do so, we should be able to make a successful sally from the camp and beat him in detail.

cc) When we can count on relief, like the Saxons at Pirna in 1756, and as was fundamentally the case after the battle of Prague in 1757, because Prague itself could only be regarded as an entrenched camp in which Prince Charles would never have allowed himself to be shut up, if he had not known that the Moravian army could liberate him.

One of these three conditions is, therefore, absolutely necessary to justify the choice of a strong position for the main body of an army. We must, nevertheless, admit that the last two border closely upon being extremely dangerous for the defender.

But if it is a question of exposing a subordinate corps to the risk of being sacrificed if necessary, for the benefit of the whole, then these conditions disappear, and the only point to decide is whether by such a sacrifice a greater evil can really be avoided. This will, perhaps, but seldom be the case; nevertheless, it is certainly not inconceivable. The entrenched camp at Pirna prevented Frederick the Great from attacking Bohemia, as he would have done in 1756. The Austrians at that time were so little prepared that the loss of that kingdom would appear to have been beyond doubt, and perhaps also a greater loss of men would have been involved in it than the 17,000 allied troops who capitulated in the Pirna camp.

c. If none of the possibilities in favor of the assailant specified in *a* and *b* exist, and therefore the conditions are fulfilled which we have there set out as in favor of the defender, there remains, indeed, nothing for the assailant to do but to stand fast before the position like a setter before a covey of birds; at most to spread himself as much as possible over the country by detachments and, contenting himself with this small and indecisive advantage, leave to the future the real decision as to the possession of the tract of country. In this case, the position has completely fulfilled its purpose.

3. *Entrenched camps near fortresses.* They belong, as we have said, to the class of entrenched positions generally, in so far as they have for their object to cover not a tract of territory but a military force against a hostile attack. They actually differ from the others only in the fact that with the fortress they form an inseparable whole by which they naturally acquire much greater strength.

From this, however, result the following characteristics:

a. That they may also have the particular object of rendering the siege of the fortress either quite impossible or very difficult. This object may be worth a great sacrifice of troops if the place is a port which cannot be blockaded, but in any other case it is to be feared that the place would fall through famine too soon to be quite worth the sacrifice of a considerable number of troops.

b. Entrenched camps can be formed near fortresses for smaller bodies of troops than those in the open field. Four or five thousand men may be undefeatable under the walls of a fortress, whereas in the open field in the strongest camp in the world they would be lost.

c. They may be used for the assembly and training of forces which have still too little stability to be trusted in contact with the enemy without the support afforded by the works of the fortress—recruits, for example, militia, national levies, and so forth.

They might, therefore, be recommended as a very useful expedient if they had not the immense disadvantage of being more or less prejudicial to the fortress in case they cannot be garrisoned, while to keep the fortress always provided with a garrison sufficient to some extent for the camp as well would be a much too onerous condition.

We are, therefore, inclined to consider them only to be recommended for places on the coast and in all other cases as more harmful than helpful.

In conclusion, to summarize our opinion in a general view, fortified and entrenched positions are:

1. The less to be dispensed with, according as the country is smaller and there is less space for evading the enemy.
2. The less dangerous, according as they can count more certainly on being reinforced and relieved either by other forces, or the season of bad weather, or popular insurrection, or famine, and so forth.
3. The more effective, according as the elemental force of the enemy's onset is weaker.

CHAPTER 14

FLANK POSITIONS

We have only allotted to this conception, so prominent in the world of ordinary military theory, a special chapter, in dictionary fashion, for greater facility of reference, for we do not believe that anything independent in itself is denoted by the term.

Every position which is to be held, even if the enemy passes it by, is a flank position, for from the moment he does so it can have no other effect but that upon the enemy's strategic flank. Therefore, necessarily, all *fortified positions* are, at the same time, flank positions, for as they cannot be attacked and the enemy is accordingly obliged to pass them by, they can only get their value by their effect on his strategic flank. Whatever the actual front of the fortified position may be, whether it runs parallel to the enemy's strategic flank, as at Colberg, or perpendicular to it, as at Bunzelwitz and Drissa, is quite immaterial, for a fortified position must front every way.

But it may still be desirable to maintain a position that is *not* impregnable, even if the enemy passes it by, should its situation, for instance, give us so preponderant an advantage in relation to the lines of retreat and communication that not only can we make an effective attack on the strategic flank of the advancing enemy, but also that the enemy, alarmed for his own line of retreat, is unable to get entire possession of ours. For if this last were not the case, our position not being a fortified, that is to say, an *unassailable*, one, we should run the risk of having to fight without a line of retreat.

The year 1806 affords us an example that illustrates this. The disposition of the Prussian army on the right bank of the Saale might in relation to Bonaparte's advance by Hof have become in every sense a flank position, if, that is, the army had been drawn up parallel to the Saale and in this position had awaited further developments.

If in this case there had not been such a disproportion between the physical and moral forces, if only a Daun had been at the head of the French army, the Prussian position would have shown its effectiveness by a most brilliant result. To pass it by was quite impossible; that was acknowledged even by Bonaparte by his resolution to attack it. Even Bonaparte did not *entirely* succeed in cutting off its line of retreat, and had the disproportion between the physical and moral forces been less serious, to do so would have been just as impracticable as to pass it by. For the Prussian army ran less risk from the overpowering of their left wing than did the French from that of their own left wing. Even with the existing disproportion of the forces, resolute and judicious leadership would still have afforded great hopes of a victory. There would have been nothing to hinder the Duke of Brunswick on the 13th from making such arrangements that at daybreak on the morning of the 14th he might oppose 80,000 men to the 60,000 with which Bonaparte had crossed the Saale near Jena and Dornburg. Though this preponderance and the steep valley of the Saale in the rear of the French would not have been enough to give a decisive victory, yet it must be admitted that it was in itself an extremely advantageous result. If, with such a result, no favorable decision could be won, no decision at all could be expected in that district and instead we should have retreated farther, to gain reinforcements thereby and weaken the enemy.

The Prussian position on the Saale, therefore, though assailable, might have been regarded as a flank position to the road coming by way of Hof, only this description could not be applied to it absolutely, any more than to any assailable position, because it only became applicable if the enemy did not dare to attack.

Still less would it be consistent with clear ideas if even those positions which *cannot* be maintained after the enemy has passed them by, and from which consequently the defender seeks to attack the assailant's flank, were called *flank positions* merely because his attack is directed against a flank. For this flank attack has hardly anything to do with the position itself, or at all events does not essentially arise from its properties, as is the case with the operation against a strategic flank.

It is obvious from this that there is nothing new to establish about the properties of a flank position. A few words only on the character of this expedient may conveniently be introduced here. Really fortified positions, we disregard entirely because we have already said enough about them.

A flank position which is not impregnable is an extremely effective instrument, but at the same time on that very account, also an extremely dangerous one. If the assailant is checked by it, we have obtained a great effect with a small expenditure of force; it is the touch of the rider's finger on the long lever of a sharp bit. But if the effect is too slight, if the assailant is not stopped, the defender has more or less sacrificed his retreat and must either still try to escape in haste and by detours, therefore under very unfavorable circumstances, or he is in danger of having to fight without a line of retreat. Against a bold and morally superior adversary who is seeking a thorough decision, this expedient therefore is extremely hazardous and entirely out of place, as the example from 1806 quoted above shows. On the other hand, against a cautious opponent and in a war of mere observation, it can rank as one of the best means to which the talent of the defender can have recourse. The Duke Ferdinand's defense of the Weser by his position on the left bank and the well-known positions of Schmotseifen and Landshut are examples of this, but the latter, certainly, at the same time, in the catastrophe which befell Fouqué's corps in 1760, illustrates the danger of a wrong employment of it.

CHAPTER 15

DEFENSE OF MOUNTAINS

The influence of mountain ranges on the conduct of war is very great; the subject, therefore, is very important for theory. As this influence introduces into action a retarding principle, it belongs first of all to the defensive. We shall, therefore, treat it here without confining ourselves to the narrower conception as defense of mountains. As we shall discover in our consideration of the subject results which run counter to general opinion in many points, we shall be obliged to enter into a rather elaborate analysis of it.

We shall first examine the tactical nature of the subject, in order to find its point of contact with strategy.

The endless difficulty attending the march of large columns on mountain roads, the extraordinary strength which a small post obtains by a steep slope covering its front, and by ravines right and left supporting its flanks, are unquestionably the two principal causes why such efficacy and strength are universally attributed to the defense of mountains that nothing but peculiarities in armament and tactics at certain periods have prevented large masses of military forces from engaging in it.

When a column, winding like a serpent, toils its way through narrow ravines up to the top of a mountain, and passes over it at a snail's pace, when artillery- and transport-drivers, with oaths and shouts, flog their over-driven horses through the rugged narrow roads, when each broken wagon has to be gotten out of the way with indescribable trouble, while all those behind are detained, cursing and blaspheming, everyone then thinks to himself: "Now, if the enemy came up with but a few hundred men he might scatter the whole lot." From this has originated the expression used by historical writers when they describe a narrow pass as a place where "a handful of men could keep a whole army in check." At the same time everyone who has had any experience in war knows, or ought to know, that such a march through mountains has little or nothing in common with *the attack* on them, and that therefore to infer from *this* difficulty that the difficulty of attacking them must be much greater still is a false conclusion.

It is natural enough that an inexperienced person should argue thus, and it is almost as natural that the art of war itself of a certain age should have been entangled in the same error. The phenomenon was almost as new at that time to those accustomed to war as to the uninitiated. Before the Thirty Years' War, owing to the deep order of battle, the numerous cavalry, the primitive firearms, and other peculiarities, it was quite unusual to make use of formidable obstacles of ground in war, and an actual defense of mountains, at least by regular troops, was almost impossible. It was not until a more extended order of battle was introduced, and infantry and their firearms became the chief part of an army, that the use which might be made of hills and valleys was thought of. But it was not until a hundred years afterward, i.e., about the middle of the eighteenth century, that the idea became fully developed.

The second circumstance, namely, the great power of resistance which a small post acquires by being placed on a point difficult of access, was still more suited to lead to an exaggerated idea of the strength of mountain defenses. The opinion arose that it was only necessary to multiply such a post by a certain number to make an army out of a battalion, a mountain range out of a mountain.

It is undeniable that a small post acquires an extraordinary strength by selecting a good position in a mountainous country. A small detachment, which would be put to flight in level country by a couple of squadrons, and would think itself lucky to save itself from rout or capture by a hasty retreat, can in the mountains, we might say, stand up before a whole army, and, with a kind of tactical effrontery, exact from it the military honor of a regular attack, of having its flank turned, and so forth. How it obtains this power of resistance by obstacles to approach, *points d'appui* for its flanks, and new positions which it finds on its retreat, is a subject for tactics to explain; we accept that as an established fact.

It was very natural to believe that a great number of such posts placed in a line would provide a very strong, almost unassailable front, and all that remained to be done was to prevent the position from being turned by extending it right and left until either flank supports were found, commensurate with the importance of the whole, or until the extent of the position itself might be supposed to give security against turning movements. A mountainous country especially invites such a course by presenting such a number of positions, each one apparently better than the other, that for this very reason one does not know where to stop; and therefore finally each and every approach to the mountains within a certain distance is occupied and defended. Ten or fifteen single posts, thus spread over a space of about fifty miles or more, were supposed to bid defiance to that odious turning movement. Now as the connection between these posts was considered sufficiently secure by the intervening spaces being ground of an impassable nature (because except on roads it is impossible to march with columns), it was thought that a wall of iron had thus been placed before the enemy. As an extra precaution, a few battalions, some horse artillery and a dozen squadrons of cavalry, formed a reserve to provide against the case of an unexpected break-through actually occurring at any point.

No one will deny that this picture is accurately historical, and it cannot be said that we have absolutely got rid of such absurd ideas.

The course of improvement in tactics since the Middle Ages, with armies ever increasing in numbers, likewise contributed to bring mountainous districts in this sense more within the scope of military action.

The chief characteristic of mountain defense is its most decisive passivity; in this light the tendency toward the defense of mountains was very natural before armies attained their present mobility. But armies were constantly becoming larger, and, on account of the effect of firearms, began to extend more and more into long, thin lines connected with great ingenuity, and very difficult, often almost impossible, to move. To dispose, in order of battle, such an ingenious machine, was often a half day's work, and half the battle; and almost everything that must now be attended to in the preliminary plan of the battle was included in this first disposition. After this work was done, it was difficult to make any modifications when new circumstances arose. From this it followed that the assailant, being the last to form his line of battle, had to adapt it to the position chosen by the enemy, without the latter being able in turn to modify his accordingly. The attack thus acquired a general superiority, and the defense had no other means to make up for this than by seeking protection from impediments of ground, and for this nothing was so favorable and general as mountainous ground. Thus it became an object to couple, as it were, the army with a formidable obstacle of ground, and the two united then made common cause. The battalion defended the mountain, and the mountain the battalion; thus the passive defense through the aid of mountain ground acquired a high degree of strength, and the only disadvantage in the thing itself was that it entailed an even greater loss of freedom of movement, a quality of which in any case no one knew how to make any particular use.

When two antagonistic systems act upon each other, the exposed side, that is, the weak point, of the one always draws upon itself the blows from the other. If the defensive stands motionless and as it were spellbound in posts which are in themselves strong, and cannot be taken, the aggressor is thereby encouraged to be bold in turning movements, because he has no apprehension about his own flanks. This is what took place, and *turning*, as it was called, soon became the order of the day. To counteract this, positions were extended more and more; they were thus weakened in front, and the offensive suddenly turned upon that part. Instead of trying to outflank by extending, the assailant now concentrated his masses for attack at some point, and the line was broken. This is roughly what took place in regard to mountain defense in the most recent period of military history.

The offensive had once again gained an absolute preponderance through the continuous advance being made in mobility, and it was only from the same means that the defense could seek help. But mountain ground, by its nature, is opposed to mobility, and thus the whole mountain defense experienced, if we may use the expression, a defeat not unlike that which the armies engaged in it in the Revolutionary War so often suffered.

But that we may not empty the baby with the bath, and allow ourselves to be carried away by the stream of commonplaces to assertions which, in actual experience, will be refuted a thousand times by the force of circumstances, we must distinguish the effects of mountain defense according to the nature of the cases.

The principal question to be decided here, and that which throws the greatest light upon the whole subject, is whether the resistance which the mountain defense is intended to offer is to be *relative* or *absolute*—whether it is only to last for a time or to end in a decisive victory. For resistance of the first kind, mountain ground is in a high degree suitable, and introduces into it a very powerful element of strength; for one of the latter kind, on the contrary, it is in general not at all suitable, or only so in some special cases.

In mountains every movement is slower and more difficult, costs therefore more time, and more men as well, if executed within the sphere of danger. But the loss to the assailant in time and men is the standard by which the resistance offered is measured. As long as the movement is all on the side of the offensive, the defensive has a marked advantage; but as soon as the defensive also applies this principle of movement, that advantage ceases. Now, naturally, on tactical grounds, a relative resistance permits a much greater degree of passivity than one which is intended to lead to a decisive result, and it allows this passivity to be carried to the extreme limit, that is, to the end of the engagement, which in the other case may never happen. The impeding element of mountain ground, which as a medium of greater density weakens all positive activity, is therefore completely suited to the passive defense.

We have said that a small post in a mountain range acquires an extraordinary strength by the nature of the ground; but although this tactical result requires no further proof, we must add some further explanation. It is here especially that a distinction must be drawn between what is relatively and what is absolutely small. If a small body of troops, whatever its size may be, isolates a part of itself in a position, this part may possibly be exposed to the attack of the whole body of the enemy's troops, therefore of a superior force in comparison to which it is itself small. In this case, as a rule, no absolute, but only a relative, resistance can be the object. The smaller the post in relation to the whole body of the enemy, the more this applies.

But also a post which is small in an absolute sense, that is, one which is not opposed by an enemy superior to itself and which, therefore, may contemplate an absolute resistance, a real victory, will be infinitely better off in mountains than a large army and can derive more advantage from the ground, as we shall show later.

Our conclusion, therefore, is that a small post in mountains possesses great strength. How this may be of decisive use in all cases where everything depends on a *relative* resistance is obvious. But will it be of the same decisive use for the *absolute* resistance of an army? This is the question which we now propose to examine.

First of all, we further ask whether a front line composed of several such posts has, as has hitherto been assumed, the same strength proportionally as each separate post. This is certainly not the case, and to suppose so would involve us in one or other of two errors.

In the first place, a country *without roads* is often mistaken for one which is *impassable*. Where a column, or where artillery and cavalry cannot *march*, infantry may still, in general, be able to advance, and even artillery may often be brought there as well, for the movements, very exhausting but brief, in an engagement are not to be compared with those on a march. The secure connection of the single posts with one another rests therefore on an illusion, and their flanks are consequently threatened.

Or, next, it is supposed, that the line of small posts, which are very strong in front, are also equally strong on their flanks, because ravines, precipices, and so forth, form excellent supports for a small post. But why are they so? Not because they make the turning of the post impossible, but because, owing to them, the operation demands an expenditure of time and of force proportionate to the effective action of the post. The enemy who, in spite of the difficulties of the ground, wishes, and in fact is obliged, to turn such a post, because the front is unassailable, requires, perhaps, a half day to execute his purpose, and nevertheless cannot accomplish it without some loss of men. Now if such a post can depend on support, or if it is only designed to resist for a certain length of time, or, lastly, if it is a match for the enemy in strength—then the flank supports have done their part, and we may say the position had not only a strong front, but strong flanks as well. But such is not the case if it is a question of a line of posts, forming part of an extended mountain position. None of these three conditions is realized in that case. The enemy attacks one point with an overwhelming force, the support from the rear is perhaps slight, and yet everything depends on an absolute resistance. Under such circumstances the flank supports of such posts are worth nothing.

Upon a weak point like this the attack usually directs its blows. The assault with concentrated, and therefore very superior, forces upon a point of the front *may certainly be met by a resistance which is very violent as regards that point, but which is very insignificant as regards the whole*. After it is overcome, the line is pierced, and the object of the attack attained.

From this it follows that the *relative* resistance in mountain warfare is, in general, greater than in a level country, that it is comparatively greatest in small posts, but that it does not increase in the same measure as the masses increase.

Let us now turn to the real object of great engagements generally—to the *positive* victory which may also be the object in the defense of mountains. If the whole mass, or the main force is employed for that purpose, then the defense of mountains changes itself by that very fact into a *defensive battle in the mountains*. A battle, that is, the application of all our military forces to the destruction of the enemy, is now the form, a victory the object, of the engagement. The defense of mountains which takes place in this engagement appears now a subordinate consideration, for it is no longer the object; it is only the means. Now, in this case, how does the ground in mountains suit the object?

The character of a defensive battle is a passive reaction in front, and an increased active reaction in rear, but for this the ground in mountains is a paralyzing influence. There are two reasons for this: first, lack of roads affording means of moving rapidly in all directions, from the rear toward the front—even the sudden tactical surprise is hampered by the unevenness of ground—second, lack of a free view over the country and the enemy's movements. The ground in mountains, therefore, in this case offers the enemy the same advantages which it gave to us in front, and paralyzes the entire better half of the resistance. To this is to be added a third objection, namely, the danger of being cut off. Much as a mountainous country is favorable to a retreat made under a pressure exerted along the whole front, and great as may be the loss of time to an enemy who wants to turn us, still these again are only advantages in the case of a *relative defense*, advantages which have no connection with the decisive battle, the holding out to the last extremity. The resistance will last, it is true, somewhat longer, that is, until the enemy has reached with his flank columns points which menace or completely block our retreat. Once he has gained such points, however, relief is hardly any longer possible. No offensive action made from the rear can drive him out again from the points where he threatens us; no desperate assault with our whole mass can clear the passage which he *blocks*. Whoever sees in this a contradiction, and believes that the advantages which the assailant has in mountain warfare, must also be profitable to the defensive in an attempt to cut his way through, forgets the difference in the circumstances. The corps which contests the passage is not engaged in an *absolute* defense; a few hours' resistance will probably be sufficient. It is, therefore, in the situation of a small post. Besides this, the original defender is no longer in full possession of all his fighting means; he is thrown into disorder, lacks ammunition, and so forth. Therefore, in any case, the prospect of success is very small, and this is the danger that the defensive fears above all; this fear is at work even during the battle, and weakens every fiber of the struggling athlete. A morbid sensibility arises on the flanks, and any handful of men which the aggressor displays on a wooded slope in our rear becomes for him another aid to victory.

These disadvantages, would, for the most part, disappear, leaving all the advantages, if the defense of a mountain district consisted in the concentrated disposition of the army on an extensive mountain plateau. There we might imagine a very strong front, flanks very difficult to approach, and yet the most perfect freedom of movement, both within and in the rear of the position. Such a position would be one of the strongest possible, but it is little more than an illusory conception, for although most mountains are somewhat more easily accessible along their crests than on their slopes, yet most mountain plateaux are either too small for such a purpose, or they have no proper right to be called plateaux, and are so termed more in a geological than in a geometrical sense.

For smaller bodies of troops, the disadvantages of a defensive position in mountains diminish as we have indicated before. The reason for this is that such bodies take up less space, require fewer roads for retreat, and so forth. A single mountain is not a mountain range, and has not the same disadvantages. The smaller the force, the more easily it can confine itself to single ridges and mountains, and the less it will be necessary for it to get entangled in the intricacies of countless steep mountain gorges.

CHAPTER 16

Defense of Mountains

(continued)

We now turn to the strategic use of the tactical results analyzed in the preceding chapter.

We distinguish the following points:

1. A mountain range as a battlefield.
2. The influence which the possession of it exercises on other parts of the country.
3. Its effect as a strategic barrier.
4. The attention which it demands in respect of the subsistence of troops.

The first and most important of these points, we must again subdivide as follows:

a. A main battle.
b. Minor engagements.

1. A MOUNTAIN RANGE AS A BATTLEFIELD

We have shown in the preceding chapter how little favorable *mountainous ground* is to the defensive in a *decisive battle*, and, consequently, how much it favors the assailant. This runs exactly counter to common opinion; but then how many things there are which common opinion confuses; how little does it draw distinctions between things which are of the most opposite nature! From the extraordinary resistance which small, subordinate bodies of troops may offer in a mountainous country, common opinion gets the impression that all defense is extremely strong, and is astonished when any one denies this great strength in the case of the greatest act of all defense, the defensive battle. On the other hand, general opinion is instantly ready to perceive in every battle lost by the defensive in mountain warfare the inconceivable error of a cordon war, without having regard to the nature of things and its inevitable influence. We do not hesitate to be diametrically opposed to such an opinion, and at the same time we must mention, to our great satisfaction, that we have found our views supported in the works of an author whose opinion must have great weight for us in this matter in more than one respect; we allude to the history of the campaigns of 1796 and 1797 by the Archduke Charles, a good historian, a good critic and, above all, a good general—all in *one* person.

We can only characterize it as a lamentable situation when the weaker defender, who has laboriously and with the greatest effort assembled all his forces, in order to make the assailant feel the effect of his patriotism, of his enthusiasm and his wise discretion in a decisive battle—when he on whom every eye is fixed in anxious expectation, having stationed himself in the obscurity of thickly covered mountain land, and hampered in every movement by the obstinate ground, stands exposed to the thousand possible forms of attack which his powerful adversary can use against him. Only in one single direction is there still left a wide field for his intelligence—that is, in making all possible use of every obstacle of ground; but this leads close to the borders of the disastrous war of cordons, which, under all circumstances, is to be avoided. Therefore far from seeing in a mountainous country, a refuge for the defensive, when a decisive battle is sought, we should rather advise a general to avoid it by every possible means at his disposal.

It is true, however, that this is sometimes impossible; but the battle will then necessarily have a very different character from one in a level country; the disposition of the troops will be much more extended—in most cases twice or three times the length—the resistance more passive, the counter-blow much less effective. These are influences of mountain ground which cannot be avoided; still, in such a battle the defensive is not to change into a mere defense of mountains; its predominating character should be only a concentrated disposition of the military forces in the mountains, where everything unites into *one* engagement and is carried on to a great extent under the eye of *one* commander and where there are sufficient reserves to make the decision something more than a mere warding off, a mere thrusting of our shield before us. This condition is indispensable, but very difficult to realize, and the drifting into what is really a defense of mountains comes so naturally that we cannot be surprised if it happens so often. The danger in this is so great that theory cannot too urgently raise a warning voice.

So much for the decisive battle with the main body of the army.

For engagements of minor significance and importance, mountain country, on the other hand, may be very useful, because the main point in these cases is not absolute defense, and because no decisive results are connected with them. We may make this plainer by enumerating the objects of this reaction.

a. Merely to gain time. This motive occurs a hundred times: always in the case of a defensive line formed to obtain information; further, in all cases in which reinforcement is expected.

b. The repulse of a mere demonstration or minor enterprise of the enemy. If a province is protected by mountains which are defended by troops, then this defense, however weak, will always suffice to prevent attacks of enemy raiding parties and other small expeditions intended to plunder the province. Without the mountains, such a weak chain of posts would be useless.

c. To make demonstrations on our own part. It will be some time yet before opinion with respect to mountains has come to the right point. Until then, there will always be opponents who are afraid of them and shrink back from them in their operations. In such a case, therefore, the principal body may also be used for the defense of a range of mountains. In wars carried on with little energy or movement, this state of affairs will often occur; but it must always be a condition that we neither intend to accept a main battle in this mountain position nor can be compelled to do so.

d. In general, a mountainous country is suited for all positions in which we do not intend to accept any main engagement, for each of the separate parts of the army is stronger there, and it is only the whole as such that is weaker; besides, in such a position, it is not so easy to be suddenly attacked and forced into a decisive engagement.

e. Lastly, mountains are the native element of national risings. But while national risings should always be supported by small bodies of regular troops, on the other hand, the proximity of a great army seems to have an unfavorable effect upon them. This reason, therefore, as a rule, will never give occasion for taking to the mountains with the whole army.

This much for mountains in connection with the battle positions which may be taken up there.

2. THE INFLUENCE OF MOUNTAINS ON OTHER PARTS OF THE COUNTRY

As we have seen, it is so easy on mountain ground to secure a considerable tract of territory by small posts, weak in numbers, which in an easily accessible district could not maintain themselves and would be continually exposed to danger. Every advance in mountains which have been occupied by the enemy is much slower than in a level country and therefore cannot keep pace with that enemy. For these reasons the question of who is in possession is also much more important in the case of mountains than in that of any other tract of country of equal extent. In an open country, the possession may change from day to day. The mere advance of strong detachments compels the enemy to give up the country we need. But it is not so in mountains. Here even in the case of much inferior forces a marked resistance is possible, and for that reason, if we need a section of the country which includes mountains, particular operations, specially adapted thereto, and often necessitating a considerable expenditure of time as well as of men, are always required in order to put us in possession of it. Even if, therefore, a mountain range is not the theater of principal operations, we cannot, as in the case of more accessible country, look upon it as dependent on those operations, and we cannot regard its capture and possession as a necessary consequence of our advance.

A mountainous district has therefore a much greater independence, and the possession of it is much more decisive and less liable to change. If we add to this the fact that a mountain ridge naturally affords a good view over the adjacent open country from its crests, while itself remaining as if in the darkness of night, then we may understand that when we are close to mountains, without being in actual possession of them, they are to be regarded as an inexhaustible source of unfavorable influences—a sort of secret workshop of hostile forces; and this will be the case in a still greater degree if the mountains are not only occupied by the enemy, but also form part of his territory. The smallest bodies of adventurous irregulars find shelter there if pursued, and can then appear again suddenly at other points with impunity; the largest bodies, under their cover, can approach unperceived, and our forces must always keep at a sufficient distance if they would avoid getting within reach of their dominating influence, and being exposed to disadvantageous engagements and sudden attacks which they cannot return.

In this manner every range of mountains exercises a great influence over the lower and more level country adjacent to it, up to a certain distance. Whether this influence will become effective at once, for instance in a battle (as at Maltsch on the Rhine, 1796), or only after some time, upon the lines of communication, depends on the geographical situation. Whether it can be overcome or not, through some decisive event happening in the valley or in the level country, depends on the relations of the armed forces to each other respectively.

Bonaparte, in 1805 and 1809, advanced as far as Vienna without troubling himself much about the Tyrol; but Moreau had to leave Swabia in 1796, chiefly because he was not master of the higher parts of the country, and too many troops were required to watch them. In campaigns, where there is a balanced interplay of forces, we shall not expose ourselves to the constant disadvantage of mountains remaining in possession of the enemy. We shall, therefore, only endeavor to seize and retain possession of that portion of them which is required on account of the direction of the principal lines of our attack. For this reason it is usually found that in such cases the mountains are the arena of separate minor engagements which take place between the forces of each side. But we must be careful of overrating the importance of this circumstance and being led to consider a range of mountains as in every case the key to the whole and its possession as the main point. When a victory is the object sought, then *this victory* is the principal object, and if it is gained, other things can be regulated according to the paramount needs of the situation.

3. MOUNTAINS CONSIDERED AS A STRATEGIC BARRIER

We must divide this subject into two parts.

The first is again that of a decisive battle. We can consider the mountain range as a river, that is, as a barrier with certain points of passage, which may afford us an opportunity of gaining a victory, because the enemy will be compelled by it to divide his forces in advancing and is tied down to certain roads which will enable us with our forces concentrated behind the mountains to fall upon separate parts of his force. As the assailant on his advance through mountains, irrespective of all other considerations, cannot march in a single column because he would thus expose himself to the danger of getting engaged in a decisive battle with only one line of retreat, this method of defense, no doubt, rests on very essential circumstances. But mountains and mountain outlets are very indefinite terms, and in adopting this measure everything depends entirely on the nature of the country itself. It can only, therefore, be indicated as a possible measure, which however must be conceived of as having two disadvantages. The first is that if the enemy has received a severe blow, he soon finds shelter in the mountains; the second, that he is in possession of the higher ground, which, although not decisive, must still always be regarded as a disadvantage for the pursuer.

We know of no battle fought in such circumstances unless the battle with Alvinzi in 1796 can be counted as such. But that the case *may* occur is plain from Bonaparte's passage of the Alps in the year 1800, when Melas might and should have fallen on him with his whole force before Napoleon had united his columns.

The second influence which mountains may have as a barrier is that which they have upon the enemy's lines of communication if they cross those lines. Without taking into account what may be effected by erecting forts at the points of passage and by arming the people, bad roads in bad seasons of the year may of themselves prove destructive to an army. They have frequently compelled a retreat after having sucked all the marrow and blood out of an army. If, in addition, there are frequent patrols of active irregulars, or even a national rising, then the enemy is obliged to make large detachments and at last is driven to form strong posts in the mountains and thus he becomes involved in one of the most disadvantageous situations that can exist in an offensive war.

4. MOUNTAINS IN THEIR RELATION TO THE SUBSISTENCE OF THE ARMY

This is a very simple subject, easy to understand. The best use the defender can make of mountains in this respect is when the assailant is either obliged to remain in them, or at least to leave them in his rear.

These considerations on the defense of mountains, which, fundamentally include the whole of mountain warfare, and, by their reflection, throw also the necessary light on offensive war, must not be deemed incorrect or impracticable because we can neither make plains out of mountains, nor hills out of plains, and the choice of a theater of war is determined by so many other things that it appears as if there was little margin left for considerations of this kind. In operations on a large scale it will be found that this margin is not so small. If it is a question of the disposition and effective employment of the main force at the moment of a decisive battle, by a few marches more to the front or rear an army can be brought out of mountain ground into the level country, and a resolute concentration of the main force in the plain will neutralize the adjoining mountains.

We shall now once more collect the light which has been thrown on the subject, and bring it to focus in one distinct picture.

We maintain, and believe we have shown, that mountains, both tactically and strategically, are in general unfavorable to the defensive, meaning thereby, that kind of defensive which is *decisive,* on the result of which the question of the possession or loss of the country depends. They limit the view and prevent movements in every direction; they force us to passivity, and make it necessary to block every way of access, which always leads more or less to a war of cordons. We should, therefore, if possible, avoid mountains with the principal mass of our force, and leave them on one side, or keep them before or behind us.

On the other hand, we think that, for minor operations and objects, there is an element of increased strength to be found in mountain ground, and after what has been said, we shall not be accused of inconsistency in maintaining that such a country is the real place of refuge for the weak, that is, for those who no longer dare seek an absolute decision. This claim which secondary objects have on mountain ground once more excludes the main army from it.

Still all these considerations will hardly counteract the impressions of the senses. The imagination not only of the inexperienced but also of all those accustomed to bad methods of war will still receive in the concrete case such an overpowering impression of the difficulties which the inflexible and retarding nature of mountainous ground oppose to all the movements of an assailant that they will hardly be able to look upon our opinion as anything but a most singular paradox. With those, however, who take a general view, the history of the last century (with its characteristic form of war) will take the place of sense impressions, and therefore there will be but few who will bring themselves to believe that Austria, for example, should not more easily be able to defend her states against Italy than against the Rhine. On the other hand, the French who carried on war for twenty years under a leadership both energetic and reckless, and have constantly before their eyes the successful results thus obtained, will, for some time to come, distinguish themselves in this as well as in other cases by the instinct resulting from a trained judgment.

Does it follow from this that a state would be better protected by an open country than by mountains, that Spain would be stronger without the Pyrenees, Lombardy less accessible without the Alps, and a level country such as North Germany more difficult to conquer than a mountainous country? To these false deductions we shall devote our concluding remarks.

We do not assert that Spain would be stronger *without* its Pyrenees than *with* them, but we say that a Spanish army, feeling itself strong enough to engage in a decisive battle, would do better by concentrating itself in a position behind the Ebro than by distributing itself among the fifteen passes of the Pyrenees. The influence of the Pyrenees on the war is very far from being suppressed on that account. We say the same respecting an Italian army. If it divided itself in the High Alps it would be vanquished by every resolute opponent it encountered, without even having the alternative of victory or defeat; while in the plains of Turin it would have the same chance as any other army. But still no one will on that account suppose that it is desirable for an aggressor to have to march over a mass of mountains such as the Alps, and to leave them in his rear. Moreover, the acceptance of a great battle in the plains by no means excludes a preliminary defense of the mountains by subordinate forces, an arrangement very advisable in respect to such masses of mountains as the Alps and Pyrenees. Lastly, it is far from our intention to argue that the conquest of a level country is easier than that of a mountainous one, unless a single victory sufficed to disarm the enemy completely. After this victory a state of defense ensues for the conqueror, during which the mountainous ground must be as disadvantageous to the assailant as it was to the defensive, and even more so. If the war continues, if foreign assistance arrives, if the people take up arms, this reaction will gain strength from a mountainous country.

It is here, as in dioptrics: the image represented becomes more luminous when moved in a certain direction, not, however, as far as one pleases, but only until the focus is reached, beyond which the effect is reversed.

If the defensive is weaker in the mountains, that would seem to be a reason for the assailant to direct his line of operation preferably in the direction of the mountain range. But this will seldom occur, because the difficulties of subsisting an army, and those arising from the roads, the uncertainty as to whether the enemy will accept a main battle in the mountains, and even whether he will take up his position there with his principal force, tend fully to neutralize that possible advantage.

CHAPTER 17

DEFENSE OF MOUNTAINS
(CONTINUED)

In Chapter 15 we spoke of the nature of engagements in mountains, and in Chapter 16 of the use to be made of them by strategy. In so doing we often came upon the idea of a real *mountain defense*, without stopping to consider the form and arrangement of such a measure. We shall now examine it in greater detail.

Mountain ranges frequently extend like bands or belts over the surface of the earth, separating the streams flowing in one direction from those flowing in another, and thus forming the divides of the whole water system. This form of the whole is repeated in the parts of it, inasmuch as these go off from the main chain in branches and ridges and so constitute the divides for smaller water systems. The idea of a mountain defense has naturally, therefore, been founded in the first instance on the primary conception of an obstacle longer than it is broad, and consequently extending like a great barrier; and out of this conception the idea has been developed. Although geologists are not yet agreed as to the origin of mountains and the laws of their formation, still in every case the water courses indicate in the shortest and surest way the form of the system, whether their action has contributed to its formation (by the process of erosion) or they are themselves the result of it. It was, therefore, natural in devising a mountain defense, to take the water courses as a guide. Not only are they to be regarded as a natural series of levels from which we can learn both the general height and profile of the mountains, but the valleys formed by the waters are also to be considered as the best means of access to the highest points, because in every case this much may be said of erosive action, that it tends to smooth out the inequalities of the slopes into one regular curve. The theory of mountain defense would accordingly lay down that a mountain range running almost parallel to the defensive front should be regarded as a great obstacle to approach, a sort of rampart, the gates in which are formed by the valleys. The real defense would therefore have to be made on the crest of this rampart—that is, on the edge of the plateau that crowns the heights of the range—and cut the main valleys transversely. If the main direction of the range were more perpendicular to the defensive front, one of its principal branches would have to be defended—one running up parallel to one of the main valleys right to the main ridge, the latter having to be regarded as its termination.

We have here indicated this abstract scheme for the defense of a mountain range in accordance with its geological structure because it has actually for some time held a vague but prominent place in theory, and in the so-called "theory of terrain" has amalgamated the laws of erosion with the conduct of war.

But all this is so full of false hypotheses and incorrect substitutions that in reality too little is left remaining of such a view for anyone to be able to construct any sort of a helpful system out of it.

The principal ridges of real mountain ranges are far too inhospitable and inaccessible to have large masses of troops placed on them; it is often the same with the minor ridges; often, too, they are too short and irregular. Plateaux do not exist on all mountain ranges, and where they are found, they are mostly narrow and on that account very unfit to accommodate many troops. Indeed, there are few mountain ranges that, more closely examined, form an unbroken main ridge with sides at such an angle that they may be regarded as inclined planes or, at all events, as terraced slopes. The main ridge winds, bends and branches. Mighty offshoots stretch away into the country in curved lines, and often, quite at their extremities, rise to more considerable heights than the main ridge itself. Promontories join on to them and form deep valleys, which do not fit into the system. Furthermore, where several lines of mountains cross one another, or at the point from which several branch out, the conception of a narrow band or belt comes completely to an end and gives place to mountain and water lines radiating like the rays of a star.

From this it follows—and anyone who has examined mountains in this way will feel it still more clearly—that the idea of a systematic disposition of troops fades away and that it would be extremely unpractical for anyone to try to make it the basis of his arrangements. There is, however, still another important point to notice in the matter of practical application.

If we look closely at mountain warfare in its tactical aspects, it is clear that these include two main elements, of which the first is the defense of steep slopes, and the second that of narrow valleys. Now the latter, which often, indeed in most cases, give the greater efficacy in resistance, do not admit of being easily combined with the disposition on the main ridge, for the occupation of the valley *itself* is often required, and that more where the valley issues from the mass of mountains and not at its beginning, for there its sides are more precipitous. Furthermore, this valley defense offers a means of defending mountain districts even in cases where on the ridge itself no position can possibly be taken up. The part it plays is usually, therefore, greater in proportion as the mass of mountains is higher and more inaccessible.

From these considerations it follows that the idea of a more or less regular defensive line, coinciding with one of the basic geological lines, must be entirely abandoned. A mountain range must be regarded merely as a plain covered with inequalities and obstacles of many kinds, and our aim is to make as good a use of the parts of it as the circumstances permit. Although, therefore, the geological features of the ground are indispensable for a clear insight into the form of the mass of mountain ranges, nevertheless in considering measures of defense they can count for little.

Neither in the War of the Austrian Succession, nor in the Seven Years' War nor in the War of the French Revolution do we find dispositions which included a whole system of mountains and in which the defense had been organized according to the main features of it. Never do we find the armies on the main ridges, but always on the slopes; sometimes higher, sometimes lower; sometimes in one direction, sometimes in another; parallel, perpendicular and oblique; with the watercourse, and against it; in loftier ranges, such as the Alps, even extended often in a valley; in less lofty, such as the Sudetics—and this is the strangest anomaly—half-way up the slope that faced the defender, and thus with the main ridge in front, like the position in which Frederick the Great in 1762 covered the siege of Schweidnitz and had the Hohe Eule before the front of his camp.

The famous positions of Schmotseifen and Landshut in the Seven Years' War are for the most part in the bottoms of valleys; the same is the case with the position of Feldkirch in Vorarlberg. In the campaigns of 1799 and 1800 the chief posts both of the French and the Austrians were always right in the valleys, not merely across them so as to block them, but also lengthwise, while the ridges were either not occupied at all or were held only with a few single posts.

The ridges of the higher Alps in particular are so inaccessible and inhospitable that it becomes impossible to occupy them with any considerable bodies of troops. Now if we must positively have forces in mountains to keep possession of them, the only thing to be done is to place them in the valleys. At first sight this appears to be a mistake, because according to the prevalent theoretical ideas it would be said that the mountains "command" the valleys. But it is not so. Mountain ridges are only accessible by a few paths and tracks, and with rare exceptions, only for infantry, because the carriage roads follow the valleys. The enemy, therefore, could only appear at particular points of them and only with infantry. But in these mountain masses the distances are too great for any effective fire of small arms, and therefore a position in a valley is less dangerous than it seems. Such a defensive position in a valley is certainly, however, exposed to another and serious danger, that of being cut off. The enemy, it is true, can only descend into the valley with infantry at certain points, slowly and with great effort. He cannot, therefore, take us by surprise. But none of our positions defends the outlet of such a path into the valley, and the enemy can, therefore, bring down superior numbers of troops, then spread out and burst through the thin and, from that moment, weak line, which perhaps has no more protection than the rocky bed of a shallow mountain stream. But now retreat, which in a valley must always be made piecemeal until an outlet from the mountains has been found, is for many parts of the line impossible. This is the reason why the Austrians in Switzerland have almost always lost one-third or one-half of their troops in prisoners.

Now a few words as to the extent of the division to which such defense forces are usually subjected.

Every such subordinate position proceeds from a position taken up by the main force more or less in the middle of the whole line, on the principal road of approach. From this central position other corps are detached right and left to occupy the most important points of approach, and thus the whole is disposed in from three to six posts or more, roughly in one line. How far this extension may, or can, be carried depends on the requirements of the individual case. A couple of days' march, that is, from thirty to forty miles, is a very moderate length for it, and cases have been seen in which it has been increased to a hundred miles, or a hundred and fifty.

Between the separate posts, which are one or two hours' march from one another, there may easily be found other less important points of approach to which attention is given later. Excellent separate posts, very well suited for communications with the chief posts, are found for a couple of battalions, and these also are occupied. It is easy to see that the division of forces may be carried still further and come down to single companies and squadrons. Such cases have often occurred. There are, therefore, in this no general limits to the process of splitting. On the other hand, the strength of the single posts depends on the strength of the whole, and so we can say nothing as to the possible or natural extent of the strength which the principal posts will possess. We shall only give as a guide some propositions drawn from experience and common sense.

1. The more lofty and inaccessible the mountains are, the further the process of division may be carried, and the further it *must* be carried. For the less any district can be kept secure by plans which depend on movements, the more must that security be obtained by immediate covering. The defense of the Alps requires a much greater division of forces and makes it approximate much more to a cordon than does the defense of the Vosges or the Riesengebirge.
2. Hitherto, whenever a defense of mountains has been undertaken, the forces have been divided to such an extent that the principal posts have had only one line of infantry and, in the second line, some squadrons of cavalry. Only the chief post, stationed in the middle, has had at best a few battalions as well in the second line.
3. A strategic reserve in the rear to strengthen the points attacked has very seldom been kept, because with such an extended front it was felt that the line was everywhere already too weak. On that account the support which the post attacked could receive has mostly been taken from other posts in the line, themselves not attacked.
4. Even when the division of the forces was still comparatively moderate and the strength of the single posts still considerable, the main resistance of the latter has always consisted in a local defense, and if the enemy was once in complete possession of a post, no help was any longer to be expected from newly arrived support.

How much, according to this, is to be expected from a mountain defense, in what cases this means may be employed, how far we can and may go in the splitting up of our forces—all this, theory must leave to the instinct of the general. It is enough if theory tells him what this means really is and what part it may play in the operations of the army in war.

A general who gets himself into disaster in an extended mountain position deserves to be court-martialed.

CHAPTER 18

DEFENSE OF RIVERS AND STREAMS

Rivers and large streams, in so far as we speak of their defense, belong, like mountains, to the category of strategic barriers. But they differ from mountains in two respects. The one concerns their relative, the other their absolute defense.

Like mountains, they strengthen the relative defense; but their peculiarity is that they are like implements of hard and brittle material; they either stand every blow without bending, or their defense breaks and then ends altogether. If the river is very large, and the other conditions are favorable, then the passage may be absolutely impossible. But if the defense of any river is forced at one point, then there cannot still be, as in mountain warfare, a persistent defense afterward. The matter is settled with that one act, unless the river itself runs between mountains.

The other peculiarity of rivers in relation to the engagement is that in some cases they admit of very good, and in general of better, combinations for a decisive battle than do mountains.

Both again have this property in common: they are dangerous and seductive objects which have often led to wrong measures and placed generals in awkward situations. We shall point out these results in examining more closely the defense of rivers.

History shows rather few examples of rivers defended with success, and therefore the opinion is justified that rivers and streams are no such formidable barriers as was supposed at the time when an absolute defensive system seized upon all means of strengthening itself which the country offered. Still the influence which they exercise to the advantage of the engagement and the defense of the country in general cannot be denied.

In order to survey the matter as a whole, we wish to summarize the different points of view from which we propose to examine it.

First and foremost, the strategic results which rivers and streams produce through their defense must be distinguished from the influence which they have on the defense of a country, even when not themselves specially defended.

Further, the defense itself may have three different intentions:

1. An absolute resistance with the main body.
2. A mere demonstration of resistance.
3. A relative resistance by subordinate bodies of troops, such as outposts, covering lines, minor corps, etc.

Lastly, we must, in respect of its form, distinguish three different degrees or kinds of defense, namely:

1. A direct defense by preventing the passage.
2. A less direct one, in which the river and its valley are only used as a means toward a better combination for the battle.
3. An absolutely direct one, by holding an unassailable position on the enemy's side of the river.

We shall subdivide our observations in accordance with these three degrees, and after we have made ourselves acquainted with each of them in its relation to the first and the most important intention, we shall then in conclusion consider the other two. Therefore, first, the direct defense, that is, such a defense as is to prevent the passage of the enemy's army itself.

This can only come into the question in the case of large rivers, that is, great bodies of water.

The combinations of space, time and force, which must be looked upon as elements of this theory of defense, make the subject somewhat complicated, so that it is not easy to get a fixed point of view. The following is the result at which everyone will arrive upon more careful consideration.

The time required to build a bridge determines the distance from each other at which the detachments charged with the defense of the river should be posted. If we divide the whole length of the line of defense by this distance, we get the number of corps; if with that number we divide the mass of troops disposable, we shall get the strength of each detachment. If we now compare the strength of each corps with the number of troops which the enemy, by using other means, can have passed over during the construction of the bridge, we shall be able to judge whether we can count upon a successful resistance. We can only assume that a crossing cannot be forced when it is possible for the defender to attack the troops which have crossed over with *a considerable numerical superiority,* say, double, before the bridge is completed. An example will make this plain.

If the enemy requires twenty-four hours for the construction of a bridge, and if he cannot by other means pass over more than 20,000 men in those twenty-four hours, while the defender within twelve hours can appear at any given point with 20,000 men, the passage cannot be forced, for the defender will arrive when the enemy has passed over only half of those 20,000 men. Now since in twelve hours, the time for conveying intelligence included, we can march twenty miles, every forty miles 20,000 men would be required, which would make 60,000 for the defense of a length of 120 miles of river. These would be sufficient for the appearance of 20,000 men at any given point, even if the enemy attempted the passage at two points at the same time; and of twice that number, if that should not be the case.

Here, then, three circumstances are decisive:

1. the breadth of the river;
2. the means of passage, for the two determine both the time required to construct the bridge and the number of troops that can cross during the time the bridge is being built;
3. the strength of the defender. The strength of the enemy's force itself does not as yet come into consideration. According to this theory we may say that there is a point at which the possibility of crossing ceases completely, and that no numerical superiority on the part of the enemy would enable him to force a passage.

This is the simple theory of the direct defense of a river, that is, of a defense intended to prevent the enemy from finishing his bridge and from making the passage itself; in this there is as yet no notice taken of the effect of demonstrations which those crossing may use. We shall now consider particulars in detail, and measures required for such a defense.

Setting aside, in the first place, geographical peculiarities, we have only to say that the detachments as determined by the present theory must be posted directly on the river and each kept concentrated. They must be directly on the river because every position farther back lengthens the distances unnecessarily and uselessly; for since the waters of the river give security against any important action on the part of the enemy, it is not necessary to maintain a reserve, as in the case of land defense. Besides, the roads running parallel to rivers, up and down, are generally better than transverse roads from behind leading to any particular point on the river. Lastly, the river is unquestionably better watched by means of such a position than by a mere chain of posts, more particularly since the commanders are all close at hand. Each of these bodies must be kept concentrated, because otherwise all the calculation as to time would be different. He who knows the loss of time in effecting a concentration will easily comprehend that just in this concentrated disposition lies the greatest effectiveness of the defense. No doubt, at first sight, it is very tempting to make the crossing, even in boats, impossible for the enemy from the first by a line of posts. But with a few exceptions of points specially favorable for crossing, such a measure would be extremely dangerous. To say nothing of the objection that the enemy can generally drive off such a post by a superior rifle fire from the opposite side, it is, as a rule, a waste of strength; that is to say, the most that can be achieved by such a post is that the enemy chooses another point of passage. If, therefore, we are not so strong that we can treat and defend the river as if it were the moat of a fortress, a case for which no further rules are required, such a method of directly defending the bank of a river necessarily leads away from the proposed object. Besides these general principles for positions, we have further to deal with: first, the consideration of the special peculiarities of the river; second, the removal of all means of passage; and third, the influence of any fortresses situated on the river.

A river, considered as a line of defense, must have at extreme points, right and left, *points d'appui*, such as, for instance, the sea or a neutral territory; or there must be other conditions which make it impracticable for the enemy to cross the extreme point of the line of defense. Now since neither such *points d'appui* nor such conditions occur except at wide intervals, we see that the defense of a river must be spread out over very considerable distances, and that, therefore, the possibility of a defense by placing a large body of troops behind a relatively short length of the river vanishes from the class of possible cases, to which we must always confine ourselves. We say *a relatively short length of the river*, by which we mean a length which does not very much exceed that which the same number of troops would usually occupy on an ordinary position without a river. Such cases, we say, do not occur, and every direct defense of a river always becomes a kind of cordon system, at least as far as regards the extension of the position, and therefore is not at all suited to oppose a turning movement on the part of the enemy in the way which is the natural one in the case of a concentrated position. Therefore, where such turning movement is possible, the direct defense of the river, however promising its results in other respects, is a highly dangerous measure.

Now, as regards the portion of the river between its extreme points, it is obvious that all points are not equally well suited for crossing. This matter can, no doubt, be determined somewhat more precisely on general principles but not positively settled by them, for the very smallest local peculiarity often decides more than all that looks large and important in books. However, such a determination is also wholly unnecessary, for the appearance of the river and the information to be obtained from those residing near it are clear enough indications, and there is no necessity for referring to books.

Going into more detail, we may observe that roads leading to a river, its tributaries, the great cities through which it passes, and lastly, above all, its islands are generally most favorable for a passage; that on the other hand, the elevation of one bank above another, and the bend in the course of the river at the point of passage, which usually play such an important role in books, have seldom been of any consequence. The reason for this is that the influence of these two things rests on the limited idea of absolute defense of the river bank—a case which seldom or never happens in connection with great rivers.

Now, whatever may be the nature of the circumstances which make it easier to cross a river at particular points, they must have an influence on the position of the troops, and modify the general geometrical law; but it is not advisable to deviate too far from that law, relying on the difficulties of the passage at many points. The enemy then chooses exactly those spots which are naturally least favorable for crossing, if he can hope that there he will be least likely to meet us.

In any case, the strongest possible occupation of islands is a measure to be recommended, because a serious attack on an island indicates in the surest way the intended point of passage.

The troops stationed close to the river must be able to move either up or down along its banks, according as circumstances require. If, therefore, there is no road parallel to the river, one of the most essential preparatory measures for the defense is to put the nearest small roads running in a direction parallel to the river into suitable order or to construct new ones for short distances.

The second point of which we must speak is the removal of the means of crossing. Even on the river itself the thing is no easy matter and at least requires considerable time; but on the tributaries which fall into the river on the enemy's side the difficulties are in most cases insurmountable, since these branch rivers are generally already in the hands of the enemy. For that reason it is important to block the mouths of such rivers by fortifications.

Since the equipment for crossing rivers which the enemy brings with him, that is, his pontoons, are rarely sufficient for the passage of great rivers, much depends on the means which he finds on the river itself and its tributaries and in the great cities on his side of it, and, finally, on the timber for building boats and rafts in forests near the river. There are cases in which all these circumstances are so unfavorable that the crossing of the river becomes almost impossible because of this.

Lastly, the fortresses, which lie on both sides, or on the enemy's side of the river, serve both to prevent crossing at any points near them, up or down the river, and as a means of blocking the mouths of tributaries and of quickly taking in the means of crossing.

So much for the direct defense of a river which presupposes a great volume of water. If a deep valley with precipitous sides or if marshy banks are added to the barrier of the river itself, the difficulty of passing and the strength of the defense are certainly increased; but the volume of water is not compensated for by such obstacles, for such circumstances constitute no absolute natural barrier, which is an *indispensable* condition of direct defense.

If we are asked what role such a direct river defense can play in the strategic plan of the campaign, we must admit that it can never lead to a decisive victory, partly because the object is not to let the enemy pass over to our side at all or to crush the first considerable force he has put across and partly because the river prevents our being able to convert the advantages gained into a decisive victory by means of a strong attack.

On the other hand, the defense of a river in this way may produce a great gain in time, which is generally important for the defender. Collecting the means of crossing often involves much time; if several attempts fail, a good deal more time is gained. If the enemy, on account of the river, gives his forces an entirely different direction, still further advantages may probably be gained by that means. Lastly, whenever the enemy is not in downright earnest about advancing, a river will cause him to stop his movements and thereby afford a lasting protection to the country.

A direct defense of a river, therefore, when the masses of troops engaged are considerable, the river large and other circumstances favorable, may be regarded as a very good defensive means, and may yield results to which commanders in modern times, thinking only of unfortunate attempts to defend rivers, which failed because of insufficient means, have paid too little attention. For if, in accordance with the supposition just made (which may easily be realized in connection with such rivers as the Rhine or the Danube), an efficient defense of 120 miles of river is possible by 60,000 men in face of a very considerably superior force, we may well say that this is a remarkable result.

We say, in opposition to a *considerably superior force*, and we must come back to this point again. According to the theory which we have given, all depends on the means of crossing, and nothing on the numerical strength of the force seeking to cross, always supposing it is not smaller than the force which defends the river. This appears very extraordinary, and yet it is true. But we must take care not to forget that most defenses of rivers, or, more correctly speaking, all of them, have no absolute *points d'appui*, and therefore may be turned, and that this turning movement will be very much easier if there is great superiority of numbers.

If we now reflect that such a direct defense of a river, even if overcome by the enemy, is still not to be compared to a lost battle, and has very little chance to lead to a complete defeat, since only a part of our force has been engaged, and the enemy, detained by the tedious crossing over of his troops on a single bridge, cannot immediately gain great results from his victory over ours, we shall be the less disposed to underestimate this means of defense.

In all the practical affairs of human life it is important to hit the right point, and so also in the defense of a river, it makes a great difference whether we correctly grasp all the circumstances. An apparently insignificant detail may essentially alter the case and make a measure which is very wise and effective in one instance a disastrous mistake in another. This difficulty of forming a correct judgment and of avoiding the notion that "a river is a river" is perhaps greater here than anywhere else. We must therefore especially guard against mistaken applications and interpretations; but having done so, we do not hesitate to declare plainly that we do not think it worth while to listen to the clamor of those who, under the influence of some dim feeling or vague idea, expect everything from attack and movement, and think they see the truest picture of war in a hussar at full gallop brandishing his sword over his head.

Such ideas and feelings are not always sufficient (we shall only instance here the once famous Dictator Wedel at Zullichau in 1759); but the worst of all is that they seldom hold their ground, but forsake the general at the last moment if great, complex cases with a thousand ramifications bear heavily upon him.

We therefore believe that a direct defense of a river in the case of large bodies of troops, under favorable conditions, can lead to successful results if we content ourselves with the moderate negative object; but this does not hold good in the case of smaller masses. Although 60,000 men on a certain length of river could prevent an army of 100,000 or more from passing, a body of 10,000 on the same length would not be able to oppose the passage of 10,000 men, indeed, probably not even of one half that number, if such a body chose to run the risk of placing itself on the same side of the river with an enemy so much superior in numbers. This matter is clear, since the means of crossing do not alter.

We have as yet said little about feigned crossings, as they do not essentially come into consideration in the direct defense of a river, partly because such a defense does not depend on a concentration of the army at one point, but each corps has in any case the defense of a portion of river distinctly allotted to it; partly because such feigned crossings are very difficult even under the circumstances we have supposed. If, as it is, the means of crossing are limited, that is, not in such abundance as the assailant must desire to ensure the success of his undertaking, he will then scarcely be able or willing to use a large share of them for a mere demonstration. At all events, the mass of troops to be passed over at the true point of crossing must consequently be smaller, and the defender gains again in time what he might have lost through uncertainty.

This direct defense, as a rule, seems only suitable to large rivers, and on the last half of their course.

The second form of defense is suitable for smaller rivers with deep valleys, often even for very unimportant ones. It consists in a position taken up farther back from the river at such a distance that the enemy's army at the time of crossing may be found either divided, if it passes at several points at the same time, or near the river and confined to *one* bridge and road, if the crossing is made at one point. An army with its rear squeezed against a river or a deep valley, and confined to one line of retreat, is in a most disadvantageous position for battle. The most effective defense of rivers of moderate size, running in deep valleys, lies in making proper use of this circumstance.

The disposition of an army in large detachments close to a river, which we consider the best in a direct defense, presupposes that the enemy cannot pass the river unexpectedly and in great force, because otherwise, by making such a disposition, there would be great danger of being beaten in detail. If, therefore, the circumstances which favor the defense are not sufficiently advantageous, if the enemy has in hand ample means of crossing, if the river has many islands or even fords, if it is not broad enough, if we are too weak, and so forth, then that method is out of the question. The troops must be drawn back a little from the river, for the sake of more secure connection with each other, and all that now remains to be done is to ensure the most rapid concentration possible upon that point where the enemy is attempting to cross, so as to attack him before he has gained so much ground that he has command of several passages. In this case the river or its valleys must be watched and weakly defended by a chain of outposts while the army is disposed in several corps at suitable points and at a certain distance (usually a few hours) from the river.

The greatest difficulty lies here in the passage through the defile formed by the river and its valley. In this case it does not merely depend on the volume of water but on the whole of the defile, and, as a rule, a deep rocky valley is a much greater impediment to pass than a river of considerable breadth. The difficulty of the march of a large body of troops through a long defile is in reality much greater than it appears from mere contemplation. The time required is very considerable; and the danger that the enemy may make himself master of the surrounding heights during the march is very disquieting. If the troops in front advance too far, they encounter the enemy too soon, and are in danger of being overpowered by a superior force; if they remain near the point of passage, then they fight in the worse situation. The passage across such an obstacle of ground in order to try one's strength against the enemy on the other side is, therefore, a bold undertaking, or implies a great superiority of numbers and great confidence on the part of the commander.

Such a defensive line cannot, certainly, be extended to such a length as in the direct defense of a great river, for it is intended to fight with the whole force united, and the passages, however difficult, cannot be compared with those over a large river; it is, therefore, much more likely that the enemy will make a turning movement against us. But at the same time, such a movement carries him out of his natural direction (for we suppose, as is obvious, that the valley crosses that direction at about right angles), and the disadvantageous effect of confined lines of retreat only disappears gradually, not all at once, so that the defender will still have some advantage over the advancing foe, even if the latter has not been caught by him exactly at the crisis of the passage, but by means of his turning movement has already gained a little more room to move.

Since we are not speaking of rivers merely in respect of the volume of their waters, but have almost more in view the deep cleft formed by their valleys, we must explain that under this term we do not mean a regular mountain gorge, because in that case all that has been said about mountains would be applicable. But, as everyone knows, there are many level districts where even the smallest rivers form deep and precipitous clefts; besides these, marshy banks or other difficulties of approach belong to the same class.

Under these conditions, therefore, an army on the defensive, posted behind a large river or deep valley with steep sides, is in a very excellent position, and this sort of river defense is to be counted among the best strategic measures.

Its defect (the point on which the defender is very apt to err) is the over-extension of the forces. It is so natural in such a case to be drawn on from one point of passage to another, and to miss the right point at which we ought to stop; but if we do not succeed in fighting with the whole army united, the intended effect is lost; a lost engagement, the necessity of retreat, confusion in many ways and losses reduce the army nearly to ruin, even if it does not stand its ground to the last.

In saying that the defender, under the above conditions, should not extend his forces widely, that he must in any case have all his forces assembled on the evening of the day on which the enemy passes, enough is said, and it may stand in place of all further combinations of time, power and space—things which, in this case, must depend on many local circumstances.

The battle brought on in these circumstances must have a special character—that of the greatest impetuosity on the part of the defender. The feigned crossings, by which the enemy may have kept him for some time in uncertainty, will, as a rule, cause him to appear only at the last moment. The peculiar advantages of the situation of the defender consist in the disadvantageous situation of the enemy's troops just immediately in his front; if other corps, having passed at other points, menace his flank, he cannot, as in a defensive battle, counteract such movements by vigorous blows from his rear, for that would be to sacrifice the advantage of his situation. He must, therefore, settle affairs in his front before such other corps can become dangerous; that is, he must attack what he has before him as swiftly and vigorously as possible, and decide all by its defeat.

But the object of this form of river defense can never be resistance to a very greatly superior force, as is conceivable in the direct defense of a large river; for, as a rule, we have really to deal with the bulk of the enemy's force, and although we do so under favorable circumstances, still it is easy to see that the relative strength of the forces becomes at once a serious consideration.

This is the nature of the defense of rivers of moderate size and deep valleys when the principal masses of the army are concerned, where the considerable resistance which can be offered on the ridges of the valley cannot compensate for the disadvantages of a scattered position, and a decisive victory is a matter of necessity. But if it is only a question of the reinforcement of a secondary line of defense, which is intended to hold out for a short time and counts on being reinforced, certainly a direct defense of the ridges of the valley, or even of the river bank, may be made. Although advantages similar to those in mountain positions are not to be expected here, still the resistance will always last longer than in ordinary country. Only one circumstance makes this measure very dangerous, if not impossible: it is when the river has many windings and sharp turnings, which is just what is often the case when a river runs in a deep valley. One needs merely to consider the course of the Moselle in Germany. In the case of its defense, the corps in advance of the salients of the bends would almost inevitably be lost in the event of a retreat.

That a great river allows the same defensive means and the same form of defense, which we have pointed out as best suited for rivers of a moderate size, in connection with the mass of an army and even under much more favorable circumstances, is obvious. It will come into use more especially when it is important for the defender to gain a decisive victory (Aspern).

The case of an army drawn up with its front close on a river or stream or deep valley, in order by that means to command a tactical obstacle to the approach or to strengthen its front, is quite a different one, the detailed examination of which belongs to tactics. Of the effect of this we shall say only this much: that fundamentally it rests on a complete delusion. If the cleft in the ground is very considerable, the front of the position becomes absolutely unassailable. Now, as there is no more difficulty in passing round such a position than any other, it is actually just the same as if the defender had himself gone out of the way of the assailant, yet that could hardly be the object of the position. A position of this kind, therefore, can only be useful when, as a consequence of its location, it threatens the communications of the assailant, so that every deviation from the direct road would involve consequences altogether too serious.

In this second form of defense feigned crossings are much more dangerous, for the assailant can make them more easily, while, on the other hand, the defender has the problem of assembling his whole army at the right point. The defender in this case, however, is not quite so limited as to time, because his advantage lasts until the assailant has massed his whole force and has taken several crossings; on the other hand, the feigned attack has not yet the same degree of effect here as in the defense of a cordon, where all must be held, and where, therefore, in the employment of the reserve, the question is not merely, as in our problem, where the enemy has his principal force, but the much more difficult one: which is the point he will first seek to force?

With respect to both forms of defense of large and small rivers, we must furthermore observe generally that if they are undertaken in the haste and confusion of a retreat, without preparation, without the removal of means of passage, and without an exact knowledge of the country, they cannot, of course, fulfill what has been here supposed. In most such cases, nothing of the kind is to be reckoned on, and therefore it will always be a great error for an army to divide itself over extended positions.

Just as everything usually miscarries in war, if it is not done with clear knowledge and with the whole will and energy, so a *river defense* will generally end badly when it is only resorted to because we have not the heart to meet the enemy in the open field, and hope that the broad river or the deep valley will stop him. In this case, there is so little confidence in the actual situation that both the general and his army are usually filled with anxious forebodings, which in most cases will be realized quickly enough. A battle in the open field does not presuppose a perfectly equal state of circumstances, like a duel; and the defender who does not know how to gain for himself any advantages in such a battle, either through the special nature of the defense, through rapid marches, or by knowledge of the country and freedom of movement, is one whom nothing can save, and least of all will a river or its valley be able to help him.

The third form of defense—by a fortified position taken up on the enemy's side of the river—founds its efficiency on the danger in which it places the enemy of having his communications cut by the river, and being limited to a few bridges only. It follows, as a matter of course, that we are only speaking of great rivers with a great volume of water, since these alone make that case possible, while a river which is merely in a deep ravine usually affords such a number of passages that all such danger disappears.

But the position of the defender must be very well fortified, almost unassailable; otherwise he would just meet the enemy half way, and give up his advantages. But if it is of such strength that the enemy will not decide to attack it, he will, under certain circumstances, be confined thereby to the same bank with the defender. If the assailant crossed, he would expose his communications; but he would threaten ours, of course, at the same time. Here, as in all cases in which one opponent passes by the other, it depends on whose communications, by number, situation and other circumstances, are the best secured, and who has also, in other respects, most to lose, and therefore can be outbid by his opponent; lastly, on who still possesses in his army the greater power of victory upon which he can depend in an extreme case. The influence of the river merely amounts to this: that it increases the danger of such a movement for both parties, since both are limited to bridges. Now, in so far as we can assume that, according to the usual course of things, the passages of the defender, as well as his depots of all kinds, are better secured by fortresses than those of the assailant, such a defense is, indeed, conceivable, and would then be substituted for the direct defense in those cases where other circumstances are not favorable enough to that form. Certainly in this case the river is not defended by the army, nor the army by the river, but by the connection between the two is the country defended, which is the main point.

At the same time we must admit that this mode of defense, without a decisive blow, resembling the state of tension of the two kinds of electricity when their poles are not in contact, cannot stop any very powerful attack. It might be applicable against a cautious, irresolute general, who never pushes forward with energy, even if his forces are greatly superior; it might also answer when a kind of equilibrium between the contending forces has previously arisen, and nothing but small advantages are looked for on either side. But if we have to deal with superior forces, led by a bold general, we are upon a dangerous course and very close to disaster.

This form of defense looks so bold, and at the same time so scientific, that it might be called the elegant form; but since elegance easily borders on fatuity, and that is not so easily excused in war as in society, we have few instances of this elegant form. From it a special means of assistance for the first two forms is developed, which is, by holding a bridge and a bridgehead, to keep up a constant threat of crossing.

Besides the object of an absolute resistance with the main body, each of the three forms of defense may also have that of a *feigned defense*.

This show of resistance, which is not intended really to be offered, is an act which is of course combined with many other measures, and fundamentally with every position which is something more than a mere bivouac; but the feigned defense of a great river becomes a true demonstration through the adoption of a number of more or less complicated measures, and through its effect being usually on a greater scale and of longer duration than that of any other. For the act of passing a great river in sight of an army is always an important step for the assailant, one over which he will ponder long and many times, or which he will postpone to a more favorable moment.

For such a feigned defense it is therefore requisite that the main army should divide and post itself along the river (much in the same way as for a real defense); but since the intention of a mere feigned defense shows that circumstances are not favorable enough for a real defense, such a measure, which always demands a more or less extended and scattered disposition, would very easily cause danger of serious loss, if the detachments should become engaged in a real resistance, even if only on a moderate scale. It would then be in the true sense a half-hearted measure. In a feigned defense, therefore, everything must be arranged so as to ensure concentration of the army at a point considerably, perhaps several days' march, in the rear, and the defense should not be carried beyond what is consistent with this arrangement.

In order to make our views clear, and to show the importance such a feigned defensive may have, let us call to mind the end of the campaign of 1813. Bonaparte repassed the Rhine with forty or fifty thousand men. To attempt to defend this river with such a force along the whole extent within which the Allies, according to the direction of their forces, could easily pass, that is, between Mannheim and Nimwegen, would have been an impossibility. Bonaparte could therefore only think of offering his first serious resistance somewhere on the French Meuse, where he could make his appearance with the army in some measure reinforced. Had he at once withdrawn his forces to that point, the Allies would have followed close at his heels; had he placed his army in rest-quarters behind the Rhine, the same thing could scarcely have failed to take place a moment later, for in spite of their most faint-hearted caution, the Allies would have sent over swarms of Cossacks and other light troops in pursuit, and, if that measure produced good results, other corps would have followed. The French corps were therefore forced to take steps to defend the Rhine in earnest. As it could be foreseen that nothing could come of this defense as soon as the Allies seriously undertook to cross the river, it was therefore to be regarded as a mere demonstration, in which the French corps incurred hardly any danger, since their point of concentration lay on the Upper Moselle. Only Macdonald, who, as is known, stood at Nimwegen with twenty thousand men, committed the mistake of deferring his retreat until he was actually compelled to retire. This delay prevented his joining Bonaparte before the battle of Brienne, as the retreat was not forced on him until after the arrival of Winzingerode's corps in January. This feigned defense of the Rhine, therefore, was sufficient to check the Allies in their advance, and to induce them to postpone the crossing of the river until their reinforcements arrived, which did not take place for six weeks. These six weeks were bound to be of infinite value to Bonaparte. Without this feigned defense of the Rhine the victory at Leipzig would have led directly to Paris, and it would have been absolutely impossible for the French to have given battle this side of their capital.

In a river defense of the second class, that is, in the case of rivers of a smaller size, such a demonstration may also be used, but it will generally be less effective, because mere attempts to cross are in such a case easier, and therefore the spell is sooner broken.

In the third kind of river defense a demonstration would in all probability be still less effective and produce no more result than that of the occupation of any other temporary position.

Lastly, the first two forms of defense are very well suited to give a chain of outposts, or any other defensive line (cordon) established for some secondary object, or to a corps of observation, much greater and more certain strength than it would have without the river. In all these cases we can only speak of a relative resistance and that must naturally be considerably strengthened by such a natural obstacle. At the same time, we must think not merely of the relatively considerable gain in time, which the resistance in the engagement itself can produce, but also of the many anxieties which such an undertaking usually excites in the mind of the enemy, and which in ninety-nine cases out of a hundred lead to its not being carried out unless there are urgent reasons for it.

CHAPTER 19

Defense of Rivers and Streams
(continued)

We have still to add something in regard to the effect of rivers and streams on the defense of a country, even when they are not themselves defended.

Every important river, with its main valleys and its adjacent valleys, forms a very considerable obstacle in a country, and is in that way advantageous to defense in general; but its peculiar influence does not admit of being more closely determined in its principal relations.

First we must distinguish whether it flows parallel to the frontier, that is, the general strategical front, or at an oblique, or a right, angle to it. In the case of the parallel direction we must observe the difference between having our own army or that of the assailant behind it, and in both cases again the distance between it and the army.

An army on the defensive, having a large river close behind it, but not less than a day's march, and on that river an adequate number of secure crossings, is unquestionably in a much stronger position than it would be without the river; for if it loses a little in freedom in all its movements by having to consider the crossings, still it gains much more by the security of its strategic rear, that is, essentially, its lines of communication. In all this we allude to a defense in *our own country*; for in the enemy's country, although his army might be before us, we should still always have to fear more or less his appearance behind us on the other side of the river, and then the river, involving as it does narrow defiles in roads, would be more disadvantageous than advantageous in its effect on our situation. The farther the river is behind the army, the less useful it will be, and at certain distances its influence disappears altogether.

If an attacking army, in its advance, has to leave a river in its rear, this can only be disadvantageous to its movements, for it restricts the communications of the army to a few single passages. When Prince Henry in 1760 marched against the Russians on the right bank of the Oder near Breslau, he had obviously a *point d'appui* in the Oder, a day's march behind him; on the other hand, when the Russians under Czernitschef passed the Oder later, they were in a very embarrassing situation, just through the danger of losing their line of retreat, which was limited to one bridge.

If a river crosses the theater of war more or less at a right angle with the strategic front, then the advantage is again on the side of the defensive. In the first place, there are generally a number of good positions leaning on the river, and using the transverse valleys running down to it to strengthen the front (like the Elbe for the Prussians in the Seven Years' War); secondly, the assailant will have to leave one side of the river or the other unoccupied, or he must divide his forces; and such division cannot fail to be in favor again of the defensive, because he will be in possession of more well-secured passages than the assailant. One need only regard the Seven Years' War as a whole to be convinced that the Oder and Elbe were very useful to Frederick the Great in the defense of his theater of war (namely, Silesia, Saxony and the Mark), and consequently a great impediment to the conquest of these provinces by the Austrians and Russians, although there was not one real defense of those rivers in the whole Seven Years' War, and their course is mostly, with regard to the enemy, at an oblique or a right angle to the front rather than parallel with it.

It is only the convenience of a river as a means of transport, when its course is more or less at right angles to the front, which can, in general, be advantageous to the assailant; this may be so for the reason that as the assailant has the longer line of communication, and therefore the greater difficulty in the transport of all he requires, transport by water may relieve him of a great deal of trouble and prove very useful. The defender, it is true, has the advantage here also of being able to close the navigation within his own frontier by fortresses. Still, even by that means the advantage which the river affords the assailant will not be lost so far as its course up to that frontier is concerned. But many rivers are often not navigable, even where they are otherwise of no unimportant breadth from the military point of view, and others not navigable at all seasons. Navigation against the stream is tedious, and the winding of a river often doubles the length of the way. The chief communications, moreover, between countries now are high roads, and lastly, now more than ever the needs of an army are supplied from the nearest provinces, and not by transport from distant parts. Bearing all this in mind, we can well see that the use of a river does not generally play such a prominent part in the subsistence of troops as is usually presented in books and that its influence on the course of events is, therefore, very remote and uncertain.

CHAPTER 20

A. DEFENSE OF SWAMPS

Very large, extensive swamps, such as the Bourtang Moor in North Germany, are so uncommon that it would not be worth while to lose time on them; but we must not forget that certain lowlands and marshy banks of rivers are more common and they form very considerable obstacles of ground which can be and often are used for defensive purposes.

Measures for their defense are naturally much the same as in the case of rivers; there are, however, some peculiarities to be specially noticed. The first and principal one is that a marsh, which except on the dykes is quite impracticable for infantry, is much more difficult to cross than any river. In the first place, a dyke is not built so quickly as a bridge, and second, there are no provisional means by which the troops covering the construction of it can be put across. No one would begin to build a bridge without using some of the boats to take over an advance guard, but in the case of a morass no similar assistance can be employed. For infantry alone, the easiest way to get them across a swamp would be just planks, but if the morass is of some breadth this process takes incomparably longer than the crossing of the first boats on a river. If now in the middle of the morass there is a river which cannot be passed without a bridge, the crossing of the first detachment of troops becomes still more difficult, for though single men can very well get across on mere planks, it is impossible by this means to transport heavy materials such as are required for building bridges. This difficulty in some circumstances may be found insurmountable.

A second peculiarity of a swamp is that the means for crossing it cannot be completely removed, like those of a river. Bridges can be broken off or so completely destroyed as to be made absolutely unusable, but the most that can be done with dykes is to cut them, which does not mean much. If there is a small river in the middle, the bridge across it can, of course, be removed, but the possibility of crossing will not thereby be so wholly put an end to as it is in the case of a large river. The consequence is that the existing dykes must always be occupied with a fairly strong force and strenuously defended if we want to derive any advantage at all from the morass.

We are, therefore, on the one hand, obliged to adopt a local defense, while, on the other, such a defense is made easier by the difficulty of getting across elsewhere. The result of these two peculiarities is that the defense of swamps must be more local and passive than that of rivers.

It follows that we must be relatively stronger than in the direct defense of a river and, consequently, the line of defense cannot be made so long, especially in cultivated countries, where the number of ways across, even under the most favorable circumstances, is usually still very great.

In this respect, therefore, swamps are inferior to great rivers, and it is a very important respect, for all local defense has in it something extremely insidious and dangerous. But such swamps and lowlands usually have a breadth with which that of the largest European rivers is not to be compared, and consequently a post stationed for the defense of a passage is never in danger of being overpowered by the fire from the other side. The effect of its own fire is immensely increased by a quite narrow, long dyke, and the time required to pass such a defile, a mile or two long, is incomparably greater than that needed to cross a bridge. Bearing all this in mind, we must admit that such lowlands and morasses, if the means of crossing them are not too numerous, are among the strongest lines of defense that can be found.

An indirect defense, such as we have become acquainted with in the case of rivers and streams, in which obstacles of ground are made use of to bring on a great battle under advantageous circumstances, can be employed equally well in the case of morasses.

The third method of river defense by means of a position on the enemy's side would be much too hazardous because of the long time required for crossing.

It is extremely dangerous to venture on the defense of such morasses, soft meadows, bogs and so forth, as are not absolutely impassable except on the dykes. A single possible crossing discovered by the enemy is enough to break the line of defense, a thing which, in case of a serious resistance, is always attended by great losses to the defender.

B. INUNDATIONS

We have still to consider inundations. Both as means of defense and as natural phenomena, they are unquestionably most like morasses.

Certainly they are not common. Holland is perhaps the only country in which they constitute a phenomenon which makes them worth notice in connection with our subject. But it is precisely that country which obliges us, on account of the remarkable campaigns of 1672 and 1787 and of its important relation to Germany and France, to devote some consideration to this matter.

The character of these Dutch inundations differs from that of ordinary swampy and impassable wet lowlands in the following respects:

1. The soil itself is dry and consists either of dry meadows or of cultivated fields.
2. For purposes of irrigation or of drainage, a number of small ditches of greater or less depth and breadth intersect the country in such a way that in certain districts they run in parallel directions.
3. Larger canals, enclosed by dykes and intended for irrigation, drainage and navigation, run through the country in all possible directions, and are such that they cannot be crossed without bridges.
4. The level of the ground throughout the whole district subject to inundations lies considerably below the level of the sea and consequently below that of the canals.
5. The consequence of this is that by cutting the dykes and closing and opening the sluices the whole country can be laid under water, so that there are no dry roads except those on the tops of the higher dykes, all others being either entirely under water or at all events so soaked with water as to be no longer usable. Now though the inundation is only three or four feet deep, so that in case of need it could be waded through for short distances, this is prevented by the small ditches mentioned in 2, which are not visible. It is only where these ditches have an appropriate direction, so that we can move between two of them without crossing one or the other, that the inundation ceases to be an obstacle to approach. It is easy to see that this can only be the case for short distances, and therefore can only be used for tactical purposes of a very special character.

From all this we deduce:

1. That the assailant is confined to a more or less small choice of approaches, which run along rather narrow dykes and usually in addi-

tion have on each side of them ditches filled with water, thus forming very long dangerous defiles.

2. That every defensive means on such a dyke may easily be strengthened to the point of being impregnable.

3. That the defender, however, just because he is so hemmed in, must at each separate point confine himself to the most passive defense, and consequently must look entirely to passive resistance for his safety.

4. That it is not a question of a single line of defense, shutting in the country like a simple barrier, but, as everywhere we have the same obstacle to approach to protect our flanks, we can continually form new posts, and in this way replace a lost section of the line of defense by a new one. We might say that the number of combinations here, like those on a chessboard, is infinite.

5. But as this general condition of a country is only conceivable on the presupposition that it is highly cultivated and thickly inhabited, it follows of itself that the number of ways through, and consequently the number of the posts which block them, will be very large as compared with that in other strategical positions. From which it follows again as a consequence that such a line of defense must not be long.

The principal line of defense in Holland runs from Naarden on the Zuyder Zee, mostly behind the Vecht, to Gorkum on the Waal, that is, actually, to the Biesbosch, and is about forty miles long. For the defense of this line in 1672 and in 1787 a force of from 25,000 to 30,000 men was employed. If we could safely reckon on an invincible resistance, the results would certainly be very great, at all events for Holland, the province lying behind that line. In 1672 the line actually withstood a considerably superior force commanded by great generals, Condé at the beginning and later Luxemburg, who could have attacked it with from 40,000 to 50,000 men. Yet they would not attempt an assault but preferred to wait for the winter, which, however, was not severe enough. In 1787, on the other hand, the resistance offered on this line amounted to nothing. Even that which was made on a much shorter line between the Zuyder Zee and the sea of Haarlem, although somewhat more serious, was overcome by the Duke of Brunswick in one day, through the mere effect of a very skilful tactical disposition, well adapted to the locality, and this though the Prussian force actually engaged in the attack was little, if at all, superior in numbers to the troops guarding the lines.

The different result in the two cases is to be attributed to the difference in the supreme command. In 1672 the Dutch were surprised by Louis XIV while everything was in a state of peace, in which, as is well known, there existed very little military spirit so far as the land forces were concerned. For that reason the greater number of the fortresses were deficient in all articles of equipment, garrisoned only by weak bodies of hired troops, and defended by governors who were native-born incapables or treacherous foreigners. Thus all the Brandenburg fortresses on the Rhine garrisoned by the Dutch, as well as all their own places situated to the east of the line of defense above described, except Groningen, very soon fell into the hands of the French, and for the most part without any real defense. And in the conquest of this great number of fortresses consisted the activity of the French army, at that time 150,000 strong.

But when, after the murder of the brothers De Witt, in August, 1672, the Prince of Orange came into power, bringing unity into the measures for defense, there was just enough time to close the above-mentioned defensive line. All measures now fitted so well together that neither Condé nor Luxemburg, who commanded the French forces left in Holland after the departure of the two armies under Turenne and Louis XIV, ventured to make any attack on the separate posts.

In 1787 the situation was quite different. It was not the government of the seven united provinces but really only the province of Holland that was to resist the invasion. Of the conquest of all the fortresses, which had been the principal object in 1672, there was, therefore, no question; the defense was confined at once to the above-mentioned line. But the assailant this time, instead of 150,000 men, had only 25,000 and was no mighty sovereign of a great adjoining country, but the deputed general of a distant prince, himself fettered by many considerations. In Holland, as everywhere else, the people were divided into two parties, but in Holland the republican party was decidedly predominant and at the same time in a state of truly enthusiastic excitement. In these circumstances, the resistance in 1787 could certainly have yielded at least as good a result as that of 1672. But there was an important difference: in 1787 the unity of command was lacking. What in 1672 had been left to the wise, skilful and energetic guidance of the Prince of Orange, was entrusted in 1787 to a so-called Defense Commission, which, although it consisted of four energetic men, was not able to introduce enough unity of measures into the whole work and inspire individuals with enough confidence to prevent the whole instrument from becoming imperfect and inefficient in use.

We have dwelt for a moment on this example to give more distinctness to the conception of this defensive measure and at the same time to show the difference in the effects produced, according as more or less unity and consistency prevail in the direction of the whole.

Although the organization and kind of resistance of such a defensive line are a subject for tactics, still in reference to the defensive line, which is somewhat more closely related to strategy, we cannot omit to make an observation to which the campaign of 1787 gives occasion. We think, namely, that however passive the defense must naturally be at individual posts, still an offensive action from some one point of the line is not impossible and may produce good results if the enemy, as was the case in 1787, is not decidedly superior. For although such an assault must be carried out on dykes, and on that account certainly will have no great freedom of movement or great impulsive force, nevertheless it is impossible for the assailant to occupy all the dykes and roads on which he is not himself advancing. For the defender, therefore, who knows the country and is in possession of the strong points, there should still always be means in this way either of effecting a real flank attack against the columns of the assailant, or of cutting them off from their sources of supply. If, on the other hand, we consider in what a very constrained position the assailant is placed, how much more dependent he is on his communications than in almost any other conceivable case, we shall well understand that any assault by the defender which has but a remote possibility of success in its favor, must at once, as a demonstration, be extremely effective. We doubt very much if the prudent and cautious Duke of Brunswick would have ventured to approach Amsterdam if the Dutch had only made a single demonstration of that kind, from Utrecht, for instance.

CHAPTER 21

DEFENSE OF FORESTS

Above all things we must distinguish thick, impassable, virgin forests from cultivated, extensive plantations which are partly quite open and partly intersected by many paths.

The latter kind, whenever it is a question of a defensive line, are either to be left in the rear or else avoided altogether. The defender has more need than the assailant of an unimpeded view all round him, partly because, as a rule, he is the weaker, partly because the natural advantages of his position cause him to develop his plans later than the assailant. If he chose to leave a wooded district before him, he would be fighting like a blind man against one who can see. If he should place himself in the middle of the wood, both, of course, would be blind, but it is just this equality that would not answer the natural requirements of the defender.

Such a wooded country cannot, therefore, be brought into any sort of favorable relation to the defender's engagements, unless he keeps it in the rear of him so as to conceal whatever is taking place behind him from the enemy, and at the same time to make use of it for covering and facilitating his retreat.

At present, however, we are speaking only of forests in level country, for where a decidedly mountainous character prevails, its influence on tactical and strategic measures becomes predominant, and of this we have already spoken elsewhere.

But impassable forests, that is, such as can only be traversed on certain roads, afford, of course, advantages for an indirect defense, similar to those derived from mountains, for bringing on a battle under favorable circumstances. Behind the forest the army can await the enemy in a more or less concentrated position in order to fall upon him the moment he debouches from the defiles. Such a forest resembles a mountain range more than it does a stream in its effects, for it affords, it is true, only a very long and difficult defile; but in regard to the retreat it is rather advantageous than dangerous.

A direct defense of forests, however impassable they may be, is a very hazardous piece of work for even the thinnest chain of outposts. For entanglements are only imaginary barriers, and no forest is so impassable that it cannot be penetrated in a hundred places by small detachments. These, in the case of a chain of defensive posts, are comparable to the first drops of water that ooze through a dyke, to be presently followed by a general burst.

Much more important is the influence of great forests of every kind in connection with a national rising. They are undoubtedly the true element for such irregular forces. If, therefore, the strategic plan of defense can be so arranged that the enemy's line of communication passes through great forests, by that means another mighty lever is brought into use in the work of defense.

CHAPTER 22

THE CORDON

The term cordon is used to denote every defensive arrangement which is intended directly to protect a whole district of country by a line of posts connected with one another. We say "directly," for several corps of a great army posted in line with each other might protect a large tract of country from incursion of the enemy without constituting a cordon. This protection, however, would not be direct, but through the effect of combinations and movements.

It is evident that a defensive line, long enough to cover an extensive tract of country directly, can only have a very slight capacity for resistance. Even when very large bodies of troops occupy the lines, this would still be the case if similar bodies were acting against them. The object of a cordon can, therefore, only be to protect against a weak blow, whether that weakness lies in the will power behind it or in the man power with which it may be delivered.

With this view the wall of China was built as a protection against Tartar raids. This is the intention of all lines and frontier defenses of the European states bordering on Asia and Turkey. Used in this way the cordon system is neither absurd nor does it appear unpractical. Certainly it is not sufficient to stop all raids, but still they will become more difficult and consequently less frequent, and in circumstances such as prevail among the peoples of Asia, where the state of war is almost permanent, that is very important. Next to this class of cordons come the lines which in the wars of modern times have been formed between European states, such as the French lines on the Rhine and in the Netherlands. These were originally formed only with a view to protect a country against attacks made merely for the purpose of levying contributions or living at the expense of the enemy. They are, therefore, only intended to check minor operations, and consequently should be defended with small bodies of troops. But of course in the event of the enemy's main force taking its direction against these lines, the defender is forced to occupy them with his own main force, a course by no means productive of the best defensive arrangements. On account of this disadvantage, and because the protection against raids in a transient war is an object of very minor importance, for which through the existence of these lines we may easily be compelled to incur an excessive expenditure of force, they have in our day come to be looked upon as a harmful expedient. The more fiercely the war rages, the more useless and dangerous does such a means become.

Lastly, all very extended lines of outposts covering the quarters of an army and intended to offer a certain amount of resistance are to be regarded as really cordons.

This kind of resistance is chiefly designed for protection against raids and other small operations directed against the security of single quarters, and for this purpose it may be quite sufficient if favored by the ground. Against the advance of the main body of the enemy the resistance can only be relative, that is, intended to gain time. But this gain in time will not in most cases be very considerable and can therefore less be regarded as the object of the cordon of outposts. The assembling and advance of the enemy's army can never take place so unobservedly that the first information the defender got of it would be through his outposts, and in such a case he would be very much to be pitied.

Consequently, in this case also, the cordon is only intended to resist the attack of a weak force, and the object, therefore, in this as in the other two cases is not at variance with the means.

But that the main force intended for the defense of a country against the main force of the enemy should disperse itself into a long series of defensive posts, that is to say, into a cordon, seems so absurd that we must try to discover the particular circumstances which lead to and accompany such a proceeding.

In mountainous country every position, although taken up with a view to a battle with all forces united, may and must necessarily be more extended than it would be on level ground. It *may* be so, because the aid of the ground increases the capacity for resistance; it *must* be so, because a wider basis of retreat is required, as we have already shown in the chapter "Defense of Mountains." But suppose there is no near prospect of a battle, and that the enemy will probably remain in his position opposite to us for some time without undertaking anything unless tempted by some favorable opportunity—the usual state of things in most wars formerly. In this case it is natural for us not to limit ourselves to the occupation of only so much country as is absolutely necessary, but to remain masters of as much ground right and left of us as the security of our own army permits, which will give us advantages of many kinds, as we shall presently show in more detail. In open and accessible country by means of a war of *movement* this object can be attained to a greater extent than in mountains, the extension and dispersion of our forces being in the former case less necessary to secure it. But it would also be much more dangerous because each part has less capacity for resistance.

But in mountains, all possession of ground depends on its local defense. A threatened point cannot be so easily reached, and if the enemy has got there first, he cannot easily be driven out again by a superior force. In mountains, therefore, in such circumstances we shall always come to a form of disposition which though not actually a cordon, yet, as a series of defensive posts, comes very near it. From such a disposition, broken up into several posts, to the cordon, there is certainly a wide step, but nevertheless generals take it, often without themselves being aware of the fact, being drawn on from one stage to another. At the beginning the object of the dispersion of the forces is the covering and possession of the country; later on it becomes the security of the forces themselves. Every commander of a post calculates the advantages he would get from the occupation of this or that point of approach on the right or left side of his post, and so the whole passes imperceptibly from one degree of dispersion to another.

A cordon war, therefore, carried on by the main body, if it occurs, is not to be considered as a form intentionally chosen, for the purpose of arresting every onset of the enemy forces, but as a situation into which the army has fallen in the pursuit of quite another object, which is the holding and covering of the country against an enemy who has no decisive operation in view. Such a situation must always be a mistake, and the reasons which have gradually coaxed one small post after another out of the general must be called trivial in relation to the purpose of a main army. But this point of view at least shows the possibility of such an aberration. The fact that it is such an aberration, this failure to understand the enemy's and one's own position, is overlooked, and all that is spoken of is the faulty *system*. But this system is quietly approved when it has been pursued with advantage, or, at all events, without loss. Everyone praises the *faultless* campaigns of Prince Henry in the Seven Years' War, because the king has so pronounced them, although these campaigns afford the most glaring and incomprehensible examples of a chain of posts so extended that no other has a better claim to be called a cordon. We may completely justify the prince by saying that he knew his opponents, and was aware that he had no decisive operations to fear. As the object, moreover, of his dispositions was always to occupy as large a tract of country as possible, he went as far as the circumstances in any way would permit. If the prince had once come to grief in spinning such a cobweb and sustained a serious loss, it would have had to be said not that the prince pursued a faulty system, but that he had been mistaken in his methods and had applied them to a case to which they were not suited.

We are trying in this way to explain how a so-called cordon system may arise with the principal force in a theater of war, and how it may even be a judicious and useful measure and no longer in that case seem an absurdity. At the same time we must acknowledge that there actually have been cases where generals or their staffs have overlooked the real meaning of a cordon system and assumed its relative value to be a general one. They have believed that it was actually suited for a protection against any sort of hostile attack, which implies not a misapplication of the measure, but a complete misunderstanding of it. We admit that one instance of this truly absurd belief seems to have been afforded by the Prussian and Austrian army during the defense of the Vosges in 1793 and 1794.

CHAPTER 23

KEY OF THE COUNTRY

There is no theoretical idea in the art of war which has played such a part in criticism as that which is here our subject. It is the show piece in all descriptions of battles and campaigns, the most frequent point of view in all military arguments, and one of those scraps of scientific form with which critics make a show of learning. And yet the conception embodied in it has never been established nor has it ever been clearly formulated.

We shall try to ascertain its real meaning and see what value it will still have for practical action.

We treat of it here because the defense of mountains and rivers, as well as the conceptions of fortified and entrenched camps, with which it is closely connected, had to take precedence.

The indefinite, confused conception which is concealed behind this ancient military metaphor has sometimes signified the most exposed part of a country, at other times the strongest.

If there is a district without the *possession of which no one dare venture to penetrate into an enemy's country,* that may with propriety be called the key of that country. But this simple, but certainly not very fruitful, notion has not satisfied the theorists. They have amplified it and imagined under the term "key of the country" *points which decide the possession of the country.*

When the Russians wanted to advance into the Crimean peninsula they were obliged to make themselves masters of the isthmus of Perekop and its lines, not so much to gain an entrance at all—for Lascy turned it twice, in 1737 and 1738—but so as to be able to establish themselves in the Crimea with tolerable security. That is very simple, but at the same time we do not gain much through the conception of a key point. But if it might be said: Whoever has possession of the district of Langres commands all France as far as Paris—that is to say, it only rests with himself to take possession of it—that is plainly a very different thing, something of much higher importance. According to the first kind of conception, the possession of the country cannot be thought of without the possession of the point which is to be called the key, which is intelligible to mere common sense. But according to the second kind of conception, the point which is to be called the key cannot be imagined without the possession of the country following as a necessary consequence, which is plainly something marvelous. Common sense is no longer sufficient to grasp this; the magic of occult science is required. This cabala actually originated in books published about fifty years ago, and reached its zenith at the end of the last century. Notwithstanding the irresistible force, certainty and lucidity with which Bonaparte's conduct of war carried away men's convictions that cabala has nevertheless still succeeded in spinning out in books a thin thread of tenacious existence.

Setting aside *our* conception of a key point, it is obvious that in every country there are points of *commanding* importance, points at which many roads meet, in which our subsistence can conveniently be obtained, and from which we can conveniently move in various directions—points, in short, the possession of which satisfies many needs and affords many advantages. Now if generals have wanted a word to denote such a point and have therefore called it the *key of the country*, it would be a piece of pedantry to take offense at their doing so; the expression is, on the contrary, in that case very applicable and acceptable. But to propose out of this mere flower of speech to produce a seed out of which a whole system with many branches, like a tree, is to be developed, is a challenge to common sense to reduce the expression to its real value.

For the practical but certainly very indefinite meaning which the conception of a key of the country has in the narratives of generals when they speak of their military enterprises, it was necessary to substitute one more definite and therefore more narrow. Among all its references the one chosen was "high ground."

When a road traverses a mountain ridge, we thank heaven when we get to the top and now begin to descend. This feeling, so natural in the case of the individual traveler, is still more so in the case of an army. All difficulties seem to have been overcome, and so, indeed, in most cases they are. The descent is easy; we feel our superiority over anyone who would try to stop us; we see the whole country spread out before us; and we command it in anticipation with our glance. Thus the highest point of a mountain pass has always been regarded as the decisive one, and so in most cases it really is, though by no means in all. In the historical narratives of generals, such points have therefore very frequently been called key points, again, no doubt, in a slightly different sense and for the most part with a more limited reference. This idea has been the main starting point of the false theory of which Lloyd may perhaps be regarded as the founder, and on this account high points, from which several roads descend into the country to be entered, have been looked upon as key points of that country—points which *command* that country. It was natural that this view should combine with one closely related to it, namely, that of a *systematic defense of mountains*, and that the matter should be pushed still further into the realm of illusion. For now a number of tactical elements, important in the defense of mountains, came into play as well, and thus the conception of the highest *point of the road* was soon abandoned and the highest point generally of the whole mountain system, that is, the point of the *watershed* was looked upon as the key of the country.

Now just about that time, that is, in the latter half of the last century, more definite ideas on the forms given to the surface of the earth by the action of water became current. Thus natural science with this geological system lent a hand to the theory of war, and every barrier of practical truth was now broken through and every discussion of such subjects floated vaguely in the illusory system of a geological analogy. In consequence of this, at the end of the last century we heard, or rather we *read*, of nothing but the sources of the Rhine and the Danube. This nuisance, it is true, prevailed for the most part only in books, as in such cases it is only a small fraction of book wisdom that ever passes over into the real world, and all the less so, the more foolish the theory; but the theory of which we are speaking, unfortunately for Germany, has not remained without influence upon practice. We are, therefore, not tilting at windmills, and in proof of this we will quote two examples: first, the important but scientific campaigns of the Prussian army in 1793 and 1794 in the Vosges, to which the books of Gravert and Massenbach serve as a theoretical key, and, second, the campaign of 1814, in which the Bohemian army of 200,000 men allowed itself, as a result of this theory, to be led by the nose through Switzerland on to the so-called plateau of Langres.

But a high point in a country from which all its waters flow is as a rule nothing more than a high point, and all that at the end of the eighteenth century and the beginning of the nineteenth was written about its influence on military events, in exaggeration and wrong application of an idea in itself true, is sheer imagination. If the Rhine and the Danube and all the six rivers of Germany chose to honor *one* mountain with their common source, that mountain would not on this account have a claim to any greater military value than as something on which to erect a trigonometrical landmark. For a signal tower it would be less useful, for a vedette, still less so, and for an army, of absolutely no use at all.

To seek for a *key position*, therefore, in a so-called *key district* of a country, that is, where the different mountain ranges branch out from a common point and the highest river sources lie, is a mere book idea, opposed by Nature herself, who does not make the ridges and valleys so accessible from above as the hitherto so-called theory of terrain assumes, but distributes peaks and gorges as she pleases and not infrequently surrounds the lowest-lying sheets of water with the loftiest masses of mountain. If we question military history on the subject, we shall realize how little influence as a rule the leading geological features of a district have on the use made of it in war, and how greatly, on the other hand, that little is outweighed by other local circumstances and other requirements, so that the line of positions will often run quite close to one of those points and yet not be attracted by it.

We have dwelt so long on this false idea only because a whole—very pretentious—system has been built upon it. We now leave it and return to our own view.

We say then that if the expression, *key position*, is to represent in strategy an independent conception, it can only be that of a district without the possession of which we may not dare to force our way into a country. But if we choose to denote by the term every convenient entrance to a country, or every convenient central point in it, the term loses its specific meaning, that is, its value, and denotes something which must be found more or less anywhere. It then becomes merely a pleasing figure of speech.

But positions such as the term conveys here to us are certainly very rarely to be found. In general the best key to a country lies in the enemy's army, and if the concept of ground is to predominate over that of military force, exceptionally favorable conditions must prevail. These, according to our opinion, may be recognized by two principal effects. In the first place, the force occupying a position, by the aid of the ground, is capable of a strong tactical resistance; in the second place, the position effectively threatens the enemy's line of communications sooner than he threatens ours.

CHAPTER 24

OPERATING AGAINST A FLANK

We need hardly observe that we speak of the strategic flank, that is, of the side of the theater of war, and that the attack from one side in battle, or the tactical movement against a flank, must not be confounded with it. Even in cases in which the strategic operation against a flank, in its last stage, ends in the tactical operation, they can quite easily be kept separate, because the one never follows necessarily out of the other.

These flanking operations, and the flanking positions connected with them, belong also to the show pieces of theory, which are seldom met with in actual war. Not that the means itself is either ineffectual or illusory, but because both sides generally seek to guard themselves against its effects, and cases in which this would be impossible are rare. Now in these rare cases this means has often proved highly efficient, and for this reason, as well as on account of the constant watching against it which is required in war, it is important that it should be clearly explained in theory. Although the strategic operation against a flank can naturally be imagined, not only on the part of the defensive, but also on that of the offensive, still it has much more affinity with the first, and therefore finds its place among the defensive means.

Before we enter into the subject, we must establish the simple principle, which must never be lost sight of afterward in its consideration, that troops which are to act against the rear or flank of the enemy cannot be employed against his front, and that, therefore, whether it be in tactics or strategy, it is an entirely erroneous notion to consider that *coming in the rear* of the enemy is an advantage in itself. In itself, it is as yet nothing; but it will become something in connection with other things, and something either advantageous or disadvantageous, according to the nature of these other things, with the examination of which we are now primarily concerned.

First, in the action against the strategic flank, we must make a distinction between its two objectives—between the action merely against the communications and that against the line of retreat, with which, at the same time, an action upon the communications may be combined.

When Daun, in 1758, sent raiding parties to seize the convoys on their way to the siege of Olmütz, he had plainly no intention of impeding the king's retreat into Silesia; he rather wished to bring about that retreat, and would willingly have opened the way for him.

In the campaign of 1812, the object of all raiding parties that were detached from the Russian main army in the months of September and October was only to interrupt the communications, not to stop the retreat; but the latter was quite plainly the design of the Moldau army which, under Tschitschagof, advanced against the Beresina, as well as of the attack which General Wittgenstein was instructed to make on the French troops stationed on the Dwina.

These examples are merely to make the ideas clear.

The action against the lines of communication is directed against the enemy's convoys, against small detachments following in rear of the army, against couriers and travelers, small depots of the enemy, etc.; therefore against all the means which the enemy requires to keep his army in a vigorous and healthy condition. Its object is, therefore, to weaken the condition of the army in this way, and by this means to cause it to retreat.

The action against the enemy's line of retreat is to cut his army off from that line. It cannot effect this object unless the enemy really decides to retreat; but it may certainly cause him to do so by threatening him, and, therefore, it may have the same effect as the action against the line of communication, by acting as a demonstration. But as we said before, none of these effects is to be expected from the mere turning movement, from the mere geometrical form given to the disposition of the troops; they only result from the conditions suitable to this purpose.

In order to recognize these conditions more distinctly, we shall separate completely the two actions against the flank, and first consider that which is directed against the communications.

Here we must first establish two principal conditions, one or the other of which must always be forthcoming.

The first is that the forces used for this action against the communications of the enemy must be so insignificant in number that their absence is hardly felt at the front.

The second, that the enemy's army must be at the end of its advance, and therefore be unable to make further use of a fresh victory over our army, or to pursue us if we evade an engagement by moving out of the way.

This last case, which is by no means so uncommon as might be supposed, we shall not consider for the moment, but occupy ourselves with the further prerequisites of the first.

The first of these is that the enemy's communications have a certain length, and can no longer be protected by a few good posts; the second point is that the situation of the line is such as exposes it to our action.

This exposure of the line may arise in two ways: either by its direction, if it is not perpendicular to the strategic front of the enemy's army, or because his lines of communication pass through our territory; if both of these circumstances are combined the line is so much the more exposed. Both relations require a closer examination.

One would think that when it is a question of covering a line of communication 200 or 250 miles long, it would be of little consequence whether the position occupied by an army standing at the one end of this line is at an oblique angle or a right angle to it, since the breadth of this position is little more than a mere point in comparison to the line; and yet it is not so. When an enemy's army is posted at right angles to its communications, it is difficult, even with a considerable superiority, to interrupt the communications by raiding parties sent out for the purpose. If we think only of the difficulty of covering absolutely a certain space, we should not believe this, but rather suppose, on the contrary, that it must be very difficult for an army to protect its rear (that is, the country behind it) against all expeditions which an enemy superior in numbers may undertake. It certainly would be, if we could survey everything in war as easily as if it were on paper. In that case the party covering the line, in its uncertainty as to the point where raiding parties will appear, would be, so to speak, blind, and only the raiders would see. But if we think of the uncertainty and incompleteness of all the intelligence we get in war, and know that both parties are constantly groping in the dark, then we easily perceive that a raiding party sent round the enemy's flank to gain his rear is in the position of a man who has to contend with many in a dark room. In the end this man must fall; and it is just the same with the raiding parties which get around an army occupying a perpendicular position, and therefore find themselves near to the enemy, and completely separated from their own army. Not only is there danger of losing considerable forces in this way; but the instrument itself is blunted at once. For the very first misfortune which befalls one such party will make all the others timid, and instead of bold assaults and insolent tricks, we shall only see constant running away.

Through this difficulty, therefore, an army occupying a perpendicular position covers the nearest points on its line of communications for a distance of two or three days' march, according to the strength of the army; but these nearest points are just those which are most in danger, as they are also the nearest to the army of the enemy.

On the other hand, in the case of a decidedly oblique position, no such part of the line of communication is covered; the smallest pressure, the least dangerous attempt on the part of the enemy, leads at once to a vulnerable point.

But now, what is it which determines the front of a position, if it is not just the direction perpendicular to the line of communication? The front of the enemy. But then, again, this may be equally well supposed to be dependent on our front. Here a reciprocal action comes in, the origin of which we must find.

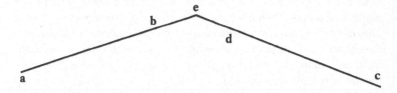

If we suppose the lines of communication of the assailant, *a b,* so situated with respect to those of the defender, *c d,* that the two lines form a considerable angle with each other, it is evident that if the defender wished to take up a position at *e,* where the two lines intersect, the assailant from *b,* by the mere geometrical relation, could compel him to form front opposite to him, and thus to expose his communications. The case would be reversed if the defender took up his position on this side of the point of junction, about *d;* then the assailant would have to make a front toward him, provided that his line of operations, which closely depends on geographical conditions, cannot be arbitrarily changed, and moved, for instance, to the direction *a d.* From this it would follow that the defender has an advantage in this system of reciprocal action, because he only has to take up his position on this side of the intersection of the two lines. But very far from attaching great importance to this geometrical element, we only traced our examination back to this point to make ourselves perfectly clear; and we are, on the contrary, convinced that local and individual relations in general have much more to do with determining the position of the defender; that, therefore, it is quite impossible to lay down in general which of two belligerents will be obliged to expose his communications most.

If the lines of communication of both sides lie in one and the same direction, then whichever of the two parties takes up an oblique position in respect to it will certainly compel his adversary to do the same. But then there is nothing gained geometrically by this, and both parties attain the same advantages and disadvantages.

We shall therefore confine our further consideration to the instance in which the line of communications is exposed only on one side.

Now as far as the second disadvantageous relation of a line of communication is concerned, that is to say, when it runs through an enemy's country, it is clear in itself how much the line is exposed by that circumstance, if the inhabitants of the country have taken up arms. Consequently the affair must be looked at as if a force of the enemy were posted all along the line. This force, it is true, is in itself weak without solidity or intensive strength, but we must also consider what the close contact and influence of such a hostile force may nevertheless effect through the number of points which offer themselves, one next to the other, on long lines of communication. That requires no further explanation. But even if the enemy's subjects have *not* taken up arms, and even if there is no militia in the country, or other military preparations, indeed even if the people are very unwarlike in spirit, still the mere fact that the people are subjects of a hostile government is a noticeable disadvantage for the communications of the other side. The assistance which raiding parties enjoy merely through a better understanding with the people, through a knowledge of the country and its inhabitants, through good information, through the support of official functionaries, is, for them, of decisive value; and this assistance will be accorded to all such parties without any special effort on their part. In addition to this, within a certain distance there will never be lacking fortresses, rivers, mountains or other places of refuge, which ordinarily belong to the enemy, if they have not been formally taken possession of and occupied by our troops.

Now in such a case, especially if accompanied by other favorable circumstances, it is possible to act against the enemy's communications, although their direction is perpendicular to the position of that army; for the raiding parties do not then need to fall back always on their own army, because they will find sufficient protection by merely escaping into their own country.

We have, therefore, now ascertained that

1. a considerable length,
2. an oblique direction, and
3. a hostile environment,

are the principal circumstances under which the lines of communication of an army may be interrupted by relatively slight enemy forces. In order to make this interruption effective, a fourth condition is still required, which is a certain duration of time. Respecting this point, we refer to what has been said in Book V, Chapter 15.

But these four conditions are only the principal points relating to the subject; a number of local and individual circumstances are attached to these, which are often a more decisive and important influence than the principal ones themselves. To recall only the most essential, we mention the character of the roads, the nature of the country through which they pass, the means of cover which can be afforded by rivers, mountains and morasses, the seasons and weather, the importance of particular convoys, such as siege trains, the number of light troops, etc., etc.

On all these circumstances, therefore, will depend the success with which a general can act on his opponent's communications; and by comparing the result of the whole of these circumstances on the one side with the result of the whole on the other, we obtain a just estimate of the relative advantages of the two systems of communication, on which will depend which of the two generals can outdo the other in this respect.

What here seems so lengthy in the explanation is often decided in the concrete case at first sight; but nevertheless the instinct of a practical judgment is required and a person must have thought over all the cases analyzed here in order to realize the absurdity of those critical writers who think they have settled something by the mere words "turning" and "acting on a flank," without furnishing more detailed reasons.

We now come to the *second main condition,* under which the strategic action against the enemy's flank may take place.

If the enemy's army is hindered from advancing by any other cause than the resistance of our own, no matter what this cause may be, then our army should no longer hesitate to weaken itself by sending out considerable detachments; for if the enemy actually proposed to retaliate on us by an attack, all we need to do would be to decline the engagement. Such was the case with the chief Russian army at Moscow in 1812. But it is not at all necessary that everything should be of the same great dimensions and proportions as in that campaign to produce such a case. In the first Silesian wars Frederick the Great was each time in this situation, on the frontiers of Bohemia and Moravia and in the complex conditions of the generals and their armies, many causes of different kinds, particularly political ones, may be imagined, which make further advance an impossibility.

Since in this case more considerable forces may be employed against the enemy's flank, the other conditions need not be quite so favorable: even the nature of our own communications in relation to those of the enemy need not be to our advantage, as an enemy who is not in a condition to make any particular use of our further retreat is not likely to use his right to retaliate, but on the contrary will be more concerned about the direct covering of his own line of retreat.

Such a situation is therefore very well suited to obtain for us, by means less brilliant and effective than a victory but at the same time less dangerous, those results which it would be too great a risk to seek to obtain by a battle.

As in such a case we feel little anxiety about exposing our own line of communications, by taking up a position on one or the other flank, and as the enemy by that means may always be compelled to form front at an oblique angle to his line of communications, therefore *this one* of the conditions above named will seldom be lacking. The more the rest of the conditions, as well as other favorable circumstances, co-operate, so much the more certain are we of success from the means now in question; but the less such favorable circumstances are present, the more will everything depend on superior skill in the plans adopted and on rapidity and precision in their execution.

Here is the proper field for strategic maneuvers, such as are to be found so frequently during the Seven Years' War, in the campaigns of 1760 and 1762 in Silesia and Saxony. If, in many wars in which only a moderate amount of genuine energy is displayed, such strategic maneuvering very often appears, this is not because a commander on each occasion found himself at the end of his course, but because lack of resolution, of courage, of an enterprising spirit and a dread of responsibility have often taken the place of real impediments. For a case in point, we have only to call to mind Field-Marshal Daun.

As a summary of the main results of our considerations, we may say that the action against a flank will be most effective

On War · 759

1. in the defensive;
2. toward the end of a campaign;
3. preferably in a retreat into the heart of the country; and
4. in connection with a general arming of the people.

On the mode of executing this action against communications, we have only a few words to say.

The enterprises must be conducted by skilful irregulars, who, at the head of small bodies, by bold marches and attacks, fall upon the enemy's weak garrisons and convoys, and on small detachments on the march here and there, encourage national levies, and sometimes join with them in particular undertakings. These parties must be more numerous than strong, and so organized that it may be possible to unite several of them for a greater undertaking without any obstacle from the vanity or caprice of the single leaders.

We have still to speak of the action against the enemy's line of retreat.

Here we must keep in view, above all things, the principle which we laid down at the very beginning: that forces which are to operate in the rear cannot be used in front; that, therefore, the action against the rear or flanks is not an increase of the forces in itself; it is only to be regarded as a more effective employment of them, increasing the degree of success in prospect, but also increasing the degree of risk.

Every opposition offered with the sword which is not of a direct and simple nature has a tendency to raise the result at the cost of its certainty. An operation against the enemy's flank, whether with one concentrated force or with separate bodies converging from several quarters, belongs to this category.

But now, if cutting off the enemy's retreat is not to be a mere demonstration, but is seriously intended, the real solution is a decisive battle, or, at least, the conjunction of all the conditions for it, and in this very solution we find again the two elements mentioned above—the greater result and the greater danger. If a general, therefore, is to consider himself justified in adopting this method of action, his motives for it must be supplied by favorable conditions.

In this method of resistance we must distinguish the two forms we have mentioned before. The first is: if a general intends to attack the enemy in rear with his whole force, either from a position taken up on the flank for that purpose, or by an actual turning movement; the second is: if he divides his forces, and, by an enveloping position with one part threatens the enemy's rear, with the other part his front.

The result is intensified in both cases alike: either there is a real interception of the retreat, and as a result a great part of the enemy's army is taken prisoner or scattered, or there may be a long and hasty retreat of the enemy's force to prevent this danger.

But the intensified risk is different in the two cases.

If we turn the enemy with our whole force, the danger lies in the exposure of the rear of our own army, and hence all depends again on the relation of the lines of retreat to each other, just as in the action against the lines of communications, in a similar case, it depended on the relation between those lines.

The defender, if he is in his own country, is of course less restricted than the assailant, both as to his lines of retreat and of communication, and is accordingly in a better position to turn his adversary strategically; but this general relation is not of a sufficiently decisive character to be used as the foundation of an effective method. Therefore, nothing but the sum total of the circumstances in each individual case can decide.

Only so much we may add: that favorable conditions are naturally more common in large areas than in small; more common, also, with independent states than with weak ones waiting for foreign aid, and whose armies therefore must above all things have their attention centered on the point of junction with the auxiliary army. Lastly, they become most favorable for the defender toward the close of the campaign, when the impulsive force of the assailant is exhausted; very much, again, in the same manner as in the case of the lines of communication.

Such a flank position as the Russians took up with so much advantage on the road from Moscow to Kaluga in 1812, when Bonaparte's impulsive force was spent, would have brought them into great difficulties at the beginning of the campaign in the camp of Drissa, if they had not been wise enough to change their plan in good time.

The other method of turning the enemy, and cutting off his retreat by dividing our force, entails the risk attending a division of our own forces while the enemy, having the advantage of the interior lines, retains his forces united, and therefore has the power of acting with very superior numbers against one of our divisions. This is a disadvantage which nothing can remove, and exposing ourselves to it can only be justified by one of three principal reasons:

1. The original division of the forces which makes such a method of action necessary, if we do not want to expose ourselves to a great loss of time.

2. A great physical and moral superiority, which justifies the adoption of a decisive method.
3. The lack of impulsive force in the enemy as soon as he has arrived at the end of his advance.

When, in 1757, Frederick the Great invaded Bohemia on converging lines, he certainly had not intended to combine an attack in front with one on the strategic rear; at all events, this was by no means his principal object, as we shall explain more fully elsewhere. But in any case, it is evident that there never could have been any question of a concentration of forces in Silesia or Saxony before the invasion, as he would thereby have sacrificed all the advantages of a surprise.

When the Allies arranged their plan for the second part of the campaign of 1813, with their great physical superiority they could very well at that time entertain the idea of attacking Bonaparte's right flank on the Elbe with their main force, and of thus shifting the theater of war from the Oder to the Elbe. The fact that they fared so badly at Dresden is to be ascribed not to this general plan but to their faulty dispositions both strategic and tactical. They could have concentrated 220,000 men at Dresden against Bonaparte's 130,000, a proportion of numbers eminently favorable to them (at Leipzig, at least, the proportion was as 285:157). It is true that Bonaparte had distributed his forces too evenly for the proper system of a defense on one line (in Silesia 70,000 against 90,000, in the Mark-Brandenburg—70,000 against 110,000), but at all events it would have been difficult for him, without completely abandoning Silesia, to assemble on the Elbe a force which could deal the principal army of the Allies the decisive blow. The Allies could also have easily ordered the army of Wrede to advance to the Maine, and thereby employed it to try to cut Bonaparte off from the road to Mayence.

Lastly, in 1812, the Russians could direct their army of Moldavia upon Volhynia and Lithuania in order to move it forward afterward in the rear of the principal French army, because it was quite certain that Moscow had to be the culminating point of the French line of operations. For any part of Russia beyond Moscow there was nothing to fear in that campaign; therefore the Russian main army had no cause to consider itself too weak.

This same scheme underlay the disposition of the forces suggested in the first defensive plan proposed by General Phul, according to which Barclay's army was to occupy the camp at Drissa, while Bagration's was to press forward in the rear of the main army. But what a difference of circumstances in the two cases! In the first case, the French were three times as strong as the Russians; in the second, the Russians were decidedly superior. In the first, Bonaparte's main army has in it an impulsive force which carries it to Moscow, four hundred miles beyond Drissa; in the second, it is unfit to make a day's march beyond Moscow; in the first plan, the line of retreat on the Niemen would not have exceeded 150 miles; in the second it was 560. The same action against the enemy's retreat, therefore, which was so successful in the second case, would, in the first, have been the wildest folly.

Since the action against the enemy's line of retreat, if it is more than a demonstration, becomes an actual attack from the rear, there still remains a good deal to be said on the subject, but it will come in more appropriately in the book on "The Attack"; we shall therefore discontinue our discussion at this point and content ourselves with having indicated the conditions under which this kind of reaction may take place.

Usually the intention of causing the enemy to retreat by menacing his line of retreat is understood to imply a mere demonstration rather than the actual execution of the threat. If it were necessary that every effective demonstration should be founded on the complete feasibility of the real action, which at first sight seems a matter of course, it would coincide with the action in all respects. But this is not the case. On the contrary, in the chapter on demonstrations we shall see that these, of course, are connected with conditions which are somewhat different, and therefore we must refer our readers to that chapter (Book VII, Chapter 13).

CHAPTER 25

RETREAT INTO THE INTERIOR

OF THE COUNTRY

We have regarded the voluntary retreat into the interior of the country as a special indirect form of defense through which the enemy is expected to be destroyed, not so much by the sword as by exhaustion from his own efforts. In this case, therefore, a great battle is either not supposed at all, or it is assumed to take place so late that the enemy's forces have previously been considerably reduced.

Every assailant in advancing diminishes his military force by the advance. We shall consider this in greater detail in Book VII. Here we must assume that result as a fact, which we may the more readily do, as it is clearly shown by military history to take place in every campaign in which there has been a considerable advance.

This weakening in the advance is increased if the enemy has not been beaten and withdraws of his own accord with his forces intact, offering a constant, steady resistance, and selling every step of ground at the cost of blood, so that the advance is a constant pushing forward and not a mere pursuit.

On the other hand, the losses which the defender suffers on a retreat are much greater if this has been preceded by a defeat in battle than if he retreats voluntarily. For even if he were able to offer the pursuer the daily resistance which we expect on a voluntary retreat, his losses would be *at least* the same in that way, over and above which those sustained in the battle have still to be added. But what an unnatural supposition this would be! The best army in the world, if obliged to retreat far into the interior of the country after the loss of a battle, will suffer losses *quite out of proportion*; and if the enemy is considerably superior, as we suppose him to be in the case of which we are now speaking, if he pursues with great energy, as has almost always happened in modern wars, there will be the highest probability that a regular flight will take place by which the army will usually be completely annihilated.

A *regulated* daily resistance, that is, one which each time only lasts as long as the equilibrium of the combat can be kept doubtful and in which we secure ourselves from defeat by giving up at the right moment the ground which has been contested—such a combat will cost the assailant at least as many men as the defender. The loss which the latter by retreating must now and again unavoidably suffer in prisoners will be balanced by the losses of the other under fire, as the assailant must always fight against the advantages of the ground. It is true that the retreating side loses entirely all those men who are badly wounded, but the assailant likewise suffers the loss of his wounded temporarily, since they usually remain in hospital for several months.

The result will be that the two armies will wear each other away in nearly equal proportions in these perpetual collisions.

It is quite different in the pursuit of a beaten army. Here the troops lost in battle, the general disorganization, the broken courage and the anxiety about the retreat make such a resistance on the part of the retreating army very difficult, in many cases impossible; and the pursuer who, in the first case, advances extremely cautiously, even hesitatingly, like a blind man, always groping about, presses forward in the second case with the firm step of a victor, with the exuberance of one who has met with success, with the confidence of a demi-god, and the more daringly he presses forward so much the more he hastens things on in the direction which they have already taken, because here is the proper field for the moral forces which intensify and multiply themselves without being restricted to the rigid calculations and measurements of the physical world.

It is, therefore, very plain how different the relations of two armies will be, according as it is by the first or the second of the above ways that they arrive at that point which may be regarded as the end of the assailant's advance.

This is merely the result of the mutual destruction; to this must now be added the reductions which the advancing party suffers in yet other ways, and respecting which, as has been said, we refer to Book VII. On the other hand, we have to take into account the reinforcements which the retreating party receives in the great majority of cases, by forces subsequently joining him either in the form of help from abroad or through persistent efforts at home.

Lastly, there is, in the means of subsistence, such a disproportion between the retreating side and the advancing that the first not infrequently lives in affluence when the other perishes in want.

The army in retreat has the means of storing provisions everywhere, and it marches toward them, while the pursuer must have everything brought after him, which, as long as he is in motion, even with the shortest lines of communication, is difficult, and on that account begets scarcity from the very beginning.

All that the country itself has to offer will first be used by the retreating army and will be mostly consumed. Nothing remains but wasted villages and cities, fields stripped of their crops and trampled down, empty wells and muddy brooks.

The pursuing army, therefore, from the very first day, frequently has to contend with the most pressing wants. On taking the enemy's supplies he cannot reckon at all. It is only through accident, or some unpardonable blunder on the part of the enemy, that here and there something falls into his hands.

Thus there can be no doubt that in countries of considerable dimensions, and when there is not too great a difference in strength between the belligerent powers, a relation may be produced in this way between the military forces which promises the defender an infinitely greater chance for a final success in his favor than he would have had if there had been a great battle on the frontier. But not only does the probability of gaining a victory become greater through this alteration in the proportion of power, but the prospects of great results from the victory are increased as well through the change of situation. What a difference between a battle lost close to the frontier of our country and one in the middle of the enemy's country! Indeed, the situation of the assailant is often such at the end of his advance that even a battle *gained* may force him to retreat, because he neither has enough impulsive power left to complete and make use of the victory, nor is he in a condition to replace the forces he has lost.

There is, therefore, an immense difference between a decisive blow at the beginning of the attack and one at the end of it.

To the great advantages of this method of defense are opposed two drawbacks. The first is the loss which the country suffers through the advance of the enemy; the other is the moral impression.

To protect the country from loss can certainly never be looked upon as the object of the whole defense. That object is an advantageous peace. To obtain that as surely as possible is the endeavor, and for this no momentary sacrifice must be considered too great. At the same time, this loss, although it should not be the decisive factor, must still be carefully weighed, for it always affects our interests.

This loss does not affect our military forces directly; it only acts upon them in a more or less roundabout way, while the retreat itself directly reinforces them. It is, therefore, difficult to compare the advantage and disadvantage in this case; they are things of a different kind which have no common sphere of action. We must, therefore, content ourselves with saying that the loss is greater when we have to sacrifice fertile and well-populated provinces and large commercial cities; but it arrives at a maximum when at the same time we lose war materials either ready for use or in course of production.

The second drawback is the moral impression. There are cases in which the commander must be above regarding such a thing, in which he must quietly follow his plans and expose himself to the disadvantages which a short-sighted despondency calls forth; but nevertheless, this impression is no phantom to be treated with contempt. It is not like a force which acts upon one single point, but like a force which, with the speed of lightning, penetrates every fiber, and paralyzes all the powers which should be in full activity, both in the nation and in its army. There are indeed cases in which the cause of the retreat into the interior of the country is quickly understood by both nation and army, and trust, as well as hope, can even be increased by the step; but such cases are very rare. Usually, the people and the army will not even distinguish whether it is a voluntary movement or a precipitate retreat, and still less whether the plan is one wisely adopted, with the prospect of assured advantages, or the result of fear of the enemy's sword. The people will have a mingled feeling of compassion and dissatisfaction when they see the fate of the provinces which were sacrificed; the army will easily lose confidence in its leader, or even in itself, and the constant engagement of the rear-guard during the retreat tends always to give new strength to its fears. *These are consequences* of the retreat about which we must never deceive ourselves. And it certainly is—considered in itself—more natural, simpler, nobler and more in accordance with the moral being of a people to take up the challenge at once, so that the enemy may not cross the frontiers of a nation without being opposed by its genius, which calls him to bloody account.

These are the advantages and disadvantages of this kind of defense; now a few words on its conditions and the circumstances in their favor.

A country of great extent, or, at all events, a long line of retreat, is the first and fundamental condition; for an advance of a few marches will naturally not weaken the enemy noticeably. Bonaparte's center, in the year 1812, at Witebsk, was 250,000 strong, at Smolensk, 182,000, and it was only at Borodino that it had diminished to 130,000, that is to say, had become equal in numbers to the Russian center. Borodino is 450 miles from the frontier; but it was not until they came near Moscow that the Russians reached that decided superiority in numbers, which of itself made the reversal so inevitable, that the French victory at Malo-Jaroslawitz could not essentially alter it again.

No other European state has such dimensions as Russia, and in very few is a line of retreat 500 miles long conceivable. But neither is a force such as that of the French in 1812 likely to appear under different circumstances, still less such a superiority in numbers as existed at the beginning of the campaign, when the French army had more than double the numbers of its adversary, and, furthermore, a decisive moral superiority. Therefore, what was here only effected at the end of the 500 miles, may, perhaps, in other cases, be attained at the end of 250 or 150 miles.

Among the circumstances which favor this mode of defense are:

1. A country only little cultivated.
2. A loyal and warlike people.
3. The season of bad weather.

All these things increase the difficulty for the enemy of maintaining his army, render great convoys necessary, many detachments, harassing duties, cause the spread of sickness and make operations against the flanks easier for the defender.

Lastly, we must yet speak of the absolute quantity of the military forces, which influences this method of defense.

In itself it is natural that, irrespective of the relation of the opposing forces to each other, a small force is sooner exhausted than a larger, and, therefore, that its course cannot be so long, nor its theater of war so wide. There is, therefore, to a certain extent, a constant relation between the absolute size of an army and the space which that army can occupy. It is out of the question to try to express this relation by a figure, and furthermore, it will always be modified by other circumstances. It is sufficient for our purpose to say that these things fundamentally and essentially have this connection. We may be able to march upon Moscow with 500,000 but not with 50,000, even if the relation of the invader's army to that of the defender in point of numbers were much more favorable in the latter case.

Now if we assume this relation of absolute strength to space to be the same in two different cases, it is certain that the effect of our retreat into the interior in weakening the enemy will increase with the masses.

1. Subsistence and quartering of troops become more difficult for the enemy, since even if the space which an army covers increases in proportion to the size of the army, still the subsistence for the army will never be obtained from this space alone, and everything which has to be brought after an army is subject to greater loss; nor is the whole space occupied ever used for quartering, but only a small part of it, and this does not increase in the same proportion as the masses.

2. The advance becomes slower in proportion as the masses increase; consequently, the time is longer before the course of aggression has come to an end, and the sum total of the daily losses becomes greater.

   Three thousand men driving 2,000 before them in an ordinary country will not allow them to retreat by short marches of five, ten or at most fifteen miles a day, and from time to time make a few days' halt. To reach them, to attack them and disperse them is the work of a few hours; but if we multiply these masses by 100, the case is altered. Operations for which a few hours sufficed in the first case require now perhaps a whole day, or even two. The contending forces can no longer remain together near one point; thereby, the diversity of movements and combinations increases, and, consequently, also the time required. But this places the assailant at a disadvantage, because, his difficulty with subsistence being greater, he is obliged to extend his force more than the pursued. Therefore, he is always in danger of being overpowered by the latter at some particular point, as the Russians tried to do at Witepsk.

3. The greater the masses, the more severe are the exertions demanded from each individual for the daily duties required strategically and tactically. A hundred thousand men who have to march to and from the point of assembly every day, who have now to be halted, and then set in movement again, now to be called to arms, then to cook a meal or receive their rations—100,000 men who must not enter camp until the necessary reports are delivered from all quarters—these men, as a rule, require for all these minor exertions connected with the actual march twice as much time as 50,000 would require, but there are only twenty-four hours in the day for both. How much the time and fatigue of the march itself differs according to the size of the body of troops to be moved has been shown in Chapter 10 of the preceding book. Now, the retreating army, it is true, shares these fatigues with the advancing party, but they are much greater for the latter.

1, Because the mass of his troops is greater on account of the superiority which we presuppose.

2, Because the defender, by being always the party to yield ground, purchases by this sacrifice the right of the initiative, and, therefore, the right always to determine the action of his opponent. He forms his plan beforehand, and, in most cases, nothing interferes with it, but the aggressor, on the other hand, can only make his plans after his adversary has taken up his position, which he has always first to find out.

We must, however, remind our readers that we are speaking of the pursuit of an enemy who has not suffered a defeat, who has not even lost a battle. It is necessary to mention this, in order that we may not be supposed to contradict what was said in Book IV, Chapter 12.

But this privilege of determining the action of the enemy makes a difference in saving of time and strength, as well as in respect of other minor advantages, which, in the long run, becomes very important.

3, Because the retreating force, on the one hand, does all it can to make its own retreat easy, repairs roads and bridges, chooses the most convenient places for encampment, etc., and, on the other hand, it does just as much to throw impediments in the way of the pursuer. It destroys bridges, makes bad roads worse by the mere act of marching over them, deprives the enemy of the best places for encampment with the best water supply by occupying them himself, and so on.

Lastly, we must still add, as a specially favorable circumstance, the war made by the people. This does not require further examination here, since we shall allot a special chapter to the subject (Book VI, Chapter 26).

Hitherto we have been discussing the advantages which such a retreat ensures, the sacrifices which it requires and the conditions which must exist. We shall now say something concerning the mode of executing it.

The first question which we have to propose to ourselves is with reference to the direction of the retreat.

It should be made into the *interior* of the country, therefore, if possible, toward a point where the enemy will be surrounded on both sides by our provinces. In that case he will be exposed to their influence, and we shall not be in danger *of being driven out of the main mass of our territory*, which might happen, if we chose a line too near the frontier, as it would have happened to the Russians in 1812, if they had proposed to retreat to the south instead of to the east.

This is the condition which the object of the measure itself imposes. Which point in the country is best, how far the choice of that point will accord with the intention of covering the capital or any other important point directly, or drawing the enemy away from the direction of such important places depends on circumstances.

If the Russians, in 1812, had considered their retreat beforehand, and, therefore, made it completely in conformity with a regular plan, they might easily, from Smolensk, have taken the direction to Kaluga, which they only took on leaving Moscow. It is very possible that under these circumstances Moscow would have been entirely spared.

That is to say, the French were about 130,000 strong at Borodino, and there is no reason for assuming that they would have been any stronger if this battle had been accepted by the Russians half way to Kaluga. Now, how many of these men could they have spared to detach against Moscow? Plainly, very few; but it is not with a few troops that an expedition can be sent a distance of 250 miles (the distance from Smolensk to Moscow) against such a place as Moscow.

Let us suppose that Bonaparte at Smolensk, where after the engagement he was still 160,000 strong, had thought he could venture to send a part of his army against Moscow *before* engaging in a great battle, and had used 40,000 men for that purpose, leaving 120,000 to face the principal Russian army. In that case, these 120,000 men would not have been more than 90,000 in the battle of Borodino, that is 40,000 fewer than the number which actually fought there. The Russians, therefore, would have had a superiority of 30,000 men. Taking the course of the battle of Borodino as a standard, we may very well assume that with such a superiority they would have been victorious. At all events, the relative strength would have been more favorable for the Russians than it was at Borodino. But the retreat of the Russians was not the result of a well-matured plan; they retreated as far as they did because each time they were on the point of accepting battle they did not consider themselves strong enough yet for the battle that should decide the campaign. All means of subsistence and reinforcements had been despatched on the road from Moscow to Smolensk, and it could not occur to anyone in Smolensk to leave this road. Aside from that, however, a victory between Smolensk and Kaluga would never have excused, in the eyes of the Russians, the offense of having left Moscow uncovered, and having exposed it to the possibility of being captured.

Bonaparte, in 1813, would have secured Paris with still greater certainty from an attack if he had taken up a position at some distance in a lateral direction, somewhere behind the canal of Burgundy, leaving in Paris only a few thousand regular troops, together with the city's large force of National Guards. The Allies would never have had the courage to march a corps of 50,000 or 60,000 against Paris while they knew that Bonaparte stood at Auxerre with 100,000 men. If the situation were reversed, no one would have advised an allied army in Bonaparte's situation to leave open the way to their own capital, if *he,* Bonaparte, were their opponent. With such a preponderance he would not have hesitated a moment to march on the capital. So different will be the result even under the same circumstances but with a different morale.

As we shall return later to this subject when treating of the plan of a war, we shall now only add that, when such a lateral position is taken, the capital or place which by this means it is desired to keep out of the conflict, must, in every case, be capable of making some resistance, so that it may not be occupied and laid under contribution by every raiding party.

But we have still to consider another peculiarity in respect to such a line of retreat, that is, a sudden *change of direction.* After the Russians had kept the same direction as far as Moscow they left that direction, which would have taken them to Waladimir, went first farther in the direction of Riazan, and then branched off in that of Kaluga. If they had been obliged to continue their retreat they could easily have done so in this new direction, which would have led them to Kiev, therefore much nearer again to the enemy's frontier. It is obvious that, even though at that time the French should still have preserved a large numerical superiority over the Russians, they could not have maintained their line of communication by way of Moscow. They would have been forced to give up not only Moscow but, in all probability, Smolensk also, that is, all the conquests obtained with so much toil, and to content themselves with a theater of war on this side of the Beresina.

Now, no doubt, the Russian army would thus have gotten into the same difficulty to which it would have exposed itself by taking the direction of Kiev in the first place, namely, that of being separated from the mass of its own territory; but this disadvantage would now have become almost illusory, for how far different would have been the condition of the French army if it had marched straight upon Kiev without making the detour by Moscow!

It is evident that such a sudden *change of direction* in a line of retreat, which is very practicable in the case of long distances, ensures remarkable advantages.

1. It makes it impossible for the enemy (the advancing force) to maintain his old line of communication; but the organization of a new one

is always a difficult matter, and, furthermore, since the change is made gradually, he has more than once to find a new line.

2. Both parties in this manner approach the frontier again; the position of the aggressor no longer covers his conquests, and he must in all probability give them up. Russia, with its enormous dimensions, is a country in which two armies can, in this way, really play tag with each other.

But such a change in the line of retreat is also possible in smaller countries, when other conditions are favorable, a thing which can only be gathered from all the circumstances of each individual case.

Once the direction in which the enemy is to be drawn into the country is determined, then it naturally follows that our principal army should take that direction, for otherwise the enemy with his own principal army would not take it in his advance, and even if he did, we should not then be able to impose upon him all the conditions above supposed. The only question remaining is whether we shall take this direction with our forces undivided, or whether considerable portions should spread out laterally and give the retreat a divergent form.

To this we answer that this form in itself is to be rejected:

1. Because it divides our forces, while their concentration on one point is precisely one of the chief difficulties for the assailant.

2. Because the enemy gets the advantage of operating on interior lines, he can remain more concentrated than we are and consequently can be so much the more superior at any one point. Now this superiority is certainly less to be dreaded when we are following a system which for the moment consists of constantly giving way; but the very condition of this giving way will always be to remain formidable to the enemy and not to allow him to beat us in detail, which might easily happen. A further condition of such a retreat is to bring our principal force gradually to a superiority of numbers, in order, with this superiority, to be able to deal a decisive blow; which with our forces divided would remain uncertain.

3. Because as a general rule a convergent action against the enemy is not suitable for the weaker forces.

4. Because many disadvantages of the weak points of the assailant disappear when the defender's army is divided into separate parts.

The weakest features in a long advance on the part of the aggressor are: the length of the lines of communication and the exposure of the strategic flanks. By the divergent form of retreat, the assailant is compelled to let a portion of his force make front to the flank, and this portion, originally intended only to neutralize our force immediately in his front, now effects to a certain extent something else in addition: that is, it covers a part of the lines of communication.

For the mere strategic effect of the retreat, the divergent form is therefore not favorable; but if it is to prepare subsequent action against the enemy's line of retreat, we must refer to what has been said about that in the last chapter.

There is *only one* object which can cause a divergent retreat: that is, when we can by that means protect provinces which otherwise the enemy would occupy.

What tracts of country the advancing foe will occupy right and left of his course can usually be discerned with fair probability by the point of assembly of his forces and the direction given to them and by the situation of his provinces, fortresses, etc., with respect to ours. To place troops in those tracts which the enemy will in all probability leave unoccupied would be dangerous waste of our forces. But whether *by any disposition of our forces we shall be able to hinder him* from occupying tracts, which he will probably desire to occupy, is more difficult to decide, and it is therefore a point, the solution of which depends much on instinctive judgment.

When the Russians retreated in 1812, they left 30,000 men under Tormassow in Volhynia, to oppose the Austrian force which was expected to invade that province. The size of the province, the numerous obstacles of ground which the country presents, the lack of superiority in the forces likely to attack them, justified the Russians in their expectations that they would be able to keep the upper hand in that quarter, or at least to maintain themselves near their frontier. From this, very important advantages could result later, with which we shall not concern ourselves here; besides this, it was almost impossible for these troops to have joined the main army in time even if they had wished to. For these reasons, the determination to leave the army in Volhynia to carry on a separate war of its own there was right. If, on the other hand, according to the plan of campaign proposed by General Phul, only Barclay's army (80,000 men) was to retire to Drissa and Bagration's army (40,000) was to remain on the right flank of the French, with a view to falling upon them in the rear subsequently, it is evident at once that this corps could not possibly have maintained itself in South Lithuania so near to the rear of the main French army, and would soon have been destroyed by their overwhelming masses.

That fundamentally it is the defender's interest to give up as few provinces as possible to the assailant is intelligible enough, but this is always a secondary consideration; that the attack is also made the more difficult the smaller, or rather the narrower, the theater of war to which we can confine the enemy, is likewise obvious; but all this is subject to the condition that in so doing we have the probability of success in our favor, and that the main force will not be too much weakened thereby, for upon that force we must chiefly depend for the final decision, because distress suffered by the main body of the enemy is most likely to call forth his determination to retreat, and increases in the greatest degree the loss of physical and moral forces connected with it.

The retreat into the interior of the country should therefore as a rule be made directly before the enemy, and as slowly as possible, with an army which has not suffered defeat and is undivided; and by its incessant resistance it should force the enemy to a constant state of readiness for battle, and to a ruinous expenditure of forces in tactical and strategical measures of precaution.

When both sides have in this manner reached the end of the course of attack, the defender should then dispose his army in a position, if possible, forming an oblique angle with the route of his opponent, and from now onward operate against the enemy's rear with all the means at his command.

The campaign of 1812 in Russia shows all these measures on a great scale, and their effects, as it were, in a magnifying glass. Although it was not a voluntary retreat, we may nevertheless conveniently consider it from that point of view. If the Russians, with the experience they now have of the results thus produced, had to undertake the defense of their country over again, under exactly the same circumstances, they would do voluntarily and systematically what in great part was done without a definite plan in 1812. But it would be a great mistake to suppose that there neither is nor can be any instance of the same mode of action where the dimensions of the Russian empire are lacking.

Wherever a strategic attack, without coming to the issue of a battle, founders merely on the difficulties encountered, and the aggressor is compelled to make a more or less disastrous retreat, the chief condition and the chief effect of this sort of resistance are found, whatever modifying circumstances may otherwise accompany it. Frederick the Great's campaign of 1742 in Moravia, of 1744 in Bohemia, the French campaign of 1743 in Austria and Bohemia, the Duke of Brunswick's campaign of 1792 in France and Massena's winter campaign of 1810–11 in Portugal are examples of cases similar but much less imposing in scale and circumstances. There are, besides, innumerable partial operations of this kind, the success of which, although not wholly, is still partly to be ascribed to the principle which we here uphold. These we do not bring forward because it would necessitate an analysis of circumstances which would lead us into too wide a field.

In Russia, and in the other cases cited, the crisis or turn of affairs took place without any successful battle having given the decision at the culminating point; but even when such an effect is not to be expected, it is always a matter of sufficient importance to bring about by this mode of defense, such a relation between the forces as makes the victory possible, and through that victory, as through a first blow, to cause a movement which usually goes on increasing in its disastrous effects according to the laws which govern falling bodies.

CHAPTER 26

ARMING THE NATION

A people's war in civilized Europe is a phenomenon of the nineteenth century. It has its advocates and its opponents, the latter either considering it in a political sense as a revolutionary means, a state of anarchy declared lawful, as dangerous to the social order at home as to the enemy; or on military grounds, believing that the result is not commensurate with the expenditure of force. The first point does not concern us here, for we are considering a people's war merely as a means of fighting, therefore, in its connection with the enemy; but with reference to the latter point, we must observe that a people's war in general is to be regarded as a consequence of the way in which in our day the elemental violence of war has burst its old artificial barriers; as an expansion and strengthening, therefore, of the whole ferment which we call war. The requisition system, the enormous increase in the size of armies by means of that system and of universal conscription and the employment of militia are all things which lie in the same direction, if we make the limited military system of former days our starting point; and the *levée en masse*, or arming of the people, lies also in the same direction. If the first named of these new aids to war are the natural and necessary consequences of barriers thrown down, and if they have so enormously increased the power of those who first used them that the enemy was carried along in the current and obliged to adopt them likewise, such will also be the case with national wars. In the majority of cases the nation which makes judicious use of this means will gain a proportionate superiority over those who despise its use. If this is so, then the only question is whether this new intensification of the violence of war is, on the whole, salutary for humanity or otherwise, a question which would be about as easy to answer as the question of war itself. We leave both to the philosophers. But the opinion might be advanced that the resources which a people's war costs might be more profitably employed, if used in providing other military means; no very deep investigation, however, is necessary to be convinced that these forces are for the most part not at our disposal, and cannot be utilized at will. One essential part of them, that is, the moral element, is indeed only called into existence by its employment in this kind of way.

Therefore, we no longer ask: how much does the resistance which the whole nation in arms is capable of offering cost that nation? But: what is the influence which such resistance can have? What are its conditions and how is it to be used?

Naturally a resistance so widely distributed is not suited to great blows requiring concentrated action in time and space. Its action, like the process of evaporation in physical nature, depends on the extent of the surface exposed. The greater this is, the greater the contact with the enemy's army, and the more that army spreads out, so much the greater will be the effects of arming the nation. Like a slow, gradual fire, it destroys the foundation of the enemy's army. As it requires time to produce its effects, there is, while the hostile elements are working on each other, a state of tension which either gradually subsides, if the people's war is extinguished at some points, and burns slowly down at others, or leads to a crisis, if the flames of this general conflagration engulf the enemy's army and compel it to evacuate the country before it is utterly destroyed. That this result should be produced by a people's war alone presupposes either a surface extent of the invaded state exceeding that of any country in Europe except Russia, or a disproportion between the strength of the invading army and the extent of the country, such as never occurs in reality. Therefore, to avoid following a phantom, we must imagine a people's war always in combination with a war carried on by a regular army, and both carried on according to a plan embracing the operations of the whole.

The conditions under which alone the people's war can become effective are the following:

1. That the war is carried on in the interior of the country.
2. That it is not decided by a single catastrophe.
3. That the theater of war embraces a considerable extent of country.

4. That the national character supports the measures.
5. That the country is of a broken and inaccessible nature, either from being mountainous, or by reason of woods and marshes, or from the peculiar mode of cultivation in use.

Whether the population is numerous or not is of little consequence, as there is less likelihood of a lack of men than of anything else. Whether the inhabitants are rich or poor is also a point not absolutely decisive; at least it should not be. But it must be admitted that a poor population accustomed to hard work and privation usually shows itself more vigorous and better suited to war.

One peculiarity of country which greatly favors the action of a people's war is the scattered distribution of homesteads, such as we find in many parts of Germany. The country is thus more intersected and covered; the roads become worse, although more numerous; the quartering of troops is attended with endless difficulties, but especially that peculiarity repeats itself on a small scale, which a people's war possesses on a great one, namely, that the spirit of resistance exists everywhere, but is nowhere tangible. If the inhabitants live together in villages, the most troublesome have troops quartered upon them, or as a punishment they are plundered, their houses burnt, and so forth, a system which could not be carried out very easily with the peasant community of Westphalia.

National levies and masses of armed peasants cannot and should not be employed against the main body of the enemy's army, or even against any considerable forces; they must not attempt to crunch the core; they must only nibble at the surface and the edges. They should rise in the provinces situated at the sides of the theater of war, and in which the assailant does not appear in force, in order to draw these provinces entirely from his influence. Where there is as yet no enemy, there is no lack of courage to oppose him, and the mass of the neighboring population is gradually kindled at this example. Thus the fire spreads as it does in heather, and reaches at last that stretch of ground on which the aggressor is based; it seizes his lines of communication and preys upon the vital thread by which his existence is supported. For even if we entertain no exaggerated ideas of the omnipotence of a people's war, even if we do not consider it an inexhaustible, unconquerable element, over which the mere force of an army has as little control as the human will has over the wind or the rain; in short, although our opinion is not founded on rhetorical pamphlets, still we must admit that we cannot drive armed peasants before us like a body of soldiers who keep together like a herd of cattle, and usually follow their noses. Armed peasants, on the contrary, when scattered, disperse in all directions, for which no elaborate plan is required. Thereby the march of every small body of troops in a mountainous, thickly wooded or otherwise very difficult country becomes very dangerous, for at any moment the march may become an engagement. If in point of fact no armed bodies have even been heard of for some time, yet the same peasants already driven off by the head of a column may at any hour make their appearance in its rear. If it is a question of destroying roads and blocking narrow defiles, the means which outposts or raiding parties of an army can apply to that purpose bear about the same relation to those furnished by a body of insurgent peasants as the movements of an automaton do to those of a human being. The enemy has no other means to oppose to the action of national levies except that of detaching numerous parties to furnish escorts for convoys, to occupy military stations, defiles, bridges, etc. In proportion as the first efforts of the national levies are small, so the detachments sent out by the enemy will be weak in numbers, because he is afraid to divide his forces much. It is on these weak bodies that the fire of the national war now kindles itself even more; the enemy is overpowered by numbers at some points, courage rises, the love of fighting gains strength and the intensity of this struggle increases until the culminating point approaches, which is to decide the issue.

According to our idea of a people's war, it should, like a kind of nebulous vapory essence, nowhere condense into a solid body; otherwise the enemy sends an adequate force against this core, crushes it and takes a great many prisoners; then courage sinks; everyone thinks the main question is decided, any further effort useless and the weapons fall from the people's hands. However, on the other hand, it is necessary that this mist should gather at some points into denser masses and form threatening clouds from which now and again a formidable flash of lightning may burst forth. These points are chiefly on the flanks of the enemy's theater of war, as we have said before. There, the national rising must be organized into larger and more ordered units, supported by a small force of regular troops, so as to give it the appearance of a regular force and to make it fit to venture upon enterprises on a larger scale. Starting from these points the organization of the people in arms must become more irregular in proportion as they are to be employed more in the direction of the rear of the enemy, where he is exposed to their hardest blows. Those better-organized masses are for the purpose of falling upon the larger garrisons which the enemy leaves behind him. Besides, they serve to create a feeling of uneasiness and dread, and increase the moral impression of the whole; without them the total effect would be lacking in force, and the whole condition of the enemy would not be made sufficiently uncomfortable.

The easiest way for a general to produce this more effective form of a national rising is to support the movement by small detachments sent from the army. Without such a support of a few regular troops as an encouragement, the inhabitants generally lack the impulse and the confidence to take up arms. The stronger the bodies are, detached for this purpose, the greater will be their power of attraction, the greater will be the avalanche which is to come hurtling down. But this has its limits, partly because it would be detrimental to split up the whole army for this secondary objective, to dissolve it, as it were, into a body of irregulars, and form with it an extended and weak defensive line, by which proceeding we may be sure both regular army and national levies alike would become completely ruined; partly because experience seems to tell us that when there are too many regular troops in a district, the people's war loses in vigor and effectiveness. The causes of this are, in the first place, that too many of the enemy's troops are thus drawn into the district; in the second place, that the inhabitants then rely on their own regular troops; and, thirdly, that the presence of large bodies of troops makes too great demands on the powers of the people in other ways, that is, in providing quarters, transport, contributions, etc.

Another means of preventing any too serious reaction on the part of the enemy against the people's war constitutes, at the same time, a leading principle in the method of using such levies; this is the rule, that with this great strategic means of defense, a tactical defense should seldom or never take place. The character of *engagements fought by national levies* is the same as that of all engagements of troops of an inferior quality—great impetuosity and fiery ardor at the start, but little coolness or tenacity if the combat is prolonged. Further, it is a matter of little consequence whether a force of the national levy is defeated or dispersed, since that is what it is made for, but a body of this description must not be broken up by too great losses in killed, wounded and prisoners; a defeat of that kind would soon cool their ardor. But both of these peculiarities are entirely opposed to the nature of a tactical defensive. In the defensive engagement a persistent, slow, systematic action is required, and great risks must be run. A mere attempt, from which we can desist as soon as we please, can never lead to results in the defensive. If, therefore, the national levy is to take over the defense of any particular natural obstacle, things must never come to a first-rate, decisive defensive engagement; for however favorable the circumstances may be, the national levy will then be defeated. It may, and should, therefore, defend the approaches to mountains, the dykes of a swamp, the passages over a river, as long as possible; but when it is once broken, it should disperse, and continue its defense by unexpected attacks, rather than concentrate and allow itself to be shut up in some narrow last refuge in a regular defensive position. However brave a nation may be, however warlike its habits, however intense its hatred of the enemy, however favorable the nature of the country, it is an undeniable fact that a people's war cannot be kept alive in an atmosphere too thick with danger. If, therefore, its combustible material is anywhere to be fanned into a considerable flame, it must be at remote points where it has air, and where it cannot be extinguished by one great blow.

These reflections are more a feeling-out of the truth than an objective analysis, because the subject has as yet really been too little in evidence and has been too little treated by those who have long observed it with their own eyes. We have only to add that the strategic plan of defense can include the co-operation of a general levy in two different ways, either as a last resource after a lost battle or as a natural assistance before a decisive battle has been fought. The latter case supposes a retreat into the interior of the country and that indirect kind of action of which we have treated in Chapters 8 and 24 of this Book. Therefore, we need only say a few words here on the summoning of the national levy after a battle has been lost.

No state should believe its fate, that is, its entire existence, to be dependent upon one battle, no matter how decisive it may be. If it is beaten, the calling forth of fresh forces and the natural weakening which every offensive undergoes in the long run may bring about a turn of fortune, or assistance may come from abroad. There is always still time to die; and just as it is a natural impulse for a drowning man to seize at a straw, so it is in the natural order of the moral world that a people tries the last means of deliverance when it sees itself thrown on the brink of an abyss.

However small and weak a state may be in comparison with its enemy, if it foregoes a last supreme effort, we must say that there is no longer any soul left in it. This does not exclude the possibility of its saving itself from complete destruction by a peace fraught with sacrifice; but neither does such an intention do away with the usefulness of fresh measures for defense; they will make peace neither more difficult nor worse, but easier and better. They are still more necessary if there is an expectation of assistance from those who are interested in maintaining our political existence. Any government, therefore, which, after the loss of a great battle thinks only of quickly permitting its people to enjoy the comforts of peace, and overpowered by the feeling of disappointed hope, no longer feels in itself the courage and the desire to give the spur to its every force, is guilty in any case through weakness of a gross inconsistency, and shows that it did not deserve the victory, and, perhaps, just on that account was utterly incapable of gaining it.

Therefore, however decisive the defeat experienced by a state may be, still by the retreat of the army into the interior, its fortresses and its national levies must be brought into operation. In connection with this it is advantageous if the flanks of the principal theater of war are limited by mountains or tracts of country which are otherwise very difficult. These now stand forth as bastions, whose strategic flanking fire the assailant has to endure.

If the victorious enemy is engaged in siege works, if he has left strong garrisons behind him everywhere to secure his communications, or even detached troops to gain elbow room and keep the adjacent provinces in order, if he is already weakened by his various losses in man power and material of war, then the moment has arrived when the defensive army must again enter the lists, and by a well-directed blow make the assailant stagger in his disadvantageous position.

CHAPTER 27

Defense of a Theater of War

Having treated the *most important defensive means*, we might perhaps be contented to leave the manner in which these means are related to the plan of defense as a whole to be discussed in the last Book, which will be devoted to the "Plan of a War"; for every secondary plan, either of attack or defense, originates from this and is determined by it in its leading features; and moreover in many cases the plan of the war itself is nothing more than the plan of the attack or defense of the principal theater of war. But in no case have we been able to begin with war as a whole, although in war, more than anywhere else, the parts are determined by the whole, imbued with and essentially altered by its character; instead, we have been obliged to make ourselves thoroughly acquainted, in the first instance, with each single subject as a separate part. Without this progress from the simple to the complex, a host of vague ideas would have overpowered us, and the reciprocal activities, of which there are so many in war, in particular would have constantly confused our conceptions. We shall, therefore, take one step more in our approach to the whole; that is, we shall consider the defense of a theater of war in itself, and look for the thread by which the subjects treated connect themselves with it.

The defensive, according to our conception, is nothing *but the stronger form of combat*. The preservation of our own forces and the destruction of those of the enemy—in a word, the victory—is the object of this combat, but at the same time not its ultimate object.

That object is the preservation of our own political state and the overthrow of that of the enemy; or again, in one word, *the intended peace*, because only in that does this conflict adjust itself and end in a comprehensive result.

But just what is the enemy's political state in connection with war? Above all things, its military force is important, then its territory; but of course there are still many other things, which, through particular circumstances, may obtain a predominant importance. To these belong, primarily, foreign and domestic political relations, which sometimes decide more than all the rest. But although the military force and the territory of the enemy are not the state itself, nor do they exhaust all the connections which the state may have with the war, still these two things are always those which predominate, for the most part infinitely surpassing all other connections in importance. The military force is to protect the territory of the state, or to conquer that of an enemy; the territory on the other hand, constantly supports and renews the military force. The two, therefore, depend on each other, mutually support each other, are equally important to each other. But, nevertheless, there is a difference in their mutual relations. If the military force is destroyed, that is, completely overthrown, rendered incapable of further resistance, the loss of the territory follows of itself; but on the other hand, the destruction of the military force by no means follows from the conquest of the country, because that force may of its own accord evacuate the territory, in order to conquer it the more easily afterwards. Indeed, not only does the *complete* defeat of its army decide the fate of a country, but even every *considerable weakening* of its military force leads regularly to a loss of territory. On the other hand, every considerable loss of territory does not cause a proportionate diminution of military force; in the long run it will do so of course, but not always within the interval of time in which the decision of the war takes place.

From this it follows that the preservation of our own military power, and the weakening or destruction of that of the enemy, takes precedence in importance over the possession of territory, and, therefore, is the *first object* which a general should strive for. The possession of territory forces itself into prominence *as an object* only if that means (weakening or destruction of the enemy's military force) has not yet effected it.

If the whole of the enemy's military force were united in *one* army, and the whole war consisted of *one* engagement, the possession of the country would depend on the issue of that engagement; destruction of the enemy's military forces, conquest of his country and safety of our own would follow from that result, and, in a certain sense, be identical with it. Now the question is: what can induce the defender to deviate from this simplest form of the act of warfare and to distribute his power in space?

The answer is: the insufficiency of the victory which he might gain with all his forces united. Every victory has its sphere of influence. If this extends over the entire state of the enemy, consequently over the whole of his military force and his territory, that is, if all the parts are swept away at the same time in the movement which we have given to the core of his power, such a victory is all we require, and a division of our forces would not be justified by sufficient grounds. But if there are portions of the enemy's military force, and of country belonging to either party, over which our victory would have no effect, we must give particular attention to those parts; and as we cannot unite territory like a military force in one point, therefore we must divide our forces for the purpose of attacking or defending those portions.

It is only in small, well-developed states that it is possible and probable to have such a unity of military force, that all depends upon a victory over *that force*. In the case of larger tracts of territory which border on our own for a great distance, or in the case of an alliance of such states as surround us on several sides, such a unity is practically impossible. Here, therefore, divisions of force must necessarily take place, giving occasion to different theaters of war.

The sphere of influence of a victory will naturally depend on its magnitude, and this in turn on *the number of troops conquered*. Therefore *the blow* which, if successful, will produce the greatest effect must be made against *that part* of the country where the greatest number of the enemy's forces are concentrated. The greater the number of our own forces which we use for this blow, the more certain we are of this success. This natural sequence of ideas leads us to an illustration from which we shall see this truth more clearly; it is the nature and effect of the center of gravity in mechanics.

As the center of gravity is always situated where the greatest mass of matter is concentrated, and as a blow given to the center of gravity of a body is the most effective; and, further, as the strongest blow is that struck with the center of gravity of the power used, so it is also in war. The armed forces of every belligerent, whether it is a single state or an alliance of states, have a certain unity and, by means of this, cohesion; but wherever there is cohesion, analogies drawn from the center of gravity are applicable. There are, therefore, in these armed forces certain centers of gravity, the movement and direction of which decide that of the other points, and these centers of gravity are situated where the greatest bodies of troops are assembled. But just as, in the world of inanimate matter, the action against the center of gravity has its measure and limits in the cohesion of the parts, so it is in war, and here as well as there the force exerted may easily be greater than the resistance requires, and then there is a blow in the air, a waste of force.

What a difference there is between the cohesion of an army under *one* flag, led into battle under the personal command of *one* general, and that of an *allied army* extended over 250 or 500 miles, or even with its bases scattered in quite different directions! There we see cohesion in the strongest degree, unity most complete; here unity is very far away, often only existing in the common political intention, and even there in a poor and insufficient degree, the cohesion of parts mostly very weak, often quite an illusion.

Therefore, if on the one hand, the violence with which we wish to strike the blow demands the greatest concentration of force, then on the other hand, we have to fear every excess as a real disadvantage, because it entails a waste of power, and that in turn a *deficiency of power* at other points.

To recognize these *centra gravitatis* in the enemy's military force, to discern their spheres of action is, therefore, one of the principal functions of strategic judgment. We must constantly ask ourselves what effect the advance or retreat of one part of the forces on either side will produce on the other parts.

We do not by this lay claim in any way to the discovery of a new method; we have only sought to base the method of all generals and all ages on ideas which should show more clearly how natural it is.

How this conception of the center of gravity of the enemy's force affects the whole plan of war, we shall consider in the last Book, for that is the proper place for the subject, and we have only borrowed it from there to avoid leaving a gap in the sequence of ideas. In this examination we have seen the general conditions which determine a distribution of forces. They consist fundamentally of two interests which are in opposition to each other; the one, *the possession of territory,* strives to divide the forces; the other, *the blow against the center of gravity of the enemy's military power,* up to a certain point once more unites them.

Thus it is that theaters of war or separate army regions originate. They are in fact such boundaries of the tract of country and of the forces distributed on it that every decision given by the main force of such a region affects the whole *directly,* and carries it along in its own direction. We say *directly,* because a decision on one theater of war must naturally also have a more or less indirect influence over those adjoining it.

Although it is really quite natural, we must again remind our readers expressly that, here as well as everywhere else, our definitions are only concerned with the central points of certain groups of ideas, the limits of which we neither desire, nor are able, to define by sharp lines.

We think, therefore, a theater of war, whether large or small, with its military force, of whatever size this may be, represents a unity which may be reduced to *one* center of gravity. At this center of gravity the decision must be made, and to be victor here means to defend the theater of war in the widest sense.

CHAPTER 28

DEFENSE OF A THEATER OF WAR

(CONTINUED)

Defense, however, consists of two different elements, namely, the *decision* and the *awaiting*. The combination of these two elements is to be the subject of this chapter.

First, we must observe that the state of awaiting is not, in point of fact, the complete defense; it is, however, that province of defense in which it proceeds to its aim. As long as a military force has not abandoned the portion of territory entrusted to it, the tension of forces created by the attack on both sides continues, and this lasts until there is a decision. This decision can only be regarded as having actually taken place when either the assailant or defender has left the theater of war.

As long as an armed force maintains itself within its section of the country, its defense of that section continues, and in this sense the defense of the theater of war is identical with the defense *in* the theater. Whether the enemy has for the time being obtained possession of much or little of that section of country is not essential, for it is only lent to him until the decision.

But this kind of conception by which we wish to settle the proper relation of the state of awaiting to the whole is only correct when a decision is really to take place and is regarded by both parties as inevitable. For it is only by that decision that the centers of gravity of the respective forces and the theater of war determined through them become *effective principles*. Whenever the idea of a decision disappears, the centers of gravity are neutralized; indeed, in a certain sense, the whole of the armed forces become so too, and now the possession of territory, which forms the second principal component of the whole theater of war, comes forward as the direct object. In other words, the less a decisive blow is sought for by both sides in a war, and the more the war is merely a mutual observation, so much the more important becomes the possession of territory, so much the more the defensive seeks to cover everything directly, and the assailant to extend his forces in his advance.

Now we cannot conceal from ourselves the fact that the majority of wars and campaigns approach much more to a state of observation than to a struggle for life or death, that is, a contest in which one at least of the combatants seeks in every way to bring about a decision. Only the wars of the nineteenth century have possessed this character to such a degree that a theory founded on this point of view can be made use of in relation to them. But as it is unlikely that all future wars will have this character, as we must rather assume that most of them will once more tend to that of a state of observation, any theory to be of use in practical life must take this into account. Hence we shall first concern ourselves with the case in which the desire for a decision permeates and guides the whole, therefore with *real*, or, if we may use the expression, *absolute war*; then in another chapter we shall examine those modifications which arise through the approach, in a greater or less degree, to the state of observation.

In the first case, therefore, whether the decision is sought by the assailant or by the defender, the defense of a theater of war will consist in the defender establishing himself in it in such a manner that he can offer the decision advantageously at any moment. This decision may be either a battle, or a series of great engagements, but it may also consist in the result of mere relations which arise from the position of the opposing forces, that is, in the result of *possible engagements*.

Even if the battle were not the most powerful, the most usual and most effective means to a decision, as we think we have shown on several previous occasions that it is, still the mere fact of its being one of the means of reaching this solution would be sufficient to demand *the greatest concentration of our forces* which circumstances will at all permit. A great battle upon the theater of war is the blow of the one center of gravity against the other; the more forces can be assembled in the one or the other, the more certain and greater will be the effect. Therefore every separation of forces which is not called for by an object (which either cannot itself be attained by the successful issue of a battle or which itself is necessary to the successful issue of the battle) is *objectionable*.

But the greatest concentration of the forces is not only a fundamental condition; it is also necessary that they should have such a position and be so situated that the battle may be fought under favorable circumstances.

The different types of defense which we have become acquainted with in the chapter "Methods of Resistance" (Book VI, Chapter 8), are wholly in keeping with these fundamental conditions; there will therefore be no difficulty in connecting them with it, according to the requirements of each individual case. But there is one point which seems at first sight to involve a contradiction, and which, as one of the most important in the defense, is the more in need of explanation. It is the hitting upon the center of gravity of the enemy's force.

If the defender ascertains in time the roads by which the enemy will advance, and upon which in particular the flower of his forces is without fail to be found, he may march against him on that road. This will be the most usual case, for although the defense precedes the attack in measures of a general nature, in the establishment of strong places, great arsenals and in the peace arrangements of the army, and thus gives a line of direction to the assailant in his preparations, still, when the campaign really opens, the defender, in relation to the aggressor, has the advantage peculiar to the whole defense of playing the last hand.

To advance with a considerable force into a foreign country demands extensive preparations, in the way of collecting provisions and supplies of articles of equipment of all kinds. The time required for this is enough to give the defender time to prepare accordingly, and we must not forget that the defensive generally requires less time, because in every state things are prepared rather for the defensive than the offensive.

But although this may be perfectly true for the majority of cases, there is always a possibility that, in particular cases, the defensive may remain in uncertainty as to the principal line by which the enemy intends to advance; and this case is more likely to occur when the defense is based on measures which themselves take a good deal of time, as for example, the preparation of a strong position and so forth. Further, although the defender actually places himself on the aggressor's line of advance, then in every instance in which the aggressor is not offered an offensive battle he can avoid the position which the defender has taken up, merely by slightly altering his original line of attack, for in the cultivated parts of Europe we can never be so situated that there are not roads to the right or left by which any position may be avoided. Plainly, in such a case the defender could not wait for his enemy in a position, or at least could not wait there with the intention of giving battle.

But before entering on the means remaining to the defensive in this case, we must inquire more particularly into the nature of such a case, and the probability of its occurrence.

Naturally there are in every state, and also in every theater of war (of which alone we are at present speaking), objects and points upon which an attack is likely to be specially effective. About this we think it will be better to speak when we come to the attack. Here we shall confine ourselves to observing that, if the most advantageous object and point of attack becomes the motive for the direction of the assailant's blow, this motive also reacts on the defensive, and must be his guide in cases in which he knows nothing of the intentions of his adversary. If the assailant were not to take this direction favorable to him, he would forgo part of his natural advantages. It is evident that, if the defender has taken up a position in that direction, the evading of his position, or passing it by, is not to be done for nothing; it costs a sacrifice. From this it follows that there is not on the side of the defender such a risk of *missing the direction of his enemy;* neither, on the other hand, is it so easy for the assailant *to pass his adversary by* as appears at first sight, because there exists beforehand a very distinct, and in most cases preponderating, motive in favor of one or the other direction, and consequently the defender, although his preparations are bound to one place, will not fail in most cases to come into contact with the main strength of the enemy's forces. In other words, *if the defender has taken up the right position, he may be almost sure that the assailant will march to meet him.*

But by this we shall not and cannot deny the possibility that the defender with all these arrangements might on some occasion not meet with the assailant, and therefore the question arises, what he should do in that case, and how much of the advantages proper to his position still remain available to him.

If we ask ourselves what means still remain generally to the defender when the assailant passes by his position, they are the following:

1. To divide his forces from the beginning, so as to be certain to find the assailant with one portion, and then to hurry to the support of that portion with the other.
2. To take up a position with his force united, and in case the assailant passes him by, to push on rapidly in front of him by a lateral movement. In most cases there will not be time to make such a movement directly to a flank; it will, therefore, be necessary to take up the new position somewhat farther back.
3. With his whole force to attack the enemy in flank.
4. To operate against his communications.

5. By a counterattack on *his* theater of war, to do exactly what the enemy has done to us in passing us by.

We mention this last measure here because it is possible to imagine a case in which it would be effective; but as it is in contradiction to the object of the defense, that is, the grounds on which that form has been chosen, it can only be regarded as an abnormal procedure, which can only be adopted because the enemy has made some great mistake, or because there are other peculiarities in a particular case.

Operating against the enemy's communications implies that our own are superior, which is indeed one of the fundamental requirements of a good defensive position. But although on that ground this action may promise the defender a certain amount of advantage, still, in the defense of a theater of a war, it is seldom an operation suited to lead to a *decision*, which we have supposed to be the object of the campaign.

The dimensions of a single theater of war are seldom so large that the assailant's lines of communications are exposed to much danger by their length, and even if they are, still the time which the assailant requires for the execution of his blow is usually too short for its progress to be arrested by the slow effects of the action against his communications.

Therefore this means (that is, the action against the communications) will prove quite ineffective in most cases against an enemy determined upon a decision, and also in case the defender seeks this decision.

The object of the three other means which remain for the defender is a direct decision—a meeting of centers of gravity; they correspond better, therefore, to the task. But we shall at once say that we decidedly prefer the third to the other two, and without quite rejecting the latter, we hold the former to be in the majority of cases the true means of defense.

In a position where our forces are divided, there is always a danger of getting involved in a war of posts, from which, if our adversary is resolute, under the best of circumstances, there can only follow a *considerable relative defense,* never a decision such as we desire. Even if by a correct instinct we have been able to avoid this error, nevertheless our attack will have been decidedly weakened by the temporarily divided opposition, and we can never be certain that the troops first advanced will not suffer disproportionate losses. To this has to be added the fact that the resistance of this corps, which usually ends in its falling back on the main body coming to its assistance, appears to the troops in the light of lost engagements, or unsuccessful plans, and in this manner noticeably weakens the moral forces.

The second means, that of placing our army concentrated in one position in front of the enemy, in whichever direction he may turn his march, involves the risk of our arriving too late and being caught between two measures. Besides, a defensive battle requires calm and consideration, a knowledge, indeed an intimate knowledge, of the country, and all that is not to be expected in a hasty oblique movement to a flank. Lastly, positions suitable for a good defensive battlefield are too rare to reckon upon them at every point of every road.

On the other hand, the third means, namely, to attack the enemy in flank, therefore to give battle with a changed front, is attended with great advantages.

First, there is always in this case, as we know, an exposure of the lines of communication, here the lines of retreat, and in this respect the defender will have the advantage first of all through his general circumstances as defender, and next and chiefly through the strategic properties we have claimed for his position.

Second—and this is the main thing—every assailant who attempts to pass his opponent by is caught between two tendencies which are entirely opposed to one another. His first desire is to advance to obtain the object of his attack; but the possibility of being attacked in flank at any moment creates a desire to deliver a blow in that direction at any moment, and furthermore a blow with his united forces. These two tendencies contradict each other, and engender such a confusion in his own conditions, such a difficulty in taking measures to suit every possibility, that strategically there can hardly be a more abominable position. If the assailant knew with certainty the moment when he would be attacked, he might prepare with skill and ability; but in his uncertainty on this point, and pressed by the necessity of advancing, it is almost certain that if the battle ensues, it finds him with preparations hurriedly made and utterly inadequate, in a situation therefore certainly not advantageous.

If then there are favorable moments for the defender to deliver an offensive battle, they are to be found most surely at such a moment as this. If we consider, further, that the knowledge of the country and choice of ground are on the side of the defender, that, moreover, he can prepare his movements, and can time them, no one can doubt that he possesses even in such a situation still a decided superiority, strategically, over his adversary.

We think, therefore, that a defender occupying a well-chosen position, with his forces united, may quietly wait for the enemy to pass him by. Should the enemy not attack him in his position, and if an operation against the enemy's communications does not suit the circumstances, there still remains for the defender an excellent means of bringing about a decision by resorting to a flank attack.

If cases of this kind are almost entirely lacking in military history, the reason is, partly, that the defender has seldom had the courage to remain in such a position, but has either divided his forces or rashly thrown himself in front of his enemy by a march in a lateral or an oblique direction, or that no assailant dares pass by the defender under such circumstances, and that for this reason his movement usually comes to a stand-still.

The defender is in this case compelled to resort to an offensive battle. The further advantages of *awaiting the enemy, of a strong position, of good entrenchments,* etc., he must give up; in most cases the situation in which he finds the advancing enemy will not quite make up for these advantages, for it is just to evade their influence that the assailant has put himself in this situation; but it always offers him a *certain compensation,* and theory is therefore not in the position of seeing a quantity disappear suddenly from the calculation and the *pro and contra* mutually cancel each other, as so often happens when critical historians introduce a little bit of theory.

It must not, in fact, be supposed that we are now dealing with logical subtleties; on the contrary, the more this subject is practically considered, the more it appears as an idea embracing the entire system of defense, everywhere dominating and regulating it.

It is only by the determination on the part of the defender to assail his opponent with all his force, the moment he passes by him, that he can safely avoid the two pitfalls, so close to which he is led by the defensive form, namely, a division of his forces and a hasty march to a flank. In both he accepts the law of the assailant; in both he resorts to measures of a very critical nature, and with a most dangerous degree of haste; and wherever a resolute adversary, thirsting for victory and a decision, has encountered such a system of defense, he has shattered it. But when the defender has assembled his forces at the right point to fight a united action, if he is determined with this force, if it comes to the worst, to attack his enemy in flank, he is and remains *in the right,* and he is supported by all the advantages which the defense can offer in his situation. His action will then bear the stamp of *good preparation, coolness, certainty, unity and simplicity.*

At this juncture, we cannot avoid mentioning a great historical event, which has a close analogy with the ideas here developed; we do so to anticipate its being used in a wrong application. When the Prussian army, in October, 1806, was waiting in Thuringia for the French under Bonaparte, the former was posted between the two main roads on which the latter could advance, that is, the road to Berlin by Erfurt, and that by Hof and

Leipzig. The original intention of breaking into Franconia straight through the Thuringian Forest, and afterward, when that plan was abandoned, the uncertainty as to which of the roads the French would choose for their advance, caused the occupation of this intermediate position. As *such* it was bound to lead to the measure of a hasty lateral movement.

This was, in fact, the intention in case the enemy had marched by Erfurt, for the roads in that direction were perfectly accessible; on the other hand, the idea of a lateral movement in the direction of the road by Hof could not be entertained, partly because the army was two or three days' march away from that road, partly because the deep valley of the Saale lay in between; nor had this ever been the intention of the Duke of Brunswick, so that there was no kind of preparation made for carrying it into effect. But it was always contemplated by Prince Hohenlohe, that is, by Colonel Massenbach, who exerted all his influence to draw the duke into his plan. Still less could the idea be entertained of leaving the position which had been taken on the left bank of the Saale to proceed to an offensive battle against Bonaparte on his advance, that is, to such an attack in flank as we have been considering. For if the Saale was an obstacle to intercepting the enemy at the last moment it would have been a still greater obstacle to assuming the offensive at a moment when the enemy would be in possession of the opposite side of the river, at least partially. The duke, therefore, decided to wait behind the Saale to see what would happen, that is to say, if we can call anything a decision which emanated from this many-headed headquarters, and in this time of confusion and utter indecision.

Whatever may have been the truth about this waiting, the resulting situation of the army was this:

1. That the enemy might be attacked if he crossed the Saale to march against the Prussian army.
2. That if he did not march against that army, operations might be begun against his communications, or
3. If it should be found practicable and advisable, he might still be intercepted near Leipzig by a rapid flank march.

In the first case, the Prussian army possessed a great strategic and tactical superiority because of the deep valley of the Saale. In the second, the purely strategic advantage was just as great, for the enemy had only a very narrow base between our position and the neutral territory of Bohemia, while ours was extremely broad; even in the third case, our army, covered by the Saale, was still in no disadvantageous situation. All these three measures, in spite of the confusion and lack of any clear perception at headquarters, *were really discussed;* but of course we cannot wonder that, although a right idea may have been entertained, it was necessarily bound to fail *in the execution* by the complete indecision and the confusion everywhere prevailing.

In the two first cases, the position on the left bank of the Saale was regarded as a real flank position, and as such it had undoubtedly very great qualities; but, of course, against a very superior enemy, *against a Bonaparte,* a flank position with an army that is not very sure of its success, *is a very bold measure.*

After long indecision, the duke on the 13th adopted the last of the three plans specified, but it was too late. Bonaparte had already begun to cross the Saale, and the battles of Jena and Auerstädt were inevitable. The duke, through his indecision, had fallen between two stools; he quitted his first position too late for *a lateral movement,* and too soon for *a battle suited to the object.* Nevertheless, the natural strength of this position proved itself so far that the duke could have destroyed the right wing of the enemy's army at Auerstädt, while Prince Hohenlohe, by a bloody rearguard action, would have still been able to back out of the snare; but at Auerstädt they did not venture to realize the victory, which was *quite certain;* and at Jena they thought they might reckon upon one which was *quite impossible.*

In any case, Bonaparte felt the strategic importance of the position on the Saale so much, that he did not venture to pass it by, but decided to cross the Saale in sight of the enemy.

By what we have now said we think we have sufficiently specified the relations between defense and attack in the case of decisive action, and shown the threads to which, according to their situation and connection, the different subjects of plans of defense attach themselves. To go through the different arrangements in greater detail cannot be our intention, for it would lead us into a boundless field of particular cases. If a general has laid down a definite point for his direction, he will see how far it agrees with the geographical, statistical and political circumstances and the material and personal conditions of his own army and that of the enemy, and how far these things must modify his plan in one way or another in execution.

But in order to go more definitely here into the gradations of defense specified in the chapter on "Methods of Resistance," and examine them again more closely, we shall here state what strikes us as being, generally speaking, important with regard to them.

1. Reasons for marching against the enemy with a view to an offensive battle may be as follows:

a) If we know that the assailant is advancing with his forces very much divided and we, therefore, even though very weak, have still the prospect of a victory.

But such an advance on the part of the assailant is in itself very improbable, and consequently unless we know of it on certain information, this plan is inadmissible; for to reckon upon it and to rest all our hopes on it, on a mere *supposition* and without sufficient motive, usually leads to a disadvantageous situation. Things do not in such a case turn out as we expected them to do; we have to give up the offensive battle, and, not being prepared beforehand for a defensive one, are obliged to begin with a forced retreat and to leave almost everything to chance.

This is more or less what happened in the defense conducted by the army under Dona against the Russians in the campaign of 1759, which under General Wedel ended in the disastrous battle of Züllichau.

This measure settles matters so quickly that planmakers are only too ready to propose it, without troubling to ask whether the presuppositions on which it rests are well founded.

b) If we are generally strong enough for a battle, and

c) if an awkward, irresolute adversary specially invites attack.

In this case the effect of surprise may be worth more than any assistance furnished by the ground through a good position. It is the truest essence of good generalship thus to bring into play the power of the moral forces, but theory can never insist strongly enough and often enough that there must be an *objective foundation* for these presuppositions. Without such a *foundation in fact*, to be always talking of surprises and the superiority of an unusual kind of attack, and to found thereon plans, opinions and criticisms, is a wholly inadmissible and unwarranted method of proceeding.

d) If the nature of our army makes it specially suitable for the offensive.

It was certainly no baseless or false idea if Frederick the Great believed that in his mobile, courageous army, full of confidence in him, obedient by habit, trained to precision, animated and elevated by pride, and well practiced in his oblique attack, he possessed an instrument which, in his firm and daring hand, was more suited to attack than to defense. All these qualities were lacking in his opponents, and it was just in this respect that he had the most decisive superiority. To make use of this was worth more to him in most cases than the assistance of entrenchments and natural obstacles. But such a superiority will always be rare. A well-trained army, thoroughly practiced in great movements is only a part of it. If Frederick the

Great maintained that the Prussian army was particularly adapted for attack—and if this has been incessantly repeated since his time—still we must not attach too much weight to such an assertion. In most cases in war we feel more exhilarated, more courageous, when acting offensively than defensively, but this is a feeling that all troops have in common, and there is hardly an army of which its generals and leaders have not expressed the same opinion. We should not, therefore, carelessly rely upon the semblance of a superiority and neglect real advantages for that.

A very natural and weighty reason for resorting to an offensive battle may be the composition of the army as regards the three arms, that is to say, a numerous cavalry and little artillery.

We continue our enumeration of reasons.

e) When we can nowhere find a good position.

f) When we must hasten with the decision.

g) Lastly, the combined influence of several or all of these reasons.

2. The waiting for the enemy in a locality in which we ourselves intend then to attack him (Minden, 1759) most naturally proceeds from

a) there being no such disproportion of force to our disadvantage as to make it necessary to seek a strong and strengthened position;

b) the finding of a terrain particularly suitable for the purpose. The properties which determine this belong to tactics; we shall only observe here that they must consist chiefly in an easy approach on the side of the defender and in all kinds of obstacles on the side of the enemy.

3. A position will be taken up with the express intention of actually awaiting in it the attack of the enemy:

a) If the disproportion of strength compels us to seek cover from natural obstacles or behind entrenchments.

b) If the country affords a specially good position of the kind.

The two modes of resistance (2) and (3) will come more into consideration according as we do not seek the decision itself but content ourselves with the negative result and can expect our opponent to hesitate, be irresolute and finally get stuck with his plans.

4. An entrenched, unassailable position only fulfils its purpose:

a) If it is situated at a specially important strategic point.

The characteristic of such a position consists in the fact that we cannot possibly be overpowered in it. The enemy is, therefore, obliged to try every other means, that is, to pursue his object regardless of the position, or blockade it and reduce it by starvation. If it is impossible for him to do this, the strategic qualities of this position must be very great.

b) If we have reason to expect aid from abroad.

Such was the case with the Saxon army in its position at Pirna. Whatever may have been said against this measure after its ill success, still it remains certain that 17,000 Saxons could never have neutralized 40,000 Prussians in any other way. If the Austrian army made no better use of the superiority it gained at Lowositz, that only proves how bad was their whole method of making war and their whole military organization. There can be no doubt that if the Saxons, instead of going into the camp at Pirna, had retired into Bohemia, Frederick the Great would have driven both Austrians and Saxons beyond Prague and taken that place in the same campaign. Whoever does not admit the value of this advantage but continues to think only of the capture of the whole Saxon army is really incapable of making any calculation of such matters, and without calculation no certain result is possible.

But as the cases *a* and *b* very rarely occur, the entrenched camp is a measure which requires to be very well considered and which is very seldom successfully employed. The hope of *impressing* the enemy by such a camp and crippling thereby his whole activity is attended with too great danger, the danger, that is, of being obliged to fight without the possibility of retreat. When Frederick the Great gained his object in this way at Bunzelwitz, we must admire the correct judgment he formed of his adversary, but we must certainly also lay more stress than is in other cases permissible on the means which he would have found at the last moment to clear a way for himself with the ruins of his army, and also on the independence enjoyed by a king.

5. If there happens to be a fortress or several fortresses near the frontier, the great question arises whether the defender is to seek his decision before or behind them. The latter recommends itself

a) by the superiority of the enemy in numbers, which forces us to break his strength before we come to grips with it;

b) by those fortresses being near, so that the sacrifice of country is not greater than we are compelled to make;

c) by the fitness of the fortresses for defense.

One of the principal purposes of fortresses undoubtedly is, or ought to be, to break the enemy's strength in its advance and seriously to weaken that part of his forces from which we are demanding the decision. If we so seldom see this use made of fortresses, the reason is that a case in which the decision is sought by one of the two parties is so rare. But it is only with this case that we are here dealing. We, therefore, look upon it as a principle equally simple and important in all cases in which the defender has one or more fortresses near him, that he should keep them before him and give the decisive battle behind them. We admit that a battle lost behind our fortresses will force us to retreat somewhat farther into the interior of the country than one lost in front of them, tactical results being in both cases the same, although the causes of this difference are based rather on imagination than on material things. Neither do we forget that a battle may be fought in front of the fortresses in a well-chosen position, whilst behind them the battle in many cases must be an offensive one, particularly if the enemy is laying siege to a fortress which is in danger of being lost. But what are these fine shades compared with the advantage of finding the enemy weaker in the decisive battle by a quarter or a third of his strength, or, if several fortresses are concerned, even perhaps by a half?

We think, therefore, in all cases of an *inevitable decision*, whether sought by the offensive or the defensive, and when the latter is not tolerably sure of a victory, or if the nature of the country does not offer some most decisive reason for giving battle in a position farther forward—in all these cases, we say, when a fortress is near at hand and capable of resistance, the defender should by all means withdraw from the very beginning behind it and let the decision take place on this side of it and, consequently, with its co-operation. If he takes up his position so near the fortress that the assailant can neither form the siege nor blockade the place without first driving him off, he furthermore forces the assailant to attack him in his position. To us, therefore, of all defensive measures in a critical situation none appears so simple and effective as the choice of a good position near to and behind a strong fortress.

It would, of course, be quite another question if the fortress lay far back, for then it would be necessary to abandon a considerable part of our theater of war, a sacrifice which, as we know, is only made if urgent circumstances demand it. In such a case this measure is more like a retirement into the interior of the country.

Another condition is the fitness of the place for resistance. It is well known that there are fortified places, especially larger ones, which should not be brought into any contact with the enemy's army because they are not equal to resisting a vigorous attack by a considerable body of troops. In that case, our position would have to be at all events so close behind as to be able to support the garrison.

6. Lastly, the retirement into the interior of the country is only a proper expedient in the following circumstances:

a) when owing to the physical and moral relation in which we stand to the enemy, the idea of a successful resistance on the frontier cannot be entertained;

b) when it is a principal object to gain time;

c) when the condition of the country is favorable to such a measure, a subject with which we have already dealt in Chapter 25.

We thus close the chapter on the defense of a theater of war if a decision is sought by one side or the other, and is therefore inevitable. But we must, of course, remember that events in war do not present themselves in such a pure abstract form, and that, therefore, if our propositions and arguments are applied in reasoning to actual war, we must keep Chapter 30 in mind and think of the general being faced in most cases by two courses and urged *more* to the one or the other according to circumstances.

CHAPTER 29

DEFENSE OF A THEATER OF WAR

(CONTINUED)

SUCCESSIVE RESISTANCE

We have proved in Chapters 12 and 13 of Book III that in strategy a successive resistance is unnatural, and that all forces available should be used simultaneously.

As regards all movable forces, this requires no further demonstration; but if we consider the seat of war itself, with its fortresses, its natural obstacles of ground, and even the mere extent of its surface, as also one of our forces, this is immovable and we can, therefore, only bring it gradually into play, or we must at once retire so far that all parts which are to come into operation are in our front. Everything which the country occupied by the enemy can contribute to weaken him, then comes into operation. The assailant must at least blockade the defender's fortresses; he must make his occupation of the country secure with garrisons and other posts; he must make long marches; he must fetch everything he needs from long distances, and so forth. All these effects come into operation against the assailant whether he advances *before the decision or after it*, but in the former case they will be somewhat stronger than in the latter. From all this it follows that if the defender wishes to postpone the decision, he certainly has thereby the means to bring into play all these immovable forces together.

On the other hand, it is clear that this *postponing of the decision* has no influence upon the sphere of influence of a victory gained by the assailant. In treating of the attack we shall examine in greater detail the sphere of influence; here we shall only observe that it reaches to the exhaustion of the superiority (the product, that is to say, of the moral and physical factors). But this superiority exhausts itself, in the first place, by the employment of forces to secure the theater of war, and, in the second place, by losses in engagements. The weakness produced by these two things cannot be essentially altered whether the engagements take place at the beginning or at the end, near the frontier or in the interior. We believe, for example, that a victory gained by Bonaparte over the Russians at Vilna, in 1812, would have carried him just as far as that of Borodino—assuming it had been equally great—and that a victory at Moscow would not have carried him any farther either; Moscow was in any case the limit of this sphere of victory. Indeed, it cannot be doubted for a moment that a decisive battle on the frontier (for other reasons) would have resulted in a far greater victory, and then perhaps in a wider sphere of influence for that victory as well. From this point of view, therefore, the postponement of the decision makes no difference for the defender.

In the chapter on "Methods of Resistance" we have already become acquainted with that postponement of the decision which may be regarded as the extreme form, under the name of *retreat into the interior* of the country and as a special kind of resistance in which the object is rather that the assailant should wear himself out than that he should be destroyed by the sword on the field of battle. But it is only when such an intention predominates that the postponement of the decision can be looked upon as a special *kind of resistance,* for otherwise it is clear that an infinite number of gradations may be conceived in this method and that these may be combined with every means of defense. We, therefore, look upon the greater or less co-operation of the theater of war, not as a special kind of resistance, but only as an introduction at our pleasure of the immovable means of resistance, according as the circumstances and the situation require it.

But now, if a defender does not think he requires any assistance from these immovable forces for his decision, or if the sacrifice in other ways which the use of them involves is too great for him, they are kept in reserve for the future and form, so to speak, gradual reinforcements. These perhaps ensure the possibility of keeping the movable forces strong enough to be able to follow up the first favorable decision with a second, and perhaps with yet a third. In other words, a *successive* use of forces will in this way become possible.

If the defender has lost a battle on the frontier, which is not exactly a complete defeat, we may very well imagine that, by placing himself behind his nearest fortress, he may be at once in a position to fight a second battle. Indeed, if he is only dealing with an adversary who has not much resolution, perhaps some considerable obstacle of ground is in itself enough to stop him.

There is, therefore, in strategy's use of the theater of war, as in everything else, *a proper management of resources.* The less we can make suffice, the better, but we must be sure it does suffice, and here, as in commerce, there is naturally something else to be thought of besides mere niggardliness.

In order, however, to prevent a great misconception, we must draw attention to the fact that the thing we are here considering is not at all what after a lost battle we can still do or undertake in the matter of defense, but merely how much success we can expect *beforehand* from this second kind of resistance and how highly, therefore, we can estimate it in our plan. Here there is hardly more than one point that the defender has to look to, namely, his opponent and his opponent's character and situation. An opponent of weak character, lacking in self-confidence, destitute of noble ambition or in a very cramped situation will content himself, in case of success, with a moderate advantage, and at every fresh decision which the defender ventures to offer him will timidly hold back. In this case the defender may count upon gradually making successful use of the means of resistance afforded by his theater of war, in constantly fresh, though in themselves minor, acts of decision, in which the prospect is continually revived for him of a final decision in his favor.

But who does not feel that we are here on the way to campaigns without decision, which are much more the field of a successive application of force. Of these we shall speak in the next chapter.

CHAPTER 30

DEFENSE OF A THEATER OF WAR
WHEN NO DECISION IS SOUGHT

Whether and how far there can be a war in which neither party is the assailant, and in which neither, therefore, has a *positive aim*, we shall consider more closely in the last Book. Here it is not necessary for us to occupy ourselves with this contradiction, because on a single theater of war we can easily suppose reasons for such a defensive on both sides in accordance with the relations which these parts bear to the whole.

But not merely have single campaigns taken place without the focus of a necessary decision, as history shows us, there were very many in which there was certainly no lack of an assailant, and therefore none of a *positive will* on one side. That will, however, was so weak that instead of striving to attain its object at any cost and forcing the *necessary* decision, it contented itself with such advantages as arose, so to speak, spontaneously out of the circumstances. Or the assailant pursued *no* self-selected aim *at all*, but made it depend on circumstances, in the meanwhile gathering such fruits as from time to time presented themselves.

Such an offensive, which departs from the strict logical necessity of an advance against its object and saunters through the campaign almost like an idler, looking out right and left for a cheap bargain, differs very little from the defense itself, which also allows the general to pick up such bargains. Nevertheless, we shall postpone the more detailed, philosophical consideration of this way of conducting a war to Book VII, "The Attack." Here we shall confine ourselves to the conclusion that in such a campaign neither by the assailant nor the defender can everything be related to the decision, and that, therefore, this decision ceases to be the keystone to which all the lines of the strategic arch converge. Campaigns of this kind (as the military history of all times and lands teaches us) are not only numerous, but form so overwhelming a majority that the others only appear as exceptions. Even if this proportion should change in the future, yet it is certain that there will always be many such campaigns, and that in dealing with the defense of a theater of war, we must therefore take them into consideration. We shall try to specify the qualities which characterize them. Real war will in most cases fall between the two tendencies, lying nearer now to the one, now to the other, and consequently we can only see the practical effect of these qualities in the modification they produce in the *absolute form* of war by their reciprocal action. We have already, in Chapter 3 of this Book, said that the *state of awaiting* is one of the greatest advantages which the defensive has over the offensive; as a general rule it is seldom in life, and least of all in war, that *everything* happens which, according to the circumstances, should have happened. The imperfection of human insight, the fear of evil results, the accidents that befall our plan of action when put into execution bring it about that of all the plans of action dictated by the circumstances there are always very many that never get executed. In war, where the imperfection of our knowledge, the danger of catastrophe, the number of accidents are incomparably greater than in any other human activity, the number of fallings short, if we wish to call them so, must, on that account, be also necessarily much greater. This is the rich field in which the defense gathers fruits which have grown for it spontaneously. If we add to this experience the independent importance of the possession of the surface of the ground in war, then that principle derived from experience, which has become a proverb, *beati sunt possidentes* (fortunate are those in possession), holds good no less in war than in peace. It is *this principle* that here takes the place of the decision, which in every war directed to *mutual destruction* is the focus of the whole affair. It is extraordinarily fruitful, not, certainly, in actions called forth by it, but in motives for inaction and for the sort of action that takes place in the interest of inaction. When no decision can be sought or expected, there is no reason for giving up anything, for that could only be done to gain advantages thereby for the decision. The consequence is that the defender's object is to keep everything, or as much as he can (as much, that is, as he can cover), and the assailant's to take possession of all he can (that is, to extend his occupation as far as possible), without a decision. We have only, in this place, to deal with the defender.

Wherever the defender is not present with his military forces, the assailant can take possession, and then the advantage of the state of awaiting is *on his side;* hence the endeavor to cover the country everywhere directly, and to take the chance of the assailant attacking the troops posted for this purpose.

Before we now deal in greater detail with the characteristics of the defense we must borrow from Book VII, "The Attack," those objects which the assailant usually aims at when a decision (by battle) is not seriously sought. They are as follows:

1. The seizure of a considerable tract of territory, as far as that can be done without a decisive engagement.
2. The capture of an important magazine under the same condition.
3. The capture of a fortress not covered. No doubt a siege is more or less a great operation, often demanding great efforts; but it is an undertaking which involves nothing in the nature of a catastrophe. If it comes to the worst, the siege can be raised without any considerable positive loss being thereby sustained.
4. Lastly, a successful engagement of some importance, but in which there is not much risked, and consequently not much to be gained, an engagement which takes place not as the climax, rich in results, of an entire strategic scheme, but on its own account for the sake of trophies or the honor of the army. For such an object, of course, the engagement is not fought *at all costs;* we either wait for the chance of a favorable opportunity or seek to bring one about by skill.

These four objects of attack give rise to the following efforts on the part of the defense:

1. To cover the fortresses by keeping them behind us.
2. To cover the country by extending the troops over it.
3. Where the extension is not sufficient, to throw the army rapidly in front of the enemy by a flank march.
4. To guard against engaging at a disadvantage.

It is clear that the object of the first three measures is to force on the enemy the initiative, and to derive the utmost advantage from the state of awaiting, and this object is so natural that it would be great folly to despise it *prima facie.* It must necessarily be the more likely to be adopted the less a decision is to be expected, and it is the ruling principle in all such campaigns, though superficially, in small, indecisive actions, a tolerably lively play of activity may take place.

Hannibal as well as Fabius, and Frederick the Great as well as Daun, have done homage to this principle whenever they did not either seek or expect a decision. The fourth effort serves as a corrective to the three others: it is their *conditio sine qua non.*

We shall now discuss these four objects more in detail.

At first sight it appears somewhat preposterous to protect a fortress from the enemy's attack by placing an army *in front* of it. Such a measure looks like a sort of pleonasm, as fortifications are built for the express purpose of themselves resisting a hostile attack. Yet it is a measure which we see resorted to thousands and thousands of times. But thus it is with the conduct of war: the most ordinary things often seem to be the most incomprehensible. Who would presume on the ground of this apparent inconsistency to pronounce these thousands and thousands of instances to be just so many blunders? The constant recurrence of this measure shows that there must be some deep-lying reason for it. This reason, however, is none other than that pointed out above, emanating from moral sluggishness and inactivity.

If the defender places himself before his fortress, the enemy cannot attack it unless he first beats the army in front of it. But a battle is a decision, and if that is *not* the enemy's object, there will be no battle, and the defender will remain in possession of his fortress without striking a blow. Consequently, whenever we do not believe the enemy intends to seek a decision, we must venture on the chance of his not making up his mind to do so, especially as in most cases we still have the power to retire behind the fortress the moment that the enemy, contrary to our expectation, advances to the attack. The position before the fortress thus becomes free from risk, and the probability of maintaining the *status quo* without any sacrifice is then not even attended with a risk *later on.*

If the defender places himself behind the fortress, he offers the assailant an object exactly suited to his situation. If the fortress is not of great strength and if the assailant is not quite unprepared, he will begin the siege. In order that this may not end in the fall of the place, the defender must march to its relief. The positive action, the initiative, is thus now laid on him, and his opponent who with his siege is to be regarded as advancing toward his object is in possession. Experience teaches that this is always the turn that the matter takes, and that is also quite natural. A siege, as we have said before, is not necessarily linked with a catastrophe. Even a general devoid of enterprise and energy, who would never have made up his mind to a battle, will proceed to undertake a siege with perhaps nothing but field artillery when he can approach a fortress without risk. At the worst, he can abandon his undertaking without any positive loss. There is still to be considered the risk to which most fortresses are more or less exposed, of being taken by storm or in some other irregular manner, and this circumstance should certainly not be overlooked by the defender in his calculation of probabilities.

In weighing the different chances, it seems natural that the defender considers the advantage of fighting *under favorable* circumstances inferior to that of very probably not being obliged to fight at all. And so it seems to us that the practice of placing our army in the field before our fortress is very natural and easily understood. Frederick the Great almost always followed it, as, for instance, at Glogau against the Russians, and at Schweidnitz, Neisse and Dresden against the Austrians. This measure, however, brought misfortune on the Duke of Bevern at Breslau; *behind* Breslau he could not have been attacked; the superiority of the Austrians in the king's absence was bound to cease soon on his approach, and by taking up a position *behind* Breslau the Duke could have avoided a battle until Frederick's arrival. No doubt he would have preferred this course, but it would have exposed that important city and its great stores to a bombardment, and for this the king, whose judgment in such cases was anything but mild, would have taken him very severely to task. *An attempt made* by the duke to protect Breslau, by taking up an entrenched position before it, cannot after all be regarded with disapproval, for it was very possible that Prince Charles of Lorraine, contented with the capture of Schweidnitz and threatened by the approach of the king, would have let himself be prevented thereby from advancing farther. The best thing would have been not to let matters come to a battle, but to retire through Breslau the moment the Austrians advanced to the attack. In this way he would have gotten all the advantages of a state of awaiting without paying for them with a great risk.

If we have here explained and justified on higher and conclusive grounds the defender's taking up a position *before* his fortresses, we have still to observe that there exists in addition a subordinate ground, which, however, though more obvious, is not in itself alone sufficient, because it is not conclusive. We refer to the use which the army commonly makes of the nearest fortress as a depot of provisions and munitions. This is so convenient and offers so many advantages that a general will not easily make up his mind to draw his supplies from more distant fortresses or to store them in open cities. But if a fortress is the magazine of an army the position before it is in many cases a matter of absolute necessity and in most cases is very natural. It is, however, easy to see that this obvious reason, which can easily be overvalued by people who are not as a rule much interested in reasons less obvious, neither suffices to explain all the cases that occur, nor is it important enough in its relations for the supreme decision to be entrusted to it.

The capture of one or more fortresses without risking a battle is such a very natural object of all attacks which do not aim at a great decision that the defender makes the frustration of them his chief task. Thus it is that on theaters of war which contain many fortresses, we see that almost all movements are pivoted on these, the assailant seeking to approach one of them unexpectedly and employing various feints for the purpose, while the defender tries by well-prepared movements quickly to get before him again. This is the general character of almost all campaigns in the Netherlands from Louis XIV to Marshal Saxe.

So much for the covering of fortresses.

The covering of a country by an extended disposition of the military forces is only conceivable in combination with considerable obstacles of ground. The great and small posts which must be formed for the purpose can only by means of strong positions acquire a certain capacity for resistance; and as the natural conditions are seldom found sufficient, the art of entrenchment comes to our aid. We must, however, observe that the power of resistance which is thus obtained at any point, is always only *relative* (see the chapter "On the Significance of the Engagement," Book IV, Chapter 5) and can never be regarded as *absolute*. It may, of course, happen that such a post resists all attacks, and that thus in a single instance an absolute result occurs; but from the great number of posts, any single one, in comparison with the whole, still appears but weak and exposed to the possible attack of a greatly superior force, and consequently it would be unreasonable to make our whole safety depend on the resistance of any single post. With such an extended disposition, therefore, we can only count upon a comparatively long resistance, and not upon a real victory. This value of single posts is, however, sufficient for the object and intention of the whole. In campaigns in which no great decision, no impetuous advance toward the subjugation of the whole force, is to be feared, engagements in which a post is involved, even though they end in the loss of the post, are less dangerous. There is seldom any further result in connection with them except just the loss of this post and a few trophies. The victory has no further influence upon the state of affairs; it breaks down no foundation that will bring down with it a mass of ruins. In the worst case, that is, if the whole defensive system has been disorganized by the loss of single posts, the defender will always have time to concentrate his corps and with his united strength *offer* the decision which the assailant, according to our supposition, does not desire. It usually happens, therefore, that with this concentration of strength the act is closed and the further advance of the assailant stopped. A bit of ground, a few men and cannon are all that the defender loses and all that the assailant gains, but enough to satisfy him.

To such a risk, we say, in the event of things going wrong the defender may very well expose himself, if he has, on the other hand, the possibility or rather the probability, in his favor, that the assailant from timidity or prudence will halt before his posts without attacking them. Only with regard to this we must not lose sight of the fact that we are now supposing an assailant who has no wish to venture a great stroke. A moderate or strong post will very well serve to stop such an adversary, for although he can undoubtedly make himself master of it, still he asks himself what price he will have to pay for it, and whether this price is not too high for what use, in his position, he can make of the victory.

In this way we may see how the powerful relative resistance which the defender can obtain from an extended disposition, consisting of a number of posts in juxtaposition with one another, may constitute a satisfactory result in the balance sheet of his whole campaign. In order to direct at once to the right point the glance which the reader, with his mind's eye, will here cast upon military history, we must observe that these extended positions appear more frequently in the latter half of a campaign, because by that time the defender has become thoroughly acquainted with his adversary, and with his intentions and conditions; and the small quantity of the spirit of enterprise with which the assailant started is usually exhausted.

In this defensive in an extended position, by which *the country, the supplies* and *the fortresses* are covered, all great natural obstacles, such as rivers, streams, mountains, woods and morasses, must naturally play a great part, and acquire a predominant influence. Concerning their use, we refer the reader to what we have already said about it.

It is through this predominating importance of the topographical element that the knowledge and activity which are looked upon as the specialty of the General Staff are particularly required. Now as the General Staff is usually that part of the army which writes and publishes most, it follows that these parts of campaigns are recorded more fully in history; and, furthermore, from this there arises a not unnatural tendency to systematize them, and out of the historical solution of one case to construct general solutions for all succeeding cases. But this is a futile, and, therefore, a mistaken endeavor. Even in this more passive kind of war, which is more closely connected with locality, each case is different and must be differently dealt with. The ablest memoirs of a critical character respecting these subjects are therefore only suited to make us familiar with them, but never to serve as instructions for action.

However necessary and estimable the activity of the General Staff may be, which, following the common view, we have distinguished as most peculiarly its own, we must still raise a warning voice against the usurpations which often spring from it to the prejudice of the whole. The authority acquired by those leading members of it who are strongest in this branch of the service often gives them a sort of general dominion over people's minds, beginning with the general himself, and from this arises a habit of thinking which leads to onesidedness. At last the general sees nothing but mountains and passes, and that which should be a measure of free choice guided by circumstances becomes a mannerism, a second nature.

Thus in 1793 and 1794, Colonel Grawert of the Prussian army, who was the animating spirit of the General Staff of the day and well known as great on mountains and passes, led two generals of the most opposite personal characteristics, the Duke of Brunswick and General Möllendorf, into exactly the same ways of carrying on war.

That a defensive line parallel to the course of a formidable natural obstacle may lead to a cordon war is evident. It would, in most cases, necessarily lead to that if really the whole extent of the theater of war could be directly covered in that manner. But most theaters of war have such an extent that the normal tactical extension of the troops destined for their defense would be far too thin. Since, however, the assailant, by his own dispositions and other circumstances, is confined to certain principal directions and roads, and any great deviation from these, even if he is only opposed to a very inactive defender, would be attended with great embarrassment and disadvantage, generally all the defender has to do is to cover the country for a certain number of miles or marches right and left of these principal lines of direction. To effect this covering, we may again be contented with defensive posts on the principal roads and means of approach, and merely watch the country in between by small posts of observation. The consequence of this is, of course, that the assailant may pass a column between two of these posts and thus make his intended attack upon one of these two posts from several quarters at once. Now these posts are in some measure arranged to meet this, partly by their having supports for their flanks, partly by the formation of flank defenses (called crochets), partly by their being able to receive assistance from a reserve posted in the rear, or by some troops detached from adjoining posts. In this manner the number of posts is reduced still more and the result is that an army engaged in a defense of this kind usually is divided into four or five principal posts.

For important points of approach, beyond a certain distance, and yet in some measure threatened, special centers are established which, in a certain sense, form small theaters of war within the principal one. In this manner the Austrians, during the Seven Years' War, generally placed the main body of their army in four or five posts in the mountains of Lower Silesia; while a small, more or less independent, detachment organized for itself a similar system of defense in Upper Silesia.

Now the further such a defensive system diverges from direct covering, the more it must seek the assistance of mobility (active defense) and even of offensive means. Certain bodies are regarded as reserves; besides which, one post hastens to send to the help of another all the troops it can spare. This assistance may be rendered either by actually hastening up from the rear to reinforce and re-establish the passive defense, or by attacking the enemy in flank, or even by threatening his line of retreat. If the assailant threatens the flank of a post not by an attack, but only by a position through which he tries to act upon the communications of this post, either the troops which have been advanced for this purpose must be actually attacked, or the way of reprisal must be resorted to by an attempt to act upon the communications of the enemy.

We see then that however passive in character the defense may be fundamentally, it must still include many active means and can in many ways be organized to meet the complex conditions. Usually those defenses are reckoned the best which make the most use of active and even offensive means, but this depends very much on the nature of the ground, the composition of the forces, and even on the talent of the general. We are also liable, in general, to expect altogether too much from mobility and other auxiliary means of an active nature and to place altogether too little confidence in the local defensive power of a formidable natural obstacle. We think we have thus sufficiently explained what we understand by an extended line of defense, and we now turn to the third auxiliary means, the placing ourselves in front of the enemy by a rapid march to a flank.

This means necessarily belongs to the set-up of that kind of defense of a country which we are now considering. In the first place, even with the most extended positions, the defender often cannot occupy all the approaches to his country which are threatened. Secondly, he must in many cases be ready to betake himself with the main strength of his forces to those posts against which that of the enemy's forces is meant to be thrown, as otherwise those posts would be too easily overpowered. Lastly, a general who does not like to have his forces condemned to passive resistance in an extended position must, as a rule, seek all the more to attain his object, the protection of the country, by rapid movements, well planned and well directed. The greater the spaces he leaves exposed, the greater must be his mastery in movement in order everywhere to interpose at the right moment.

The natural consequence of striving to do this is that everywhere positions are sought out, which in such a case are occupied and which offer enough advantages to banish from our adversary's mind the idea of attacking any of them from the moment that our own army or even only a part of it has reached it. As these positions continually recur and everything depends on their being reached at the right moment, they are, so to speak, the keynotes of the whole of this way of carrying on war, which on that account has sometimes been called a *war of posts*.

Just as an extended position and the relative resistance it offers in a war *without a great decision* do not present the dangers originally inherent in them, so to throw ourselves in front of the enemy by a flank march is not so hazardous as it would be in a moment of great decisions. With a determined adversary, able and willing to deal heavy blows and not afraid therefore of a considerable expenditure of forces, to propose at the last moment in great haste to thrust an army in front of him would be going half way to meet a most decisive defeat, for against a ruthless blow delivered with full force such a running and stumbling into a position would be futile. But against an adversary who instead of going to work with his whole fist does so only with the tips of his fingers, who is incapable of taking advantage of a great result, or rather of the preparation for it, who only seeks a trifling advantage at small expense—against such an adversary, this kind of resistance can certainly be employed with success.

A natural consequence of this is that this means is generally more in evidence in the latter half of a campaign than at its opening.

Here also the General Staff has an opportunity of displaying its topographical knowledge in framing a system of connected measures with reference to the choice and preparation of positions and the roads leading to them.

When, finally, the whole object of the one side is to gain a certain point, and that of the other is to prevent it, both sides are often obliged to make their movements each under the eyes of the other; for this reason these movements must be made with a degree of precaution and precision not otherwise required. Formerly, when the main army was not divided into independent divisions and even on the march was regarded as an indivisible whole, this precaution and precision was attended with many formalities and a great expenditure of tactical skill. On such occasions, certainly, single brigades had to hurry forward from the line of battle, to secure certain points and play an independent part until the army could arrive; but these were, and continued to be, *anomalous proceedings;* and in general, the aim in the order of march was always to move the army about as a whole with its formation undisturbed and to avoid such exceptional proceedings as much as possible. Nowadays, when the parts of the main army are again divided into independent bodies and these bodies may venture to enter into an engagement even with the whole of the enemy's army, provided the others are near enough to carry it on and finish it—nowadays such a flank march even under the eyes of the enemy offers less difficulty. What formerly could only be effected by the actual mechanism of the order of march can now be effected by starting single divisions at an earlier hour, by hastening the march of others and by greater freedom in the employment of the whole.

By the means of defense just considered, the assailant is to be prevented from taking a fortress, from occupying any important tract of country or from capturing a magazine. He will be so prevented if everywhere engagements are offered him in which he finds either too small probability of success, too great danger of a reaction in case of failure or, in general, an expenditure of force too great for his object and his situation.

If now the defender succeeds in this triumph of his skill and contrivance, and the assailant, wherever he turns his eyes, sees himself deprived by judicious precautions of all prospect of realizing his modest wishes, the offensive principle often seeks an escape in the mere satisfaction of military honor. The winning of any sort of important engagement gives to the arms of the victor the semblance of a superiority, satisfies the vanity of the general, the court, the army and the people, and therewith to some extent the expectations which naturally are associated with every attack.

A successful engagement of some importance merely for the sake of the victory and the trophies is thus the assailant's last hope. Let no one suppose that we are here involving ourselves in a contradiction, for by our *own hypothesis,* which still covers us, the sound measures of the defender have deprived the assailant of all prospect of achieving one of those other objects by means of *a successful engagement.* Such a prospect would depend on two conditions; namely, *advantageous conditions in the engagement* and, next, that *the success should also actually lead to the attainment of one of those objects.*

The first condition can very well exist without the second, and therefore separate detachments and posts of the defender's will much more frequently find themselves in danger of being involved in disadvantageous engagements if the assailant is merely aiming at *the honor of the battlefield* than if he couples with it the condition that it should bring further advantages as well.

If we place ourselves entirely in Daun's position and mental attitude, we can understand that he could venture on the surprise of Hochkirch without inconsistency, provided he was aiming at nothing but winning the trophies of the day. But a victory rich in results, which would have compelled the king to abandon to him Dresden and Neisse, was quite another task, on which he had no wish to venture.

Let no one think that these are trifling or quite idle distinctions; on the contrary, we have here to do with one of the deepest-rooted, fundamental characteristics of war. The significance of an engagement is the soul of it for strategy, and we cannot too often repeat that in strategy all leading events invariably proceed from the final intention of the two parties, as from the final conclusion of the whole train of thought. This is why there can be such a difference strategically between one battle and another that they can no longer be regarded as at all the same means.

Now a fruitless victory of the assailant's can scarcely be regarded as doing any essential damage to the defense. But still, the defender will not willingly concede even *this* advantage to his adversary, especially as one never knows what may still happen to be involved in it. So it is necessary that he should keep incessant watch on the situation of his important corps and posts. Here, no doubt, very much depends on the judicious arrangements made by the corps commanders, though these may also be involved in inevitable catastrophes by injudicious instructions from the general. Who is not reminded here of Fouqué's corps at Landshut and of Fink's at Maxen?

In both cases Frederick the Great had reckoned too much on the effect of traditional ideas. He could not possibly think that in the position of Landshut 10,000 men could successfully engage 30,000, or that Fink could resist overwhelmingly superior numbers pouring upon him from all sides. But he thought the strength of the position of Landshut would be accepted, like a bill of exchange, as heretofore, and that Daun would see in the demonstration against his flank sufficient reason to exchange his uncomfortable position in Saxony for the more comfortable one in Bohemia. There he misjudged Laudon and here for once Daun, and in that lies the error of these measures.

But apart from such errors, into which even generals may fall who are not so proud, daring and obstinate as Frederick the Great may perhaps in some of his proceedings be reproached with being, there always, with regard to the subject we are now considering, lies a great difficulty in the fact that the general cannot always expect all he desires from the sagacity, good will, courage and firmness of character of his corps commanders. He cannot, therefore, leave all to their judgment; he must on many points give them orders by which their course of action, being restricted, may easily fail to suit the circumstances of the moment. This is an unavoidable drawback. Without an imperious, commanding will, the influence of which penetrates the whole army, war cannot be well conducted, and whoever insisted on following the practice of always expecting the best from his subordinates would for that very reason be quite unfit to be a good commander of an army.

The conditions of every single corps and post must, therefore, be always kept clearly in view to prevent any of them being involved in a catastrophe.

The aim of all these four efforts is to preserve the *status quo*. The more fortunate and successful they are, the longer will the war linger at the same point; but the longer the war continues at one point, the more serious becomes the problem of subsistence.

The place of requisitions and contributions from the country is taken from the very beginning, or at all events very soon, by a system of subsistence from magazines; that of local vehicles, collected for each occasion, more or less by the formation of a permanent transport service, composed of local vehicles or of such as belong to the army itself. In short, there arises an approach to that regular system of feeding troops from magazines, which we have already treated in Book V, Chapter 14, "Subsistence."

At the same time, it is not this which exercises a great influence on this mode of conducting war, for as this mode, by its object and character, is tied down to a limited space, the question of subsistence may very well have a part in determining its course—and will indeed do so for the greater part—but without changing the character of the whole war. On the other hand, the action of each side on the other's line of communications gains a much greater importance for two reasons. First, through the lack in such campaigns of more important and conclusive measures, the generals must turn their energies to feebler ones of this sort; and second, because there is in this case no lack of the time required to wait for the effect of this means. The security of his own line of communication will therefore appear to the defender specially important, for though, it is true, its interruption cannot be an object of the enemy's attack, yet it might force the defender to retreat and to give up other objects.

All measures for the protection of the area of the theater of war itself must naturally also have the effect of covering the lines of communication. Their security therefore is in part provided for in that way, and we have only to observe that it is a principal condition in choosing a position.

A *special* means of security consists in small or fairly considerable bodies of troops escorting the separate convoys. For one thing, the most extended positions are not always sufficient to secure the lines of communication; for another, such an escort becomes particularly necessary when the general has wanted to avoid a very extended disposition. Consequently we find in Tempelhoff's *History of the Seven Years' War* instances without end in which Frederick the Great caused his bread and flour wagons to be escorted by single regiments of infantry or cavalry, sometimes even by whole brigades. On the side of the Austrians, we nowhere find mention of the same thing, which certainly may be partly due to the fact that they had no such circumstantial historian on their side, but partly also because they always took up much more extended positions.

Having mentioned the four efforts which form the basis of a defensive *that does not aim at a decision*, and which are fundamentally free from all offensive elements, we must now say something of the offensive means with which they may become more or less mixed up, or, so to speak, flavored. These offensive means are chiefly:

1. Operating against the enemy's communications, under which we likewise include enterprises against his places of supply.
2. Diversions and raids within the enemy's territory.
3. Attacks on the enemy's detachments and posts and even upon his main body, under favorable circumstances, or even merely the threat of such attacks.

The first of these means is incessantly in action in all campaigns of this kind, but, so to speak, very quietly without actually making an appearance. Every suitable position for the defender derives a great part of its efficacy from the disquietude which it causes the assailant with regard to his communications; and as the question of subsistence in such a war becomes, as we have observed, one of vital importance, affecting the assailant just as much as the defender, a great part of the strategic web is determined by this apprehension of possible offensive action proceeding from the enemy's positions, as we shall again note in dealing with the attack.

Not only this general influence, proceeding from the choice of positions, which, like pressure in mechanics, acts *invisibly,* but also an actual offensive advance with part of the forces against the enemy's line of communication, comes within the compass of such a defensive. But if it is to be done advantageously, *the situation of the lines of communication, the nature of the country or the peculiar quality of the troops* must always be specially suitable for the purpose.

Incursions into the enemy's country which have as their object reprisals or the levying of contributions for the sake of profit cannot properly be regarded as defensive means. On the contrary, they are true offensive means, but are usually combined with the object of what is really a diversion. This has for its object the weakening of the enemy force opposing us and may be regarded as a true defensive means. But as it may be used just as well in the attack and is in itself actual attack, we think it more suitable to leave its further examination for the next Book. Accordingly, we shall only list it here in order to render a full account of the arsenal of small offensive means belonging to the defender of a theater of war. For the present we merely add that in extent and importance it may reach such a point as to give the whole war the *appearance* and, along with that, the honor, of the offensive. Of this nature are Frederick the Great's enterprises in Poland, Bohemia and Franconia before the opening of the campaign of 1759. His campaign itself is plainly a pure defense, but these incursions into the enemy's territory have given it the character of an offensive, which perhaps has a special value on account of the moral weight it carries.

An attack on one of the enemy's detachments or on his main army must always be kept in view as a necessary complement of the whole defense whenever the assailant takes the matter too easily, and on that account lays himself very open to attack at particular points. Under this tacit condition the whole action takes place. But here also the defender, in the same way as in operating against the communications of the enemy, may go a step further into the province of the offensive, and, like his adversary, make lying in wait for *a favorable stroke* an object of his efforts. In order to expect some result in this field, he must either be decidedly superior in strength to his adversary—which certainly in general is contrary to the nature of the defensive but still can happen—or he must possess the method and the talent of keeping his forces more concentrated, and make up by activity and mobility for what, on the other hand, he must sacrifice in such a case.

The former was Daun's case in the Seven Years' War, the latter the case of Frederick the Great. We hardly ever see Daun's offensive make its appearance except when Frederick the Great invited it by excessive boldness and a display of contempt for him (Hochkirch, Maxen, Landshut). On the other hand, we see Frederick the Great almost constantly on the move in order to beat one or other of Daun's corps with his main body. He seldom succeeded; at least the results were never great, because Daun, in addition to his great superiority in numbers, had also a rare degree of prudence and caution; but we must not suppose that, on this account, the king's attempts were altogether fruitless. In these attempts lay rather a very effectual resistance, for the carefulness and exertion imposed upon his adversary in order to avoid fighting at a disadvantage neutralized those forces which otherwise would have aided in advancing the attack. Let us only call to mind the campaign of 1760 in Silesia, where Daun and the Russians, out of sheer apprehension of being attacked and beaten by the king, now here and now there, could not succeed in advancing one step.

We believe we have now gone through all the subjects which form the predominant ideas, the principal aims and therefore the mainstay of the whole action in the defense of a theater of war if no decision is intended. Our chief object in putting them side by side has been merely to provide a general view of how the whole strategic action hangs together; the individual measures by which they are realized in practice—marches, positions, and so forth—we have hitherto considered in detail.

If we now cast another glance at the whole of the subject as we here conceive it, the idea must strike us that with such a weak offensive principle, with so little desire for a decision on either side, with so little positive motive, with so many subjective counteracting influences which stop us and hold us back, the essential difference between attack and defense must more and more tend to disappear. At the opening of a campaign, no doubt, one party will enter the other's theater of war and thereby his action will assume the form of attack. But it may very well be and frequently happens that he must soon enough employ all his forces to defend his own country on the enemy's ground. Thus both really stand opposite each other in a state of mutual observation, both of them intent on losing nothing, both also perhaps equally intent on making some positive gain. Indeed it may happen, as with Frederick the Great, that the real defender aims even higher in that way than his adversary.

Now the more the assailant gives up the attitude of advance, and the less the defender is threatened by him and confined by the urgent need of security to a strict defensive, so much the more there arises an equality of circumstances, in which the energies of each party are directed to winning an advantage from its adversary and protecting itself against any disadvantage, therefore to a true strategic *maneuvering*. And indeed this is more or less the character of all campaigns when conditions or political intentions do not allow of any great decision.

In the following Book we have allotted a chapter specially to the subject of strategic maneuvers; but as this equipoised play of forces has frequently acquired in theory a false importance, we find ourselves obliged here, while treating of the defense, to examine it more in detail, for it is more particularly in the defensive that this false importance has been given to it.

We call it an *equipoised play of forces* because when there is no movement of the whole there is a state of equipoise. Where no great object impels, there is no movement of the whole; therefore, in such a case, however unequal the two parties may be, they are still to be regarded as in equipoise. From this state of equipoise now come forth the particular motives to actions on a minor scale and for more trifling objects. They can here develop because they are no longer kept down by the pressure of a great decision and a great danger. What can be lost or won at all consists of smaller counters, and the whole action is broken up into smaller operations. With these smaller operations at this slighter cost there now arises between the two generals a contest of skill. But as in war, chance and, consequently, luck, can never be quite denied entrance, this contest will never cease to be a *gamble.* Meanwhile two other questions here arise, namely, whether in this maneuvering, chance will not have a smaller share in the decision than when everything is concentrated into a single great act, and whether calculating intelligence will not have a greater. The latter of these two questions we must answer in the affirmative. The more intricate the whole becomes and the oftener time and space—the former at single moments, the latter at single points—come into consideration, so much the greater obviously becomes the field of calculation and consequently the dominance of reasoning intelligence. What the reasoning intelligence gains is partly taken from chance, but not necessarily altogether so, and therefore we are not obliged to give an affirmative answer to the first question as well. For we must not forget that a reasoning intelligence is not the only mental quality of a general. Courage, force, resolution, presence of mind and so forth are qualities which will again count for more when everything depends on a single great decision. They will, therefore, count for somewhat less in an equipoised play of forces, and the predominating importance of clever calculation increases not merely at the expense of chance but also at the expense of these qualities. On the other hand, these same brilliant qualities in the moment of a great decision may rob chance of a great part of its dominance, and therefore, so to speak, bind fast what calculating cleverness in this case had to let go free. We see by this that here a conflict takes place between several forces, and that we cannot positively assert that there is a greater field left open to chance in the case of a great decision than there is in the total result when that equipoised play of forces takes place. If, therefore, we see in this play of forces more particularly a contest of mutual skill, that must be taken to refer only to clever calculation and not to the sum total of military genius.

Now it is just from this aspect of strategic maneuvering that the whole has been given that false importance of which we have spoken above. In the first place, this skill has been assumed to constitute the whole of a general's mental endowment; but this is a great mistake, for, as we have said, it is not to be denied that in moments of great decisions his other qualities, mental and moral, can make him master of the force of events. If this mastery proceeds more from the impulse of great emotions and of those flashes of genius which come into being almost unconsciously and not therefore as the conclusion of a long train of thought, it has none the less on that account full right of citizenship in the art of war, for that art is neither a mere act of intelligence nor are the activities of the intelligence the highest that have part in it. Further, it has been believed that every ineffectual activity in a campaign must be due to that sort of skill on the part of one or even of both of the two generals, while nevertheless its usual and chief reason was always to be found in the general circumstances which made of war such a gamble.

As most wars between civilized states have had for their object rather the observation of the enemy than his destruction, the majority of campaigns must naturally have assumed the character of strategic maneuvering. Those of them which had no famous general to show have attracted no attention; but where there was a great general on whom all eyes were fixed, or even two, opposed to each other, like Turenne and Montecuculi, there the seal of perfection has been stamped upon this whole art of maneuvering through the names of those generals. A further consequence was that this game was looked upon as the acme of art, the manifestation of its high perfection, and consequently also as the source at which the art of war must chiefly be studied.

This view was universal in the world of theory before the Wars of the French Revolution. But these wars suddenly opened to view a wholly different world of military phenomena, which at first appeared somewhat rough and wild, but afterwards systematized under Bonaparte into a method on a grand scale, produced results which roused the astonishment of young and old. Old models were abandoned and it was thought that all this was the result of new discoveries, magnificent ideas, and so forth, but also, of course, of the changes in the state of society. It was now thought that the old methods were of no further use whatever and would never be seen again. But in such revolutions in opinions, parties always arise and in this case also the old views have found their champions, who look upon the new phenomena as rude blows of brute force, a general decadence of the art, and who cherish the belief that it is precisely the equipoised, fruitless, nugatory war game which must be the goal of perfection. There lies at the bottom of this last view such a want of logic and philosophy that we can only term it a desperate confusion of ideas. But at the same time, the opposite opinion, that nothing like the past will ever appear again, is very rash. Of the new phenomena in the field of war very few indeed are to be ascribed to new discoveries or new tendencies of thought. Most of them are due to new social conditions and circumstances. But these must not be taken as a norm, either, belonging as they do just to the crisis of a process of fermentation, and we cannot doubt that a great part of the earlier conditions of war will once more reappear. This is not the place to enter further into these matters; it is enough for us that by directing attention to the position which this equipoised play of forces occupies in the conduct of war as a whole, we have shown with regard to its significance and its internal connection with other matters that it is always the product of limitations imposed by circumstances on both parties and a very attenuated form of war's essential violence. In this play, one general may show himself more skilful than his opponent, and consequently, if he is a match for him in forces, may also gain many advantages over him; or if he is the weaker, he may maintain the equipoise with him by his superior ability; but it is manifestly absurd to look here for the highest honor and glory of a general. Such a campaign is, on the contrary, always an infallible sign that neither of the generals has any great military talent, or that he who has talent is prevented by circumstances from venturing on a great decision. But where this is the case, the opportunity of winning the highest military glory no longer exists.

We have so far spoken of the general character of strategic maneuvering; we must now proceed to a special influence which it has on the conduct of war, the fact, namely, that it frequently leads the military forces away from the principal roads and places into out-of-the-way or, at all events, unimportant districts. Where trifling interests, which exist for a moment and then disappear, are paramount, the great features of a country have less influence on the conduct of the war. We therefore find that forces are often sent off to points at which, having regard to the great and simple requirements of the war, we should never expect to find them, and consequently the details of the war's course are here much more subject to change and variation than they are in wars that seek a great decision. Let us only look at the last five campaigns of the Seven Years' War. In spite of the relations in general remaining unchanged, each of these campaigns took a different form, and, on close examination, no single measure ever appears twice. And yet in these campaigns the offensive principle manifests itself on the side of the Allied armies much more decidedly than in most other earlier wars.

In this chapter on the defense of a theater of war if no great decision is in view, we have only shown the tendencies which the action will have, together with their connection, their relations to one another, and their nature; the particular measures involved in it have been described in detail earlier in this work. Now the question arises whether for these different tendencies no general comprehensive principles, rules or methods can be given. To this we reply that if we keep to history, we are decidedly not led to any deductions of the kind through constantly recurring forms; and at the same time for a whole subject of so manifold and variable a character, we could hardly admit any theoretical rule, except one founded on experience. A war with great decisions is not only much simpler but also much more natural; it is freer from inconsistencies, more objective, more governed by a law of inherent necessity. Hence reason can prescribe forms and laws for it. But for a war without a decision, this appears to us to be far more difficult. Even the two fundamental principles of the theory of war on a large scale developed only in our own time, the *breadth of the base* by Bülow, and the *position on interior lines* by Jomini, if applied to the defense of a theater of war have actually in no instance shown themselves absolute and effective. But this is just where as mere forms they should prove most effective, because forms must become more and more effective and acquire more and more preponderance over the other factors in the product, the more the action extends in time and space. Nevertheless we find that they are nothing but particular aspects of the subject and certainly anything but decisive advantages. That the particular character of the means and circumstances must have a great influence, defying all general principles, is very obvious. What Daun did by the extension and prudent choice of positions, the king did by keeping his main force always concentrated, always hugging the enemy close, and constantly ready to act on the spur of the moment. Both methods proceeded not only from the nature of the armies but also from the circumstances in which they were placed. To act on the spur of the moment is much easier for a king than it is for a general responsible to higher authority. We shall here once more call special attention to the fact that the critic has no right to regard the different manners and methods that may arise as different degrees of perfection, treating one as inferior to another. They exist side by side and it must be left to the judgment to appraise their use in each particular case.

To enumerate the different manners which may arise from the particular character of the army, the country or the situation is, of course, not our intention here. We have already pointed out the general influence of these things.

We acknowledge, therefore, that in this chapter we are unable to give any principles, rules or methods, because history offers us nothing of the kind. On the contrary, at almost every single moment we stumble upon peculiarities which are very frequently quite inexplicable and often even startle us with their oddity. But it is not on that account unprofitable to study history in connection with this subject also. Where there is no system, no contrivance for finding truth, still a truth is there, and this truth will usually be discovered only by a practiced judgment and the instinct that comes of long experience. Even if history does not here furnish any formula, we may be certain that here as everywhere else it will give us *exercise for the judgment*.

We shall only set up one comprehensive general principle, or rather we shall repeat and put more vividly before the eye, in the form of a separate principle, the natural presupposition of all that has now been said.

All the means here enumerated have only a *relative* value; they are under the legal ban of a certain disability on both sides; over this region a higher law prevails, and in that a totally different world of phenomena exists. The general must never forget this; he must never move in imaginary security within the narrow sphere as if in something absolute; never look upon the means which he employs here as the *necessary* and the *only* means, and never *still adhere to them even when he himself is trembling at their insufficiency*.

From the point of view which we have here taken up, such an error may appear to be almost impossible; but it is not impossible in the real world, because there things do not appear in such sharp contrast.

We must again remind our readers that, for the sake of giving clearness, distinctness and force to our ideas, we have taken as the subject of our treatment the complete antithesis of the extremes of each side, but that the concrete case in war generally lies between these two extremes and is only influenced by either of these extremes, according to the degree in which it approaches one or the other.

Therefore, quite commonly everything depends on the general making up his mind above all as to whether his adversary has the inclination and the means of outbidding him by the use of some greater and more decisive measure. As soon as he has reason to apprehend this, he must give up the small measures intended to ward off small disadvantages. The course which in that case remains to him is to put himself by a voluntary sacrifice in a better position to be equal to a greater decision. In other words, the first requisite is that the general should apply the right standard in laying out his work.

In order to give these ideas still more distinctness by the aid of actual life, we shall briefly review some cases in which in our opinion a wrong standard was used, that is, in which one of the generals in his calculations very much underestimated the decisive action intended by his adversary. We begin with the opening of the campaign of 1757, in which the Austrians showed by the disposition of their forces that they had not counted upon so thorough an offensive as that adopted by Frederick the Great. Even the lingering of Piccolomini's corps on the Silesian frontier, while Duke Charles of Lorraine was in danger of having to surrender with his whole army, is such a case of complete misconception of the situation.

In 1758, the French were not only completely deceived as to the effects of the convention of Kloster Seeven (a fact, of course, with which we have nothing to do here), but two months afterward they were also entirely mistaken in their judgment of what their opponent might undertake, which, very shortly after, cost them the country between the Weser and the Rhine. That Frederick the Great in 1759, at Maxen, and in 1760, at Landshut misjudged his enemies in not supposing them capable of such decisive measures has been mentioned before.

But in all history we can hardly find a greater error in the standard adopted than that in 1792. It was then imagined possible to turn the tide in a national war by a moderate-sized auxiliary army, which brought down on those who attempted it the enormous weight of the whole French people, at that time completely unhinged by political fanaticism. We only call this error a great one because it has proved so since, and not because it would have been easy to avoid. As far as regards the conduct of the war itself, it cannot be denied that the foundation of all the disastrous years which followed was laid in the campaign of 1794. On the side of the Allies in that campaign itself, the powerful nature of the enemy's attack was quite misunderstood, by opposing to it a miserable system of extended positions and strategic maneuvers; and further, in the political differences between Prussia and Austria and the foolish abandonment of Belgium and the Netherlands, we may also see how little presentiment the cabinets of the day had of the force of the torrent which had just broken loose. In

1796, the separate acts of resistance offered at Montenotte, Lodi, and so on, show sufficiently how little the Austrians understood what it meant to be confronted with a Bonaparte.

In 1800 it was not by the direct effect of the surprise, but by the false view which Melas took of the possible consequences of this surprise, that his catastrophe was brought about.

Ulm, in 1805, was the last knot of a loose network of scientific but extremely feeble strategic schemes, good enough to stop a Daun or a Lascy but not a Bonaparte, the Emperor of the Revolution.

The indecision and confusion of the Prussians in 1806 proceeded from antiquated, petty, impracticable views and measures being mixed up with some lucid ideas and a true feeling of the immense importance of the moment. If there had been a clear consciousness and a complete appreciation of the situation of the country, how could they have left 30,000 men in Prussia and entertained the idea of forming a special theater of war in Westphalia and of gaining any results from a trivial offensive such as that for which Rüchel's and the Weimar corps were intended? How could they have still gone on talking of danger to magazines and loss of this or that piece of territory in the last moments left for deliberation?

Even in 1812, in that grandest of all campaigns, there was no want at first of unsound tendencies proceeding from the use of an erroneous standard. In the headquarters at Vilna there was a party of men of high repute who insisted on a battle on the frontier in order that no hostile foot should tread on Russian ground with impunity. That this battle on the frontier *might* be lost, nay, that it *would* be lost, these men probably admitted; for, although they did not know that there would be 300,000 French to meet 80,000 Russians, still they knew that a considerable superiority on the part of the enemy had to be assumed. The chief error consisted in the value which they ascribed to this battle; they thought it would be a lost battle like many others, whereas it may almost certainly be asserted that this supreme decision on the frontier would have produced a very different series of events from those following other battles. Even the camp at Drissa was a measure at the root of which there lay a completely erroneous standard with regard to the enemy. If the Russian army had intended to remain here they were bound to be completely isolated and cut off from every quarter, and then the French army would not have been at a loss for means to compel the Russians to lay down their arms. The designer of that camp never thought of power and will on such a scale as that.

But even Bonaparte sometimes used a false standard. After the armistice of 1813 he believed he could hold in check the subordinate armies of the Allies under Blücher and the Crown Prince of Sweden by forces which, although not able to offer any effectual resistance, might still give a cautious opponent sufficient cause to refuse to risk anything, as had so often been seen in earlier wars. He did not take enough account of the reaction of a deep-rooted hatred and sense of pressing danger such as animated Blücher and Bülow.

In general he underestimated the enterprising spirit of old Blücher. At Leipzig Blücher alone wrested from him the victory; at Laon Blücher might have entirely ruined him, and if he did not do so, the cause lay in circumstances completely out of Bonaparte's calculation. Finally, at Waterloo, the penalty of this mistake reached him like a thunderbolt.

Book VII

The Attack

CHAPTER 1[1]

THE ATTACK IN RELATION
TO THE DEFENSE

If two ideas form an exact logical antithesis, that is to say, if the one is the complement of the other, then, fundamentally, each one is implied in the other. Even when the limited power of our mind is insufficient to apprehend both at once, and by the mere antithesis to recognize the totality of the one in that of the other as well, still, at all events, the one will always throw on the other a strong and, in many parts, sufficient light. Thus we think the first chapters on the defense throw a sufficient light on all the points of the attack which they touch upon. But it will not be so throughout in respect of every point. It has never been possible to exhaust all aspects of the subject, and it is, therefore, natural that where the antithesis does not lie so immediately in the root of the conception as in the first chapters, all that can be said about the attack does not follow immediately from what has been said of the defense. A change in our point of view brings us nearer to the subject, and it is, therefore, natural, from this nearer point of view to bring under observation what we overlooked from the more distant. What we perceive here will thus be the complement of the set of ideas we had there collected, and it will not unfrequently happen that what is said of the attack will throw new light on the defense as well. Thus in treating of the attack we shall, for the most part, have the same subjects before us which we had in treating of the defense. But it would be contrary to our own view and to the nature of the facts to do as is done in most textbooks of fortification and ignore or upset all that we

---

[1] Book VII, "The Attack," is a rough draft only.—Ed.

have found of positive value in the defense, by proving that against every means of defense there is an infallible means of attack. The defense has its strong points and its weak ones; the first are not unsurmountable, yet they cost a disproportionate price, and that must remain true from whatever point of view we look at it, or we get involved in a contradiction. Further, it is not our intention exhaustively to review the reciprocal action of the various means. Each means of defense suggests a means of attack, but this is often so evident that there is no necessity to pass over from the standpoint of the defense to that of the attack in order to become aware of it; the one follows from the other of itself. Our intention in each subject is to set forth the special characteristics of the attack in so far as they are not directly implied by the defense, and this mode of treatment must necessarily lead us also to many chapters to which there are none corresponding in the defense.

CHAPTER 2

NATURE OF THE STRATEGIC ATTACK

We have seen that the defensive in war generally—therefore, also the strategic defensive—is no absolute state of waiting and warding off, therefore no completely passive state, but that it is a relative state and consequently permeated more or less with offensive elements. In the same way the offensive is no homogeneous whole, but incessantly mingled with the defensive. But there is this difference between the two: that a defensive, without a return blow, cannot be conceived; that this return blow is a necessary constituent of the defensive, while in the attack, the blow or act of attack is in itself one complete conception. The defense in itself is not necessarily a part of the attack; but time and space, to which it is bound, introduce the defense into it as a necessary evil. For in the *first* place, the attack cannot be continued uninterruptedly up to its conclusion; it must have stages of rest, and in these stages, when its action is neutralized, the state of defense intervenes of itself. In the *second* place, the space which a military force, in its advance, leaves behind it, and which is essential to its existence, cannot always be covered by the attack itself, but must be specially protected.

The act of attack in war, but particularly in strategy, is therefore a perpetual alternating and combining of attack and defense; but the latter is not to be regarded as an effective preparation for attack, as a means by which its force is heightened; that is to say, not as an active principle, but purely as a necessary evil; as the retarding weight produced by the sheer weight of the mass; as its original sin, its seed of mortality. We say a *re-*

tarding weight, because if the defense does not contribute to strengthen the attack, it must tend to diminish its effect by the very loss of time which it represents. But now, may not this defensive element, which is contained in every attack, have even a *positively disadvantageous* effect upon it? If we say to ourselves that *the attack is the weaker, the defense the stronger form of war,* it seems to follow that the latter cannot have a positively disadvantageous effect upon the former; for as long as we still have enough forces for the *weaker* form, we should have more than enough for the *stronger.* In general—that is as regards the main point—this is true, and how this works out in detail we shall set forth more precisely in Chapter 22, "The Culminating Point of Victory." But we must not forget that the superiority of *strategic defense* is partly founded on the very fact that the attack itself cannot be without an admixture of defense, and indeed of defense of a much weaker kind. The admixture of defense with which the attack is encumbered consists of its worst elements. With respect to these, that which holds good of the whole in a general sense can no longer be maintained, and therefore we can understand that these elements of the defensive may even positively become a weakening influence for the attack. It is just in these moments of weak defensive in the attack that the effective energy of the offensive principle in the defense should come into action. During the twelve hours' rest which usually succeeds a day's work, what a difference there is between the situation of the defender in his chosen, well-known and prepared position, and that of the assailant, occupying a bivouac into which—like a blind man—he has groped his way; or during a longer period of rest, which may be required to obtain provisions and await reinforcements, etc., when the defender is close to his fortresses and supplies, while the situation of the assailant, on the other hand, is like that of a bird on a bough. Every attack must end with a defense. What the nature of this will be depends on circumstances; these may be very favorable if the enemy's forces are destroyed, but they may also be very difficult, if that is not the case. Although this defensive does not belong to the attack itself, yet its nature must react upon the attack and help to determine its value.

The result of this examination is that in every attack regard must be had to the defensive, which is a necessary component of it, in order to be able to get a clear view of the drawbacks to which it is subject and to be prepared for them.

On the other hand, in another respect, the attack is always in itself completely one and the same. But the defensive has its gradations according to the degree in which the fundamental principle of awaiting the enemy is being made exhaustive use of. This produces forms which differ essentially from one another, as has been developed in the chapter on methods of resistance.

As the attack has only *one* active principle, the defensive element it contains being only a dead weight, hanging upon it, it does not present the same variety as does the defense. An immense difference is certainly found in the energy of the attack, and in the rapidity and force of the blow, but only a difference of *degree* and not of *kind*. We might well think that the assailant also might on occasion choose the defensive form, the better to attain his end; that he might, for instance, occupy a strong position in order to be attacked there. But such instances are so rare that we do not think it necessary to dwell upon them in our classification of principles and facts, which is always based upon practice. In the attack no such gradations are found as the methods of resistance present.

Lastly, the means of attack, as a rule, consist only of the armed force; to this we must, of course, add the fortresses, for if these are in the vicinity of the enemy's theater of war, they have a noticeable influence on the attack. But this influence gradually diminishes as the attack progresses, and it is obvious that, in the attack, our own fortresses can never play such an important part as in the defense, in which they often become the principal factor. The assistance of the people may be supposed to be in co-operation with the attack in those cases in which the inhabitants are more attached to the assailant than to their own army. Finally, the assailant may also have allies, but these are then only the result of special or accidental circumstances, not an assistance proceeding from the nature of attack. If, therefore, in the case of the defense, we have included fortresses, popular insurrections and allies among the means of resistance available, we cannot do the same in the case of the attack. In defense they are essential elements; in attack they only appear rarely and then for the most part accidentally.

CHAPTER 3

On the Objects of the
Strategic Attack

The overthrow of the enemy is the aim in war, destruction of the enemy's military force the means, both in attack and defense. By the destruction of the enemy's forces the defensive is led on to the offensive; the offensive, to the conquest of territory. Territory is, therefore, the object of the attack; but that need not be a whole country; it may be confined to a part—a province, a strip of country or a fortress. All these things may have a sufficient value as political counters in settling the terms of peace, whether they are retained or exchanged.

The object of the strategic attack is, therefore, conceivable in an infinite number of gradations, from the conquest of the whole country down to that of the most insignificant spot. As soon as this object is attained and the attack ceases, the defensive begins. We could, therefore, represent to ourselves the strategic attack as a distinctly limited unit. But it is not so if we consider the matter practically, that is, in accordance with actual phenomena. Practically the elements of the attack, that is, its intentions and measures, often end in defense as vaguely as the plans of the defense end in attack. It is seldom, or at all events not always, that the general lays down positively for himself what he wishes to conquer; he leaves that dependent on the course of events. His attack often leads him further than he had intended; after a more or less short rest, he often gets new strength, without our having occasion to make out of this two entirely separate stages of action. At another time he is brought to a standstill sooner than he had expected, without, however, giving up his intentions and changing to a real defensive. We see, therefore, that if the successful defense may change imperceptibly into the offensive, an attack may, on the other hand, change into a defense. These gradations must be kept in view in order to avoid making a wrong application of what we have to say of the attack in general.

CHAPTER 4

DECREASING FORCE

OF THE ATTACK

This is one of the principal subjects of strategy; on its right evaluation in the individual case depends our being able to judge correctly what we are able to do.

The decrease of absolute power arises:

1. Through the object of the attack, the occupation of the enemy's country; this generally begins only after the first decision, but the attack does not cease with the first decision.
2. Through the necessity imposed on the attacking armies to occupy the country in their rear in order to be able to secure their lines of communication and subsistence.
3. Through losses in action, and through sickness.
4. Distance of the various sources of supplies and reinforcements.
5. Sieges and blockades of fortresses.
6. Relaxation of efforts.
7. Secession of allies.

But against these grounds for the attack growing weaker must also be set some which may strengthen it. It is clear, however, that only the balancing of these different elements determines the general result. Thus, for instance, the weakening of the attack may be in part completely compensated or more than compensated by the weakening of the defense. This last is seldom the case; we only have to be careful to bring into the comparison not all the forces in the field but only those at the point of contact or facing one another at decisive points. Different examples: the French in Austria and Prussia, and in Russia; the Allies in France; the French in Spain.

CHAPTER 5

CULMINATING POINT

OF THE ATTACK

The success of the attack is the result of a present superiority of force, it being understood that the moral as well as the physical forces are included. In the preceding chapter we have shown that the force of the attack gradually exhausts itself; possibly at the same time the superiority may increase, but in the great majority of cases it diminishes. The assailant purchases advantages to be turned to account in the subsequent negotiations for peace; but he has to pay cash for them on the spot with his forces. If he maintains this daily diminishing preponderance in favor of the attack till peace is concluded, his end is attained. There are strategic attacks which have led to an immediate peace, but such instances are very rare; the majority, on the contrary, lead only to a point at which the forces remaining are just sufficient to maintain a defensive and to wait for peace. Beyond this point comes the turn of the tide, the counterstroke. The violence of such a counterstroke is usually much greater than the force of the original blow. This we call the culminating point of the attack. As the object of the attack is to get possession of the enemy's territory, it follows that the advance must continue till the superiority is exhausted; this, therefore, pushes us to our goal and may easily lead us beyond it. If we reflect on the number of elements which have to be taken into account in comparing the forces operating, we can understand how hard it is in many cases to make out which of the two sides has the superiority. Often all hangs on the silken thread of imagination.

Thus everything depends on a delicate instinctive judgment that intuitively recognizes the culminating moment. Here we come upon a seeming contradiction. The defense being stronger than the attack, we should thus suppose that the latter can never lead us too far, for so long as the weaker form remains strong enough, we are all the more so for the stronger.[1]

---

[1] Here follows in the manuscript the words: "Development of this subject after Book III, in the section about the culminating point of victory."

Under this title is now found in an envelope inscribed, "Separate Essays as Materials," a section which appears to be a draft of the chapter here indicated and which has been printed at the end of Book VII.—Note of Frau Marie von Clausewitz, the author's wife.

CHAPTER 6

DESTRUCTION OF THE ENEMY'S MILITARY FORCES

The destruction of the enemy's military forces is the means to the end. What is meant by this and what is the price it costs? The different points of view possible with regard to the subject are:

1. Only to destroy as much as the object of the attack demands.
2. Or as much as is at all possible.
3. The sparing of our own forces in the process as the principal point of view.
4. This, again, can go so far that only when a favorable opportunity offers does the attack make any attempt at the destruction of the enemy's forces. How this can also be the case with the object of the attack has already been mentioned in Chapter 3.

The only means of destroying the enemy's forces is the engagement but, of course, in two ways: 1. directly; 2. indirectly, through a combination of engagements. If, therefore, the battle is the chief means, it is not the only one. The capture of a fortress or of a portion of territory is in itself a destruction of the enemy's forces, and it may lead to a still greater destruction and thus become also an indirect means.

The occupation of an undefended strip of territory, therefore, apart from the value which it had as a direct attainment of the object, may also count as a destruction of the enemy's forces as well. Maneuvering the enemy out of a district occupied by him is something not very different and can, therefore, only be regarded from the same point of view, and not as a real success of arms. The value of these means is for the most part rated too high; they seldom have the value of a battle; besides which, it is always to be feared that the disadvantageous position to which they lead will be overlooked. They are seductive through the low price they cost.

We must always consider means of this description as small investments, which only lead to small profits and are suited only for more limited circumstances and weaker motives. Then they are obviously better than battles without a purpose. Victories the results of which cannot be realized to the full.[1]

---

[1] This is the heading of a paragraph never written. These chapters must be regarded merely as notes to be subsequently expanded.—Ed.

CHAPTER 7

THE OFFENSIVE BATTLE

What we have said about the defensive battle throws great light upon the offensive also.

We there had in view the battle in which the defense is most strongly pronounced, in order that we might convey a more vivid impression of its nature. But only very few are of that kind; most battles are *demi-rencontres* in which the defensive character to a great extent disappears. It is otherwise with the offensive battle; it preserves its character under all circumstances and may keep up that character the more boldly when the defender has departed from his own. For this reason, in the battle which is not definitely defensive and in the real *rencontres* there always remains something of the difference in the character of the battle, both on the one side and the other. The chief characteristic of the offensive battle is the maneuver to outflank or envelop, and therefore to gain the initiative as well.

The engagement in lines formed to outflank has obviously in itself great advantages; this is, however, a subject of tactics. The attack cannot give up these advantages because the defense has a means of counteracting them, for it cannot itself employ this means inasmuch as it is too closely connected with the other circumstances of the defense. In order successfully to outflank the outflanking enemy, it is necessary to have a well-chosen and well-prepared position. But what is much more important is that all the advantages which the defensive possesses cannot actually be employed. Most defenses are poor makeshifts; the greater number of defenders find themselves in a very harassing and critical position, in which, expecting the worst, they meet the attack half way. The consequence of this is that battles formed with outflanking lines, or even with an oblique front, which should properly result from an advantageous situation of the lines of communication, are commonly the result of a moral and physical preponderance (Marengo, Austerlitz, Jena). Besides, in the first battle fought, the base of the assailant, if not superior to that of the defender, is yet very wide on account of the proximity of the frontier; he can, therefore, afford to take some risk. The flank attack, that is, the battle with oblique front, is, by the way, generally more effective than the outflanking form. It is an erroneous idea that an outflanking strategic advance from the very beginning must be connected with it, as at Prague. That strategic measure has seldom anything in common with it, and is very hazardous. We shall speak further of this in the chapter, "Attack on a Theater of War." As it is the object of the commander in a defensive battle to delay the decision as long as possible and gain time, because a defensive battle undecided at sunset is commonly one gained, so it is the object of the commander in an offensive battle to hasten the decision; but, on the other hand, there is a great risk in too much haste, because it leads to a waste of forces. One characteristic of the offensive battle is the uncertainty, in most cases, as to the position of the enemy; it is an actual groping about among things that are unknown (Austerlitz, Wagram, Hohenlinden, Jena, Katzbach). The more this is the case, the more is concentration of forces required and the more is enveloping to be preferred to outflanking. That the chief fruits of victory are only gathered in the pursuit, we have already learned in Book IV, Chapter 12. The pursuit is naturally more an integral part of the whole action in the offensive, than in the defensive, battle.

CHAPTER 8

CROSSING OF RIVERS

1. A large river which crosses the direction of the attack is always very embarrassing for the assailant; for when he has crossed it, he is generally limited to one point of passage, and therefore unless he wants to remain close to the river, he becomes very much hampered in his movements. If, moreover, he intends to bring on a decisive engagement after crossing or may expect the enemy to do so, he exposes himself to great dangers; therefore without a decided superiority, both moral and physical, a general will not place himself in such a position.

2. The difficulty caused by the mere fact of the assailant having a river in his rear makes it possible actually to defend it much oftener than would otherwise be the case. If we suppose that this defense is not considered the only means of salvation but is so planned that even if it fails, still a stand can be made near the river, then the assailant in his calculations must add to the resistance which he may experience through the defense of the river all the advantages for the defender mentioned in 1. The effect of the two together is that attacking generals usually show great respect to a defended river.

3. But in the preceding Book we have seen that under certain conditions the real defense of a river promises very good results. If we refer to experience, we must allow that such results follow in reality much more frequently than theory promises. In theory we take account only of the actual circumstances, as we find them, while in the execution all circumstances commonly appear to the assailant more difficult than they actually are, and so become a stronger brake upon his action.

Now if it is a question of an attack which does not aim at an important decision and is not delivered with determined energy, we may be sure that in carrying it out a multitude of little obstacles and accidents, quite impossible for theory to take account of, will show themselves to the disadvantage of the assailant, because he is the acting party and thus is the first to come into conflict with them. Let us think how often the rivers of Lombardy, in themselves inconsiderable, have been successfully defended! If, on the other hand, in military history defenses of rivers also occur which have not produced the results expected of them, the reason lies in the wholly exaggerated effect which at times has been required of such operations, an effect not based in the least on their tactical character but merely on their effectiveness as known from experience, which people insisted on exaggerating beyond all due limits.

4. It is only when the defender commits the mistake of resting his whole salvation upon the defense of a river and exposes himself to the risk, if it is forced, of falling into great difficulties and a sort of catastrophe—it is only then that the defense of a river can be looked upon as a form of resistance favorable to the attack, for it is certainly easier to force the passage of a defended river than to win an ordinary battle.

5. It follows naturally from what has been just said that the defense of a river may become of great value if no great decision is sought; but where this is to be expected from the superior numbers or the energy of the enemy, then this operation if wrongly undertaken may be of positive value to the assailant.

6. There are very few river lines of defense which cannot be turned either in respect of the whole line of defense or of one particular point of it. Therefore an assailant superior in numbers and out for serious blows always has the means of making a demonstration at one point and crossing at another, and then by his superiority in numbers and by relentlessly pushing forward, of making good the first untoward circumstances he may have met with in the engagement; for he can achieve this by means of his superiority in numbers. An actual tactical forcing of the passage of a defended river—by one of the enemy's principal posts being dislodged by superior fire and superior valor—rarely or never occurs. The expression *forcing a passage* is always to be understood only in a strategic sense, in so far as the assailant by his passage at an undefended, or only slightly defended, point within the line of defense braves all the disadvantages which, according to the intention of the defender, should accrue to him from his crossing. But the worst an assailant can do is attempt the real passage at several points, unless they lie very close together and make a joint

blow possible. For as the defender must necessarily have his forces separated, the assailant throws away his natural advantage by dividing his own forces. In this way Bellegarde lost the battle on the Mincio in 1814, where by chance both armies passed at the same time at different points, and the Austrians were more divided than the French.

7. If the defender remains on this side of the river, it necessarily follows that there are two ways to gain a strategic advantage over him: either to pass at some point regardless of him, and thus outbid him in the use of the same means, or to give battle. In the first case the conditions of the base and of the lines of communication should be the chief deciding factors, but we often, certainly, see that special arrangements influence the decision more than do the general conditions. Ability to select better outposts and make better dispositions, better discipline, quicker marching—things like these can contend successfully against general circumstances. As regards the second alternative, it presupposes on the part of the assailant the means, the conditions and the resolution to fight a battle; but when these are presupposed, the defender will not be likely to venture upon this mode of defending a river.

8. As a final result we must therefore assert that though the passage of a river in itself rarely presents great difficulties, yet in all cases not concerned with a great decision, so many apprehensions of the consequences, immediate and more remote, are bound up with it that certainly in many cases the assailant can be brought to a standstill thereby. So that he either leaves the defender on this side of the river, or, at most, passes, but then remains close to the river. For it rarely happens that two armies remain for any length of time facing each other on opposite sides of the river.

But in cases also of a great decision, a river is an important object; it always weakens and upsets the offensive, and the most fortunate thing in such a case is if the defender is induced by this to look upon the river as a tactical barrier and to make the defense of that barrier the centerpiece of his resistance, so that the assailant obtains the advantage of striking the decisive blow in a very easy manner. This blow, it is true, will never in the first moment be a complete overthrow of the enemy, but it will consist of several advantageous engagements, and these will bring about very bad general conditions on the side of the enemy, as happened to the Austrians on the Lower Rhine in 1796.

CHAPTER 9

ATTACK OF DEFENSIVE POSITIONS

In the Book on defense it has been sufficiently explained how far defensive positions compel the assailant either to attack them or to give up his advance. Only those which can effect this are useful for our object and suited to wear out or neutralize the forces of the assailant, either wholly or in part. With regard to these the attack can do nothing against them; that is to say, there is no means at its disposal to counterbalance the advantage they give. But not all defensive positions are actually of this kind. If the assailant sees that he can pursue his object without attacking them, to attack them would be an error. If he cannot pursue his object, the question arises whether he cannot maneuver the enemy out of his position by threatening his flank. It is only if such means are ineffectual that a commander determines on the attack of a good position, and then an attack from the side usually offers somewhat less difficulty. The choice of which side to attack is decided by the situation and direction of the lines of retreat open to each party, so as to threaten those of the enemy and to cover our own. Between these two objects a conflict may arise, in which case the former is naturally entitled to the preference, as it is itself of an offensive nature, and thus in keeping with the attack, while the other is of a defensive nature. But it is certain and must be regarded as a truth of the first importance that *to attack a war-seasoned enemy in a good position is a dangerous thing.* No doubt instances are not wanting of such battles, and of successful ones too, as Torgau and Wagram. (We do not say Dresden because we may not call the enemy there "war-seasoned.") But on the whole the danger to the defender is very slight and disappears when we consider the infinity of cases in which we see the most resolute commanders making their bow to such positions.

We must not, however, corfuse the subject now before us with ordinary battles. Most battles are real *rencontres* in which one party, it is true, stands its ground but not in a prepared position.

CHAPTER 10

ATTACK OF ENTRENCHED CAMPS

It was for a time the fashion to speak slightingly of entrenchments and their effects. The cordon-like lines of the French frontier, which had often been forced; the entrenched camp at Breslau in which the Duke of Bevern was defeated, the battle of Torgau and several other cases led to this opinion of their value. The victories of Frederick the Great, won by rapid movement and the use of the offensive, had cast a reflection upon all defense, all stationary engagements and more particularly on all entrenchments, which still further increased this contempt. Certainly when a few thousand men are to defend several miles of country or when entrenchments are nothing more than reversed communication trenches, they are worth nothing, and the confidence placed in them thus creates a dangerous gap. But is it not inconsistent, or rather, nonsensical, to extend this contempt, in the spirit of a common braggart, to the idea of entrenchment itself (as Tempelhoff does)? What would be the point of having entrenchments at all if they were useless for strengthening the defense? No, not only reason but experience in hundreds and thousands of cases shows that a well-designed, well-manned and well-defended entrenchment is *as a rule* to be regarded as *an impregnable point*, and is also so regarded by the assailant. Starting from this point of the efficiency of a single entrenchment, we argue that there can be no doubt as to the attack of an entrenched camp being a most difficult operation, and one in which in most cases it will be impossible for the assailant to succeed.

It is natural that an entrenched camp should be weakly manned; but with good natural obstacles and well-made entrenchments, it can be defended against a great numerical superiority. Frederick the Great considered the attack on the camp of Pirna as impracticable, although he had at his command double the force of the garrison; and although it has since been here and there asserted that it would have been quite possible to take it. The only proof in favor of this assertion is founded on the very bad condition of the Saxon troops, an argument which certainly proves nothing against the efficacy of entrenchments. But it is a question whether those who have since contended not only for the feasibility, but also for the facility, of the attack would have made up their minds to it at the moment when it was to be carried out.

We think, therefore, that the attack of an entrenched camp is not at all one of the usual means for the offensive to adopt. It is only if the entrenchments have been thrown up in haste and not completed, much less strengthened with obstacles to prevent their being approached, or when, as is often the case, the whole camp is altogether only an outline of what it should be, a half-finished ruin, that an attack on it can be advisable and even become a way to gain an easy victory over the enemy.

CHAPTER 11

ATTACK OF A MOUNTAIN RANGE

From the fifteenth and following chapters of Book VI may be deduced sufficiently the general strategic relations of a mountain range, both as regards defense and even attack. We have there also endeavored to explain the part which a mountain range plays as a line of defense, properly so called, and from that naturally follows how it is to be looked upon in this significance from the side of the assailant. There remains, therefore, little for us to say here on this important subject. Our main conclusion there was that the defense must take an entirely different point of view in the case of a subordinate engagement from that which it takes in the case of a main battle. In the former, the attack of a mountain can only be regarded as a necessary evil, because it has all the conditions against it; in the latter, however, the advantages are on the side of the attack.

An attack, therefore, armed with the forces and the resolution for a battle, will meet the enemy on the mountains and certainly will reap advantage by so doing.

But we must here once more repeat that it will be difficult to obtain respect for this conclusion, because it runs counter to appearances, and is also, at first sight, contrary to the experience of war. It has been observed in most cases hitherto that an army pressing forward to the attack (whether seeking a decisive battle or not) reckoned it a rare piece of good fortune if the enemy had not occupied the intervening mountains, and that it then hastened to forestall him. No one will find this forestalling of the enemy inconsistent with the interests of the assailant; in our view, too, this is quite admissible, only we must here distinguish more carefully between the circumstances.

An army advancing against the enemy to bring him to a decisive battle, if it has to traverse an unoccupied mountain range, will naturally have to apprehend that just those passes of which it wishes to avail itself may at the last moment be blocked by the enemy. In that case the assailant would no longer have the same advantages which the occupation by the enemy of an ordinary mountain position had offered him. The enemy is no longer unduly extended, no longer in ignorance of the route which his assailant is taking; the assailant has not been able to choose his route with reference to the enemy's position, and therefore this battle in the mountains no longer has for him all the advantages of which we have spoken in Book VI. Under such circumstances the defender might be found in an impregnable position. According to this, the defender would after all have the means at his command of making an advantageous use of the mountains for his decisive battle. This would indeed be possible; but if we consider the difficulties the defender would have in establishing himself at the last moment in a good position in the mountains, especially if he had hitherto left them entirely unoccupied, we may well consider this means of defense entirely inadmissible, and the case which the assailant had to fear extremely *improbable*. But extremely improbable though the case may be, yet it is natural to fear it, for in war it often happens that an anxiety is very natural and yet tolerably superfluous.

But another measure which the assailant has here to fear is the provisional defense of the mountains by an advance guard or a chain of outposts. This means will also seldom accord with the interests of the defender, but the assailant is not well able to distinguish whether this will be the case or not, and so he fears the worst.

Further, our view by no means excludes the possibility of a position becoming quite unassailable on account of the mountainous character of the terrain. There are such positions which are not necessarily in the mountains (Pirna, Schmottseifen, Meissen, Feldkirch), and just because they are not in the mountains, they are the more suited for defense. We may, however, very well conceive that such positions may also be found in mountains themselves, where the defender can avoid the usual disadvantages of mountain positions as, for instance, on high plateaux. Yet they are extremely rare, and here we could only keep in view the majority of cases.

How little mountains are suited for decisive defensive battles, we see clearly from military history, for great generals have always preferred a position in the plains if they wanted to fight such a battle. Throughout the whole range of military history there are no examples of decisive engagements in the mountains except in the Revolutionary Wars, in which obviously a false application and analogy led to the use of mountain positions even in cases in which a decisive battle had to be reckoned upon (1793 and 1794 in the Vosges, and 1795, 1796 and 1797 in Italy). Melas has been generally blamed for not having occupied the Alpine passes in 1800; but such criticisms are those of first impressions, of mere childish judging by appearances, we might say. Bonaparte, in Melas's position, would no more have occupied them than Melas did.

The dispositions for the attack of mountain positions are mostly of a tactical nature; but we think it necessary to indicate here the first outlines for those parts, that is to say, that lie next to strategy and are coincident with it.

1. In the mountains we cannot, as in other districts, leave the road and make out of one column two or three if the exigency of the moment requires the troops to be divided, but for the most part we move slowly in long defiles. The advance, therefore, must generally be made on several roads or, rather, upon a somewhat wider front.

2. Against a mountain defense line of wide extent, the attack naturally is made with concentrated forces; to outflank the whole cannot in such a case be thought of, and if an important victory is the result sought, it must be attained by bursting through the enemy's lines and forcing the wings apart, rather than by outflanking the force and so cutting it off. A rapid continuous advance upon the enemy's main line of retreat is there the natural endeavor of the assailant.

3. But if the enemy is to be attacked in a more or less concentrated formation in the mountains, enveloping movements are a very essential part of the attack, for frontal thrusts will fall on the defense where it is strongest. But the enveloping movements must again aim rather at an actual cutting off than at a tactical assault on the flank or rear. Mountain positions are capable of a prolonged resistance even in rear if forces are not wanting, and the quickest result is invariably to be expected from the apprehension excited in the enemy of losing his line of retreat. This apprehension arises sooner in the mountains and operates more strongly, because if it comes to the worst, it is not so easy to cut one's way out with cold steel. A mere demonstration is here no adequate means; it would at most maneuver the enemy out of his position, but would not ensure any special result. The aim must therefore be actually to cut him off.

CHAPTER 12

ATTACK ON CORDON LINES

If a supreme decision is to lie in their defense or attack, they give the assailant a real advantage, for their excessive length is still more opposed to all the requirements of a decisive battle than is the direct defense of a river or a mountain range. Eugene's lines at Denain in 1712 may be instanced here, for the loss of them was fully equivalent to a lost battle. Villars would hardly have gained such a victory over Eugene in a concentrated position. If the offensive side does not possess the means required for a decisive battle, then even cordon lines are treated with respect, that is, if they are occupied by the enemy's main army; for instance, those of Stolhofen, held by Louis of Baden in 1703, were respected even by Villars. But if they are only held by a secondary force, then it is merely a question of the strength of the detachment which we can spare for their attack. The resistance in such cases is seldom great, but at the same time the result of the victory is seldom worth much.

The circumvallation lines of a besieger have a peculiar character, of which we shall speak in the chapter on the attack of a theater of war.

All cordonlike positions, as for instance, strengthened lines of outposts, etc., etc., have always the characteristic of being easily broken through; but if they are not forced with a view to going farther and bringing about a decision, they yield for the most part only a feeble result, not worth the trouble that has been spent upon it.

CHAPTER 13

MANEUVERING

1. We have already touched upon this subject in Chapter 30 of Book VI. It is one which concerns the defense and the attack alike; nevertheless, it certainly always has in it more of the nature of the offensive than of the defensive. We shall therefore examine it here more closely.

2. Maneuvering is opposed not to the carrying out of the attack by force and by means of great engagements, but to every such carrying out as proceeds directly from the offensive means, even if it is an operation against the enemy's communications or line of retreat, a diversion, etc.

3. If we adhere to the ordinary use of the word, there is in the conception of maneuvering an efficacy *called forth,* so to speak, out of nothing, that is to say, out of a state of *equilibrium,* only by the mistakes which the enemy is lured into committing. It is like the first moves in a game of chess. It is a play of evenly balanced forces in order to produce a favorable opportunity for success, and then to use this success as a superiority over the enemy.

4. The interests which in this connection must be considered partly as the object, partly as the basis, of action, are chiefly:

a) The subsistence which we are seeking to cut off from the enemy or to restrict.

b) The junction with other detachments.

c) The threatening of other communications with the interior of the country or with other armies or detachments.

d) Threatening the retreat.

e) Attack of special points with superior forces.

These five interests can find a lodgment in the most minute details of a particular situation and these thereby become the object round which everything for a time revolves. A bridge, a road or an entrenchment then plays the principal part. It is easy to show in each case that it is only the relation which any such object bears to one of the above interests which gives it importance.

f) The result of a successful maneuver is then for the assailant, or rather for the active party (which certainly can also be the defender), a piece of land, a magazine, etc.

g) In the strategic maneuver two antitheses appear, which look like different maneuvers and have sometimes served for the derivation of false maxims and rules. But all four of their members are in reality necessary constituents of the matter and are to be regarded as such. The first antithesis is between the outflanking of the enemy and the operating on interior lines; the second, between the concentration of forces and the spreading of them out in a number of posts.

h) As regards the first antithesis, we certainly cannot say that one of its members deserves a general preference over the other. For one thing, it is natural that one kind of effort calls out the other kind as its natural counterpoise, its proper antidote; for another, outflanking is in keeping with the attack while remaining on the inner lines is in keeping with the defense, and therefore the former will for the most part suit the assailant and the latter the defender. That form will maintain the upper hand which is the best handled.

i) The members of the other antithesis can just as little be classed the one above the other. The stronger force can afford to spread itself over several posts and by that means will in many respects obtain for itself a convenient strategic position and liberty of action and spare the energies of the troops. The weaker, on the other hand, must keep itself more concentrated, and seek by rapidity of movement to counteract the disadvantage that would otherwise be created for him thereby. This greater mobility assumes a higher degree of skill in marching. The weaker must, therefore, put a greater strain on his physical and moral forces—a final result which we must naturally meet everywhere if we have always been consistent, and which, therefore, may to a certain extent be regarded as the logical test of our reasoning. The campaigns of Frederick the Great against Daun in 1759 and 1760, against Laudon in 1761, and Montecuculi's against Turenne in 1673 and 1675 have always been regarded as the most skilful movements of this kind, and from them we have chiefly derived our views.

j) Just as the four members of the two antitheses above supposed must not be abused by being made the foundation of false maxims and rules, so we must also enter a warning against attaching to other general conditions, such as base, terrain, etc., an importance and a decisive influence which they do not in reality possess. The smaller the interests at stake, so much the more important become the details of time and place, so much the more does what is general and great retire into the background, as having, so to speak, no place in small calculations. Looked at from a general point of view, is there to be found a more absurd situation than that of Turenne in 1675, when he stood with his back to the Rhine, his army extended in a line fifteen miles long and with his bridge of retreat at the extremity of his right wing? But nevertheless his measures attained their object, and it is not without reason that they are acknowledged to show a high degree of skill and intelligence. We can, however, only understand this success and this skill when we look more closely into details and judge of them according to the value which they were bound to have in this particular case.

We are convinced that there are no rules of any kind for strategic maneuvering; that no method, no general principle can determine the mode of action; but that superior energy, precision, order, obedience, intrepidity in the most special and trifling circumstances can find the means to create for themselves signal advantages, and that thus victory in this competition will chiefly depend on these qualities.

CHAPTER 14

ATTACK OF MORASSES,
INUNDATIONS, WOODS

Morasses, that is, impassable meadows intersected only by a few dykes, present special difficulties to the tactical attack, as we have stated in treating of the defense. Their breadth hardly ever admits of the enemy being driven from the opposite bank by artillery and of the construction of a roadway across. The strategic consequence is that one tries to avoid and go round them. Where the level of cultivation is so high, as in many low countries, that the means of passing are innumerable, the resistance of the defender is still strong enough relatively, but it is proportionately weakened for an absolute decision, and therefore wholly unsuitable for it. On the other hand, if the low land (as in Holland) is aided by an inundation, the resistance it offers may increase till it becomes absolute and every attack upon it will end in failure. This was shown in Holland in 1672, when, after the conquest and occupation of all the fortresses outside the inundation line, 50,000 French troops were still left over but nevertheless—first under Condé and then under Luxemburg—were not able to force the inundation line, though perhaps not more than 20,000 men were defending it. The campaign of the Prussians in 1787 under the Duke of Brunswick against the Dutch ended, it is true, in quite an opposite result, the inundation lines being forced with almost no superiority in numbers and very trifling loss. The reason of this, however, must be sought in the dissensions among the defenders due to political animosities and in a want of unity in the command. And, nevertheless, nothing is more certain than that the success of the campaign—the advance through the last line of inundation up to the walls of Amsterdam—rested on so fine a point that no general deduction can possibly be drawn from it. This point was the unguarded Sea of Haarlem. By means of this the duke turned the line of defense and got in rear of the post of Amselvoen. If the Dutch had had an armed vessel or two on this sea, the duke would never have got to Amsterdam, for he was at the end of his tether. What influence that might have had on the conclusion of peace does not concern us here, but it is certain that a forcing of the last line of inundation would have been out of the question.

The winter is, no doubt, the natural enemy of this means of defense, as the French have shown in 1794 and 1795, but it must be a *severe* winter.

Woods that are scarcely passable we have also included among the means which afford the defense powerful assistance. If they are of no great depth, the assailant may force his way through by several roads running near one another, and thus reach better ground. The tactical strength of single points will not be great because a wood can never be considered so absolutely impassable as a river or a morass. But when, as in Russia and Poland, a very large tract of country is nearly everywhere covered with forest, and the assailant has not the power of getting beyond it, his situation will be a very difficult one. We have only to think of the many difficulties of supply with which he has to contend and how little he can do in the obscurity of the forest to make his ubiquitous adversary feel his superiority in numbers. Certainly this is one of the worst situations in which the offensive can be placed.

CHAPTER 15

ATTACK ON A THEATER OF WAR
WHEN A DECISION IS SOUGHT

Most of the subjects have already been touched upon in Book VI, and by their mere reflection, throw sufficient light on the attack.

In any case, the conception of a closed theater of war has a nearer relation to the defense than to the attack. Many of the leading points, *the object of attack, the sphere of action of victory,* etc., have been already treated of in this Book, and that which is most decisive and essential in the nature of the attack cannot be explained until we get to the "Plan of a War." Still there remains a good deal to say here, and we shall again begin with the kind of campaign *in which a great decision is intended.*

1. The first aim of the attack is a victory. To all the advantages which the defender finds in the nature of his situation the assailant can only oppose superior numbers, and perhaps, in addition, the slight advantage which the feeling of being the offensive and advancing side gives an army. The importance of this feeling, however, is generally overrated, for it does not last long and will not hold out against real difficulties. Of course, we assume that the defender acts just as faultlessly and judiciously as the assailant. Our object in this remark is to rule out those vague ideas of sudden attack and surprise, which in the attack are commonly thought to be rich sources of victory, and which yet in reality do not occur except in special circumstances. The problems of the real strategic surprise we have already dealt with elsewhere. If, then, the attack is inferior in physical power, it must have a superiority in morale in order to make up for the disadvantages of the offensive form. When this also is lacking, there are no good grounds for the attack and it will not succeed.

2. As prudence should be the guardian genius of the defender, so boldness and confidence should animate the assailant. We do not mean that the opposite qualities in each case may be altogether wanting, but that the qualities named have a greater affinity in the one case with the defense, in the other with the attack. All these qualities are after all only necessary because action in war is no mere mathematical calculation, but an activity carried on in the dark, or at best, in a feeble twilight, in which we must trust ourselves to that leader who is best suited to carry out the aim we have in view. The weaker the defender shows himself morally, the bolder the assailant should become.

3. For victory it is necessary that there should be a meeting between the enemy's principal force and our own. This is less doubtful as regards the attack than in regard to the defense, for the assailant seeks out the defender in his position. But we have maintained (in treating of the defensive) that the assailant should not seek the defender out if the latter has placed himself in a *wrong* position, because he may be sure that the defender will seek *him* out and he will then have the advantage of finding the defender unprepared. Here all depends on which road and direction has the greatest importance. This is a point which we did not examine in treating of the defense, reserving it for the present chapter. We shall therefore say what is necessary about it here.

4. We have already pointed out those objects to which the attack can be more immediately directed, and which, therefore, are the *objects* to be attained by victory. Now if these are within the theater of war which is attacked, and within the probable sphere of victory, the road to them is the natural direction of the blow to be struck. But we must not forget that the object of the attack does not generally acquire its importance till after the victory, and that therefore the victory must always be thought of in connection with it. Consequently the chief point for the assailant is not so much merely to attain his object; it is rather to gain it as a conqueror. The direction of the blow must therefore be not so much on the object itself as on the road the enemy's army has to take to reach it. This road is the immediate object of attack. To meet the enemy before he has reached this object, to cut him off from it, and in that position to beat him—the victory this gives is victory raised to a higher power. If, for example, the enemy's capital is the object of the attack, and the defender has not placed himself between it and the assailant, the latter would be wrong in marching direct upon the capital. He does better to make for the line connecting the defender's army with the capital and to seek there the victory which is to place the capital in his hands.

If there is no great object within the assailant's sphere of victory, the enemy's line of communication with the next great object is the point of paramount importance. The question then for every assailant to ask himself is: If I am successful in the battle, what am I to do with the victory? The object to be gained which this indicates to him is then the natural mark at which to direct his blow. If the defender has placed himself in that direction, he has done right and nothing further remains but for the assailant to seek him there. If his position is too strong, the assailant must try to pass it by, thus making a virtue of necessity. But if the defender has not placed himself on this right spot, the assailant chooses this direction, and as soon as he draws level with the defender, if the latter has not meanwhile made a lateral movement, turns in the direction of the defender's line of communications with that major object in order to seek out the enemy's army there. If the latter were to remain quite stationary, the assailant would have to wheel round upon it and attack it from the rear.

Of all the roads among which the assailant has a choice, the great commercial roads are always the best and the most natural to choose. In places, certainly, where these make too sharp a bend, the straighter roads, even if smaller, must be chosen, for a line of retreat which deviates much from the straight line is always subject to grave danger.

5. When the assailant sets out with a view to a great decision, he has no reason whatever for dividing his forces, and if, notwithstanding this, he does so, it must usually be regarded as an error proceeding from lack of clear views. He should, therefore, only advance with his columns in such a width of front as to admit of them all going into action together. If the enemy himself has divided his forces, so much the better for the assailant, only in that case, no doubt, small demonstrations may be made. These are, so to speak, strategic sham attacks and their intention is to preserve the advantages gained. The division of the forces, if done *for this purpose,* would then be justifiable.

Such division into several columns, as in any case is necessary, must be made use of to effect the dispositions required for the outflanking form of the tactical attack, for this form is natural to the attack and must not be disregarded without good reason. But it must remain only of a tactical nature, for a strategic outflanking while a great blow is being delivered is a complete waste of force. It can only be excused when the assailant is so strong that there can be no doubt at all about the result.

6. But the attack also demands prudence, for the assailant has also a rear, and communications which must be protected. This protection must be provided as far as possible by the manner in which the army advances,

that is, *eo ipso*, by the army itself. If a force must be specially detailed for this duty and therefore a division of the forces is required, this naturally cannot but weaken the force of the blow itself. As a large army is always in the habit of advancing with a front of at least a day's march in breadth, the covering of the lines of retreat and communications, if they do not deviate too much from the perpendicular, is in most cases accomplished by the front of the army.

Dangers of this description, to which the assailant is exposed, must be measured chiefly by the situation and character of the adversary. When everything lies under the pressure of an imminent great decision, there is little room for the defender to engage in undertakings of this kind, and the assailant has, therefore, in ordinary circumstances not much to fear. But when the advance is over, when the assailant himself is gradually changing into the defensive, then the covering of the rear becomes more and more necessary, more and more a thing of the first importance. For the rear of the assailant being naturally weaker than that of the defender, the latter, long before he passes over to the real offensive, and even at the time he is still yielding ground, may have begun to operate against the communications of the assailant.

CHAPTER 16

ATTACK ON A THEATER OF WAR WHEN A DECISION IS NOT SOUGHT

1. Even if there is neither the will nor the power sufficient for a great decision, there may still exist the decided intention of a strategic attack, but it is directed upon some secondary object. If the attack succeeds, with the attainment of this object the whole comes into a state of rest and equilibrium. If, to some extent, difficulties present themselves, the general progress of the attack comes to a standstill before this. Then in its place begins a mere occasional offensive or strategic maneuvering. This is the character of most campaigns.

2. The objects which may be made the aim of an offensive of this kind are:

a) *A piece of territory.* Gain in means of subsistence, perhaps contributions, sparing our own territory, equivalents in negotiations for peace—such are the advantages to be derived from this object. Sometimes also the idea of military honor is associated with it, as constantly occurs in the campaigns of the French marshals under Louis XIV. It makes a very important difference whether a piece of territory can be kept or not. In general, the first is the case only when the territory is on the border of our own theater of war and forms a natural complement to it. Only such pieces come into consideration as an equivalent in negotiating a peace, others are usually taken possession of for the duration of a campaign and are to be evacuated in the winter.

b) *One of the enemy's principal magazines.* If it is not one of considerable importance, it can hardly be looked upon as the object of an offensive determining a whole campaign. It certainly in itself is a loss to the defender and a gain to the assailant; its main advantage, however, for the latter is that the loss may compel the defender to retire a little and give up a piece of territory which he would otherwise have kept. The capture of the magazine is therefore in reality more a means, and is only spoken of here as an object, because it becomes the immediate definite aim of action.

c) *The capture of a fortress.* We have made the siege of fortresses the subject of a separate chapter, to which we refer our readers. For the reasons there explained we can understand how fortresses always constitute the best and most desirable objects in those offensive wars and campaigns in which the intention cannot be the complete overthrow of the enemy or the conquest of a considerable part of his territory. This explains why in the wars in the Low Countries, where fortresses are so abundant, everything always turned on the possession of one or another of them—so much so that gradual conquest of the whole province *does not even appear as the leading feature* of the campaign. On the contrary, each fortress was looked upon as a separate entity, which would have a value in itself, and more attention was paid to the convenience and facility with which it could be attacked than to the value of the place itself.

At the same time the siege of a place of some importance is always a formidable undertaking, because it involves great expenditure of money, and in wars in which the whole issue is not always at stake, this must be very carefully considered. Such a siege, therefore, must here be included among the important objects of a strategic attack. The less important the place or the less in earnest we are with the siege of it, and the less the preparation for it and the more everything is done casually, so much the smaller the strategic object becomes, and so much more an affair for feeble forces and motives. The whole thing then often sinks into a mere sham fight, to carry off the campaign with honor, because as assailant one does want to do something.

d) *A successful engagement, encounter or even battle,* for the sake of trophies, or merely for military honor, and even at times for the mere ambition of the commander—that such things do occur no one could doubt who knew anything at all of military history. In the campaigns of the French during the reign of Louis XIV, most of the offensive battles were of this kind. But what is more necessary for us to observe is that these things are not without objective importance, not a mere pastime of vanity. They have a very definite influence on the peace and thus lead fairly directly to the end in view. Military honor, the moral superiority of the army and its commander are things the influence of which, although invisible, never ceases to affect the whole course of the war.

The aim of such an engagement, of course, presupposes:

aa) That there is a fair prospect of victory.

bb) That in the event of the engagement being lost, too much is not being staked on the issue.

Such a battle, fought in confined circumstances and with a limited object, must naturally not be confused with victories which mere moral weakness has failed to turn to account.

3) With the exception of the last of these objects (d), they may all be attained without any considerable engagement and are usually so obtained by the assailant. Now the means which the assailant has at his command without resorting to a decisive engagement are derived from all the interests which the defender has to protect in his theater of war. They consist in threatening his lines of communication with points of supply, like magazines, fertile provinces, watercourses, etc., or with another corps, or with strong points like bridges, passes, etc.; in the occupation of strong positions from which the defender cannot drive us out again and which are so situated as to embarrass him; in the occupation of important cities, fertile lands and disturbed districts which could be seduced into rebellion; in the threatening of weaker allies, etc., etc. By actually cutting the said communications, and that in such a way that the defender cannot re-establish them without considerable sacrifice, and by setting himself to occupy the said points, the assailant forces the defender to take up another position more to the rear or to a flank, in order to cover those objectives even at the sacrifice of less important ones. Thus a strip of territory is left open, a magazine or a fortress is exposed, the one to conquest, the others to investment. In the process, engagements, greater or less, may occur, but they are not then sought and treated as ends but as a necessary evil, and can never exceed a certain degree of magnitude and importance.

4) The operation of the defense on the communication lines of the offensive is a sort of resistance which in wars aiming at a great decision can only occur if the lines of operation become very long. In wars, on the contrary, which do not aim at a great decision this kind of resistance is more natural. In such a case the enemy's lines of communication will, of course, rarely be very long; but then neither is it here so much a question of inflicting great losses of this kind on the enemy. A mere impeding and cutting short his means of subsistence are often effective, and what the lines want in length is made up for in some degree by the length of time which can be spent in this kind of contest with the enemy. For this reason the covering of his strategic flanks becomes an important object for the assailant. If, therefore, a contest or rivalry of this description takes place be-

tween the assailant and the defender, the assailant must seek to compensate by his numbers for his natural disadvantages. If he still retains power and resolution enough occasionally to venture a decisive stroke against one of the enemy's detachments, or against his main army itself, the danger which he thus holds over the head of his opponent is his best means of covering himself.

5) In conclusion we must notice another great advantage which in wars of this kind the assailant certainly has over the defender, which is that of being better able to judge of the intentions and means of his adversary than the latter can of his. It is much more difficult to discover in what degree an assailant is enterprising and bold than to decide whether the defender is meditating some great stroke. Viewed practically, there usually lies already in the choice of the defensive form of war a kind of guaranty that nothing positive is intended; besides this, the preparations for a great counterstroke differ much more from the ordinary preparations for defense than the preparations for an attack differ according as its intentions are more or less important. Finally, the defender is obliged to take his measures sooner than is the assailant, who thus has the advantage of playing the last hand.

CHAPTER 17

ATTACK OF FORTRESSES

We cannot here, of course, concern ourselves with the attack on fortresses from the point of view of the military engineer who constructs them. We shall deal with it, first, in relation to the strategic object with which it is connected, second, in relation to the choice among several fortresses, and, third, in relation to the way in which a siege should be covered.

That the loss of a fortress weakens the defense, especially when it has formed an essential part of that defense; that many conveniences accrue to the assailant from gaining possession of one, inasmuch as he can use it for magazines and depots and by means of it can cover strips of territory, quarters, etc.; that if his offensive should finally have to be changed into a defense, it can afford the strongest support for this defense—all these relations that fortresses bear to the theater of war as the war proceeds can be sufficiently apprehended from what has been said about fortresses in Book VI, "Defense," the reflection from which throws all the light required on the attack.

In relation to the taking of strong places, too, there is a great difference between campaigns which aim at a great decision and others. In the first, a conquest of this kind is always to be regarded as a necessary evil. As long as there is yet a decision to be made we undertake no sieges but such as are positively unavoidable. Only when the decision has already taken place, when the crisis and the tension of forces have for some time subsided, and when, therefore, a state of rest has begun, should the capture of strong places be undertaken. It then serves to consolidate the conquests made, and it can then usually be carried out, if not without effort and expenditure of force, yet still without danger. In the crisis itself the siege of a fortress heightens the intensity of the crisis to the prejudice of the offensive. It is evident that nothing so much weakens the force of the offensive, and therefore there is nothing so certain to rob it for some time of its preponderance. But there are cases in which the capture of this or that fortress is quite unavoidable, if the offensive is to be continued, and in such a case a siege is to be considered as an intensive advance of the attack. The less has previously been decided, the greater then becomes the crisis. All that now remains for consideration on this subject belongs to the Book on "Plan of a War."

In campaigns with a limited object, a fortress is generally not the means but itself the object; it is regarded as a small independent conquest, and as such has the following advantages over every other:

1. That a fortress is a small, distinctly defined conquest, which does not demand a major expenditure of force and which therefore gives no cause to fear a reaction.

2. That in negotiating for peace, its value in exchange can be turned to such good account.

3. That a siege is an intensive advance of the attack, or at least seems so, without constantly diminishing our forces as does every other advance of the attack.

4. That the siege is an enterprise without a catastrophe.

The result of these things is that the capture of one or more of the enemy's strong places is very frequently the object of those strategic attacks which cannot aim at any greater object.

The grounds for deciding the choice of the fortress to be besieged, in case this may be at all doubtful, are:

a) That it is one which can be easily kept, therefore stands high in value as an exchange in case of negotiations for peace.

b) That the means of taking it are available. Small means are only sufficient to take small places; but it is better to take a small one than to fail before a large one.

c) The strength of its defenses, which obviously are not always in proportion to the importance of the place. Nothing would be more foolish than to waste forces before a very strong place of little importance, if a place of less strength may be made the object of attack.

d) The strength of the armament and, therefore, of the garrison as well. If a fortress is weakly armed and insufficiently garrisoned, its capture must naturally be easier; but here we must observe that the strength of the garrison and armament are at the same time to be reckoned among those things which help to make up the *importance* of the place, because garrison and armaments are directly parts of the enemy's military strength, which cannot be said in the same measure of works of fortification. The conquest of a fortress with a strong garrison can, therefore, much more readily repay the sacrifice it costs than one with very strong works.

e) The facility of moving the siege-train. Most sieges fail for want of means, and the means are generally wanting from the difficulty of transport. Eugene's siege of Landreci in 1712 and Frederick the Great's siege of Olmütz in 1758 are the most remarkable examples.

f) Finally, there remains the facility of covering the siege, a point still to be considered.

There are two essentially different ways by which a siege may be covered: by entrenching the besieging force, that is, by a line of circumvallation, and by what are called lines of observation. The first of these methods has gone quite out of fashion, although evidently one important point speaks in its favor, namely, that by this method the force of the assailant does not at all suffer by division that weakening which is so generally found a great disadvantage at sieges. But we grant there is still a weakening in another way, to a very considerable degree, because:

1) The position round the fortress, as a rule, requires too great an extent for the strength of the army.

2) The garrison, the strength of which, added to that of the relieving army, would only make up the force originally opposed to us, *under these circumstances* is to be looked upon as an enemy's corps in the middle of our camp, which, protected by its walls, is *invulnerable*, or at least not to be overpowered. Hereby its power is greatly increased.

3) The defense of a line of circumvallation admits of nothing but the most absolute defensive, because the circular position, facing outwards, is the weakest and most disadvantageous of all possible orders of battle, and is particularly unfavorable to any advantageous sorties. There is, therefore, no alternative but to defend ourselves to the utmost in our entrenchments. That these circumstances may entail a much greater weakening of the defense than the diminution of the army by one-third of its strength, which perhaps would occur in an army of observation, is easily understood. If now we think of the general preference which has existed since the time of Frederick the Great for the offensive, as it is called (but which in reality is not always so), for movements and maneuvering, and the aversion to entrenchments, we shall not be surprised at lines of circumvallation having gone out of fashion. But this weakening of the tactical resistance is by no means their only disadvantage, and we have only men-

tioned the prejudices which would force themselves into the judgment on the lines of circumvallation, together with that disadvantage because they are nearly akin to each other. A line of circumvallation really only covers that portion of the theater of war which it actually encloses; all the rest is more or less abandoned to the enemy unless special detachments are detailed to cover it, which, however, would entail the very division of our forces that we want to avoid. Thus the besieging army will always be in anxiety and embarrassment on account of the convoys it requires. The covering of these by lines of circumvallation is not to be thought of if the army and the siege supplies required are considerable and the enemy is in the field in strong force. Such covering is only possible under such conditions as are found in the Netherlands, where a whole system of fortresses lying close to one another and connected by intermediate lines covers all the rest of the theater of war and considerably shortens the transport lines. In the time of Louis XIV the conception of a theater of war had not yet been associated with the position of a military force. In the Thirty Years' War particularly, the armies moved here and there sporadically, before this or that fortress in the neighborhood of which there happened to be no enemy force, and besieged it as long as the siege equipment they had brought with them lasted, and till a hostile army approached to relieve the place. At that time lines of circumvallation had their justification in the nature of the circumstances.

In future it is not likely that they will often be used again except when the enemy in the field is very weak, and when the conception of the theater of war is to some extent replaced by that of the siege. Only then will it be natural to keep all the forces concentrated on the siege itself, as a siege by that means unquestionably gains in energy in a high degree.

The lines of circumvallation in the reign of Louis XIV, at Cambray and Valenciennes, were of little use when the former were stormed by Turenne, opposed to Condé, and the latter by Condé, opposed to Turenne. But we must not overlook the endless number of cases in which they were respected, even when there existed in the place the most urgent need for relief and the commander on the defensive was a man of great enterprise, as in 1708, when Villars did not venture to attack the Allies in their lines at Lille. Frederick the Great also at Olmütz in 1758 and at Dresden in 1760, although he had no regular lines of circumvallation, had a system which, essentially, amounted to the same thing: he used the same army both for the siege and for covering it. The remoteness of the Austrian army induced him to adopt this plan at Olmütz, but the loss of his convoy at Domstädtel made him repent it. At Dresden in 1760 the motives which led him to this mode of proceeding were his contempt for the imperial army and his desire to take Dresden as soon as possible.

Lastly, it is a disadvantage of lines of circumvallation that in case of a reverse it is more difficult to save the siege artillery. But if the defeat has been sustained at a distance of one or more days' march from the place besieged, the siege can be raised before the enemy arrives, and the heavy transport may probably get the start of a day's march.

In taking up a position for an army of observation, an important question to be considered is the distance at which it should be stationed from the besieged place. This question will in most cases be decided by the nature of the terrain or by the position of other armies or corps with which the besiegers want to remain in communication. In other respects, it is easy to see that, with a greater distance, the siege is better covered, but that by a smaller distance, not exceeding a few miles, the two armies are better able to afford each other mutual support.

CHAPTER 18

ATTACK OF CONVOYS

The attack and defense of a convoy form a subject of tactics. We should, therefore, have nothing to say about it here, if it were not necessary first to demonstrate generally, so to speak, its possibility, which can only be done by reference to strategic principles and circumstances. We should have had to speak of it in this connection before when treating of the defense, had it not been that the little which can be said about it was applicable to the attack and the defense, and the former in this matter played the chief part.

A moderate convoy of three or four hundred wagons, let the load be what it may, takes up a couple of miles; a large convoy may be several miles in length. Now how is it possible to expect that the few troops usually allotted to a convoy will suffice for its defense? If to this difficulty we add the unwieldy nature of this mass, which can only advance at the slowest pace and which, besides, is always liable to be thrown into disorder, and, lastly, that every part of a convoy must be equally protected, because the moment that one part is attacked by the enemy, the whole is brought to a stop and thrown into a state of confusion, we may well ask: How can the covering and defense of such a train be possible at all? Or, in other words, why are not all transport trains taken when they are attacked and why are not all attacked which require an escort, or, which is the same thing, all that come within reach of the enemy? It is plain that all tactical expedients, such as Tempelhoff's most impracticable scheme of shortening the train by a process of continuous stopping and starting, and the much better plan of Scharnhorst's, of breaking up the convoy into several columns, only afford feeble remedies for a radical disease.

The explanation consists in this; that by far the greater number of transport trains derive more security from the strategic situation in general than any other parts exposed to the attacks of the enemy, and this gives their limited means of defense a very much increased efficacy. Convoys generally move more or less in the rear of their own army, or, at least, at a great distance from that of the enemy. The consequence is that only weak detachments can be sent to attack them, and these are obliged to cover themselves by strong reserves. Added to this, the very unwieldiness of the vehicles used makes it very difficult to carry them off; the assailant must, therefore, usually content himself with cutting the traces, driving off the horses, blowing up the powderwagons, etc., by which the whole is certainly detained and thrown into disorder, but not completely lost. By all this we perceive still more clearly that the security of such trains lies more in these general circumstances than in the defensive power of its escort. If now, to all this we add the defense by the escort, which, though it cannot, certainly, protect its convoy directly by striking resolutely at the enemy, is still able to derange the plan of his attack, it finally appears that the attack of a convoy, instead of being easy and sure of success, is fairly difficult and, in its results, uncertain.

But there still remains a very important point, which is the danger of the enemy's army or one of its corps taking revenge on the assailant and punishing him ultimately for the undertaking by defeating him. The apprehension of this causes the abandonment of many such undertakings, without the reason coming to light; so that the safety of the convoy is attributed to the escort, and people wonder how a pitiful arrangement, such as an escort, should meet with such respect. In order to realize the truth of this observation we have only to think of the famous retreat which Frederick the Great made through Bohemia after the siege of Olmütz in 1758, when half of his army was broken up into a column of companies to cover a convoy of 4,000 carriages. What prevented Daun from falling on this monstrosity? The fear that Frederick would throw himself upon him with the other half of his army and involve him in a battle which Daun did not desire. What prevented Laudon, who was constantly at the side of that convoy, from falling upon it at Zischbowitz sooner and more boldly than he did? The fear that he would get a rap over the knuckles. Fifty miles from his main army and completely separated from it by the Prussian army, he thought himself in danger of a serious defeat if the king, who was in no way interfered with by Daun, should march against him with the bulk of his forces.

It is only if the strategic situation of an army involves it in the unnatural necessity of receiving its convoys quite from the flank or even quite from the front that these convoys are really in great danger and become an advantageous object of attack for the enemy, if his position allows him to detach troops for that purpose. The same campaign of 1758 affords an instance of the most complete success of an undertaking of this description in the capture of the convoy at Domstädtel. The road to Neisse lay on the left flank of the Prussian position, and the king's forces were so neutralized by the siege and by the troops watching Daun that the irregulars had no reason whatever to be anxious on their own account and were able to make their attack completely at their ease.

When Eugene besieged Landreci in 1712, he drew his supplies for the siege from Bouchain via Denain; therefore, actually from the front of the strategic position. It is well known what means he used to overcome the difficulty of protecting his convoys in these circumstances and in what embarrassments he involved himself, ending in a complete reversal of the situation.

The conclusion we draw, therefore, is that however easy an attack on a convoy may appear in its technical aspect, still it has not so much in its favor on strategic grounds but only promises important results in the exceptional instances where the lines of communication are very much exposed.

CHAPTER 19

ATTACK ON THE ENEMY'S ARMY
IN ITS QUARTERS

We have not treated this subject in the defense because a line of quarters is not to be regarded as a defensive means but as a mere condition of the army, and one which implies little readiness for battle. In respect to this readiness for battle we therefore did not go beyond what we had to say about this condition of an army in Book V, Chapter 13.

But here, in considering the attack, we have to think of an enemy's army in quarters as a special object; for, in the first place such an attack is of a very peculiar kind in itself, and, in the next place, it may be considered as a strategic means of particular efficacy. Here we have before us, therefore, not the question of an onslaught on a single enemy billet or a small corps dispersed amongst a few villages, as the arrangements for that are entirely of a tactical nature. Here it is the question of an attack on a large army, distributed in quarters more or less extensive—an attack of which the object is not merely to surprise a single billet but to prevent the assembly of the army.

The attack on an enemy's army in quarters is, therefore, the surprise of an army not assembled. If this surprise is to be regarded as successful, then the enemy's army must be prevented from reaching its appointed place of assembly, and therefore be compelled to choose another, more in the rear. As this change of the point of assembly to the rear in a state of such emergency can seldom be effected in less than a day's march, but will generally require several days, the loss of ground which this causes is by no means insignificant. This is the first advantage gained by the assailant.

But now this surprise in respect of circumstances in general can certainly at the beginning be at the same time a surprise of certain single quarters, only certainly not of all and not of very many, because this would presuppose a spreading and scattering of the attacking army which would never be advisable. Therefore only the most advanced quarters, only those which lie in the direction of the attacking columns, can be surprised, and even this will seldom be quite successful with many of them, as large forces cannot easily approach unobserved. However, this element of the attack is by no means to be overlooked, and we reckon the success which may be thus obtained as the second advantage of such a surprise.

A third advantage consists in the partial engagements forced upon the enemy in which his losses may be considerable. A great body of troops does not assemble by single battalions at the general assembly point. They usually first form by brigades, divisions or corps, and these bodies cannot then rush at full speed to the rendezvous, but if an enemy column comes into collision with them they have to accept the engagement. Now it is certainly conceivable that they may come off victorious in it, in the event of the attacking column not having been strong enough, but even in the victory they lose time, and generally, as is easy to understand, in such circumstances, when all are trying to get to a point that lies to the rear, a corps can make no particular use of its victory. On the other hand, they may be beaten, and that is in itself more probable because they have no time to organize a good resistance. We may, therefore, very well suppose that in a well-planned and executed surprise attack, the assailant through these partial engagements will pick up substantial trophies, which will then be a leading feature in the general success.

Lastly, the fourth advantage, and the keystone of the whole, is a certain momentary disorganization and discouragement on the side of the enemy, which, when the force is at last assembled, seldom allows of its being immediately brought into action, and generally obliges the party attacked to abandon still more ground and, as a rule, to make a change in his intended operations.

Such are the characteristic good results of a successful surprise of the enemy in quarters, that is, of one in which he has been prevented from assembling his army without loss at the point fixed in his plan. But naturally, the success will have very many gradations, and the results will in one case be very considerable, and in another hardly worth mentioning. But even when, through the complete success of the enterprise, these results are considerable, they will still seldom yield the success yielded by victory in a decisive battle. In the first place, the trophies are seldom as great, and, second, the moral effect cannot be estimated so highly.

This general result must always be kept in view, so that we may not promise ourselves more from an enterprise of this kind than it can give. Many consider it to be the *non plus ultra* of offensive activity, but it is not so by any means, as we may see by this analysis, as well as from military history.

One of the most brilliant surprises in history is that made by the Duke of Lorraine in 1643 on the quarters of the French under General Ranzau, at Duttlingen. The corps was 16,000 strong and it lost the commanding general and 7,000 men. It was an absolute rout. The complete lack of outposts was responsible for this result.

The surprise of Turenne at Mergentheim (Mariendal, as the French call it) in 1644 is in like manner to be regarded as equal to a defeat in its effects, for he lost 3,000 men out of 8,000, principally owing to his having been led into making an untimely stand after he had got his men assembled. Such effects we can seldom reckon upon. It was rather the result of an ill-judged encounter than of the actual surprise, for Turenne might easily have avoided the engagement and effected a union somewhere else with those of his troops in the more distant quarters.

A third famous surprise is that which Turenne made on the Allies under the Great Elector, the Imperial General Bournonville and the Duke of Lorraine, in Alsace in 1674. The trophies were very small and the loss of the Allies did not exceed 2,000 or 3,000 men, which with a force of 50,000 could not be decisive. Nevertheless, they thought that they could not venture to offer any further resistance in Alsace, and retired across the Rhine. This strategic result was all that Turenne wanted, but it is not in the actual surprise that we must seek the reason. Turenne surprised the plans of his opponent rather than his troops. The want of unanimity among the allied generals and the proximity of the Rhine did the rest. This event altogether deserves a closer examination, as it is generally interpreted in a wrong way.

In 1741 Neiperg surprised Frederick in his quarters. The only result was that the king was obliged to fight the battle of Mollwitz before he had collected all his forces and with a changed front.

In 1745, Frederick the Great surprised the Duke of Lorraine in his quarters in Lusatia. The chief success came about through the real surprise of one of the most important quarters, that of Hennersdorf, by which the Austrians suffered a loss of 2,000 men. The general result was that the Duke of Lorraine retreated to Bohemia by Upper Lusatia, which, however, did not prevent his returning into Saxony by the left bank of the Elbe, so that without the battle of Kesselsdorf there would have been no important success.

In 1758 Duke Ferdinand surprised the French quarters. The immediate result was that the French lost some thousands of men and were obliged to take up their position behind the Aller. The moral effect may have had some influence on the subsequent evacuation of the whole of Westphalia.

If from these different examples we seek for a conclusion as to the efficacy of this kind of attack, only the first two can be considered equal to a battle gained. But the forces engaged were only small, and the want of outposts in the warfare of those days was a circumstance greatly in favor of surprises. Although the four other cases must be reckoned completely successful operations, they obviously, in their results, cannot be considered as equal to battles gained. The general result could not have taken place in any of them except with an adversary weak in will and character, and therefore, in the case of 1741, it did not take place at all.

In 1806 the Prussian army contemplated surprising the French in this manner in Franconia. The case promised well for a satisfactory result. Bonaparte was not present; the French corps were in widely extended quarters; and under these circumstances the Prussians, acting with great resolution and activity, might very well have reckoned on driving the French back across the Rhine with more or less loss. But this was all. If they had reckoned upon more, upon following up, for instance, their advantages beyond the Rhine, or of gaining such a moral preponderance that the French would not again have ventured to appear on the right bank of the river in the same campaign, such an expectation would have had no sufficient grounds.

In the beginning of August, 1812, the Russians from Smolensk meant to fall upon the quarters of the French when Napoleon halted his army in the neighborhood of Witebsk. But they lost courage while carrying out the operation, and it was fortunate for them that they did, for the French commander with his center was not only more than twice as strong numerically as they were but also the most resolute leader that ever lived. Further the loss of a few miles of ground could have decided nothing, and there was absolutely no natural obstacle near enough for them to pursue their success up to it and thereby to some extent make it secure. Lastly, the war of 1812 was not in any way a campaign of the kind which drags languidly to its conclusion, but the serious plan of an assailant who had made up his mind utterly to overthrow his opponent. The trifling advantages to be expected from a surprise of the enemy in his quarters seem out of all proportion to the task to be performed. They could not justify the hope of making up by their means for the great inequality of forces and circumstances. But this attempt serves to show how a confused idea of the effect of this means may lead to an entirely wrong application of it.

What has been said on the subject hitherto, analyzes it as a *strategic means*. But its execution also is, naturally, not purely tactical, but in part belongs again to strategy, in so far, that is, as such an attack is usually made on a front of considerable width, and the army which carries it out can, and generally will, come into action before it is concentrated, so that the whole is an agglomeration of separate engagements. We must, therefore, now add a word or two on the most natural organization of such an attack.

The first condition is:

1. To attack the front of the enemy's quarters in a certain width of front. That is the only means by which we can really surprise several quarters, cut off others and create generally that disorganization in the enemy's army which is intended. The number of the columns and the intervals between them must depend on circumstances.

2. The direction of the different columns must converge upon one point, where it is intended that they should unite, for the enemy ends, more or less, with a concentration of his force, and therefore we must do the same. The point should, if possible, be the enemy's point of assembly or lie on his line of retreat, best, of course, where that line is cut by a natural obstacle.

3. The separate columns when they come in contact with the enemy's forces must attack them with great determination, with boldness and dash, for they have the general circumstances in their favor, and in such a case daring is always in its right place. From this it follows that the commanders of the separate columns must be allowed great freedom and full discretion.

4. The tactical plan of attack against those of the enemy's troops that are the first to offer resistance must always be directed to turn a flank. The greatest result is always to be expected by separating the several corps and cutting them off.

5. Each of the columns must be composed of portions of the three arms, and must not be too weak in cavalry. It may even sometimes be well to divide among them the whole of the reserve cavalry, for it would be a great mistake to suppose that, as such, this could play any great part in an enterprise of this sort. The first village, the smallest bridge, the most insignificant thicket would bring it to a halt.

6. Although in a surprise the assailant should naturally not send his advance guard very far in front, that principle only applies to the first approach. If the fight has begun in the enemy's line of quarters and thus all

that was to be expected from actual surprise has been gained, the columns must then push on as advance guards of all arms as far as possible, for they may greatly increase the confusion on the side of the enemy by their more rapid movement. It is only by this means that it becomes possible to carry off here and there baggage and artillery, and the stragglers who usually follow quarters that are being suddenly broken up. These advance guards must also become the chief instruments in turning and cutting off the enemy.

7. Finally the retreat in case of ill success must be thought of and a rallying point be assigned beforehand.

CHAPTER 20

DIVERSION

According to the ordinary use of language, by the term "diversion" is understood such an attack on the enemy's country as draws off forces from the principal point. It is only when this is the chief intention, and not the gaining of the object attacked on the occasion, that it is an enterprise of a special character; otherwise it is only an ordinary attack.

Naturally the diversion must at the same time always have an object of attack, for it is only the value of this object that will induce the enemy to send troops. Besides, in case the operation does not succeed as a diversion, such objects are a compensation for the forces expended on it.

These objects of attack may be fortresses, or important magazines, or rich and large cities, especially capital cities, contributions of all kinds, and, lastly, assistance to be offered to discontented subjects of the enemy.

It is easy to conceive that diversions may be useful, but they certainly are not so always; on the contrary, they are frequently even injurious. The chief condition is that they should draw off from the principal theater of war more of the enemy's troops than we employ on the diversion; for if they only succeed in drawing off just the same number, then their efficacy as diversions, properly so called, ceases, and the undertaking becomes a subordinate attack. Even where a secondary attack is arranged because, on account of circumstances, we have a good chance of attaining with small forces a disproportionately great result, as, for instance, to make an easy capture of an important fortress, that must no longer be called diversion. When a state is resisting another, and is at the same time attacked by a third state, such an event is commonly also called a diversion; but such an attack differs in nothing from an ordinary attack except in its direction. There is, therefore, no reason for giving it a particular name, for in theory one should designate by special terms only such things as also have a special meaning.

But if small forces are to attract large ones, special circumstances must obviously be the cause, and, therefore, for the object of a diversion it is not sufficient merely to detach some troops to a point not hitherto touched.

If the assailant with a small detachment of 1,000 men overruns one of his enemy's provinces, not belonging to the main theater of war, in order to levy contributions, etc., it is of course to be expected that the enemy cannot stop this by detaching 1,000 men, but if he means to protect the province from invaders, he must, of course, send out a larger force. But it must be asked: cannot a defender, instead of protecting his own province, restore the balance by sending a similar detachment to plunder a province in our country? Therefore, if an advantage is to be obtained by an aggressor in this way, it must first be ascertained that there is more to be gotten or to be threatened in the defender's provinces than in his own. If this is the case, then no doubt a weak diversion will occupy more enemy forces than our own so employed amount to. On the other hand, this advantage naturally diminishes as the masses increase, for 50,000 men can defend a province of moderate extent not only against equal, but even against somewhat superior numbers. The advantage of large diversions is, therefore, very doubtful, and the greater they become, the more decisive the other circumstances must be which favor a diversion if any good at all is to come of it.

Now these favorable circumstances may be:

a) Forces which the assailant can make available for a diversion without weakening his main attack.

b) Points belonging to the defender which are of vital importance to him and can be threatened by a diversion.

c) Discontented subjects of the defender.

d) A rich province which can supply a considerable means for war.

If such a diversion is to be undertaken, which, when tested by these different considerations, promises results, it will be found that an opportunity for it is not frequent.

But now comes another important point. Every diversion brings war into a district into which it would not otherwise have penetrated; for that reason it will always call forth some enemy forces which would otherwise have remained idle. It will do this in an extremely perceptible manner if the enemy is prepared for war by means of an organized militia and national armament. It is quite natural and amply shown by experience that if a district is suddenly threatened by a detachment of the enemy, and nothing has been prepared beforehand for its defense, all the most capable officials in the district call up all conceivable extraordinary means and set them in motion, in order to ward off the impending danger. Thus, new powers of resistance spring up, such as border upon a people's war, and may easily excite one.

This is a point which should be kept well in view in every diversion, in order that we may not dig our own graves.

The expeditions to North Holland in 1799, and to Walcheren in 1809, regarded as diversions, are only to be justified in that there was no other way of employing the English troops; but there is no doubt that the total means of resistance used by the French was thereby increased, and every landing in France itself would have had just the same effect. To threaten the French coast certainly offers great advantages, because by that means an important body of troops becomes neutralized in watching the coast, but a landing with a large force can never be justifiable unless we can count on the assistance of a province against its government.

The less a great decision is intended in war the more will diversions be permissible, but of course so much the smaller also will be the gain to be derived from them. They are only a means of setting the stagnant masses in motion.

EXECUTION

1. A diversion may include in itself a real attack; then the execution has no special character except boldness and expedition.
2. It may also have the intention of appearing more than it really is, being in fact, a demonstration as well. What particular means are to be employed in such a case can only be specified by a subtle mind, which is well acquainted with the character of the populace and with the existing state of circumstances. Naturally, there must be a great dispersion of forces on such occasions.
3. If the forces employed are quite considerable, and if the retreat is restricted to certain points, a reserve on which the whole rallies is an essential condition.

CHAPTER 21

INVASION

Almost all that we have to say on this subject consists in an explanation of the term. We find the expression very frequently used by modern authors and they even pretend to denote by it something particular. *Guerre d'invasion* occurs constantly among French authors. They use it as a term for every attack which enters deep into the enemy's country, and would like, at all events, to use it as the antithesis to a regulation attack, that is, to one which only nibbles at the frontier. But this is a very unphilosophical confusion of language. Whether an attack is to be confined to the frontier or carried into the heart of the country, whether it is to make the seizure of the enemy's strong places the chief object or to seek out the core of the enemy's power, and pursue it unremittingly, is the result of circumstances, and does not depend on a fashion. In some cases, to push forward may be more regular, and at the same time more prudent, than to stay on the frontier, but in most cases it is nothing else than just the fortunate result of a vigorous *attack*, and consequently does not differ from it in any respect.

CHAPTER 22

On the Culminating Point

of Victory[1]

The conqueror in a war is not always in a condition to defeat his adversary completely. Often, in fact mostly, there is a culminating point of victory. Experience shows this sufficiently; but as the subject is one especially important for the theory of war, and the foundation of almost all plans of campaign, while, at the same time, there hovers on its surface as with iridescent colors a flicker of apparent contradictions, we wish to examine it more closely and concern ourselves with its inherent causes.

Victory, as a rule, arises from a preponderance of the sum of all physical and moral powers; undoubtedly it increases this preponderance, or it would not be sought for and purchased at a great price. Victory *itself* does so unhesitatingly; so too do its consequences, but not to the final end—generally only up to a certain point. This point may be very near at hand, and is sometimes so near that all the results of a victorious battle can be confined to an increase of moral superiority. How this comes about we have now to examine.

In the progress of action in war, the military force is incessantly meeting with elements which increase it, and others which decrease it. Hence it is a question of the preponderance of one or the other. As every diminution of power on one side is to be regarded as an increase on the side of the enemy, it follows, of course, that this double current, this ebb and flow, takes place alike whether troops are advancing or retiring.

[1] Compare Chapters 4 and 5.

It is only necessary to find out the principal cause of this alteration, in the one case, to have determined the other along with it.

In advancing, the most important causes of the *increase of strength* on the part of the assailant are:

1) The loss which the enemy's military force suffers, because his loss is usually greater than that of the assailant.
2) The loss which the enemy suffers in the way of material military resources, such as magazines, depots, bridges, etc., and which the assailant does not share with him at all.
3) That from the moment the assailant enters the enemy's territory, there is a loss of provinces to the defense, consequently of the sources of new military force.
4) That the advancing army gains a portion of those resources, in other words, gains the advantage of living at the expense of the enemy.
5) The loss of internal organization and of the regular working of all its parts on the side of the enemy.
6) That the allies of the enemy abandon him, and others join the conqueror.
7) Lastly, the discouragement of the enemy who lets the weapons to some extent drop out of his hands.

The causes of *decrease of strength* in an advancing army are:

1) That it is compelled to lay siege to the enemy's fortresses, to blockade them or observe them; or that the enemy, who did the same before the victory, in his retreat draws in these troops to the main body.
2) That from the moment the assailant enters the enemy's territory, the nature of the theater of war is changed; it becomes hostile; we must occupy it, for it belongs to us only in so far as we occupy it and yet it everywhere presents difficulties to the whole machine which must necessarily tend to weaken its effects.
3) That we are moving farther away from our resources, while the enemy is drawing nearer to his; this causes a delay in the replacement of expended forces.
4) That the danger which threatens the state rouses other powers to its protection.
5) Lastly, the greater efforts of the adversary, in consequence of the increased danger; on the other hand, a relaxation of effort on the side of the winning state.

All these advantages and disadvantages can exist together, meet each other, so to speak, and pursue their way in opposite directions. Only the last ones meet as real opposites, cannot pass each other and, therefore, mutually exclude each other. This alone shows how infinitely different the effect of a victory may be according as it stuns the vanquished or stimulates him to greater exertions.

We shall now try with a few remarks to characterize each of these points separately.

1) The loss of the enemy's forces when defeated may be at the greatest in the first moment of defeat, and then daily diminish in amount until it arrives at a point where it balances ours; but it also may grow every day in geometrical progression. The difference of situations and conditions determines this. We can only say that, in general, with a good army the first will be the case, with a bad army the second; next to the spirit of the troops, the spirit of the government is here the most important thing. It is very important in war to distinguish between the two cases, in order not to stop just at the point where we should really begin, and vice versa.

2) The loss which the enemy sustains in the way of natural resources may increase and decrease in just the same manner, and this will depend on the accidental position and nature of the depots. This subject, however, in the present day, cannot be compared in importance with the others.

3) The third advantage must necessarily increase as the army advances; indeed, it may be said that it does not come into consideration until an army has penetrated deep into the enemy's country; that is to say, until a third or fourth of his country has been left behind. Moreover, the intrinsic value which a province has in connection with the war also comes into consideration.

In the same way the fourth advantage must increase with the advance. But with respect to these two last, it is also to be observed that their influence on the military forces actually engaged in the struggle is seldom felt immediately; they only work slowly and in a roundabout way. Therefore we should not bend the bow too much on their account, that is to say, not place ourselves in too dangerous a position.

The fifth advantage, again, only comes into consideration when we have made a considerable advance, and when by the form of the enemy's country some provinces can be separated from the principal mass, as these, like limbs compressed by ligatures, then usually soon die off.

As to 6 and 7, it is at least probable that they increase with the advance; furthermore, we shall return to them later.

Let us now pass on to the causes of weakness.

1) The besieging, attacking and blockading of fortresses will generally increase as the army advances. This weakening influence alone acts so powerfully on· *the immediate condition of the military forces* that it may easily counterbalance all the advantages gained. No doubt, in modern times, a system has been introduced of attacking fortresses with a small number of troops or of watching them with a still smaller number; and also the enemy must keep garrisons in them. Nevertheless, they remain a great element of security. The garrisons usually consist half of men who have taken no part in the war previously. Before those places which are situated near the line of communication, it is necessary for the assailant to leave a force at least double the strength of the garrison; and if it is desirable to lay formal siege to one single considerable place, or starve it out, a small army is required for the purpose.

2) The second cause, the establishment of a theater of war in the enemy's country, increases necessarily with the advance, and has a still greater effect on the permanent state of the military forces, though not on their condition at the moment.

We can only regard as our theater of war as much of the enemy's country as we occupy; that is to say, where we have left either small detachments in the open field or here and there garrisons in the most important cities or stations along the roads, etc. Now, however small the garrisons may be which we leave behind, they still weaken the military force considerably. But this is the smallest evil.

Every army has strategic flanks, that is, the country which borders both sides of its lines of communications; because, however, the army of the enemy likewise has such flanks, the weakness of these parts is not perceptible. But that can only be the case as long as we are in our own country; as soon as we find ourselves in the enemy's country, the weakness of these parts is felt very much, because the smallest operation promises some result when directed against a very long line only feebly covered or not at all; and these attacks may be made from any direction in an enemy's country.

The farther we advance, the longer these flanks become, and the danger arising from them grows in geometrical progression, for not only are they difficult to cover, but the spirit of enterprise is also first roused in the enemy chiefly by long insecure lines of communication, and the consequences which their loss may entail in case of a retreat are extremely serious.

All this helps to impose a fresh burden upon an advancing army at every step of its progress; so that if it has not started out with an extraordinary superiority, it will gradually feel itself more and more cramped in its plans, more and more weakened in its power of attack, and at last in a state of uncertainty and anxiety as to its situation.

3) The third cause, the distance from the source from which the incessantly diminishing military force must also be incessantly reinforced increases with the advance. A conquering army is like the light of a lamp in this respect; the more the oil which feeds it sinks in the reservoir and recedes from the center of light, the smaller the light becomes, until at length it is quite extinguished.

The wealth of the conquered provinces may, of course, diminish this evil very much, but can never entirely remove it, because there are always a number of things which we must procure from our own country—men in particular, because the supplies furnished by the enemy's country are, in most cases, neither so promptly nor so surely forthcoming as those furnished by our own; because the means of meeting any unexpected necessity cannot be so quickly procured; because misunderstandings and mistakes of all kinds cannot so soon be discovered and remedied.

If a prince does not lead his army in person, as became the custom in the last wars, if he is not anywhere near it, then another and very great inconvenience arises in the loss of time occasioned by communications backwards and forwards; for the fullest powers conferred on a commander of an army are never sufficient to meet every case in the wide expanse of his activity.

4) The change in political alliances. If these changes, produced by a victory, should be such as are disadvantageous to the conqueror, they will probably be so in a direct relation to his progress, just as is the case if they are of an advantageous nature. Everything depends here on the existing political alliances, interests, customs and tendencies, on princes, ministers, favorites, mistresses, etc. In general we can only say that when a great state which has smaller allies is conquered, these usually secede very soon from their alliance, so that the victor, in this respect, becomes stronger with every blow; but if the conquered state is small, protectors arise much sooner when its existence is threatened, and others, who have helped to shake its stability, will turn round to prevent its complete downfall.

5) The increased resistance which is called forth on the part of the army. At one time, terror-stricken and stupefied, the enemy lets the weapons fall from his hands; at another an enthusiastic paroxysm seizes him, everyone hastens to arms, and the resistance after the first defeat is

much stronger than it was before. The character of the people and of the government, the nature of the country and its political alliances, are here the data from which the probable effect must be conjectured.

How infinitely different these two last points alone render the plans which may and should be made in war in the one case and the other! While in the one through scrupulousness and so-called methodical procedure we trifle away our best chance of success, in the other through rashness we fall head and ears into destruction.

In addition, we must mention the slackness which frequently appears in the victor in his own country when the danger is removed; while, on the contrary renewed efforts would be necessary in order to follow up the victory. If we cast a general glance on these different and antagonistic principles, the deduction doubtless is that the following up of the victory, the onward march in a war of aggression, in the majority of cases, diminishes the preponderance with which the assailant set out, or which has been gained by the victory.

Here the question must necessarily strike us: If this is so, what is it then that impels the conqueror to pursue his course of victory, to continue the offensive? And can this really still be called following up the victory? Would it not be better to stop where as yet there is hardly any diminution of the preponderance gained?

To this we must naturally answer: the preponderance of military forces is only the means, not the end. The end is either to overthrow the enemy or at least to take from him a part of his lands, in order thereby to place ourselves in a position to make the advantages gained count in the conclusion of peace. Even if our aim is to overthrow the enemy completely, we must put up with the fact that, perhaps, every step we advance, reduces our preponderance. It does not, however, necessarily follow from this that it must be *nil* before the fall of the enemy. The fall of the enemy may take place before that, and if it is to be obtained by the last minimum of preponderance, it would be a mistake not to use it for that purpose.

The preponderance which we have or acquire in war is, therefore, only the means, not the object, and it must be staked to gain the latter. But it is necessary to know how far it will reach, in order not to go beyond that point and, instead of fresh advantages, to reap disgrace.

It is not necessary to adduce special examples from experience in order to prove that this is the way in which the strategic preponderance exhausts itself in the strategic attack; it is rather the multitude of instances which has forced us to investigate the causes of it. It is only since the appearance of Bonaparte that we have known campaigns between civilized nations in which the preponderance has led without interruption to the fall of the enemy. Before his time, every campaign ended with the victorious army seeking to win a point where it could simply maintain itself in a state of equilibrium. At this point, the movement of victory stopped, if, indeed, a retreat did not become necessary. Now this culminating point of victory will also appear in the future, in all wars in which the overthrow of the enemy cannot be the military object of the war; and the majority of wars will still be of this kind. The natural goal of every single plan of campaign is the point at which the offensive changes into the defensive.

To go beyond this goal is more than simply a *useless* expenditure of power, yielding no further result; it is a *ruinous* one which causes reactions, and these reactions, according to universal experience, have always disproportionate effects. This last fact is so common, and appears so natural and easy to understand, that we need not enter circumstantially into the causes. Lack of organization in the conquered land and the violent revulsion of feeling that results when a serious loss takes the place of the looked-for new success are the chief causes in every case. The moral forces, courage on the one side rising often to *bravado*, and extreme depression on the other, now generally begin to come into very active play. The losses on the retreat are increased thereby, and the hitherto successful party thanks heaven if he escapes with only the surrender of all his gains, without losing some of his own territory.

We must now clear up an apparent contradiction.

One would, of course, think that as long as the progress in the attack continues, there must still be a preponderance, and that as the defensive, which will begin at the end of the victorious advance, is a stronger form of war than the offensive, therefore, there is so much the less danger of the victor becoming unexpectedly the weaker party. But nevertheless this danger does exist, and keeping history in view, we must admit that the greatest danger of a reverse often does not arrive until the moment when the offensive ceases and passes into the defensive. We shall try to find the cause of this.

The superiority which we have attributed to the defensive form of war consists:

1) In the use of ground.
2) In the possession of a prepared theater of war.
3) In the support of the people.
4) In the advantage of awaiting the enemy.

It is evident that these advantages cannot always be forthcoming and active in a like degree; that, consequently, one defense is not always like another; and therefore, also, that the defense will not always have this same superiority over the offensive. This must be particularly the case in a defensive, which begins after the exhaustion of an offensive, and has its theater of war usually situated at the apex of an offensive triangle thrust far forward. Of the four advantages named above, this defensive retains only the first unaltered—the use of the ground. The second generally disappears altogether, the third becomes negative, and the fourth is very greatly weakened. A word or two more, by way of explanation, with regard to the last point only.

Under the influence of an imagined equilibrium whole campaigns often pass without any result, because the side which should assume the initiative is wanting in the necessary resolution. It is just in this, as we conceive, that the advantage of the state of awaiting lies. But if this equilibrium is disturbed by an offensive act, if the enemy's interests are damaged, and his will stirred to action, then the probability of his remaining in a state of indolent irresolution is greatly diminished. A defense, which is organized on conquered territory, has a much more challenging character than one upon our own soil; the offensive principle is engrafted on it, so to speak, and its nature is thereby weakened. The peace which Daun granted Frederick II in Silesia and Saxony, he would never have permitted him in Bohemia.

Thus it is clear that the defensive, which is interwoven with an offensive undertaking, is weakened in all its chief advantages, and, therefore, will no longer have the superiority which originally is due to it.

As no defensive campaign is composed of purely defensive elements, so likewise no offensive campaign is made up entirely of offensive elements: because, besides the short intervals in every campaign, in which both sides are on the defensive, every attack which does not lead to a peace must necessarily end in a defensive.

In this manner it is the defensive itself which contributes to the weakening of the offensive. This is so far from being an idle subtlety, that on the contrary, we consider it the chief disadvantage of the attack that we are afterwards reduced through it to a very disadvantageous defensive.

And this explains how the difference which originally exists between the strength of the offensive and defensive forms in war is gradually reduced. We shall now further show how it may completely disappear, and the advantage for a short time may change into the reverse.

If we may be allowed to make use of a concept from nature to explain our point, we will be able to express ourselves more briefly. It is the time which every force in the material world requires to produce its effect. A force, which if applied slowly and by degrees would be sufficient to bring to rest a body in motion, will be overcome by it if time is lacking. This law of the material world is a striking image of many of the phenomena in our inner life. If we are once roused to a certain trend of thought, not every motive, sufficient in itself, is capable of changing or stopping that current of thought. Time, tranquillity and durable impressions on our consciousness are required. So it is also in war. When once the mind has taken a decided trend toward an object or is, on the other hand, turned back toward a harbor of refuge, it may easily happen that the motives which compel one man to stop and which challenge another to dare are not felt at once in their full force; and as the progress of action in the meantime continues, they are carried along by the stream of movement beyond the limits of equilibrium, beyond the culminating point, without being aware of it. Indeed, it may even happen that, in spite of the exhaustion of force, the assailant, supported by the moral forces which chiefly lie in the offensive, finds it less difficult to advance than to stop, like a horse drawing a load uphill. By this, we believe, we have now shown, without inconsistency, how the assailant may pass that point which, at the moment of stopping and assuming the defensive, still promises him good results, that is, equilibrium. Rightly, to determine this point is, therefore, important in framing a plan of campaign, both for the assailant, that he may not undertake what is beyond his powers and, so to speak, incur debts, and for the defender, that he may perceive and profit by this error if committed by the assailant.

If now we look back at all the points which the commander should bear in mind in making his decision, and remind ourselves that he can only estimate the tendency and value of the most important of them through the consideration of many other near and distant circumstances, that he must to a certain extent *guess* them—guess whether the enemy's army, after the first blow, will show a stronger core and a steadily increasing solidity, or, like a Bologna phial, will crumble into dust as soon as the surface is injured—guess the extent of weakness and paralysis which the drying up of certain sources, the interruption of certain communications will produce on the military condition of the enemy—guess whether the enemy, from the burning pain of the blow which has been dealt him, will collapse powerless, or whether, like a wounded bull, he will be roused to a state of fury—lastly, guess whether other powers will be terrified or enraged and what political alliances will be dissolved or formed. When we say that in all this, and much more, he must hit the mark every time with an instinctive judgment, as the rifleman hits the bull's-eye, it must be admitted that such a feat of the human mind is no trifle. A thousand by-paths running this way and that, present themselves to the judgment; and whatever the number, the confusion and complexity of objects left undone is completed by the sense of danger and responsibility.

Thus it happens that the great majority of generals prefer to remain far short of the goal rather than to approach too close; and thus it happens that a fine courage and great spirit of enterprise often go beyond it, and therefore fail to attain their object. Only he who does great things with small means has successfully reached the goal.

Book VIII

PLAN OF A WAR

CHAPTER 1[1]

Introduction

In the chapter on the essence and object of war, we have sketched, in a certain sense, its general conception, and pointed out its relations to surrounding things, in order to start out with a sound fundamental idea. We hinted at the manifold difficulties which the mind encounters in the consideration of this subject, while we postponed the closer examination of them and stopped at the conclusion that the overthrow of the enemy, consequently the destruction of his military forces, is the chief object of the whole act of war. This put us in a position to show in the following chapter that the means which the act of war employs is the engagement alone. In this manner we think we have obtained for the time being a correct point of view.

We then went separately through all the principal relations and forms which appear in military action, but are extraneous to the engagement, in order that we might determine their value more distinctly, partly through the nature of the thing, partly from the experience which military history affords. We did this, furthermore, in order to purify them from those vague, ambiguous ideas which are generally mixed up with them, and also to bring forth in them the real object of the act of war—the destruction of the enemy's military force—as the primary object. We now return to war as a whole, inasmuch as we propose to speak of the plan of war and of campaigns, and that obliges us to revert to the ideas in Book I.

[1] Book VIII, "Plan of a War," is a rough draft only.—Ed.

In these chapters, which are to deal with the problem as a whole, is contained the very essence of strategy, in its most comprehensive and important features. We enter this innermost part of its domain, where all other threads meet, not without some diffidence.

Indeed this diffidence is amply justified.

On the one hand, we see how extremely simple the operations of war appear. We hear and read how the greatest generals speak of it in the plainest and simplest manner, how on their lips the regulating and managing of this ponderous machine, with its hundred thousand parts, seems just as if it were only a question of their own persons, so that the whole tremendous act of war is individualized into a kind of duel. We find the motives of their action explained now by a few simple ideas, now by the impulse of some emotion. We see the easy, sure, we might almost say, indifferent manner in which they treat the subject. And now see, on the other hand, the immense number of circumstances which present themselves for consideration to the investigating mind; the long, often indefinite distances into which the threads of the subject spin out and the number of combinations which lie before us. If we reflect that it is the duty of theory to embrace all this systematically, that is, with clearness and comprehensiveness, and always to trace the action back to the necessity of a sufficient cause, then there comes upon us an overpowering dread of being dragged down to a pedantic dogmatism, to crawl about in the lower regions of clumsy conceptions, where we shall never meet the great general, with his easy *coup d'oeil*. If the result of an effort at theory is to be of this kind, it would have been as well, or rather, it would have been better, not to have made the attempt at all. It could only bring down on theory the contempt of genius, and would soon be forgotten. And on the other hand, this easy *coup d'oeil* of the general, this simple way of thinking, this personification of the whole action of war, is so absolutely the very essence of every sound conduct of war, that in no other than this broad way is it possible to conceive that freedom of the mind which is indispensable if the mind is to dominate events, and not to be overpowered by them.

With some fear we proceed again; we can only do so by pursuing the way which we have prescribed for ourselves from the beginning. Theory serves to throw a clear light on the mass of objects, that the mind may the more easily find its bearings; theory serves to pull up the weeds which error has sown everywhere; it is to show the relations of things to each other and separate the important from the trifling. Where ideas resolve themselves spontaneously into such a core of truth as is called principle, when they of themselves keep such a line as forms a rule, theory shall indicate this.

What the mind brings away with it from this wandering among the fundamental ideas of things, the rays of light that are quickened in it, *that is the assistance which theory affords it.* Theory can give no formulas with which to solve problems; it cannot confine the mind's course to the narrow line of necessity by principles set up on both sides. It gives the mind a glance into the mass of objects and their relations, and then dismisses it again into the higher regions of action, there to act according to the measure of its natural gifts, with the combined energy of the whole of those forces, and to grasp *the true* and *the right*, as one single clear idea, which, driven forth under the united pressure of all these forces, would seem to be rather a product of feeling than of thought.

CHAPTER 2

ABSOLUTE AND REAL WAR

The plan of war comprehends the whole military operation; through it the operation becomes a single act, which must have one final definitive object, in which all particular objects have been merged. No war is begun, or at least, no war should be begun, if people acted wisely, without first finding an answer to the question: what is to be attained by and in war? The first is the final object; the other is the intermediate aim. By this dominant idea the whole course of the war is prescribed, the extent of the means and the measure of energy are determined; its influence manifests itself down to the smallest details of action.

We said in the first chapter that the overthrow of the enemy is the natural aim of the act of war, and that if we would keep within the strictly philosophical limits of the conception, there can fundamentally be no other.

As this idea must apply to both the belligerent parties, it would follow that there can be no suspension in the military act, and a suspension cannot take place until one or the other of the parties concerned is actually overthrown.

In the chapter on the suspension of action in warfare, we have shown how the abstract principle of hostility, applied to its agent, i.e., man, and to all circumstances out of which war is made up, is subject to delays and limitations from causes which are inherent in the apparatus of war.

But this modification is not nearly sufficient to carry us from the original conception of war to the concrete form in which it almost everywhere appears. Most wars appear only as mutual anger, under the influence of which each side takes up arms to protect itself and put fear into its adversary, and—occasionally—to strike a blow. They are, therefore, not like two mutually destructive elements brought into collision, but like tensions of two elements still apart which discharge themselves in small separate shocks.

But what is now the non-conducting medium which hinders the complete discharge? Why is the philosophical conception not fulfilled? That medium consists in the great number of interests, forces and circumstances in the existence of the state which are affected by the war. Through their infinite windings the logical conclusion cannot be traced out as it would be on the simple thread of one or two inferences. In these windings it is caught fast, and man, who in great things as well as in small, usually acts more on particular prevailing ideas and emotions than according to strictly logical conclusions, is hardly conscious of his confusion, onesidedness and inconsistency.

But even if the intelligence from which war originates could have gone through all these circumstances, without for a moment losing sight of its aim, still all other intelligences in the state which are concerned would not be able to do the same. Thus an opposition would arise, and consequently a force would become necessary, capable of overcoming the inertia of the whole mass—a force which will mostly be inadequate to the task.

This inconsistency is found on one or the other of the two sides, or it may be on both sides, and becomes the cause of the war being something quite different from what it should be, according to the conception of it—a half-hearted affair, a thing without inner cohesion.

This is how we find it almost everywhere, and we might doubt whether our notion of its absolute nature had any reality, if we had not seen real warfare make its appearance in this absolute completeness right in our own times. After a short introduction performed by the French Revolution, the ruthless Bonaparte quickly brought it to this point. Under him war was carried on without slackening for a moment until the enemy was laid low, and the counterstrokes followed almost with as little remission. Is it not natural and necessary that this phenomena should lead us back to the original conception of war with all its rigorous deductions?

Shall we now rest satisfied with this and judge all wars accordingly, however much they may differ from it—and deduce therefrom all the requirements of theory?

We must decide upon this point, for we can say nothing intelligent on the plan of war until we have made up our minds whether war is to be only of this kind, or whether it may be of yet another kind.

If we give an affirmative answer to the first question, then our theory will, in all respects, come nearer to logical necessity; it will be a clearer and more settled thing. But what are we to say then of all wars from Alexander and certain campaigns of the Romans down to Bonaparte? We would have to reject them in a lump, and yet we could not, perhaps, do so without being ashamed of our presumption. But the worst of it is that we must say to ourselves that in the next ten years there may perhaps be a war of that same kind again, in spite of our theory, and that this theory, with its rigorous logic, is still quite powerless against the force of circumstances. We must, therefore, be prepared to construe war as it is to be, not from pure conception, but by allowing room for everything of a foreign nature which is involved in it and attaches itself to it—all the natural inertia and friction of its parts, the whole of the inconsistency, the vagueness and timidity of the human mind; we shall have to admit that war, and the form which we give it, proceeds from ideas, emotions and circumstances prevailing for the moment; indeed, if we would be perfectly candid we must admit that this has even been the case where it has taken its absolute character, that is, under Bonaparte.

If we must do so, if we must grant that war originates and takes its form not from a final adjustment of all the innumerable relations which it affects, but from some among them which happen to predominate, then it follows, as a matter of course, that it rests upon a play of possibilities, probabilities, good fortune and bad, in which rigorous logical deduction often gets altogether lost, and in which it is in general an unhelpful, inconvenient instrument for the brain to work with. Then it also follows that war may be a thing which is sometimes war in a greater, sometimes in a lesser, degree.

All this, theory must admit, but it is its duty to give the foremost place to the absolute form of war, and to use that form as a general point of direction, that he who wishes to learn something from theory may accustom himself never to lose sight of it, to regard it as the fundamental standard of all his hopes and fears, in order to approach it *where he can or where he must.*

That a leading idea, which lies at the root of our thoughts and actions, gives them a certain tone and character, even when the immediate reasons for a decision come from totally different regions, is just as certain as that the painter can give this or that tone to his picture by the colors which he uses for his ground.

Theory is indebted to the last wars for being able to do this effectually now. Without these warning examples of the destructive force of the unrestrained element, it would have talked itself hoarse to no purpose; no one would have believed possible what all have now lived to see realized.

Would Prussia have ventured to invade France in the year 1798 with 70,000 men, if she had foreseen that the reaction in case of failure would be so strong as to overthrow the old balance of power in Europe?

Would Prussia in 1806 have made war upon France with 100,000 men, if she had considered that the first pistol shot would be a spark to fire the mine which was to blow her into the air?

CHAPTER 3

A. INTERDEPENDENCE OF
THE PARTS IN WAR

According as we have in view the absolute form of war, or one of the real forms deviating more or less from it, so likewise two different notions of its result will arise.

In the absolute form, where everything is the effect of necessary causes, one thing swiftly affects another; there is, if we may use the expression, no neutral space; there is—on account of the manifold reciprocal effects which war contains in itself,[1] on account of the connection in which, strictly speaking, the whole series of engagements[2] follow one after another, on account of the culminating point which every victory has, beyond which the period of losses and defeats begins[3]—on account of all these natural circumstances of war there is, I say, only one result, namely, the *final result*. Until it takes place nothing is decided, nothing won, nothing lost. Here we must constantly say: the end crowns the work. In this conception, therefore, war is an indivisible whole, the parts of which (the individual results) have no value except in their relation to this whole. The conquest of Moscow, and of half Russia in 1812, was of no value to Bonaparte unless it procured for him the peace which he had in view. But it was only a part of his plan of campaign; to complete that plan, one part was still lacking, the destruction of the Russian army. If we suppose this added to the other success, then the peace was as certain as is possible with

[1] Book I, Chapter 1.
[2] Book I, Chapter 2.
[3] Book VII, Chapter 5 ("Culminating Point of the Attack").

things of this kind. This second part Bonaparte could no longer attain because he had failed to do so earlier, and so the whole of the first part was not only useless, but fatal to him.

To this view of the connection of results in war, which may be regarded as extreme, stands opposed another extreme, according to which war is composed of single independent results, in which, as with different rounds in a game, the preceding result has no influence on those following; here, therefore, everything depends only on the sum total of the results, and we can set aside each separate result like a counter in a game.

Just as the first kind of conception derives its truth from the nature of the thing, so we find that of the second in history. There are cases without number in which it has been possible to gain a small moderate advantage without any very onerous condition being attached to it. The more the element of war is modified, the more common these cases become; but as little as the first of the views was ever completely realized in any war, just as little is there any war in which the last is true in all respects, and the first can be dispensed with.

If we keep to the first of these two views, we must perceive the necessity of every war being looked upon as a whole from the very outset, and that at the very first step forward, the commander should have the end in view to which every line must converge.

If we admit the second view, subordinate advantages may be pursued for their own sake, and the rest left to subsequent events.

As neither of these views is entirely without result, theory cannot, therefore, dispense with either. But it makes this difference in the use of them: it requires the first to be laid as a fundamental idea at the root of everything, and the latter is only to be used as a modification justified by circumstances.

When Frederick the Great in the years 1742, 1744, 1757 and 1758 thrust out from Silesia and Saxony a fresh offensive point into the Austrian Empire, which he knew very well could not lead to a new and permanent conquest like that of Silesia and Saxony, he had in view not the overthrow of the Austrian Empire, but a subordinate object, namely, to gain time and strength, and he could pursue that subordinate object without being afraid that he should risk his whole existence.[1] But if Prussia in 1806, and Austria

---

[1] Had Frederick the Great gained the battle of Kollin, and consequently captured the chief Austrian army with its two field marshals in Prague, it would have been such a tremendous blow that he might then have entertained the idea of marching to Vienna to shake the Austrian monarchy, and gain a peace directly. This, in those times, unparalleled success,

in 1805 and 1809, proposed to themselves a still more moderate object, that of driving the French over the Rhine, they would not have acted in a reasonable manner if they had not first carefully reviewed the whole series of events which, either in the case of success or of the reverse, would probably follow the first step and lead up to peace. This was quite indispensable, in order that they might determine not only how far on their side they could pursue victory without danger, but also how and where they would be able to check the course of victory on the side of the enemy.

An attentive consideration of history shows wherein the difference of the two cases consists. At the time of the Silesian Wars in the eighteenth century, war was still a mere cabinet affair, in which the people participated only as a blind instrument; at the beginning of the nineteenth century the peoples on each side weighed in the scale. The commanders opposed to Frederick the Great were men who acted on commission, and just on that account men in whom caution was a predominant characteristic; the opponent of the Austrians and Prussians was, to put it bluntly, the God of War himself.

Did not these different circumstances necessarily give rise to quite different views? Did they not necessarily in the years 1805, 1806 and 1809 point to the uttermost disaster as a very close possibility, nay, even a great probability, and consequently should they not have led to widely different plans and measures from those that merely aimed at the conquest of a couple of fortresses or a paltry province?

They did not do so to the extent necessary, although both Austria and Prussia, at the time of their armament, felt that storms were brewing in the political atmosphere. It was impossible, because at that time those circumstances had not yet been so clearly exposed by history. It is just those very campaigns of 1805, 1806, 1809, and the following ones, which have made it easier for us to form a conception of modern absolute war in its smashing energy.

Theory demands, therefore, that at the beginning of every war its character and main outline shall be defined according to what the political

---

which would have been quite like what we have seen in our day, only still more wonderful and brilliant—being a contest between a little David and a great Goliath—might very probably have taken place after the gain of this one battle; but that does not contradict the assertion made above, for it only refers to what the king originally intended with his offensive. The surrounding and capture of the enemy's army was an event which was beyond all calculation, and which the king never thought of, at least not until the Austrians laid themselves open to it by the awkward position in which they placed themselves at Prague.

conditions and relations lead us to anticipate as probable. The more nearly, according to this probability, its character approaches the form of absolute war, the more its outlines embrace the mass of the belligerent states and draw them into the vortex—so much the more closely its events will be connected and so much the more necessary it will also be not to take the first step without thinking what may be the last.

B. OF THE MAGNITUDE OF THE MILITARY OBJECT AND THE EFFORTS TO BE MADE

The compulsion which we must use toward our enemy will be regulated by the magnitude of our own and his political demands. In so far as these are mutually known they would give the measure of the efforts on each side; but they are not always quite so evident, and this may be a first reason for a difference in the means used by each.

The situation and conditions of the states are not like each other; this may become a second cause.

The strength of will, the character and capacities of the governments are just as little alike; this is a third cause.

These three elements cause an uncertainty in the calculation of the resistance to be expected, consequently an uncertainty as to the means to be employed and as to the aim we may set before ourselves.

As in war insufficient efforts may result not only in lack of success, but in positive loss; therefore, the two sides respectively seek to outdo each other, which produces a reciprocal action.

This might lead to the utmost aim of effort, if it were possible to define such a point. But then regard for the magnitude of the political demands would be lost, the means would lose all relation to the end, and in most cases this intention for an extreme effort would be wrecked by the opposing weight of circumstances inherent in itself.

In this manner, he who undertakes war is brought back again into a middle course, in which he acts to a certain extent upon the principle of employing only such forces, and having only such a war aim in mind, as are just sufficient for the attainment of his political object. To make this principle practicable he must renounce every absolute necessity of a result, and exclude remote contingencies from the calculation.

Here the activity of the intellect leaves the province of strict science, of logic and mathematics, and becomes, in the wider sense of the term, *art,* that is, the skill to pick out, by instinctive judgment, from an infinite multitude of objects and circumstances the most important and decisive. This instinctive judgment consists unquestionably more or less in some intuitive comparison of things and relations by which the remote and unimportant are more quickly eliminated, and the more immediate and important are sooner discovered than they could be by strictly logical deduction.

In order to ascertain what amount of means we have to call up for the war, we must consider the political object both on our own side and on that of the enemy; we must consider the power and conditions of the enemy's state, as well as of our own; we must consider the character of the enemy's government and of his people, and the capacities of both. These same factors must be considered on our own side; we must take into account the political connections of other states and the effect which the war will produce on those states. That the determination of these diverse circumstances and their diverse connections is an immense problem, that it is a true flash of genius which, confronted with them, quickly picks out the right course, while it would be quite impossible to become master of their complexity by mere academic study, is easily understood.

In this sense Bonaparte was quite right when he said that it would be a problem in algebra before which even a Newton might stand aghast.

If the diversity and magnitude of the circumstances and the uncertainty as to the right measure increase in a high degree the difficulty of obtaining a favorable result, we must not overlook the fact that although the incomparable *importance* of the matter does not increase the complexity and difficulty of the problem, it does nevertheless increase the merit of its solution. In ordinary men freedom and activity of mind are reduced, not increased, by the sense of danger and responsibility; but where these things give wings to strengthen the judgment, there undoubtedly must be unusual greatness of mind.

First of all, therefore, we must admit that the judgment on an approaching war, on the aim which it may have, and on the means which are required, can only be formed after a general survey of all the circumstances, among which the most characteristic features of the moment must thus be included; next, that this decision, like all in military life, can never be purely objective, but must be determined by the mental and moral qualities of princes, statesmen and generals, whether they are united in the person of one man or not.

The subject becomes general and better suited to an abstract treatment if we look at the general relations imposed upon states by their time and circumstances. We must here allow ourselves a passing glance at history.

Half-civilized Tartars, republics of ancient times, feudal lords and commercial cities of the Middle Ages, kings of the eighteenth century, and, lastly, princes and people of the nineteenth century, all carry on war in their own ways, carry it on differently, with different means and with different aims.

The Tartar hordes sought new abodes. They marched out as a whole nation with their wives and children; they were, therefore, greater than any other army in point of numbers, and their aim was to make the enemy submit or to expel him altogether. By these means they would have soon overthrown everything before them, if a high degree of civilization could have been made compatible with such a condition.

The old republics, with the exception of Rome, were of small extent; still smaller were their armies, for they excluded the great mass, the populace. There were too many of them and they lay too close together not to find an obstacle to great enterprises in the natural equilibrium into which small separate parts always settle according to a quite general law of nature; therefore their wars were confined to devastating the open country and taking single cities, in order to secure for themselves by means of these things a certain degree of influence for the future.

Rome alone is an exception to this, but not until the later period of her history. For a long time, by means of small bands, she carried on the usual warfare with her neighbors for booty and alliances. She became great more through the alliances which she formed, and through which neighboring peoples by degrees amalgamated with her into one whole, than through actual conquests. It was only after having spread in this manner all over Southern Italy that she began to advance as a really conquering power. Carthage fell, Spain and Gaul were conquered, Greece subdued, and Rome's dominion extended to Egypt and Asia. At this period her military forces were immense, without her efforts being equally so. These forces were kept up by her riches; she no longer resembles the ancient republics, nor her own former self. She stands alone.

Just as distinctive in their way were the wars of Alexander. With a small army, but one distinguished for its perfect organization, he overthrew the rotten structures of the Asiatic states. Restlessly and ruthlessly, he penetrated the wide Asiatic continent and pushed on as far as India. No republic could have done this. Only a king, who, so to speak, was his own *condottiere,* could have done it quickly.

The great and small monarchies of the Middle Ages carried on their wars with feudal levies. Everything was restricted to a short period of time; whatever could not be done in that time had to be regarded as impracticable. The feudal force itself consisted of an organization of vassaldom; the bond which held it together was partly legal obligation, partly voluntary alliance; the whole formed a real confederation. The armament and tactics were based on the right of might, on single combat, and therefore little suited to large bodies. In fact, at no period has state union been so lax and the individual citizen so independent. All this influenced the character of the wars at that period in the most distinct manner. They were carried out with comparative haste; there was little time spent idly in the field, but the object was generally only punishing, not subduing, the enemy. They carried off his cattle, burned his castles and then went home again.

The great commercial towns and small republics introduced the *condottieri.* That was an expensive, and therefore, in point of numbers a very limited military force; in point of intensive strength, it was of still less value; of extreme energy and effort in the field it showed so little that its combats became for the most part only sham fights. In a word, hatred and enmity no longer roused a state to personal activity, but had become an article of its trade. War had lost a great part of its danger, altered completely its nature, and nothing that can be deduced from this nature was applicable to it.

The feudal system condensed itself by degrees into a definite territorial sovereignty; the ties binding the state together became closer; personal obligations were transformed into material ones, money gradually became the substitute in most cases, and the feudal levies were turned into armies of mercenaries. The *condottieri* formed the connecting link in the change, and were, therefore, for a time, the instrument of the more powerful states; but this had not lasted long when the soldier, hired for a limited term, was turned into a *standing mercenary,* and the military force of states now became the standing army, supported by the public treasury.

Naturally the slow advance to this stage brought about many different combinations of all the three kinds of military force. Under Henry IV we find the feudal contingents, *condottieri* and standing army all employed together. The *condottieri* prolonged their existence up to the period of the Thirty Years' War; indeed there are some slight traces of them even in the eighteenth century.

The other relations of the states of Europe at these different periods were quite as peculiar as their military forces. Fundamentally, this continent had split up into a mass of petty states, partly republics in a state of internal dissension, partly small monarchies in which the power of the government was very limited and insecure. Such a state could not be considered a real unity; it was rather an agglomeration of loosely connected forces. Neither, therefore, could such a state be considered an intelligence, acting in accordance with simple logical rules.

It is from this point of view that we must look at the foreign politics and wars of the Middle Ages. Let us only think of the continual expeditions of the German emperors to Italy for five centuries, without any substantial conquest of that country resulting from them, or even having been intended. It is easy to look upon this as a blunder repeated over and over again—as a false view which had its root in the nature of the times, but it is more logical to regard it as the consequence of a hundred important causes which we can at any rate partially understand, but which it is impossible for us to realize as vividly as those people could who were brought into actual conflict with them. As long as the great states which have risen out of this chaos required time to consolidate and organize themselves, their whole power and energy are directed solely *to that point;* their foreign wars are few, and those that took place bear the stamp of an immature political union.

The wars between France and England are the first that appear, and yet at that time France is not to be considered as really a monarchy, but as an agglomeration of dukedoms and countships; England, although bearing more the semblance of a unity, still fought with the feudal organization, and was hampered by many domestic troubles.

Under Louis XI, France made its greatest step toward internal unity; under Charles VIII it appeared in Italy as a power bent on conquest; and under Louis XIV it had brought its political state and its standing army to the highest perfection.

Spain was united under Ferdinand the Catholic; through accidental marriage connections, the great Spanish monarchy under Charles V suddenly arose, composed of Spain, Burgundy, Germany and Italy united. What this colossus lacked in the way of unity and internal political cohesion, it made up for in money, and its standing army came first into collision with the standing army of France. The great Spanish colossus on the abdication of Charles V fell into two pieces, Spain and Austria. The latter, strengthened by the acquisition of Bohemia and Hungary, now appears on the scene as a great power, towing the German Confederation like a tender behind her.

The end of the seventeenth century, the time of Louis XIV, is to be regarded as the point in history at which the standing military power, such as it existed in the eighteenth century, reached its height. That military power was based on recruiting and money. States had organized themselves into complete unities; and the governments, by commuting the personal obligations of their subjects into taxes, had concentrated their whole power in their treasuries. Through the rapid progress in social improvements, and a constantly developing system of government, this power had become very great in comparison to what it had been. France appeared in the field with a standing army of 200,000 men or more, and the other powers in proportion.

The other relations of states had likewise altered. Europe was divided among a dozen kingdoms and a few republics; it was now conceivable that two of these powers might fight with each other without ten times as many others being involved, as would certainly have been the case formerly. The possible combinations in political relations were still extremely various, but they could be surveyed and determined from time to time according to probability.

Internal relations had almost everywhere been simplified into a plain monarchical form; the rights and influence of privileged estates had gradually died out, and the cabinet had become a complete unity, representing the state in all its external relations. The time had therefore come when a suitable instrument and an independent will could give war a form in accordance with its theoretical conception.

And at this epoch appeared three new Alexanders—Gustavus Adolphus, Charles XII and Frederick the Great—whose aim was, by means of small but highly perfected armies, to raise little states to the rank of great monarchies, and to throw down everything in their way. Had they only had to deal with Asiatic states they would have more closely resembled Alexander in the parts they acted. In any case, we may look upon them as the precursors of Bonaparte in respect to what may be risked in war.

But what war gained on the one side in force and consistency was lost again on the other side.

Armies were supported out of the treasury, which the sovereign regarded partly as his private purse, or at least as a resource belonging to the government, and not to the people. Relations with other states, except for a few commercial matters, mostly concerned only the interests of the treasury or of the government, not those of the people; at least ideas tended everywhere in that direction. The cabinets, therefore, looked upon themselves as the owners and administrators of large estates, which they were continually seeking to increase without the tenants on these estates being particularly interested in this improvement. The people, who in the Tartar invasions were everything in war, who in the old republics and in the Middle Ages (if we properly restrict the idea to those actually possessing the rights of citizens) were of great consequence, in the eighteenth century were absolutely nothing directly, only retaining an indirect influence on the war, through their general virtues and faults.

In this manner, in proportion as the government separated itself from the people and regarded itself as the state, war became exclusively a business of the government, which it carried on by means of the money in its coffers and the idle vagabonds it could pick up in its own and neighboring countries. The consequence of this was that the means which the government could command had fairly well-defined limits which could be mutually estimated, both as to their extent and duration; this robbed war of its most dangerous feature, namely, the effort toward the extreme and the obscure series of possibilities connected therewith.

The financial means, the contents of the treasury and the state of credit of the enemy were approximately known as well as the size of his army. Any large increase at the outbreak of a war was not feasible. Inasmuch as the limits of the enemy's powers were thus recognized, a state felt fairly secure against complete subjugation, and as the state was conscious at the same time of the limits of its own means, it saw itself restricted to a moderate aim. Protected from an extreme, there was no necessity to venture on an extreme. Necessity no longer giving an impulse in that direction, that impulse could only now be given by courage and ambition. But these found a powerful counterpoise in the circumstances of the state. Even kings in command were obliged to use the instrument of war with caution. If the army was disbanded, no new one could be obtained, and apart from the army there was nothing. This imposed as a necessity great prudence in all undertakings. It was only when a decided advantage seemed to present itself that they made use of the costly instrument; to bring about such an opportunity was the general's master stroke; but until it was brought about, everything floated, so to speak, in an absolute vacuum; there was no reason for action, and all forces, that is, all motives, seemed to rest. The original motive of the aggressor faded away in prudence and circumspection.

Thus war became essentially a regular game in which time and chance shuffled the cards; but in its significance, it was only diplomacy somewhat intensified, a more forceful way of negotiating, in which battles and sieges were the diplomatic notes. To obtain some moderate advantage in order to make use of it in negotiations for peace was the aim even of the most ambitious.

This restricted, shriveled-up form of war proceeded, as we have said, from the narrow basis on which it rested. But that distinguished generals and kings, like Gustavus Adolphus, Charles XII and Frederick the Great, at the head of armies no less distinguished, could not emerge more prominently from the mass of things in general—that even these men were obliged to be content to remain at the general level of moderate achievement is to be attributed to the balance of power in Europe. Now that states had become greater, and their centers farther apart from each other, that which had formerly in the multitude of small states been done through direct and perfectly natural interests—proximity, contact, family connections, personal friendship—to prevent any one single state from becoming suddenly great was now effected by a higher cultivation of the art of diplomacy. Political interests, attractions and repulsions had developed into a very refined system, so that no cannon shot could be fired in Europe without all the cabinets having an interest in it.

A new Alexander, therefore, in addition to his good sword, had to wield also a good pen, and yet he never went very far with his conquests.

But although Louis XIV intended to overthrow the balance of power in Europe, and at the end of the seventeenth century had already got to such a point as to trouble himself little about the general feeling of animosity, he carried on war in the traditional manner, for while his army was certainly that of the greatest and richest monarch in Europe, it was just like the others in its nature.

Plundering and devasting the enemy's country, which play such an important part with Tartars, with ancient nations and even in the Middle Ages, were no longer in accordance with the spirit of the age. They were justly looked upon as unnecessary barbarity, which might easily induce reprisals, and which did more injury to the enemy's subjects than the enemy's government, producing, therefore, no effect and only serving indefinitely to retard the progress of national civilization. War, therefore, confined itself more and more, both as regards means and end, to the army itself. The army, with its fortresses and some prepared positions, constituted a state in a state, within which the element of war slowly consumed itself. All Europe rejoiced at its taking this direction, and held it to be the necessary consequence of the spirit of progress. Although there was an error in this, inasmuch as the progress of the human mind can never lead to what is absurd, can never make five out of twice two, as we have already said and shall have to repeat again, still on the whole this change had an effect beneficial to the people; only it is not to be denied that it had a tendency to make war still more an affair of the government, and to separate it still more from the interests of the people. The plan of war on the part of the state assuming the offensive in those times consisted generally in the conquest of one or another of the enemy's provinces; the plan of the defender was to prevent this. The particular plan of campaign was to take one or another of the enemy's fortresses, or to prevent one of our own from being taken. It was only when a battle became unavoidable for this purpose that it was sought for and fought. Whoever fought a battle without this unavoidable necessity, from mere innate desire of gaining a victory, was reckoned a daring general. Generally the campaign was over with one siege, or, if it was a very active one, with two sieges, and winter quarters, which were regarded as a necessity. During these the bad organization of the one party could never be taken advantage of by the other, and the mutual contacts of the two parties almost entirely ceased. They formed a distinct limit to the activity which was to take place in a campaign.

If the forces opposed were too much on a par, or if the aggressor was decidedly the weaker of the two, then neither battle nor siege took place, and the whole of the operations of the campaign pivoted on the maintenance of certain positions and magazines, and the regular devastation of particular districts of country.

As long as war was universally conducted in this manner, and the natural limits of its force were so close and obvious, no one found anything contradictory in it. All was considered to be in the finest order; and criticism, which in the eighteenth century began to turn its attention to the art of war, directed itself to details without troubling itself much about beginning or end. Thus there was eminence and perfection of every kind, and even Field-Marshal Daun—who is chiefly responsible for Frederick the Great having completely attained his object, and for Maria Theresa having completely failed in hers—could still pass for a great general. Only now and again did a penetrating judgment make its appearance, that is, did sound common sense recognize that with superior numbers something positive must be attained or else war was being badly conducted, whatever art might be displayed.

Thus matters stood when the French Revolution broke out; Austria and Prussia tried their diplomatic art of war; this soon proved insufficient. While, according to the usual way of seeing things, all hopes were placed on a very limited military force, in 1793 such a force as no one had had any conception of made its appearance. War had again suddenly become an affair of the people, and that of a people numbering thirty millions, every one of whom regarded himself as a citizen of the state. Without entering here into the details of circumstances, by which this great phenomenon was attended, we shall confine ourselves to the results which interest us at present. By this participation of the people in war, instead of a cabinet and an army, a whole nation with its natural weight entered the scale. Henceforward, the means available—the efforts which might be called forth—had no longer any definite limits; the energy with which the war itself could be conducted had no longer any counterpoise, and consequently the danger for the adversary had risen to the extreme.

If the whole war of the Revolution ran its course without all this making itself felt in its full force and becoming quite evident; if the generals of the Revolution did not advance irresistibly up to the final aim and lay in ruins the monarchies of Europe; if the German armies now and again had the opportunity of resisting with success and checking the torrent of victory—the cause really lay in that technical imperfection with which the French had to contend, which showed itself first among the common soldiers, then in the generals, lastly, at the time of the Directory, in the government itself.

After everything had been perfected by the hand of Bonaparte, this military power, based on the strength of the whole nation, marched shattering over Europe with such confidence and certainty that wherever it only encountered the old-fashioned armies the result was never even for a moment doubtful. A reaction, however, awoke in due time. In Spain, the war became of itself an affair of the people. In Austria, in the year 1809, the government made extraordinary efforts, by means of reserves and *Landwehr,* which came nearer to the end in view, and surpassed anything that this state had hitherto conceived possible. In Russia, in 1812, the example of Spain and Austria was taken as a model. The enormous dimensions of that empire, on the one hand, allowed the preparations, although too long deferred, still to produce an effect; and, on the other hand, intensified the effect produced. The result was brilliant. In Germany, it was Prussia who pulled herself together first, made the war a national cause, and without either money or credit, and with a population reduced by one-half, took the field with an army twice as strong as that of 1806. The rest of Germany sooner or later followed the example of Prussia, and Austria, although less energetic than in 1809, still came forward with unusual strength. Thus it was that Germany and Russia, in the years 1813 and 1814, including all who took an active part or were killed in these two campaigns, appeared against France with about a million men.

Under these circumstances, the energy also thrown into the conduct of the war was quite different; and, although not quite on a par with that of the French, although at some points timidity still prevailed, the course of the campaigns, upon the whole, may be said to have been in the new, not in the old, style. In eight months the theater of war was removed from the Oder to the Seine. Proud Paris had to bow its head for the first time; and the redoubtable Bonaparte lay fettered on the ground.

Since the time of Bonaparte, war, through being first on one side, then on the other, again an affair of the whole nation, has assumed quite a new nature, or rather it has approached much nearer to its real nature, to its absolute perfection. The means then called forth had no visible limit, the limit lost itself in the energy and enthusiasm of the governments and their subjects. By the extent of the means and the wide field of possible results, as well as by the powerful excitement of feeling, energy in the conduct of war was immensely increased; the object of its action was the overthrow of the foe; and not until the enemy lay powerless on the ground was it supposed to be possible to stop and to come to any understanding with respect to the mutual objects of the contest.

Thus, the primitive violence of war, freed from all conventional restrictions, broke loose with all its natural force. The cause was the participation of the people in this great affair of state, and this participation arose partly from the effects of the French Revolution on the internal affairs of countries, partly from the threatening attitude of the French toward all nations.

Now, whether this will always be the case, whether all future wars in Europe will be carried on with the whole power of the states, and, consequently, take place only on account of great interests closely affecting the people, or whether a separation of the government from the people will gradually arise again, would be a difficult point to settle; least of all shall we take it upon ourselves to settle it. But everyone will agree with us, that bounds, which only existed in the non-consciousness, so to speak, of what is possible, when once thrown down, are not easily built up again; and that, at least, whenever great interests are in question, mutual hostility will discharge itself in the same manner as it has done in our time.

We here bring our historical survey to a close, for it was not our design hastily to assign to every age some principles of the conduct of war, but only to show how each age has had its own peculiar forms of war, its own restrictive conditions and its own prejudices. Each, therefore, would also keep its own theory of war, even if everywhere, in early times as well as in later, there had been an inclination to work it out on philosophical principles. The events in each age must, therefore, be judged with due regard to the peculiarities of the time, and only he who, less by an anxious study of minute details than by a shrewd glance at the main features, can place himself in each particular age is able to understand and appreciate its generals.

But this conduct of war, conditioned by the peculiar relations of states and of military power, must nevertheless always contain in itself something more general, or rather something quite general, with which theory is above all concerned.

The period just elapsed, in which war reached its absolute strength, contains most of what is universally valid and necessary. But it is just as improbable that wars henceforth will all have this grand character as that the wide barriers which have been opened to them will ever be completely closed again. Therefore, by a theory which only dwells upon this absolute war, all cases in which external influences alter the nature of war would be excluded or condemned as errors. This cannot be the object of theory, which ought to be the science of war, not under ideal, but under real, circumstances. Theory, therefore, while casting a searching, discriminating and classifying glance at objects, should always have in view the diversity of causes from which war may proceed, and will, therefore, so trace out the great features of war as to leave room for the needs of the age and the moment.

Accordingly, we must say that the object which everyone who undertakes war proposes to himself and the means which he calls forth are determined entirely according to the particular details of his position. On that very account they will also partake of the character of the age and of its *general* circumstances. Lastly, *they are always subject to the general conclusions which must be deduced from the nature of war.*

CHAPTER 4

AIM OF WAR MORE
PRECISELY DEFINED

OVERTHROW OF THE ENEMY

The aim of war according to its conception is always supposed to be the overthrow of the enemy; this is the fundamental idea from which we set out.

Now, what is this overthrow? It does not always necessarily imply the complete conquest of the enemy's country. If the Germans had reached Paris in 1792, there—in all human probability—the war with the Revolutionary party would have been brought to an end temporarily. It was not even necessary to beat their armies beforehand, for those armies were not yet to be looked upon as the only effective power. On the other hand, in 1814, the Allies would not have gained everything by taking Paris if Bonaparte had still remained at the head of a considerable army; but since his army had been for the most part annihilated, the capture of Paris decided everything both in the years 1814 and 1815. If Bonaparte in the year 1812, either before or after taking Moscow, had been able to destroy the Russian army of 120,000 on the Kaluga road completely, as he did the Austrian in 1805, and the Prussian army in 1806, the possession of that capital would most probably have brought about a peace, although an enormous tract of land still remained to be conquered. In the year 1805 it was the battle of Austerlitz that was decisive; and, therefore, the previous possession of Vienna and two-thirds of the Austrian states was not sufficient to gain a peace. On the other hand, however, even after that battle, the fact that all

Hungary was still untouched was not of sufficient weight to prevent the conclusion of peace. The defeat of the Russian army was the last blow required; the Emperor Alexander had no other army near at hand, and, therefore, peace was the indubitable consequence of victory. If the Russian army had been on the Danube along with the Austrian in 1805, and had shared in its defeat, then probably the conquest of Vienna would not have been necessary at all and peace would have been concluded in Linz.

In other cases the complete conquest of a country does not suffice, as in the year 1807, in Prussia, when the blow against the Russian auxiliary army, in the doubtful battle of Eylau, had not been decisive enough, and the undoubted victory of Friedland had to turn the scale, as the victory at Austerlitz had done a year before.

We see that here, too, the result cannot be determined from general causes; the individual causes, which no one knows who is not on the spot, and many of a moral nature which are never heard of, even the smallest features and accidents, which only appear in history as anecdotes, are often decisive. All that theory can say here is that the main point is to keep the predominant conditions of both parties in view. Out of them a certain center of gravity, a center of power and movement, will form itself, upon which everything depends; and against this center of gravity of the enemy the concentrated blow of all the forces must be directed.

The little always depends on the great, the unimportant on the important, and the accidental on the essential. This must guide our view.

Alexander had his center of gravity in his army, so had Gustavus Adolphus, Charles XII and Frederick the Great, and the career of any one of them would soon have been brought to an end by the destruction of his fighting force. In states torn by internal dissensions, this center generally lies in the capital; in small states dependent on greater ones, it lies generally in the army of these allies; in a confederacy, it lies in the unity of interests; in a national insurrection, in the person of the chief leader and in public opinion; against these points the blow must be directed. If the enemy hereby loses his balance, no time must be allowed for him to recover it. The blow must be persistently repeated in the same direction, or, in other words, the conqueror must always direct his blows upon the whole, but not against a part of the enemy. It is not by conquering one of the enemy's provinces, with ease and with superior numbers, and preferring the more secure possession of this unimportant conquest to great results, but by seeking out constantly the nucleus of hostile power, and staking the whole thing in order to gain the whole, that we can actually strike the enemy to the ground.

But whatever may be the central point of the enemy's power against which we are to direct our operations, still the conquest and destruction of his arm is the surest beginning, and in all cases the most essential.

Hence we think that, according to the majority of experiences, the following circumstances chiefly bring about the overthrow of the enemy:

1. Dispersion of his army if it forms, in some degree, an effective power.
2. Capture of the enemy's capital, if it is not merely the center of the state powers, but also the seat of political bodies and parties.
3. An effective blow against the principal ally, if he is in himself more powerful than the enemy.

Hitherto, we have always thought of the enemy in war as a unity, which was permissible for considerations of a very general nature. But having said that the subjugation of the enemy lies in the overcoming of his resistance, concentrated in the center of gravity, we must lay aside this supposition and discuss the case in which we have to deal with more than one opponent.

If two or more states combine against a third, this constitutes, politically speaking, only *one* war. However, this political union also has its degrees.

The question is whether each state in the coalition possesses an independent interest in, and an independent force with which to prosecute, the war, or whether there is one among them on whose interests and forces the others lean for support. The more the latter is the case, the easier it is to regard the different enemies as one alone, and the more readily we can simplify our principal enterprise to one great blow; and as long as this is in any way possible, it is the most thorough and complete means of success.

We would, therefore, establish it as a principle that if we can defeat all our enemies by defeating one of them, the defeat of that one must be the aim of the war, because in that one our blow strikes the common center of gravity of the whole war.

There are very few cases in which this kind of conception is not permissible, and where this reduction of several centers of gravity to one cannot be made. But if this cannot be done, then indeed there is no alternative but to look upon the war as two or more separate wars, each of which has its own aim. As this case presupposes the independence of several enemies, consequently the great superiority of all together, the overthrow of the enemy will be entirely out of the question.

We now turn more particularly to the question: When is such an aim possible and advisable?

In the first place, our military forces must be sufficient:

1. To gain a decisive victory over those of the enemy.
2. To make the expenditure of force which is necessary, when we follow up the victory to the point where the establishment of an equilibrium is no longer conceivable.

Next, we must feel sure that our political situation is such that this result will not excite against us new enemies, who may at once compel us to turn away from the first enemy.

France, in the year 1806, was able to conquer Prussia completely, although in doing so it brought down upon itself the whole military power of Russia, because it was in a condition to defend itself against the Russians in Prussia.

France was able to do the same in Spain in 1808 with respect to England, but not with respect to Austria. It had to weaken itself considerably in Spain in 1809, and would have been forced to give up the contest in that country altogether if it had not already had too great a superiority, both physically and morally, over Austria.

These three instances must therefore be carefully studied, so that we may not lose in the last the cause which we have won in the former ones, and then be condemned in costs.

In estimating the strength of forces, and that which may be effected by them, the idea very often suggests itself to look upon time, by analogy with dynamics, as a factor of the forces, and to assume accordingly that half the efforts, or half the number of forces, would accomplish in two years what could only be effected in one year by the whole force united. This view, which lies at the bottom of military plans, sometimes clearly, sometimes less plainly, is completely wrong.

A military operation, like everything else on earth, requires its time; as a matter of course we cannot walk from Vilna to Moscow in eight days; but there is no trace to be found in war of any reciprocal action between time and force, such as takes place in dynamics.

Time is necessary to both belligerents, and the only question is: Which of the two, judging by his position, has most reason to expect *special advantages* from time first? This, however, the peculiarities of the one case being weighed against those of the other is, obviously, the vanquished; of course not according to dynamic, but according to psychological, laws.

Envy, jealousy, anxiety and maybe even sometimes magnanimity are the natural mediators for the unfortunate. On the one hand they create friends for him, and on the other hand weaken and dissolve the coalition among his enemies. Therefore, by delay something advantageous is more likely to happen for the conquered than for the conqueror. Further, we must recollect that to make right use of a first victory, as we have already shown, a great expenditure of force is necessary. This expenditure is not merely to be made, but it must be maintained, like a big household; the forces which have been sufficient to give us possession of an enemy province are not always sufficient to meet this additional outlay; gradually the strain on our resources becomes greater, until at last they become insufficient. Thus time of itself may bring about a change.

Could the contributions which Bonaparte levied from the Russians and Poles, in money and in other ways, in 1812, have procured the hundreds of thousands of men that he would have had to send to Moscow in order to retain his position there?

But if the conquered provinces are sufficiently important, if there are in them points which are essential to the well-being of those parts which are not conquered, so that the evil, like a cancer, eats onward of itself, then it is possible that the conqueror, although nothing further is done, may gain more than he loses. Now in this case, if no help comes from without, time may complete the work thus begun; what still remains unconquered will, perhaps, fall of itself. Thus time may also become a factor of his forces, but this can only take place if a return blow from the conquered is no longer possible, a change of fortune in his favor no longer conceivable, when, therefore, this factor of his forces is no longer of any value to the conqueror, for he has accomplished the chief object, the danger of the crisis is past, in short, the enemy has already been overthrown.

Our object in the above reasoning has been to show clearly that no conquest can be finished too soon, that spreading it over a *greater space of time* than is absolutely necessary for its completion, instead of *facilitating* it, *makes it more difficult*. If this assertion is true, then it is also true that if we are at all strong enough to effect a certain conquest, we must also be strong enough to do it at one stretch without intermediate stations. Of course we do not mean by this the insignificant halts to concentrate the forces, and to make one or another arrangement.

By this view, which makes the character of a speedy and irresistible decision essential to offensive war, we think we have completely set aside all grounds for *that* theory which, in place of the unrestrained and continued following up of victory, would substitute a slow, and what is called methodical, system as being more sure and prudent. But even for those who have readily followed us thus far, our assertion has, after all, so much the appearance of a paradox—is at first sight so much opposed and offensive to an opinion which, as an old prejudice, has taken such deep root and has been repeated a thousand times in books—that we consider it advisable to examine more closely the apparent reasons which are against us.

Of course, it is easier to reach an object near us than one at a distance, but when the near one does not suit our purpose it does not follow that a pause, a resting-point, will enable us to get over the second half of the road more easily. A short jump is easier than a long one, but no one on that account, wishing to cross a wide ditch, would first jump into the middle of it.

If we look more closely at what underlies the conception of a so-called methodical offensive war, we shall find that it is generally the following things:

1. Conquest of those fortresses belonging to the enemy which we meet.
2. Accumulation of necessary supplies.
3. Fortifying important points, as *magazines, bridges, positions,* etc.
4. The resting of troops in winter and rest quarters.
5. Waiting for the reinforcements of the ensuing year.

If for the attainment of all these objects we make a formal halt in the course of the offensive action, a resting point in the movement, it is supposed that we gain a new base of operation and renewed strength, as if our own state were following up in the rear of the army, and as if the latter with every new campaign acquired renewed vigor.

All these praiseworthy motives may make the offensive war more comfortable, but they do not make its results more certain, and are mostly only pretenses to cover certain counteracting forces in the temperament of the commander and in the indecision of the cabinet. We shall try to roll them up from the left flank.

1. Waiting for reinforcements is just as much, and we may very well say, even more on the side of the enemy and in his favor. Besides, it is natural that a state can muster pretty well as many military forces in one year as in two; for all the actual increase of military force in the second year is but trifling in relation to the whole.
2. The enemy rests at the same time that we do.
3. The fortification of cities and positions is not the work of the army, and therefore no ground for a delay.

4. In the present system of subsisting armies, magazines are more necessary when the troops are in quarters than when they are advancing. As long as we advance with success, we continually come into possession of some of the enemy's provision depots, which assist us when the country itself is poor.

5. The capture of the enemy's fortresses cannot be regarded as a suspension of the attack; it is an intensified progress, and therefore the seeming suspension which is caused thereby is not properly a case such as we are speaking of; it is neither a suspension nor a mitigation of force. But whether a regular siege, blockade or a mere observation of one or the other is most to the purpose is a question which can only be decided according to particular circumstances. We can only say this in general: that the answer to this question must be entirely decided by the further question, whether by mere blockading and further advance we would not be taking too great a risk. Where this is not the case, and when there is still ample room to extend our forces, it is better to postpone the formal siege till the end of the whole offensive movement. We must, therefore, take care not to be led astray by the idea of immediately making secure that which is conquered, and in doing so neglect something more important.

No doubt it seems as if, by further advance, we at once risk again what has been already won. Our opinion is, therefore, that no pause, no resting point, no intermediate stations are in accordance with the nature of offensive war, and that when they are unavoidable, they are to be regarded as an evil which makes the result not more certain, but, on the contrary, more uncertain; and further, that, keeping strictly to the general truth, if from weakness or any cause we have been obliged to stop, a second attempt at the object we have in view is, as a rule, impossible; but if such a second attempt is possible, then the stoppage was unnecessary, and that when an object at the very beginning is beyond our strength, it will always remain so.

We say that this appears to be the general truth, by which we only wish to eliminate the idea that time of itself can do something for the advantage of the assailant. But as the political relations may change from year to year, on that account alone many cases may happen which are exceptions to this general truth.

It may appear, perhaps, as if we had lost our general point of view, and had nothing in sight except offensive war; but this is not at all the case. Certainly, he who can make the complete overthrow of the enemy his object will not be easily reduced to take refuge in the defensive, the immediate object of which is only the preservation of possessions. But we must maintain throughout that a defensive without any positive principle is to be regarded as a self-contradiction in strategy as well as in tactics, and therefore we always come back to the fact that every defensive, according to its strength, will seek to change to the attack as soon as it has exhausted the advantages of the defensive. However great, therefore, or however small the defense may be, we must also include in it, if possible, the overthrow of the enemy as an object which this attack may have, and which is to be considered as the proper object of the defensive. We say that there may be cases in which the assailant, notwithstanding that he has in view such a great object, may still prefer at first to make use of the defensive form. That this idea is not unsupported by facts is easily shown by the campaign of 1812. The Emperor Alexander in engaging in the war did not perhaps think of destroying his enemy completely, as was done afterward. But would such an idea have been impossible? And would it not still have been very natural that the Russians began the war on the defensive?

CHAPTER 5

AIM OF WAR MORE PRECISELY DEFINED (CONTINUED)

LIMITED AIM

In the preceding chapter we have said that by the expression "overthrow of the enemy" we understand the real absolute aim of the act of war. Now we shall consider what remains to be done when the conditions under which this aim might be attained do not exist.

These conditions presuppose a great physical or moral superiority, or a great spirit of enterprise, a predilection for great risks. Now where all this is not forthcoming, the aim in the act of war can only be of two kinds; either the conquest of some small or moderate portion of the enemy's country or the defense of our own until better times. This last is the usual case in defensive war.

Whether the one or the other of these aims is right in a given case can always be settled by calling to mind the expression used in reference to the last. *The waiting till more favorable times* implies that we have reason to expect such times hereafter, and this waiting, that is, defensive war, is always based on this prospect; on the other hand, offensive war, that is, taking advantage of the present moment, is always imperative when the future holds out a better prospect, not to ourselves, but to our adversary.

The third case, which is probably the most common, is when neither party has anything definite to expect from the future, when therefore it furnishes no motive for decision. In this case offensive war is plainly imperative for him who is politically the aggressor, that is, who has the positive motive, for he has taken up arms with that object, and every moment of time which is lost without any good reason is so much lost time *for him*.

We have here decided for offensive or defensive war on grounds which have nothing to do with the relative strength of the belligerents, and yet it might appear much more natural to make the choice of the offensive or defensive depend chiefly on the relative strength. However, our opinion is that just in so doing we should go astray. The logical correctness of our simple argument no one will dispute; we shall now see whether it leads to absurdity in concrete cases.

Let us suppose a small state which is involved in a conflict with a very superior power, but foresees that with each year its position will become worse. If it cannot avoid war, must it not make use of the time when its situation is not yet so bad? It must, therefore, attack, not because the attack *in itself* ensures any advantages—it will rather increase the disparity of forces still more—but because such a state is under the necessity of either bringing the matter to a final issue before the worst time arrives or of gaining at least in the meantime some advantages on which it may hereafter maintain itself. This theory cannot appear absurd. But if this small state were quite certain that the enemy will advance against it, then, certainly, it can and may make use of the defensive against its enemy to procure a first advantage. There is then, at any rate, no danger of losing time.

Further, if we suppose a small state to be engaged in war with a greater, and the future to have no influence whatever on their decisions, still, if the small state is politically the assailant, we must demand of it also that it should advance toward its object.

If it has had the audacity to propose to itself the positive object against a more powerful opponent, then it must also act, that is, attack the foe, if the latter does not save it the trouble. Waiting would be an absurdity; unless at the moment of execution it has altered its political decision, a case which very frequently occurs, and contributes not a little to giving to wars an indefinite character, a fact of which the philosopher does not know what to make.

Our consideration of the limited aim leads us to offensive war with such an aim, and to defensive war. We wish to discuss both in special chapters, but we must first turn our attention in another direction.

Hitherto we have deduced the modification of the aim of war solely from intrinsic reasons. The nature of the political intention we have only taken into consideration, in so far as it is or is not directed at something positive. Everything else in the political intention is fundamentally something extraneous to war; but in Book I, Chapter 2, "End and Means in War," we have already admitted that the nature of the political object, the extent of our own or the enemy's demand, and our whole political relation have in reality a most decisive influence on the conduct of the war, and we shall therefore devote the following chapter to that subject especially.

CHAPTER 6

A. Influence of the Political Object on the Military Aim

We never find that a state joining in the cause of another state takes it as seriously as its own. An auxiliary army of moderate strength is sent ahead; if it is not successful, then the ally looks upon the affair as in a manner ended, and tries to get out of it on the cheapest terms possible.

In European politics it is an established thing for states to pledge themselves to mutual assistance by an offensive and defensive alliance. Not to such an extent that one shares in the interests and quarrels of the other, but only so far as to promise each other beforehand the assistance of a fixed, generally very moderate, contingent of troops without regard to the object of the war or the extent of efforts made by the foe. In a treaty of alliance of this kind the ally does not look upon himself as engaged with the enemy in a war, properly speaking, which would necessarily have to begin with a declaration of war and end with a peace treaty. Still, this idea is nowhere fixed with any distinctness, and usage varies.

The thing would have a kind of consistency and the theory of war would have less difficulty in regard to it, if this promised contingent of ten, twenty or thirty thousand men were handed over entirely to the state engaged in war, so that it might be used as required; it might then be regarded as a hired force. But the usual practice is widely different. Usually the auxiliary force has its own commander, who depends only on his government, which prescribes to him an object such as best suits the half-hearted measures it has in view.

But even if two states really go to war with a third, they do not always both look in like measure upon this common enemy as one that they must destroy or be destroyed by him. The affair is often settled like a commercial transaction; each, according to the amount of risk he incurs or the advantage to be expected, takes a share in the concern to the extent of 30,000 or 40,000 men, and acts as if he could not lose more than the amount of his investment.

Not only is this the point of view taken when a state comes to the assistance of another in a cause which is rather foreign to it; but even when both have a common and great interest at stake, nothing can be done without diplomatic support, and the contracting parties usually only agree to furnish a small stipulated contingent, in order to employ the rest of their military forces on the special ends to which policy might happen to lead them.

This way of regarding wars entered into because of alliances was very prevalent, and was forced to give way to the natural point of view only in very modern times, when the extremest danger drove men's minds into natural pathways (as *against* Bonaparte) and when boundless power compelled them to it (as *under* Bonaparte). It was a half-hearted thing, an anomaly, for war and peace are ideas which fundamentally can have no gradations. Nevertheless, it was no mere diplomatic tradition which reason could disregard, but deeply rooted in the natural limitation and weakness of human nature.

Lastly, even in wars carried on without allies, the political cause of a war has a great influence on the method in which it is conducted.

If we wish from the enemy only a small sacrifice, we are satisfied with winning by means of the war only a small equivalent, and we expect to attain that by moderate efforts. The enemy reasons in very much the same way. Now if one or the other finds that he has deceived himself in his calculation—that, in place of being slightly superior to his enemy, as he supposed, he is, if anything, somewhat weaker, still, at that moment, money and all other means, as well as sufficient moral impulse for great exertions, are very often deficient. In such a case he manages as best he can, and hopes for favorable events from the future, although he has not the slightest foundation for such hope, and the war in the meantime drags itself feebly along, like a body worn out with sickness.

Thus it comes to pass that the reciprocal action, the effort to outbid, the violence and irresistibleness of war are lost in the stagnation of weak motives, and that both parties move with a certain kind of security in very reduced spheres.

If this influence of the political object on war is once permitted, as it must be, there is no longer any limit, and we must put up with descending to such warfare as consists in a *mere threatening of the enemy* and in *negotiating*.

It is evident that the theory of war, if it is to be and remain a philosophical study, finds itself in difficulty here. All that is inherent in the concept of what is essential to war seems to flee from it, and theory is in danger of being left without any point of support. But the natural solution soon appears. According as a modifying principle gains influence over the act of war, or rather, the weaker the motives of action become, so much the more action turns into passive resistance, so much the less occurs, and the less it requires guiding principles. All military art then changes into mere prudence, the principal object of which will be to prevent the wavering balance from suddenly turning to our disadvantage, and the half-hearted war from turning into a real one.

B. WAR AS AN INSTRUMENT OF POLICY

Up to this point we have had to consider, now from this side, now from that, the state of antagonism in which the nature of war stands with relation to the other interests of men individually and in a social group, in order not to neglect any of the opposing elements—an antagonism which is founded in our own nature, and which, therefore, no philosophy can unravel. We shall now look for that unity to which, in practical life, these antagonistic elements attach themselves by partly neutralizing each other. We would have brought forward this unity at the very beginning if it had not been necessary to emphasize these very contradictions, and also to look at the different elements separately. Now this unity is *the conception that war is only a part of political intercourse, therefore by no means an independent thing in itself.*

We know, of course, that war is only caused through the political intercourse of governments and nations; but in general it is supposed that such intercourse is broken off by war, and that a totally different state of things ensues, subject to no laws but its own.

We maintain, on the contrary, that war is nothing but a continuation of political intercourse with an admixture of other means. We say "with an admixture of other means," in order thereby to maintain at the same time that this political intercourse does not cease through the war itself, is not changed into something quite different, but that, in its essence, it continues to exist, whatever may be the means which it uses, and that the main lines along which the events of the war proceed and to which they are bound are only the general features of policy which run on all through the war until peace takes place. And how can we conceive it to be otherwise? Does the cessation of diplomatic notes stop the political relations between different nations and governments? Is not war merely another kind of writing and language for their thought? It has, to be sure, its own grammar, but not its own logic.

Accordingly, war can never be separated from political intercourse, and if, in the consideration of the matter, this occurs anywhere, all the threads of the different relations are, in a certain sense, broken, and we have before us a senseless thing without an object.

This way of looking at the matter would be indispensable even if war were entirely war, entirely the unbridled element of hostility. All the circumstances on which it rests, and which determine its leading features, viz., our own power, the enemy's power, allies on both sides, the characteristics of the people and the governments respectively, etc., as enumerated in Book I, Chapter 1—are they not of a political nature, and are they not so intimately connected with the whole political intercourse that it is impossible to separate them from it? But this view is doubly indispensable if we reflect that real war is no such consistent effort tending to the last extreme, as it should be according to abstract theory, but a half-hearted thing, a contradiction in itself; that, as such, it cannot follow its own laws, but must be looked upon as part of another whole—and this whole is policy.

Policy in making use of war avoids all those rigorous conclusions which proceed from its nature; it troubles itself little about final possibilities, confining its attention to immediate probabilities. If there ensues much uncertainty in the whole transaction because of this, if war thereby becomes a sort of game, the policy of each cabinet cherishes the confident belief that in this game it will surpass its opponent in skill and discernment.

Thus policy makes out of the all-overpowering element of war a mere instrument, out of the fearsome battle-sword, which should be lifted with both hands and the whole power of the body to strike once and not more, a light, handy dagger, which is even sometimes nothing more than a rapier, which it uses in turn for thrusts, feints and parries.

Thus the contradictions in which man, naturally timid, becomes involved by war may be solved, if we choose to accept this as a solution.

If war belongs to policy, it will naturally take on its character. If policy is grand and powerful, so also will be war, and this may be carried to the height at which war attains *its absolute form.*

In this way of conceiving it, therefore, we need not lose sight of the absolute form of war, rather its image must constantly hover in the background.

Only through this way of conceiving it does war once more become a unity; only thus can we regard all wars as things of *one* kind; and only thus can judgment obtain the true and exact basis and point of view from which great plans are to be made and judged.

It is true the political element does not penetrate deeply into the details of war. Vedettes are not planted, patrols are not sent round on political considerations. But its influence is all the more decisive in regard to the plan of a whole war, or campaign, and often even for a battle.

For this reason we were in no hurry to establish this view at the beginning. While engaged with particulars, it would have given us little help, and, rather would have distracted our attention to a certain extent; in the plan of a war or campaign it is indispensable.

There is, on the whole, nothing more important in life than to find out exactly the point of view from which things must be regarded and judged, and then to keep to it, for we can only apprehend the mass of events in their unity from *one* standpoint, and it is only the keeping to one point of view that can save us from inconsistency.

If, therefore, in drawing up a plan of war, it is not permissible to have two or three points of view, from which things might be regarded, now with a soldier's eye, now with an administrator's, now with a politician's, and so on, then the next question is whether *policy* is necessarily paramount and everything else subordinate to it.

It is assumed that policy unites and reconciles within itself all the interests of internal administration and also those of humanity and of whatever else the philosophical mind might bring up, for it is nothing in itself but a mere representative of all these interests toward other states. That policy may take a wrong direction, and prefer to promote ambitious ends, private interests or the vanity of rulers, does not concern us here, for under no circumstances can the art of war be considered as its tutor, and we can only regard policy here as the representative of all the interests of the whole community.

The only question, therefore, is whether in forming plans for a war the political point of view should give way to the purely military (if such a point of view were conceivable), that is to say, should disappear altogether, or subordinate itself to it, or whether the political must remain the ruling point of view and the military be subordinated to it.

That the political point of view should end completely when war begins would only be conceivable if wars were struggles of life or death, from pure hatred. As wars are in reality, they are, as we said before, only the manifestations of policy itself. The subordination of the political point of view to the military would be unreasonable, for policy has created the war; policy is the intelligent faculty, war only the instrument, and not the reverse. The subordination of the military point of view to the political is, therefore, the only thing which is possible.

If we reflect on the nature of real war, and call to mind what has been said in the third chapter of this book, that *every war should be understood according to the probability of its character and its leading features as they are to be deduced from the political forces and conditions,* and that often—indeed we may safely affirm, in our days *almost always*—war is to be regarded as an organic whole, from which the single members cannot be separated, in which therefore every individual activity flows into the whole and also has its origin in the idea of this whole, then it becomes perfectly certain and clear that the highest standpoint for the conduct of war, from which its leading features proceed, can be no other than that policy.

From this point of view our plans come out as from a mold; our comprehension and judgment become easier and more natural, our convictions gain in force, motives are more satisfactory and history more intelligible.

At all events from this point of view there no longer exists a natural conflict between the military and political interests, and where it does appear, it is to be regarded merely as imperfect knowledge. That policy makes demands on war which it cannot fulfill would be contrary to the presupposition that it knows the instrument which it is going to use, contrary therefore to a presupposition that is natural and quite indispensable. But if policy judges correctly of the course of military events, it is entirely its affair to determine what events and what direction of events correspond to the aim of the war.

In a word, the art of war in its highest point of view becomes policy, but, of course, a policy which fights battles instead of writing notes.

According to this view, it is an unpermissible and even harmful distinction, according to which a great military event or the plan for such an event should admit a *purely military judgment;* indeed, it is an unreasonable procedure to consult professional soldiers on the plan of war, that they may give a *purely military* opinion, as is frequently done by cabinets; but still more absurd is the demand of theorists that a statement of the available means of war should be laid before the general, that he may draw up a purely military plan for the war or for the campaign in accordance with them. General experience also teaches us that in spite of the great diversity and development of the present system of war, still the main outlines of a war have always been determined by the cabinet, that is, if we would use technical language, by a purely political and not a military organ.

This is perfectly natural. None of the principal plans which are necessary for a war can be made without insight into the political conditions, and when people speak, as they often do, of the harmful influence of policy on the conduct of the war, they really say something very different from what they intend. It is not this influence, but the policy itself, which should be found fault with. If the policy is right, that is, if it achieves its end, it can only affect the war favorably—in the sense of that policy. Where this influence deviates from the end, the cause is to be sought only in a mistaken policy.

It is only when policy promises itself a wrong effect from certain military means and measures, an effect opposed to their nature, that it can exercise a harmful effect on war by the course it prescribes. Just as a person in a language which he has not entirely mastered sometimes says what he does not intend, so policy will often order things which do not correspond to its own intentions.

This has very often happened and shows that a certain knowledge of military affairs is essential to the management of political intercourse.

But before going further, we must guard ourselves against a wrong interpretation, which readily suggests itself. We are far from holding the opinion that a war minister, buried in official papers, a learned engineer or even a soldier who has been well tried in the field would, any of them, necessarily make the best minister of state in a country where the sovereign does not act for himself. In other words, we do not mean to say that this acquaintance with military affairs is the principal qualification for a minister of state; a remarkable, superior mind and strength of character—these are the principal qualifications which he must possess; a knowledge of war may be supplied in one way or another. France was never worse advised in its military and political affairs than by the two brothers Belleisle and the Duke of Choiseul, although all three were good soldiers.

If war is to correspond entirely with the intentions of policy, and policy is to accommodate itself to the means available for war, in a case in which the statesman and the soldier are not combined in one person, there is only one satisfactory alternative left, which is to make the commander-in-chief a member of the cabinet, that he may take part in its councils and decisions on important occasions. But then, again, this is only possible when the cabinet, that is, the government itself, is near the theater of war, so that things can be settled without noticeable waste of time.

This is what the Emperor of Austria did in 1809, and the allied sovereigns in 1813, 1814, 1815, and the arrangement proved perfectly satisfactory.

The influence in the cabinet of any military man except the commander-in-chief is extremely dangerous; it very seldom leads to sound vigorous action. The example of France in 1793, 1794 and 1795, when Carnot, while residing in Paris, managed the conduct of the war, is thoroughly objectionable, because a system of terror is not at the command of any but a revolutionary government.

We shall now conclude with some reflections derived from history.

In the last decade of the past century, when that remarkable change in the art of war in Europe took place by which the best armies saw a part of their method of war become ineffective, and military successes occurred on a scale of which no one had up to then had any conception, it certainly seemed that all wrong calculations were to be laid to the charge of the art of war. It was plain that, while confined by habit within a narrow circle of conceptions, Europe had been surprised by possibilities which lay outside of this circle, but, certainly, not outside of the nature of things.

Those observers who took the most comprehensive view ascribed the circumstance to the general influence which policy had exercised for centuries on the art of war, to its very great disadvantage, and as a result of which it had sunk into a half-hearted affair, often into mere sham-fighting. They were right as to the fact, but wrong in regarding it as an avoidable condition arising by chance.

Others thought that everything was to be explained by the momentary influence of the particular policy of Austria, Prussia, England, etc.

But is it true that the real surprise by which men's minds were seized was due to something in the conduct of war, and not rather to something in policy itself? That is: did the misfortune proceed from the influence of policy on the war or from an intrinsically wrong policy?

The tremendous effects of the French Revolution abroad were evidently brought about much less through new methods and views introduced by the French in the conduct of war than through the change in statecraft and civil administration, in the character of government, in the condition of the people, and so forth. That other governments took a mistaken view of all these things, that they endeavored, with their ordinary means, to hold their own against forces of a novel kind and overwhelming strength—all that was a blunder of policy.

Would it have been possible to perceive and correct these errors from the standpoint of a purely military conception of war? Impossible. For if there had been a philosophical strategist, who merely from the nature of the hostile elements had foreseen all the consequences, and prophesied remote possibilities, still it would have been quite impossible for such a wholly theoretical argument to produce the least result.

Only if policy had risen to a just appreciation of the forces which had been awakened in France, and of the new relations in the political state of

Europe, could it have foreseen the consequences which were bound to follow in respect to the great features of war, and only in this way could it be led to a correct view of the extent of the means required, and the best use to make of them.

We may, therefore, say that the twenty years' victories of the Revolution are chiefly to be ascribed to the faulty policy of the governments by which it was opposed.

It is true that these faults were first displayed in the war, and the events of war completely disappointed the expectations which policy entertained. But this did not take place because policy neglected to consult its military advisers. The art of war in which the politician of the day could believe, namely, that derived from the reality of that time, that which belonged to the policy of the day, that familiar instrument which had been hitherto used—*that* art of war, I say, was naturally involved in the same error as policy, and therefore could not teach it better. It is true that war itself has undergone important alterations, both in its nature and forms, which have brought it nearer to its absolute form; but these changes were not brought about by the French government having, so to speak, freed itself from the leading-strings of policy; they arose from an altered policy which proceeded from the French Revolution not only in France but in the rest of Europe as well. This policy had called forth other means and other forces by which it became possible to conduct war with a degree of energy which could not have been thought of before.

Also the actual changes in the art of war are a consequence of alterations in policy, and, far from being an argument for the possible separation of the two, they are, on the contrary, very strong evidence of the intimacy of their connection.

Therefore, once more: war is an instrument of policy; it must necessarily bear the character of policy; it must measure with policy's measure. The conduct of war, in its great outlines, is, therefore, policy itself, which takes up the sword in place of the pen, but does not on that account cease to think according to its own laws.

CHAPTER 7

LIMITED AIM—OFFENSIVE WAR

Even if the complete overthrow of the enemy cannot be the aim, there may still be one which is directly positive, and this positive aim can be nothing else than the conquest of a part of the enemy's country.

The use of such a conquest is that we weaken the enemy's national forces and consequently his military forces, while we increase our own; that we therefore carry on the war, to a certain extent, at his expense; further, that in negotiations for peace the possession of the enemy's provinces may be regarded as net gain, because we can either keep them or exchange them for other advantages.

This view of a conquest of the enemy's provinces is very natural, and would be open to no objection if it were not that the state of defense, which must follow the offensive, might often cause uneasiness.

In the chapter "Culminating Point of the Attack" (Book VII, Chapter 5), we have sufficiently explained the manner in which such an offensive weakens the military forces, and that it may be followed by a situation which causes apprehension of dangerous consequences.

The weakening of our military force by the conquest of a portion of the enemy's territory has its degrees, and these depend chiefly on the geographical position of this portion of territory. The more it is an annex of our own country, being contiguous or within it, and the more it lies in the direction of our principal forces, by so much the less will it weaken our military force. In the Seven Years' War Saxony was a natural complement of the Prussian theater of war, and Frederick the Great's army, instead of being weakened, was strengthened by the possession of that province, because it lies nearer to Silesia than to the Mark, and yet at the same time covers the latter.

Even in 1740 and 1741, after Frederick the Great had once conquered Silesia, it did not weaken his military forces, because, owing to its form and situation, as well as the contour of its frontier line, it offered only a narrow point to the Austrians, as long as they were not masters of Saxony, and, besides that, this small point of contact lay in the line which the attacks on both sides had to take.

If, on the other hand, the conquered territory is a strip running between hostile provinces and has an eccentric position and unfavorable configuration of ground, the weakening increases so visibly that a victorious battle becomes not only much easier for the enemy, but it may even become unnecessary.

The Austrians have always been obliged to evacuate Provence without a battle when they have made attempts on it from Italy. In 1774 the French were glad enough to get out of Bohemia without having lost a battle. In 1758 Frederick the Great could not hold his position in Bohemia and Moravia with the same force with which he had obtained such brilliant successes in Silesia and Saxony in 1757. Examples of armies not being able to keep possession of conquered territory, merely because their military force was being so much weakened thereby, are common occurrences, and it is therefore not worth the trouble to cite any more of them.

Therefore, the question whether we should aim at such an object depends on whether we can expect to hold possession of the conquest or whether a temporary occupation (invasion, diversion) would repay the expenditure of force required, especially on whether we have not to apprehend such a vigorous counterstroke as would completely upset our equilibrium. In the chapter "Culminating Point of the Attack," we have treated the manifold consideration due to this question in each particular case.

There is just one point which we must yet add.

An offensive of this kind will not always compensate us for what we lose upon other points. While we are engaged in making a partial conquest, the enemy may be doing the same at other points, and if our enterprise is not of superior importance then it will not compel the enemy to give up his. It is, therefore, a question for serious consideration whether we do not lose more on the one side than we gain on the other.

Even if we suppose two provinces (one on each side) to be of equal value, we shall always lose more by the one which the enemy takes from us than we can gain by the one we take, because a number of our resources, as *faux frais*, so to speak, become ineffective. But since the same thing occurs on the enemy's side also, one would suppose that in reality there is no reason to attach more importance to the maintenance of what is our own than to the conquest. And yet it is so. The retaining possession of our own country is always a matter which concerns us more deeply, and the suffering inflicted on our own state cannot be outweighed, nor, so to speak, neutralized by what we gain in return, except when the latter promises considerable profits, that is, if it is much greater.

The consequence of all this is that such a strategic attack, with only a moderate aim, is much less able to free itself from the defense of other points which it does not directly cover than one which is directed against the center of gravity of the enemy's state. Consequently, in such an attack, the concentration of forces in time and space cannot be carried out to the same extent. In order that it may take place, at least in respect of time, it becomes necessary for the advance to be made offensively from every suitable point at the same time, and therefore this attack loses the other advantage of being able to make shift with a much smaller force by acting on the defensive at particular points. In this way the effect of aiming at such a minor object is to bring all things more to one level; the whole act of the war can no longer be concentrated into one principal action which can be conducted according to main points of view; it is more dispersed; the friction becomes everywhere greater, and there is everywhere more room for chance.

This is the natural tendency of the thing. The commander is dragged down by it, finds himself more and more powerless. The more he is conscious of his own powers, and the more internal and external resources he has, the more he will seek to free himself from this tendency in order to give to some one point a preponderant importance, even if that should only be possible through greater daring.

CHAPTER 8

LIMITED AIM–DEFENSE

The ultimate aim of a defensive war can never be an absolute negation, as we have observed before. Even for the weakest there must be something by means of which he can hurt his opponent and threaten him.

We could, no doubt, say that this object can consist in exhausting our opponent, for since he has a positive object, every one of his undertakings which fails, even if it has no other result than the loss of the force employed, still may be considered *fundamentally* a retrograde step, while the loss which the defensive suffers was not in vain, because his object was keeping possession, and that he has achieved. This would be tantamount to saying that the defender has his positive object in merely keeping possession. Such reasoning might be valid, if it were a fact that the offensive must tire and give up after a certain number of vain attempts. But just this necessity is lacking. If we look at the actual exhaustion of forces, the defender is at a disadvantage. The attack weakens, but only in the sense that there may be a turning point; where this is no more to be thought of, the weakening is certainly greater on the defensive side than on the offensive, for, in the first place, he is the weaker, and, therefore, if the losses on both sides are equal, relatively he loses more than the other, and, in the second place, he is generally deprived of a portion of territory and of his resources. From this we cannot deduce any reason for our opponent to relax his efforts and we can only conclude that if the assailant repeats his blows, while the defensive does nothing but ward them off, the latter cannot by any counteraction meet the danger that, sooner or later, one of the attacks may succeed.

Although actually the exhaustion, or rather the tiring out of the stronger, has often brought about a peace, that is to be attributed to the half-heartedness with which war is usually conducted, and cannot be regarded philosophically as the general and ultimate aim of any defensive war whatever. There is, therefore, no alternative but that the defense should find its aim in the conception of awaiting the enemy, which is, moreover, its real character. This conception includes an alteration of circumstances, an improvement of the situation, which, therefore, when it cannot possibly be brought about by internal means, that is, by the defensive itself, can only be expected from without. Now, this improvement from without can be nothing else than a change in political relations; either new alliances spring up in favor of the defender or old ones directed against him collapse.

This is, therefore, the aim of the defender, in case his weakness does not permit him to think of any important counterstroke. But this is not the nature of every defensive according to the concept of it which we have given. According to that concept, it is the stronger form of war, and because of that strength it can also be employed when a more or less strong counterstroke is intended.

These two cases must be kept distinct from the very first, as they have an influence on the defense.

In the first case, the defender tries to keep possession of his own country as long as possible, because in that way he gains most time, and gaining time is the only way to his aim. The positive aim which he can in most cases attain, and which shall give him the opportunity of carrying out his intentions in the negotiations for peace, he cannot yet include in his plan of war. In this state of strategic passivity, the advantages which the defender can gain at certain points consist in merely repelling separate attacks; the preponderance gained at those points he carries over to others, for he is usually hard-pressed at all points. If he has no opportunity of doing this, then there often only remains for him the small advantage that the enemy will leave him alone for a time.

If the defender is not altogether too weak, small offensive operations directed less toward permanent possession than toward a temporary advantage to cover losses, which may be sustained later, invasions, diversions or enterprises against a single fortress, may have a place in this defensive system without altering its aim or essence.

But in the second case, where the defensive has already conceived a positive intention, it assumes also a more positive character, and indeed so much the more, the greater the counterstroke that is warranted by circumstances. In other words, the more the defense has been adopted voluntarily, in order to make the first blow sure, the bolder may be the snares which the defender sets for his opponent. The boldest, and if it succeeds, the most effective, is the retreat into the interior of the country, and this means is then at the same time the one which differs most widely from the other system.

Let us think only of the difference between the position in which Frederick the Great found himself in the Seven Years' War and that of Russia in 1812.

When the war began, Frederick, because of his readiness for battle, had a kind of superiority; this gave him the advantage of being able to make himself master of Saxony, which was, by the way, such a natural complement to his theater of war that the possession of it did not diminish but increased his military forces.

At the opening of the campaign of 1757, the king endeavored to proceed with his strategic attack, which seemed not impossible as long as the Russians and French had not yet reached the theater of war in Silesia, the Mark and Saxony. But the attack failed, and Frederick was thrown back on the defensive for the rest of the campaign, was obliged to evacuate Bohemia and to rescue his own theater from the enemy, in which he only succeeded by turning with one and the same army first upon the French, and then upon the Austrians. This advantage he owed entirely to the defensive.

In the year 1758, when his enemies had drawn round him in a closer circle, and his forces were dwindling down to a very unequal proportion, he wanted to try an offensive on a small scale in Moravia. His plan was to take Olmütz before his enemies were prepared, not with the hope of keeping possession of it or of making it a base for further advance, but to use it as an outwork, a counter-approach against the Austrians, who would be obliged to devote the rest of the present campaign, and perhaps even a second, to recovering it. This attack also failed. Frederick now gave up the thought of a real offensive, because he saw that it only increased the disproportion of his forces. A compact position in the heart of his own provinces in Saxony and Silesia, the use of short lines, that he might be able rapidly to increase his forces at any point which might be menaced, a battle when unavoidable, small invasions when opportunity offered, and along with this a patient state of awaiting, a saving of his means for better times, became now his general plan. Gradually the execution of it became more and more passive. Since he saw that even the victories cost him too much, he tried to manage at still less expense; everything depended on gaining time and on keeping what he had; he became more and more averse to yielding any ground, and did not hesitate to adopt a veritable cordon system. The positions of Prince Henry in Saxony, as well as those of the king in the Silesian mountains, deserve this designation. In his letters to the Marquis d'Argens we see the impatience with which he looks forward to winter quarters and how glad he is of being able to take them up again without having suffered any serious loss.

Whoever would blame Frederick for this, and see in it a sign of fallen spirit, would, we think, pass judgment without much reflection.

If the entrenched camp at Bunzelwitz, the positions taken up by Prince Henry in Saxony, and by the king in the Silesian mountains no longer appear to us as measures on which a general can place his last hope, because a Bonaparte would soon have thrust his sword through such tactical cobwebs, we must not forget that times have changed, that war has become a totally different thing, quickened by different powers and that, therefore, positions could be effective at that time, which are effective no longer, and that the character of the opponent also deserves attention. Against the imperial army, against Daun and Butterlin, the use of means which Frederick would have despised, if used against himself, could be the highest wisdom.

The result justified this view. By quietly awaiting the enemy Frederick attained his aim and evaded difficulties in collision with which his forces would have been destroyed.

The relation in point of numbers between the Russian and French armies opposed to each other at the opening of the campaign in 1812 was still much more unfavorable to the former than that between Frederick and his enemies in the Seven Years' War. But the Russians had the prospect of strengthening themselves considerably in the course of the campaign. All Europe was in secret hostility to Bonaparte, his power had been screwed up to the last pitch, a consuming war occupied him in Spain, and the vast extent of Russia permitted the weakening of the enemy's military powers to be carried to the last extremity in a retreat five hundred miles long. Under these splendid circumstances a tremendous counterstroke was not only to be expected, if the French enterprise failed (and how could it succeed if the Russian Emperor would not make peace, or his subjects did not rise in insurrection against him?), but this counterstroke could also bring about the complete destruction of the enemy. The most profound sagacity could, therefore, not have devised a better plan of war than that which the Russians followed unintentionally.

That this was not the opinion of the time, and that such a view would then have been looked upon as preposterous, is no reason for our now denying it to be the right one. If we are to learn from history, we must look upon things which have actually happened as also possible in the future, and everyone who can claim the right to an opinion in such matters will admit that the series of great events which followed the march upon Moscow is not a series of accidents. If it had been possible for the Russians to put up some scanty defense of their frontier, it is probable that in such a case the power of the French would still have given way, and they would have suffered a reverse of fortune; but the reverse would certainly not have been so violent and so decisive. By sufferings and sacrifices (which certainly for any other country would have been greater, and for most countries, impossible) Russia purchased this enormous advantage.

Thus a great positive success can never be obtained except through positive measures, planned not with a view to merely awaiting the enemy, but with a view to a *decision*. In short, even on the defensive, there is no great gain to be won except by a great stake.

CHAPTER 9

PLAN OF WAR WHEN THE DESTRUCTION OF THE ENEMY IS THE AIM

Having characterized in detail the different aims to which war may be directed, we shall go through the organization of the whole war for each of the three separate gradations corresponding to these aims.

In conformity with all that has been said on the subject up to now, two fundamental principles will comprehend the whole plan of war and serve as a guide for everything else.

The first is: to trace the weight of the enemy's power back to as few centers of gravity as possible, to one if it can be done; again, to confine the attack against these centers of gravity to as few principal undertakings as possible, to one if possible; lastly, to keep all secondary undertakings as subordinate as possible. In a word, the first principle is: *to act with as much concentration as possible.*

The second principle is: *to act as swiftly as possible*; therefore to permit no delay or detour without sufficient reason.

The determination of the center of gravity of the enemy's power depends (1) On how that power is politically constituted. If it consists of armies of one state, there is generally no difficulty; if of allied armies, of which one is acting simply as an ally without much interest of its own, then the difficulty is not much greater; if of a coalition for common objects, then it depends on the cordiality of the alliance. We have mentioned this before. (2) On the situation of the theater of war upon which the different hostile armies make their appearance.

If the enemy's forces are concentrated in one army upon one theater of war, they constitute a real unity, and we need not inquire further; if they are upon one theater of war but in separate armies, which belong to different powers, there is no longer absolute unity. There is, however, still a sufficient connection of the parts for a decisive blow upon *one* part to carry away the other along with it. If the armies are posted in theaters of war adjoining each other, and not separated by any great natural obstacles, there is in such a case still a decided influence of the one upon the other; but if the theaters of war are far apart, if there is neutral territory, great mountains, etc., intervening between them, the influence is very doubtful, therefore improbable; if they are even on quite opposite sides of the state against which the war is being waged, so that operations directed against them must diverge on eccentric lines, then almost every trace of connection has disappeared.

If Prussia was attacked by France and Russia at the same time, it would be in respect of the conduct of war much the same as if there were two separate wars; at best the unity would appear in the negotiations.

The Saxon and Austrian military powers in the Seven Years' War were, on the contrary, to be regarded as one; what the one suffered the other felt too, partly because the theaters of war lay in the same direction for Frederick the Great, partly because Saxony had no political independence.

Numerous as were the enemies which Bonaparte had to contend with in Germany in 1813, they all lay somewhat in the same direction for him, and the theaters of war for their armies were in close connection, and reciprocally influenced each other very powerfully. If by a concentration of all his forces he had been able to overpower the main army, such a defeat would have had a decisive effect on all the parts. If he had beaten the main Bohemian army, and marched upon Vienna by way of Prague, Blücher, however willing, could not have remained in Saxony, because he would have been called upon to co-operate in Bohemia, and the Crown Prince of Sweden would have been unwilling even to remain in the Mark.

On the other hand, Austria, if carrying on war against the French on the Rhine and Italy at the same time, will always find it difficult to gain a decision upon one of those theaters by means of a successful stroke on the other. This is due partly to the fact that Switzerland, with its mountains, forms too strong a barrier between the two theaters, and partly to the fact that the direction of the roads on each side is divergent. France, on the contrary, can much sooner reach a decision in the one by means of a decisive success in the other, because the direction of its forces in both converges upon Vienna and the center of gravity of the whole Austrian empire; we may add, further, that a decisive blow in Italy will have more effect on the Rhine theater than a success on the Rhine would have in

Italy, because the blow from Italy strikes nearer to the center, and that from the Rhine more upon the flank of the Austrian power.

It proceeds from what we have said that the conception of separated or connected hostile power extends also through all degrees of relationship, and that therefore only in the individual case can one discover the influence which events in one theater may have upon the other, according to which we may afterwards settle how far the different centers of gravity of the enemy's power may be reduced to one.

There is only one exception to the principle of directing all our strength against the center of gravity of the enemy's power, that is, if secondary expeditions promise *extraordinary advantages*. Still we assume in such a case that a decisive superiority puts us in a position to undertake such enterprises without risking too much in respect to the main theater of war.

When General Bülow marched into Holland in 1814, it was to be foreseen that the 30,000 men composing his corps would not only neutralize the same number of Frenchmen, but would, besides, give the English and Dutch an opportunity of entering the field with forces which otherwise would never have been brought into activity.

Thus the first consideration in a plan of war is to determine the centers of gravity of the enemy's power, and, if possible, to reduce them to one. The second is to unite the forces which are to be employed against this center of gravity into one great action.

Now at this point the following grounds for dividing and separating our forces may present themselves:

(1) The original disposition of the military forces, therefore also the situation of the states engaged in the offensive.

If the concentration of the forces would occasion detours and loss of time, and the danger of advancing by separate lines is not too great, the division may be justifiable on those grounds, for to effect an unnecessary concentration of forces, with great loss of time, by which the freshness and rapidity of the first blow is diminished, would be contrary to the second leading principle we have laid down. In all cases in which there is a hope of surprising the enemy in some measure, this deserves particular attention.

But the case becomes still more important if the attack is undertaken by allied states which are not situated on a direct line toward the state attacked—not one behind the other—but situated side by side. If Prussia and Austria undertook a war against France, it would be a forced measure, one costing time and strength, if the armies of both powers wanted to proceed from one point, since the natural line for an army operating from Prussia against the heart of France is from the lower Rhine, and that of the Austrians is from the Upper Rhine. Concentration, therefore, in this case, could only be effected by a sacrifice; consequently, in any particular instance, the question to be decided would be whether the necessity for concentration is so great that this sacrifice must be made.

(2) The attack by separate lines may offer greater results.

Since we are now speaking of advancing by separate lines against *one* center of gravity, we are, therefore, supposing an advance by *concentric lines*. A separate advance on parallel or eccentric lines comes under the head of *secondary undertakings*, of which we have already spoken.

Now, every concentric attack has, in strategy as well as in tactics, the tendency to *greater* successes, for if it succeeds, the consequence is not simply a defeat, but more or less the cutting off of the enemy's armies. The concentric attack is, therefore, always that which may lead to the greatest results, but on account of the separation of the parts of the force, and the enlargement of the theater of war, it involves also the greater risk. It is the same here as with attack and defense: the weaker form offers a prospect of greater results.

The question therefore is whether the assailant feels strong enough to strive for this great aim.

When Frederick the Great advanced upon Bohemia, in 1757, he set out from Saxony and Silesia with his forces divided. The two principal reasons for his doing so were, first, that his forces were so arranged in the winter that a concentration of them at one point would have divested the attack of all the advantages of a surprise, and, second, that by this concentric advance each of the two Austrian theaters of war was threatened in the flank and rear. The danger to which Frederick the Great exposed himself on that occasion was that one of his two armies might have been completely defeated by superior forces; should the Austrians *not see this*, then they could either accept battle only in the center or they ran the risk of being completely thrown off their line of retreat on one side or the other and suffering a catastrophe. This was the great result which this advance promised the king. The Austrians preferred the battle in the center, but Prague, where they took up their position, was still in a situation too much under the influence of the enveloping attack, which, as they remained perfectly passive in their position, had time to develop its efficacy to the utmost. The consequence of this was that when they lost the battle, it was a real catastrophe; the fact that two-thirds of the army with the commanding general had to let themselves be shut up in Prague vouches for this.

This brilliant success of the campaign was due to the daring venture of the concentric attack. If Frederick considered the precision of his own movements, the energy of his generals, the moral superiority of his troops, on the one side, and the sluggishness of the Austrians on the other, as sufficient to assure the success of his plan, who can blame him? But we cannot leave these moral quantities out of consideration and ascribe the success solely to the mere geometrical form of the attack. We have only to think of the no less brilliant campaign of Bonaparte in the year 1796, when the Austrians were so notably punished for their concentric march into Italy. The means which the French general had at his command on that occasion, the Austrian general would have had also at his disposal in 1757 (with the exception of the moral). Indeed, he had rather more, for he was not, like Bonaparte, weaker than his adversary. Therefore, when it is to be feared that the advance on separate converging lines may afford the enemy the means of counteracting the inequality in number by using interior lines, such a form of attack is not advisable, and if, on account of the situation of the belligerents, it must be resorted to, it can only be regarded as a necessary evil.

If, from this point of view, we consider the plan which was adopted for the invasion of France in 1814, it is impossible to give it approval. The Russian, Austrian and Prussian armies were concentrated at a point near Frankfurt-on-the-Main, on the most natural and most direct line to the center of gravity of the French monarchy. These armies were separated, so that one might penetrate into France from Mayence, the other from Switzerland. Since the enemy's force was so reduced that a defense of the frontier was out of the question, the whole advantage to be expected from this concentric advance, if it succeeded, was that while Lorraine and Alsace were conquered by one army, Franche-Comté would be taken by the other. Was this trifling advantage worth the trouble of marching to Switzerland? We know very well that there were other, but just as insufficient grounds, which caused this march, but we confine ourselves here to the element which we are just now considering.

On the other hand, Bonaparte was a man who thoroughly understood the defensive against a concentric attack, as he had shown in his masterly campaign of 1796, and although the Allies were very considerably superior in numbers, his great superiority as a general was generally recognized. He joined his army too late near Châlons, and had altogether too low an opinion of his opponents. Still he was very near hitting the two armies separately, and in what a weakened condition did he find them at Brienne! Blücher had only 27,000 of his 65,000 men with him, and the main army, out of 200,000, had only 100,000. It was impossible to give the opponent a better chance. And from the moment that action began, nothing was felt to be more needed than reunion.

After all these reflections, we think that although the concentric attack is in itself a means of obtaining greater results, still it should generally only proceed from a previous separation of military powers, and there will be few cases in which we act rightly in giving up for the sake of it the shortest and most direct line of operation.

(3) The breadth of a theater of war can be a motive for attacking on separate lines.

If an army on the offensive in its advance from any point penetrates with success farther into the interior of the enemy's country, then, certainly, the space which it commands is not restricted exactly to the line of road by which it marches, but it will extend somewhat on each side; still that will depend very much, if we may use the figure, on the solidity and cohesion of the opposing state. If the hostile state is only loosely united, if its people are an effeminate race unaccustomed to war, then, without our taking much trouble, a considerable extent of country will remain open behind our victorious armies; but if we have to deal with a brave and loyal population, the space behind our army will form a more or less acute triangle.

In order to prevent this evil the assailant desires to arrange his advance on a certain width of front. If the enemy's force is concentrated at a particular point, this breadth of front can only be retained so long as we are not in contact with the enemy, and must be contracted as we approach his position; that is easy to understand.

But if the enemy himself has taken up a position with a certain extent of front, then there would be nothing unreasonable in a corresponding extension on our part. We speak here of one theater of war, or of several which, however, lie near to one another. Obviously this is the case when, according to our view, the chief operation is, at the same time, to decide on subordinate points.

But, now, can we *always* take this chance? And may we expose ourselves to the danger which must arise if the influence of the main point is not sufficient to decide at minor points? Does not the need of a certain breadth for a theater of war deserve special consideration?

Here as everywhere else it is impossible to exhaust the number of combinations which *may* take place; but we maintain that, with few exceptions, the decision on the main point will carry with it the decision on all minor points. Therefore, the action should be arranged in conformity with this principle, in all cases in which the contrary is not evident.

When Bonaparte invaded Russia, he had good reason to believe that by conquering the main body of the Russian army he would be able to sweep away at the same time their forces on the upper Dwina. He left at first only the corps of Oudinot to oppose them, but Wittgenstein assumed the offensive, and Bonaparte was then obliged to send also the Sixth Corps to that quarter.

On the other hand he had originally directed a part of his forces against Bagration; but that general was swept away by the retreating movement of the center, and Bonaparte was enabled then to recall that part of his forces. If Wittgenstein had not had to cover the second capital he would also have followed the retreating movement of the main army under Barclay.

In 1805 and 1809 Bonaparte's victories at Ulm and Ratisbon decided matters in Italy and also in the Tyrol, although the first was a rather distant theater and independent. In the year 1806 his victories at Jena and Auerstädt were decisive in respect to everything that might have been attempted against him in Westphalia and Hesse or on the Frankfurt road.

Among the number of circumstances which may have an influence on the resistance of the secondary points there are two which are the most prominent.

The first is when a country of vast extent and also of relatively great strength, like Russia, can put off the decisive blow at the chief point for some time and is not obliged to concentrate all its forces there in a hurry.

The second is when a secondary point (like Silesia in the year 1806), through a great number of fortresses, possesses an extraordinary degree of independence. Yet Bonaparte treated that point with great contempt, inasmuch as, when he had to leave it right in his rear on the march to Warsaw, he only detached 20,000 men under his brother Jerome against it.

If it happens that the blow at the chief point, in all probability, will not shake such secondary points, or has not actually done so, and if the enemy still has troops at these points, adequate forces must be opposed to these—a necessary evil—because we absolutely cannot surrender our line of communication from the outset.

But prudence may go a step farther; it may require that the advance upon the chief point keep pace with that on the secondary points, and consequently the principal undertaking be delayed whenever the secondary points will not succumb.

This principle would not directly contradict our principle of uniting all action as far as possible in one great undertaking, but the spirit from which it springs is diametrically opposed to the spirit in which ours is conceived. By following such a principle there would be such a measured pace in the movements, such a paralyzing of the attacking force, such a play of chance and such a loss of time as would be practically inconsistent with an offensive directed to the overthrow of the enemy.

The difficulty becomes still greater if the forces stationed at these minor points can retire on divergent lines. What would then become of the unity of our attack?

We must, therefore, declare ourselves completely opposed in principle to the dependence of the chief attack on secondary points, and we maintain that an attack directed toward the overthrow of the enemy which has not the boldness to shoot, like the point of an arrow, direct at the heart of the enemy's power, can never hit the mark.

(4) Lastly, there is still a fourth ground for a separate advance in the facility which it may afford for subsistence.

It is certainly much pleasanter to march with a small army through an opulent province than with a large army through a poor one; but by suitable measures and with an army accustomed to privations, the latter is not impossible, and, therefore, the first should never have such an influence on our plans as to expose us to great danger.

We have now done justice to the grounds for a separation of forces which divides a main operation into several, and if the separation takes place on any of these grounds, with a distinct conception of the object, and after due consideration of the advantages and disadvantages, we shall not venture to criticize.

But if, as usually happens, the plan is made in this way by a learned general staff, merely according to routine; if different theaters of war, like the squares on a chess board, must each be occupied by its piece before the moves begin, if these moves approach the aim with an imagined skill in combination in complicated lines and relations, if the armies are to separate today in order to display all their skill in reuniting at the greatest risk in two weeks—then we detest this abandonment of the direct, simple, common-sense road in order to plunge intentionally into sheer confusion. This folly happens more easily the less it is the commander-in-chief who directs the war and conducts it, in the sense which we have pointed out in the first chapter, as a simple action of his own individuality, invested with extraordinary powers; the more the whole plan is concocted by an unpractical general staff and from the ideas of a dozen smatterers.

We must still consider the third part of our first principle; that is, to keep the subordinate parts as subordinate as possible.

By endeavoring to reduce all operations of the war to a *single* aim, and trying to attain this as far as possible by *one* great action, we deprive the other points of contact of the belligerent states of a part of their independence; they have become subordinate actions. If we could concentrate everything absolutely into a single action, those points of contact would be completely neutralized; but this is seldom possible, and, therefore, it is a matter of keeping them so far within bounds that they shall not draw off too much force from the main issue.

Next, we maintain that the plan of war must have this tendency even when it is not possible to reduce the whole resistance of the enemy to one center of gravity. Therefore, in case we are placed in the position, as we have already put it, of carrying on two almost quite separate wars at the same time, the one must always be regarded as the *main issue* to which our forces and activity are to be chiefly devoted.

According to this view, it is advisable to advance *offensively* only toward that one principal point, and to remain on the defensive at all others. Only where unusual circumstances invite an attack would it be justified.

Further, we will seek to carry on this defensive, which takes place at minor points, with as few troops as possible, and to avail ourselves of every advantage which this form of resistance can give.

This view applies with still more force to all theaters of war on which, though armies of different powers also appear, they are still such as will be hit when the general center of gravity is hit.

But against *that* enemy at whom the main blow is aimed there can no longer, according to this, be any defensive on minor theaters of war. The main attack itself and the secondary attacks, which are brought about by other considerations, make up this blow, and make every defensive, on points not directly covered by it, superfluous. All depends on the main decision; by it every loss will be compensated. If the forces are sufficient to make it reasonable to seek such a main decision, then the *possibility of failure* can no longer be used as a ground for guarding oneself in any event against injury at other points; for *that very course* makes this failure much more probable, and therefore a contradiction here arises in our action.

This predominance of the principal action over the minor ones must also be the principle observed even in separate branches of the whole attack. But since there are usually other motives which determine which forces shall advance from one theater of war and which from another against the common center of gravity, we only mean here that there must be an *effort to make the principal action predominate*, for everything will become simpler and less subject to the influence of chance events the nearer this state of predominance can be attained.

The second principle concerns the rapid use of the forces.

Every unnecessary expenditure of time, every unnecessary detour, is a waste of power, and therefore contrary to the principles of strategy.

It is extremely important always to bear in mind that the attack generally possesses almost its only advantage in the surprise with which the opening of the action can be effective. Suddenness and irresistibility are its strongest pinions, and when the object is the overthrow of the enemy, it can rarely dispense with them.

Theory demands, therefore, the shortest way to the object, and completely excludes from consideration endless discussions over right and left, hither and yon.

If we call to mind what was said in the chapter concerning the object of the strategic attack (Book VII, Chapter 3) respecting the weakest point of the state, and, further, what appears in Chapter 4 of this Book on the influence of time, we believe no further argument is required to prove that the influence which we claim for that principle really belongs to it.

Bonaparte never acted otherwise. The shortest main road from army to army, from one capital to another, was his favorite.

And in what will now consist the principal action to which we have referred everything and for which we have demanded a swift and straightforward execution?

In Chapter 4 we have explained as far as it is possible in a general way what the overthrow of the enemy means, and it is unnecessary to repeat it. Whatever it may finally depend on in individual cases, the beginning is everywhere the same, namely: *the destruction of the enemy's military force,* that is, *a great victory over it, and its dispersion.* The sooner, which means the nearer our own frontier, this victory is sought, the *easier* it is; the later, that is, the deeper in the heart of the enemy's country, it is gained, the *more decisive* it is. Here, as everywhere, the facility of success and its magnitude balance each other.

If we are not so superior to the enemy that the victory is beyond doubt, then we must, when possible, seek him out, that is, his main force. We say *"when possible,"* for if this seeking out led to great detours, wrong directions and loss of time for us, it might very likely turn out to be a mistake. If the enemy's main force is not on our road and our interests otherwise prevent our searching for it, we may be sure we shall meet it later, for it will not fail to place itself in our way. We shall then, as we have said, fight under less advantageous circumstances—an evil to which we must submit. If we gain the battle in spite of that, it will be so much the more decisive.

From this it follows that, in the case now assumed, it would be an error to pass by the enemy's main force purposely, if it places itself in our way, at least if we expected thereby to facilitate the victory.

On the other hand, it follows from what precedes, that if we have a decided superiority over the enemy's main force, we may purposely pass it by in order at a future time to deliver a more decisive battle.

We have been speaking of a complete victory, therefore of an overthrow of the enemy, and not of a mere battle gained. But such a victory requires an enveloping attack, or a battle with an oblique front, for these two forms always give the result a decisive character. It is therefore an essential part of a plan of war to make arrangements for this movement, both as regards the mass of forces required and the direction to be given them, of which more will be said in the chapter on the plan of campaign.[1]

It is, of course, not impossible that battles fought with parallel fronts may also lead to complete overthrows, and cases in point are not wanting in military history; but such an event is uncommon and will be still more so the more armies become on a par as regards training and skill. We no longer capture twenty-one battalions in one village, as they did at Blenheim.

Once the great victory is gained, there should be no talk of rest, of getting breath, of considering, or of consolidating, etc., but only of pursuit, of fresh blows wherever necessary, of the capture of the enemy's capital, of attacking the enemy's auxiliary forces or of whatever else appears to be the point of support of the hostile state.

If the tide of victory carries us past the enemy's fortresses, it will depend upon our strength whether we lay siege to them or not. If we have a great superiority of force it would be a loss of time not to take them as soon as possible; but if we are not certain of the further success ahead of us, we must keep the fortresses in check with as few troops as possible, and that excludes an extensive siege. The moment that the siege of a fortress compels us to suspend our advance, that advance, *as a rule*, has reached its culminating point. We demand, therefore, that the main force should press forward rapidly in pursuit without any rest; we have already condemned the idea of allowing the advance on the principal point to be made dependent on success at secondary points; the consequence of this will be that in all ordinary cases our chief army only keeps behind it a nar-

[1] This chapter was never written.—Ed.

row strip of territory which it can call its own, and which therefore constitutes its theater of war. How this weakens the momentum at the head, and the dangers for the offensive arising therefrom, we have shown earlier. Will not this difficulty, will not this inner counterpoise come to a point which impedes further advance? That may be, no doubt. But just as we have insisted above that it would be a mistake to try to avoid this contracted theater of war from the start, and for the sake of that object to rob the advance of its impetus, so we also now maintain, that as long as the commander has not yet overthrown his opponent, as long as he considers himself strong enough to attain that aim, so long must he also pursue it. He does so perhaps at an increasing risk, but also with the prospect of an increasingly greater success. If there comes a point where he cannot venture to proceed, where he believes it necessary to protect his rear and extend himself right and left—well, then, this is most probably his culminating point. His momentum is then spent, and if the enemy is not overthrown, it becomes extremely probable that nothing will come of it.

All that the assailant does to intensify his attack by conquest of fortresses, defiles and provinces is no doubt still a slow advance, but it is only a relative one; it is no longer absolute. The enemy is no longer in flight; he is perhaps preparing a renewed resistance; and it is therefore quite possible that, although the assailant still advances with all his force, the position of the defense is improving every day. In short, we come back to this, that, as a rule, there is no second assault after a halt has once been necessary.

Therefore theory only requires that, as long as there is an intention of overthrowing the enemy, we proceed without halt against him. If the commander gives up this object because he finds the danger involved too great, he is right in stopping and extending his force. Theory only objects to this when he does it with a view to more readily overthrowing the enemy.

We are not so foolish as to maintain that no instance can be found of states having been gradually reduced to the last extremity. In the first place, the principle we now maintain is no absolute truth, to which an exception would be impossible, but one founded only on the ordinary and probable result; next, we must make a distinction between cases in which the downfall of a state has actually been effected by a slow, gradual process, and those in which the event was the result of the first campaign. We are here only treating the latter case, for it is only in this that there is that tension of forces which either overcomes the center of gravity of the burden or is in danger of being overcome by it. If in the first year we gain a moderate advantage, to which in the following we add another, and thus gradually advance toward our object, there is nowhere very imminent danger, but on that account it is distributed over many points. Each pause between one success and another gives the enemy fresh chances. The effects of the first successes have very little influence on those which follow, often none, often a negative one, because the enemy recovers, or is perhaps excited to increased resistance, or obtains foreign aid; whereas, where everything happens in one campaign, the success of yesterday carries today's with it, one fire lights itself from another. If there are cases in which states have been overcome by successive blows—in which, consequently, *time,* the patron of the defensive, has proved fatal—how infinitely more numerous are the instances in which the intention of the aggressor has by that means utterly failed! We need only to think of the result of the Seven Years' War, in which the Austrians sought to attain their aim with such ease, caution and prudence that they completely missed it.

Taking this view, therefore, we cannot possibly be of the opinion that the concern for the proper arrangement of a theater of war and the impulse which urges us onward should always go together, and that the former must, to a certain extent, be a counterbalance to the latter; but we look upon the disadvantages which spring from the forward movement as an unavoidable evil which only deserves attention when, ahead of us, there is no longer any hope for us.

Bonaparte's example in 1812, far from shaking our opinion, has rather confirmed us in it.

His campaign did not fail because he advanced too swiftly or too far, as is commonly believed, but because the only means to secure success failed. The Russian Empire is no country which can be really conquered, that is to say, which can be held in occupation, at least not by the forces of the present states of Europe, nor by the 500,000 men with which Bonaparte invaded the country. Such a country can only be subdued by its own weaknesses, and by the effects of internal dissension. In order to strike these vulnerable points in its political existence, the country must be shaken to its very center. It was only by reaching Moscow with the force of his blow that Bonaparte could hope to shake the courage of the government and the loyalty and steadfastness of the people. In Moscow he expected to conclude the peace treaty, and this was the only reasonable object which he could set before himself in undertaking such a campaign.

He therefore led his main force against that of the Russians, which staggered back before him past the camp at Drissa and did not stop until it reached Smolensk. He carried Bagration along in his movement, beat the

Russian main army and took Moscow. He acted on this occasion as he had always done, it was only in that way that he had made himself the arbiter of Europe, and only in that way that it was possible for him to do so.

Therefore, he who admires Bonaparte in all his earlier campaigns as the greatest of generals ought not to look down upon him in this instance.

It is permissible to judge an event according to the result, as that is the best criticism upon it (see Book II, Chapter 5), but this judgment, derived merely from the result, must not be passed off as evidence of human wisdom. Seeking out the causes for the failure of a campaign is not going so far as to make a criticism of it; it is only if we show that these causes should neither have been overlooked nor disregarded that we make a criticism and place ourselves above the general.

Now we maintain that anyone who pronounces the campaign of 1812 an absurdity, merely on account of the tremendous counterstroke with which it ended, and who, if it had been successful, would look upon it as a most splendid scheme, shows an utter incapacity of judgment.

If Bonaparte had remained in Lithuania, as most of his critics think he should, in order first to get possession of the fortresses (of which, by the way, there are scarcely any, except Riga, situated quite at one side, because Bobruisk is a small insignificant place), he would have involved himself for the winter in a dismal defensive system; then the same people would have been the first to exclaim: This is not the old Bonaparte! Why has he not got even so far as a first great battle—he who used to place the final seal to his conquests on the last ramparts of the enemy's states by such victories as Austerlitz and Friedland? Has he faint-heartedly failed to take the enemy's capital, the defenseless Moscow, ready to open its gates, and thus left a nucleus round which new elements of resistance could gather? He had the singular luck to take this far-off and enormous colossus by surprise, as easily as one would surprise a neighboring city, or as Frederick the Great surprised the little state of Silesia, lying at his door, and he makes no use of his good fortune, halts in the middle of his victorious course, as if an evil spirit had laid itself at his heels! This is the way he would have been judged, for this is the fashion of critics' judgments in general.

In opposition to this, we say, the campaign of 1812 did not succeed because the enemy's government remained firm, the people loyal and steadfast, because it therefore could not succeed. Bonaparte may have made a mistake in undertaking such an expedition; at all events, the result has shown that he deceived himself in his calculations, but we maintain that, supposing it necessary to seek the attainment of this aim, in all general points it could not have been done in any other way.

Instead of burdening himself with an interminable costly defensive war in the East, such as he had on his hands in the West, Bonaparte tried the only means to achieve his object: by one bold stroke to extort a peace from his astonished adversary. The destruction of his army was the danger to which he exposed himself in the venture; it was the stake in the game, the price of his great expectations. If this destruction of his army was more complete through his own fault than it need have been, this fault was not in his having penetrated too far into the heart of the country, for that was his object and unavoidable, but in the late period at which the campaign opened, the sacrifice of life occasioned by his tactics, the lack of due care for the supply of his army, and for his line of retreat, and, lastly, in his somewhat delayed march from Moscow.

That the Russian armies were able to place themselves in his way on the Beresina, in order to cut off his retreat absolutely, is no strong argument against us. For, in the first place, the failure of that attempt just shows how difficult it is really to cut off an army, as the army which was intercepted in this case, under the most unfavorable circumstances that can be conceived, still managed to cut its way through. Although this act contributed certainly to increase the catastrophe, still it was not essentially the cause of it. Second, it was only the unusual character of the country which afforded the means to carry things so far, for if it had not been for the marshes of the Beresina, with their wooded impassable borders lying across the great road, the cutting off would have been still less possible. Third, there are no means whatsoever of guarding against such a possibility except by making the forward movement with the army in a certain width, a thing of which we have previously disapproved, for if once we proceed on the plan of advancing with the center and covering the wings with armies, which one leaves behind to the left and right, then if either of these armies meets with misfortune, we would have to fall back with the van, and then probably very little could be gained by the attack.

Moreover, it cannot be said that Bonaparte neglected his wings. A superior force remained fronting Wittgenstein; a proportionate siege corps stood before Riga, which was really superfluous there; and in the South Schwarzenberg had 50,000 men with which he was superior to Tormasoff and even almost equal to Tschitschagow; in addition, there were 30,000 men under Victor covering the rear of the center. Even in the month of November, that is, at the decisive moment when the Russian armies had been reinforced, and the French were very much reduced, the superiority of the Russians in the rear of the Moscow army was not so very extraordinary. Wittgenstein, Tschitschagow and Sacken made up together a force of 110,000. Schwarzenberg, Regnier, Victor, Oudinot and St. Cyr had still 80,000 usable men. The most cautious general, in advancing, would hardly devote a greater proportion of his force to the protection of his flanks.

If out of 600,000 men who crossed the Niemen in 1812 Bonaparte had brought back 250,000 instead of the 50,000 who recrossed it under Schwarzenberg, Regnier and Macdonald, which would have been possible, by avoiding the mistakes with which we have reproached him, the campaign would still have been an unfortunate one, but theory could have had nothing to object to, for the loss of more than half an army in such a case is not at all unusual, and only appears so to us in this instance because of the enormous scale of the whole enterprise.

So much for the main operation, its necessary tendency and its unavoidable risks. In respect to the subordinate operations, we say above all things: there must be a common aim for all of them, but this aim must be such as not to paralyze the action of the individual parts. If we invade France from the Upper and Middle Rhine and Holland with the intention of uniting at Paris, but neither of the armies is supposed to risk anything on the advance and is to keep itself intact until the concentration is effected, that is what we call a *ruinous* plan. There must be a balancing of the threefold movement, which causes delay, indecision and timidity in the advance of each of the armies. It is better to assign to each part its mission, and only to place unity at that point where these several activities become a unity of themselves.

Therefore, when the military forces advance on separate theaters of war to the attack, a separate object should be assigned to each army, on which it can direct its power of attack. The thing that matters is that *this attack* should take place from all sides, and not that all should gain proportionate advantages.

If the task assigned to one army is found too difficult because the enemy has made a disposition of his force different from that which was expected, if it sustains reverses, this must not and should not have any influence on the action of the others, or else we should turn the probability of the general success against ourselves at the very outset. It is only the unsuccessful issue of the majority of enterprises or of the principal one which can and must have an influence upon the others, for then it comes under the head of a plan which has failed.

This same rule applies to those armies and detachments which originally were intended for the defensive, and, owing to a success gained, can assume the offensive, unless we prefer to use such spare forces for the principal offensive, a point which will chiefly depend on the geographical situation of the theater of war.

But under these circumstances, what becomes of the geometrical form and unity of the whole attack, what of the flanks and rear of detachments when the part next to them is beaten?

That is precisely what we wish chiefly to contest. This gluing down of a great attack to a geometrical square is a straying into a false system of thought.

In Book III, Chapter 15, we have shown that the geometrical element has less influence in strategy than in tactics, and we shall only repeat here the conclusion reached there: that in the attack especially, the actual results at the various points deserve much more attention than the geometrical figure, which may gradually be formed through the diversity of results.

But in any case it is quite certain that, in view of the vast spaces with which strategy has to deal, the considerations and decisions which create the geometrical situation of the parts can properly be left to the commander-in-chief; that, therefore, no subordinate general has a right to ask what his neighbor is doing or leaving undone, but can be directed peremptorily to pursue his object. If any serious incongruity arises from this, a remedy can always be prescribed in time by the supreme authority. Thus, then, may be obviated the chief evil of this separate mode of action, which is, that in the place of realities a cloud of apprehensions and suppositions insinuates itself into the course of events, that every accident affects not only the part to which it happens, but also the whole, by the communication of impressions, and that a wide field is opened for the personal failings and personal animosities of subordinate commanders.

We think that these views will appear paradoxical only to those who have not studied military history long enough or with sufficient attention, who have not distinguished the important from the unimportant, nor made proper allowances for the influence of human weaknesses.

If even in tactics it is difficult, as all experienced soldiers admit it is, for an attack in several separate columns to succeeed by the perfectly harmonious working of all parts, how much more difficult, or rather how impossible, must this be in strategy, where the separation is so much wider? Therefore, if a constant connection of all parts were a necessary condition of success, such a strategic plan of attack would have to be given up altogether. But on the one hand, it is not left to our option to discard it completely, because circumstances over which we have no control may determine in favor of it; on the other hand, even in tactics, this constant harmonious working of all parts at every moment of the execution is not necessary, and it is still less so in strategy. Therefore, in strategy we should disregard it so much the more, and insist the more that an independent piece of work be assigned to each part.

To this we must add one more important observation; it relates to the proper allotment of duties.

In 1793 and 1794 the main Austrian army was in the Netherlands, that of the Prussians on the upper Rhine. The Austrians marched from Vienna to Condé and Valenciennes, crossing the line of march of the Prussians from Berlin to Landau. It is true the Austrians had to defend their Belgian provinces in that quarter, and any conquests made in French Flanders would have been acquisitions conveniently situated for them, but that interest was not strong enough. After the death of Prince Kaunitz, the minister, Thugut, put through a measure for giving up the Netherlands entirely, for the better concentration of the Austrian forces. In fact, Austria is almost twice as far from Flanders as from Alsace, and at a time when military resources were very limited, and everything had to be paid for in cash, that was no trifling consideration. Still, Thugut plainly had something else in view; his object was, through the urgency of the danger, to compel Holland, England and Prussia, the powers interested in the defense of the Netherlands and the Lower Rhine, to make greater efforts. He deceived himself in his calculations because nothing could be done with the Prussian cabinet at that time, but this occurrence nevertheless shows the influence of political interests on the course of a war.

Prussia had neither anything to conquer nor to defend in Alsace. In 1792 she had undertaken the march through Lorraine into Champagne in a sort of chivalrous spirit. But as that enterprise ended in nothing, through the unfavorable course of circumstances, she continued the war with very little interest. If the Prussian troops had been in the Netherlands, they would have been in direct communication with Holland, which they might have looked upon almost as their own country, having conquered it in 1787; they would then have covered the Lower Rhine, and consequently that part of the Prussian monarchy which lay next to the theater of war. Prussia, on account of subsidies, also had a closer alliance with England, which, under these circumstances, could not so easily have degenerated into the deceitfulness of which the Prussian cabinet was guilty at that time.

A much better result, therefore, might have been expected if the Austrians had appeared with their main force on the Upper Rhine, the Prussians with their whole force in the Netherlands and the Austrians had left there only a force of proportionate strength.

If, instead of the enterprising Blücher, General Barclay had been placed at the head of the Silesian army in 1814, and Blücher had been kept with the main army under Schwarzenberg, the campaign would perhaps have turned out a complete failure.

If the enterprising Laudon, instead of having his theater of war at the strongest point of the Prussian dominions, namely, in Silesia, had been in the position of the German Imperial army, perhaps the whole Seven Years' War would have taken a different turn. In order to examine this subject more closely, we must look at these cases according to their chief distinctions.

The first is if we carry on war in conjunction with other powers, who not only take part as our allies, but also have an independent interest.

The second is if the army of an ally has come to our assistance.

The third is when it is only a question of the personal characteristics of the generals.

In the first two cases, the point may be raised whether it is better to mix up the troops of the different powers completely, so that each separate army is composed of corps of different powers, as was done in the wars of 1813 and 1814, or to keep them separate as much as possible, so that each may act more independently.

Obviously, the first is the most advantageous plan; but it supposes a degree of friendly feeling and community of interests which is seldom found. When there is this close union of the military forces, it is much more difficult for the cabinets to separate their respective interests; and as regards the prejudicial influence of the egotistical views of commanders, it can show itself under these circumstances only among the subordinate generals, therefore, only in the province of tactics, and even here not so freely or with such impunity as when there is a complete separation. In the latter case, it passes over into strategy and therefore operates in decisive features. But, as we have said, a rare self-effacement on the part of the governments is required for this. In 1813, the exigencies of the time forced all governments in that direction, and yet we cannot sufficiently praise the conduct of the Emperor of Russia, who, though he entered the field with the strongest army and it was he by whom the change of fortune was chiefly brought about, set aside all pride about appearing at the head of an independent Russian army and placed his troops under the Prussian and Austrian commanders.

If such a fusion of forces cannot be effected, a complete separation of them is certainly better than a semi-separation; the worst of all is when two independent commanders of armies of different powers are on the same theater of war, as frequently happened in the Seven Years' War with the armies of Russia, Austria and the German states. When there is a complete separation of forces, the burdens which must be borne are also more divided, and each suffers only from what is his own, consequently is more impelled to activity by the force of circumstances; but if they find themselves in close connection, or even on the same theater of war, this is not the case, and, besides that, the ill-will of one paralyzes also the powers of the other as well.

In the first of the three supposed cases, there will be no difficulty in the complete separation, as the natural interest of each state generally indicates to it a separate direction of employing its forces. This may not be so in the second case, and then, as a rule, there is nothing to be done but to place oneself completely under the auxiliary army, if its strength is in any way proportionate to that measure, as the Austrians did in the latter part of the campaign of 1815 and the Prussians in the campaign of 1807.

With regard to the personal characteristics of generals, it becomes an entirely individual matter; but we must not omit to make one general remark, which is, that we should not, as is generally done, place at the head of subordinate armies the most prudent and cautious commanders, but *the most enterprising*, for we repeat once more that in strategic operations conducted separately, there is nothing more important than that every part should develop its powers to the full, in which case faults committed at one point may be compensated for by successes at others. This complete activity of all parts, however, is only to be expected when the commanders are spirited, enterprising men, who are driven forward by an inner urge, by their own hearts, because a mere objective, coolly reasoned-out conviction of the necessity of action seldom suffices.

Lastly, we have to remark that, if circumstances in other respects permit, the troops and their commanders, as regards their purpose and the nature of the country, should be employed in accordance with their qualities—that is, standing armies, good troops, numerous cavalry, old prudent intelligent generals in an open country; militia, national levies, young, enterprising commanders in wooded country, mountains and defiles; auxiliary forces in rich provinces where they like to be.

What we have said up till now concerning a plan of war in general, and in this chapter concerning that in particular which is directed to the overthrow of the enemy, was intended to give special prominence to the aim of the plan, and in addition to indicate principles which shall serve as guides in the preparation of ways and means. Our desire has been in this way to give a clear conception of what one wants and ought to do in such a war. We wanted to emphasize the necessary and general, and to leave a margin for the play of the particular and accidental, but to exclude all that is *arbitrary, unfounded, trifling, fantastic or sophistic.* If we have succeeded in this object, we look upon our task as accomplished.

Now, if anyone wonders at finding nothing here about turning rivers, about commanding mountains from their highest points, about avoiding strong positions, and about the keys of a country, he has not, in our opinion, understood us; neither does he as yet understand war in its general relations.

In the preceding Books we have characterized these subjects in general, and we there arrived at the conclusion that they are usually much more insignificant in their nature than we should think from their high repute. So much the less is it possible or proper that they should play a great role in a war, whose aim is the overthrow of the enemy—that is, such a role as would have an influence on the whole plan of war.

At the end of this Book we shall devote a special chapter to the arrangement of the supreme command;[1] the present chapter we shall close with an example.

If Austria, Prussia, the German Confederation, the Netherlands and England determine on a war with France, but Russia remains neutral—a case which has frequently happened during the last 150 years—they are able to carry on an offensive war, having for its object the overthrow of the enemy. For powerful and great as France is, it is still possible for her to see more than half her territory overrun by the enemy, her capital occupied and herself reduced to inadequate sources of supply, without there being any power, except Russia, which could support her with great effectiveness. Spain is too distant and too disadvantageously situated; the Italian States are at present too brittle and powerless.

The countries we have named have, exclusive of their possessions outside Europe, more than 75,000,000 inhabitants, while France has only 30,000,000; and the army which they could call out for a war against France, really meant in earnest, would be as follows, without exaggeration:

Austria	250,000	
Prussia	200,000	
The rest of Germany	150,000	
Netherlands	75,000	
England	50,000	
Total		725,000

[1] This chapter was never written.—Ed.

Should this force actually be placed on a war footing it would, in all probability, very much exceed that which France could oppose to it, for under Bonaparte the country never raised troops of the like strength. Now, if we take into account the deductions required as garrisons for fortresses and depots, to watch the coasts, etc., there can be no doubt that the Allies would have a great superiority on the principal theater of war, and upon that the aim of overthrowing the enemy is chiefly founded.

The center of gravity of the French power lies in its military force and in Paris. To defeat the former in one or more great battles, to take Paris and to drive the rest of the French across the Loire must be the object of the Allies. The weakest point of the French monarchy is between Paris and Brussels; on that side, the frontier is only 150 miles from the capital. Part of the Allies—the English, Netherlanders, Prussians and North German States—have their natural point of assembly there, as these states lie partly in the immediate vicinity, partly in a direct line behind it. Austria and South Germany can only carry on their war conveniently from the Upper Rhine. Their most natural direction is upon Troyes and Paris, or it may be Orleans. Both blows, therefore, that from the Netherlands and the other from the Upper Rhine, are quite direct and natural, short and powerful, and both strike the center of gravity of the enemy's power. Between these two points, therefore, the whole attacking army should be divided.

There are only two considerations which interfere with the simplicity of this plan.

The Austrians will not lay bare their Italian dominions; they will want to remain master of events there in any case. Therefore they will not let things reach the point at which, because of an attack on the heart of France, Italy would be only indirectly covered. In view of the political state of the country, this secondary intention is not to be treated with contempt; but it would be a decided mistake if the old and oft-tried plan of an attack from Italy, directed against the South of France, was bound up with it, and if on that account the force in Italy was increased to a size not required for mere security against contingencies in the first campaign. Only the number needed for that security should remain in Italy; only that number may be withdrawn from the main undertaking, if we would not be unfaithful to that first maxim, *unity of plan, concentration of force.* To think of conquering France by the Rhone would be like trying to lift a musket by the point of its bayonet; but even as an auxiliary enterprise, an attack on the South of France is to be condemned, for it only raises new forces against us. Whenever an attack is made on a distant province, interest and activities are aroused, which would otherwise have lain dormant. Only if it is shown that the forces left in Italy were too large for the mere security of the country and would therefore have to remain idle is an attack on the South of France from that quarter justified.

We therefore repeat that the force left in Italy must be kept down as low as circumstances will permit, and it will be quite large enough if it will suffice to prevent the Austrians from losing the whole country in one campaign. Let us suppose that number to be 50,000 men for the purpose of our illustration.

The other consideration is the circumstance that France has a coast line. Since England has the upper hand at sea, it follows that France must, on that account, be very susceptible with regard to the whole of her Atlantic coast, and consequently must protect it with more or less strong garrisons. Now, however weak this coast defense may be, still the French frontiers are tripled by it, and considerable forces, on that account, cannot fail to be withdrawn from the French army on the theater of war. The twenty or thirty thousand troops disposable to effect a landing, with which the English threaten France, would probably absorb twice or three times the number of French troops; and, further, we must not think only of troops, but also of the money, artillery, etc., required for ships and coast batteries. Let us suppose that the English devote 25,000 men to this object.

Our plan of war would then consist simply in this:

(1) That in the Netherlands

> 200,000 Prussians
> 75,000 Netherlanders
> 25,000 English
> 50,000 North German Confederation
> ————
> total 350,000 be assembled,

of whom about 50,000 should be set aside to garrison frontier fortresses, and the remaining 300,000 should advance against Paris, and engage the French army in a great battle.

(2) That 200,000 Austrians and 100,000 South German troops should assemble on the Upper Rhine to advance at the same time as the army of the Netherlands, their direction being toward the Upper Seine, and thence toward the Loire, likewise with a view to a great battle. Those two attacks would, perhaps, unite in one on the Loire.

By this the chief point is determined. What we have to add is mainly intended to remove false conceptions, and it is as follows:

(1) To seek the prescribed great battle, and to deliver it with such a relation in point of numerical strength and under such circumstances as promise a decisive victory, must be the object of the chief commanders' efforts; to this object everything must be sacrificed, and as few men as possible should be employed in sieges, blockades, garrisons, etc. If, like Schwarzenberg in 1814, they spread out in eccentric rays as soon as they enter the enemy's provinces, all is lost, and the Allies in 1814 had only the powerlessness of France to thank that all was not actually lost in the first fourteen days. The attack should be like a wedge well driven home, not like a soap-bubble, which distends itself until it bursts.

(2) Switzerland must be left to its own forces. If it remains neutral, it forms a good *point d'appui* on the Upper Rhine; if it is attacked by France, let it stand up for itself, which in more than one respect it is very well able to do. Nothing would be more absurd than to attribute to Switzerland a predominant geographical influence upon the events of a war because it is the highest land in Europe. Such an influence only exists under certain very restricted conditions which do not prevail here. When the French are attacked in the heart of their country they can undertake no strong offensive from Switzerland, either against Italy or Swabia, and, least of all, can the elevated situation of this country come into consideration as a decisive circumstance. The advantage of dominating strategically is, in the first place, chiefly important in the defensive, and any importance which it has in the offensive can only manifest itself in a single encounter. Whoever does not know this has not thought over the matter and arrived at a clear conception of it, and in case that at any future learned council of potentates and generals some learned officer of the General Staff should be found who, with an anxious brow, shows off such wisdom, we now declare it beforehand to be mere folly, and wish that in the same council some true soldier, some man of sound common sense, may be present, to silence him.

(3) The space between the two attacks we consider of very little consequence. When 600,000 men assemble 150 to 200 miles from Paris to march against the heart of France, would any one think of covering the Middle Rhine as well as Berlin, Dresden, Vienna and Munich? There would be no sense in such a thing. Are we to cover the communications? That would not be unimportant; but then we might soon be led into giving this covering the strength and importance of an attack, and thus, instead of advancing on two lines, as the situation of the states positively requires, we should be led to advance upon three, which is not required. These three would then, perhaps, become five, or perhaps seven, and in that way the old rigmarole would once more become the order of the day.

Our two attacks have each their object; the forces employed on them are very probably noticeably superior to the enemy in numbers. If each pursues his march with vigor, they cannot fail to react advantageously upon each other. If one of the two attacks is unfortunate because the enemy has not divided his force equally, we may fairly expect that the result of the other will of itself repair this disaster, and this is the true interdependence between the two. An interdependence extending to the events of individual days is impossible on account of the distance; neither is it necessary, and, therefore, the immediate or rather the direct connection is of no great value.

Besides, the enemy attacked in the very center of his dominions will have no forces worth speaking of to employ in interrupting this connection; all that is to be apprehended is that this interruption may be attempted by the inhabitants, supported by raiding parties, so that this object does not cost the enemy any actual troops. To prevent that, it is sufficient to send a body of 10,000 or 15,000 men, particularly strong in cavalry, in the direction from Treves to Rheims. It will be able to disperse all raiders, and keep in line with the grand army. This corps should neither invest nor watch fortresses, but march between them, depend on no fixed basis, but give way before superior forces in any direction. No great misfortune could happen to it, and if such did happen, it would again be no serious misfortune for the whole. Under these circumstances, such a force might probably serve as an intermediate link between the two attacks.

(4) The two subordinate undertakings, that is, the Austrian army in Italy and the English army for landing on the coast, might follow their object as seems best. If they do not remain idle, their mission is fulfilled as regards the chief point, and on no account should either of the two great attacks be made dependent in any way on them.

We are quite convinced that in this way France may be overthrown and chastised whenever she thinks fit to put on that insolent air with which she has oppressed Europe for 150 years. It is only on the other side of Paris, on the Loire, that those conditions can be wrung from her which are necessary for the peace of Europe. In this way alone the natural relation between 30,000,000 men and 75,000,000 will quickly make itself known, but not if that country from Dunkirk to Genoa is to be surrounded in the way it has been for 150 years by a girdle of armies, while fifty different small objects are aimed at, not one of which has the power to overcome the inertia, friction and extraneous influences which everywhere, but more especially in allied armies, are generated and forever born anew.

How little the provisional organization of the German Federal armies is adapted to such a disposition will strike the reader. By that organization the federal part of Germany forms the nucleus of the German power, and Prussia and Austria, thus weakened, lose their natural weight. But a federate state is a very brittle nucleus in war; there is in it no unity, no energy, no rational choice of a commander, no authority, no responsibility.

Austria and Prussia are the two natural centers of force of the German Empire; they form the pivot of the balance, the forte of the sword; they are monarchical states, used to war; they have well-defined interests, independence of power; they are predominant over the others. The organization should follow these natural lines, and not a false notion of unity, which is an impossibility in such a case. And he who neglects the possible in quest of the impossible is a fool.

INDEX

A NOTE ON THE TYPE

The principal text of this Modern Library edition was set in a
digitized version of Janson, a typeface that dates from about
1690 and was cut by Nicholas Kis, a Hungarian working in
Amsterdam. The original matrices have survived and are held by
the Stempel foundry in Germany. Hermann Zapf
redesigned some of the weights and sizes for Stempel,
basing his revisions on the original design.